Handbook of Big Data Technologies

Albert Y. Zomaya · Sherif Sakr
Editors

Handbook of Big Data Technologies

Foreword by Sartaj Sahni, University of Florida

Springer

Editors
Albert Y. Zomaya
School of Information Technologies
The University of Sydney
Sydney, NSW
Australia

Sherif Sakr
The School of Computer Science
The University of New South Wales
Eveleigh, NSW
Australia

and

King Saud Bin Abdulaziz University
of Health Science
Riyadh
Saudi Arabia

ISBN 978-3-319-84138-0   ISBN 978-3-319-49340-4   (eBook)
DOI 10.1007/978-3-319-49340-4

Printed on acid-free paper

This Springer imprint is published by Springer Nature
The registered company is Springer International Publishing AG
The registered company address is: Gewerbestrasse 11, 6330 Cham, Switzerland

To the loving memory of my Grandparents.

Albert Y. Zomaya

To my wife, Radwa,
my daughter, Jana,
and my son, Shehab
for their love, encouragement, and support.

Sherif Sakr

Foreword

Handbook of Big Data Technologies (edited by Albert Y. Zomaya and Sherif Sakr) is an exciting and well-written book that deals with a wide range of topical themes in the field of Big Data. The book probes many issues related to this important and growing field—processing, management, analytics, and applications.

Today, we are witnessing many advances in Big Data research and technologies brought about by developments in big data algorithms, high performance computing, databases, data mining, and more. In addition to covering these advances, the book showcases critical evolving applications and technologies. These developments in Big Data technologies will lead to serious breakthroughs in science and engineering over the next few years.

I believe that the current book is a great addition to the literature. It will serve as a keystone of gathered research in this continuously changing area. The book also provides an opportunity for researchers to explore the use of advanced computing technologies and their impact on enhancing our capabilities to conduct more sophisticated studies.

The book will be well received by the research and development community and will be beneficial for researchers and graduate students focusing on Big Data. Also, the book is a useful reference source for practitioners and application developers.

Finally, I would like to congratulate Profs. Zomaya and Sakr on a job well done!

Sartaj Sahni
University of Florida
Gainesville, FL, USA

Preface

We live in the era of Big Data. We are witnessing radical expansion and integration of digital devices, networking, data storage, and computation systems. Data generation and consumption is becoming a main part of people's daily life especially with the pervasive availability and usage of Internet technology and applications. In the enterprise world, many companies continuously gather massive datasets that store customer interactions, product sales, results from advertising campaigns on the Web in addition to various types of other information. The term *Big Data* has been coined to reflect the tremendous growth of the world's digital data which is generated from various sources and many formats. Big Data has attracted a lot of interest from both the research and industrial worlds with a goal of creating the best means to process, analyze, and make the most of this data.

This handbook presents comprehensive coverage of recent advancements in Big Data technologies and related paradigms. Chapters are authored by international leading experts in the field. All contributions have been reviewed and revised for maximum reader value. The volume consists of twenty-five chapters organized into four main parts. Part I covers the fundamental concepts of Big Data technologies including data curation mechanisms, data models, storage models, programming models, and programming platforms. It also dives into the details of implementing Big SQL query engines and big stream processing systems. Part II focuses on the semantic aspects of Big Data management, including data integration and exploratory ad hoc analysis in addition to structured querying and pattern matching techniques. Part III presents a comprehensive overview of large-scale graph processing. It covers the most recent research in large-scale graph processing platforms, introducing several scalable graph querying and mining mechanisms in domains such as social networks. Part IV details novel applications that have been made possible by the rapid emergence of Big Data technologies, such as Internet-of-Things (IOT), Cognitive Computing, and SCADA Systems. All parts of the book discuss open research problems, including potential opportunities, that have arisen from the rapid progress of Big Data technologies and the associated increasing requirements of application domains. We hope that our readers will benefit from these discussions to enrich their own future research and development.

This book is a timely contribution to the growing Big Data field, designed for researchers and IT professionals and graduate students. Big Data has been recognized as one of leading emerging technologies that will have a major contribution and impact on the various fields of science and varies aspect of the human society over the coming decades. Therefore, the content in this book will be an essential tool to help readers understand the development and future of the field.

Sydney, Australia   Albert Y. Zomaya
Eveleigh, Australia; Riyadh, Saudi Arabia   Sherif Sakr

Contents

Part I
Fundamentals of Big Data Processing

Big Data Storage and Data Models

Dongyao Wu, Sherif Sakr and Liming Zhu

Abstract Data and storage models are the basis for big data ecosystem stacks. While storage model captures the physical aspects and features for data storage, data model captures the logical representation and structures for data processing and management. Understanding storage and data model together is essential for understanding the built-on big data ecosystems. In this chapter we are going to investigate and compare the key storage and data models in the spectrum of big data frameworks.

The growing demand of storing and processing large scale data sets has been driving the development of data storage and databases systems in the last decade. The data storage has been improved and enhanced from that of local storage to clustered, distributed and cloud-based storage. Additionally, the database systems have been migrated from traditional RDBMS to the more current NoSQL-based systems. In this chapter, we are going to present the major storage and data models with some illustrations of related example systems in big data scenarios and contexts based on taxonomy of data store systems and platforms which is illustrated in Fig. 1.

1 Storage Models

A storage model is the core of any big-data related systems. It affects the scalability, data-structures, programming and computational models for the systems that are built on top of any big data-related systems [1, 2]. Understanding about the under-

D. Wu (✉) · S. Sakr · L. Zhu
Data61, CSIRO, Sydney, Australia
e-mail: Dongyao.Wu@data61.csiro.au

D. Wu · S. Sakr · L. Zhu
School of Computer Science and Engineering, University of New South Wales,
Sydney, Australia

S. Sakr
National Guard, King Saud Bin Abdulaziz University for Health Sciences,
Riyadh, Saudi Arabia

© Springer International Publishing AG 2017
A.Y. Zomaya and S. Sakr (eds.), *Handbook of Big Data Technologies*,
DOI 10.1007/978-3-319-49340-4_1

Storage Models   Data Models

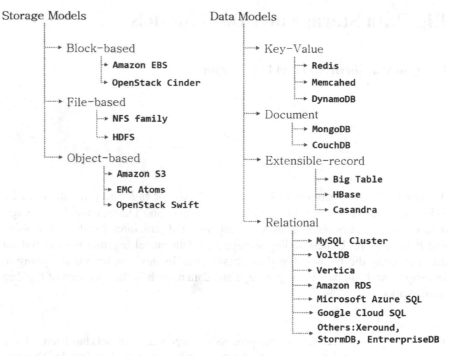

Fig. 1 Taxonomy of data stores and platforms

lying storage model is also the key of understanding the entire spectrum of big-data frameworks. For addressing different considerations and focus, there has been three main storage models developed during the past a few decades, namely, Block-based storage, File-based Storage and Object-based Storage.

1.1 Block-Based Storage

Block level storage is one of the most classical storage model in computer science. A traditional block-based storage system presents itself to servers using industry standard Fibre Channel and iSCSI [3] connectivity mechanisms. Basically, block level storage can be considered as a hard drive in a server except that the hard drive might be installed in a remote chassis and is accessible using Fibre Channel or iSCSI. In addition, for block-based storage, data is stored as blocks which normally have a fixed size yet with no additional information (metadata). A unique identifier is used to access each block. Block based storage focus on performance and scalability to store and access very large scale data. As a result, block-based storage is usually used as a low level storage paradigm which are widely used for higher level storage systems such as File-based systems, Object-based systems and Transactional Databases, etc.

Fig. 2 Block-based storage
model

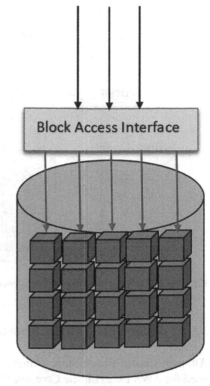

Block Storage Device

1.1.1 Architecture

A simple model of block-based storage can be seen in Fig. 2. Basically, data are
stored as blocks which normally have a fixed size yet with no additional information
(metadata). A unique identifier is used to access each block. The identifier is mapped
to the exact location of actual data blocks through access interfaces. Traditionally,
block-based storage is bound to physical storage protocols, such as SCSI [4], iSCSI,
ATA [5] and SATA [6].

With the development of distributed computing and big data, block-based storage
model are also developed to support distributed and cloud-based environments. As
we can see from the Fig. 3, the architecture of a distributed block-storage system
is composed of the block server and a group of block nodes. The block server is
responsible for maintaining the mapping or indexing from block IDs to the actual
data blocks in the block nodes. The block nodes are responsible for storing the actual
data into fixed-size partitions, each of which is considered as a block.

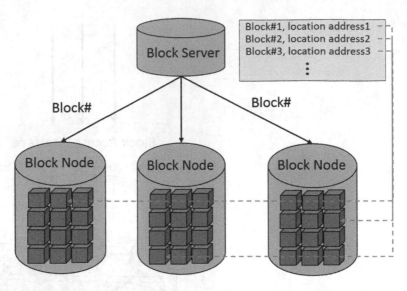

Block#1, location address1
Block#2, location address2
Block#3, location address3

Fig. 3 Architecture of distributed Block-based storage

1.1.2 Amazon Elastic Block Store (Amazon EBS)

Amazon Elastic Block Store (Amazon EBS) [7] is a block-level storage service used for AWS EC2 (Elastic Compute Cloud) [8] instances hosted in Amazon Cloud platform. Amazon EBS can be considered as a massive SAN (Storage Area Network) in the AWS infrastructure. The physical storage could be hard disks, SSDs, etc. under the EBS architecture. Amazon EBS is one of the most important and heavily used storage services of AWS, even the building blocks component offerings from AWS like RDS [9], DynamoDB [10] and CloudSearch [11], rely on EBS in the Cloud. In Amazon EBS, block volumes are automatically replicated within the availability zone to protect against data loss and failures. It also provides high availability and durability for users. EBS volumes can be used just as traditional block devices and simply plugged into EC2 virtual machines. In addition, users can scale up or down their volume within minutes. Since the Amazon EBS lifetime is separate from the instance on which it is mounted, users can detach and later attach the volumes on other EC2 instances in the same availability zone.

1.1.3 OpenStack Cinder and Nova

In the open-source cloud such as OpenStack [12], the block storage service is provided by the Nova [13] system working with the Cinder [14] system. When you start a Nova compute instance, it should come configured with some block storage devices by default, at the very least to hold the read/write partitions of the running OS. These block storage instances can be "ephemeral" (the data goes away when

Fig. 4 File-based storage model

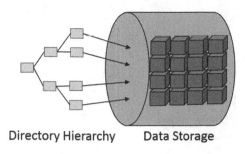

Directory Hierarchy   Data Storage

the compute instance stops) or "persistent" (the data is kept, can be used later again after the compute instances stops), depending on the configuration of the OpenStack system you are using.

Cinder manages the creation, attaching and detaching of the block devices to instances in OpenStack. Block storage volumes are fully integrated into OpenStack Compute and the Dashboard allowing for cloud users to manage their own storage on demand. Data in volumes are replicated and also backed up through snapshots. In addition, snapshots can be restored or used to create a new block storage volume.

1.2 File-Based Storage

File-based storage inherits from the traditional file system architecture, considers data as files that are maintained in a hierarchical structure. It is the most common storage model and is relatively easy to implement and use. In big data scenario, a file-based storage system could be built on some other low-level abstraction (such as Block-based and Object-based model) to improve its performance and scalability.

1.2.1 Architecture

The file-based storage paradigm is shown in Fig. 4. File paths are organized in a hierarchy and are used as the entries for accessing data in the physical storage. For a big data scenario, distributed file systems (DFS) are commonly used as basic storage systems. Figure 5 shows a typical architecture of a distributed file system which normally contains one or several name nodes and a bunch of data nodes. The name node is responsible for maintaining the file entries hierarchy for the entire system while the data nodes are responsible for the persistence of file data.

In a file based system, a user would need to know of the namespaces and paths in order to access the stored files. For sharing files across systems, the path or namespace of a file would include three main parts: the protocol, the domain name and the path of the file. For example, a HDFS [15] file can be indicated as: "[hdfs://][ServerAddress:ServerPort]/[FilePath]" (Fig. 6).

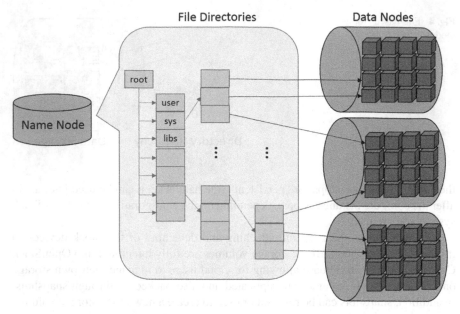

Fig. 5 Architecture of distributed file systems

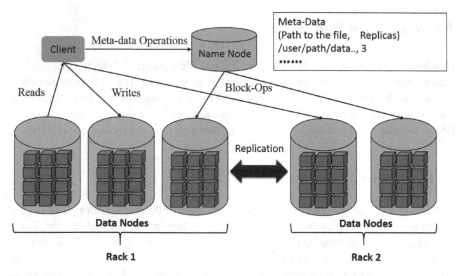

Fig. 6 Architecture of Hadoop distributed file systems

For a distributed infrastructure, replication is very important for providing fault tolerance in file-based systems. Normally, every file has multiple copies stored on the underlying storage nodes. And if one of the copies is lost or failed, the name node can automatically find the next available copy to make the failure transparent for users.

1.2.2 NFS-Family

Network File System (NFS) is a distributed file system protocol originally developed by Sun Microsystems. Basically, A Network File System allows remote hosts to mount file systems over a network and interact with those file systems as though they are mounted locally. This enables system administrators to consolidate resources onto centralized servers on the network. NFS is built on the Open Network Computing Remote Procedure Call (ONC RPC) system. NFS has been widely used in Unix and Linux-based operating systems and also inspired the development of modern distributed file systems. There have been three main generations (NFSv2, NFSv3 and NFsv4) for the NFS protocol due to the continuous development of storage technology and the growth of user requirements.

NFS consists of a few servers and more clients. The client remotely accesses the data that is stored on the server machines. In order for this to function properly, a few processes have to be configured and running. NFS is well-suited for sharing entire file systems with a large number of known hosts in a transparent manner. However, with ease-of-use comes a variety of potential security problems. Therefore, NFS also provides two basic options for access control of shared files:

- First, the server restricts which hosts are allowed to mount which file systems either by IP address or by host name.
- Second, the server enforces file system permissions for users on NFS clients in the same way it does for local users.

1.2.3 HDFS

HDFS (Hadoop Distributed File System) [15] is an open source distributed file system written in Java. It is the open source implementation of Google File System (GFS) and works as the core storage for Hadoop ecosystems and the majority of the existing big data platforms. HDFS inherits the design principles from GFS to provide highly scalable and reliable data storage across a large set of commodity server nodes [16]. HDFS has demonstrated production scalability of up to 200 PB of storage and a single cluster of 4500 servers, supporting close to a billion files and blocks. Basically, HDFS is designed to serve the following goals:

- Fault detection and recovery: Since HDFS includes a large number of commodity hardware, failure of components is expected to be frequent. Therefore, HDFS have mechanisms for quick and automatic fault detection and recovery.
- Huge datasets: HDFS should have hundreds of nodes per cluster to manage the applications having huge datasets.
- Hardware at data: A requested task can be done efficiently, when the computation takes place near the data. Especially where huge datasets are involved, it reduces the network traffic and increases the throughput.

As shown in Fig. 6, the architecture of HDFS consists of a name node and a set of data nodes. Name node manages the file system namespace, regulates the access to files and also executes some file system operations such as renaming, closing, etc. Data node performs read-write operations on the actual data stored in each node and also performs operations such as block creation, deletion, and replication according to the instructions of the name node.

Data in HDFS is seen as files and automatically partitioned and replicated within the cluster. The capacity of storage for HDFS grows almost linearly by adding new data nodes into the cluster. HDFS also provides an automated balancer to improve the utilization of the cluster storage. In addition, recent versions of HDFS have introduced a backup node to solve the problem caused by single-node failure of the primary name node.

1.3 Object-Based Storage

The object-based storage model was firstly introduced on Network Attached Secure devices [17] for providing more flexible data containers objects. For the past decade, object-based storage has been further developed with further investments being made by both system vendors such as EMC, HP, IBM and Redhat, etc. and cloud providers such as Amazon, Microsoft and Google, etc.

In the object-based storage model, data is managed as objects. As shown in Fig. 7, every object includes the data itself, some meta-data, attributes and a globally unique object identifier (OID). Object-based storage model abstracts the lower layers of storage away from the administrators and applications. Object storage systems can be implemented at different levels, including at the device level, system level and interface level.

Data is exposed and managed as objects which includes additional descriptive meta-data that can be used for better indexing or management. Meta-data can be anything from security, privacy and authentication properties to any applications associated information.

Fig. 7 Object-based storage model

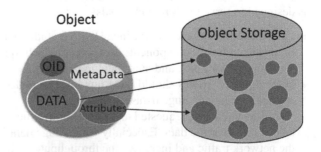

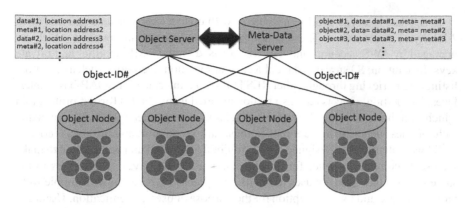

Fig. 8 Architecture of object-based storage

1.3.1 Architecture

The typical architecture of an object-based storage system is shown in Fig. 8. As we can see from the figure, the object-based storage system normally uses a flat namespace, in which the identifier of data and their locations are usually maintained as key-value pairs in the object server. In principle, the object server provides location-independent addressing and constant lookup latency for reading every object. In addition, meta-data of the data is separated from data and is also maintained as objects in a meta-data server (might be co-located with the object server). As a result, it provides a standard and easier way of processing, analyzing and manipulating of the meta-data without affecting the data itself.

Due to the flat architecture, it is very easy to scale out object-based storage systems by adding additional storage nodes to the system. Besides, the added storage can be automatically expanded as capacity that is available for all users. Drawing on the object container and meta-data maintained, it is also able to provide much more flexible and fine-grained data policies at different levels, for example, Amazon S3 [18] provides bucket level policy, Azure [19] provides storage account level policy, Atmos [20] provides per-object policy.

1.3.2 Amazon S3

Amazon S3 (Simple Storage Service) [18] is a cloud-based object storage system offered by Amazon Web Services (AWS). It has been widely used for online backup and archiving of data and application programs. Although the architecture and implementation of S3 is not published, it has been designed with high scalability, availability and low latency at commodity costs.

In S3, data is stored as arbitrary objects with up to 5 terabytes data size and up to 2 kilobytes of meta-data. These data objects are organized into buckets which are managed by AWS accounts and authorized based on the AMI identifier and private keys. In addition, S3 supports data/objects manipulation operations such as creation, listing and retrieving through either RESTful HTTP interfaces or SOAP-based interfaces. In addition, objects can also be downloaded using the BitTorrent protocol, in which each bucket is served as a feed. S3 claims to guarantee 99.9% SLA by using technologies such as redundant replications, failover support and fast data recovery.

S3 was intentionally designed with a minimal feature set and was created to make web-scale computing easier for developers. The service gives users access to the same systems that Amazon uses to run its own Web sites. S3 employs a simple web-based interface and uses encryption for the purpose of user authentication. Users can choose to keep their data private or make it publicly accessible and even encrypt data prior to writing it out to storage.

1.3.3 EMC Atmos

EMC Atmos [20] is a object-based storage services platform developed by EMC Corporation. Atmos can be deployed as either a hardware appliance or a software in a virtual environment such as cloud. Atmos is designed based on the object storage architecture aiming to manage petabytes of information and billions of objects across multiple geographic locations yet be used as a single system. In addition, Atmos supports two forms of replication: synchronous replication and asynchronous replication. For a particular object, both types of replication can be specified, depending on the needs of the application and the criticality of the data.

Atmos can be used as a data storage system for custom or packaged applications using either a REST or SOAP data API, or even traditional storage interfaces like NFS and CIFS. It stores information as objects (files + metadata) and provides a single unified namespace/object-space which is managed by user or administrator-defined policies. In addition, EMC has recently added support for the Amazon S3 application interfaces that allow for the movement of data from S3 to any Atmos public or private cloud.

1.3.4 OpenStack Swift

Swift [21] is a scalable, redundant and distributed object storage system for the OpenStack cloud platform. With the data replication service of OpenStack, objects and files in Swift are written to multiple nodes that are spread throughout the cluster in the data center. Storage in Swift can scale horizontally simply by adding new servers. Once a server or hard drive fails, Swift automatically replicates its content from other active nodes to new locations in the cluster. Swift uses software logic to ensure data replication and distribution across different devices. In addition, inexpensive commodity hard drives and servers can be used for Swift clusters (Fig. 9).

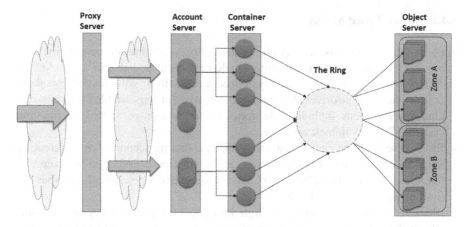

Fig. 9 Architecture of swift object store

The architecture of Swift consists of several components including proxy server, account servers, container servers and object servers:

- The Proxy Server is responsible for tying together the rest of the Swift architecture. It exposes the Swift API to users and streams objects to and from the client based on requests.
- The Object Server is a simple blob storage server which handles storage functions such as the retrieval and deletion of objects stored on local devices.
- The Container Server is responsible to handle the listings of objects. Objects in Swift are logically organized in specific containers. The listings relations are stored as sqlite database files and replicated across the cluster.
- The Account Server is similar to the Container Server except that it is responsible for the listings of containers rather than objects.

Objects in Swift are accessed through the REST interfaces, and can be stored, retrieved, and updated on demand. The object store can be easily scaled across a large number of servers. Swift uses rings to keep track of the locations of partitions and replicas for objects and data.

1.4 Comparison of Storage Models

In practice, there is no perfect model which can suit all possible scenarios. Therefore, developers and users should choose the storage models according to their application requirements and context. Basically, each of the storage model that we have discussed in this section has its own pros and cons.

1.4.1 Block-Based Model

- Block-based storage is famous for its flexibility, versatility and simplicity. In a block level storage system, raw storage volumes (composed of a set of blocks) are created, and then the server-based system connects to these volumes and uses them as individual storage drives. This makes block-based storage usable for almost any kind of applications, including file storage, database storage, virtual machine file system (VMFS) volumes, and more.
- Block-based storage can be also used for data-sharing scenarios. After creating block-based volumes, they can be logically connected or migrated between different user spaces. Therefore, users can use these overlapped block volumes for sharing data between each other.
- Block-based storage normally has high throughput and performance and is generally configurable for capacity and performance. As data is partitioned and maintained in fix-sized blocks, it reduces the amount of small data segments and also increases the IO throughput due to more sequential reading and writing of data blocks.
- However, block-based storage is complex to manage and not easy to use due to the lack of information (such as meta-data, logical semantics and relation between data blocks) when compared with that of other storage models such as file-based storage and object-based storage.

1.4.2 File-Based Model

- File storage is easy to manage and implement. It is also less expensive to use than block-storage. It is used more often on home computers and in smaller businesses, while block-level storage is used by larger enterprises, with each block being controlled by its own hard drive and managed through a server-based operating system.
- File-based storage is usually accessible using common file level protocols such as SMB/CIFS (Windows) and NFS (Linux, VMware). At the same time, files contain more information for management purposes, such as authentication, permissions, access control and backup. Therefore, it is more user-friendly and maintainable.
- Due to the hierarchical structure, file-based storage is less scalable the the number of files becomes extremely huge. It becomes extremely challenging to maintain both low-latency and scalability for large scale distributed file systems such as NFS and HDFS.

1.4.3 Object-Based Model

- Object-based storage solves the provisioning management issues presented by the expansion of storage at very large scale. Object-based storage architectures can be scaled out and managed simply by adding additional nodes. The flat name space

Table 1 Comparison of storage models

Storage Model	Data model	Indexing	Scalability	Consistency
Block-based	Blocks with fixed size	Block Id	Flat	Strong
File-based	Files	File path	Hierarchy	Configurable
Object-based	Objects and meta data size not fixed	Block Id or URI	Flat	Configurable

organization of the data, in combination with the expandable metadata functionality, facilitate this ease of use. Object storage are commonly used for the storage of large scale unstructured data such as photos in Facebook, songs on Spotify and even files in Dropbox.

- Object storage facilitates the storage for unstructured data sets where data is generally read yet not written-to. Object storage generally does not provide the ability of incrementally editing one part of a file (as block storage and file storage do). Objects have to be manipulated as a whole unit, requiring the entire object to be accessed, updated then re-written into the physical storage. This may cause some performance implications. It is also not recommended to use object storage for transactional data because of the eventual consistency model.

1.4.4   Summary of Data Storage Models

As a result, the main features of each storage model can be summarized as shown in Table 1. Generally, block-based storage has a fixed size for each storage unit while file-based and object-based models can have various sizes of storage unit based on application requirements. In addition, file-based models use the file-based directory to locate the data whilst block-based and object-based models both reply on a global identifier for locating data. Furthermore, both block-based and object-based models have flat scalability while file-based storage may be limited by its hierarchical indexing structure. Lastly, block-based storage can normally guarantee a strong consistency while for file-based and object-based models the consistency model is configurable for different scenarios.

2   Data Models

A data model illustrates how the data elements are organized and structured. It also represents the relations among different data elements. A data model is at the core for data storage, analytic and processing of contemporary big data systems. According to different data models, current data storage systems can be categorized into two big families: relational-stores (SQL) and NoSQL stores.

For past decades, relational database management systems (RDBMS) have been considered as the dominant solution for most of the data persistence and management service. However, with the tremendous growth of the data size and data variety, the traditional strong consistency and pre-defined schema for relational databases have limited their capability for dealing with large scale and semi/un-structured data in the new era. Therefore, recently, a new generation of highly-scalable, more flexible data store systems has emerged to challenge the dominance of relational databases. This new groups of systems are called NoSQL (Not only SQL) systems. The principle underneath the advance of NoSQL systems is actually a trade-off between the CAP properties of distributed storage systems.

As we know from the CAP theorem [22], a distributed system can only guarantee at most two out of the three properties: Consistency, Availability and Partition tolerance. Traditional RDBMS normally provide a strong consistency model based on their ACID [23] transaction model while NoSQL systems try to sacrifice some extent of consistency for either higher availability or better partition tolerance. As a result, data storage systems can be categorized into three main groups base on their CAP properties:

- CA systems, which are consistent and highly available yet not partition-tolerant.
- CP systems, which are consistent and partition-tolerant but not highly available.
- AP systems, which are highly available and partition-tolerant but not strictly consistent.

In the remaining of this section, we will discuss about major NoSQL systems and scalable relation databases, respectively.

2.1 NoSQL (Not only SQL)

Rational databases management systems (such as MySQL [24], Oracle [25], SQL Server [26] and PostgreSQL [27]) have been dominating the database community for decades until they face the limitation of scaling to very large scale datasets. Therefore, recently a group of database systems which abandoned the support of ACID transactions (Atomicity, Consistency, Isolation and Durability, which are key principles for relational databases) has emerged to tackle the challenge of big data. The group of these database systems are named as NoSQL (Not only SQL) systems which aims to provide horizontal scalability towards any large scale of datasets. A majority of NoSQL systems are originally designed and built to support distributed environments with the need to improve performance by adding new nodes to the existing ones. Recall that the CAP theorem states that a distributed system can only choose at most two of the three properties: Consistency, Availability and Partition tolerance. One key principle of NoSQL systems is to compromise the consistency to trade for high availability and scalability. Basically the implementation of a majority NoSQL systems share a few common design features as below:

- High scalability, which requires the ability to scale up horizontally over a large cluster of nodes;
- High availability and fault tolerance, which is supported by replicating and distributing data over distributed servers;
- Flexible data models, with the ability to dynamically define and update attributes and schemas;
- Weaker consistency models, which abandoned the ACID transactions and are usually referred as BASE models (Basically Available, Soft state, Eventually consistent) [28];
- Simple interfaces, which are normally single call-level interfaces or protocol in contrast to the SQL bindings.

For different scenarios and focus of usage, more NoSQL systems have developed in both industry and academia. Based on different data model ported, these NoSQL systems can be classified as three main groups: Key-Value Stores, Document stores and Extensible-Record/Column-based stores.

2.1.1 Key-Value Stores

Key-value stores use a simple data model, in which data are considered as a set Key-Value pairs, in which, keys are unique IDs for each data and also work as indexes during accessing the data (Fig. 10). Values are attributes or objects which contains the actual information of data. Therefore, these systems are called key-value stores. The data in key-value stores can be accessed using simple interfaces such as insert, delete and search by key. Normally, secondary keys and indexes are not supported. In addition, these systems also provide persistence mechanism as well as replication, locking, sorting and other features.

- Redis

Redis [29] is an open source Key-Value store system written in C. It supports a fairly rich data model for stored values. Values can be lists, sets and structures in addition to basic types (Integer, String, Double and Boolean). Apart from ordinary operations such as reading and writing, Redis also provides a few atomic modifier such as increment of a numeric value by one, adding an element to a list, etc. Redis mainly stores data in memory, which ensures high performance. To provide persistence, data

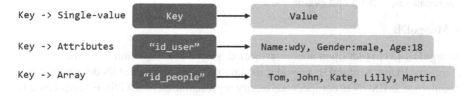

Fig. 10 Data model of Key-value stores

snapshots and modification operations are written out to disk for failure tolerance. Redis can scale out by distributing data (normally achieved at client side) among multiple Redis servers and providing asynchronous data replication through master-slaves.

- Memcached family

Memcached [30] is the first generation of Key-Value stores initially working as cache for web servers then being developed as a memory based Key-value store system. Memcahed has been enhanced to support features such as high availability, dynamic growth and backup. The original design of Memcached does not support persistence and replication. However, its follow-up variation: Membrain and Membase do include these features which make them more like storage systems.

- DynamoDB

DynamoDB [10] is a NoSQL store service provided by Amazon. Dynamo supports a much more flexible data model especially for key-value stores. Data in Dynamo are stored as tables each of which has a unique primary ID for accessing. Each table can have a set for attributes which are schema free and scalar types and sets are supported. Data in Dynamo can be manipulated by searching, inserting and deletion based on the primary keys. In addition, conditional operation, atomic modification and search by non-key attributes are also supported (yet inefficient), which makes it also closer to that of a document store. Dynamo provides a fast and scalable architecture where sharding and replication are automatically performed. In addition, Dynamo provides support for both eventually consistency and strong consistency for reads while strong consistency degrades the performance. E.g. Redis, Memcached, DynamoDB (also support document store).

2.1.2 Document Stores

Document stores provide a more complex data structure and richer capabilities than Key-Value systems. In document stores, the unit of data is called a document which is actually an object that can contains an arbitrary set of fields, values and even nested objects and arrays (Fig. 11). Document stores normally do not have predefined schemas for data and support search and indexing by document fields and attributes. Unlike key-value stores, they generally support secondary indexes, nested objects and lists. Additionally, some of them can even support queries with constraints, aggregations, sorting and evaluations.

- MongoDB

MongoDB [31] is an open source project developed in C++ and supported by the company 10gen. MongoDB provides its data model based on JSON documents and maintained as BSON (a compact and binary representation of JSON). Each document in MongoDB has a unique identifier which can be automatically generated by the

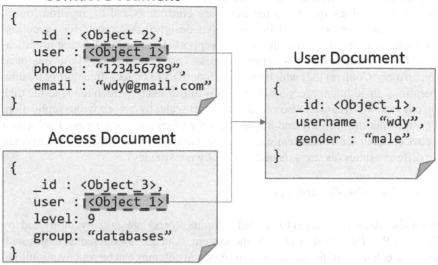

Fig. 11 Data model of document stores

server or manually created by users. A document contains an arbitrary set of fields which can be either arrays or embedded documents. MongoDB is schema free and even documents in the same collection can have completely different fields. Documents in MongoDB are manipulated based on the JSON representation using search, insertion, deletion and modification operations. Users can find or query documents by writing them as expressions of constraints of fields. In addition, complex operations such as sorting, iteration and projecting are supported. Moreover, users can perform MapReduce-like program and aggregation paradigms on documents, which makes it possible to execute more complicated analytic queries and programs. Documents can be completely replaced and any parts of their fields can also be manipulated and replaced.

Indexes of one or more fields in a collection are supported to speed up the searching queries. In addition, MongoDB scales up by distributing documents of a collection among nodes based on a sharding key. Replication between master and slaves with different consistency models depending on whether reading from secondary nodes are allowed and how many nodes are required to reach a confirmation.

- CouchDB

CouchDB [32] is an Apache open source project written in Erlang. It is a distributed documents-based store that manipulates JSON documents. CouchDB is schema free, documents are organized as collections. Each document contains a unique identifier and a set of fields which can be scalar fields, arrays and embedded documents.

Queries on CouchDB documents are called views which are MapReduce-based JavaScript functions specifying the matching constraints and aggregation logics. These functions are structured into so-called designed documents for execution. For these views, B-Tree based indexes are supported and updated during modifications. CouchDB also supports optimistic locks based on MVCC (Multi-Versioned Concurrency Control) [33] which enables CouchDB to be lock-free during reading operations. In addition, every modification is immediately written down to the disk and old versions of data are also saved. CouchDB scales by asynchronous replication, in which both master-slave and master-master replication is supported. Each client is guaranteed to see a consistent state of the database, however, different clients may see different states (as strengthened eventually consistency).

2.1.3   Extensible-Record Stores

Extensible-Record Stores (also called column stores) are initially motivated by Google's Big Table project [34]. In the system, data are considered as tables with rows and column families in which both rows and columns can be split over multiple nodes (Fig. 12). Due to this flexible and loosely coupled data model, these systems support both horizontal and vertical partitioning for the scalability purposes. In addition, correlated fields/columns (named as column families) are located on the same partition to facilitate query performance. Normally column families are predefined before creating a data table. However, this is not a big limitation as new columns and fields can always be dynamically added to the existing tables.

- BigTable

BigTable [34] is introduced by Google in 2004 as a column store to support various Google services. Big Data is built on Google File System (GFS) [35] and can be easily scaled up to hundreds and thousands of nodes maintaining Terabytes and Petabytes scale of data.

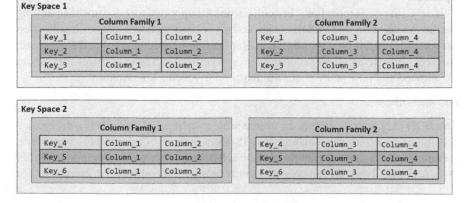

Fig. 12 Data model of extensible-record stores

BigTable is designed based on an extended table model which maintains a three dimensional mapping from row key, column key and timestamps to associated data. Each table is divided into a set of small segments called tablets based on row keys and column keys. Tablets are also the unit for performing load balancing when needed. Columns are grouped as column families which are collocated in the disk and optimized for reading correlated fields in a table. Each column family may contain an arbitrary set of columns and each column of a record in the table can have several versions of data marked and ordered by timestamps.

BigTable supports operations including writing and deleting values, reading rows, searching and scanning a subset of data. In addition, it supports creation and deletion of tables and column families and modification of meta-data (such as access rights). BigTable also supports asynchronous replication between clusters and nodes to ensure an eventually consistency.

- Hbase

HBase [36] is an Apache open source project and is developed in Java based on the principles of Google's BigTable. HBase is built on the Apache Hadoop Framework and Apache Zookeeper [37] to provide a column-store database. As HBase is inherited from BigTable, they share a lot of features in both data model and architecture. However, HBase is built on HDFS (Hadoop Distributed File System) instead of GFS and it uses ZooKeeper for cluster coordination compared with using Chubby in BigTable. HBase puts updates in the memory and periodically writes them to disk. Row operations are atomic with the support of row-level transactions. Partitions and distributions are transparent to users and there is no client-side hashing like some of the other NoSQL systems. HBase provides multiple master nodes to tackle the problem of single-point failure of the master node. Compared with BigTable, HBase does not have location groups but only that of column families. In addition, HBase does not support secondary indexing, therefore, queries can only be performed based on primary keys or by fully scanning the table. Nevertheless, additional indexes can be manually created using extra tables.

- Casandra

Casandra [38] is an open source NoSQL database initially developed by Facebook in Java. It combines the ideas of both BigTable and Dynamo and it is now open sourced under the Apache license. Casandra shares the majority of the features as other extensible record stores (column stores) in both data model and functionality. It has column groups and updates are cached in the memory first then flushed to disk. However, there still some differences:

- Casandra have columns which are the minimum unit for storage and super columns which contains a set of columns to provide additional nestedness.
- Casandra is fully decentralized of which every node in the cluster is considered equal and performs identical functions. In Casandra, a leader is selected based on the Gossip Protocol; failures are detected by using the phi accrual algorithm

and scalability is achieved by Consistent Hashing. All the process that have been mentioned before: leader selection, failure detection and recovery are performed automatically.

- Casandra only supports the eventually consistency model. It provides quorum reads to ensure clients get the latest data from majority of the replicas. Writes in Casandra are atomic within a column family and some extent of versioning and conflict resolution are supported.

2.1.4 Summary of Major Data Store Systems

Table 2 shows the comparison of existing data store systems. As we can see from the table, Key-Value stores generally trade-off consistency for availability and partition-tolerance while Document stores normally provide different levels of consistency based on different requirements of availability and partition-tolerance. In addition, we can see that the majority of NoSQL data stores provide at least eventual consistency and use MVCC for concurrent controlling. Most of the NoSQL data stores still use master-slave architecture while some more advanced systems (Casandra, etc.) are built on a decentralized, share-nothing architecture.

2.2 Relational-Based

Traditional DBMSs are designed based on the relational paradigm in which all data is represented in terms of tuples and grouped into relations. The purpose of the relational model is to provide a declarative method for specifying data and queries (SQL). Unlike NoSQL systems, these databases have a complete pre-defined schema and SQL interfaces with the support of ACID transactions. However, the ever increasing need for scalability in order to store very large datasets have brought about some key challenges for traditional DBMSs. Therefore, further performance improvements have been made to relational databases to provide comparable scalability with NoSQL databases. Those improvements are based on two main provisos:

- Small-scope operations: As large scale relational operations like Join cannot scale well with partitioning and sharding, these operations are limited to smaller scopes to achieve better performance.
- Small-scope transactions: Note that, transaction is also one key reason to cause the scalability problem for relational databases. Therefore, limiting the scope of transactions can significantly improve the scalability of DBMS clusters.

In terms of product systems, based on their model of usage, they can be classified into two groups: Scalable Rational Systems and Database-as-a-service (DaaS).

Table 2 Comparison of major data store systems

System	Data model	CAP	Consistency	Concurrent control	Partitioning	Data storage	Distributed arch
Redis	Key-value	AP	Eventual/Strong	Locks	Client-side partitioning	RAM + Disk	Multi-data nodes
Memcached	Key-value	AP	Eventual	Locks	None	RAM	Master-slave
Dynamo	Key-value or document	AP	Eventual/Strict	MVCC	Key-based sharding	RAM + Disk	Master-slave (multiple masters)
Voldemort	Key-value	AP	Read-repair	MVCC	Consistent hashing	RAM or BDB	Master-slave
MongoDB	Document	CP	Eventual/Strong	Multi-granularity locking	Key-based sharding	Disk	Master-save
CouchDB	Document	CP	Eventual	MVCC	Consistent hashing	Disk	Master-save
Terrastore	Document	CP	Eventual	MVCC + Locks	Consistent hashing	RAM	Master-save
BigTable	Column-family	CP	Eventual	Locks and stamps	Range-based	Disk	Master-save
HBase	Column-family	CP	Eventual	Optimistic locking with MVCC	Range-based	RAM + Disk	Master-save (backup Master)
Casandra	Column-family	AP/CP	Eventual/Strong	MVCC	Consistent hashing	Disk	Decentralized
MySQL cluster	Relational tables	CP	Strong	ACID	Row-based hashing	Disk	Master-slave
VoltDB	Relational tables	CP	Strong	ACID	Horizontal partitioning	RAM + Disk	Share-nothing

2.2.1 Scalable Rational Systems

With the requirement for dealing with large scale datasets, optimizations and improvements have been done on traditional DBMS systems such as MySQL. And several new products have also come out with the promise to have good per-node performance as well as scalability.

- MySQL Cluster

MySQL Cluster [39] has been part of the mainline MySQL releases as an extension that supports distributed, Multi-master and ACID compliant databases. MySQL Cluster automatically shards data across multiple nodes to scale out read and write operations on large datasets. It can be accessed through both SQL and NoSQL APIs.

The synchronous replication MySQL Cluster is based on a two-phase commit mechanism to guarantee the data consistency on multiple replicas. It also automatically creates node groups among the replicas to protect against data loss and provide support for swift failover.

MySQL Cluster is implemented as fully distributed databases with multi-master, each of them can accept write operations and updates are instantly visible for all the nodes within the cluster. Tables in MySQL Cluster are automatically partitioned among all the data nodes based on a Hashing algorithm of the primary key of each table. In addition, sharding, load balancing, failing over and recovery in MySQL Cluster are transparent to users, so it is generally easy to setup.

- VoltDB

VoltDB [40] is an open source SQL-based in-memory database which is designed for high performance as well as scalability. VoltDB is also ACID-compliant and built on a shared nothing architecture. Tables are partitioned over multiple nodes and data can be accessed through any server. Partitions are replicated among different nodes and data snapshots are also supported to provide fast failover and data recovery. VoltDB is designed as a database that can be fit into distributed RAM on the servers, so generally operations do not need to wait for the disk IO. All VoltDB SQL calls are made through stored procedures each of which is considered as one transaction. VoltDB is fully ACID compliant of which data is durable on the disk and ensured by continuous snapshots and command logging. VoltDB has been further developed in recent releases to be able to be integrated with Big Data ecosystems such as Hadoop, HDFS, Kafka, etc. And it is also extended to support geo-spatial query and data models.

- Vertica Analytics Platform

Vertica Analytics Platform (Vertica for short) [41] is a cloud-based, column-oriented, distributed database management system. It is designed for the management of large, fast-growing volumes of data as well as supporting highly optimized query performance for data warehouses and other query-intensive applications. Vertica claims to dramatically improve query performance over traditional relational database systems along with high-availability and petabyte-scalability on commodity enterprise servers. The design features of Vertica include:

- Column-oriented store: Vertica leverages the columnar data store model to offer significant improvement on the performance of sequential record access at the expense of common transactional operations such as single record retrievals, updates, and deletes. The column-oriented data model also improves the performance of I/O, storage footprint and efficiency when it comes to analytic workloads due to the lower volume of data during loading.
- Real-time loading and query: Vertica is designed with a novel time travel transactional model that ensures extremely high query concurrency. Vertica is able to load data up to 10x faster than traditional row-stores by leveraging on its design of simultaneously loading data in the system. In addition, Vertica is purposely built with a hybrid in-memory/on-disk architecture to ensure near-real-time availability of information.
- Advanced database analytics: Vertica offers a set of Advanced In-Database Analytics functionality so that users can conduct their analytics computations within the database rather than extracting data to a separate environment for processing. The in-database analytics mechanism is especially critical for applying computation on large scale data sets with the size range from terabytes to petabytes and beyond.
- Data compression: Vertica operates on encoded data which dramatically improves analytic performance by reducing CPU, memory, and disk I/O at processing time. Due to the aggressive data compression, Vertica can reduce the original data size to 1/5th or 1/10th its original size even with high-availability redundancy.
- Massively Parallel Processing (MPP) [42] support: Vertica delivers a simple, but highly robust and scalable MPP solution which offers linear scaling and native high availability on industry standard parallel hardware.
- Shared nothing architecture: Vertica is designed in a shared nothing architecture which, on one hand, reduces system contention for shared resources and on the other hand allows gradual degradation of performance when the system encounters both software or hardware failures.

2.2.2 Database-as-a-Service (DaaS)

Database-as-a-service is a service model where a third party service provider hosts scalable relational databases as services by applying the multi-tenancy technology on database systems. Those services relieve their users from the need to purchase and maintain expensive hardware and software for provisioning database functionality. Three main approaches including Shared Server, Shared Process and Shared Table are used to avoid the problem of under-utilization of data center resources. In practice, the shared-server model is most commonly used by DaaS providers as it is considered the most efficient way for providing isolation for the data of each tenant.

- Amazon RDS

Amazon RDS (Relational Database Service) [9] is a DaaS service provided by Amazon Web Services (AWS). It is a cloud service to simplify setup, configuration, operation and auto-scaling of relational databases for use by applications. It also helps in the sake of backing up, patching and recovery of users database instances. Amazon RDS provides asynchronous replication of data across multiple nodes to improve the scalability of reading operations for relational databases. It also provisions and maintains replicas across availability zones to enhance the availability of database services. For flexibility considerations, Amazon RDS supports various types of databases including MySQL, Oracle, PostgreSQL and Microsoft SQL, etc.

- Microsoft Azure SQL

Microsoft also released their SQL Azure [43] as a cloud based service for relational databases. Azure SQL is, namely, built on the Azure cloud infrastructure with Microsoft SQL Server as its databases backend. It provides highly available, multi-tenant database service with the support of T-SQL, native ODBC and ADO.NET for data access. Azure SQL provides high availability by storing multiple copies of databases with elastic scaling and rapid provisioning. It also provides self-management functions for database instances and predictable performance during scaling.

- Google Cloud SQL

Google Cloud SQL [44] is another fully managed DaaS service hosted on Google Cloud Platform. It provides easy setup, management, maintenance and administrations for MySQL databases in cloud environments. Google Cloud SQL provides automated replication, patch management, and database management with effortless scaling based on users' demand. For reliability, Google Cloud SQL also replicates databases across multiple zones with automated failover and provides backups and point-in-time recovery automatically.

- Other DaaS Platforms

Following the main stream of cloud-based solutions, more and more database software providers have been migrating their products as cloud services. There are various DaaS provided by different venders including:

- Xeround [45] offers its own elastic database service based on MySQL across a variety of cloud providers and platforms. The Xeround service allows for high availability and scalability and it can work across a variety of cloud providers including AWS, Rackspace, Joyent, HP, OpenStack and Citrix platforms.
- StormDB runs its fully distributed, relational database on bare-metal servers, meaning there is no virtualization of machines. Despite running on bare metal servers, customers still share clusters of servers with promises of isolation among customer databases. StormDB also automatically shards databases in its cloud environments.

Table 3 Comparison for different data models

Name	Data model	CAP	Consistency	Scalability	Schema	Transaction
Key-value stores	Key-values	AP	Loose	High	Key with scalar value	BASE
Record stores	Column families	AP/CP	Loose	High	Rows with scalar columns	BASE
Document stores	Documents JSON-like	AP/CP	Loose	High	Schema free	BASE
Rational databases	Relational tables	C/CP	Strict	Low	Row-based predefined schema	ACID

- EnterpriseDB [46] provides its cloud database service mainly based on the open source PostgreSQL databases. The Management Console in its cloud service provisions PostgreSQL databases with database compatibility with Oracle. Users can choose to deploy their database in single instances, high availability clusters, or development sandboxes for Database-as-a-Service environments. With EnterpriseDB's Postgres Plus Advanced Server, enterprise users can deploy their applications written for Oracle databases through EnterpriseDB, which runs in cloud platforms such as Amazon Web Services and HP.

2.3 Summary of Data Models

A comparison of the different data models is shown in Table 3. Basically, NoSQL data models:Key-Value, Column families and Document-based models has looser consistency constraints as a trade-off for high availability and/or partition-tolerance in comparison with that of relational data models. In addition, NoSQL data models have more dynamic and flexible schemas based on their data models while relational databases use predefined and row-based schemas. Lastly, NoSQL databases apply the BASE models while relational databases guarantee ACID transactions.

References

1. S. Sakr, M. Medhat Gaber (eds.), *Large Scale and Big Data - Processing and Management* (Auerbach Publications, Boston, 2014)
2. S. Sakr, A. Liu, A.G. Fayoumi, The family of mapreduce and large-scale data processing systems. ACM Comput. Surv. **46**(1), 11 (2013)
3. J. Satran, K. Meth, Internet small computer systems interface (iscsi) (2004)

4. SCSI Protocol. Information technologyscsi architecture model5 (sam-5). *INCITS document*, 10
5. S. Hopkins, B. Coile, Aoe (ata over ethernet). *The Brantley Coile Company, Inc., Technical report AoEr11*, 2009
6. ATA Serial. High-speed serialized at attachment. Serial ATA working group, available at www.sata-io.org (2001)
7. EBS Amazon. Elastic block store has launched all things distributed (2008). https://aws.amazon.com/ebs/
8. EC2 Amazon. Amazon elastic compute cloud (amazon ec2), *Amazon Elastic Compute Cloud (Amazon EC2)* (2010)
9. RDS Amazon. Amazon relational database service (amazon rds). https://aws.amazon.com/rds/. Accessed 27 Feb 2016
10. S. Sivasubramanian, Amazon dynamodb: a seamlessly scalable non-relational database service. in *Proceedings of the 2012 ACM SIGMOD International Conference on Management of Data* (ACM, New York, 2012), pp. 729–730
11. Amazon. Amazon cloudsearch service. https://aws.amazon.com/cloudsearch/. Accessed 27 Feb 2016
12. O. Sefraoui, M. Aissaoui, M. Eleuldj, Openstack: toward an open-source solution for cloud computing. Intern. J. Comput. Appl. **55**(3), 38–42 (2012)
13. K. Pepple, Openstack nova architecture. Viitattu **25**, 2012 (2011)
14. OpenStack. Openstack block storage cinder. https://wiki.openstack.org/wiki/Cinder. Accessed 27 Feb 2016
15. K. Shvachko, H. Kuang, S. Radia, R. Chansler, The Hadoop distributed file system. in *IEEE MSST* (2010)
16. S. Sakr, *Big Data 2.0 Processing Systems* (Springer, Switzerland, 2016)
17. K. Goda, Network attached secure device. in *Encyclopedia of Database Systems* (Springer, New York, 2009), pp. 1899–1900
18. S3 Amazon. Amazon simple storage service(amazon s3). https://aws.amazon.com/s3/. Accessed 27 Feb 2016
19. Azure Microsoft. Microsoft azure: Cloud computing platform and services. https://azure.microsoft.com. Accessed 27 Feb 2016
20. Atoms EMC. Atmos - cloud storage, big data - emc. http://australia.emc.com/storage/atmos/atmos.htm. Accessed 27 Feb 2016
21. Swift OpenStack. Openstack swift - enterprise storage from swiftstack. https://www.swiftstack.com/openstack-swift/. Accessed 27 Feb 2016
22. E.A. Brewer, Towards robust distributed systems. in *Proceedings of the PODC*, vol. 7 (2000)
23. J. Gray et al., The transaction concept: virtues and limitations. in *Proceedings of the VLDB*, vol. 81 (1981), pp. 144–154
24. A.B. MySQL, *MySQL: The World's Most Popular Open Source Database* (MySQL AB, 1995)
25. K. Loney, *Oracle Database 10g: The Complete Reference* (McGraw-Hill/Osborne, London, 2004)
26. Microsoft. Sql server 2014. https://www.microsoft.com/en-au/server-cloud/products/sql-server/overview.aspx. Accessed 27 Feb 2016
27. PostgreSQL Datatype. Postgresql: the world's most advanced open source database. http://www.postgresql.org. Accessed 27 Feb 2016
28. D. Pritchett, Base: an acid alternative. Queue **6**(3), 48–55 (2008)
29. J. Zawodny, Redis: lightweight key/value store that goes the extra mile. Linux Mag. **79**, (2009)
30. B. Fitzpatrick, Distributed caching with memcached. Linux J. **2004**(124), 5 (2004)
31. MongoDB Inc. Mongodb for giant ideas. https://www.mongodb.org/. Accessed 27 Feb 2016
32. Apache. Apache couchdb. http://couchdb.apache.org/. Accessed 27 Feb 2016
33. P.A. Bernstein, N. Goodman, Concurrency control in distributed database systems. ACM Comput. Surv. (CSUR) **13**(2), 185–221 (1981)
34. F. Chang, J. Dean, S. Ghemawat, W.C. Hsieh, D.A. Wallach, M. Burrows, T. Chandra, A. Fikes, R.E. Gruber, Bigtable: a distributed storage system for structured data. ACM Trans. Comput. Syst. (TOCS) **26**(2), 4 (2008)

35. S. Ghemawat, H. Gobioff, S.-T. Leung, The google file system. in *ACM SIGOPS Operating Systems Review*, vol. 37 (ACM, Bolton Landing, 2003), pp. 29–43
36. L. George, *HBase: The Definitive Guide* (O'Reilly Media, Inc., Sebastopol, 2011)
37. P. Hunt, M. Konar, F.P. Junqueira, B. Reed, Zookeeper: wait-free coordination for internet-scale systems. in *USENIX Annual Technical Conference*, vol. 8 (2010), p. 9
38. A. Lakshman, P. Malik, Cassandra: a decentralized structured storage system. ACM SIGOPS Oper. Syst. Rev. **44**(2), 35–40 (2010)
39. M. Ronstrom, L. Thalmann, Mysql cluster architecture overview. *MySQL Technical White Paper* (2004)
40. M. Stonebraker, A. Weisberg, The voltdb main memory dbms. IEEE Data Eng. Bull. **36**(2), 21–27 (2013)
41. A. Lamb, M. Fuller, R. Varadarajan, N. Tran, B. Vandiver, L. Doshi, C. Bear, The vertica analytic database: C-store 7 years later. Proc. VLDB Endow. **5**(12), 1790–1801 (2012)
42. F. Fernández de Vega, E. Cantú-Paz, *Parallel and Distributed Computational Intelligence*, vol. 269 (Springer, Berlin, 2010)
43. Microsoft. Sql database - relational database service. https://azure.microsoft.com/en-us/services/sql-database/. Accessed 27 Feb 2016
44. Google. Cloud sql - mysql relational database. https://cloud.google.com/sql/. Accessed 27 Feb 2016
45. Xeround. Xeround. https://en.wikipedia.org/wiki/Xeround. Accessed 27 Feb 2016
46. EnterpriseDB. Enterprisedb - the postgres database company. https://www.enterprisedb.com. Accessed 27 Feb 2016

Big Data Programming Models

Dongyao Wu, Sherif Sakr and Liming Zhu

Abstract Big Data programming models represent the style of programming and present the interfaces paradigm for developers to write big data applications and programs. Programming models normally the core feature of big data frameworks as they implicitly affects the execution model of big data processing engines and also drives the way for users to express and construct the big data applications and programs. In this chapter, we comprehensively investigate different programming models for big data frameworks with comparison and concrete code examples.

A programming model is the fundamental style and interfaces for developers to write computing programs and applications. In big data programming, users focus on writing data-driven parallel programs which can be executed on large scale and distributed environments. There have been a variety of programming models being introduced for big data with different focus and advantages. In this chapter, we will discuss and compare the major programming models for writing big data applications based on the taxonomy which is illustrated in Fig. 1.

1 MapReduce

MapReduce [24] the current defacto framework/paradigm for writing data-centric parallel applications in both industry and academia. MapReduce is inspired by the commonly used functions - Map and Reduce in combination with the divide-and-

D. Wu (✉) · S. Sakr · L. Zhu
Data61, CSIRO, Sydney, NSW, Australia
e-mail: Dongyao.Wu@data61.csiro.au

D. Wu · S. Sakr · L. Zhu
School of Computer Science and Engineering, University of New South Wales,
Sydney, NSW, Australia

S. Sakr
King Saud Bin Abdulaziz University for Health Sciences, National Guard,
Riyadh, Saudi Arabia

© Springer International Publishing AG 2017   31
A.Y. Zomaya and S. Sakr (eds.), *Handbook of Big Data Technologies*,
DOI 10.1007/978-3-319-49340-4_2

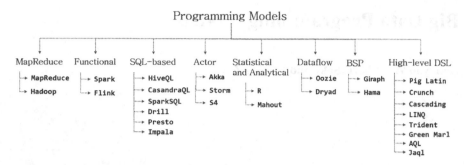

Fig. 1 Taxonomy of programming models

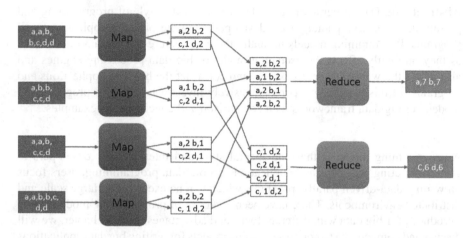

Fig. 2 MapReduce paradigm

conquer [22] parallel paradigm. For a single MapReduce job, users implement two basic procedure objects Mapper and Reducer for different processing stages as shown in Fig. 2. Then the MapReduce program is automatically interpreted by the execution engine and executed in parallel in a distributed environments. MapReduce is considered as a simple yet powerful enough programming model to support a variety of the data-intensive programs [43, 44].

1.1 Features

- *Map and Reduce functions.* A MapReduce program contains a Map function doing the parallel transformation and a Reduce function doing the parallel aggregation and summary of the job. Between Map and Reduce an implied Shuffle step is responsible for grouping and sorting the Mapped results and then feeding it into the Reduce step.
- *Simple paradigm.* In MapReduce programming, users only need to write the logic of Mapper and Reducer while the logic of shuffling, partitioning and sorting is

automatically done by the execution engine. Complex applications and algorithms can be implemented by connecting a sequence of MapReduce jobs. Due to this simple programming paradigm, it is much more convenient to write data-driven parallel applications, because users only need to consider the logic of processing data in each Mapper and Reducer without worrying about how to parallelize and coordinate the jobs.

- *Key-Value based.* In MapReduce, both input and output data are considered as Key-Value pairs with different types. This design is because of the requirements of parallelization and scalability. Key-value pairs can be easily partitioned and distributed to be processed on distributed clusters.
- *Parallelable and Scalable.* Both Map and Reduce functions are designed to facilitate parallelization, so MapReduce applications are generally linearly-scalable to thousands of nodes.

1.2 Examples

1.2.1 Hadoop

Hadoop [8] is the open-source implementation of Google's MapReduce paradigm. The native programming primitives in Hadoop are Mapper and Reducer interfaces which can be implemented by programmers with their actual logic of processing map and reduce stage transformation and processing. To support more complicated applications, users may need to chain a sequence of MapReduce jobs each of which is responsible for a processing module with well defined functionality.

Hadoop is mainly implemented in Java, therefore, the map and reduce functions are wrapped as two interfaces called Mapper and Reducer. The Mapper contains the logic of processing each key-value pair from the input. The Reducer contains the logic for processing a set of values for each key. Programmers build their MapReduce application by implementing those two interfaces and chaining them as an execution pipeline.

As an example, the program below shows the implementation of a WordCount program using Hadoop. Note that, the example only lists the code implementation of map and reduce methods but omits the signature of the classes.

Listing 1 WordCount example in Hadoop
```
1 public void map(Object key,
2                 Text value, Context context) {
3   String text = value.toString();
4   StringTokenizer itr = new StringTokenizer(text);
5   while(itr.hasMoreTokens()) {
6     word.set(itr.nextToken());
7     context.write(word, one);
8   }
9 }
10
```

```
11 public void reduce(Text key,
12      Iterable <IntWritable> values, Context context) {
13    int sum = 0;
14    for (IntWritable val : values) {
15      sum += val.get();
16    }
17    result.set(sum);
18    context.write(key, result);
19 }
```

2 Functional Programming

Functional programming is becoming the emerging paradigm for the next generation of big data processing systems, for example, recent frameworks like Spark [53], Flink [1] both utilize the functional interfaces to facilitate programmers to write data applications in a easy and declarative way. In functional programming, programming interfaces are specified as functions that applied on input data sources. The computation is treated as a calculation of functions. Functional programming itself is declarative and it avoids mutable states sharing. Compared to Object-oriented Programming it is more compact and intuitive for representing data driven transformations and applications.

2.1 Features

Functional Programming is one of the most recognized programming paradigms. It contains a set of features which facilitate the development in different aspects:

- Declarative: In functional programming, developers build the programs by specifying the semantic logic of computation rather than the control flow of the procedures.
- Functions are the first level citizens in Functional Programming. Primitives of programming are provided in functional manner and most of them can take user defined functions as parameters.
- In principle, functional programming does not allow the sharing of states, which means variables in functional programming are immutable. Therefore, there is no side effects for calling functions. This makes it easier to write functionally correct programs that are also easy to be verified formally.
- Recursive: In functional programming, many loops are normally represented as recursively calling of functions. This facilitates the optimization of performance by applying tail-recursive to reduce creating intermediate data and variables shared in different loops.
- Parallelization: As there is generally no state sharing in functional programming, it is easy and suitable for applying parallelization to multi-core and distributed computing infrastructures.

- Referential Transparent: In functional programming, as there is no states sharing and side effects. Functions are essentially re-producible. This means that re-calculation of functional results is not necessary. Therefore, once a function is calculated, its results could be cached and reused safely.

2.2 Example Frameworks

2.2.1 Spark

Spark provides programmers a functional programming paradigm with data-centric programming interfaces based on its built-in data model - resilient distributed dataset (RDD) [54]. Spark was developed in response to the limitations of the MapReduce paradigm, which forces distributed programs to be written in a linear and coarsely-defined dataflow as a chain of connected Mapper and Reducer tasks. In Spark, programs are represented as RDD transformation DAGs as shown in Fig. 3. Programmers are facilitated by using a rich set of high-level function primitives, actions and transformations to implement complicated algorithms in a much easier and compact way. In addition, Spark provides data centric operations such as sampling and caching to facilitate data-centric programming from different aspects.

Spark is well known for its support of rich functional transformations and actions, Table 1 shows the major transformations and operators provided in Spark. The code snippet in Listing 2 shows how to write a WoundCount program in Spark using its functional primitives.

Basically, programming primitives in Spark just look like general functional programming interfaces by hiding complex operations such as data partitioning, distribution and parallelization to programmers and leaving them to the cluster side.

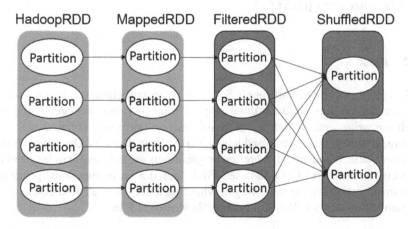

Fig. 3 RDD chain of a Spark program

Table 1 Main operations in Spark

Transformation	Description
map	Transform each element in the data set into a new type by passing them through a map function
flatMap	Transform elements in a set of collections into a plain sequence of elements by passing through a function
reduce	Aggregate elements in the data set with a function, the function needs to be commutative and associative
filter	find and return a subset of elements for the data set by checking a boolean function
groupBy	cluster the elements into different groups by mapping them with a function. The result is in (key, list of values) format
union	combine current data set with another data set to generate a new united data set
intersection	find out the overlapping elements from two data sets
distinct	find out the distinct elements from two data set
join	join two key-value based data sets and group the values for the same key. The result is in (key, (value1, value2)) format
cogroup	For two data sets of type (K, V) and (K, W), returns a grouped data set of (K, (Iterable [V], Iterable [W])) tuples

Therefore, the client side code is very declarative and simple to write by programmers. The Spark programs are essentially RDD dependency flows which will be eventually translated into parallel tasks to be executed on distributed environments.

Listing 2 WordCount example of Spark

```
1 val counts = text.flatMap{_.split("\\s+")}
2         .map{(_, 1)}
3         .reduceByKey(_ + _)
4 counts.foreach(print(_))
```

2.2.2 Flink

Flink [1] is an emerging competitor to Spark which provides functional programming interfaces that are quite similar to those of Spark. Filnk programs are regular programs which are written with a rich set of transformation operations (such as mapping, filtering, grouping, aggregating and joining) to the input data sets. Dataset in Flink is based on a table based model, therefore programmers are able to use index numbers to specify a certain field of a data set. Flink shared a lot of functional primitives and transformations in the same way as what Spark does for batch processing. The program below shows a WordCount example written in Flink.

Listing 3 WordCount example of Flink

```
1  val  counts  =  text . flatMap { _ . toLowerCase . split (" \\ s+") }
2              . map { ( _ ,  1) }
3              . groupBy (0)
4              . sum (1)
5  counts . print ( _ )
```

Apart from regular batch processing primitives, Flink is also natively designed for stream processing with the support of a rich set of functional operations. The streaming APIs of Flink is in its core bundle and users are able to write and execute stream processing and batch processing applications in the same framework. Unlike spark's use a min-batch to simulate stream processing, Flink uses the producer-consumer model for the execution of streaming programs. Therefore, it claims itself as being a more natural framework which integrates both batch and stream processing. Table 2 shows the major streaming operations provided in Flink.

In addition to normal transformations, the streaming API of Flink also provide a couple of window-based operations to apply functions and transformations on different groups of elements in the stream according to their time of arrival. A couple of window-based operations for Flink is listed in Table 3.

Table 2 Main operations for streaming processing in Flink

Transformation	Description
map	Takes one element and produces another one element with a transformation
flatMap	Takes one element and produces a collection of elements with a transformation function
KeyBy	Split input stream into different Partitions, each partition has the same key
reduce	Rolling combine elements in a keyed data stream with a aggregation function
filter	Retains a subset of elements from the input stream by evaluating a boolean function
fold	Same like reduce but provides a initial value and then combines the current element with the last folded value

Table 3 Window-based operations for streaming processing in Flink

Transformation	Description
window	Windows can be applied on Keyed stream to group the data in each key according to specific characteristics, e.g. arrival time
windowAll	Windows can be applied on the entire input of a stream to divide data into different groups based on specific characteristics
Window Reduce/Aggregation/Fold	After the input is grouped by windows, users can apply normal stream APIs such as reduce, fold and aggregation on the grouped streams

3   SQL-Like

SQL (Structured Query Language) is the most classic data query language, originally designed for rational databases based on the rational algebra. It contains four basic primitives: create, insert, update, delete for modifying the datasets considered as tables with schemas. SQL is a declarative language and also includes a few procedural elements.

SQL programs contain a few basic elements including: (1) Clauses which are constituent elements of statements and queries; (2) Expressions which can be valuated to produce a set of resulting data; (3) Predicates which specify conditions that can be used to limit the effects of statements and queries; (4) Queries which retrieve the data based on some specific criteria; (5) Statements which have a persistent effect on data, schema or even the database.

During execution of SQL, the SQLs are explained as syntax trees and then further translated into execution plans. And there are a bunch of optimizations have been developed to optimize the performance based on the syntax trees and execution plans.

3.1   Features

- *Declarative and self-interpretable*: SQL is a typical declarative language, it clearly specifies what transformation and operations are being done to which part of the data. By reading the SQL queries users can easily understand the semantics of the queries just like understand about literal descriptions.
- *Data-driven*: SQL is data-driven, all the operations and primitives are representing the transformation and manipulation of the target dataset (tables of data in SQL). This makes SQL one of the most facilitated programming model for data-centric applications for traditional databases and recent big data context.
- *Standardized and Inter-operable*: SQL is officially standardized by data communities such as IBM and W3C. Therefore, different platform provider can provide their own implementation while keeping the inter-operability between different platforms and frameworks. In Big data context, although there are some variations for SQL such as HQL (Hadoop Query Language) in Hive and CQL (Casandra Query Language) in Casandra, users can still easily understand and ship such programs into each other.

3.2   Examples

3.2.1   Hive Query Language (HQL)

Hive [47] is a query engine built on Hadoop ecosystems, it provides a SQL-like interface called Hive Query Language (HQL), which read input data based on defined schema and then transparently converts the queries into MapReduce jobs connected as a directed acyclic graph (DAG). Although based on SQL, HQL does not fully follow the SQL standard. Basically, HQL lacks support for transactions and materialized views, and only support for limited indexing and sub-queries. However, HQL also provides some extensions which are not supported in SQL, such as multi-table inserts and creating table as select. The program below shows how to write a WordCount program in Hive with HQL:

Listing 4 WordCount example of Hive

```
1 SELECT word, count(1) AS words FROM(
2 SELECT EXPLODE(SPLIT(line,'⌴') AS word FROM myinput)
3 words GROUP BY word
```

Basically, HiveQL has great extent of compatibility with SQL, the Table 4 shows the semantics and data types supported in current HiveQL v0.11.

3.2.2 Cassandra Query Language (CQL)

Apache Cassandra [36] was introduced by Facebook to power up the indexing of their in-box searching. Cassandra follows the design of Amazon Dynamo with its own query interfaces - Cassandra Query Language (CQL). CQL is an SQL based query language that is provided as the alternative to the traditional RPC interface. CQL adds an abstraction layer that hides implementation details of its query structure and provides native syntaxes for collections and common encodings. The program snippet below shows some simple operations written in CQL 3.0:

Listing 5 Code example of CQL

```
1 BEGIN BATCH
2 INSERT INTO users (userID, password, name)
3 VALUES ('user2', 'chngemb', 'second⌴user')
4 UPDATE users SET password = 'psdhds'
5 WHERE userID = 'user2'
6 INSERT INTO users (userID, password)
7 VALUES ('user3', 'ch@ngem3c')
8 DELETE name FROM users WHERE userID = 'user2'
9 INSERT INTO users (userID, password, name)
```

Table 4 Hive SQL compatibility

Data types	INT/TINYINT/SMALLINT/BIGINT, BOOLEAN, FLOAT, DOUBLE STRING, TIMESTAMP, BINARY, ARRAY, MAP, STRUCT, UNION DECIMAL
Semantics	SELECT, LOAD, INSERT from query, WHERE, HAVING, UNION GROUP BY, ORDER BY, SORT BY, CLUSTER BY, DISTRIBUTE BY LEFT, RIGHT and FULL INNER/OUTER JOIN, CROSS JOIN Sub-quires in FROM clause, ROLLUP and CUBE window functions (OVER, RANK, etc.)

```
10 VALUES ('user4', 'ch@ngem3c', 'Andrew')
11 APPLY BATCH;
```

Although CQL looks generally similar like SQL, there are some major differences:

- CQL uses KEYSPACE and COLUMNFAMILY compared to DATABASE and TABLE in SQL. And KEYSPACE requires more specifications (such as strategy and replication factor) than a standard relational database.
- There is no support for relation operations such as JOIN, GROUP BY, or FOREIGN KEY in CQL. Leaving these features out is important towards ensuring that the writing and retrieving data is much more efficient in Cassandra.
- Cheap writes, updates and inserts in CQL are extremely fast due to its Key-Value, and column family organization.
- Expiring records, CQL enables users to set expiry time for records by using the "USING TTL" (Time To Live) clause.
- Delayed Deletion, execution of DELETE queries doesn't really remove the data instantly. Basically, deleted records are marked with a tombstone (defined in TTL, which would exist for a period of time affected by the GC interval). Then, those marked data will be automatically removed during the upcoming compaction process.

3.2.3 Spark SQL

Spark introduces its rational query interfaces as Spark SQL [13], which is built on the DataFrame model and consider input data sets as table based structure. Spark SQL can be embedded into general programs of native Spark and MLlib [38] to enable interactability between different Spark modules. In addition, as Spark SQL draws on Catalyst to optimize the execution plans of SQL queries, Spark SQL can outperform native Spark APIs on most of the benchmarked APIs. The code snippet below shows how to define a DataFrame and use it to apply Spark SQL queries:

Listing 6 Code example of Spark SQL
```
1 val people = sc.textFile("people.txt").map(_.split(","))
2 .map(p => Person(p(0), p(1).trim.toInt)).toDF()
3 people.registerTempTable("people")
4
5 sqlContext.sql("SELECT name, age FROM people
6 WHERE age >= 13 AND age <= 19")
```

Basically, Spark SQL are embedded in the general programming context and supports most of the basic syntaxes of SQL as shown in Listing 7. Spark SQL is also compatible with various data sources including Hive, Avro [7], Parquet [10], ORC [6], JSON, JDBC and ODBC compatible databases and supports data set joins across these data sources.

Listing 7 Supported Syntax of Spark SQL

```
1  /* The syntax of a SELECT query */
2  SELECT [DISTINCT] [column names]|[wildcard]
3  FROM [kesypace name.] table name
4  [JOIN clause table name ON join condition]
5  [WHERE condition]
6  [GROUP BY column name]
7  [HAVING conditions]
8  [ORDER BY column names [ASC | DSC]]
9
10 /* The syntax of a SELECT query with joins. */
11 SELECT statement
12 FROM statement
13 [JOIN | INNER JOIN | LEFT JOIN | LEFT SEMI JOIN |
14 LEFT OUTER JOIN | RIGHT JOIN | RIGHT OUTER JOIN |
15 FULL JOIN | FULL OUTER JOIN]
16 ON join condition
17
18 /* Several select clauses can be combined in a
19 UNION, INTERSECT, or EXCEPT query. */
20 SELECT statement 1
21 [UNION | UNION ALL | UNION DISTINCT |
22 INTERSECT | EXCEPT]
23 SELECT statement 2
24
25 /* The syntax defines an INSERT query. */
26 INSERT [OVERWRITE] INTO [keyspace name.]
27 table name [(columns)]
28 VALUES values
29
30 /* The syntax defines an CACHE TABLE query. */
31 CACHE TABLE table name [AS table alias]
32
33 /* The syntax defines an UNCACHE TABLE query. */
34 UNCACHE TABLE table name
```

3.2.4 Apache Drill

Apache Drill [3] is the open source version of Google's Dremel system, which is a schema-free SQL Query Engine for MapReduce, NoSQL and Cloud Storage. Drill is well known for its connectivity to variety of NoSQL databases and file systems, including HBase [26], MongoDB [39], MapR-DB, HDFS [45], MapR-FS, Amazon S3 [46], Azure Blob Storage [15], Google Cloud Storage [27], Swift [41], NAS and local files. A single query can join data from multiple data stores. For example, you can join a user profile collection in MongoDB with a directory of event logs in Hadoop. The main features of Apache Drill are listed as below:

- Drill uses a JSON based data model similar to MongoDB and ElasticSearch.
- Drill supports multiple industry-standard APIs, such as JDBC/ODBC, SQL and RESTful APIs.
- Drill is designed as a pluggable architecture which supports connecting with multiple data stores, including Hadoop, NoSQL and cloud-based storages.
- Drill also supports different of data formats such as JSON, Parquet and plain text.

Drill supports standard ANSI of SQL to query data from different databases and file systems regardless of its source system or its schema and data types. Listing 8 shows an example about creating a table from a JSON file in Drill.

Listing 8 Create a table from JSON data source in Drill

```
1 CREATE TABLE dfs.tmp.sampleparquet AS
2 (SELECT trans_id ,
3 cast('date' AS date) transdate ,
4 cast('time' AS time) transtime ,
5 cast(amount AS double) amountm,
6 user_info , marketing_info , trans_info
7 FROM dfs.'/Users/drilluser/sample.json ');
```

Apart from normal SQL syntaxes, Drill also offers a couple of nested function within SQL queries as listed in Table 5.

3.2.5 Other SQL-like Query Frameworks

- Impala [21], provides high-performance, low-latency SQL queries on data stored in popular Apache Hadoop file formats. The fast response for queries enables interactive exploration and fine-tuning of analytic queries, rather than long batch jobs traditionally associated with SQL-on-Hadoop technologies. Impala integrates with the Apache Hive metastore database, to share databases and tables between both components. The high level of integration with Hive, and compatibility with the HiveQL syntax, lets you use either Impala or Hive to create tables, issue queries, load data, and so on.

Table 5 Nested functions in Drill

FLATTEN	Separate the elements in nested data from a repeated field into individual records
KVGEN	Return a repeated map generating key-value pairs for querying of complex data having unknown column names
REPEATED_COUNT	Count the values in an array
REPEATED_CONTAINS	Search for a keyword in an array

- Presto [25], is an open source distributed SQL query engine for running inter- active analytic queries against data sources of all sizes ranging from gigabytes to petabytes. Presto was designed and written from the ground up for interactive analytics and approaches the speed of commercial data warehouses while scaling to the size of organizations like Facebook.

4 Actor Model

The Actor model [29] is a programming model for concurrent computation, which consider "Actor" as the universal primitive unit for computation meaning it treats everything as an actor. An actor is responsible to react to a set of messages to trigger specific processing logics (such as making decisions, building more actors, sending more messages) for different contexts. The Actor model is also considered as a reactive programming model in which programmers write acting logic in response to events and context changes. Unlike other programming models which are normally sequential, the actor model is inherently concurrent. The reactions of an actor can happen in any order and actions for different actors are also in parallel.

4.1 Features

- *Message-driven*: The Actor model inherits the message-oriented architecture for communication. messages are the primitive and the only data carrier among the systems.
- *Stateless and isolation*: Actors are loosely coupled to each other. Therefore, there is no global state shared between different actors. In addition, actors are separate functional units which are not suppose to affect others when failures and errors are encountered.
- *Concurrent*: Actors in the actor system are in action at the same time, and there is no fixed order for sending and receiving messages. Therefore, the whole actor system is inherently concurrent.

4.2 Examples

4.2.1 Akka

Akka [49] is a distributed and concurrent programming model developed by Typesafe with inspiration drawn from Erlang. Akka has been widely used in recent distributed and concurrent frameworks such as Spark, Play [50], Flink, etc. Akka provides different programming models but it emphasizes the actor-based concurrency model.

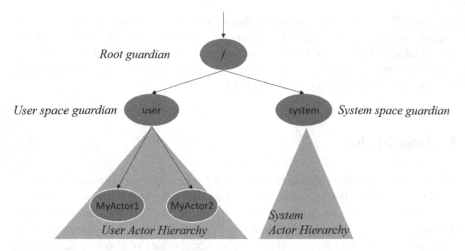

Fig. 4 Hierarchy of Akka actors

Actors in Akka communicate with each other through asynchronous messages, so typically no mutable data are shared and no synchronization primitives are used. In addition, Akka provides a hierarchical supervision model which enforces failure and fault management to the parental actors as shown in Fig. 4. Failures in Akka are also considered as messages passed to parents. Lastly, actors in Akka are portable to be executed in local and distributed environments without the need to modify existing program logics.

Akka provides a reflective way of creating a actor, in which users explicitly specify the class object and the name of an actor. In addition, Akka offers two basic message passing primitives to support communication among actors:

- ask (written as "?"), sends a message to a target actor and waits for its response as a future message.
- tell (written as "!"), sends a message to a target actor then finishes this communication, also known as fire-and-forget.

As an example, the program below shows the HelloWorld application written using Akka:

Listing 9 HelloWorld exmaple in Akka

```
1 class HelloActor extends Actor {
2   val log = Logging(context.system, this)
3   def receive = {
4     case "hello" => log.info("Hello World")
5     case _       => log.info("received unknown message")
6   }
7 }
8 \* create an HelloActor instance *\
9 val system = ActorSystem("mySystem")
```

```
10 val actor = system.actorOf(Props[HelloActor],"actor")
11 \* send a message to the actor *\
12 actor ! "hello"
```

4.2.2 Storm

Storm [48] is an open source programming framework for distributed realtime data processing. Storm inherits from the actor-model and provides two types of processing actors: Spouts and Bolts.

- Spout is the data source of a stream and is continuously generating or collecting new data for subsequent processing.
- Bolt is a processing entity within a streaming processing flow, each bolt is responsible for a certain processing logic, such as transformation, aggregation, partitioning and redirection, etc.

Jobs in Storm is defined as directed acyclic graphs (DAG) with connected Spouts and Bolts as vertices. Edges on the graph are data streams and direct data from one node to another. Unlike batch jobs being only executed once, Storm jobs are running until they are killed. The code snippet in Listing 10 shows an example about writing a Bolt to produce tuple streams.

In Storm, a complete application is built by connecting Spouts and Bolts. As shown in Listing 11 users can define the topology of the application by appending each Bolt to its predecessor.

Listing 10 Building a Bolt to Generate Tuple Streams

```
1  public class DoubleAndTripleBolt extends BaseRichBolt{
2    private OutputCollectorBase _collector;
3
4    @Override
5    public void prepare(Map conf, TopologyContext context,
6                        OutputCollectorBase collector){
7      _collector = collector;
8    }
9
10   @Override
11   public void execute(Tuple input) {
12     int val = input.getInteger(0);
13     _collector.emit(input, new Values(val*2, val*3));
14     _collector.ack(input);
15   }
16
17   @Override
18   public void declareOutputFields(
19                 OutputFieldsDeclarer declarer){
```

```
20      declarer . declare (new Fields ("double", "triple"));
21    }
22  }
```

Listing 11 Building a WordCount Topology in Storm

```
1 TopologyBuilder builder = new TopologyBuilder ();
2 builder . setSpout ("words", new TestWordSpout (), 10);
3 builder . setBolt ("exclaim1", new ExclamationBolt (), 3)
4          . shuffleGrouping ("words");
5 builder . setBolt ("exclaim2", new ExclamationBolt (), 2)
6          . shuffleGrouping ("exclaim1");
```

4.2.3 Apache S4

Apache S4 (Simple Scalable Streaming System) was introduced by Yahoo in 2008 for stream processing. S4 is a general-purpose, distributed, scalable, fault-tolerant, pluggable platform for processing continuous, unbounded streams of data. The main features of S4 are listed as below:

- *Decentralized*: All nodes in S4 are symmetric with no centralized service and no single point of failure.
- *Scalable*: The throughput of S4 increases linearly as additional nodes are added to the cluster. There is no predefined limit on the number of nodes that can be supported.
- *Extensible*: Applications can be easily written using a simple API. Building blocks of the platform such as message queues and processors, serializer and checkpointing backend can be replaced by custom implementations.
- *Fault-tolerant*: Once a server in the cluster fails, a stand-by server is automatically activated to take over the tasks.

In S4, each processing actor is called a ProcessElement (PE) which is responsible to processing in response to each data element (an event) within the input stream. An S4 application is built by connecting a couple of ProcessElement in a certain topology. Listing 12 shows a example of implementing a Hello ProcessElement in S4.

Listing 12 Implement a Process Element in S4

```
1 public class HelloPE extends ProcessingElement {
2    // PEs can maintain some state
3    boolean seen = false ;
4
5    // called upon every new Event on an incoming stream.
6    public void onEvent (Event event) {
7       System . out . println ("Hello" + event . get ("name"));
8       seen = true ;
9    }
10 }
```

5 Statistical and Analytical

In recent years significant effort was spent to offer semantically friendly environments for statistical and analytical computation, which leads to the development and revolution of statistical and analytical programming models. For example many current analytics libraries or frameworks provide a linear algebra based programming model which works with vectors, matrices and tensor data structures to deal with algebraically defined mathematical problems in machine learning, statistics and data mining, etc.

5.1 Features

Due to the mathematical nature of statistical and analytical programming, it is essentially functional with manipulations on matrix and vector-based data structures.

- *Functional*: Mathematical operations are essentially functions consuming a set of input parameters to generate an output. Also many complicated functions or models are wrapped into functional libraries so that users can directly use it without knowing the implementation details of the functions.
- *Matrix-based data structure*: A matrix is one of the most widely used data structure for representing modern analytics and statistic problems and solutions. Therefore, the majority of existing analytic programming frameworks are based on matrices, vectors and data frames to manipulate the data.
- *Declarative*: In statistical and analytical programming, the programs explicitly specify the functions and operations that have been applied on the data (matrix, vector and data frame.)

5.2 Examples

5.2.1 R

R [32] combines the S [17] programming language and Lexical Scoping inspired by Scheme [20]. It is well known for statistical programming and drawing graphics. In R, data are essentially represented as matrices which are very convenient for implementing mathematical and statistical formulas. R and its libraries implement a wide variety of statistical and graphical techniques and is easy to be extended by developers. R is recently introduced to the big data processing context (RHadoop [23], SparkR [11], RHIPE [28]) to facilitate the development of statistical and analytics programs and applications. The code snippet below shows how to implement formula: $G = BB^T - C - C^T + s_q s_q^T \xi^T \xi$ using R:

Listing 13 Writing formula of R

```
1 g <- t(b) %*% b - c - t(c)
2      + (sq %*% t(sq)) * (t(xi) %*% xi)
```

5.2.2 Mahout

Apache Mahout [9] is an open-source implementations of distributed and scalable machine learning and data mining algorithms. Mahout provides libraries that are mainly focused in the areas of collaborative filtering, clustering and classification. The initial implementation of Mahout is based on Apache Hadoop, but recently it has started to provide compatible bindings on Spark and also being able to provide matrix-based programming interfaces. For example the same formula shown in the R section can be written in Mahout as the code segment below:

Listing 14 Code example of Mahout

```
1 val g = bt.t %*% bt - c - c.t
2          + (s_q cross s_q) * (xi dot xi)
```

The Mahout project recently (since release 1.0.0) shifts its focus to building backend-independent programming framework, which is named "Samsara". The new project consists of an algebraic optimizer and an Scala DSL to unify both distributed and in-memory algebraic operators. The current version supports execution of algebraic programs on platforms including Apache Spark and H20. The support of Apache Flink operators is also in progress.

6 Dataflow-Based

The dataflow programming paradigm models programs as directed graphs with operations and dependencies as nodes and edges. Dataflow programming [35] was first introduced by Jack Dennis and his students at MIT in the 1960s. Dataflow programming emphasizes the movement of data and considers programs as a series of connections. Every operator and processor normally has explicitly defined inputs and outputs and functions like black boxes. An operation runs as soon as all of its inputs become valid. Thus, dataflow languages are inherently parallel and can work well in large, decentralized systems.

In the big data scenario, data-centric jobs can also be modeled as dataflows in which each node represents a small task while the edges represents the data dependencies between different tasks. Developers may be need to write the process logic of each node using other general programming languages such as Java, C and Python while leave the dependency and connecting logic to the dataflow.

6.1 Features

The major features of dataflow programming can be listed as follows:

- *Trackable states*: Dataflow programming consider programs as connections of tasks in combination with control logic. Therefore, unlike other programming models, it provides a inherently trackable states during execution.
- *Various representation*: Dataflow programming model could be represented in different ways for different purposes. As we have already discussed, it can be inherently represented in a graph-based manner and also can be represented in connected texts introductions and Hash tables.

6.2 Examples

6.2.1 Oozie

Apache Oozie [34] is a server side workflow scheduling system to manage complex Hadoop jobs. In Oozie workflows are directed acyclic graphs with control flow and nodes (each node as a MapReduce jobs). In Oozie, the workflow is specified as XML-based documents presenting the connection and dataflow of different MapReduce jobs. Oozie can be integrated with other Hadoop ecosystems and also support different types of jobs such as Pig, Hive, Streaming MapReduce, etc. The XML segment below shows a Fork–Join workflow defined in Oozie:

Listing 15 Fork and Join example in Oozie

```
1  <workflow−app name="sample−wf"
2      xmlns="uri:oozie:workflow:0.1">
3      ...
4      <fork name="forking">
5          <path start="firstparalleljob"/>
6          <path start="secondparalleljob"/>
7      </fork>
8      <action name="firstparallejob">
9          <map−reduce>
10             <job−tracker>foo:8021</job−tracker>
11             <name−node>bar:8020</name−node>
12             <job−xml>job1.xml</job−xml>
13         </map−reduce>
14         <ok to="joining"/>
15         <error to="kill"/>
16     </action>
17     <action name="secondparalleljob">
18         <map−reduce>
19             <job−tracker>foo:8021</job−tracker>
```

```
20                    <name—node>bar:8020</name—node>
21                    <job—xml>job2.xml</job—xml>
22              </map—reduce>
23              <ok to="joining"/>
24              <error to="kill"/>
25         </action>
26         <join name="joining" to="nextaction"/>
27         ...
28 </workflow—app>
```

With the workflow specification, each action between control logic is a MapReduce associated with its Job Tracker and job definition (in a separate xml file). Actions in Oozie are triggered by time and data availability.

6.2.2 Microsoft Dryad

Microsoft Dryad [33] is a high performance, general purpose distributed computing engine which supports writing and execution of data-centric parallel programs. Dryad allows a programmer to utilize the resources of a computer cluster or a data center to run data-parallel programs. By using Dryad, programmers write simple programs which will be concurrently executed on thousands of machines (each of which with multiple processors or cores) while hiding the complexity of concurrent programming. The code below shows an example about building a graph in Dryad:

Listing 16 Building Graphs in Dryad
```
 1 GraphBuilder XInputs = (ugriz1 >= XSet)
 2                              || (neighbor >= XSet);
 3 GraphBuilder YInputs = ugriz2 >= YSet;
 4 GraphBuilder XToY = XSet >= DSet >> MSet >= SSet;
 5 for (i = 0; i < N*4; ++i){
 6   XToY = XToY || (SSet.GetVertex(i) >= YSet.GetVertex(i/4));
 7 }
 8 GraphBuilder YToH = YSet >= HSet;
 9 GraphBuilder HOutputs = HSet >= output;
10 GraphBuilder final = XInputs || YInputs
11                              || XToY || YToH || HOutputs;
```

A Dryad job contains several sequential programs and are connected using one-way channels. The program written by programmers is structured as a directed graphs, in which, programs are vertices, while the channels are edges. A Dryad job is a graph generator which can synthesize any directed acyclic graph. These graphs can also be changed during execution to respond to important events or notifications.

7 Bulk Synchronous Parallel

The Bulk Synchronous Parallel (BSP) [51] is a computation and programming model for designing parallel algorithms. A BSP algorithm is considered as a computation proceeds in a series of global super-steps, which consists of three components:

- *Concurrent computation*: Every participating processor may perform local computations, i.e., each process can only make use of values stored in the fast local memory of the processor. The computations occur asynchronously but may overlap with communication.
- *Communication*: The processes exchange data between themselves to facilitate remote data storage capabilities.
- *Barrier synchronisation*: When a process reaches this point (the barrier), it waits until all other processes have reached the same barrier.

7.1 Features

The BSP model contains the following key features:

- *Message-based communications*: BSP considers every communication action as a message and it also considers all messages of a super-step as a unit. This significantly reduces the effort for users to handle low-level parallel communications.
- *Barrel-based Synchronization*: BSP uses barrels to guarantee the consistency when needed, although barrel is a costly operation, it provides strong consistency and can also provides support for fault tolerance in an easy and understandable way.

7.2 Examples

7.2.1 Apache Giraph and Google Pregel

Apache Giraph [4] is an iterative graph processing system built for high scalability. Giraph is inspired by Google's Pregel [37] which is based on the Bulk Synchronous Parallel (BSP) model of distributed computation. Giraph adds several features beyond the basic Pregel model, including master computation, sharded aggregators, edge-oriented input, out-of-core computation, etc.

Listing 17 Shortest Path implementation in Giraph

```
1  public void compute ( Iterable <DoubleWritable> messages ){
2    double minDist = Double .MAX_VALUE;
3    for (DoubleWritable message : messages) {
4      minDist = Math . min ( minDist , message . get ());
5    }
```

```
 6    if  (minDist < getValue().get()) {
 7      setValue(new DoubleWritable(minDist));
 8      for  (Edge<LongWritable, FloatWritable>
 9            edge : getEdges()) {
10        double distance = minDist+edge.getValue().get();
11        sendMessage(edge.getTargetVertexId(),
12                    new DoubleWritable(distance));
13      }
14    }
15    voteToHalt();
16 }
```

7.2.2 Hama

Apache Hama (stands for Hadoop Matrix) [5] is a distributed computing framework based on Bulk Synchronous Parallel computing model for massive scientific computations. Writing a Hama graph application involves inheriting the predefined Vertex class. Its template arguments define three value types, associated with vertices, edges, and messages. Hama also provides very flexible input and output options such as the ability to extract Vertex from programmers' data without any pre-processing. Hama also allows programmers to do optimizations by writing Combiner, Aggregator and Counter in data processing flows. The following code snippet shows an example of PageRank implementation in Hama:

Listing 18 PageRank implementation in Hama

```
 1 public static class PageRankVertex extends
 2         Vertex<Text, NullWritable, DoubleWritable> {
 3
 4   @Override
 5   public void compute(Iterator<DoubleWritable> messages)
 6                                      throws IOException {
 7   // initialize this vertex to 1/count
 8   if (this.getSuperstepCount() == 0) {
 9     setValue(new DoubleWritable(1.0 /
10                        this.getNumVertices()));
11   } else if (this.getSuperstepCount() >= 1) {
12     double sum = 0;
13     for (DoubleWritable msg : messages) {
14       sum += msg.get();
15     }
16     double alpha = (1.0d–DAMPING_FACTOR)
17                    / this.getNumVertices();
18     setValue(new DoubleWritable(alpha +
19                        (sum*DAMPING_FACTOR)));
20     aggregate(0, this.getValue());
```

```
21    }
22
23    // if  have  not  reached  global  error,  then  proceed.
24    DoubleWritable  globalError  =  getAggregatedValue(0);
25
26    if (globalError  !=  null  &&  this.getSuperstepCount()>2
27        && MAXIMUM_CONVERGENCE_ERROR> globalError.get()){
28      voteToHalt();
29    } else {
30      // in  each  superstep  send  a  new  rank  to  neighbours
31      sendMessageToNeighbors(new DoubleWritable(
32          this.getValue().get()/this.getEdges().size()));
33    }
34  }
35 }
```

8   High Level DSL

There is no a single programming model that can satisfy everyone and every scenario. Many frameworks provide their own Domain Specific Language (DSL, in contrast to general purpose programming language) for writing data-intensive parallel applications/programs in order to provide a better programming model in certain domains or purposes.

8.1   Pig Latin

Pig [40] is a high level platform to create data centric programs on top of Hadoop. The programming interface of Pig is called Pig Latin which is an ETL-like query language. In comparison to SQL, Pig uses extract, transform, load (ETL) as its basic primitives. In addition, in Pig Latin, it is able to store data at any point during a pipeline. At the same time, Pig supports the ability to declare execution plans as well as support for pipeline splits, thus allowing workflows to proceed along DAGs instead of strictly sequential pipelines. Lastly, Pig Latin scripts are automatically compiled to generate equivalent MapReduce jobs for execution.

Listing 19 WordCount example of Pig Latin

```
1  input_lines = LOAD '/tmp/wordcount−input'
2                   AS (line:chararray);
3  words = FOREACH input_lines
4          GENERATE flatten(TOKENIZE(line)) AS word;
5  filtered_words = FILTER words BY word MATCHES '\\w+';
6  word_groups = GROUP filtered_words BY word;
```

Table 6 Basic relational operators in Pig Latin

Operators	Description
LOAD	Load data from underlying file systems
FILTER	Select matched tuples from data set based on some conditions
FOREACH	Generate new data transformations based on each columns of a data set
GROUP	Group a data set based on some relations
JOIN	Join two or more data sets based on expressions of the values of their column fields
ORDERBY	Sort the data set based on one or more columns
DISTINCT	Remove duplicated elements from a given data set
MAPREDUCE	Execute native MapReduce jobs inside the Pig scripts
LIMIT	Limit the number of elements in the output

```
7   word_count  =  FOREACH  word_groups
8                   GENERATE  count(filtered_words)
9                        AS count,  group AS word;
```

Pig offers a bunch of operators to support transformation and manipulation on input datasets. Table 6 shows the basic relational operators provided in Pig Latin and the code snippet in Listing 19 shows a WordCount example written in Pig scripts.

8.2   Crunch/FlumeJava

Apache Crunch [2] is high-level library supports writing testing and running data-driven pipelines on top of Hadoop and Spark. The programming interface of Crunch is partially inspired by Google's FlumeJava [19]. Crunch wraps native MapReduce interface into high level declarative primitives such as parallelDo, groupByKey, combineValues and union to make it easy for programmers to write and read their applications. Crunch provides a couple of high level processing patterns (as shown in Table 7) to facilitate developers to write data-centered applications.

Listing 20 WordCount example of Crunch

```
1  Pipeline pipeline  =  new  MRPipeline(WordCount.class);
2  PCollection<String> lines  =  pipeline.readTextFile(args[0]);
3
4  DoFn<String, String> func  =  new  DoFn<String, String>(){
5    public void process(String line,
6                        Emitter<String> emitter){
7      for (String word : line.split("\\s+")) {
8        emitter.emit(word);
9      }
10   }
```

Table 7 Common data processing patterns in Crunch

Pattern	Description
groupByKey	Group and shuffle data set based on the key of the tuples
combineValues	Aggregate elements in a grouped data set based on the combination function
aggregations	Common aggregation patterns are provided as methods on the PCollection data type, including count, max, min, and length
join	Join two keyed data sets by group the elements with the same key
sorting	Sort data set based on the value of a selected column

```
11 }
12 PCollection<String> words =
13 lines.parallelDo(func, Writables.strings());
14
15 for (Pair<String, Long> wordCount : words.count())) {
16    System.out.println(wordCount);
17 }
```

In Crunch, each job is considered as a Pipeline and data are considered as Collections. Programmers write their process logic within DoFn interfaces and use basic primitives to apply transformation, filtering, aggregation and sorting to the input data sets to implement expected applications. The WordCount example of Crunch is shown in Listing 20.

8.3 Cascading

Apache Cascading [31] is a high-level development layer for building data applications on Hadoop. Cascading is designed to support the building and execution of complex data processing pipelines on a Hadoop cluster while hiding the underlying complexity of MapReduce jobs. The below code snippet shows the example of WordCount written using the Cascading API:

Listing 21 WordCount example of Cascading
```
1  Tap docTap = new Hfs( new AvroScheme(), docPath );
2  Tap wcTap = new Hfs( new TextDelimited(), wcPath, true );
3  Pipe wcPipe = new Pipe( "wordcount" );
4  wcPipe = new GroupBy( wcPipe, new Fields("count") );
5  wcPipe = new Every( wcPipe,
6                      Fields.ALL,
7                      new Count(new Fields("countcount")),
8                      Fields.ALL );
9
10 FlowDef flowDef = FlowDef.flowDef()
11   .setName( "wc" )
```

```
12   . addSource ( wcPipe , docTap )
13   . addTailSink ( wcPipe , wcTap );
14
15 Flow wcFlow = flowConnector . connect ( flowDef );
16 wcFlow . writeDOT ( "dot / wcr . dot" );
17 wcFlow . complete ();
```

As we can see from the example, a cascading job is defined as a Flow, in which it can contains multiple pipes. Each pipe is actually a function block which is responsible for a certain data process step such as GroupBy, Filtering, Joining and Sorting. Pipes are connected to construct the final Flow for execution.

8.4 Dryad LINQ

DryadLINQ [52] is a compiler which translates LINQ (Language-Integrated Query) programs to distributed computations which can be run on a cluster. The goal of LINQ is to bridge the gap between the world of objects and the world of data. LINQ uses query expressions akin to SQL statements such as select, where, join, groupBy and orderBy, etc. In addition, LINQ also defines a set of method names (called standard query operators), along with translation rules used by the compiler to translate fluent-style query expressions into expressions using these method names, lambda expressions and anonymous types. In DryadLINQ, the data queries are automatically compiled as DAG tasks to be executed on the Dryad engine to support the building and execution of large scale data-driven applications and programs.

Listing 22 WordCount example of Dryad LINQ

```
1 public  static  IQueryable <Pair> Histogram (
2                   IQueryable <string> input , int  k){
3   IQueryable <string> words =
4              input . SelectMany (x  =>  x . Split (' '));
5   IQueryable <IGrouping <string ,  string >> groups =
6              words . GroupBy (x  =>  x );
7   IQueryable <Pair> counts =
8              groups . Select (x  =>  new  Pair (x . Key ,  x . Count ()));
9   IQueryable <Pair> ordered =
10             counts . OrderByDescending (x  =>  x . count );
11  IQueryable <Pair> top = ordered . Take (k);
12  return  top ;
13 }
```

The code snippet above shows an example of the WordCount program written using DryadLINQ. As we can see from the example, LINQ actually provides ETL operations in an Object-oriented ways. Query primitives are object operations which associated to the data, and the result of queries are represented as collections with specific types.

8.5 Trident

Trident [12] is a high-level abstraction for doing realtime computing on top of Storm. It allows you to seamlessly intermix high throughput (millions of messages per second), stateful stream processing with low latency distributed querying. If you're familiar with high level batch processing tools like Pig or Cascading, the concepts of Trident will be very familiar. Trident has joins, aggregations, grouping, functions, and filters. In addition to these, Trident adds primitives for doing stateful, incremental processing on top of any database or persistence store. Trident has consistent, exactly-once semantics, so it is easy to reason about Trident topologies.

Listing 23 Code snippet of WordCount using Trident

```
1  TridentTopology topology = new TridentTopology();
2  TridentState wordCounts =
3       topology.newStream("spout1", spout)
4       .each(new Fields("sentence"),
5            new Split(),
6            new Fields("word"))
7       .groupBy(new Fields("word"))
8       .persistentAggregate(new MemoryMapState.Factory(),
9                            new Count(),
10                            new Fields("count"))
11      .parallelismHint(6);
```

8.6 Green Marl

Green Marl [30] is a DSL introduced by the Pervasive Parallelism Laboratory of Stanford University and specifically designed for graph analysis. Green Marl allows user to describe their graphs intuitively through a high level interface while inherently provide data-driven parallelism. Green Marl, provides the ability to define both directed graphs and undirected graphs and supports basic types (like Int, Long, Float, Double and Bool) and collections (like Set, Sequence and Order). Green Marl introduces its own compiler to interpret the program into C++ code for execution. The compiler of Green Marl also introduces a couple of optimizations during compile-time to improve the execution performance. The code snippet below shows an example about the Betweenness Centrality algorithm written in Green Marl.

Listing 24 Betweenness Centrality algorithm described in Green Marl

```
1  Procedure Compute_BC(
2  G: Graph, BC: Node_Prop<Float>(G)) {
3  G.BC = 0; // initialize BC
4  Foreach(s: G.Nodes) {
5  // define temporary properties
```

```
 6   Node_Prop<Float >(G)  Sigma ;
 7   Node_Prop<Float >(G)  Delta ;
 8   s . Sigma = 1;  // Initialize Sigma for root
 9   // Traverse graph in BFS-order from s
10   InBFS ( v :  G . Nodes  From  s )( v !=s )  {
11   // sum over BFS-parents
12   v . Sigma = Sum(w:  v . UpNbrs )  {w. Sigma };
13   }
14   // Traverse graph in reverse BFS-order
15   InRBFS ( v !=s )  {
16   // sum over BFS-children
17   v . Delta = Sum  (w: v . DownNbrs )  {
18   v . Sigma  /  w. Sigma  *  (1+ w. Delta )
19   };
20   v . BC += v . Delta ;  // accumulate BC
21   } } }
```

8.7 Asterix Query Language (AQL)

The Asterix Query Language (AQL) [14] is the language interface provided by Aster-ixDB which is a scalable big data management system (BDMS) with the capability for querying semi-structured data sets. AQL is based on a NoSQL style data model (ADM) which extends JSON with object database concepts. Basically, AQL is an expressive and declarative query language for querying semi-structured data with the support for a rich set of primitive types, including spatial, temporal and textual data. The code snippet below shows an example about joining two data sets in AQL.

Listing 25 Join two data sets by AQL
```
1 for $user in dataset FacebookUsers
2 for $message in dataset FacebookMessages
3 where $message . author-id = $user . id
4 return
5   {
6     "uname": $user . name ,
7     "message": $message . message
8   };
```

As we can see from the example code, AQL combines the style of an SQL query with the data model of JSON to provide a programming style that is both declarative and data-driven. The core of AQL is called FLWOR (for-let-where-orderby-return) expression which is borrowed from XQuery expressions. A FLWOR expression starts with one or more clauses which establishes the variable bindings.

- A *for* clause binds a variable incrementally to each element of its associated expression and includes an optional positional variable for counting/numbering the bindings.
- A *let* clause binds a variable to the collection of elements computed by its associated expression.
- The *where* clause in a FLWOR expression filters the preceding bindings via a boolean expression, much like a where clause does in an SQL query.
- The *order by* clause in a FLWOR expression induces an ordering on the data.
- The *return* clause defines the data expected to be sent back as the results of a query.

8.8 IBM Jaql

Jaql (or JAQL) [18] is a functional data processing and query language mostly focusing on JSON-based query processing on BigData. Jaql was originally introduced by Google and then further developed by IBM. Jaql is designed to elegantly handle deeply nested semi-structured data and even deal with heterogeneous data. Jaql can also be used in Hadoop as a expressive query language that is comparable with Pig and Hive. The code snippet below shows some basic examples of queries written in Jaql.

Listing 26 Basic opertions in Jaql

```
1  a = {name : "scott", age : 42, children : ["jake", "sam"]};
2  a.name; // returns "scott"
3  a.children [0]; // returns "jake"
4
5  // for local file system
6  read(del("file :///home/user/test.csv"));
7  // for hdfs file system
8  read(del("hdfs://localhost:9000/user/test.csv"));
9
10 recs = [ {a: 1, b: 4}, {a: 2, b: 5}, {a: −1, b: 4} ];
11 recs −> transform .a+.b; // returns [ 5, 7, 4 ]
```

9 Discussion and Conclusion

In this section, we summarize the main features and compare the various programming models presented in this chapter.

Due to its declarative feature, functional programming is natural fit for data-driven programs and applications. As pure functions are stateless and have no side effects, functional programs are easier to be parallelized and proofed for correctness. In addition, functional programs are easier to debug and test as functions are a better isolation of functionalities without the uncertainties caused by state sharing and other side-effects. However, programming in a functional way is much different from programming in the imperative programming. Developers need to shift from imperative and procedure-based thinking to a functional way of thinking when writing the programs. This may require considerable efforts from the developer to learn and practice in order to gain sufficient mastery.

MapReduce is considered as an easy way of writing data-driven parallel programs. The emergence of MapReduce significantly eased the task for developing data-parallel applications on large scale data sets. Although the paradigm is simple, it can still cover the majority of the algorithms in practice. MapReduce is not guaranteed to be fast as its focus is more on scalability and fault tolerance. In addition, MapReduce is criticized for lacking the novelty of more recent developments and its restricted programming paradigm which does not support iterative and streaming algorithms.

SQL is considered as having limited semantics and not sufficiently expressive. Basically, SQL is not a Turing-complete language, it is more towards a query rather than a general programming language such as Java and C. As a result, it is more suitable for writing ETL (Extract, Transform and Load) or CRUD (Create, Read, Update and Delete) queries rather than general algorithms. For example, it would be a horrible choice to use SQL for writing data mining and machine learning algorithms. Traditional rational queries are slow and less scalable in Big Data scenarios, therefore, query languages such as HQL and CQL cut down the majority of the rational paradigms provided by traditional SQL in order to be more scalable in a big data context.

The first advantage of dataflow programming is that it facilitates for visualized programming and monitoring. Due to its simplified graph-based interfaces, it is easy to prototype and implement certain systems and applications. In addition, it is

Table 8 Comparison for different programming models

Model	Features	Abstraction	Semantics	Computation model
MapReduce	Non-declarative skeleton-based	low	Limited inherent parallel	MapReduce
Functional	Declarative stateless	High	Rich and general purpose	DAG or Evaluation-based
SQL-based	Declarative	High	Limited	Execution plan
Data flow	Non-declarative modularized	mediate	Rich control-logic based	DAG
Statistical	Declarative	High	Limited domain-specific	Mathemetical
BSP	Skeleton-based	Low	Rich	BSP
Actor	Event-driven message-based	Low	Rich and inherent concurrency	Reactive actors

also well known for providing end-user programming in which WYSIWYG (what you see is what you get) interfaces are required. Another point in favour of data flow programming is that, by writing programs in a dataflow manner, it actually help developers to modularize their programs as connected processing components and provide good loosely coupled structure and flexibility. However, compared to other programming models such as functions and SQLs, dataflow-based model is relatively non-declarative, unproductive for programming as it basically provides low-level programming abstractions and interfaces, which is hard to be integrated with.

To sum up, the comparison of basic programming models are listed in Table 8. Basically, MapReduce and BSP models are programming skeletons, functional, SQL and statistical models are declarative, while data-flow model is inherently modularized and actor model is essentially event-driven and message-based. In addition, functional, SQL and statistical models are high-level abstraction models and MapReduce, BSP and actor models are low-level abstraction. Lastly, functional, BSP and actor models are more semantically complete to support richer operations while other models are generally limited in semantics to trade off among understandability, user-convenience and productivity [16, 42].

References

1. A. Alexandrov, R. Bergmann, S. Ewen, J.-C. Freytag, F. Hueske, A. Heise, O. Kao, M. Leich, U. Leser, V. Markl, F. Naumann, M. Peters, A. Rheinländer, M.J. Sax, S. Schelter, M. Höger, K. Tzoumas, D. Warneke, The stratosphere platform for big data analytics, VLDB J. **23**(6) (2014)
2. Apache. Apache crunch (2016). https://crunch.apache.org/. Accessed 17 Mar 2016
3. Apache. Apache drill (2016). https://drill.apache.org/. Accessed 17 Mar 2016
4. Apache. Apache giraph (2016). https://giraph.apache.org/. Accessed 17 Mar 2016
5. Apache. Apache hama (2016). https://hama.apache.org/. Accessed 17 Mar 2016
6. Apache. Apache orc (2016). https://orc.apache.org/. Accessed 17 Mar 2016
7. Apache. Avro (2016). https://avro.apache.org/. Accessed 17 Mar 2016
8. Apache. Hadoop (2016). http://hadoop.apache.org/. Accessed 17 Mar 2016
9. Apache. Mahout: Scalable machine learning and data mining (2016). https://mahout.apache.org/. Accessed 17 Mar 2016
10. Apache. Parquet (2016). https://parquet.apache.org/. Accessed 17 Mar 2016
11. Apache. Spark r (2016). https://spark.apache.org/docs/1.6.0/sparkr.html. Accessed 17 Mar 2016
12. Apache Storm. Trident (2016). http://storm.apache.org/documentation/Trident-tutorial.html. Accessed 17 Mar 2016
13. M. Armbrust, R.S. Xin, C. Lian, Y. Huai, D. Liu, J.K. Bradley, X. Meng, T. Kaftan, M.J. Franklin, A. Ghodsi, M. Zaharia, Spark SQL: relational data processing in spark, in *SIGMOD* (2015), pp. 1383–1394
14. AsterixDB. Asterix query language (aql) (2016). https://asterixdb.ics.uci.edu/documentation/aql/manual.html. Accessed 17 Mar 2016
15. Azure Microsoft. Microsoft azure: Cloud computing platform and services (2016). https://azure.microsoft.com. Accessed 27 Feb 2016

16. O. Batarfi, R. El Shawi, A.G. Fayoumi, R. Nouri, S.-M.-R. Beheshti, A. Barnawi, S. Sakr, Large scale graph processing systems: survey and an experimental evaluation. Clust. Comput. **18**(3), 1189–1213 (2015)

17. R.A. Becker, J.M. Chambers, *S: An Interactive Environment for Data Analysis and Graphics* (CRC Press, New York, 1984)

18. K.S. Beyer, V. Ercegovac, R. Gemulla, A. Balmin, M. Eltabakh, C.-C. Kanne, F. Ozcan, E.J. Shekita, Jaql: a scripting language for large scale semistructured data analysis, in *Proceedings of VLDB Conference* (2011)

19. C. Chambers, A. Raniwala, F. Perry, S. Adams, R.R. Henry, R. Bradshaw, N. Weizenbaum, FlumeJava: easy, efficient data-parallel pipelines, in *PLDI* (2010)

20. W. Clinger, J. Rees, Ieee standard for the scheme programming language, in *Institute for Electrical and Electronic Engineers* (1991), pp. 1178–1990

21. Cloudera. Apache impala (2016). http://impala.io/. Accessed 17 Mar 2016

22. T.H. Cormen, *Introduction to Algorithms* (MIT press, New York, 2009)

23. S. Das, Y. Sismanis, K.S. Beyer, R. Gemulla, P.J. Haas, J. McPherson, Ricardo: integrating r and hadoop, in *Proceedings of the 2010 ACM SIGMOD International Conference on Management of data* (ACM, 2010), pp. 987–998

24. J. Dean, S. Ghemawat, MapReduce: simplified data processing on large clusters. Commun. ACM **51**(1) (2008)

25. Facebook. Presto (2016), https://prestodb.io/. Accessed 17 Mar 2016

26. L. George, *HBase: The Definitive Guide* (O'Reilly Media, Inc., 2011)

27. Google. Cloud sql - mysql relational database (2016). https://cloud.google.com/sql/. Accessed 27 Feb 2016

28. S. Guha, R. Hafen, J. Rounds, J. Xia, J. Li, B. Xi, W.S. Cleveland, Large complex data: divide and recombine (d&r) with rhipe. Stat **1**(1), 53–67 (2012)

29. C. Hewitt, P. Bishop, R. Steiger, A universal modular actor formalism for artificial intelligence, in *Proceedings of the 3rd International Joint Conference on Artificial Intelligence* (Morgan Kaufmann Publishers Inc., 1973), pp. 235–245

30. S. Hong, H. Chafi, E. Sedlar, K. Olukotun, Green-marl: a dsl for easy and efficient graph analysis, in *ACM SIGARCH Computer Architecture News*, vol. 40 (ACM, 2012), pp. 349–362

31. Inc Concurrent. Cascading - application platform for enterprise big data (2016). http://www.cascading.org/. Accessed 17 Mar 2016

32. R. Ihaka, R. Gentleman, R: a language for data analysis and graphics. J. Comput. Graph. Stat. **5**(3), 299–314 (1996)

33. M. Isard, M. Budiu, Y. Yu, A. Birrell, D. Fetterly, Dryad: distributed data-parallel programs from sequential building blocks, in *ACM SIGOPS Operating Systems Review*, vol. 41 (ACM, 2007), pp. 59–72

34. M. Islam, A.K. Huang, M. Battisha, M. Chiang, S. Srinivasan, C. Peters, A. Neumann, A. Abdelnur, Oozie: towards a scalable workflow management system for hadoop, in *SIGMOD Workshops* (2012)

35. W.M. Johnston, J.R. Hanna, R.J. Millar, Advances in dataflow programming languages. ACM Comput. Surv. (CSUR) **36**(1), 1–34 (2004)

36. A. Lakshman, P. Malik, Cassandra: a decentralized structured storage system. ACM SIGOPS Oper. Syst. Rev. **44**(2), 35–40 (2010)

37. G. Malewicz, M.H. Austern, A.J.C. Bik, J.C. Dehnert, I. Horn, N. Leiser, G. Czajkowski, Pregel: a system for large-scale graph processing, in *SIGMOD Conference* (2010)

38. X. Meng, J. Bradley, B. Yavuz, E. Sparks, S. Venkataraman, D. Liu, J. Freeman, D.B. Tsai, M. Amde, S. Owen, et al., Mllib: machine learning in apache spark (2015). arXiv preprint, arXiv:1505.06807

39. MongoDB Inc. Mongodb for giant ideas (2016). https://www.mongodb.org/. Accessed 27 Feb 2016

40. C. Olston, B. Reed, U. Srivastava, R. Kumar, A. Tomkins, Pig latin: a not-so-foreign language for data processing, in *SIGMOD* (2008)

41. Swift OpenStack. Openstack swift - enterprise storage from swiftstack (2016). https://www.swiftstack.com/openstack-swift/. Accessed 27 Feb 2016
42. S. Sakr, *Big Data 2.0 Processing Systems* (Springer, Berlin, 2016)
43. S. Sakr, M.M. Gaber (eds.) *Large Scale and Big Data - Processing and Management* (Auerbach Publications, 2014)
44. Sherif Sakr, Anna Liu, Ayman G. Fayoumi, The family of mapreduce and large-scale data processing systems. ACM Comput. Surv. **46**(1), 11 (2013)
45. K. Shvachko, H. Kuang, S. Radia, R. Chansler, The hadoop distributed file system, in *IEEE MSST* (2010)
46. S3 Amazon. Amazon simple storage service (amazon s3) (2016). https://aws.amazon.com/s3/. Accessed 27 Feb 2016
47. A. Thusoo, J.S. Sarma, N. Jain, Z. Shao, P. Chakka, S. Anthony, H. Liu, P. Wyckoff, R. Murthy, Hive: a warehousing solution over a map-reduce framework. Proc. VLDB Endow. **2**(2), 1626–1629 (2009)
48. A. Toshniwal, S. Taneja, A. Shukla, K. Ramasamy, J.M. Patel, S. Kulkarni, J. Jackson, K. Gade, M. Fu, J. Donham, et al., Storm@ twitter, in *Proceedings of the 2014 ACM SIGMOD international conference on Management of data* (ACM, 2014), pp. 147–156
49. Typesafe. Akka (2016). http://akka.io/. Accessed 17 Mar 2016
50. Typesafe. Play framework - build modern & scalable web apps with java and scala (2016). https://www.playframework.com/. Accessed 17 Mar 2016
51. L.G. Valiant, A bridging model for parallel computation. Commun. ACM **33**(8), 103–111 (1990)
52. Y. Yu, M. Isard, D. Fetterly, M. Budiu, Ú. Erlingsson, P.K. Gunda, J. Currey, Dryadlinq: a system for general-purpose distributed data-parallel computing using a high-level language, in *OSDI*, vol. 8 (2008), pp. 1–14
53. M. Zaharia, M. Chowdhury, M.J. Franklin, S. Shenker, I. Stoica, Spark: cluster computing with working sets, in *HotCloud* (2010)
54. M. Zaharia, M. Chowdhury, T. Das, A. Dave, J. Ma, M. McCauly, M.J. Franklin, S. Shenker, I. Stoica, Resilient distributed datasets: a fault-tolerant abstraction for in-memory cluster computing, in *NSDI* (2012)

Programming Platforms for Big Data Analysis

Jiannong Cao, Shailey Chawla, Yuqi Wang and Hanqing Wu

Abstract Big data analysis imposes new challenges and requirements on programming support. Programming platforms need to provide new abstractions and run time techniques with key features like scalability, fault tolerance, efficient task distribution, usability and processing speed. In this chapter, we first provide a comprehensive survey of the requirements, give an overview and classify existing big data programming platforms based on different dimensions. Then, we present details of the architecture, methodology and features of major programming platforms like MapReduce, Storm, Spark, Pregel, GraphLab, etc. Last, we compare existing big data platforms, discuss the need for a unifying framework, present our proposed framework MatrixMap, and give a vision about future work.

Keywords Big data analysis · Programming platforms · Unifying framework · Data parallel · Graph parallel · Task parallel · Stream processing

1 Introduction

The necessity of increased computing speed and capacity offered by big data programming platforms has led to constantly evolving system architectures, novel development environments, and multiple third-party software libraries and application packages. Now, we are in an era where businesses, government sectors, small and big organizations have all realized the potential of big data analysis. The great demand

J. Cao (✉) · S. Chawla · Y. Wang · H. Wu
Department of Computing, Hong Kong Polytechnic University, King's Park, Hong Kong
email: csjcao@comp.polyu.edu.hk
URL: http://www4.comp.polyu.edu.hk/~csjcao/

S. Chawla
e-mail: csschawla@comp.polyu.edu.hk

Y. Wang
e-mail: csyqwang@comp.polyu.edu.hk

H. Wu
e-mail: cshwu@comp.polyu.edu.hk

© Springer International Publishing AG 2017   65
A.Y. Zomaya and S. Sakr (eds.), *Handbook of Big Data Technologies*,
DOI 10.1007/978-3-319-49340-4_3

for big data analysis systems is giving a thrust to the research and development in this area. Large amounts of data have to be handled in a parallel and distributed way wherein, and the computations have to be distributed across many machines in order to be finished in a reasonable amount of time. The issue of how the computation can be parallelized, how data is distributed and how failures are handled in such a wide distribution are compelling, and call for special programming platforms for big data analysis.

In recent years, a lot of programming platforms have emerged for big data analysis. Figure 1 shows the time line of systems that handle large scale data. The timeline clearly indicates the increasing amount of interest in these systems recently.

Big data processing can be done on either distributed clusters or high performance computing machines like Graphical Processing Units [10].

In this section of the chapter, we provide an overview of existing programming platforms for big data analysis, which gives the readers a brief impression on existing big data platforms. The remaining part of the chapter is organized as follows. First, we discuss the special requirements and features of programming platforms for large scale data analysis in the next section. We then present in Sect. 3, a classification schema for big data programming platforms based on different dimensions, which would give insights on types of existing systems and their suitability to different kinds of applications. In Sect. 4, we will introduce the details of major existing programming platforms. The programming platforms are described with respect to their specific purpose, programming model, implementation details and important features. We discuss our unifying framework and our proposed framework called MatrixMap [15], as well as summarize the big data programming platforms according to the essential requirements in Sect. 5. Finally, we conclude this chapter by giving our understanding and vision on programming platforms. The chapter is intended to benefit anyone who is new to big data analysis by presenting details and features of popular big data programming platforms, analysts to choose appropriate program-

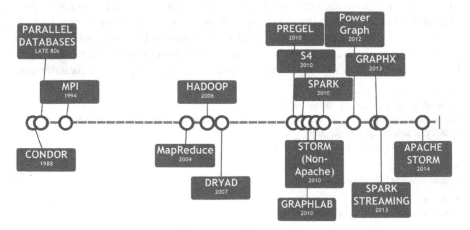

Fig. 1 Timeline of programming platforms for big data analysis

ming platforms for their specific applications by offering comparison across them, and also interested researchers by showing our current research work and vision on future direction.

2 Requirements of Big Data Programming Support

Programming platforms constitute of systems and language environments that can run on commodity, inexpensive hardware or software and can be programmed and operated by programmers and analysts with average, mainstream skills. Big data analysis need to have some essential requirements so as to deal with specific issues related to vast data and large scale computations, they also need to support distributed and local processing (data copies) and support ease of use, data abstraction, data flow and data transformations. In traditional programming platforms, the key feature is performance, but for systems with large scale data, there are many more features essential for smooth functioning of the system and being useful.

Scalability
Scalability is the ability of a system, network, or process to handle a growing amount of work in a capable manner or its ability to be enlarged to accommodate that growth [5]. Scaling can be done by either scaling up the system, which means adding additional resources on a single computer/node to improve the performance or scaling out the system, which refers to addition of more computers/nodes to the system in a distributed software system.

Support Multiple Data Types
Big data systems should be able to support multiple data types, e.g., record, graph or stream. Different common data types have been briefly explained in following text.

- *Record data* can be split into independent elements and thus data can be processed independently. Independent results can be summed up to get the final result.
- *Graph data* cannot be split into independent elements like in the case of record data. Elements may have relations with each other and thus the processing of one element depends on other elements. Graph data not only include real graphs but also other data which can be viewed as graph. The data can also be in form of stream which would require fast processing in memory.
- *Stream data* arrive at a rate that makes it infeasible to store everything in active storage. If it is not processed immediately or stored, then it is lost forever or we lose the opportunity to process them at all. Thus, stream-processing algorithm is executed in main memory, without or with only rare access to secondary storage.

Fault Tolerance
Fault tolerance is the property that enables a system to continue operating properly in the event of the failure of (or one or more faults within) some of its components. In a distributed framework with large scale data, it is imperative that some nodes

carrying data can fail. For a fault tolerant system, when a server in the cluster fails, a stand-by server is automatically activated to take over the tasks, there are also check pointing and recovery to minimize state loss.

Efficiency
Massive computation capability is required for big data analysis and hence efficiency is very critical when programming platforms are scaled up or scaled out for handling large amounts of data. Efficiency means faster speed with respect to usage of certain resources like memory or number of nodes.

Data I/O Performance
Data I/O performance refers to the rate at which the data is transferred to/from a peripheral device. In the context of big data analytics, this can be viewed as the rate at which the data is read and written to the memory (or disk) or the data transfer rate between the nodes in a cluster. The systems should have low latency to minimize the time taken for reading and writing to the memory, and high throughput for data transmission.

Iterative Task Support
This is the ability of a system to efficiently support iterative tasks. Since many of the data analysis tasks and algorithms are iterative in nature, it is an important metric to compare different platforms, especially in the context of big data analytics. The systems must be suitable for iterative algorithms so that the result of one iteration can be easily used in the next iteration, and all the parameters can be stored locally. Processes can reside and can keep running as long as the machine is running.

The properties described above are very significant for description of programming platforms. In the next section, we propose a classification of programming system based on different dimensions. We have classified the programming platforms based on the processing techniques and data sources.

3   Classification of Programming Platforms

The existing programming platforms for big data analysis have numerous special features as discussed in the previous section. It is important to realize what kinds of systems encompass what features so that it is easier to make a choice of programming system with respect to the application. We classify the existing programming platforms based on different dimensions as that have been described in the following subsections.

3.1   Data Source

Data analysis is done for different kinds of source data. The data can arrive for processing either in batches or continuous stream. Hence based on how the data arrives, various systems can be classified into Batch Processing Systems and Stream Processing Systems. Many big data analysis applications work on batch-wise input, and there are many like twitter or stock markets dealing with multiple data streams. There have been much development in this regard, and specific programming platforms have been fostered to deal with streaming data.

Batch Processing Systems are the systems that execute a series of programs which take a set of data files as input, processes the data and produce a set of output data files. It is termed as batch processing because the data is collected in batches as sets of records and processed as a unit. Output is another batch that can be reused as input if required.

Batch processing systems have existed for very long and they have various advantages. These systems utilize computing resources in an optimum and efficient manner based on the priority of other jobs. Batch processing techniques are likely to avoid system overhead.

Many distributed programming platforms like MapReduce [8], Spark [32], GraphX [12], Pregel [22] and HTCondor [14] are batch processing systems. They analyze large scale data in batches in a distributed and parallel fashion.

Stream Processing Systems are the systems that process continuous input of data. These systems should have faster rate of processing than rate of incoming data. So an input dataset coming at time t needs to be processed before dataset arrives at time $t + 1$. The stream processing systems work under a very strict time constraint. They are important in applications, which need continuous output from incoming data like stock market, twitter etc. Big data programming platforms like Storm [30], Spark Streaming [32] and S4 [24] are used for processing stream data.

3.2 Processing Technique

Programming platforms can also be classified based on the processing techniques. Large scale processing can be done using different techniques like data parallel, task parallel or graph parallel techniques. We have classified the programming platforms according to the techniques they employ for processing data.

Data parallel programming platforms focus on distributed data across parallel computing nodes. In data parallelism, each node executes the same task on different pieces of distributed data. It emphasizes that data is distributed and executed in parallel on different computing nodes, and then the result from different nodes is consolidated and processed further. The data parallel systems tend to be very fault tolerant as they can have redundancy. Also, this kind of arrangement makes processing of large-scale data simpler by breaking down data into smaller units. MapReduce, Spark, Hadoop are data parallel systems and have been very popular in big data programming community. We also proposed MatrixMap to efficiently support matrix computations.

Task parallel platforms are systems that process data in a parallel manner across multiple processors. Task parallelism focuses on distributing execution processes

Fig. 2 Classification
schema of big data
programming platforms
(stream processing platforms
are mentioned in italics)

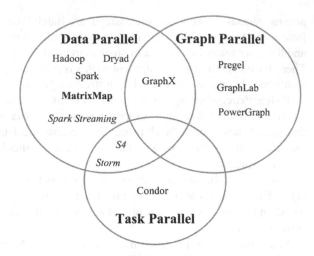

across different parallel computing nodes. HTCondor programming system is an example of task parallel system.

Graph parallel platforms are systems that encode computation as vertex programs which run in parallel and interact along edges in the graph. Graph-parallel abstractions rely on each vertex having a small neighborhood to maximize parallelism, and effective partitioning to minimize communication. Formally, a graph-parallel abstraction consists of a sparse graph $G = \{V, E\}$, and a vertex-program Q which is executed in parallel on each vertex v belongs to set V, and can interact (e.g., through shared-state in GraphLab, or messages in Pregel) with neighboring instances $Q(u)$ where (u, v) belongs to E. In contrast to more general message passing models, graph-parallel abstractions constrain the interaction of vertex-program to a graph structure enabling the optimization of data-layout and communication [11]. Pregel, Graphlab, GraphX are graph parallel systems popular for social network analysis.

Figure 2 depicts the classification schema of various programming platforms for big data analysis. The Figure presents the classification in the form of a Venn diagram, and the programming platforms are placed according to their matching criterion. The systems in italics are stream processing systems, while the remaining are batch processing systems. In the next section we describe the major existing programming platforms in detail.

4 Major Existing Programming Platforms

In this section we describe in detail some major programming platforms that are prominent in big data analysis. The programming platforms have been described according to the prominent processing techniques used in their programming models.

4.1 Data Parallel Programming Platforms

Data parallel programming platforms are the systems that distribute data over parallel computing nodes [6]. In distributed systems, data parallelism is achieved by dividing the data into a smaller size and each parallel computing node performing the same task over small sized data. The intermediate result is then integrated to achieve the final outcome of processing.

4.1.1 Hadoop

Hadoop is based on MapReduce programming model [8] which is the most popular paradigm for big data analysis till date, and brought a breakthrough in big data programming. In this model, data-parallel computations are executed on clusters of unreliable machines by systems, that automatically provide locality-aware scheduling, fault tolerance, and load balancing. Hadoop MapReduce is an open source form of Google MapReduce.

MapReduce is useful in a wide range of applications, including distributed pattern-based searching, distributed sorting, web link-graph reversal, web access log stats, inverted index construction, document clustering, machine learning, and statistical machine translation. At Google, MapReduce was used to completely regenerate Google's index of the World Wide Web. It has replaced the old ad hoc programs that updated the index and ran various analyses.

The MapReduce abstraction allows expressing simple computations without revealing the complicated details of parallelization. There are two main primitives in this abstraction called the *Map* and *Reduce* operations. The computation is expressed in form of these two functions, wherein it takes a set of input *key/value* pairs and produces a set of output *key/value* pairs.

Map, written by the user, takes an input pair and produces a set of intermediate *key/value* pairs. The MapReduce library groups together all intermediate values associated with the same intermediate key and passes them to the *Reduce* function.

Reduce function, written by user too, accepts intermediate key and a set of values for that key. It merges these values together to form a possibly smaller set of values. This is done in an iterative fashion, so that list of values that are too large can fit in memory. This is the key concept of the MapReduce paradigm that enables it to handle large scale data in an efficient way.

The MapReduce framework transforms a list of (key, value) pairs into a list of values. This behavior is different from the typical functional programming, *Map* and *Reduce* combination, which accepts a list of arbitrary values and returns one single value that combines all the values returned by map. Figure 3 [4] depicts the architecture of MapReduce programming model.

MapReduce framework for processing parallelizable problems across huge datasets using a large number of computers (nodes), collectively referred to as a cluster (if all nodes are on the same local network and use similar hardware) or a

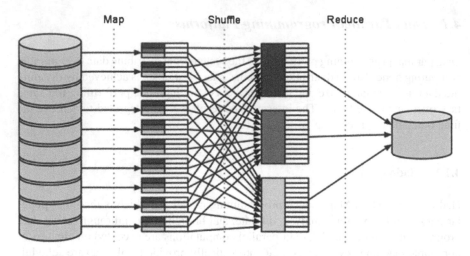

Fig. 3 Architecture of MapReduce model

grid (if the nodes are shared across geographically and administratively distributed systems, and use more heterogeneous hardware). Processing can occur on data stored either in a file system (unstructured) or in a database (structured). MapReduce can take advantage of locality of data, processing it on or near the storage assets in order to reduce the distance over which it must be transmitted.

"Map" step: Each worker node applies the "map()" function to the local data, and writes the output to a temporary storage. A master node orchestrates that for redundant copies of input data, only one is processed.

"Shuffle" step: Worker nodes redistribute data based on the output keys (produced by the "map()" function), such that all data belonging to one key are located on the same worker node.

"Reduce" step: Worker nodes now process each group of output data, per key, in parallel.

The parallelism also offers some possibility of recovering from partial failure of servers or storage during the operation: if one mapper or reducer fails, the work can be rescheduled - assuming the input data is still available.

Hadoop has following important features:

Scalability: Hadoop is highly scalable and it can be scaled out instead of scaling up. The main feature of Hadoop is that the machines with normal functioning capacity can also be used for big data analysis. Multi-node clusters of Hadoop system can be set up in distributed master slave architecture and scalability can be achieved for thousands of nodes.

Fault tolerance: Fault tolerance is the most significant feature of MapReduce programming model that makes it a robust and reliable programming system for large scale data processing. Fault tolerance is achieved in MapReduce by redundancy of data. Each dataset is duplicated in 3–4 places. Even when a node fails, the same dataset can be retrieved from other nodes.

Performance: MapReduce programming model is very efficient for large amounts of data. However, the performance is not good when the dataset is small. The time lag of Hadoop model is compromised because of its efficient fault tolerance and high scalability.

4.1.2 Spark

Spark is an efficient and iterative processing model for big data processing. At its core, Spark provides a general programming model that enables developers to write applications by composing arbitrary operators, such as *mappers*, *reducers*, *joins*, *group-bys*, and *filters*. This composition makes it easy to express a wide array of computations, including iterative machine learning, streaming, complex queries, and batch processing.

Spark programming model focuses on applications that reuse a working set of data across multiple parallel operations. This includes many iterative machine learning algorithms, as well as interactive data analysis tools. It keeps track of the data that each of the operators produces, and enables applications to reliably store this data in memory. This feature enables efficient iterative algorithms and low latency computations.

Spark provides two main abstractions for parallel programming: resilient distributed datasets and parallel operations on these datasets. Spark programming model is shown in Fig. 4 [27]. It describes two kinds of computations, iterative and

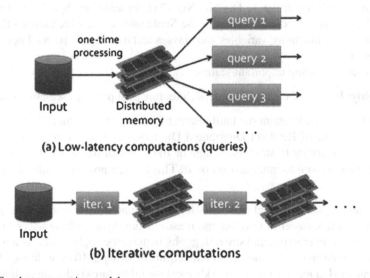

Fig. 4 Spark programming model

non-iterative. The main abstraction in Spark is that of a Resilient Distributed Dataset (RDD), which represents a read-only collection of objects partitioned across a set of machines that can be rebuilt if a partition is lost. The elements of an RDD need not exist in physical storage; instead, a handle to an RDD contains enough information to compute the RDD starting from data in reliable storage. This means that RDDs can always be reconstructed if nodes fail. In Spark, each RDD is represented by a Scala [25] object. Spark lets programmers construct RDDs in various ways like from a file in a shared file system, by "parallelizing" a Scala collection (e.g., an array) in the driver program, by transforming an existing RDD and by changing the persistence of an existing RDD. Several parallel operations like *reduce, collect, foreach* etc. can be performed on RDDs.

Spark also lets programmers create two restricted types of shared variables to support two simple but common usage patterns. Programmer can create a "broadcast variable" object that wraps the value and ensures that it is only copied to each worker once. Also, Accumulators can be defined for any type that has an "add" operation and a "zero" value. Due to their "add-only" semantics, they are easy to make fault-tolerant.

Spark is built on top of Mesos [13], a "cluster operating system" that lets multiple parallel applications share a cluster in a fine-grained manner and provides an API for applications to launch tasks on a cluster. This allows Spark to run alongside existing cluster computing frameworks, such as Mesos ports of Hadoop and MPI [26], and share data with them. In addition, building on Mesos greatly reduced the programming effort that had to go into Spark.

The two types of shared variables in Spark, broadcast variables and accumulators, are implemented using classes with custom serialization formats. Spark is implemented in Scala (Scala programming language.), a statically typed high-level programming language for the Java Virtual Machine, and exposes a functional programming interface similar to DryadLINQ [31]. In addition, Spark can be used interactively from a modified version of the Scala interpreter, which allows the user to define RDDs, functions, variables and classes and use them in parallel operations on a cluster.

Spark has following important features:

Scalability: It is based on MapReduce architecture so it provides scalability feature.

Fault tolerant: Spark retains the fault tolerant feature of map reduce. Also, its novel feature is the use of Resilient Distributed Datasets (RDD). The main property of RDD is the capability to store its lineage or the series of transformations required for creating it as well as other actions on it. This lineage provides fault tolerance to RDDs.

Easy to use: Spark's parallel programs look very much like sequential programs, which make them easier to develop and reason about. Spark allows users to easily combine batch, interactive, and streaming jobs in the same application. As a result, a Spark job can be up to 100 times faster and requires writing 2–10 times less code than an equivalent Hadoop job. One of Spark's most useful features is the interactive shell, bringing Spark's capabilities to the user immediately - no Integrated Development Environment (IDE) and code compilation required. The shell can be used as the primary tool for exploring data interactively, or as means to test portions of an application you're developing. Spark can read and write data from and to Hadoop Distributed File System (HDFS).

Better Performance: Spark can outperform Hadoop by 10x in iterative machine learning jobs, and can be used to interactively query a 39 GB dataset with sub-second response time.

4.1.3 Dryad

Dryad [17] was a research project at Microsoft Research for writing parallel and distributed programs to scale from a small cluster to a large data-center. From 2007, Microsoft made several preview releases of this programming model technology available as add-ons to Windows HPC Server 2008 R2. However, Microsoft dropped Dryad processing work and focused on Apache Hadoop in October 2011. Dryad allows a programmer to use the resources of a computer cluster or a data center for running data-parallel programs. A Dryad programmer can use thousands of machines, each of them with multiple processors or cores, without knowing anything about concurrent programming.

A Dryad programmer writes several sequential programs and connects those using one-way channels. The computation of an application written for Dryad is structured as a Directed Acyclic Graph (DAG). The DAG defines the dataflow of the application, and the vertices of the graph define the operations that are to be performed on the data. The "computational vertices" are written using sequential constructs, devoid of any concurrency or mutual exclusion semantics. A Dryad job is a graph generator which can synthesize any directed acyclic graph. The structure of Dryad jobs is shown in Fig. 5 [28]. These graphs can even change during execution, in response

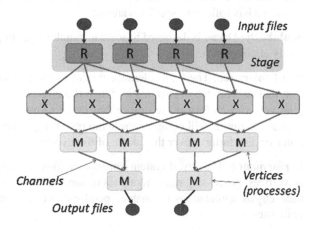

Fig. 5 The structure of Dryad jobs

to important events in the computation. The Dryad runtime parallelizes the dataflow graph by distributing the computational vertices across various execution engines. Scheduling of the computational vertices on the available hardware is handled by the Dryad runtime, without any explicit intervention by the developer of the application or administrator of the network.

The flow of data between one computational vertex to another is implemented by using communication "channels" between the vertices, which in physical implementation is realized by TCP/IP streams, shared memory or temporary files. A stream is used at runtime to transport a finite number of structured items.

Dryad defines a domain-specific language, implemented via a C++ library, that is used to create and model a Dryad execution graph. Computational vertices are written using standard C++ constructs. To make them accessible to the Dryad runtime, they must be encapsulated in a class that inherits from the GraphNode base class. The graph is defined by adding edges; edges are added by using a composition operator that connects two graphs with an edge. A lot of operators are defined to help building a graph, including Cloning, Composition, Merge and Encapsulation. Managed code wrappers for the Dryad API can also be written.

Dryad's architecture includes components that do resource management as well as the job management. A Dryad job is coordinated by a component called the Job Manager. Tasks of a job are executed on cluster machines by a Daemon process. Communication with the tasks from the job manager happens through the Daemon, which acts like a proxy. In Dryad, the scheduling decisions are local to an instance of the Dryad Job Manager C i.e., it is decentralized. The logical plan for a Dryad DAG results in each vertex being placed in a "Stage". The stages are managed by a "Stage manager" component that is part of the job manager. The Stage manager is used to detect state transitions and implement optimizations like Hadoop's speculative execution.

Overall, Dryad is quite expressive. It completely subsumes other computation frameworks, such as Google's MapReduce, or the relational algebra. Moreover, Dryad handles job creation and management, resource management, job monitoring and visualization, fault tolerance, re-execution, scheduling, and accounting.

Dryad has following special features:

Scalability: Dryad is designed to scale to much larger implementations, up to thousands of computers.

Fault tolerance: The fault tolerance model in the Dryad comes from the assumption that vertices are deterministic. Since the communication graph is acyclic, it is relatively straightforward to ensure that every terminating execution of a job with immutable inputs will compute the same result, regardless of the sequence of computer or disk failures over the course of the execution.

Performance: The Dryad system can execute jobs containing hundreds of thousands of vertices, processing many terabytes of input data in minutes. Microsoft routinely uses Dryad applications to analyze petabytes of data on clusters of thousands of computers.

Flexibility: Programmers can easily use thousands of machines and create large-scale distributed applications, without requiring them to master any concurrency programming beyond being able to draw a graph of the data dependencies of their algorithms.

4.2 Graph Parallel Programming Platforms

Graph parallel systems are systems that encode computation as vertex programs which run in parallel and interact along edges in the graph. Graph-parallel abstractions rely on each vertex having a small neighborhood to maximize parallelism and effective partitioning to minimize communication.

4.2.1 Pregel

Pregel [22] is a programming model for processing large graphs in distributed environment. It is a vertex-centric model, which defines serials of actions on an angle of a single vertex, and then the program will run such vertices through a graph and finally get the result.

Pregel has been created for solving large scale graph computations that is required in modern systems like social networks and web graphs. Many graph computing problems like shortest path, clustering, page rank, connected components etc. need to be implemented for big graphs hence the requirement of the system.

Vertices iteratively process data and send messages to neighboring vertices. Edges do not have any associated computation in this programming model. The computations consist of a sequence of iterations, called *supersteps*. Within each *superstep*, the vertices compute in parallel, each executing the same user defined function that expresses the logic of a given algorithm. A vertex can modify its state or that of its outgoing edges, receive messages sent to it in the previous *superstep*, send messages to other vertices (to be received in the next *superstep*), or even mutate the topology of the graph. The state machine of vertex is shown in Fig. 6 [22].

The input to a Pregel computation is a directed graph in which each vertex is uniquely identified by a string vertex identifier. Each vertex is associated with a modifiable, user defined value. The directed edges are associated with their source vertices, and each edge consists of a modifiable, user defined value and a target vertex identifier. A typical Pregel computation consists of input, when the graph is

Fig. 6 State machine for a vertex

initialized, followed by a sequence of *supersteps* separated by global synchronization points until the algorithm terminates, and finishing with output. Algorithm termination is based on every vertex voting to halt. The output of a Pregel program is the set of values explicitly output by the vertices. It is often a directed graph isomorphic to the input, but this is not a necessary property of the system because vertices and edges can be added and removed during computation. A clustering algorithm, for example, might generate a small set of disconnected vertices selected from a large graph.

The Pregel library divides a graph into partitions, each consisting of a set of vertices and all of those vertices' outgoing edges. Assignment of a vertex to a partition depends solely on the vertex ID, which implies it is possible to know which partition a given vertex belongs to even if the vertex is owned by a different machine, or even if the vertex does not yet exist. The default partitioning function is just hash (ID) mod N, where N is the number of partitions, but users can replace it. The execution of Pregel is depicted in Fig. 7 [16]. In the absence of faults, the execution of a Pregel program consists of several stages. First, many copies of the user program begin executing on a cluster of machines. One of these copies acts as the master. It is not assigned any portion of the graph, but is responsible for coordinating worker activity. The workers use the cluster management system's name service to discover the master's location, and send registration messages to the master. Then, the master determines how many partitions the graph will have, and assigns one or more partitions to each worker machine. Having more than one partition per worker allows parallelism among the partitions and better load balancing, and will usually improve performance. Each worker is given the complete set of assignments for all workers.

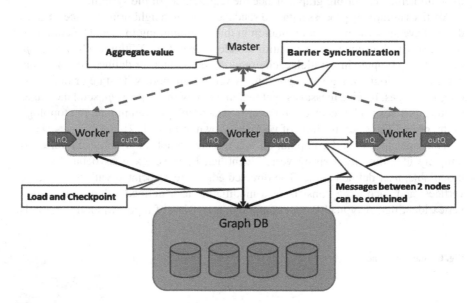

Fig. 7 Implementation of Pregel

After this stage, the master assigns a portion of the user's input to each worker. The input is treated as a set of records, each of which contains an arbitrary number of vertices and edges. The division of inputs is orthogonal to the partitioning of the graph itself, and is typically based on file boundaries. If a worker loads a vertex that belongs to that worker's section of the graph, the appropriate data structures are immediately updated. Otherwise the worker enqueuers a message to the remote peer that owns the vertex. After the input has finished loading, all vertices are marked as active.

Later, the master instructs each worker to perform a *superstep*. The worker loops through its active vertices, using one thread for each partition. The worker calls *Compute()* for each active vertex, delivering messages that were sent in the previous *superstep*. When the worker is finished it responds to the master, telling the master how many vertices will be active in the next *superstep*. This step is repeated as long as any vertices are active, or any messages are in transit. After the computation halts, the master may instruct each worker to save its portion of the graph.

Pregel has following special features:

Scalability: Pregel has very good scalability. It can work for large sized graphs with millions of vertices.

Fault tolerance: Fault tolerance is achieved through check pointing. At the beginning of a *superstep*, the master instructs the workers to save the state of their partitions to persistent storage, including vertex values, edge values, and incoming messages; the master separately saves the aggregator values. Worker failures are detected using regular "ping" messages that the master issues to workers. If a worker does not receive a ping message after a specified interval, the worker process terminates. When one or more workers fail, the current state of the partitions assigned to these workers is lost. The master reassigns graph partitions to the currently available set of workers, and they all reload their partition state from the most recent available checkpoint at the beginning of a *superstep S*.

Performance: Pregel is very fast compared to non-graph based frameworks. But during implementation it waits for the slow workers that decrease its speed.

Flexibility: Pregel provides flexibility to implement different algorithms. The Pregel implementation is easy to understand and implementation of varied algorithms can be done on it. Programming complexity is simplified by using the *supersteps*.

4.2.2   GraphX

GraphX [12] is an efficient, resilient, and distributed graph processing framework that provides graph-parallel abstractions and supports a wide range of iterative graph algorithms. Existing specialized graph processing systems, such as Pregel and GraphLab, are sufficient to process only graph data. Thus, using specialized graph processing systems in large-scale graph analytics pipeline, requires extensive data movement and duplication across file system, and even network. Moreover, users have to learn and manage multiple systems, such as Hadoop, Spark, Pregel and GraphLab. Overall, having separate systems in entire graph analytics pipeline is difficult to use and inefficient.

GraphX addresses the above challenges by providing both table view and graph view on the same physical data. On one hand, GraphX views physical data as graphs so that it can naturally express and efficiently execute iterative graph algorithms. On the other hand, graphs in GraphX are distributed as tabular data-structures so that GraphX also provides table operations on physical data. By exploiting this unified data representation, GraphX enables users to easily and efficiently express the entire graph analytics pipeline. Since graph can be composed by tables in GraphX, tabular data preprocessing and transformation between table and graph are directly realized within one system. Meanwhile, GraphX provides APIs similar to specialized graph processing systems for naturally expressing and efficiently executing iterative graph algorithms. Moreover, GraphX can leverage in-memory computation and fault-tolerance by being embedded in Spark, a general-purpose distributed dataflow framework.

Programmers can implement iterative graph algorithms without caring much about the iterations and only need to define a vertex program. However, the foundation of GraphX' graph-parallel abstractions is different from the common one that is iterative local transformation [12]. GraphX further decomposes iterative local transformation into specific dataflow operators, which are a sequence of join stages and group-by stages punctuated by map operations. The join operation and group-by operation are in the context of relational database, and the map operation is to perform update. GraphX realizes the partitioning of graphs in its representation of physical data, called distributed graph representation. Figure 8 [12] illustrates how a graph is represented by horizontally partitioned vertex and edge collections and their indices. The edges are divided into three edge partitions by applying a partition function (e.g., 2D Partitioning), and the vertices are partitioned by vertex id. Partitioned with the vertices, GraphX maintains a routing table encoding the edge partitions for each vertex.

GraphX is built as a library on top of Spark [32], which is a general-purpose distributed dataflow framework. The architecture of Spark with GraphX is illustrated by Fig. 9 [12]. As seen from the architecture, there is one more layer called Gather Apply Scatter (GAS) Pregel API between GraphX and some graph algorithms. The GAS Pregel API is implementation of Pregel abstraction of graph-parallel using GraphX dataflow operations. It is claimed that GraphX can implement Pregel abstractions in less than 20 lines of codes. Data structure of GraphX, the distributed graph representation, is built on Spark RDD abstraction, and GraphX API is expressed on top of Spark standard dataflow operators. GraphX can also exploited Scala foundation of Spark, which enables GraphX to interactively load, transform, and compute on massive graphs. GraphX requires no modifications to Spark. As a result, GraphX can also be seen as a general method to embed graph computation within distributed dataflow frameworks and distill graph computation to a specific join-map-group-by dataflow pattern. Being embedded in Spark allows GraphX to inherit many good

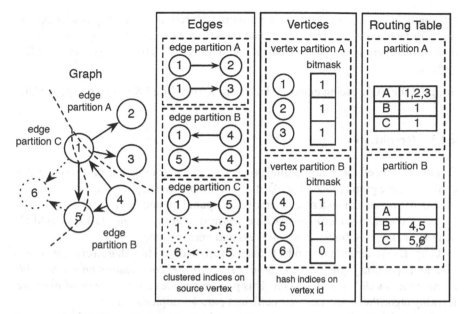

Fig. 8 Distributed graph representations

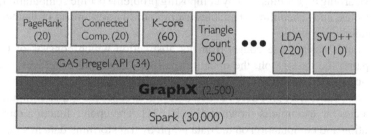

Fig. 9 Spark with GraphX

features of Spark, such as in-memory computation and fault-tolerance. Compared with Pregel and GraphLab, GraphX can achieve these features with a smaller cost.

GraphX has following important features:

Scalability: Being embedded in Spark allows GraphX to inherit Spark scalable property.

Fault tolerance: Being embedded in Spark allows GraphX to inherit Spark fault tolerance. Different from checkpoint-based fault tolerance, which is adopted by other graph systems, fault tolerance of GraphX is based on lineage. Compared with checkpoint fault tolerance, lineage-based fault tolerance produces smaller performance overhead and optimal dataset replication.

Efficient for graph analytics pipeline: Similar to specialized graph processing systems, such as Pregel and GraphLab, GraphX enables users to naturally express and efficiently execute iterative graph algorithms. Moreover, GraphX provides operations for tabular data preprocessing, and transformation between graph and tabular data so that there is no data movement and duplication across the network and file system.

Support for SQL: Being embedded in Spark allows GraphX to inherit Spark SQL.

4.2.3   GraphLab

GraphLab is an efficient and parallel processing model for big data processing especially for large graph processing. As its core, GraphLab supports the representation of structured data dependencies, iterative computation, and flexible scheduling. By targeting common patterns in machine learning algorithms and tasks, GraphLab achieves notable usability, expressiveness and performance.

GraphLab programming model focuses on applications that share a coherent computational pattern: *asynchronous iterative and parallel computation on graphs with a sequential model of computation*. This pattern encodes a broad range of machine learning algorithms, and facilitates efficient parallel implementations.

GraphLab exploits the *sparse structure* and common *computational patterns* of machine learning algorithms, and by composing problem specific computation, data-dependencies, and scheduling, it enables users to easily design and implement efficient parallel algorithms.

GraphLab's ease-of-use comes from its abstraction which consists of the following parts: the data graph, the update function, scheduling primitives, the data consistency model, and the sync operation. The data graph represents user modifiable program state, stores the user-defined data and encodes the sparse computational dependencies, an example is shown in Fig. 10 [21]. The update function represents the operation and computation on the data graph by transforming data in small overlapping contexts called scopes. Scheduling primitives determine the computation order. The data consistency model expresses how much computation can overlap. Last, the sync operation concurrently keeps track of global states.

The GraphLab is implemented in the shared memory setting [20] and distributed in-memory setting [21]. In the shared memory setting, the GraphLab abstraction uses *PThreads* for parallelism. The data consistency models have been implemented using race-free and deadlock-free ordered locking protocols. To attain maximum performance, issues related to parallel memory allocation, concurrent random number generation, and cache efficiency are addressed in [20]. The shared memory setting is extended to the distributed setting by refining the execution model, relaxing the scheduling requirements, and introducing a new distributed data-graph, execution engines, and fault-tolerance systems [21].

The GraphLab API serves as an interface between the machine learning and systems communities. Parallel machine learning algorithms built on the GraphLab API benefit from developments in parallel data structures. As new locking protocols and parallel scheduling primitives are incorporated into the GraphLab API, they become

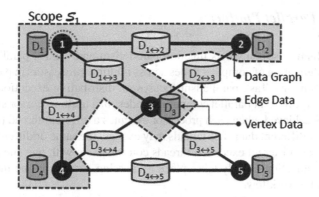

Fig. 10 The GraphLab data graph and scope S_1 of vertex 1 are illustrated in this figure. Each *gray cylinder* represents a block of user defined data and is associated with a vertex or edge. The scope of vertex 1 is illustrated by the region containing vertices $\{1, 2, 3, 4\}$. An update function applied to vertex 1 is able to read and modify all the data in S_1 (vertex data D_1, D_2, D_3, D_4 and edge data $D_{1\to2}$, $D_{1\to3}$, and $D_{1\to4}$)

immediately available to the machine learning community. On the other hand, Systems experts can use machine learning algorithms to new parallel hardware more easily by porting the GraphLab API. Actually, on top of GraphLab, several implemented libraries of algorithms in various application domains are already provided including topic modeling, graph analytics, clustering, collaborative filtering, computer vision etc.

GraphLab has following important features:

Scalability: GraphLab scales very well in various machine learning and data mining tasks, and scaling performance improves with higher computation to communication ratio.

Expressivity: Unlike many high-level abstractions (i.e., MapReduce), GraphLab is able to express complex computational dependencies with the data graph and provides sophisticated scheduling primitives which can express iterative parallel algorithms with dynamic scheduling.

Better Performance: GraphLab can outperform Hadoop by 20–60x in iterative machine learning and data mining tasks, and is competitive with tailored MPI implementation. The C++ execution engine is optimized to leverage extensive multithreading and asynchronous IO.

Powerful Machine Learning Toolkits: GraphLab has a large selection of machine learning methods already implemented. Users can also implement their own algorithms on top of the GraphLab programming API.

4.3   Task Parallel Platforms

Task parallelism (also known as function parallelism and control parallelism) is a form of parallelization of computer codes across multiple processors in parallel computing environments. Task parallelism focuses on distributing execution processes (threads) across different parallel computing nodes. In a multiprocessor system, task parallelism is achieved when each processor executes a different thread (or process) on the same or different data. The threads may execute the same or different code. In the general case, different execution threads communicate with one another as they work. Communication usually takes place by passing data from one thread to the next as part of a workflow.

4.3.1   HTCondor

HTCondor has been derived from Condor that is a batch system for harnessing idle cycles on personal workstations [19]. Since then, it has matured to become a major player in the compute resource management area and renamed HTCondor in 2012. HTCondor (HTCondor) is a high throughput computing system for compute-intensive jobs. Like other full-featured batch systems, HTCondor provides a job queueing mechanism, scheduling policy, priority scheme, resource monitoring, and resource management.

HTCondor is able to transparently produce a checkpoint and migrate a job to a different machine which would otherwise be idle when it detects that a machine is no longer available. It does not require a shared file system across machines - if no shared file system is available, it can transfer the job's data files on behalf of the user, or it may be able to transparently direct all the job's I/O requests back to the submit machine. As a result, it can be used to seamlessly combine all of an organization's computational power into one resource.

HTCondor programming model has several logical entities, as shown in Fig. 11 [23]. The central manager acts as a repository of the queues and resources. A process called the "collector" acts as an information dashboard. A process called the "startd" manages the computes resources provided by the execution machines (worker nodes in the diagram). The *startd* gathers the characteristics of compute resources such as CPU, memory, system load, etc. and publishes it to the collector. A process called the "schedd" maintains a persistent job queue for jobs submitted by the users. A process called the "negotiator" is responsible for matching the computer resources to user jobs.

The communication flow in Condor is fully asynchronous. Each *startd* and each *schedd* advertise the information to the collector asynchronously. Similarly, the negotiator starts the matchmaking cycle using its own timing. The negotiator periodically queries the *schedd* to get the characteristics of the queued jobs and matches them to available resources. All the matches are then ordered based on user priority and communicated back to the *schedds* that in turn transfer the matched user jobs to

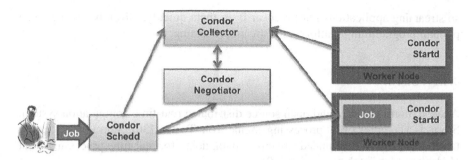

Fig. 11 Condor architecture overview

the selected *startds* for execution. To fairly distribute the resources among users, the negotiator tracks resource consumption by users and calculates user priorities accordingly.

Condor supports the transferring of input files to a worker node (startd) before a job is launched and of output files to the submit node (schedd) after the job is finished. Using a flexible plugin architecture, HTCondor can easily be extended to support domain specific protocols, such as GridFTP and Globus Online.

HTCondor has following important features:

Flexibility: The *ClassAd* mechanism in HTCondor provides an extremely flexible and expressive framework for matching resource requests (jobs) with resource offers (machines). Jobs can easily state both job requirements and job preferences. Likewise, machines can specify requirements and preferences about the jobs they are willing to run.

Efficiency: HTCondor is a high throughput computing system. Also, it utilizes the computing resources in a very efficient way.

4.4   Stream Processing Programming Platforms

Much of "big data" is received in real time, and is most valuable at its time of arrival. For example, a social network may wish to detect trending conversation topics in minutes; a search site may wish to model which users visit a new page; and a service operator may wish to monitor program logs to detect failures in seconds. To enable these low-latency processing applications, there is need for streaming computation models that scale transparently to large clusters, in the same way that batch models like MapReduce simplified offline processing.

Designing such models is challenging, however, because the scale needed for the largest applications can be hundreds of nodes. At this scale, two major problems are faults and stragglers (slow nodes). Both problems are inevitable in large clusters, so streaming applications must recover from them quickly. Given below are some popular programming platforms for stream processing.

4.4.1 Storm

Apache Storm is a free and open source distributed real-time computation system. Storm is a complex event processing engine from Twitter [30]. Storm makes it easy to reliably process unbounded streams of data, doing for real-time processing what Hadoop did for batch processing [29].

It has been used by various companies for many purposes like real time analytics, online machine learning, continuous computation, distributed RPC, ETL, and more. The fundamental concept in Storm is that of a stream, which can be defined as an unbounded sequence of tuples. Storm provides ways to transform the stream in various ways in decentralized and fault tolerant manner [1].

The storm topology lays down the architecture for processing of streams. The topology comprises of a spout, which is a reader or source of streams and a bolt, which is a processing entity and wiring together of spouts and bolts as shown in Fig. 12 [2].

Clients submit topologies to a master node, which is called the *Nimbus*. Nimbus is responsible for distributing and coordinating the execution of the topology. The actual work is done on worker nodes. Each worker node runs one or more worker processes. At any point in time a single machine may have more than one worker processes, but each worker process is mapped to a single topology. Note more than one worker process on the same machine may be executing different part of the same topology. The high level architecture of Storm is shown in Fig. 13 [22].

Each worker process runs a JVM, in which it runs one or more executors. Executors are made of one or more tasks. The actual work for a bolt or a spout is done in

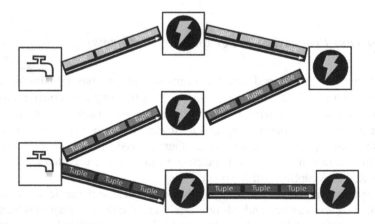

Fig. 12 Storm topology

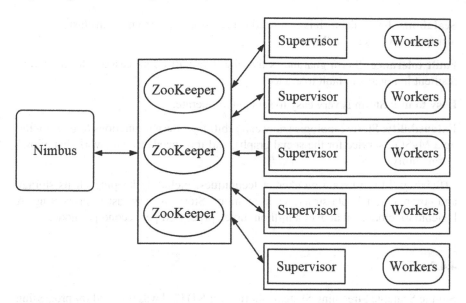

Fig. 13 High level architecture of Storm

the task. Thus, tasks provide intra-bolt/intra-spout parallelism, and the executors provide intra-topology parallelism. Worker processes serve as containers on the host machines to run Storm topologies. Spouts can read streams from Kafka (distributed publish-subscribe system from LinkedIn), Twitter, RDBMS etc.

Storm supports the following types of partitioning strategies. *Shuffle* grouping randomly partitions the tuples. *Fields* grouping hashes on a subset of the tuple attributes/fields. *All* grouping replicates the entire stream to all the consumer tasks. *Global* grouping sends the entire stream to a single bolt. *Local* grouping sends tuples to the consumer bolts in the same executor. The partitioning strategy is extensible and a topology can define and use its own partitioning strategy.

Each worker node runs a Supervisor that communicates with Nimbus. The cluster state is maintained in Zookeeper [3], and Nimbus is responsible for scheduling the topologies on the worker nodes and monitoring the progress of the tuples flowing through the topology.

Storm currently runs on hundreds of servers (spread across multiple datacenters) at Twitter. Several hundreds of topologies run on these clusters, some of which run on more than a few hundred nodes. Many terabytes of data flows through the Storm clusters every day, generating several billions of output tuples. Storm topologies are used by a number of groups inside Twitter, including revenue, user services, search, and content discovery. These topologies are used to do simple things like filtering and aggregating the content of various streams at Twitter (e.g., computing counts), and also for more complex things like running simple machine learning algorithms (e.g., clustering) on stream data.

Storm has following important features:

Scalability: It is scalable. It is easy to add or remove nodes from storm cluster without disrupting existing data flows.

Fault tolerance: Storm guarantees that the data will be processed. Storm is very resilient in regards to fault tolerance.

Easy to use: Storm is very easy to set up and operate.

Extensibility: Storm topologies may call arbitrary external functions (e.g., Looking up a MySQL service for the social graph), and thus needs a framework that allows extensibility.

Efficiency: Storm uses a number of techniques, including keeping all its storage and computational data structures in memory. Storm is very fast in processing. A benchmark clocked it at over a million tuples processed per second per node.

4.4.2 S4

Simple Scalable Streaming System (shorted for S4) [24] was released for processing continuous, unbounded streams of data by Yahoo. S4 is a general-purpose, distributed, scalable, fault-tolerant, pluggable platform that allows programmers to easily develop applications for processing continuous unbounded streams of data.

S4 is designed to solve real-world problems in the context of search applications that use data mining and machine learning algorithms. Compared with current processing systems, S4, a low latency, scalable stream processing engine, is developed. The stream throughput is improved by 1000% (200 k + messages /s /stream) in S4 [18].

The design goal of S4 is developing a high performance computing platform that can hide the complexity inherent in a parallel processing system from the application programmer. Simple programming interfaces for processing data streams are provided in S4. A cluster with high availability is designed; the cluster can scale using commodity hardware. Latency is minimized by using local memory in each processing node, and the disk I/O bottlenecks are avoided as well. A symmetric and decentralized architecture is used in S4. Because all nodes in S4 share the same functionality and responsibilities, there is no central node with specialized responsibilities. Thus, the deployment and maintenance of S4 are greatly simplified. The design is friendly and easy to program and flexible by using a pluggable architecture. The gap between complex proprietary systems and batch-oriented open source computing platforms is filled in S4 [18].

S4 provides a runtime distributed platform that handles communication, scheduling and distribution across containers. The nodes are the distributed containers, which are deployed in S4 clusters. The size of clusters is fixed in S4, the size of an S4 cluster corresponds to the number of logical partitions (tasks). The key concepts are shown in Fig. 14 [18].

In S4, computation is executed by Processing Elements (PEs) and messages are transmitted between them in the form of data events. The stream is defined as a

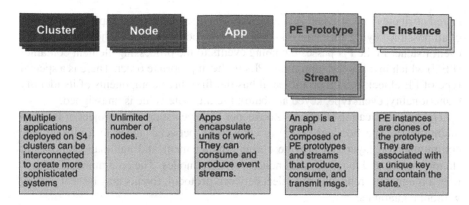

Fig. 14 Key concepts in S4 (Incubator)

sequence of elements (events). The only mode of interaction between PEs is event emission and consumption. PE cannot access to the state of other PEs. The framework provides the capability to route events to appropriate PEs and to create new instances of PEs [24].

PEs are assembled into applications using the Spring Framework. Processing Elements (PEs) are the basic computational units in S4. Each instance of a PE is uniquely identified by four components (the functionality, the types of events, the keyed attribute and the value of keyed attribute).

Processing nodes (PNs) are the logical hosts to PEs. They are responsible for listening to events, executing operations on the incoming events, dispatching events with the assistance of the communication layer, and emitting output events (Fig. 15 [24]). S4 routes each event to PNs based on a hash function of the values of all known

Fig. 15 Processing node

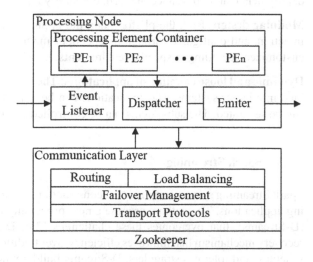

keyed attributes in that event. A single event may be routed to multiple PNs. The set of all possible keying attributes is known from the configuration of the S4 cluster. An event listener in the PN passes incoming events to the processing element container (PEC) which invokes the appropriate PEs in the appropriate order. There is a special type of PE object: the PE prototype. It has the first three components of its identity (functionality, event type, keyed attribute); the attribute value is unassigned.

The communication layer uses Zookeeper (an open source subproject of Hadoop maintained) (Apache ZooKeeper) to coordinate between nodes in a cluster. The communication layer can provide cluster management and automatic failover to standby nodes and maps physical nodes to logical nodes. The communication layer uses a pluggable architecture to select network protocol. Events may be sent with or without a guarantee.

The core platform is written in Java. The implementation is modular and pluggable, and S4 applications can be easily and dynamically combined for creating more sophisticated stream processing systems. Every PE consumes exactly those events which correspond to the value on which it is keyed. It may produce output events. Two primary handlers are implemented by developers: an input event handler *processEvent()* and an output mechanism *output()*. The *output()* method is optional and is set to be invoked in a variety of ways. The *output()* method implements the output mechanism for the PE, typically to publish internal state of the PE to some external system [24].

S4 has following important features:

Fault tolerance: When a server in the cluster fails, a stand-by server is automatically activated to take over the tasks. Check pointing and recovery mechanism are used to minimize state loss.

Flexible deployment: Application packages and platform modules are standard jar files (suffixed.s4r). The keys are homogeneously sparsed over the cluster, the flexible deployment can help balance the load, especially for fine grained partitioning.

Modular design: Both the platform and the applications are built by dependency injection, and configured through independent modules. The system is easy to be customized according to specific requirements.

Dynamic and loose coupling of applications: The subsystems are easy to be assembled into larger systems. The applications can be reused in S4, and pre-processing can be separated. The subsystems can be controlled and updated independently.

4.4.3   Spark Streaming

Spark Streaming system simplifies the construction of scalable fault-tolerant streaming applications. The authors propose a new processing model, discretized streams (D-Streams), that overcomes these challenges [33]. D-Streams enable a parallel recovery mechanism that improves efficiency over traditional replication and backup schemes, and tolerates stragglers. D-Streams build applications through high-level

operators and make efficient fault tolerance while combining streaming with batch
and interactive queries.

Existing streaming models use replication or upstream backup for fault tolerance.
This mechanism costs much time on fault tolerance and stragglers. Also their event
driven programming interface does not directly support parallel processing. The purposes of Spark Streaming are to directly support parallel processing, fault tolerance
and efficient stragglers.

Unlike stateful programming model, Spark Streaming use batch processing
method to process continuous streaming and cut streaming into discretized intervals.
It can take advantage of batch operations in Spark and also provide typical streaming
operations. Spark Streaming uses short stateless, deterministic tasks instead of continues, stateful operators. The state stored in memory across tasks into RDD. Spark
Streaming runs a streaming computation as a series of very small, deterministic batch
jobs. When the streaming data is coming, Spark Streaming chops up the live stream
into batches of 0.5–1 second. It treats each batch of data as RDDs and processes them
using RDD operations. In this way, it has potential for combining batch processing
and streaming processing in the same system.

For fault-tolerance, RDDs remember the operations that created them and replicated batches of input data in memory for fault-tolerance. So data lost due to worker
failure can be recomputed from replicated input data via RDD. Therefore, all data is
fault-tolerant. The lineage graph of RDD is shown in Fig. 16 [33].

Spark Streaming can easily be composed with batch and query model. It provides
both batch operation in Spark and standard streaming systems (Das) [7]. Batch API
in Spark includes *Map, Reduce, GroupBy, Join* operations. Streaming API in Spark
supports *Windowing, Incremental Aggregation* operations.

Spark Streaming consists of three components, shown in Fig. 17 [7]. A master,
that tracks the D-Stream lineage graph and schedules tasks to compute new RDD

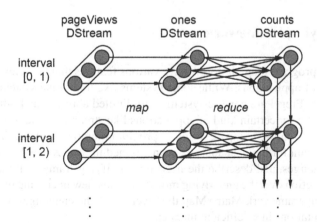

Fig. 16 Lineage graph for RDDs

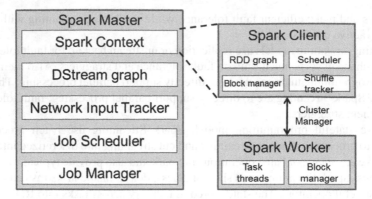

Fig. 17 Components of Spark Streaming architecture

partitions. Worker nodes that receive data, store the partitions of input and computed RDDs, and execute tasks. A client library used to send data into the system.

In the Spark Master, network Input Tracker keeps track of the data received by each network receiver and maps them to the corresponding input *DStreams*. Job Scheduler periodically queries the *DStream* graph to generate Spark jobs from received data, and hands them to Job Manager for execution. Job Manager maintains a job queue and executes the jobs in Spark.

Spark Streaming has many important features that make it a desirable programming platform. It scales to 100 s of nodes and achieves second scale latencies. It enables efficient and fault-tolerant stateful stream processing while integrating with Spark's batch and interactive processing. Spark provides a simple batch-like API for implementing complex algorithms.

5 A Unifying Framework

The existing programming platforms have various features that are relevant for particular kinds of applications. While some systems like traditional systems, MapReduce, Hadoop, Pregel are generic systems with limited abilities, and other systems are very specific for certain kind of applications like streaming data or graph based data. In this section, we compare different programming platforms against features that are important for big data analysis as mentioned in Sect. 2. We then discuss the existing challenges, and describe the need for a unifying framework that allows a generic abstraction over the underlying models and any new upcoming models. Then we present our framework MatrixMap that overcomes the challenges of supporting matrix computations in an efficient manner.

5.1   Comparison of Existing Programming Platforms

The comparison of various programming platforms with respect to some important features as discussed in the corresponding sections is summarized in Table 1. Most of the data parallel programming platforms have high scalability. Hadoop derivatives like Spark and Spark Streaming inherit similar characteristics for high scalability with distributed processing. Real time processing is supported by Storm, S4 and Spark Streaming in an efficient manner. Fault tolerance in big data analytics is a critical feature because of dependency on multiple systems and size of application. It is observed that Hadoop, Spark, Spark Streaming, GraphX and Storm are highly fault tolerant as they use redundancy and special data structures called RDDs. GraphLab, Pregel and S4 use checkpointing for fault tolerance.

The newer programming platforms like Storm, Spark Streaming have most attributes required for efficient big data analysis. Much research is being carried out to develop all machine learning algorithms for newer systems. For MapReduce based systems, not all the machine learning algorithms can be formulated as map and reduce problems. For interactive analysis, Storm, S4 and Spark Streaming can be used as programming platforms.

Table 1 Comparison of Programming platforms for big data analysis

Processing Techniques		Features/ Platforms	Scalability	Fault Tolerance	Efficiency	Usability	Real-time Processing	Iterative Task Support
Task Parallel		HTCondor	Medium	Low	High	Medium		Yes
		Storm	High	High	High	Medium	Yes	Yes
		S4	High	Medium	Medium	High	Yes	
	Data Parallel	Hadoop	High	High	Medium	Medium		
		Spark	High	High	High	High		Yes
		Spark Streaming	High	High	High	High	Yes	Yes
		Dryad	Medium	Medium	Low	High		Yes
		MatrixMap	High	Medium	High	Medium		Yes
	Graph	GraphX	Medium	High	High	High		Yes
		GraphLab	Low	Medium	High	High		Yes
		Pregel	Low	Medium	Low	Medium		Yes

5.2 Need for Unifying Framework

One of the existing challenges in big data programming is that no single programming model or framework can excel at every problem. Different big data programming platforms address different requirements, e.g., some platforms support graph based processing and some systems are specifically designed for streaming data. Programmers need to spend much time learning individual models and their corresponding language, and there are always tradeoffs between simplicity, expressivity, fault tolerance, performance etc.

Therefore, there is need of a unifying framework that allows for a generic abstraction on top of the underlying models and upcoming new models like MatrixMap as shown in Fig. 18. Such an abstraction would integrate different programming platforms so that the programmers only need to learn a single language and techniques for diverse big data applications. Integration of big data platforms would require unifying the interface so that data and operations supported by different models can be abstracted, and mapping each data processing stage to underlying models. In addition, both inter-model and intra-model tasks need to be scheduled on processing units for better efficiency. The cloud resources also have to be allocated dynamically after analyzing the different computation requirements. Integrating data storage systems such as file systems and special databases are another issue. There are various open challenges in it calling for future research efforts.

Besides this, there are still many problems for the existing platforms when performing big data analysis in different application scenarios. Thus, designing new programming platform is another challenge that attracts much attention in the research communities. We will present our proposed platform MatrixMap in the next section.

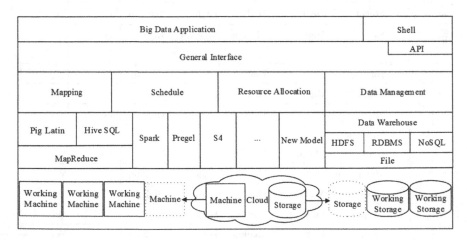

Fig. 18 Integration of diverse big data programming platforms

5.3 *MatrixMap Framework*

Machine learning and graph algorithms play vital important roles in big data analytics. Most algorithms are formulated based into matrix computations. That is they apply matrix operations on values and perform various manipulations of values according to their labels. However, existing big data programming platforms do not provide efficient support for matrix computations.

Most programming platforms provide separate models for machine learning and graph algorithms, e.g., in Spark has different interfaces: GraphX for graph algorithms and Spark for machine learning. The existing systems do not have direct support for important matrix operations, e.g., in MapReduce, matrix multiplication must be formulated into a series of map and reduce operations. The support is mostly limited to matrix multiplication, but not other popular machine learning and graph algorithms, e.g., Presto. Systems besides Spark save temporal data in secondary storage, slow to load data for operations. The cache memory uses LRU algorithm (e.g., Spark), which may not be efficient for all operations. These challenges have led us to develop a model and framework for handling matrix based computations for big data analysis.

MatrixMap [15] is a new model and framework to support data mining and graph algorithms. It provides matrix as language-level construct. The data is loaded into key matrices and then powerful and simple matrix patterns are provided that support basic operations for machine learning and graph algorithms. This model unifies data-parallel and graph-parallel models by abstracting matrix computations into graph patterns.

The framework implements parallel processing of matrix operations and data manipulations invoked by user defined functions. MatrixMap supports high-volume data with pattern-specific fetching and caching across memory and secondary storage.

Algorithms are formulated as a series of matrix patterns, which define sequences of operations on each element. Unary Operator: Map, Reduce; Binary Operator: Plus, Multiply; Mathematical matrix operations are special cases of matrix patterns filled with specific pre-defined lambda functions; User defined lambda functions according to matrix patterns to support various algorithms.

The data is loaded into Bulk Key Matrix (BKM) which is suitable for large volume data. BKM is a shared distributed data structure which spreads data into whole clusters. It can keep data across matrix patterns. It is constant and cannot be changed, after initiation. BKM is row-oriented or column-oriented. It cannot slice concrete matrix element. BKM use key (string or digit) to index row or column. MatrixMap adopts BSP model, while supporting asynchronous pipeline of IO and processing with data partitioning as shown in Fig. 19 [15].

There are many applications of matrix patterns like logistic regression, alternating least squares, all pairs shortest path, Pagerank among other applications. When compared with Spark, it achieved 20% improvement on execution time - the more iterations, the better as shown in Fig. 20.

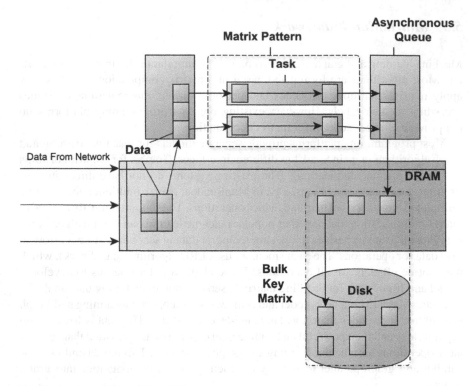

Fig. 19 Implementation of MatrixMap

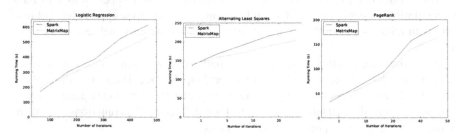

Fig. 20 MatrixMap performance w.r.t. Spark

MatrixMap provides powerful yet simple abstraction, consisting of a distributed data structure called bulk key matrix and a computation interface defined by matrix patterns. Users can easily load data into bulk key matrices and program algorithms into parallel matrix patterns. MatrixMap outperforms current state-of-the-art systems by employing three key techniques: matrix patterns with lambda functions for irregular and linear algebra matrix operations, asynchronous computation pipeline with optimized data shuffling strategies for specific matrix patterns and in-memory data structure reusing data in iterations. Moreover, it can automatically handle the parallelization and distribute execution of programs on a large cluster.

6 Conclusion and Future Directions

The purpose of this chapter is to survey various existing programming platforms for big data analysis. We have enumerated various essential features that a programming environment should possess for big data analysis. The prominent programming platforms have been discussed in brief to give an insight into their purpose, programming model, implementation and features. The comparisons of existing programming platforms against various features have been summarized as well as the need for a unifying framework and our proposed MatrixMap framework that implements machine learning and graph based algorithms using matrices as language constructs, which can handle large data in an efficient manner. In future, we would investigate more in unifying framework for different big data platforms, and improve the MatrixMap framework so that multiple machine learning algorithms can be implemented for different kinds of data. In summary, we can say that research and development of big data programming platforms are driven by real world applications and key industrial stakeholders and it's a challenging but compelling task. Programming platforms for handling big data specially streaming data are still evolving. Samza [9] is a recent addition to programming platforms for streaming data. The concept of "Lambda Architecture" that integrates batch processing and real time processing together in a harmonious way in terms of batch, speed and serving is also an area of interest for the researchers. The integration of different big data programming platforms is an open challenge with various issues related to task scheduling, resource allocation and model mapping to be resolved; while designing new platforms to better perform big data analysis in different application scenarios is another one. Developing a higher-level programming support on top of multiple models can help ease and shorten the development of big data applications.

Acknowledgements This work was partially supported by the funding for Project of Strategic Importance provided by The Hong Kong Polytechnic University (1-ZE26) and HK RGC under GRF Grant (PolyU 5104/13E).

References

1. V. Agneeswaran, *Big Data Analytics Beyond Hadoop: Real-Time Applications with Storm, Spark, and More Hadoop Alternatives*, 1st edn. (Pearson FT Press, USA, 2014)
2. Apache storm documentation, https://storm.apache.org/documentation/Home.html
3. Apache zookeeper, http://zookeeper.apache.org
4. Architecture of mapreduce model, https://cloud.google.com/appengine/docs/-python/images/mapreduce_mapshuffle.png
5. A.B. Bondi, Characteristics of scalability and their impact on performance, in *Workshop on Software and Performance* (2000), pp. 195C203
6. W. Daniel Hillis, G.L. Steele, Jr., Data parallel algorithms. Commun. ACM, **29**(12), 1170C1183 (1986)
7. T. Das, Deep dive into spark streaming. http://spark.apache.org/-documentation.html (2013)

8. J. Dean, S. Ghemawat, Mapreduce: simplified data processing on large clusters. Commun. ACM 51(1):107C113 (2008)
9. T. Feng, Z. Zhuang, Y. Pan, H. Ramachandra, A memory capacity model for high performing data-filtering applications in samza framework, in *2015 IEEE International Conference on Big Data, Big Data 2015*, Santa Clara, CA, USA, October 29 - November 1, 2015, p. 2600C2605
10. A. Fernández, S. del Ró, V. López, A. Bawakid, M. José del Jesús, J. Manuel Bentez, F. Herrera, Big data with cloud computing: an insight on the computing environment, mapreduce, and programming frameworks. Wiley Interdisc. Rew.: Data Min. Knowl. Discov. 4(5), 380C409 (2014)
11. J.E. Gonzalez, Y. Low, H. Gu, D. Bickson, C. Guestrin, Powergraph: distributed graph-parallel computation on natural graphs, in *10th USENIX Symposium on Operating Systems Design and Implementation, OSDI 2012*, Hollywood, CA, USA, October 8-10, 2012, p. 17C30
12. J.E. Gonzalez, R.S. Xin, A. Dave, D. Crankshaw, M.J. Franklin, I. Stoica, Graphx: graph processing in a distributed dataflow framework, in *11th USENIX Symposium on Operating Systems Design and Implementation, OSDI 14*, Broomfield, CO, USA, October 6–8, 2014, p. 599C613
13. B. Hindman, A. Konwinski, M. Zaharia, A. Ghodsi, A.D. Joseph, R.H. Katz, S. Shenker, I. Stoica, Mesos: A platform for fine-grained resource sharing in the data center, in *Proceedings of the 8th USENIX Symposium on Networked Systems Design and Implementation, NSDI 2011*, Boston, MA, USA (2011)
14. Htcondor, http://research.cs.wisc.edu/htcondor/description.html
15. Y. Huangfu, J. Cao, H. Lu, G. Liang, Matrixmap: programming abstraction and implementation of matrix computation for big data applications, in *21st IEEE International Conference on Parallel and Distributed Systems, ICPADS 2015*, Melbourne, Australia (2015), p. 19C28
16. Implementation of pregel, http://people.apache.org/~edwardyoon/documents/-pregel.pdf
17. M. Isard, M. Budiu, Y. Yu, A. Birrell, D. Fetterly, Dryad: distributed data-parallel programs from sequential building blocks, in *Proceedings of the 2007 EuroSys Conference*, Lisbon, Portugal, March 21–23, 2007, p. 59C72
18. Key concepts in s4 (incubator), https://incubator.apache.org/s4/doc/0.6.0/-overview
19. M. J. Litzkow, M. Livny, M.W. Mutka, Condor - a hunter of idle workstations, in *Proceedings of the 8th International Conference on Distributed Computing Systems*, San Jose, California, USA, June 13–17, 1988, p. 104C111
20. Y. Low, J. Gonzalez, A. Kyrola, D. Bickson, C. Guestrin, J.M. Hellerstein, Graphlab: a new framework for parallel machine learning, in *UAI 2010, Proceedings of the Twenty-Sixth Conference on Uncertainty in Artificial Intelligence*, Catalina Island, CA, USA, July 8–11, 2010, p. 340C349
21. Y. Low, J. Gonzalez, A. Kyrola, D. Bickson, C. Guestrin, J.M. Hellerstein, Distributed graphlab: a framework for machine learning in the cloud. PVLDB 5(8), 716C727 (2012)
22. G. Malewicz, M.H. Austern, A.J.C. Bik, J.C. Dehnert, I. Horn, N. Leiser, G. Czajkowski, Pregel: a system for large-scale graph processing, in *Proceedings of the ACM SIGMOD International Conference on Management of Data, SIGMOD 2010*, Indianapolis, Indiana, USA (2010), p. 135C146
23. P. Mhashilkar, Z. Miller, R. Kettimuthu, G. Garzoglio, B. Holzman, C. Weiss, X. Duan, L. Lacinski, End-to-end solution for integrated workload and data management using glideinwms and globus online. J. Phys. Conf. Ser. 396(3), 032076 (2012)
24. L. Neumeyer, B. Robbins, A. Nair, A. Kesari, S4: distributed stream computing platform, in *ICDMW 2010, The 10th IEEE International Conference on Data Mining Workshops*, Sydney, Australia, 13 Dec 2010, p. 170C177
25. Scala programming language, http://www.scala-lang.org
26. M. Snir, S. Otto, S. Huss-Lederman, D. Walker, J. Dongarra, MPI-The Complete Reference, vol. 1: The MPI Core, 2nd (revised) edn. (MIT Press, Cambridge 1998)
27. Spark programming model, http://blog.cloudera.com/blog/2013/11/-putting-spark-to-use-fast-in-memory-computing-for-your-big-data-applications
28. The structure of dryad jobs, http://research.microsoft.com/en-us/projects/dryad

29. M. Tim Jones, Process real-time big data with twitter storm. Technical Report pp. 1-9, IBM Developer Works (2013)
30. A. Toshniwal, S. Taneja, A. Shukla, K. Ramasamy, J.M. Patel, S. Kulkarni, J. Jackson, K. Gade, M. Fu, J. Donham, N. Bhagat, S. Mittal, D.V. Ryaboy, Storm@twitter, in *International Conference on Management of Data, SIGMOD 2014*, Snowbird, UT, USA, June 22–27, 2014, p. 147C156
31. Y. Yu, M. Isard, D. Fetterly, M. Budiu, Ú. Erlingsson, P. Kumar Gunda, J. Currey, Dryadlinq: a system for general-purpose distributed data-parallel computing using a high-level language, in *8th USENIX Symposium on Operating Systems Design and Implementation, OSDI 2008*, San Diego, California, USA, Proceedings (2008), p. 1C14
32. M. Zaharia, M. Chowdhury, M.J. Franklin, S. Shenker, I. Stoica, Spark: cluster computing with working sets, in *2nd USENIX Workshop on Hot Topics in Cloud Computing*, HotCloud10, Boston, MA, USA (2010)
33. M. Zaharia, T. Das, H. Li, T. Hunter, S. Shenker, I. Stoica, Discretized streams: fault-tolerant streaming computation at scale, in *ACM SIGOPS 24th Symposium on Operating Systems Principles, SOSP 13*, Farmington, PA, USA (2013), p. 423C438

Big Data Analysis on Clouds

Loris Belcastro, Fabrizio Marozzo, Domenico Talia
and Paolo Trunfio

Abstract The huge amount of data generated, the speed at which it is produced, and its heterogeneity in terms of format, represent a challenge to the current storage, process and analysis capabilities. Those data volumes, commonly referred as Big Data, can be exploited to extract useful information and to produce helpful knowledge for science, industry, public services and in general for humankind. Big Data analytics refer to advanced mining techniques applied to Big Data sets. In general, the process of knowledge discovery from Big Data is not so easy, mainly due to data characteristics, as size, complexity and variety, that require to address several issues. Cloud computing is a valid and cost-effective solution for supporting Big Data storage and for executing sophisticated data mining applications. Big Data analytics is a continuously growing field, so novel and efficient solutions (i.e., in terms of platforms, programming tools, frameworks, and data mining algorithms) spring up everyday to cope with the growing scope of interest in Big Data. This chapter discusses models, technologies and research trends in Big Data analysis on Clouds. In particular, the chapter presents representative examples of Cloud environments that can be used to implement applications and frameworks for data analysis, and an overview of the leading software tools and technologies that are used for developing scalable data analysis on Clouds.

Keywords Cloud computing · Big data · Data analytics · Data mining

L. Belcastro (✉) · F. Marozzo · D. Talia · P. Trunfio
DIMES, University of Calabria, Rende, Italy
e-mail: lbelcastro@dimes.unical.it

F. Marozzo
e-mail: fmarozzo@dimes.unical.it

D. Talia
e-mail: talia@dimes.unical.it

P. Trunfio
e-mail: trunfio@dimes.unical.it

© Springer International Publishing AG 2017   101
A.Y. Zomaya and S. Sakr (eds.), *Handbook of Big Data Technologies*,
DOI 10.1007/978-3-319-49340-4_4

1 Introduction

In the last years the ability to produce and gather data has increased exponentially. In fact, in the Internet of Things' era, huge amounts of digital data are generated by and collected from several sources, such as sensors, cams, in-vehicle infotainment, smart meters, mobile devices, web applications and services. The huge amount of data generated, the speed at which it is produced, and its heterogeneity in terms of format (e.g., video, text, xml, email), represent a challenge to the current storage, process and analysis capabilities. In particular, thanks to the growth of social networks (e.g., Facebook, Twitter, Pinterest, Instagram, Foursquare, etc.), the widespread diffusion of mobile phones, and the large use of location-based services, every day millions of people access social network services and share information about their interests and activities. Those data volumes, commonly referred as Big Data, can be exploited to extract useful information and to produce helpful knowledge for science, industry, public services and in general for humankind.

Although nowadays the term Big Data is often misused, it is very important in computer science for understanding business and human activities. As defined by Gartner[1]: "*Big Data is high volume, high velocity, and/or high variety information assets that require new forms of processing to enable enhanced decision making, insight discovery, and process optimization.*" Thus, Big Data is not only characterized by the large size of data sets, but also by the complexity, by the variety, and by the velocity of data that can be collected and processed. In fact, we can collect huge amounts of digital data from sources, at a very high rate that the volume of data is overwhelming our ability to make use of it. This situation is commonly called "data deluge".

In science and business, people are analyzing data to extract information and knowledge useful for making new discoveries or for supporting decision processes. This can be done by exploiting Big Data analytics techniques and tools. As an example, one of the leading trends today is the analysis of big geotagged data for creating spatio-temporal sequences or trajectories tracing user movements. Such kind of information is clearly highly valuable for science and business: tourism agencies and municipalities can know the most visited places by tourists, the time of year when such places are visited, and other useful information [4, 23]; transport operators can know the places and routes where is it more likely to serve passengers [58] or crowed areas where more transportation resources need to be allocated [57]; city managers may exploit social media analysis to reveal mobility insights in cities such as incident locations [24], or to study and prevent crime events [16, 26].

But it must be also considered that just Twitter and Facebook produce about 20 TB of data every day. According to a study conducted by the International Data Corporation (IDC), the whole world produced about 165 exabytes (1 exabytes is equal to 10^{18} bytes) of data in 2007, 800 exabytes in 2009, and it is estimated that in 2020 the global amount of data produced will reach the 35 zettabytes (1 zettabytes is equal to 10^{21} bytes). Then to extract value from such kind of data, novel technologies

---

[1]http://www.gartner.com/it-glossary/big-data.

and architectures have been developed by data scientists for capturing and analyzing complex and/or high velocity data. In this scenario data mining raised in the last decades as a research and technology field that provides several different techniques and algorithms for the automatic analysis of large data sets. The usage of sequential data mining algorithms for analyzing large volumes of data requires a very long time for extracting useful models and patterns. For this reason, high performance computers, such as many and multi-core systems, Clouds, and multi-clusters, paired with parallel and distributed algorithms are commonly used by data analysts to tackle Big Data issues and to reduce response time to a reasonable value.

Big Data analytics refer to advanced mining techniques applied to Big Data sets. In general, the process of knowledge discovery from Big Data is not so easy, mainly due to data characteristics, as size, complexity and variety, that require to address several issues. To overcame these problems and to get valuable information and knowledge in shorter time, high performance and scalable computing systems are used in combination with data and knowledge discovery techniques. In this context, Cloud computing is a valid and cost-effective solution for supporting Big Data storage and for executing sophisticated data analytic applications. In fact, thanks to elastic resource allocation and high computing power, Cloud computing represents a compelling solution for Big Data analytics, allowing faster data analysis, that means more timely results and then greater data value.

Actually, despite the Cloud is an affordable solution for many users, the number of analytics data solutions available is very limited. Most available solutions today are based on open source frameworks, such as Hadoop and Spark, but there are also some proprietary solutions, such as those proposed by IBM, EMC or Kognitio. Big Data analytics is a continuously growing field, so novel and efficient solutions (i.e., in terms of platforms, programming tools, frameworks, and data mining algorithms) spring up everyday to cope with the growing scope of interest in Big Data.

The remainder of the chapter is organized as follows. Section 2 introduces the main Cloud computing concepts. Section 3 describes representative examples of Cloud environments that can be used to implement applications and frameworks for data analysis in the Cloud. Section 4 provides an overview of the leading software tools and technologies used for developing scalable data analysis on Clouds. Section 5 discusses some research trends and open challenges on Big Data analysis. Finally, Sect. 6 concludes the chapter.

2   Introducing Cloud Computing

This section introduces the basic concepts of Cloud computing, which provides scalable storage and processing services that can be used for extracting knowledge from Big Data repositories. In the following we provide basic Cloud computing definitions (Sect. 2.1) and discuss the main service distribution and deployment models provided by Cloud vendors (Sect. 2.2).

2.1 Basic Concepts

In the last years, Clouds have emerged as effective computing platforms to face the challenge of extracting knowledge from Big Data repositories in limited time, as well as to provide effective and efficient data analysis environments to both researchers and companies. From a client perspective, the Cloud is an abstraction for remote, infinitely scalable provisioning of computation and storage resources. From an implementation point of view, Cloud systems are based on large sets of computing resources, located somewhere "in the Cloud", which are allocated to applications on demand [2]. Thus, Cloud computing can be defined as a distributed computing paradigm in which all the resources, dynamically scalable and often virtualized, are provided as services over the Internet. As defined by NIST (National Institute of Standards and Technology) [37] Cloud computing can be described as: "A model for enabling convenient, on-demand network access to a shared pool of configurable computing resources (e.g., networks, servers, storage, applications, and services) that can be rapidly provisioned and released with minimal management effort or service provider interaction". From the NIST definition, we can identify five essential characteristics of Cloud computing systems, which are on-demand self-service, broad network access, resource pooling, rapid elasticity, and measured service. Cloud systems can be classified on the basis of their service model and their deployment model.

2.2 Cloud Service Distribution and Deployment Models

Cloud computing vendors provide their services according to three main distribution models:

- *Software as a Service (SaaS)*, in which software and data are provided through Internet to customers as ready-to-use services. Specifically, software and associated data are hosted by providers, and customers access them without need to use any additional hardware or software. Examples of SaaS services are Gmail, Facebook, Twitter, Microsoft Office 365.
- *Platform as a Service (PaaS)*, in an environment including databases, application servers, development environment for building, testing and running custom applications. Developers can just focus on deploying of applications since Cloud providers are in charge of maintenance and optimization of the environment and underlying infrastructure. Examples of PaaS services are Windows Azure, Force.com, Google App Engine.
- *Infrastructure as a Service (IaaS)*, that is an outsourcing model under which customers rent resources like CPUs, disks, or more complex resources like virtualized servers or operating systems to support their operations (e.g., Amazon EC2, RackSpace Cloud). Compared to the PaaS approach, the IaaS model has a higher system administration costs for the user; on the other hand, IaaS allows a full customization of the execution environment.

The most common models for providing Big Data analytics solution on Clouds are PaaS and SaaS. IaaS is usually not used for high-level data analytics applications but mainly to handle the storage and computing needs of data analysis processes. In fact, IaaS is the more expensive delivery model, because it requires a greater investment of IT resources. On the contrary, PaaS is widely used for Big Data analytics, because it provides data analysts with tools, programming suites, environments, and libraries ready to be built, deployed and run on the Cloud platform. With the PaaS model users do not need to care about configuring and scaling the infrastructure (e.g., a distributed and scalable Hadoop system), because the Cloud vendor will do that for them. Finally, the SaaS model is used to offer complete Big Data analytics applications to end users, so that they can execute analysis on large and/or complex data sets by exploiting Cloud scalability in storing and processing data.

Regarding deployment models, Cloud computing services are delivered according to three main forms:

- *Public Cloud*: it provides services to the general public through the Internet and users have little or no control over the underlying technology infrastructure. Vendors manage their proprietary data centers delivering services built on top of them.
- *Private Cloud*: it provides services deployed over a company intranet or in a private data center. Often, small and medium-sized IT companies prefer this deployment model as it offers advance security and data control solutions that are not available in the public Cloud model.
- *Hybrid Cloud*: it is the composition of two or more (private or public) Clouds that remain different entities but are linked together.

As outlined in [27], users access Cloud computing services using different client devices and interact with Cloud-based services using a Web browser or desktop/mobile app. The business software and users data are executed and stored on servers hosted in Cloud data centers that provide storage and computing resources. Resources include thousands of servers and storage devices connected each other through an intra-Cloud network. The transfer of data between data center and users takes place on wide-area network. Several technologies and standards are used by the different components of the architecture. For example, users can interact with Cloud services through SOAP-based or RESTful Web services [42] and Ajax technologies allow Web interfaces to Cloud services to have look and interactivity equivalent to those of desktop applications. Open Cloud Computing Interface (OCCI)[2] specifies how Cloud providers can deliver their compute, data, and network resources through a standardized interface.

---

[2]OCCI Working Group, http://www.occi-wg.org.

3   Cloud Solutions for Big Data

At the beginning of the Big Data phenomenon, only big IT companies, such as Facebook, Yahoo!, Twitter, Amazon, LinkedIn, invested large amounts of resources in the development of proprietary or open source projects to cope with Big Data analysis problems. But today, Big Data analysis becomes highly significant and useful for small and medium-sized businesses. To address this increasing demand a large vendor community started offering highly distributed platforms for Big Data analysis. Among open-source projects, Apache Hadoop is the leading open-source data-processing platform, which was contributed by IT giants such as Facebook and Yahoo.

Since 2008, several companies, such as Cloudera, MapR, and Hortonworks, started offering enterprise platform for Hadoop, with greats efforts to improve Hadoop performances in terms of high-scalable storage and data processing. Instead, IBM and Pivotal started offering its own customized Hadoop distribution. Other big companies decided to provide only additional softwares and support for Hadoop platform developed by external providers: for example, Microsoft decided to base its offer on Hortonworks platform, while Oracle decided to resell Cloudera platform. However Hadoop is not the only solution for Big Data analytics. Out of the Hadoop box other solutions are emerging. In particular, in-memory analysis has become a widespread trend, so that companies started offering tools and services for faster in-memory analysis, such as SAP, that is considered the leading company with its Hana[3] platform. Other vendors, including HP, Teradata and Actian, developed analytical database tools with in-memory analysis capabilities. Moreover, some vendors, like Microsoft, IBM, Oracle, and SAP, stand out from their peers for offering a complete solution for data analysis, including DBMS systems, software for data integration, stream-processing, business intelligence, in-memory processing, and Hadoop platform.

In addition, many vendors decided to focus whole offer on the Cloud. Among these certainly there are Amazon Web Services (AWS) and 1010 data. In particular, AWS provides a wide range of services and products on the Cloud for Big Data analysis, including scalable database systems and solutions for decision support. Other smaller vendors, including Actian, InfiniDB, HP Vertica, Infobright, and Kognitio, focused their big-data offer on database management systems for analytics only. Following the approach in [48], the remainder of the section introduces representative examples of Cloud environments: Microsoft Azure as an example of public PaaS, Amazon Web Services as the most popular public IaaS, OpenNebula and OpenStack as examples of private IaaS. These environments can be used to implement applications and frameworks for data analysis in the Cloud.

---

[3]https://hana.sap.com.

3.1 Microsoft Azure

Azure[4] is the Microsoft Cloud proposal. It is environment providing a large set of Cloud services that can be used by developers to create Cloud-oriented applications, or to enhance existing applications with Cloud-based capabilities. The platform provides on-demand compute and storage resources exploiting the computational and storage power of the Microsoft data centers. Azure is designed for supporting high availability and dynamic scaling services that match user needs with a pay-per-use pricing model. The Azure platform can be used to perform the storage of large datasets, execute large volumes of batch computations, and develop SaaS applications targeted towards end-users. Microsoft Azure includes three basic components/services:

- *Compute* is the computational environment to execute Cloud applications. Each application is structured into roles: *Web role*, for Web-based applications; *Worker role*, for batch applications; *Virtuam Machines role*, for virtual-machine images.
- *Storage* provides scalable storage to manage: binary and text data (*Blobs*), non-relational tables (*Tables*), queues for asynchronous communication between components (*Queues*). In addition, for relational databases, Microsoft provides its own Cloud database services, called Azure SQL Database.
- *Fabric controller* whose aim is to build a network of interconnected nodes from the physical machines of a single data center. The Compute and Storage services are built on top of this component.

Microsoft Azure provides standard interfaces that allow developers to interact with its services. Moreover, developers can use IDEs like Microsoft Visual Studio and Eclipse to easily design and publish Azure applications.

3.2 Amazon Web Services

Amazon offers compute and storage resources of its IT infrastructure to developers in the form of Web services. Amazon Web Services (AWS)[5] is a large set of Cloud services that can be composed by users to build their SaaS applications or integrate traditional software with Cloud capabilities. It is simple to interact with these service since Amazon provides SDKs for the main programming languages and platforms (e.g. Java, .Net, PHP, Android).

AWS compute solution includes *Elastic Compute Cloud* (EC2), for creating and running virtual servers, and *Amazon Elastic MapReduce* for building and executing MapReduce applications. The Amazon storage solution is based on *S3 Storage Service*, with a range of storage classes designed to cope with different use cases (i.e., Standard, Infrequent Access, and Glacier for long term storage archive).

---

[4]https://azure.microsoft.com.
[5]https://aws.amazon.com.

A full set of database systems are also proposed: *Relational Database Service (RDS)* for relational tables; *DynamoDB* for non-relational tables; *SimpleDB* for managing small datasets; *ElasticCache* for caching data. Even though Amazon is best known to be the first IaaS provider (based on its EC2 and S3 services), it is now also a PaaS provider, with services like Elastic Beanstalk, that allows users to quickly create, deploy, and manage applications using a large set of AWS services, or Amazon Machine Learning, that provides visualization tools and wizards for easily creating machine learning models.

3.3 OpenNebula

OpenNebula [45] is an open-source framework mainly used to build private and hybrid Clouds. The main component of the OpenNebula architecture is the Core, which creates and controls virtual machines by interconnecting them with a virtual network environment. Moreover, the Core interacts with specific storage, network and virtualization operations through pluggable components called Drivers. In this way, OpenNebula is independent from the underlying infrastructure and offers a uniform management environment. The Core also supports the deployment of Services, which are a set of linked components (e.g., Web server, database) executed on several virtual machines. Another component is the Scheduler, which is responsible for allocating the virtual machines on the physical servers. To this end, the Scheduler interacts with the Core component through appropriate deployment commands.

OpenNebula can implement a hybrid Cloud using specific Cloud Drivers that allow to interact with external Clouds. In this way, the local infrastructure can be supplemented with computing and storage resources from public Clouds. Currently, OpenNebula includes drivers for using resources from Amazon EC2 and Eucalyptus [40], another open source Cloud framework.

3.4 OpenStack

OpenStack[6] is an open source Cloud operating system realesed under the terms of the Apache License 2.0. It allows the management of large pools of processing, storage, and networking resources in a datacenter through a Web-based interface. Most decisions about its development are decided by the community to the point that every six months there is a design summit to gather requirements and define new specifications for the upcoming release. The modular architecture of OpenStack is composed by four main components, as shown in Fig. 1.

OpenStack Compute provides virtual servers upon demand by managing the pool of processing resources available in the datacenter. It supports different virtualization

---

[6]https://www.openstack.org/.

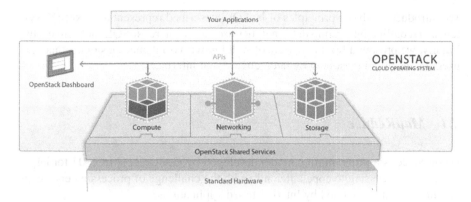

Fig. 1 OpenStack architecture (source: http://openstack.org)

technologies (e.g., VMware, KVM) and is designed to scale horizontally. *OpenStack Storage* provides a scalable and redundant storage system. It supports Object Storage and Block Storage: the former allows storing and retrieving objects and files in the datacenter. *OpenStack Networking* manages the networks and IP addresses. Finally, *OpenStack Shared Services* are additional services provided to ease the use of the datacenter, such as Identity Service for mapping users and services, Image Service for managing server images, and Database Service for relational databases.

4 Systems for Big Data Analytics in the Cloud

In this section we describe the most used tools for developing scalable data analysis on Clouds, such as MapReduce, Spark, workflow systems, and NoSQL database management systems. In particular, we discuss some frameworks commonly used to develop scalable applications that analyze big amounts of data, such as Apache Hadoop, the best-known MapReduce implementation, and Spark. We present also some powerful data mining programming tools and strategies designed to be executed in the Cloud for exploiting complex and flexible software models, such as the distributed workflows. Workflows provide a declarative way of specifying the high-level logic of an application, hiding the low-level details. They are also able to integrate existing software modules, datasets, and services in complex compositions that implement discovery processes. In this section we presented several data mining workflow systems, such as the Data Mining Cloud Framework, Microsoft Azure Machine Learning, and ClowdFlows. Moreover, we discuss about NoSQL database technology that recently became popular as an alternative or as a complement to relational databases. In the last years, several NoSQL systems have been proposed for providing more scalability and higher performance than relational databases.

We introduce the basic principles of NoSQL, described representative NoSQL systems, and outline interesting data analytics use cases where NoSQL tools are useful. Finally, we present a brief overview of well known visual analytics tools, that help users in analytical reasoning by interactive visual interfaces.

4.1 MapReduce

MapReduce is a programming model developed by Google [11] in 2004 for large-scale data processing to cope efficiently with the challenge of processing enormous amounts of data generated by Internet-based applications.

Since its introduction, MapReduce has proven to be applicable to a wide range of domains, including machine learning and data mining, social data analysis, financial analysis, scientific simulation, image retrieval and processing, blog crawling, machine translation, language modelling, and bioinformatics. Today, MapReduce is widely recognized as one of the most important programming models for Cloud computing environments, being it supported by Google and other leading Cloud providers such as Amazon, with its Elastic MapReduce service,[7] and Microsoft, with its HDInsight,[8] or on top of private Cloud infrastructures such as OpenStack, with its Sahara service.[9]

Hadoop[10] is the most used open source MapReduce implementation for developing parallel applications that analyze big amounts of data. It can be adopted for developing distributed and parallel applications using many programming languages (e.g., Java, Ruby, Python, C++). Hadoop relieves developers from having to deal with classical distributed computing issues, such as load balancing, fault tolerance, data locality, and network bandwidth saving.

The Hadoop project is not only about the MapReduce programming model (Hadoop MapReduce module), as it includes other modules such as:

- *Hadoop Distributed File System (HDFS)*: a distributed file system providing fault tolerance with automatic recovery, portability across heterogeneous commodity hardware and operating systems, high-throughput access and data reliability.
- *Hadoop YARN*: a framework for cluster resource management and job scheduling.
- *Hadoop Common*: common utilities that support the other Hadoop modules.

In particular, thanks to the introduction of YARN in 2013, Hadoop turns from a batch processing solution into a platform for running a large variety of data applications, such as streaming, in-memory, and graphs analysis. As a result, Hadoop became a reference for several other frameworks, such as: Giraph for graph analysis; Storm for streaming data analysis; Hive, which is a data warehouse software

---

[7]http://aws.amazon.com/elasticmapreduce/.

[8]http://azure.microsoft.com/services/hdinsight/.

[9]http://wiki.openstack.org/wiki/Sahara.

[10]http://hadoop.apache.org/.

for querying and managing large datasets; Pig, which is as a dataflow language for exploring large datasets; Tez for executing complex directed-acyclic graph of data processing tasks; Oozie, which is a workflow scheduler system for managing Hadoop jobs. Besides Hadoop and its ecosystem, several other MapReduce implementations have been implemented within other systems, including GridGain, Skynet, MapSharp and Twister [14]. One of the most popular alternative to Hadoop is Disco, which is a lightweight, open-source framework for distributed computing. The Disco core is written in Erlang, a functional language designed for building fault-tolerant distributed applications. Disco has been used for a variety of purposes, such as log analysis, text indexing, probabilistic modeling and data mining.

4.2 Spark

Apache Spark[11] is another Apache framework for Big Data processing. Differently from Hadoop in which intermediate data are always stored in distributed file systems, Spark stores data in RAM memory and queries it repeatedly so as to obtain better performance for some class of applications (e.g., iterative machine learning algorithms) [56]. For many years, Hadoop has been considered the leading open source Big Data framework, but recently Spark has become the more popular so that it is supported by every major Hadoop vendors. In fact, for particular tasks, Spark is up to 100 times faster than Hadoop in memory and 10 times faster on disk. Several other libraries have been built on top of Spark: *Spark SQL* for dealing with SQL and DataFrames, *MLlib* for machine learning, *GraphX* for graphs and graph-parallel computation, and *Spark Streaming* to build scalable fault-tolerant streaming applications.

For these reasons, Spark is becoming the primary execution engine for data processing and, in general, a must-have for Big Data applications. But even though in some applications Spark can be considered a better alternative to Hadoop, in many other applications it has limitations that make it complementary to Hadoop. The main limitation of Spark is that it does not provide its own distributed and scalable storage system, that is a fundamental requirement for Big Data applications that use huge and continually increasing volume of data stored across a very large number of nodes. To overcome this lack, Spark has been designed to run on top of several data sources, such as Cloud object storage (e.g., Amazon S3 Storage, Swift Object Storage), distributed filesystem (e.g., HDFS), no-SQL databases (e.g., HBase, Apache Cassandra), and others. Today an increasing number of big vendors, such Microsoft Azure or Cloudera, offer Spark as well as Hadoop, so developers can choose the most suitable framework for each data analytic application.

With respect to Hadoop, Spark loads data from data sources and executes most of its tasks in RAM memory. In this way, Spark reduces significantly the time spent in writing and reading from hard drives, so that the execution is far faster than Hadoop.

---

[11]http://spark.apache.org.

Regarding task recovering in case of failures, Hadoop flushes all of the data back to the storage after each operation. Similarly, Spark allow recovering in case of failures by arranging data in Resilient Distributed Datasets (RDD), which are a immutable and fault-tolerant collections of records which can be stored in the volatile memory or in a persistent storage (e.g., HDFS, HBase). Moreover, Spark's real-time processing capability is increasingly being used by Big Data analysts into applications that requires to extract insights quickly from data, such as recommendation and monitoring systems.

Several big companies and organizations use Spark for Big Data analysis purpose: for example, Ebay uses Spark for log transaction aggregation and analytics, Kelkoo for product recommendations, SK Telecom analyses mobile usage patterns of customers.

4.3 Mahout

Apache Mahout[12] is an open-source framework that provides scalable implementations of machine learning algorithms that are applicable on big input. Originally, the Mahout project provided implementations of machine learning algorithms executable on the top of Apache Hadoop framework. But the comparison of the performance of Mahout algorithms on Hadoop with other machine learning libraries, showed that Hadoop spends the majority of the processing time to load the state from file system at every intermediate step [44].

For these reasons, the latest version of Mahout goes beyond Hadoop and provides several machine learning algorithms for collaborative filtering, classification, and clustering, implemented not only in Hadoop MapReduce, but also in Spark, H2O.[13] Both Apache Spark and H2O process data in memory so they can achieve a significant performance gain when compared to Hadoop framework for specific classes of applications (e.g., interactive jobs, real-time queries, and stream data) [44]. In addition, the latest release of Mahout introduces a new math environment, called Samsara [29], that helps users in creating their own math providing general linear algebra and statistical operations. In the following, some examples for each algorithm's category are listed: analyzing user history and preferences to suggest accurate recommendations (collaborative filtering), selecting whether a new input matches a previously observed pattern or not (classification), and grouping large number of things together into clusters that share some similarity (clustering) [41]. In the future, Mahout will support Apache Flink,[14] an open source platform that provides data distribution, communication, and fault tolerance for distributed computations over data streams.

---

[12]http://mahout.apache.org/.
[13]http://www.h2o.ai.
[14]https://flink.apache.org/.

4.4 Hunk

Hunk[15] is a commercial data analysis platform developed by Splunk for rapidly exploring, analyzing and visualizing data in Hadoop and NoSQL data stores. Hunk uses a set of high-level user and programming interfaces to offer speed and simplicity of getting insights from large unstructured and structured data sets. One of the key components of the Hunk architecture is the *Splunk Virtual Index*. This system decouples the storage tier from the data access and analytics tiers, so enabling Hunk to route requests to different data stores. The analytics tier is based on *Splunks Search Processing Language* (SPL) designed for data exploration across large, different data sets. The Hunk web framework allows building applications on top of the Hadoop Distributed File System (HDFS) and/or the NoSQL data store.

Developers can use Hunk to build their Big Data applications on top of data in Hadoop using a set of well known languages and frameworks. Indeed, the framework enables developers to integrate data and functionality from Hunk into enterprise Big Data applications using a web framework, documented REST API and software development kits for CSharp, Java, JavaScript, PHP and Ruby. Also common development languages such as HTML5 and Python can be used by developers.

The Hunk framework can be deployed on on-premises Hadoop clusters or private Clouds and it is available as a preconfigured instance on the Amazon public Cloud using the Amazon Web Services (AWS). This public Cloud solution allows Hunk users to utilize the Hunk facilities and tools from AWS, also exploiting commodity storage on Amazon S3, according to a pay-per-use model. Finally, the framework implements and makes available a set of applications that enable the Hunk analytics platform to explore, explore and visualize data in NoSQL and other data stores, including Apache Accumulo, Apache Cassandra, MongoDB and Neo4j. Hunk is also provided in combination with the Cloudera's enterprise data hub to develop large-scale applications that can access and analyze Big Data sets.

4.5 Sector/Sphere

Sector/Sphere[16] is a Cloud framework designed at the University of Illinois-Chicago to implement data analysis applications involving large, geographically distributed datasets in which the data can be naturally processed in parallel [19]. The framework includes two components: a storage service called *Sector*, which manages the large distributed datasets with high reliability, high performance IO, and a uniform access, and a compute service called *Sphere*, which makes use of the Sector service to simplify data access, increase data IO bandwidth, and exploit wide area high performance networks. Both of them are available as open source software.[17] Sector is a

---

[15]http://www.splunk.com/en_us/products/hunk.html.

[16]http://sector.sourceforge.net/.

[17]http://sector.sourceforge.net.

distributed storage system that can be deployed over a wide area network and allows users to ingest and download large datasets from any location with a high-speed network connection to the system. The system can be deployed over a large number of commodity computers (called nodes), located either within a data center or across data centers, which are connected by high-speed networks.

In an example scenario, nodes in the same rack are connected by 1 Gbps networks, two racks in the same data center are connected by 10 Gbps networks, and two different data centers are connected by 10 Gbps networks. Sector assumes that the datasets it stores are divided into one or more separate files, called slices, which are replicated and distributed over the various nodes managed by Sector.

The Sector architecture includes a Security server, a Master server and a number of Slave nodes. The Security server maintains user accounts, file access information, and the list of authorized slave nodes. The Master server maintains the metadata of the files stored in the system, controls the running of the slave nodes, and responds to users' requests. The Slaves nodes store the files managed by the system and process the data upon the request of a Sector client. Sphere is a compute service built on top of Sector and provides a set of programming interfaces to write distributed data analysis applications. Sphere takes streams as inputs and produces streams as outputs. A stream consists of multiple data segments that are processed by Sphere Processing Engines (SPEs) using slave nodes. Usually there are many more segments than SPEs. Each SPE takes a segment from a stream as an input and produces a segment of a stream as output. These output segments can in turn be the input segments of another Sphere process. Developers can use the Sphere client APIs to initialize input streams, upload processing function libraries, start Sphere processes, and read the processing results.

4.6 BigML

BigML[18] is a system provided as a Software-as-a-Service (SaaS) for discovering predictive models from data and it uses data classification and regression algorithms. The distinctive feature of BigML is that predictive models are presented to users as interactive decision trees. The decision trees can be dynamically visualized and explored within the BigML interface, downloaded for local usage and/or integration with applications, services, and other data analysis tools. Recently, BigML launched its PaaS solution, called *BigML PredictServer*, which is a dedicated machine image that can be deployed on Amazon AWS. An example of BigML prediction model is shown in Fig. 2.

Extracting and using predictive models in BigML consists in multiple steps, as detailed as follows:

- *Data source setting and dataset creation.* A data source is the raw data from which a user wants to extract a predictive model. Each data source instance is described

---

[18]https://bigml.com.

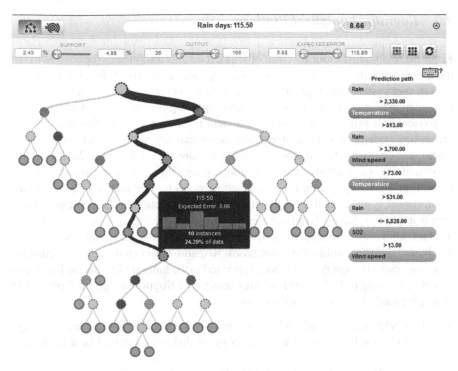

Fig. 2 Example of BigML prediction model for air pollution (source: http://bigml.com)

by a set of columns, each one representing an instance feature, or field. One of the fields is considered as the feature to be predicted. A dataset is created as a structured version of a data source in which each field has been processed and serialized according to its type (numeric, categorical, etc.).

- *Model extraction and visualization.* Given a dataset, the system generates the number of predictive models specified by the user, who can also choose the level of parallelism level for the task. The interface provides a visual tree representation of each predictive model, allowing users to adjust the support and confidence values and to observe in real time how these values influence the model.

- *Prediction making.* A model can be used individually, or in a group (the so-called ensemble, composed of multiple models extracted from different parts of a dataset), to make predictions on new data. The system provides interactive forms to submit a predictive query for a new data using the input fields from a model or ensemble. The system provides APIs to automate the generation of predictions, which is particularly useful when the number of input fields is high.

- *Models evaluation.* BigML provides functionalities to evaluate the goodness of the predictive models extracted. This is done by generating performance measures that can be applied to the kind of extracted model (classification or regression).

4.7 Kognitio Analytical Platform

Kognitio Analytical Platform,[19] available as Cloud based service or supplied as a pre-integrated appliance, allows users to pull very large amounts of data from existing data storage systems into high-speed computer memory, allowing complex analytical questions to be answered interactively. Although Kognitio has its own internal disk subsystem, it is primarily used as an analytical layer on top of existing storage/data processing systems, e.g., Hadoop clusters and/or existing traditional disk-based data warehouse products, Cloud storage, etc. A feature called *External Tables* allows persistent data to reside on external systems. Using this feature the system administrator, or a privileged user, can easily setup access to data that resides in another environment, typically a disk store such as the above-mentioned Hadoop clusters and data warehouse systems. To a final user, the Kognitio Analytical Platform looks like a relational database management system (RDBMS) similar to many commercial databases. However, unlike these databases, Kognitio has been designed specifically to handle analytical query workload, as opposed to the more traditional on-line transaction processing (OLTP) workload. Key reasons of Kognitios high performance in managing analytical query workload are:

- Data is held in high-speed RAM using structures optimized for in-memory analysis, which is different from a simple copy of disk-based data, like a traditional cache.
- Massively Parallel Processing (MPP) allows scaling out across large arrays of low-cost industry standard servers, up to thousands nodes.
- Query parallelization allows every processor core on every server to be equally involved in every query.
- Machine code generation and advanced query plan optimization techniques ensure every processor cycle is effectively used to its maximum capacity.

Parallelism in Kognitio Analytical Platform fully exploits the so-called shared nothing? distributed computing approach, in which none of the nodes share memory or disk storage, and there is no single point of contention across the system.

4.8 Data Analysis Workflows

A workflow consists of a series of activities, events or tasks that must be performed to accomplish a goal and/or obtain a result. For example, a data analysis workflow can be designed as a sequence of pre-processing, analysis, post-processing, and interpretation steps. At a practical level, a workflow can be implemented as a computer program and can be expressed in a programming language or paradigm that allows expressing the basic workflow steps and includes mechanisms to orchestrate them.

---

[19]www.kognitio.com.

Workflows have emerged as an effective paradigm to address the complexity of scientific and business applications. The wide availability of high-performance computing systems, Grids and Clouds, allowed scientists and engineers to implement more and more complex applications to access and process large data repositories and run scientific experiments in silico on distributed computing platforms. Most of these applications are designed as workflows that include data analysis, scientific computation methods and complex simulation techniques. The design and execution of many scientific applications require tools and high-level mechanisms. Simple and complex workflows are often used to reach this goal. For this reason, in the past years, many efforts have been devoted towards the development of distributed workflow management systems for scientific applications. Workflows provide a declarative way of specifying the high-level logic of an application, hiding the low-level details that are not fundamental for application design. They are also able to integrate existing software modules, datasets, and services in complex compositions that implement scientific discovery processes.

Another important benefit of workflows is that, once defined, they can be stored and retrieved for modifications and/or re-execution: this allows users to define typical patterns and reuse them in different scenarios [5]. The definition, creation, and execution of workflows are supported by a so-called Workflow Management System (WMS). A key function of a WMS during the workflow execution (or enactment) is coordinating the operations of the individual activities that constitute the workflow. There are several WMSes on the market, most of them targeted to a specific application domain. In the following we focus on some well-known software tools and frameworks designed implementing data analysis workflows on Clouds systems.

Data Mining Cloud Framework

The Data Mining Cloud Framework (DMCF) [32] is a software system that we developed at University of Calabria for allowing users to design and execute data analysis workflows on Clouds. DMCF supports a large variety of data analysis processes, including single-task applications, parameter sweeping applications, and workflow-based applications [33]. A Web-based user interface allows users to compose their applications and to submit them for execution to a Cloud platform, according to a Software-as-a-Service approach. Recently, DMCF has been extended to include the execution of MapReduce tasks [3].

The DMCFs architecture includes a set of components that can be classified as storage and compute components (see Fig. 3). The storage components include:

- A *Data Folder* that contains data sources and the results of knowledge discovery processes. Similarly, a Tool folder contains libraries and executable files for data selection, pre-processing, transformation, data mining, and results evaluation.
- The *Data Table*, *Tool Table* and *Task Table* that contain metadata information associated with data, tools, and tasks.
- The *Task Queue* that manages the tasks to be executed.

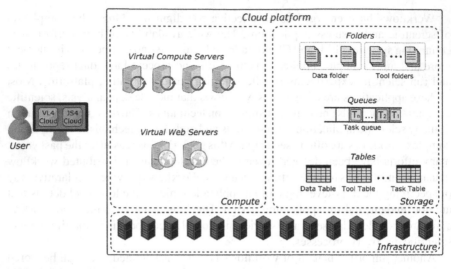

Fig. 3 DMCF architecture

The compute components are:

- A pool of *Virtual Compute Servers*, which are in charge of executing the data mining tasks.
- A pool of *Virtual Web Servers* host the Web-based user interface.

The user interface provides three functionalities:

- App submission, which allows users to submit single-task, parameter sweeping, or workflow-based applications;
- App monitoring, which is used to monitor the status and access results of the submitted applications;
- Data/Tool management, which allows users to manage input/output data and tools.

The DMCF architecture has been designed as a reference architecture to be implemented on different Cloud systems. However, a first implementation of the framework has been carried out on the Microsoft Azure Cloud platform and has been evaluated through a set of data analysis applications executed on a Microsoft Cloud data center. The DMCF framework takes advantage of Cloud computing features, such as elasticity of resources provisioning. In DMCF, at least one Virtual Web Server runs continuously in the Cloud, as it serves as user front-end. In addition, users specify the minimum and maximum number of Virtual Compute Servers. DMCF can exploit the auto-scaling features of Microsoft Azure that allows dynamic spinning up or shutting down Virtual Compute Servers, based on the number of tasks ready for execution in the DMCFs Task Queue. Since storage is managed by the Cloud platform, the number of storage servers is transparent to the user.

For designing and executing a knowledge discovery application, users interact with the system performing the following steps:

1. The Website is used to design an application (either single-task, parameter sweeping, or workflow-based) through a Web-based interface that offers both the visual programming interface and the script.
2. When a user submits an application, the system creates a set of tasks and inserts them into the Task Queue on the basis of the application requirements.
3. Each idle Virtual Compute Server picks a task from the Task Queue, and concurrently executes it.
4. Each Virtual Compute Server gets the input dataset from the location specified by the application. To this end, file transfer is performed from the Data Folder where the dataset is located, to the local storage of the Virtual Compute Server.
5. After task completion, each Virtual Compute Server puts the result on the Data Folder.
6. The Website notifies the user as soon as her/his task(s) have completed, and allows her/him to access the results.

The set of tasks created on the second step depends on the type of application submitted by a user. In the case of a single-task application, just one data mining task is inserted into the Task Queue. If users submit a parameter sweeping application, a set of tasks corresponding to the combinations of the input parameters values are executed in parallel. If a workflow-based application has to be executed, the set of tasks created depends on how many data analysis tools are invoked within the workflow. Initially, only the workflow tasks without dependencies are inserted into the Task Queue.

In DMCF workflows may encompass all the steps of discovery based on the execution of complex algorithms and the access and analysis of scientific data. In data-driven discovery processes, knowledge discovery workflows can produce results that can confirm real experiments or provide insights that cannot be achieved in laboratories. In particular, DMCF allows to program workflow applications using two languages:

- VL4Cloud (*Visual Language for Cloud*), a visual programming language that lets users develop applications by programming the workflow components graphically [33].
- JS4Cloud (*JavaScript for Cloud*), a scripting language for programming data analysis workflows based on JavaScript [34].

Both languages use two key programming abstractions:

- Data elements denote input files or storage elements (e.g., a dataset to be analyzed) or output files or stored elements (e.g., a data mining model).
- Tool elements denote algorithms, software tools or complex applications performing any kind of operation that can be applied to a data element (data mining, filtering, partitioning, etc.).

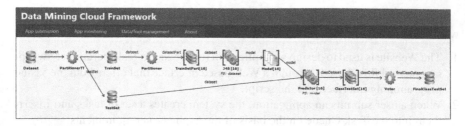

Fig. 4 Example of data analysis application designed using VL4Cloud

Another common element is the task concept, which represents the unit of parallelism in our model. A task is a *Tool*, invoked in the workflow, which is intended
to run in parallel with other tasks on a set of Cloud resources. According to this
approach, VL4Cloud and JS4Cloud implement a data-driven task parallelism. This
means that, as soon as a task does not depend on any other task in the same workflow,
the runtime asynchronously spawns it to the first available virtual machine. A task
T_j does not depend on a task T_i belonging to the same workflow (with $i \neq j$), if
T_j during its execution does not read any data element created by T_i. In VL4Cloud,
workflows are directed acyclic graphs whose nodes represent data and tools elements.
The nodes can be connected with each other through direct edges, establishing specific dependency relationships among them. When an edge is being created between
two nodes, a label is automatically attached to it representing the type of relationship
between the two nodes. Data and Tool nodes can be added to the workflow singularly
or in array form. A data array is an ordered collection of input/output data elements,
while a tool array represents multiple instances of the same tool. Figure 4 shows an
example of data analysis workflow developed using the visual workflow formalism
of DMCF [6].

In JS4Cloud, workflows are defined with a JavaScript code that interacts with
Data and Tool elements through three functions:

- Data Access, for accessing a Data element stored in the Cloud;
- Data Definition, to define a new Data element that will be created at runtime as a
  result of a Tool execution;
- Tool Execution: to invoke the execution of a Tool available in the Cloud.

Once the JS4Cloud workflow code has been submitted, an interpreter translates the
workflow into a set of concurrent tasks by analysing the existing dependencies in the
code. The main benefits of JS4Cloud are:

1. It extends the well-known JavaScript language while using only its basic functions
   (arrays, functions, loops).
2. It implements both a data-driven task parallelism that automatically spawns ready-
   to-run tasks to the Cloud resources, and data parallelism through an array-based
   formalism.
3. These two types of parallelism are exploited implicitly so that workflows can be
   programmed in a totally sequential way, which frees users from duties like work
   partitioning, synchronization and communication.

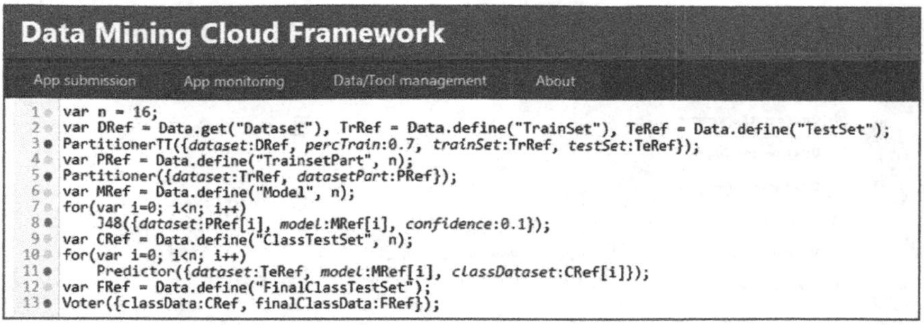

Fig. 5 Example of data analysis application designed using JS4Cloud

Figure 5 shows the script-based workflow version of the visual workflow shown in Fig. 4. In this example, parallelism is exploited in the for loop at line 7, where up to 16 instances of the J48 classifier are executed in parallel on 16 different partitions of the training sets, and in the for loop at line 10, where up to 16 instances of the Predictor tool are executed in parallel to classify the test set using 16 different classification models.

Figure 5 shows a snapshot of the parallel classification workflow taken during its execution in the DMCFs user interface. Beside each code line number, a colored circle indicates the status of execution. This feature allows user to monitor the status of the workflow execution. Green circles at lines 3 and 5 indicate that the two partitioners have completed their execution; the blue circle at line 8 indicates that J48 tasks are still running; the orange circles at lines 11 and 13 indicate that the corresponding tasks are waiting to be executed.

Microsoft Azure Machine Learning

Microsoft Azure Machine Learning (Azure ML) is a SaaS that provides a Web-based machine learning IDE (i.e., integrated development environment) for creation and automation of machine learning workflows. Through its user-friendly interface, data scientists and developers can perform several common data analysis/mining tasks on their data and automate their workflows.

Using its drag-and-drop interface, users can import their data in the environment or use special readers to retrieve data form several sources, such as Web URL (HTTP), OData Web service, Azure Blob Storage, Azure SQL Database, Azure Table. After that, users can compose their data analysis workflows where each data processing task is represented as a block that can be connected with each other through direct edges, establishing specific dependency relationships among them. Azure ML includes a rich catalog of processing tools that can be easily included in a workflow to prepare/transform data or to mine data through supervised learning (regression e classification) or unsupervised learning (clustering) algorithms. Optionally, users can include their own custom scripts (e.g., in R or Python) to extend the tools catalog. When workflows are correctly defined, users can evaluate them using some testing dataset.

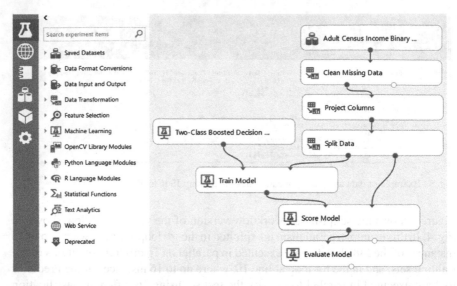

Fig. 6 Example of Azure machine learning workflow (source: http://studio.azureml.net)

Users can easily visualize the results of the tests and find very useful information about models accuracy, precision and recall. Finally, in order to use their models to predict new data or perform real time predictions, users can expose them as Web services. Always through a Web-based interface, users can monitor the Web services load and use by time. Azure Machine Learning is a fully managed service provided by Microsoft on its Cloud platform; users do not need to buy any hardware/software nor manage virtual machine manually. One of the main advantage of working with a Cloud platform like Azure is its auto-scaling feature: models are deployed as elastic Web services so as users do not have to worry about scaling them if the models usage increased. An example of workflow built on Microsoft Azure Machine Learning is shown in Fig. 6.

ClowdFlows

ClowdFlows [22] is an open source Cloud-based platform for the composition, execution, and sharing of data analysis workflows. It is provided as a software as a service that allows users to design and execute visual workflows through a simple Web browser and so it can be run from most devices (e.g., desktop PCs, laptops, and tablets). ClowdFlows is based on two software components: the workflow editor (provided by a Web browser) and the server side application that manages the execution of the application workflows and hosts a set of stored workflows. The server side consists of methods for supporting the client-side workflow editor in the composition and for executing workflows, and a relational database of workflows and data. The workflow editor includes of a workflow canvas and a widget repository. The widget repository is a list of all available workflow components that can be added to the workflow canvas. The repository includes a set of default widgets. Figure 7 shows an example of workflow built on CloudFlow.

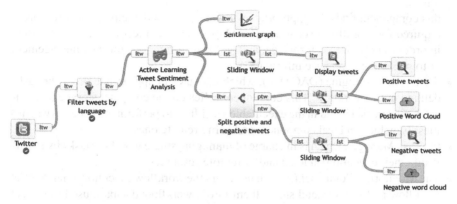

Fig. 7 Example of CloudFlow workflow (source: http://clowdflows.org)

According to this approach, the CloudFlows service-oriented architecture allows users to include in their workflow the implementations of various algorithms, tools and Web services as workflow elements. For example, the Weka's algorithms have been included and exposed as Web services and so they can be added in a workflow application. ClowdFlows is also easily extensible by importing third-party Web services that wrap open-source or custom data mining algorithms. To this end, a user has only to insert the WSDL URL of a Web service to create a new workflow element that represents the Web service in a workflow application.

Pegasus

Pegasus [12] is a workflow management system developed at the University of Southern California for supporting the implementation of scientific applications also in the area of data analysis. Pegasus includes a set of software modules to execute workflow-based applications in a number of different environments, including desktops, Clouds, clusters and grids. It has been used in several scientific areas including bioinformatics, astronomy, earthquake science, gravitational wave physics, and ocean science. The Pegasus workflow management system can manage the execution of an application expressed as a visual workflow by mapping it onto available resources and executing the workflow tasks in the order of their dependencies. In particular, significant activities have been recently performed on Pegasus to support the system implementation on Cloud platforms and manage computational workflows in the Cloud for developing data-intensive scientific applications (Juve et al. 2010; Nagavaram et al. 2011). The Pegasus system has been used with IaaS Clouds for workflow applications and the most recent versions of Pegasus can be used to map and execute workflows on commercial and academic IaaS Clouds such as Amazon EC2, Nimbus, OpenNebula and Eucalyptus [12]. The Pegasus system includes four main components:

- the Mapper, which builds an executable workflow based on an abstract workflow provided by a user or generated by the workflow composition system. To this end, this component finds the appropriate software, data, and computational resources required for workflow execution. The Mapper can also restructure the workflow in order to optimize performance, and add transformations for data management or to generate provenance information.

- the Execution Engine (DAGMan), which executes in appropriate order the tasks defined in the workflow. This component relies on the compute, storage and network resources defined in the executable workflow to perform the necessary activities. It includes a local component and some remote ones.
- the Task Manager, which is in charge of managing single workflow tasks by supervising their execution on local and/or remote resources.
- The Monitoring Component, which monitors the workflow execution, analyzes the workflow and job logs and stores them into a workflow database used to collect runtime provenance information. This component sends notifications back to users notifying them of events like failures, success and completion of workflows and jobs.

The Pegasus software architecture includes also an error recovery system that attempts to recover from failures by retrying tasks or an entire workflow, re-mapping portions of the workflow, providing workflow-level checkpointing, and using alternative data sources, when possible. The Pegasus system records provenance information including the locations of data used and produced, and which software was used with which parameters. This feature is useful when a workflow must be reproduced.

Swift

Swift [53] is a implicitly parallel scripting language that runs workflows across several distributed systems, like clusters, Clouds, grids, and supercomputers. The Swift language has been designed at the University of Chicago and at the Argonne National Lab to provide users with a workflow-based language for grid computing. Recently has been ported on Clouds and exascale systems. Swift separates the application workflow logic from runtime configuration. This approach allows a flexible development model.

As the DMCF programming interface, the Swift language allows invocation and running of external application code and allows binding with application execution environments without extra coding from the user. Swift/K is the previous version of the Swift language that runs on the Karajan grid workflow engine across wide area resources. Swift/T is a new implementation of the Swift language for high-performance computing. In this implementation, a Swift program is translated into an MPI program that uses the Turbine and ADLB runtime libraries for scalable dataflow processing over MPI. The Swift-Turbine Compiler (STC) is an optimizing compiler for Swift/T and the Swift Turbine runtime is a distributed engine that maps the load of Swift workflow tasks across multiple computing nodes. Users can also use Galaxy [17] to provide a visual interface for Swift.

The Swift language provides a functional programming paradigm where workflows are designed as a set of code invocations with their associated command-line arguments and input and output files. Swift is based on a C-like syntax and uses an implicit data-driven task parallelism [54]. In fact, it looks like a sequential language, but being a dataflow language, all variables are futures, thus execution is based on data availability. When input data is ready, functions are executed in parallel. Moreover, parallelism can be exploited through the use of the foreach statement. The Turbine runtime comprises a set of services that implement the parallel execution of Swift scripts exploiting the maximal concurrency permitted by data dependencies within a script and by external resource availability. Swift has been used for developing several scientific data analysis applications, such as prediction of protein structures, modeling the molecular structure of new materials, and decision making in climate and energy policy.

4.9   NoSQL Models for Data Analytics

With the exponential growth of data to be stored in distributed network scenarios, relational databases exhibit scalability limitations that significantly reduce the efficiency of querying and analysis [1]. In fact, most relational databases have little ability to scale horizontally over many servers, which makes challenging storing and managing the huge amounts of data produced everyday by many applications.

The NoSQL or non-relational database approach became popular in the last years as an alternative or as a complement to relational databases, in order to ensure horizontal scalability of simple read/write database operations distributed over many servers [8]. Compared to relational databases, NoSQL databases are generally more flexible and scalable, as they are capable of taking advantage of new nodes transparently, without requiring manual distribution of information or additional database management [46]. Since database management may be a challenging task with huge amounts of data, NoSQL databases are designed to ensure automatic data distribution and fault tolerance [15]. In the remainder of this section, we describe some representative NoSQL systems, and discuss some use cases for NoSQL databases, with a focus on data analytics.

NoSQL databases provide ways to store scalar values (e.g., numbers, strings), binary objects (e.g., images, videos), or more complex values. According to their data model, NoSQL databases can be grouped into three main categories [8]: Key-value stores, Document stores, Extensible Record stores.

Key-value stores provide mechanisms to store data as (key, value) pairs over multiple servers. In such kind of databases a distributed hash table (DHT) can be used to implement a scalable indexing structure, where data retrieval is performed by using key to find value [8].

Document stores are designed to manage data stored in documents that use different formats (e.g., JSON), where each document is assigned a unique key that is used to identify and retrieve the document. Therefore, document stores extend key-value stores because they provide for storing, retrieving, and managing semi-structured information, rather than single values. Unlike the key-value stores, document stores generally support secondary indexes and multiple types of documents per database, and provide mechanisms to query collections based on multiple attribute value constraints [8].

Finally, Extensible Record stores (also known as Column-oriented data stores) provide mechanisms to store extensible records that can be partitioned across multiple servers. In this type of database, records are said to be extensible because new attributes can be added on a per-record basis. Extensible record stores provide both horizontal partitioning (storing records on different nodes) and vertical partitioning (storing parts of a single record on different servers). In some systems, columns of a table can be distributed over multiple servers by using column groups, where pre-defined groups indicate which columns are best stored together.

A brief comparison of noSQL databases is shown in Table 1. For a more detailed comparison see also [20, 28, 39].

Google Bigtable

Google Bigtable[20] is a popular table store. Built above the Google File System, it is able to store up to petabytes of data and supporting tables with billions of rows and thousands of columns. Thanks to its high read and write throughput at low latency, Bigtable it is an ideal data source for batch MapReduce operations [9] and other applications oriented to the processing and analysis of large volumes of data.

Data in Bigtable are stored in sparse, distributed, persistent, multi-dimensional tables composed of rows and columns. Each row is indexed by a single row key, and a set of columns that are grouped together into sets called *column families*. Instead, a generic column is identified by a column family and a *column qualifier*, which is a unique name within the column family. Each value in the table is indexed by a tuple (row key, column key, timestamp). To improve scalability and to balance the query workload, data are ordered by row key and the row range for a table is dynamically partitioned into contiguous blocks, called *tablets*. These tablets are distributed among different Bigtable cluster's nodes (i.e., *Tablet Servers*). To improve load balancing, the Bigtable master is able to split larger and merge smaller tablets, redistributing them across nodes as needed. To ensure data durability, Bigtable stores data on Google File System (GFS) and protects it from disaster events through data replication and backup. Bigtable can be used into applications through multiple clients, including *Cloud Bigtable HBase*, a customized version of the standard client for the industry-standard Apache HBase.

Apache Cassandra

Apache Cassandra[21] is a distributed database management system providing high availability with no single point of failure. Born at Facebook and inspired by Amazon Dynamo and Google BigTable, Apache Cassandra is designed for managing large amount of data across multiple data centers and Cloud availability zones.

Cassandra uses a masterless ring architecture, where all nodes play an identical role, that allows any authorized user to connect to any node in any data center.

---

[20]https://cloud.google.com/bigtable/.
[21]http://cassandra.apache.org/.

Table 1 Comparison of some NoSQL databases. FS = File System; MEM = In-Memory

	DynamoDB	Cassandra	Hbase	Redis	CouchDB	BigTable	MongoDB	Neo4j
Type	Key-value	Column	Column	Key-value	Document	Column	Document	Graph
Data storage	MEM, FS	HDFS, CFS	HDFS	MEM, FS	MEM, FS	GFS	MEM, FS	MEM, FS
MapReduce	yes	yes	yes	no	yes	yes	yes	no
Persistence	yes	yes	yes	yes, with limits[a]	yes	yes	yes	yes
Replication	yes	yes	yes	yes	yes	yes	yes	yes
Scalability	high	high	high	high	high	high	high	high
Performance	high	high	high	high	high	high	high	high, variable
High availability	yes	yes	yes	yes	yes	yes	yes	yes
Language	Java	Java	Java	Ansi-C	Erlang	Java, Python, Go, Ruby	C++	Java
License	Proprietary	Apache 2.0	Apache 2.0	BSD	Apache 2.0	Proprietary	GNU AGPL3	GNU GPL3

[a]Last queries can be lost as explained in http://redis.io/topics/persistence

This is a really simple and flexible architecture that allows to add nodes without service downtime. The process of data distribution across nodes is very simple and no programmatic operations are needed by the developers.

Since all nodes communicate each other equally, Cassandra has no single point of failure, that ensures continuous data availability and service uptime. Moreover, Cassandra provides very customizable data replication service that allows to replicate data across nodes that participate in a ring. In this manner, in case of node failure, one or more copies of the needed data are available on other nodes.

Cassandra also provides built-in and customizable replication, which stores redundant copies of data across nodes that participate in a Cassandra ring. This means that if a node in a cluster goes down, one or more copies of data stored on that node is available on other machines in the cluster. Replication can be configured to work across one data center, many data centers, and multiple Cloud availability zones. Focusing on performance and scalability, Cassandra reaches a quite linear speedup, that means the OPS (Operations Per Second) capacity can be increased by adding new nodes (e.g., if 2 nodes can handle 10,000 OPS, 4 nodes will support 20,000 OPS, and so on).

Many companies have successfully deployed and benefited from Apache Cassandra including some large companies such as: Apple (75,000 nodes storing over 10 PB of data), Chinese search engine Easou (270 nodes, 300 TB, over 800 million requests per day), and eBay (over 100 nodes, 250 TB), Netflix (2,500 nodes, 420 TB, over 1 trillion requests per day), Instagram, Spotify, eBay, Rackspace, and many more.

Neo4j Graph Database

If we need to take into account real time data relationships (e.g. create queries using data relationships), NoSQL databases are not the best choice. In fact, relationship-based or graph databases has been created for naturally supporting operations on data that use data relationships. Graph databases provide a novel and powerful data modeling technique that does not store data in tables, but in graph models [43], with several benefits in storing and retrieving data connected by complex relationships.

There are several graph data models, such as Neo4j, OrientDB, Virtuoso, Allegro, Stardog, InfiniteGraph. Among all we focus on Neo4j. Neo4j is an open-source NoSQL graph database implemented in Java and Scala that is considered the most popular graph database used today. The Neo4j source code and issue tracking are available on GitHub, with a large support community. It is used today by a very large number of organizations working in different sectors, including software analytics, scientific research, project management, recommendations, and social networks.

In the Neo4j graph model, each node contains a list of relationship records that refer to other nodes, and additional attributes (e.g. timestamp, metadata, key-value pairs, and more). Each relationship record must have a name, a direction, a start node and an end node, and can contains additional properties. One or more labels can be assigned both to nodes and relationships. In particular, such labels can be used for representing the roles a node plays in the graph (e.g., user, address, company, and so

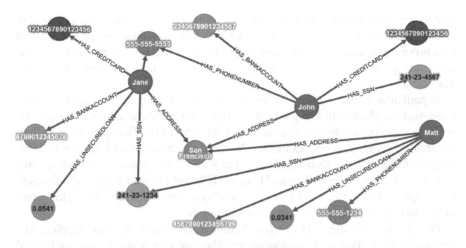

Fig. 8 Example of bank fraud graph dataset (source: http://neo4j.com)

on) or for associating indexes and constraints to groups of nodes. Figure 8 shows an example of a graph model used for detecting bank fraud.

Moreover, Neo4j clusters are designed for high availability and horizontal read scaling using master-slave replication. Focusing on performance, Neo4j is thousands of times faster than SQL in executing traversal operation. The traversal operation consists of visiting a set of nodes in the graph by moving along relationships (e.g., find potential friends in social network from user friendship). With such kind of operations, graph models allow to take into account only the data that is required, without doing expensive grouping operations as done by relational database during join operations [51]. Queries in Neo4j are written using *Cypher*, a declarative and SQL-based language for describing patterns in graphs. Cypher is a relative simple but very powerful language, that allows to execute queries in a easy way on a very complex graph database.

4.10   Visual Analytics

A primary problem in data analysis is to interpret results easily. To overcome this problem, in the last years, great progress has been made in the field of visual analytics. As defined by [50], visual analytics is the science of analytical reasoning facilitated by interactive visual interfaces. Nowadays, people use visual analytics tools and methodologies to extract synthetic information from often confusing data and use them in further analysis or business operations. The power of visual analytics techniques relies on human brain capabilities to process graphics faster than text. In particular, through a graphical data presentation, the human brain could be able to find complex and often hidden patterns and relationships in data that are difficult to discover using automatic methods. Also in the Big Data context, the tools used to visualize results and to interact with data play a key role. Thus, in order to support data presentation and interaction also in presence of Big Data, innovative methodologies (e.g., interactive charts, animations, diagrams, and much more) have been developed.

In particular, to ride the wave of visual analytics technologies, several big IT company, such as Microsoft, Google, and SAS, developed advanced data presentation and data visualization tools able to interact with existent Big Data platforms, including Hadoop-based ones. For example, Microsoft extended Excel functions to allow integration with its Big Data solution. In particular, Excel's users can be connected to Azure Storage associated to an Hadoop HDInsight cluster using the Microsoft Power Query for Excel add-in. Once data has been retrieved, users can exploit Excel functions to make more interesting charts or graphs.

Google Fusion Tables[22] is an other alternative for turning data into graphics in a very easy way. It allows to load tabular data, filter and summarize across hundreds of thousands of rows, and create geo maps, heat maps, graphs, charts, animations, and more. Also Google Charts[23] are a powerful Javascript library for making interactive charts for browsers and mobile devices. Google Charts allows to create several types of charts, from simple line charts to complex hierarchical tree maps. In the field of maps and location-based applications, advanced platforms, such as Google Maps,[24] Mapbox,[25] can be used to create interactive and dynamic maps, display additional layers on a map or generate routes. In the field of visual data analysis, several Big Data start-ups spring up in the last years. Tableau,[26] for example, is a Big Data company from Stanford with multinational operations in fifteen cities, and more than 39,000 customer accounts in 150 countries. It developed software solutions for easily creating complex charts from huge amount of data. In fact, thanks to its Cloud analytics platform, Tableau allows users to manipulate data through a simple web control panel. In this way, users can interact directly with data to find interesting insights. Among all the competitors in this field, SAS[27] probably stands out among its peers.

SAS Visual Analytics, in fact, represents a complete solution for advanced data visualization and exploratory analyses. Thanks to its drag-and-drop capabilities and no code requirements, it allows users to easily solve complex issues using several sophisticated techniques for data analysis (e.g. decision trees, network diagrams, scenario analysis, path analysis, sentiment analysis) and business intelligence. In addition, exploiting in-memory processing, SAS software makes analytic applications faster.

---

[22]https://tables.googlelabs.com.
[23]https://developers.google.com/chart.
[24]https://www.google.com/maps.
[25]https://www.mapbox.com/.
[26]http://www.tableau.com.
[27]https://www.sas.com.

4.11   Big Data Funding Projects

Open-source projects discussed in the previous sections (e.g., Hadoop, Spark, and NoSQL databases) have been widely used in several public funding projects. As examples:

- BigFoot project[28] is a cloud-based solution featuring scalable and optimized engines to store, process and interact with Big Data. It has received funding from the European Union's Horizon 2020 program.
- Optique[29] is a EU funding project with a total budget of about 14 million EUR. It is aims to provide a novel end-to-end OBDA (Ontology-Based Data Access) [7, 38] solution for improving Big Data access. In particular, Optique platform allows to quickly formulate intuitive queries exploiting user vocabularies and conceptualizations, and executing them using massive parallelism.

Also government agencies invested large amount of money on Big Data technologies in many public sector fields, such as intelligence, defense, weather forecasting, crime prediction and prevention, and scientific research.

As example, US Administration invested more that 250 million USD for Big Data research and development initiative across multiple agencies and departments. Moreover, in 2014 UK government decided to invest about 73 million GBP in Big Data and other analytics technologies with the goals of creating 58,000 new jobs in Britain by 2017, contributing 216 billion GBP to the countrys economy.

4.12   Historical Review

In this section a brief historical review of Big Data is presented. Undoubtedly, main events in Big Data evolution are due to big IT and Internet companies, like Google and Yahoo, who faced first the need of new solutions for tackling the rise of Big Data. A significant role in this context has been played by Hadoop and its related projects, that made Big Data analytics accessible also to a larger number of organizations.

Hadoop was created by Doug Cutting and it has its origins in Apache Nutch (2002), an open source web search engine, itself a part of the Lucene project (2000). After Google released the Google File System (GFS) paper (October 2003) and the MapReduce paper (December 2004), Cutting went to work with Yahoo and decided to build open source frameworks based on them: in 2006 Yahoo! created Hadoop based on GFS and MapReduce, and one year later, it started using Hadoop on a 1000 node cluster. In 2006, Yahoo Labs created Pig based on Hadoop, and then donated it to the *Apache Software Foundation* (ASF). In few years, several other projects was created around Hadoop and, in a short time, graduated to a Apache Top Level Project: HBase (2008), Hive (2008), Cassandra (2008), Storm (2011), Giraph (2011),

---

[28]http://bigfootproject.eu/.
[29]http://optique-project.eu.

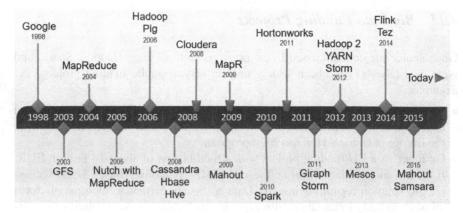

Fig. 9 A short Hadoop ecosystem's history

and so on. At the same time, many Hadoop distributor was founded, such as Cloudera (2008), MapR (2009), Hortonworks (2011). A short history of Hadoop and related project is shown in Fig. 9.

Spark represents another milestone in Big Data analytics. Spark was initially created at UC Berkeley's AMPLab in 2009, open sourced in 2010 under a BSD license, and donated to the ASF in 2013. Finally, in February 2014, Spark became a Top-Level Apache Project and declared the most active ASF project. As discussed before, Spark is nowadays considered the primary execution engine for several Big Data applications, sometimes used to complement Hadoop.

4.13 Summary

It is not easy to summarize all the features of the systems discussed till now or to do a proof comparison among them. Some of those systems have common features and, in some cases, using one rather than another is an hard choice. In fact, given a specific data analytic task, such as a machine learning application, it is possible to use several tools. Some of those are widely used commercial tools, provided through cloud services, that can be easily used by no skilled people (e.g., Azure Machine Learning or Amazon Machine Learning); other are open-source frameworks that require skilled users who prefer to program their application using a more technical approach. In addition, choosing the best solution for developing a data analytic application may depend on many other factors, such as budget (e.g., often high-level services are easy-to-use but more expensive than low-level solutions), data format, data source, the amount of data to be analyze and its velocity, and so on. Table 2 presents a brief comparison of the Big Data analytics systems.

Table 2 A brief comparison of most common big data analytics systems

Systems/Tools	Analytics				SQL	Data flow	Data processing	Workflow	Open-source	Cloud model
	Streaming	Graph	In-memory	Machine learning						
Hadoop	x	x		x		x	x	x	x	IaaS
Spark	x	x	x	x	x		x		x	IaaS
Mahout			x	x					x	IaaS
Oozie								x	x	IaaS
Tez								x	x	IaaS
Giraph		x							x	IaaS
Storm	x								x	IaaS
Hive					x				x	IaaS
Pig						x			x	IaaS
Hunk										SaaS
Sector/Sphere				x			x		x	SaaS
BigML				x						SaaS, PaaS
Kognitio analytical			x		x					PaaS
DMCF				x		x	x	x		SaaS, PaaS
Microsoft Azure ML				x			x	x		SaaS
Amazon ML	x		x	x			x			SaaS
Pegasus								x	x	IaaS
ClowdFlows								x	x	PaaS
Swift								x	x	IaaS

Table 3 Summary considerations about graph databases

Graph databases	
Horizontal scaling	Poor horizontal scaling
When to use	For storing objects without a fixed schema and linked together by relationships; when users can done naturally their reasoning about data via graph traversals instead of using complex SQL queries
CAP tradeoff	Usually prefer availability over consistency
Pros	Powerful data modeling and relationships representation; locally indexed connected data; easy to query
Cons	Highly specialized query capabilities that make them the best for graph data, but not suitable for non-graph data

Hadoop represents the most used framework for developing distributed Big Data analytics application. In fact, Hadoop-ecosystem is undoubtedly the most complete solution for any kind of problem, but at the same time it is thought for high skilled users. On the other hand, many other solutions are designed for low-skilled users or for low-medium organizations that do not want to spend resources in developing and maintaining enterprise data analytics solutions (e.g., Microsoft Azure Machine Learning, Amazon Machine Learning, Data Mining Cloud Framework, Kognitio Analytical, or BigML). Finally, other solutions have been created mainly for scientific research purposes and, for this reason, they are poorly used for developing business applications (e.g., Sector/Sphere, Pegasus).

Choosing the best database solution for creating a Big Data application is another key-step, so several aspects need to be considered. To decide what kind of database to adopt, the first aspect to be considered is probably the classes of queries will be run. So graph databases are probably the best solution for representing and querying highly connected data (e.g., data gathered from social network) or that have complex relationships and/or dynamic schema. In any other case, when non-graph data are analyzed, graph databases could result in really bad performance. About that, summary considerations on graph databases are presented in Table 3.

Another aspect to be considered in choosing the best database solution should be the CAP (Consistency, Availability, and Partition) capabilities offered, because distributed NoSQL database systems can't be fully CAP compliant. In fact, the CAP theorem, also named Brewer's theorem [18], states that a distributed system can't simultaneously guarantee all three of the following properties:

- Consistency (C), that means all nodes see the same data at the same time;
- Availability (A), that means every request will receive a response within a reasonable amount of time;
- Partition (P) tolerance, that means the system continues to function also if arbitrary network partitions occur due to failures.

Table 4 Summary considerations about Key-Value databases

Key-value databases	
Horizontal scaling	Very high scale provided via sharding
When to use	When you have a very simple data schema or extreme speed scenario (like real-time)
CAP tradeoff	Most solutions prefer consistency over availability
Pros	Simple data model; very high scalability, data can be accessed using query language like SQL
Cons	Some queries could be inefficient or limited due to sharding (e.g., join operations across shards); no API standardization; maintenance is difficult; poor for complex data

Table 5 Summary considerations about column-oriented databases.

Column-oriented databases	
Horizontal scaling	Very high scale capabilities
When to use	When you need consistency and higher scalability performance than a single machine (i.e., usually using more than 1,000 nodes), without using indexed caching front end
CAP tradeoff	Most solutions prefer consistency over availability
Pros	Higher throughput and stronger concurrency when it is possible to partition data; multi-attribute queries; data is naturally indexed by columns; support semi-structured data
Cons	More complex than the document stores; poor for interconnected data

Thus if a distributed database system guarantees Consistency and Partitioning, it can never ensure Availability. Similarly, if you need a full Availability and Partition tolerance, you can't have Consistency, anyway not immediately. In fact, on a distributed environment, data changes on one node need some time to be propagated to the other nodes. During that time the copies will be mutually inconsistent, that may lead to the possibility of reading not updated data. To try to overcome this limitation, the *Eventual Consistency* property is usually provided: it ensures that the system, sooner or later, will become consistent. This is a weak property, so if the adopted database system only provides eventual consistency, the developer must be aware that exists the possibility of reading inconsistent data. NoSQL databases usually offer a balance among CAP properties, which is the key difference among the different available solutions. For each database family, some summary considerations are also provided for Key-Value databases (Table 4), Column-oriented (Table 5), and Document-oriented databases (Table 6).

Table 6 Summary considerations about document-oriented databases

Document-oriented databases	
Horizontal scaling	Scale provided via replication or replication and sharding
When to use	When your record structure is relatively small and it is possible to store all of its related properties in a single doc
CAP tradeoff	In most cases prefer consistency over availability
Pros	High scalability and simple data model; generally support secondary indexes, multiple types of documents per database, and nested documents or lists; MapReduce support for adhoc querying.
Cons	Eventually consistent model with limited atomicity and isolation; poor for interconnected data; query model is limited to keys and indexes

5 Research Trends

Big Data analysis is a very active research area with significant impact on industrial and scientific domains where is important to analyze very large and complex data repositories. In particular, in many cases data to be analyzed are stored in Cloud platforms and elastic computing Clouds facilities are exploited to speedup the analysis. This section outlines and discusses main research trends in Big Data analytics and Cloud systems for managing and mining large-scale data repositories.

As we discussed, scalable data analytics requires high-level, easy-to-use design tools for programming large applications dealing with huge, distributed data sources. Moreover, Clouds are widely adopted by many organizations, however several existing issues remain to be addressed, so that Cloud solutions can improve their efficiency and competitiveness at each business size, from medium to large companies. This requires further research and development in several key areas such as:

- Programming models for Big Data analytics. Big Data analytics programming tools require novel complex abstract structures. The MapReduce model is often used on clusters and Clouds, but more research is needed to develop scalable higher-level models and tools. State-of-the-art solutions generated major success stories, however they are not mature and suffer several problems from data transfer bottlenecks to performance unpredictability. Several other processing models have been proposed as alternative to MapReduce, such as Dryad [21] or Pregel [30], but they have never been widely used by developers.
- Data storage scalability. The increasing amount of data generated needs even more scalable data storage systems. As discussed in the previously, traditional RDBMS systems are not the best choice for supporting Big Data applications in the Cloud, and that leads to the popularity of noSQL platforms [8]. Several noSQL solutions have been proposed, with good experimental results in term of performance gain, but several other improvements are still needed [47, 52]. In fact, RDBMS systems have been around for a long time, are quite stable and offers lots of features. In the other hand, most noSQL systems are in its early version and several additional

features have yet to be improved or implemented, such as integrating capabilities from DBMS (e.g., indexing techniques), facilities for ad-hoc queries, and more.

- Data availability. Cloud service provides have to deal with the problem of granting service and data availability. Especially in presence of huge amounts of data, granting high-quality service is an opened challenge. Several solutions have been proposed for improving exploitation, such as using a cooperative multi-Cloud model to support Big Data accessibility in emergency cases [25], but more studies are still needed to handle the continue increasing demand for more real time and broad network access to Cloud services.

- Data and tool interoperability and openness. Interoperability is a main issue in large-scale applications that use resources such as data and computing nodes. Standard formats and models are needed to support interoperability and ease cooperation among teams using different data formats and tools. The *National Institute of Standards and Technology* (NIST) just released the Big Data interoperability framework,[30] a collection of documents, organized in 7 volumes, which aim to define some standards for Big Data.

- Data quality and usability. Big Data sets are often arranged by gathering data from several heterogeneous and often not well-known sources. This leads to a poor data quality that is a big problem for data analysts. In fact, due to the lack of a common format, inconsistent and useless data can be produced as a result of joining data from heterogeneous sources. Defining some common and widely adopted format would lead to data that are consistent with data from other sources, that means high quality data. Since real-world data is highly susceptible to inconsistency, incompleteness, and noise, finding effective methodologies for data preprocessing is still an open challenge for improve data quality and the analysis results [10]. In this regard, an interesting discussion about challenges of data quality in the Big Data has been presented in [6].

- Integration of Big Data analytics frameworks. The service-oriented paradigm allows running large-scale distributed workflows on heterogeneous platforms along with software components developed using different programming languages or tools. Scalable software architectures for fine grain in-memory data access and analysis. Exascale processors and storage devices must be exploited with fine-grain runtime models. Software solutions for handling many cores and scalable processor-to-processor communications have to be designed to exploit exascale hardware [13, 36].

- Tools for massive social network analysis. The effective analysis of social network data on a large scale requires new software tools for real-time data extraction and mining, using Cloud services and high-performance computing approaches [31, 35]. Social data streaming analysis tools represent very useful technologies to understand collective behaviors from social media data. Tools for data exploration and models visualization. New approaches to data exploration and models visualization are necessary taking into account the size of data and the complexity of the knowledge extracted. As data are bigger and bigger, visualization tools will

---

[30]http://www.nist.gov/itl/bigdata/bigdatainfo.cfm.

be more useful to summarize and show data patterns and trends in a compact and easy-to-see way.

- Local mining and distributed model combination. As Big Data applications often involve several local sources and distributed coordination, collecting distributed data sources to a centralized server for analysis is not practical or in some cases possible. Scalable data analysis systems have to enable local mining of data sources and model exchange and fusion mechanisms to compose the results produced in the distributed nodes [55]. According to this approach the global analysis can be performed by distributing the local mining and supporting the global combination of every local knowledge to generate the complete model.

- In-memory analysis. Most of the data analysis tools query data sources on disks while, differently from those, in-memory analytics query data in main memory (RAM). This approach brings many benefits in terms of query speed up and faster decisions. In-memory databases are, for example, very effective in real-time data analysis, but they require high-performance hardware support and fine-grain parallel algorithms [49, 59]. New 64-bit operating systems allow to address memory up to one terabyte, so making realistic to cache very large amount of data in RAM. This is why this research area is very promising.

6   Conclusions

In the last years the ability to gather data has increased exponentially. Advances and pervasiveness of computers have been the main driver of the very huge amounts of digital data that today are collected and stored in digital repositories. Those data volumes can be analyzed to extract useful information and producing helpful knowledge for science, industry, public services and in general for humankind. However, the huge amount of data generated, the speed at which it is produced, and its heterogeneity, represent a challenge to the current storage, process and analysis capabilities. Then to extract value from such kind of data, novel technologies and architectures have been developed by data scientists for capturing and analyzing complex and/or high velocity data. In this scenario was born also the Big Data mining field as a discipline that today provides several different techniques and algorithms for the automatic analysis of large data sets. But, the process of knowledge discovery from Big Data is not so easy, mainly due to data characteristics, and to get valuable information and knowledge in shorter time, high performance and scalable computing systems are needed. In many cases, Big Data are stored and analyzed in Cloud platforms.

Clouds provide scalable storage and processing services that can be used for extracting knowledge from Big Data repositories, as well as software platforms for developing and running data analysis environments on top of such services. In this chapter we provided an overview of Cloud technologies by describing the main service models (SaaS, PaaS, and IaaS) and deployment models (public, private or hybrid Clouds) adopted by Cloud providers. We also described representative examples of Cloud environments (Microsoft Azure, Amazon Web Services, OpenNebula and OpenStack) that can be used to implement applications and frameworks for data analysis in the Cloud. The development of data analysis applications on Cloud computing systems is a complex task that needs to exploit smart software solutions and innovative technologies. In this chapter we presented the leading software tools and technologies used for developing scalable data analysis on Clouds, such as MapReduce, Spark, workflow systems, and NoSQL database management systems. In particular, we particularly focused on Hadoop, the best-known MapReduce implementation, that is commonly used to develop scalable applications that analyze big amounts of data. As we discussed, Hadoop is also a reference tool for several other frameworks, such as Storm, Hive, Oozie and Spark. Moreover, besides Hadoop and its ecosystem, several other MapReduce implementations have been implemented within other systems, including GridGain, Skynet, MapSharp, and Disco.

As such Cloud platforms become available, researchers are increasingly porting powerful data mining programming tools and strategies to the Cloud to exploit complex and flexible software models, such as the distributed workflow paradigm. Workflows provide a declarative way of specifying the high-level logic of an application, hiding the low-level details. They are also able to integrate existing software modules, datasets, and services in complex compositions that implement discovery processes. In this chapter we presented several data mining workflow systems, such as Data Mining Cloud Framework, Microsoft Azure Machine Learning, ClowdFlows.

Then we also discussed NoSQL database technology that became popular in the latest years as an alternative or as a complement to relational databases. In fact, NoSQL systems in several application scenarios are more scalable and provide higher performance than relational databases. We introduced the basic principles of NoSQL, described representative NoSQL systems, and outlined interesting data analytics use cases where NoSQL tools are useful. Finally, some research trends and open challenges on Big Data analysis has been discussed, such as scalable data analytics requirements of high-level, easy-to-use design tools for programming large applications dealing with huge distributed data sources.

Acknowledgements This work is partially supported by EU under the COST Program Action IC1305: Network for Sustainable Ultrascale Computing (NESUS).

References

1. V. Abramova, J. Bernardino, P. Furtado, Which nosql database? a performance overview. Open J. Databases (OJDB) **1**(2), 17–24 (2014)
2. R. Barga, D. Gannon, D. Reed, The client and the cloud: democratizing research computing. IEEE Internet Comput. **15**(1), 72–75 (2011)
3. L. Belcastro, F. Marozzo, D. Talia, P. Trunfio, Programming visual and script-based big data analytics workflows on clouds, in *Big Data and High Performance Computing*. Advances in Parallel Computing, vol. 26 (IOS Press, 2015), pp. 18–31
4. L. Bermingham, I. Lee, Spatio-temporal sequential pattern mining for tourism sciences. Procedia Comput. Sci. **29**, 379–389 (2014). 2014 International Conference on Computational Science

5. S. Bowers, B. Ludäscher, A.H. Ngu, T. Critchlow, Enabling scientificworkflow reuse through structured composition of dataflow and control-flow, in *22nd International Conference on Data Engineering Workshops, 2006*. Proceedings (IEEE, 2006), pp. 70–70

6. L. Cai, Y. Zhu, The challenges of data quality and data quality assessment in the big data era. Data Sci. J. **14**, 2 (2015)

7. D. Calvanese, G. De Giacomo, D. Lembo, M. Lenzerini, R. Rosati, Tractable reasoning and efficient query answering in description logics: the dl-lite family. J. Autom. Reason. **39**(3), 385–429 (2007)

8. R. Cattell, Scalable sql and nosql data stores. ACM SIGMOD Record **39**(4), 12–27 (2011)

9. F. Chang, J. Dean, S. Ghemawat, W.C. Hsieh, D.A. Wallach, M. Burrows, T. Chandra, A. Fikes, R.E. Gruber, Bigtable: a distributed storage system for structured data. ACM Trans. Comput. Syst. (TOCS) **26**(2), 4 (2008)

10. D. Che, M. Safran, Z. Peng, From big data to big data mining: challenges, issues, and opportunities, in *Database Systems for Advanced Applications: 18th International Conference, DASFAA 2013, International Workshops: BDMA, SNSM, SeCoP, Wuhan, China, 22–25 April 2013*. Proceedings (Springer, Berlin, 2013), pp. 1–15

11. J. Dean, S. Ghemawat, Mapreduce: simplified data processing on large clusters, in *Proceedings of the 6th Conference on Symposium on Opearting Systems Design & Implementation - Volume 6, OSDI'04, Berkeley, USA* (2004), p. 10

12. E. Deelman, K. Vahi, G. Juve, M. Rynge, S. Callaghan, P.J. Maechling, R. Mayani, W. Chen, R.F. da Silva, M. Livny et al., Pegasus, a workflow management system for science automation. Futur. Gener. Comput. Syst. **46**, 17–35 (2015)

13. J. Dongarra et al., The international exascale software project roadmap. Int. J. High Perform. Comput. Appl. **25**, 3–60 (2011)

14. J. Ekanayake, H. Li, B. Zhang, T. Gunarathne, S.H. Bae, J. Qiu, G. Fox, Twister: a runtime for iterative mapreduce, in *Proceedings of the 19th ACM International Symposium on High Performance Distributed Computing. HPDC '10* (ACM, New York, 2010), pp. 810–818

15. S.K. Gajendran, A survey on nosql databases. University of Illinois (2012)

16. M.S. Gerber, Predicting crime using twitter and kernel density estimation. Decision Support Syst. **61**, 115–125 (2014)

17. B. Giardine, C. Riemer, R.C. Hardison, R. Burhans, L. Elnitski, P. Shah, Y. Zhang, D. Blankenberg, I. Albert, J. Taylor et al., Galaxy: a platform for interactive large-scale genome analysis. Genome Res. **15**(10), 1451–1455 (2005)

18. S. Gilbert, N. Lynch, Brewer's conjecture and the feasibility of consistent, available, partition-tolerant web services. ACM SIGACT News **33**(2), 51–59 (2002)

19. Y. Gu, R.L. Grossman, Sector and sphere: the design and implementation of a high-performance data cloud. Philos. Trans. R. Soc. Lond. A Math. Phys. Eng. Sci. **367**(1897), 2429–2445 (2009)

20. I.A.T. Hashem, I. Yaqoob, N.B. Anuar, S. Mokhtar, A. Gani, S.U. Khan, The rise of big data on cloud computing: review and open research issues. Inf. Syst. **47**, 98–115 (2015)

21. M. Isard, M. Budiu, Y. Yu, A. Birrell, D. Fetterly, Dryad: distributed data-parallel programs from sequential building blocks. SIGOPS Oper. Syst. Rev. **41**(3), 59–72 (2007)

22. J. Kranjc, V. Podpečan, N. Lavrač, Clowdflows: a cloud based scientific workflow platform, in *Machine Learning and Knowledge Discovery in Databases* (Springer, 2012), pp. 816–819

23. T. Kurashima, T. Iwata, G. Irie, K. Fujimura, Travel route recommendation using geotags in photo sharing sites, in *Proceedings of the 19th ACM International Conference on Information and Knowledge Management. CIKM '10* (ACM, New York, 2010), pp. 579–588

24. R. Lee, S. Wakamiya, K. Sumiya, Urban area characterization based on crowd behavioral lifelogs over twitter. Personal Ubiquitous Comput. **17**(4), 605–620 (2013)

25. S. Lee, H. Park, Y. Shin, Cloud computing availability: multi-clouds for big data service, in *Convergence and Hybrid Information Technology* (Springer, 2012), pp. 799–806

26. A. Lemieux, Geotagged photos: a useful tool for criminological research? Crime Sci. **4**(1), 3 (2015)

27. A. Li, X. Yang, S. Kandula, M. Zhang, Cloudcmp: comparing public cloud providers, in *Proceedings of the 10th ACM SIGCOMM Conference on Internet Measurement* (ACM, 2010), pp. 1–14

28. J.R. Lourenço, B. Cabral, P. Carreiro, M. Vieira, J. Bernardino, Choosing the right nosql database for the job: a quality attribute evaluation. J. Big Data **2**(1), 1–26 (2015)
29. D. Lyubimov, A. Palumbo, *Apache Mahout: Beyond MapReduce* (Chapman and Hall/CRC, Boca Raton, 2016)
30. G. Malewicz, M.H. Austern, A.J. Bik, J.C. Dehnert, I. Horn, N. Leiser, G. Czajkowski, Pregel: a system for large-scale graph processing, in *Proceedings of the 2010 ACM SIGMOD International Conference on Management of Data. SIGMOD '10* (ACM, New York, 2010), pp. 135–146
31. G. Marciani, M. Piu, M. Porretta, M. Nardelli, V. Cardellini, Real-time analysis of social networks leveraging the flink framework, in *Proceedings of the 10th ACM International Conference on Distributed and Event-Based Systems. DEBS '16* (ACM, New York, 2016), pp. 386–389
32. F. Marozzo, D. Talia, P. Trunfio, A cloud framework for parameter sweeping data mining applications, in *2011 IEEE Third International Conference on Cloud Computing Technology and Science (CloudCom)* (IEEE, 2011), pp. 367–374
33. F. Marozzo, D. Talia, P. Trunfio, Using clouds for scalable knowledge discovery applications, in *Euro-Par Workshops, Rhodes Island, Greece*. Lecture Notes in Computer Science, vol. 7640 (2012), pp. 220–227
34. F. Marozzo, D. Talia, P. Trunfio, Scalable script-based data analysis workflows on clouds, in *Proceedings of the 8th Workshop on Workflows in Support of Large-Scale Science* (ACM, 2013), pp. 124–133
35. A. Martin, A. Brito, C. Fetzer, Real-time social network graph analysis using streammine3g, in *Proceedings of the 10th ACM International Conference on Distributed and Event-Based Systems. DEBS '16* (ACM, New York, 2016), pp. 322–329
36. I. Mavroidis, I. Papaefstathiou, L. Lavagno, D.S. Nikolopoulos, D. Koch, J. Goodacre, I. Sourdis, V. Papaefstathiou, M. Coppola, M. Palomino, Ecoscale: reconfigurable computing and runtime system for future exascale systems, in *2016 Design, Automation Test in Europe Conference Exhibition (DATE)* (2016), pp. 696–701
37. P.M. Mell, T. Grance, Sp 800-145. the nist definition of cloud computing. Technical report, National Institute of Standards & Technology, Gaithersburg, MD, United States (2011)
38. R. Möller, B. Neumann, Ontology-based reasoning techniques for multimedia interpretation and retrieval, in *Semantic Multimedia and Ontologies: Theory and Applications*, ed. by Y. Kompatsiaris, P. Hobson (Springer, London, 2008), pp. 55–98
39. A.B.M. Moniruzzaman, S.A. Hossain, Nosql database: new era of databases for big data analytics - classification, characteristics and comparison. CoRR abs/1307.0191 (2013)
40. D. Nurmi, R. Wolski, C. Grzegorczyk, G. Obertelli, S. Soman, L. Youseff, D. Zagorodnov, The eucalyptus open-source cloud-computing system, in *9th IEEE/ACM International Symposium on Cluster Computing and the Grid, 2009. CCGRID '09* (2009), pp. 124–131
41. S. Owen, R. Anil, T. Dunning, E. Friedman, *Mahout in Action* (Manning Publications Co., Greenwich, 2011)
42. L. Richardson, S. Ruby, *RESTful Web Services* (O'Reilly Media, Inc., Sebastopol, 2008)
43. M.A. Rodriguez, P. Neubauer, The graph traversal pattern. CoRR abs/1004.1001 (2010)
44. S. Shahrivari, Beyond batch processing: Towards real-time and streaming big data. CoRR abs/1403.3375 (2014)
45. B. Sotomayor, R.S. Montero, I.M. Llorente, I. Foster, Virtual infrastructure management in private and hybrid clouds. IEEE Internet Comput. **13**(5), 14–22 (2009)
46. M. Stonebraker, Sql databases v. nosql databases. Commun. ACM **53**(4), 10–11 (2010)
47. A. Tai, M. Wei, M.J. Freedman, I. Abraham, D. Malkhi, Replex: a scalable, highly available multi-index data store, in *2016 USENIX Annual Technical Conference (USENIX ATC 16)* (USENIX Association, Denver, 2016), pp. 337–350
48. D. Talia, P. Trunfio, F. Marozzo, *Data Analysis in the Cloud* (Elsevier, 2015). ISBN 978-0-12-802881-0
49. K.L. Tan, Q. Cai, B.C. Ooi, W.F. Wong, C. Yao, H. Zhang, In-memory databases: challenges and opportunities from software and hardware perspectives. SIGMOD Rec. **44**(2), 35–40 (2015)

50. J.J. Thomas, K.A. Cook, A visual analytics agenda. IEEE Comput. Graph. Appl. **26**(1), 10–13 (2006)
51. A. Vukotic, N. Watt, T. Abedrabbo, D. Fox, J. Partner, *Neo4j in Action* (Manning, Shelter Island, 2015)
52. Z. Wang, Y. Chu, K. Tan, D. Agrawal, A. El Abbadi, X. Xu, Scalable data cube analysis over big data. CoRR abs/1311.5663 (2013)
53. M. Wilde, M. Hategan, J.M. Wozniak, B. Clifford, D.S. Katz, I. Foster, Swift: a language for distributed parallel scripting. Parallel Comput. **37**(9), 633–652 (2011)
54. J.M. Wozniak, M. Wilde, I.T. Foster, Language features for scalable distributed-memory dataflow computing, in *2014 Fourth Workshop on Data-Flow Execution Models for Extreme Scale Computing (DFM)* (2014), pp. 50–53
55. X. Wu, X. Zhu, G.Q. Wu, W. Ding, Data mining with big data. IEEE Trans. Knowl. Data Eng. **26**(1), 97–107 (2014)
56. R.S. Xin, J. Rosen, M. Zaharia, M.J. Franklin, S. Shenker, I. Stoica, Shark: sql and rich analytics at scale, in *Proceedings of the 2013 ACM SIGMOD International Conference on Management of Data. SIGMOD '13* (ACM, New York, 2013), pp. 13–24
57. L. You, G. Motta, D. Sacco, T. Ma, Social data analysis framework in cloud and mobility analyzer for smarter cities, in *2014 IEEE International Conference on Service Operations and Logistics, and Informatics (SOLI)* (2014), pp. 96–101
58. J. Yuan, Y. Zheng, L. Zhang, X. Xie, G. Sun, Where to find my next passenger, in *Proceedings of the 13th International Conference on Ubiquitous Computing. UbiComp '11* (ACM, New York, 2011), pp. 109–118
59. H. Zhang, G. Chen, B.C. Ooi, K.L. Tan, M. Zhang, In-memory big data management and processing: a survey. IEEE Trans. Knowl. Data Eng. **27**(7), 1920–1948 (2015)

Data Organization and Curation in Big Data

Mohamed Y. Eltabakh

Abstract This chapter covers advanced techniques in Big Data analytics and query processing. As the data is getting bigger and, at the same time, workloads and analytics are getting more complex, the advances in big data applications are no longer hindered by their ability to collect or generate data. But instead, by their ability to efficiently and effectively manage the available data. Therefore, numerous scalable and distributed infrastructures have been proposed to manage big data. However, it is well known in literature that scalability and distributed processing alone are not enough to achieve high performance. Instead, the underlying infrastructure has to be highly optimized for various types of workloads and query classes. These optimizations typically start from the lowest layer of the data management stack, which is the storage layer. In this chapter, we will cover two well-known techniques for optimized storage and organization of data that have big influence on query performance, namely the *indexing*, and *data layout* techniques. However, in the cases of non-traditional workloads where queries have special execution and data-access characteristics, the standard indexing and layout techniques may fall short in providing the desired performance goals. Therefore, further optimizations specific to the workload characteristics can be applied. In this chapter, we will cover techniques addressing several of these non-traditional workloads in the context of big data. Some of these techniques rely on curating either the data or the workflows (or both) with useful metadata information. This curation information can be very valuable for both query optimization and the business logic. In this chapter, we will cover the curation and metadata management of big data in query optimization and different systems. In this chapter, we focus on the MapReduce-like infrastructures, more specifically its open-source implementation Hadoop. The chapter covers the state-of-art in big data indexing techniques, and the data layout and organization strategies to speedup queries. It will also cover advanced techniques for enabling non-traditional workloads in Hadoop. Hadoop is primarily designed for workloads that are characterized by being batch, offline, ad-hoc, and disk-based. Yet, this chapter will cover recent projects and techniques targeting non-traditional workloads such as

M.Y. Eltabakh (✉)
Computer Science Department, Worcester Polytechnic Institute,
Worcester, MA, USA
e-mail: meltabakh@cs.wpi.edu

© Springer International Publishing AG 2017   143
A.Y. Zomaya and S. Sakr (eds.), *Handbook of Big Data Technologies*,
DOI 10.1007/978-3-319-49340-4_5

continuous query evaluation, main-memory processing, and recurring workloads. In addition, the chapter covers recent techniques proposed for data curation and efficient metadata management in Hadoop. These techniques vary from being semantic specific, e.g., provenance tracking techniques, to generic frameworks for data curation and annotation.

1 Big Data Indexing Techniques

1.1 Overview

Big data infrastructures such as Hadoop are increasingly supporting applications that manage structured or semi-structured data. In many applications including scientific applications, weblog analysis, click streams, transaction logs, and airline analytics, at least partial knowledge about the data structure is known. For example, some attributes may have known types and possible domain of values, while other attributes may have little information known about them. This knowledge, even if it is partial, can enable optimization techniques that otherwise would not be possible. Query optimization in big data is fundamentally important, especially because (1) the datasets to be processed are getting very large, (2) the analytical queries are increasing in complexity and may take hours to execute if not carefully optimized, and (3) the pay-as-you-go cost models for cloud computing add additional urgency for optimized processing.

A typical query in big data applications may touch files in the order of 100s of GBs or TBs of size. These queries are typically very expensive as they consume significant resources and require long periods of time to execute. For example, in transaction log applications, e.g., transaction history of customer purchases, one query might be interested in retrieving all transactions from the last two months that exceed a certain amount of dollar money. Such query may need to scan billions of records and go over TBs of data.

Indexing techniques are well-known techniques in database systems, especially relational databases, to optimize query processing. Examples of the standard indexing techniques are the B+-Tree, R-Tree, and Hash-based indexes along with their variations. However, transforming these techniques and structures to big data is not straightforward due to the unique characteristics of both the data itself and the underlying infrastructure processing the data. At the data level, the data is no longer assumed to be stored in relational tables. Instead, the data is received and stored in the forms of big batches of flat files. In addition, the data size exceeds what relational database systems can typically handle.

On the other hand, at the infrastructure level, the processing model no longer follows the relational model of query execution, which relies on connecting a set of query operators together to form a query tree. Instead, the MapReduce computing paradigm is entirely different as it relies on two rigid phases of *map* and *reduce*.

Moreover, the access pattern of the data from the file system is also different. In relational databases, the data records are read in the form of disk pages (a.k.a disk blocks), which are very small in size (typically between 8 and 128 KBs) and usually hold few data records (10s or at most 100s of records). And thus, we assume that the database systems can support record-level access. In contrast, in the Hadoop file system (HDFS), a single data block ranges between 64 MBs and 1 GB, and usually holds many records. Therefore, the record-level access no longer holds. Even the feasible operations over the data are different from those supported in relational databases. For example, record updates and deletes are not allowed in the MapReduce infrastructure. All of these unique characteristics of big data fundamentally affect the design of the appropriate indexing and pre-processing techniques.

Plain Hadoop is found to be orders-of-magnitudes slower than distributed database management systems when evaluating queries on structured data [1, 67]. One of the main observed reasons for this slow performance is the lack of indexing in the Hadoop infrastructure. As a result, significant research efforts have been dedicated to designing indexing techniques suitable for the Hadoop infrastructure. These techniques have ranged from record-level indexing [27, 28, 46, 65] to split-level indexing [34, 38], from user-defined indexes [27, 38, 46] to system-generated and adaptive indexes [28, 34, 65], and from single-dimension indexes [27, 28, 34, 46, 65] to multi-dimensional indexes [30, 56, 58].

In Table 1, we compare several of the Hadoop-based indexing techniques with respect to different criteria. Record-level granularity techniques aim for skipping irrelevant records within each data split, but eventually they may touch all splits. In contrast, the split-level granularity techniques aim for skipping entire irrelevant splits. SpatialHadoop system provides both split-level global indexing as well as record-level local indexing, and thus it can skip irrelevant data at both granularities. Some techniques index only one attribute at a time (Dimensionality $= 1$), while others allow indexing multi-dimensional data (Dimensionality $= m$). Techniques like HadoopDB and Polybase Index inherit the multi-dimensional capabilities from the underlying DBMS. E3 technique enables indexing pairs of values (from two attributes), but only for a limited subset of the possible values. Most techniques operate only on the HDFS data (DB-Hybrid $= N$), while HadoopDB and Polybase Index have a database system integrated with HDFS to form a Hybrid system. In most of the proposed techniques, the system's admin decides on which attributes to be indexed. The only exceptions are the LIAH index, which is an adaptive index that automatically detects the changes of the workload and accordingly creates (or deletes) indexes, and the E3 index, which automatically indexes all attributes in possibly different ways depending on the data types and the workload. Finally, the index structure is either stored in HDFS along with its data as in Hadoop++, HAIL, LIAH, SpatialHadoop, and ScalaGist, in a database system along with its data as in HadoopDB, or in a database system while the data resides in HDFS as in E3 and Polybase Index. In the following, we present few of these techniques in more details.

Target Queries: Indexing techniques target optimizing queries that involve selection predicates, which is the common theme for all techniques listed in Table 1. Yet,

Table 1 Comparison of Hadoop-based indexed techniques

Technique	Granularity	Dimensionality	DB-hybrid	Definition	Index location
Hadoop++ [27]	Record	1	No	Admin	HDFS
HAIL [28]	Record	1	No	Admin	HDFS
LIAH [65]	Record	1	No	System	HDFS
E3 [34]	Split	1 and 2	No	System	DB
SpatialHadoop [30]	Record/Split	m	No	Admin	HDFS
ScalaGist [58]	Record	m	No	Admin	HDFS
HadoopDB [4]	Record	m	Yes	Admin	DB
Polybase Index [38]	Split	m	Yes	Admin	DB

they may differ on how queries are expressed and the mechanism by which the selection predicates are identified. For example, Hadoop++, HAIL, and LIAH they allow expressing the query in Java while also passing the selection predicates as arguments within the job configuration. As a result, a customized input format will receive these predicates (if any) and perform the desired filtering during execution. In contrast, E3 framework is built on top of the Jaql high-level query language [14], and thus queries are expressed, compiled, and optimized using Jaql. An example query is as follows:

```
read( hdfs("docs.json") )
    -> transform { author: $.meta.author,
                   products: $.meta.product,
                   Total: $.meta.Qty * $.meta.Price}
    -> filter $.products == "XYZ";
```

Jaql has the feature of applying *selection-push-down* during query compilation whenever possible. As a result, in the given query the `filter` operator will be pushed before the `transform` operator, with the appropriate re-writing. The E3 framework can then detect this filtering operation directly after the read operation of the base file, and thus can push the selection predicate into its customized input format to apply the filtering as early as possible.

HadoopDB provides a front-end for expressing SQL queries on top of its data, which is called SMS. SMS is an extension to Hive. In HadoopDB queries are expressed in an identical way to standard SQL as in the following example.

```
SELECT pageURL, pageRank
    FROM Rankings
    WHERE pageRank > 10;
```

SpatialHadoop is designed for spatial queries, and thus it provides a high-level language and constructs for expressing these queries and operating on spatial objects, e.g., points and rectangles. For example, a query can be expressed as follows:

```
Objects = LOAD "points" AS (id:int, Location:POINT);
Result = FILTER Objects BY
               Overlaps (Location, Rectangle(x1, y1, x2, y2));
```

ScalaGist enables building Gist indexes, e.g., B+-tree and R-tree, over HDFS data. A single query in ScalaGist can make use of multiple indexes at the same time. For example, given a table T with schema $\{x, \ldots, (a1, a2)\}$, where x is a one-dimensional column, and $(a1, a2)$ is a two-dimensional column, the following query can use both a B+-tree index (on x) and an R-tree index (on $(a1, a2)$) during its evaluation:

```
SELECT *
FROM T
WHERE x ≤ 100
     AND 10 ≤ a1 ≤ 20
     AND 30 ≤ a2 ≤ 60;
```

Finally, the Polybase system enables expressing queries using standard SQL over HDFS data that are defined as external tables. First, users need to define the external table as in the following example:

```
Create External Table hdfsLineItem
     (l_orderkey BIGINT Not Null,
     l_partkey BIGINT Not Null,
      ...)
With (Location = '/tpch1gb/lineitem.tbl',
     Data_Source = VLDB_HDP_Cluster,
     File_Format = Text_Delimited);
```

And then, a query on the external table can be expressed as follows:

```
SELECT *
FROM hdfsLineItem
WHERE l_orderkey = 1;
```

1.2 Record-Level Non-adaptive Indexing

Hadoop++ [27] is an indexing technique built on top the Hadoop infrastructure. Unlike other techniques that require extensive changes to Hadoop's execution model to offer run-time optimizations, e.g., HadoopDB [4, 5], Hadoop++ relies on augmenting the indexing structures to the data in a way that does not affect the execution mechanism of Hadoop. All processing on the indexes, e.g., creating the indexes, augmenting them to the data, and their access, are all performed through pluggable user-defined functions (UDFs) that are already available within the Hadoop framework.

The basic idea of the Hadoop++ index, which is referred to as *a Trojan Index*, is illustrated in Fig. 1a. At the loading time, the base data is partitioned using a map-reduce job. This job partitions the data based on the attribute to be indexed, i.e., if attribute X is to be indexed then depending on the X's value in each record, the record will be assigned to a specific split Id. This assignment is performed by the mapper function. On the other hand, the reducer function receives all the records belonging to a specific split, and creates the trojan index corresponding to that split. The index is then augmented to the data split to form a bigger split, referred to as *an indexed split* as depicted in the figure. Each indexed split will also have a *Split Header* (H), and a *Split Footer* (F), which together hold the metadata information about each indexed split, e.g., the split size, the number of records, the smallest and largest indexed values within this split, etc. In general, Hadoop++ can be configured to create several trojan indexes for the same data on different attributes. However, only one index can be the primary index according to which the data records are sorted within each split. This primary index is referred to as the *clustered index*, while the other additional indexes are *non-clustered indexes*.

At query time, given a query involving a selection predicate on one of the indexed attributes, the processing works as follows. First, a custom InputFormat function would read each indexed split (instead of the data splits), and consult the trojan index for that split w.r.t the selection predicate. If there are multiple indexes, then the appropriate index is selected based on the selection predicate. If none of the records satisfies the query predicate, then the entire split is skipped and the map function terminates without actually checking any record within this split. Otherwise, the trojan index will point to the data records within the split that satisfies the query. If the trojan index is clustered, then this means that the data records within the given block are ordered according to the indexed attribute, and thus the retrieval of the records will be faster and requires less I/Os.

It is worth highlighting that trojan indexes are categorized as *local* indexes meaning that a local index is created for each data split in contrast to building a single global index for the entire dataset. Local indexes have their advantages and disadvantages. For example, one of the advantages is that the entire dataset does not need to sorted, which is important because global sorting is prohibitively expensive in big data. However, one disadvantage is that each indexed split has to touched at query time. This implies that a mapper function has to be scheduled and initiated by Hadoop for each split even if many of these splits are irrelevant to the query.

Hadoop++ framework also provides a mechanism, called *Trojan Join*, to speedup the join operation between two datasets, say S and T (Refer to Fig. 1b). The basic idea is to partition both datasets (at the same time) on the join key. This partitioning can be performed at the loading time as a pre-processing step. The actual join does not take place during this partitioning phase. Instead, only the corresponding data partitions from both datasets are grouped together in bigger splits, referred to as *Co-Partition*

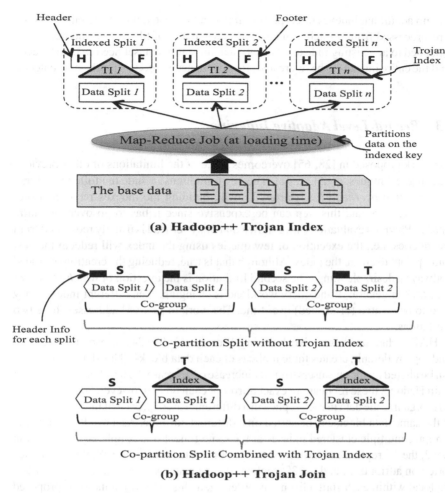

Header   Footer

Indexed Split *1*   Indexed Split *2*   Indexed Split *n*   Trojan Index

(a) Hadoop++ Trojan Index

(b) Hadoop++ Trojan Join

Fig. 1 Hadoop++ Trojan Index and Trojan Join

Splits. And then, at query time when *S* and *T* need to be joined, the join operation can take place as a map-only job, where each mapper will be assigned one complete co-partition split. As such, each mapper can join the corresponding partitions within its split. Therefore, the join operation becomes significantly less expensive since the shuffling/sorting and reduce phases have been eliminated (compared to the traditional map-reduce join operation in Hadoop). As highlighted in Fig. 1b, the individual splits from either of *S* or *T* (or both) within a single co-partitioned may or may not have a trojan index on them.

Hadoop++ framework is suitable for static indexing and joining. That is, at the time of loading the data into Hadoop, the system needs to know whether or not indexes need to be created (and on which attributes), and also whether or not co-partitioning between specific datasets need to be performed. After loading the data no additional indexes or co-partitioning can be created unless the entire dataset is re-processed from scratch. Similarly, if new batches of files arrive and need to be appended to an existing indexed dataset, then the entire dataset need to be re-loaded (and the entire indexes to be re-created) in order to accommodate for the new batches.

1.3   Record-Level Adaptive Indexing

The work proposed in [28, 65] overcomes some of the limitations of other previous indexing techniques, e.g., [27, 46]. The key limitations include the following. First, *high creation overhead for indexes*. Usually building the indexes requires a pre-processing step, and this step can be expensive since it has to go over the entire dataset. Previous evaluations have shown that this overhead is usually redeemed from few queries, i.e., the execution of few queries using the index will redeem the cost paid upfront to create the index. Although that is true, reducing the creation overhead is always a desirable thing. The second limitation is the question of *which attributes to index?* In general, if the query workload is changing, then different indexes may need to be created (or deleted) over time. The work in [28, 65] addresses these two limitations.

HAIL (Hadoop Aggressive Indexing Library) [28] makes use of the fact that Hadoop, by default, creates three replicas of each data block—This default behavior can be altered by the end-users to either increase or decrease the number of replicas. In plain Hadoop, these replicas are exact mirror of each other. However, HAIL proposes to re-organize the data in each replica in a different way, e.g., each of the three replicas of the same data block can be sorted on a different attribute. As a result, a single file can have multiple clustered indexes at the same time. For example, as illustrated in Fig. 2, the 1st replica can have each of its splits sorted on attribute X, the 2nd replica sorted on attribute Y, and the 3rd replica sorted on attribute Z. These sorting orders are local within each split. Given this ordering, a clustered trojan index as proposed in [27] can be built on each replica independently.

HAIL also proposes a *replica-aware scheduling policy*. In plain Hadoop, since all replicas are the same, the task scheduling decision does not differentiate between the replicas. In contrast in HAIL the task scheduler needs to take the query predicates into account while selecting the target replica to work on. For example referring to Fig. 2, given a query involving a selection predicate on attribute Y, then HAIL scheduler will try to assign the map tasks to the splits of the 2nd replica. Otherwise, a full scan operation has to be performed on either of the other replicas because their indexes cannot help in evaluating the given predicate.

LIAH (Lazy Indexing and Adaptivity in Hadoop) indexing framework [65] further extends the idea of HAIL by adaptively selecting the columns to be indexed under changing workloads, and also lazily building these indexes as more queries execute in the system. LIAH can incrementally build a given index starting from indexing few splits, and incrementally indexing more splits as more queries are executed until the entire index is built. This strategy is based on the idea of *piggybacking* the index

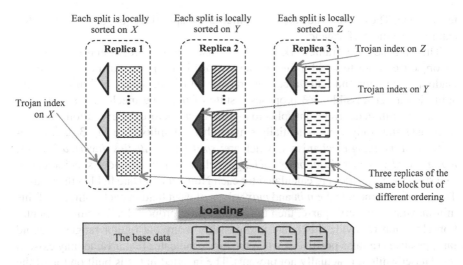

Fig. 2 HAIL indexing framework

creation task over other user's queries to reduce the overheads involved in the index creation. For example, referring to Fig. 2, in LIAH it is possible that the system starts without any indexes on the base data. And then, by automatically observing the query workload, the system decides that attributes X and Y are good candidates for indexing, e.g., many queries have selection predicates on either of these two attributes. LIAH puts a strategy to incrementally make the splits of the 1st replica sorted based on X, and for each split where its data becomes sorted, its corresponding trojan index is built. LIAH framework keeps track of which blocks have been indexed, and which blocks need to indexed (which will be done progressively and piggybacked over future users' jobs). As more user jobs are submitted to the system, and the data blocks are read anyway, additional blocks can be indexed. In this way, the overheads involved in the index creation are distributed over many user queries.

1.4   Split-Level Indexing

Most of the previous indexing techniques proposed over the Hadoop Infrastructure try to mimic the indexes in traditional databases in that they are record-level indexes. That is, their objective is to eliminate and skip irrelevant records from being processed. Although these techniques show some improvements in query execution, they still encounter unnecessary high overhead during execution. For example, imagine the extreme case where a queried value x appears only in very few splits of a given file. In this case, the indexing techniques like Hadoop++ and HAIL would encounter the overheads of starting a map task for each split, reading the split headers, searching the local index associated with the split, and then reading few data records or directly terminating. These overheads are substantial in a map-reduce job, and if eliminated can improve the performance.

The proposed $E3$ framework in [34] is based on the aforementioned insight. Its objective is not to build a fine-grained record-level index, but instead be more Hadoop-compliant and build a coarse-grained split-level index to eliminate entire splits whenever possible. $E3$ proposes a suite of indexing mechanisms that work together to eliminate irrelevant splits to a given query before the execution, and thus map tasks start only for a potentially small subset of splits (See Fig. 3a). $E3$ integrates four indexing mechanisms, which are: *split-level statistics*, *Inverted indexes*, *materialized views*, and *adaptive caching*, each is beneficial under specific cases. The split-level statistics are calculated for each Number and Date field in the dataset. These statistics include the *min* and *max* values of each field in each split, and if this min-max range is very sparse, then the authors have proposed a domain segmentation algorithm to divide this range into possibly many but tighter ranges to avoid false positives (a false positive is when the index indicates that a value may exist in the dataset while it is actually not present). The inverted index is built on top of the String fields in the dataset, i.e., each string value is added to the index and it points to all splits including this values. By combining these two types of indexes for a given query involving a selection predicate, $E3$ can identify which splits are relevant to the query, and only for those splits a set of mappers will be triggered.

The other two mechanisms, i.e., materialized views and adaptive caching, are used in the cases where indexes are mostly useless. One example provided in [34] is highlighted in Fig. 3b. This example highlights what is called *"nasty values"*, which are values that are infrequent over the entire dataset, but scattered over most of the data splits, e.g., each split has one or few records of this value. In this case, the inverted index will point to all splits, and becomes almost useless. $E3$ framework handles these nasty values by coping their records into an auxiliary materialized view. For example, the base file A in Fig. 3b will now have an addition materialized view file stored in HDFS that contains a copy of all records having the nasty value v. Identifying the nasty values and deciding on which ones have a higher priority to handle has been proven to be an NP-Hard problem, and the authors have proposed an approximate greedy algorithm to solve it [34].

The adaptive caching mechanism in $E3$ is used to optimize conjunctive predicates, e.g., $(A = x$ and $B = y)$, under the cases where each of x and y individually are frequent, but their combination in one record is very infrequent. In other words, none of x or y are nasty values, but their combination is a *nasty pair*. In this case, neither the indexes nor the materialized views are useful. Since it is prohibitively expensive to enumerate all pairs and identify the nasty ones, the $E3$ framework handles these nasty pairs by dynamically observing the query execution, and identifying on-the-fly the nasty pairs. For example, for the conjunctive predicates $(A = x$ and $B = y)$, $E3$ will consult the indexes to select a subset of splits to read. And then, it will observe the number of mappers that actually produced matching records. If the number of mappers is very small compared to the triggered ones, then $E3$ identifies (x, y) to be a nasty pair. Consequently, (x, y) will be cached along with pointers to its relevant splits.

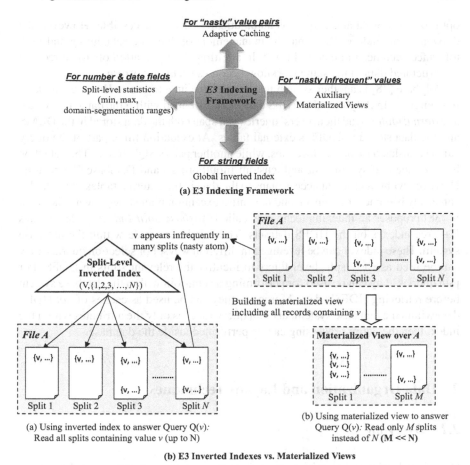

Fig. 3   E3 indexing framework [34]

1.5   Hadoop-RDBMS Hybrid Indexing

There has been a long debate on whether or not Hadoop and database system can co-exist together in a single working environment, and whether or not this strategy is beneficial. There are several successful projects that built such integration [4, 13, 36–38]. HadoopDB is one of the early projects that brings the optimizations of relational database systems to Hadoop [4]. HadoopDB proposes major changes to Hadoop's infrastructure by replacing the HDFS storage layer by a database management layer. That is, the Data Node and Task Tracker on each slave node in the Hadoop's cluster will be running an instance of a database system. This database instance replaces the HDFS layer, and thus the data on each slave node are stored and managed by the database engine. HadoopDB will push as much of work as possible to the database engine, and as a result all indexing capabilities and query optimizations of database systems automatically become accessible. However, the drawbacks of HadoopDB is that the management of dynamic scheduling and fault tolerance becomes more complicated. In addition, the integration of structured and un-structured data in the same workflow becomes tricky to perform.

Polybase [38] is another system that enables the integration of Hadoop and database engines. In Polybase, the HDFS datasets are defined within the database system as *external tables*. And then, users' queries can span both the data stored in the DBMS and the data stored in HDFS's external tables. At execution time, part of the query can be translated to map-reduce jobs, while another part is SQL-based. The data flow between the two systems through custom InputFormats and Database Connectors. However, without efficient access plans to the data in the external tables, these tables can easily become a bottleneck and the entire execution plan slows down. The work in [38] proposes an indexing technique, called *Polybase Split-Indexing*, that creates B+-Tree indexes on the HDFS datasets. These indexes reside within the database system. These indexes can be leveraged in different ways. For selection queries, they can be used as early split-level filters to identify the relevant splits in HDFS. For join queries, they can be used for performing a semi-join within the database system before retrieving HDFS's data. Moreover, they can be used as caches of *hot* HDFS data within the database system, and if a query touches only the attributes within the index, then the entire processing can be performed inside the database.

2 Data Organization and Layout Techniques

2.1 Overview

One of the dominant factors in query performance is the data layout, which determines the structure and organization of the data in the file system. Data organization is a well-known and effective strategy in boosting performance in database systems that has been studied for decades. However, not all techniques are transferable to the context of big data and the Hadoop infrastructure. Even if the same basic idea can be transferred to Hadoop, the technical details and challenges would be different because the characteristics of the data and the infrastructure are different as explained in Sect. 1.1.

At the conceptual level, both of the data organization (or re-organization) and the indexing strategies presented in Sect. 1 have the same objective, which is avoiding a full scan over the data whenever possible, and touching only a subset of the records. However, at the design level, the two strategies are different. Indexing techniques build auxiliary structures of special properties, called *indexes*, in addition to the base data, and then at query time these auxiliary structures are consulted first to identify the relevant subset of records to the query. In contrast, re-organization and data layout techniques may or may not create auxiliary structures.

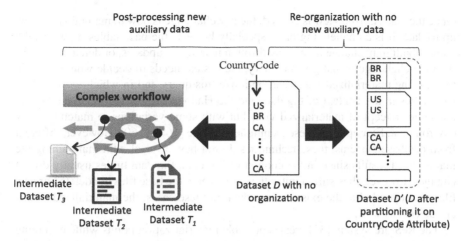

Fig. 4 Data organization with and without auxiliary data creation

In some cases as illustrated in Fig. 4, the base data is re-organized in a certain way (usually offline) to enable better query evaluation in the future. For example, the base dataset D in Fig. 4 has no specific organization or ordering for its records. As a result, a query involving a selection predicate over the CountryCode attribute would have to scan all records in D. However, if D is re-organized in the form of dataset D', where the records are partitioned based on the CountryCode values (R.H.S of Fig. 4), then the same query would execute much faster on D' since the query can now scan only the relevant partition(s). Depending on the application needs, the base data D may or may not be kept after the creation of D'.

In some other cases, data organization may involve building auxiliary datasets (in addition to the base data), where these datasets are processed or massaged in a certain way. For example, given the base dataset D in Fig. 4 and after applying a complex workflow on D, it can be beneficial to keep not only the final results of the workflow, but also the intermediate data at specific points, e.g., storing the datasets T_1, T_2, and T_3 as illustrated in L.H.S of Fig. 4. This is because these datasets may have interesting properties that may help speeding up future queries. In this case, datasets T_1, T_2, and T_3 usually do not replace D and it remains stored in the system. In the following, we present in more details several data organization techniques, which are proposed in the context of the Hadoop infrastructure.

2.2 Result Materialization and Caching Techniques

Result materialization is a known technique from relational databases where it is also referred to as *materialized views* [76]. In relational databases, materialized views are typically defined by the database admin in an explicit way through a SQL command.

Once the materialized view is created, the key challenge is how to maintain this view up to date. It is a challenging task especially because the base tables on which the view is built may change in anyway through insertions, updates, or deletes [50, 75]. Another challenge is that given a query, the system needs to decide whether or not the existing materialized views can optimize this query and in which way. Several techniques in the context of big data and the Hadoop infrastructure have inherited the same concept of materialized view but with some variations to match the new environment. Examples of these techniques include [16, 31, 60]. However, different from database systems, these techniques do not have to deal with maintaining the materialized results since in the context of big data, the data tuples usually do not change. However, they still monitor the cases where the base files got deleted or new files are appended to the existing base data. In these cases, the materialized results are deleted as well.

The ReStore system [31] creates possible materialization points while executing a workflow of map-reduce jobs expressed in Apache Pig. Similar to the L.H.S of Fig. 4, the materialization points can be at any stage within the workflow. The system allows the materialization of the results generated after each job within the workflow, or even at a finer granularity where the materialization may take place within the same job, e.g., store a map output within a map-reduce job. Since materializing everything is prohibitively expensive, ReStore deploys several heuristics to decide on which materialization points to add within a given workflow. Ultimately, the system maintains a repository of these materialized results in HDFS along with metadata information on each materialized result. This metadata includes the query or workflow structure produced the result, the dependent base files, and access statistics. And then, given a subsequent query or workflow of map-reduce jobs, the system optimizes this workflow based on the stored materialized results. As illustrated in Table 2, ReStore enables the re-usability at the workflow level, a single job, or a subset of a job.

The MRShare system [60] is different from ReStore in that the former system requires all queries targeting optimization to be submitted to the system at once in the form of a single batch. MRShare will then build a single optimized query

Table 2 Comparison of Hadoop-based sharing techniques

Technique	Granularity	Knowledge of Workload	Sharing opportunities
ReStore [31]	Across workflows	Dynamic	Workflow, one job, partial job
MRShare [60]	Single batch	Known in advance	Scan, map output, map function
Multi-query [70]	Across batches	Dynamic	Scan, map output, reduce input
HaLoop [16]	Single iterative job	Known in advance	Map input, reduce input, reduce output

plan for their execution. During the execution of this batch, intermediate results can be shared across multiple queries. Yet, these intermediate results are not kept in the system beyond the execution of the given batch. Therefore, in some sense the ReStore and MRShare systems are complementary to each other. MRShare proposes several sharing opportunities specific to the MapReduce computing paradigm as indicated in Table 2. These opportunities include: (1) *Sharing Scan*, in which two jobs accessing the same file in the same key-value pair formats can share the scan over this file, (2) *Sharing Map Output*, in which the output stream from the mapper function of two jobs can be merged into one stream if they produce the same key-value pair types. In this case, each output record will be tagged with a special tag indicating whether it belongs to either or both jobs. And (3) *Sharing Map Function*, in which the map function of both jobs can be identical and can be shared. As an extension to MRShare, the work proposed in [70] generalizes the grouping strategy of MRShare to identify better sharing opportunities, and it also enables materialization of results and re-usability across batches.

HaLoop [16] is another system that enables caching and re-usability of results in MapReduce. However, HaLoop provides such features within a single iterative workflow. These workflows are very common in data mining techniques such as Page Rank, K-means clustering, and graph analysis. These techniques are usually iterative in nature, and they may need to access the same data again and again across iterations, which creates opportunities for sharing and re-using of results from previous iterations. As presented in Table 2, HaLoop proposes extensions to the Hadoop framework to enable the caching and re-usability at different stages including: (1) *Map Input Cache*, in which a mapper retrieving remote data from another DataNode will cache this data locally to re-use it in subsequent iterations, (2) *Reduce Input Cache*, in which reducers may cache their inputs from the invariant dataset (the dataset that does not change across iterations) and re-use that in subsequent iterations, and (3) *Reduce Output Cache*, which allows reducers to compare the results from the current and previous iterations without remote data access. Not all of the three caching types are useful in all iterative jobs, and it depends on the nature of the job. For example, a PageRank job assigns a rank for every webpage depending on the ranking of the pages it points to, and the ranks of those pages iteratively depend on their references (PageRank is explained in details in [16]). In this job type of job, it can make use of the *Reduce Input Cache* and *Reduce Output Cache*, but not *Map Input Cache*. HaLoop is designed to optimize a single iterative job, and thus the caches are purged after the completion of the job.

2.3 Pre-processing and Colocation Techniques

CoHadoop [33] is a system that re-organizes the data in place without creating auxiliary datasets. The key idea of CoHadoop is to extend Hadoop's internals, more specifically the logic of the NameNode, to allow more informative decisions regarding the storage location of each data block. Typically, by default, each data block in

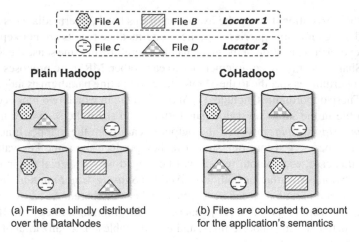

Fig. 5 Data colocation in CoHadoop

HDFS is replicated three times. The NameNode decides to which DataNode each replica of a given block should go to. This procedure is referred to as the *Data Placement Policy*. In the earlier versions of Hadoop (before Hadoop version 1.x.x), the placement policy is built-in within Hadoop and it aims for a single objective, which is *load balancing*. This means that each DataNode should have approximately equal storage load. Clearly, this policy is blind to the higher-level application semantics, and the decision may not align with the application's best interest. For example, As indicated in Fig. 5, the two files *A* and *B* can be semantically related and frequently accessed together, e.g., one file is a data file while the other one is its associated index, the two files usually join together, etc. Similarly, the two files *C* and *D* are semantically related. As shown in Fig. 5a, plain Hadoop is blind to such relationships and may randomly put un-related files together.

CoHadoop provides an interface through HDFS that enables applications to provide *hints* to HDFS regarding the relationships among the uploaded files. These hints can be viewed as simple tokens (or *locators*), where all files having the same token are assumed to be related, and now CoHadoop will try to colocate them in the same set of nodes. Referring to Fig. 5, while the four files are being uploaded (or generated), the application can assign files *A* and *B* the same token, while *C* and *D* will be assigned a different token. As a result, as shown in Fig. 5b, CoHadoop will try to colocate related files together on the same DataNotes. A typical application of CoHadoop is the join of two data files, say Customers and Transactions. Each file get partitioned on the join key using a map-reduce job, and then corresponding partition pairs from both files is assigned a specific locator, and this assignment is performed by the reduce function.

It is worth highlighting that the newer versions of Hadoop (version 1.x.x and later) enable applications to plugin an application-specific logic for the placement policy function so that, to some extent, applications can control where to store their data. One of the issues with CoHadoop is load balancing. That is, colocating many files together may negatively affect load balancing and few DataNodes may become heavily loaded. However, It has been studied in [33] that as long as new tokens are frequently introduced into the system, the overall data distribution will not be significantly affected. Moreover, the proposed colocation functionality is a best-effort approach meaning that the system does not have to enforce it if the distribution becomes skewed.

Trojan Join technique proposed in [27] is another example of data pre-processing and organization (Refer to Sect. 1.2). In Trojan Join, two datasets S and T are both pre-partitioned on the join key using a single map-reduce job. And then, each pair of corresponding partitions are stored together to form one logical split referred to as *Co-Partition Splits*. Certainly, each pair of partitions can be colocated on the same DataNode to speedup their retrieval. More details on Trojan Join technique can be found in Sect. 1.2.

The work in [46] has also studied the advantages of pre-partitioning the data on the join performance. For example, two datasets D_1 and D_2 can be both pre-partitioned on the join column, and at query time, a map-only job can join these two datasets where each mapper will read a pair of corresponding partitions; one from D_1 and one from D_2. In general, the pre-partitioning has shown to speedup join queries between 5x and 10x as studied in [27, 33, 46]. The colocation has shown to further add between 2x and 4x speedup on top of the pre-partitioning [33].

2.4 None Row-Oriented Storage Layouts

The standard layout of the data records in HDFS, or more specifically within each block of HDFS is the *Row-Oriented* layout. In this layout (See Fig. 6a), the data blocks are horizontal partitions of the data file, and each block consists of a set of entire data records. And then, within each block the records are organized

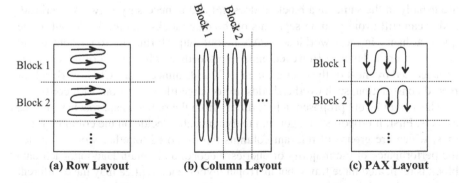

(a) Row Layout   (b) Column Layout   (c) PAX Layout

Fig. 6 Different types of data layouts

row-by-row as illustrated in Fig. 6a. This layout has a big advantage if the access pattern requests the entire records along with all their columns. This is because there is no overhead in constructing each record. However, the disadvantage is that if the data file has many columns, and typical queries are only interested in few columns each time, then there is a significant I/O waste due to reading un-needed data.

The *Column-Oriented* layout (Fig. 6b) is designed to overcome this limitation. This layout is proposed in several techniques including [12, 35, 48]. In this layout, the data blocks are vertical partitions of the data file. Depending on the number of records in the file and the HDFS block size, one HDFS block may contain sub-column, one column, or many columns. In some techniques [35], each column is stored as one file with varying number of data blocks depending on the number of values. The column-oriented layout is suitable for workloads in which each query accesses few columns while the data file originally has very large number of columns. In this case, only the columns of interest to the query are touched without wasting any I/O accessing the other columns. In addition, the column-oriented layout allows for efficient compression of the data. That is, the values within one column are of the same type, and tend to be highly compressible unlike the row-oriented layout. It is worth mentioning that writing and accessing column-oriented data only requires special output and input formats without the need for changing the internals of Hadoop. However, the disadvantage of this layout is that constructing an entire record (or many columns of each record) turns out to be an expensive process. Especially because different columns are most probably located on different DataNodes in Hadoop. Therefore, constructing records (if needed) would involved high communication and processing overheads. It has been studied in [47] that if a query is referencing a number of column around 13 or more, then the column-oriented layout performs worse than the row-oriented layout.

The *PAX* (Partition across) layout (Fig. 6c) is proposed to overcome the limitations of both previous layouts, and hopefully combining their advantages [21]. In this layout, the data blocks are horizontal partitions of the data file as in the row-oriented layout. Yet, the difference is that within each data block, the data is arranged in column-oriented layout. As a hybrid layout, PAX has several advantages including it avoids expensive tuple construction since the values contributing to one record are usually in the same data block, it does not add unnecessary network overhead, and it can still avoid reading segments of the data blocks that are irrelevant to the query at hand. In most workloads, PAX layout outperforms the row-oriented and column-oriented layouts. However, under workloads in which most of the columns are accessed by most of the queries, then PAX adds unnecessary CPU overhead for record construction, such overhead adds up for large files of billions of records.

The *Trojan* layout proposed in [47] differs from the previous approaches in three aspects. First, it proposes an algorithm to dynamically decide on the column group-ings, which are groups of relevant columns that if stored together would enhance the performance of the majority of queries. Second, as in plain Hadoop, each data block is replicated three times, but in Trojan layout, each replica may have different layout—although it is the same content. And third, a query will be routed to access specific replica depending on the query's access pattern.

In summary, data organization and layout is critical for improved query performance. The different layouts and organizations covered in this section in the context of Hadoop have shown more than 10x improvement over naive solutions under different scenarios. Certainly, there is no single optimal layout as it significantly influenced by the query workload. As the workload changes, the data layout may also need to be changed accordingly.

3 Non-traditional Workloads in Big Data

3.1 Overview

There are various ways to categorize queries (and workloads) according to different criteria. For example, one criteria considers the response time of queries, and based on that queries can be categorized as being *interactive* or *batch* queries. Another criteria considers the amount of data being touched by a query, and the granularity of the retuned results, e.g., whether it is tuple oriented or aggregation and mining oriented, and based on that queries can be categorized as *OLTP* (Online Transaction Processing), or *OLAP* (Online Analytical Processing). Another important criteria— which is our focus in this section—considers the frequency of queries' execution including how many times and how frequently the queries execute (See Fig. 7). On one end of this categorization spectrum are the ad-hoc queries, which are queries that are submitted once, get executed by the system, and then discarded (forgotten) by the system. This is the typical type of queries in big data, and more specifically the

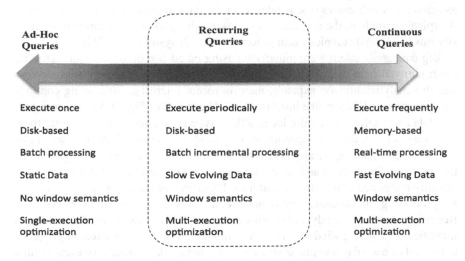

Fig. 7 Spectrum of ad-hoc, recurring, and continuous query types

Hadoop infrastructure. On the other side of the spectrum are the *continuous* queries, which are queries that are registered in the system before execution, live long in the system, and get executed frequently according to user-defined parameters. The characteristics of the ad-hoc and continuous queries are fundamentally different from each other, and thus the underlying systems and infrastructures supporting them are also significantly distinct as presented in Fig. 7.

Ad-hoc queries are queries that are submitted to the system in an ad-hoc way meaning that the system has no prior knowledge about them, and there is no mechanism to expect whether or not a given query will execute again, or what will be submitted next. When an ad-hoc query is submitted, the system tries to find the best execution plan to execute this query in isolation from any other queries. In the ad-hoc query model, the underlying data is assumed to be relatively static and does not frequently change. A typical example of the ad-hoc queries over big data infrastructures are the traditional Hadoop map-reduce jobs.

The opposite to the ad-hoc queries are the continuous queries, which are queries that are registered in the system before execution, and hence the system knows many details about these queries, and it can also learn more characteristics about these queries over time. Continuous queries are frequently executed by the system—usually in a high frequency—and the execution is initiated by a triggering mechanism, which can be a time-based (e.g., every 10 min re-executed the query), event-based (e.g., whenever a sensor's reading exceeds a specific threshold re-execute the query), or count-based (e.g., when 1000 tuples are collected re-execute the query). Since continuous queries are long lived in the system, shared execution and global execution plan(s) are typically used to efficiently execute many queries altogether [2, 3]. In the continuous query model, the underlying data is assumed to be streaming data that is arriving to the system in a very high rate and continuously changing. Each execution of a continuous query considers only a limited segment of the streaming data (known as *window*), which moves (a.k.a slides) over the data from one execution to another. A typical example of the continuous queries are the queries in stream management systems [2, 3], and complex event processing (CEP) systems [64, 74].

Big data applications were initially focusing on ad-hoc disk-based queries, e.g., the traditional map-reduce jobs. Yet, as the applications are getting more diverse and the analytics horizon expands, there is recent interest in supporting continuous queries over Hadoop-like infrastructure (The R.H.S of Fig. 7). Moreover, recent big data applications have introduced a third type of queries, which we refer to as *recurring queries*. Recurring queries are very common in most Hadoop-based applications, and big data applications in general. They appear in numerous applications that periodically generate and collect huge volumes of fresh data that must be periodically integrated into complex analytics. Examples of these applications include log processing, clickstream analysis, news feed updates, and social network services. Recurring queries are analytical queries that periodically execute over data subsets identified by a sliding window on the evolving data. For example, executing a query at the end of each day and processing the last n hours, days, weeks, or even months worth of data, depending on the granularity of interest.

As presented in Fig. 7, recurring queries have distinct characteristics from both ad-hoc and continuous query types. In fact, recurring queries combine properties from both worlds in an interesting way. For example, recurring queries are similar to continuous queries in that both are long-lived, re-execute periodically over the incoming data, have the notion of sliding windows to limit of the scope of the data to be processed, and process (possibly) large segments of overlapping data. However, they fundamentally differ in that recurring queries do not always mandate real-time millisecond processing. Instead, they tend to have a larger granularity of execution, e.g., they may execute once every hour or every day. Also, they may return the results within a certain period of time, e.g., few minutes to a couple of hours. Hence a query may remain idle for longer periods of time. Moreover, recurring queries are inherently data-intensive disk-based queries as they may process TBs of disk-resident data in each execution. In contrast, stream processing systems are optimized mostly for main-memory realtime processing.

On the other side of the spectrum, ah-hoc batch-processing systems such as plain Hadoop, are well-designed for scalability and disk-based processing—both are shared properties for recurring queries. However, these systems lack the notion of recurring execution, sliding windows, and overlapping data sets. Hence they fall short in providing backbone support for recurring queries and optimizing the execution according to their characteristics.

Example Queries: *In the following, we illustrate few examples queries under each of the three categories highlighted in Fig. 7.*

Ad-hoc Queries: Any Hadoop job given in isolation from any other job is considered as an ad-hoc query. The standard Word Count *query is a typical ad-hoc query. Transactional log processing queries that aggregate transactions by customers, items, or regions are all examples of ad-hoc queries. These queries have the characteristics of the L.H.S column in Fig. 7 including that they execute only once, they read their data from disk (the distributed file system), the input data to a query is static and does not change, and their is no notion of window semantics.*

Recurring Queries: In log processing, an aggregation query may need to execute every 12 or 24 h (the frequency of execution), and in each execution it processes and aggregates the log data from the recent past (the window of execution), e.g., the last 10 days or last month, over different dimensions such as age, country, gender to detect emerging patterns. This query has characteristics of the middle column in Fig. 7 including that the system is aware of its execution frequency, the input data is read from disk (the distributed file system), the query execution may still take long time, e.g., hours, and thus it is viewed as batch execution, and the query must have the notion of a window to define the scope of each execution. Since with each execution, the window slides over the data, there can be significant overlap between consecutive executions, and thus incremental evaluation is critical for recurring queries.

Continuous Queries: Online aggregation is an example of continuous queries where continuous real-time update of results is needed. Compared to recurring queries, in online aggregation, the window of execution is usually very small, e.g., few hours, and execution's frequency is usually higher, e.g., few minutes. Given this small granularity, the data under processing is usually kept in memory, and also passed from the mapper phase to the reducer phase (within a single Hadoop job) or even across jobs through the main memory as well.

In this section, we will cover several techniques proposed in literature for the *non-traditional workloads* in Hadoop, namely the *recurring workloads* and the *online analytics*.

3.2 Techniques for Recurring Workloads

As mentioned previously, a recurring query is a query that repeats periodically. Therefore, there are possibilities for new optimization opportunities to optimize such queries if treated as a first-class citizen. A naive approach for executing a recurring query is to manually re-issue the query every time it needs to be executed. However, this naive approach lacks both convenience and system-level optimizations. To overcome these limitations, several systems have been proposed such as Oozie [10], Nova [61], and Redoop [51, 52].

Apache Oozie [10] is a workflow scheduler that provides partial support by enabling developers to write scripts for automatic scheduling of jobs. Using this, end-users would no longer need to re-issue the recurring query over and over manually, but instead have it kicked off in an automated fashion. The Nova system [61] is also a workflow management system on top of Pig/Hadoop. It offers scheduling of job and queries such that a recurring query can be automatically triggered when an event takes place. Nova forms the execution flow as a directed graph, where the nodes represent either dataset or analytical tasks, and edges represent the flow of the data. The edges are annotated with instructions that guide the execution. For example, an analytical task may receive in each execution a delta changes (new data) or an entire data set including the changes. Similarly, the output from an analytical task can be either delta to the previous results or a complete new results. To support continuous and recurring workloads, the analytical tasks can have different types to enable incremental processing. Examples of these types are *Stateless Incremental, Stateless Incremental with Lookup Table*, or *Stateful Incremental*. Nova also provides several triggering mechanisms to trigger the execution of a given task. These mechanisms are either *data-driven*, e.g., when a new data arrives, *time-driven*, e.g., every one hour, or *cascade*, e.g., when another task finishes.

Both Oozie and Nova work on top of Hadoop without the need to change any of Hadoop's internals or its execution engine. This approach has advantages and disadvantages. The advantages are that the system design is relatively easier as it does not involve altering Hadoop's behavior, and also the portability since these systems can seamlessly work on different versions of Hadoop. However, the main disadvantage as pointed in [51] is the lack of system-level optimizations (See Fig. 8). Basically, in both Oozie and Nova, Hadoop infrastructure is unaware of the recurring nature of the submitted queries, e.g., it is unaware of the window semantics, the possible overlapping of the data being re-processed by consecutive execution, etc.

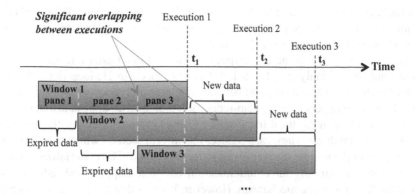

Fig. 8 Consecutive executions of a recurring query

For example, referring to Fig. 8 which shows three consecutive executions of one recurring query at times t_1, t_2, and t_3, and each execution processes a window of data configured by the query, e.g., the last 30 days of data. As the time moves forward, *Execution 2* will be triggered at time t_2 to process *Window 2*. It is clear that there is a significant amount of work that may be re-done if the overlapping segments between *Window 1* and *Window 2* is ignored. And the same applies for *Window 3*, and so on.

The Redoop system [51] has proposed extensions to Hadoop that enable registering these recurring queries inside Hadoop. And then, by analyzing the query semantics and configuration parameters, e.g., the window size, and the sliding frequency, it provides various types of query optimizations. For example, it can divide each window into smaller units, called *panes*, which become the unit of processing. The results from each pane can be shared across possibly many windows of executions, and hence redundant execution is avoided. For example, in Fig. 8, *Window 1* is divided into three panes, where *pane 1* evaluation belongs only to *Window 1*, *pane 2* evaluation is shared between *Window 1* and *Window 2*, while *pane 3* evaluation is shared between the three windows. Redoop also offers caching strategies in the local file system of the DataNotes that allows future executions to efficiently make use and build on top of previous execution results. Redoop system focuses on optimizing a single recurring query. As an extension to Redoop, the Helix system [52] proposes different mechanisms to enable efficient sharing of execution among multiple recurring queries possibly having different configuration parameters such as different window sizes, sliding frequencies, and quality of service requirements.

3.3   Techniques for Fast Online Analytics

Plain Hadoop is designed to suit a wide range of applications, and thus several of its design choices aim for flexibility and simplicity instead of performance. Therefore by default, Hadoop is suitable for fast online analytics. Several techniques and optimizations have been proposed to overcome this limitation of Hadoop and remove many of its bottleneck operations [8, 11, 23, 53].

The HOP (Hadoop Online Prototype) system modifies Hadoop to support continuous and online analytics [23, 24]. A key bottleneck in Hadoop that HOP has resolved is the materialization point between mappers and reducers. In Hadoop, the output from the mappers within a map-reduce job is an intermediate data that is materialized and stored in the local file system of the DataNodes running the mappers. And then, when reducers start, each reducer has information on which files to access and bring from the remote mapper nodes to its local node. The materialization is an important step to simplify the communication between mappers and reducers and also to the fault tolerance mechanism. However, it slows down the processing and it cannot be part of an infrastructure targeting fast online analytics.

HOP proposes to replace the built-in materialization step with a pipelining mechanism where the data flows from mappers to reducers through main memory buffers. Mappers are extended to push (pipeline) their output records to reducers using established TCP connections between each mapper and all reducers. For efficiency, mappers will not push each record as it get produced, instead they will buffer a specific number of records in main memory, possibly apply a combiner over the buffer, and once the pre-defined threshold is reached, the buffer is pushed to the reduce function. Given this change of the communication channel between the map and reduce phases, HOP has re-visited the fault tolerance mechanism to ensure seamless recovery under failures. Under these changes, HOP has shown to enable fast online analytical queries on Hadoop, e.g., online aggregations, and also enable continuous queries. That is, a set of mappers and reducers are continuous running to consume newly arriving data, pipeline the outputs from mappers to reducers, and produce continuous stream of results (Table 3).

The SOPA system proposed in [53] also enables one-pass analytics over Hadoop. The system relies more on in-memory processing, reading the input data only once, and incrementally processing new batches of arrived data. One of the built-in

Table 3 Comparison of Hadoop-based indexed techniques

Technique	Primary storage	Pipelining	Shuffling phase	Analytics
HOP [23]	HDFS	Yes	Disk-based sorting	Incremental and non-incremental
SOPA [53]	HDFS	Yes	Memory-based hashing	Incremental and non-incremental
M3 [8]	Memory	Yes	Memory-based hashing	Incremental
M3R [66]	Memory	Yes	Memory-based hashing	Incremental
C-MR [11]	Memory	Yes	Memory-based sorting	Incremental

operations in Hadoop that has been replaced is the *sort-merge* operation, which is part of the shuffle/sort phase between the mappers and the reducers. Sorting the data is performed in the shuffle/sort phase to group all map outputs records having specific key k altogether and passing that to a single reducer instance. However, it is well known that sorting is an expensive process—especially when performed over very large datasets. SOPA proposes to replace the sort-merge operation with another less-expensive operation, which is a hash-based partitioning operation. The hash-based operation would still achieve the same goal, which is grouping all records of the same key together, but without encountering the high overhead. Another advantage of the hash-based operation is that it is not a blocking operation, meaning it does not need to collect all the input before it starts producing output. Therefore, it can be easily pipelined and it can leverage in-memory processing more easily.

Several types of hash functions have been proposed in SOPA depending on whether or not the reduce function can be incrementally computed. If incremental computation is possible, e.g., in the case of simple aggregates such as $sum()$ and $count()$ functions, then reducers can receive partial inputs, incrementally update their state, and then consume more inputs. This flow enables better pipelining between mappers and reducers, yet it requires an extended interface for reducers that allows for creating and maintaining a state for each key in the reduce function. The design tries to keep as many states as possible in memory such that their incremental updates become more efficient. However, if that is not the case, the paper has presented another mechanism, referred to as *dynamic hashing*, to adaptively select a subset of key states to keep in memory while moving the other states to local disk.

Other techniques have been proposed to alter Hadoop's disk-based and batch processing nature to be streaming nature. The M3 system [8] is designed to replace the HDFS layer in data and rely only on main-memory processing. Moreover, in M3 the jobs are continuous jobs in the sense that they get registered in the system, and the mappers and reducers remain continuously running and consuming data. M3 mimics stream processing systems in its design as well as its functionalities. For example, it supports incremental processing and the query results is also computed incrementally by reporting *delta* changes over the previously reported results. The Continuous-MapReduce (C-MR) [11], and Main Memory Map Reduce (M3R) [66] systems aim for the same objective as M3. They try to entirely (or partially) replace the disk-based file system HDFS with main-memory storage and processing. In all of these techniques, especial consideration has to be given to fault tolerance to be able to recover from failures. The common strategy used by these techniques is to replicate the data multiple times in the memory of different machines. Moreover, since these techniques will manage continuous queries over possibly infinite input data, they all inherit the *window* semantics from stream processing systems to limit the scope of their computations. Certainly, main-memory processing puts limitations on the size of the data to be processed, and also puts restrictions on the type of jobs to be supported, e.g., M3 supports only jobs that can be incrementally evaluated. Yet, the availability of large memories in the modern clusters, e.g., each single machine can have 100s of GBs of memory, enable many applications to store and manage their data entirely in memory.

4 Curation and Metadata Management in Big Data

4.1 Overview

Metadata is a general term that references all auxiliary information that is related to the base data, but not really part of the data. This auxiliary information may range from execution statistics, optimization hints, users' comments, related articles or documents, to corrections and highlights of errors, provenance information, and special tagging or marking. As highlighted in Fig. 9a, the metadata information can be attached at different granularities and be related to, for example, specific table cells, row, columns or arbitrary combination of them. This type of metadata is referred to as *data-centric annotations*, where is the metadata is associated with the base data regardless of any execution workflows. Another type of metadata captures execution statistics as data goes under complex transformations and workflows, and this type is referred to as *execution-centric annotations* (Fig. 9b). In contrast, another type of metadata may capture the lineage of the data including the input records contributed to each output, the applied transformation functions, and the configuration parameters used during execution. This type is referred to as *provenance-centric annotations* (Fig. 9c).

Metadata and annotation management is relatively a new research topic in the database community. However, the concept of annotations and curating the data has been known for a long time. Historically, it goes back to the paper-based *Post-It* yellow notes that scientists and people have used in early 1970s to write down their thoughts, ideas, and exchange information [29]. Then, with the advances of data management and the increasing use of DBMSs in scientific applications, the virtue, size, and complexity of the annotation repositories have increased. Consequently,

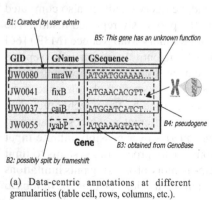

(a) Data-centric annotations at different granularities (table cell, rows, columns, etc.).

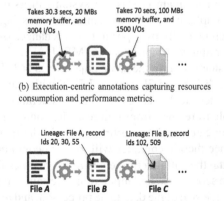

(b) Execution-centric annotations capturing resources consumption and performance metrics.

(c) Provenance-centric annotations capturing the lineage of output records in terms of the input records.

Fig. 9 Examples of metadata usage

there was a pressing need to develop annotation management techniques that can capture and query the annotations in more systematic and advanced ways. This triggered numerous projects and research initiatives to address annotation management in the context of relational database management systems.

In general, the metadata and the annotations cannot be treated as regular data. This is because they have a fundamental logical difference compared to the base data, which is that they are viewed as *auxiliary* information that should propagate (be carried) automatically with the data. For, example, since scientific data may go though complex transformations during query processing, e.g., projection, join, grouping and aggregation, and duplicate elimination, the related annotations must also go though corresponding transformations by each query operator. If annotations are modeled as regular data—which was the case before the development of annotation management engines—then the annotation management tasks are entirely delegated to end-users and higher-level applications starting from the storage and indexing of annotations and ending by explicitly encoding the propagation semantics within each of the users' queries. Both tasks have been shown to be very complex and sophisticated. For example, the storage and indexing mechanisms need to deal with the combinatorial relationship between annotations and data, e.g., annotations can be attached to single table cells (attributes), rows, columns, arbitrary sets and combinations of them, or even attached to sub-attributes [32, 42]. Moreover, manually encoding the annotations' propagation within each query is not only error-prune, and lacks optimizations, but also renders even simple queries very complex [17, 41, 68]. That is why annotation management engines have been proposed to efficiently and transparently manage such complexities across applications.

Annotation management in relational databases has ranged from developing generic frameworks [15, 22, 32, 43, 68] to developing semantic-specific techniques for annotations [19, 40, 49]. The generic frameworks, e.g., Mondrian [43], DBNotes [22], InsightNotes [44, 73], and others [54], target extending the database functionalities with annotation management capabilities. This includes efficient storage and indexing, interfaces to add and query the annotations, and new algebraic semantics—and possibly new query operators—to enable efficient annotation propagation within a query pipeline. The semantic-specific techniques have mostly focused on lineage and provenance tracking [18, 20, 26, 71], where each output record from a SQL query carries references to all input records that have contributed to the output record.

As the data management systems have evolved to the cloud-based systems and the emerging infrastructures such as Hadoop for managing big data, the annotation management techniques have also evolved to operate on these new infrastructures. The usage of annotations in these techniques varies from execution statistics [55, 61], and provenance tracking [6, 9, 45, 62] to tagging for query optimization [7] and generic frameworks [59]. In the following, we describe several of these techniques in more details.

4.2   Execution-Centric Metadata Approach

Several techniques have been proposed in the context of MapReduce to leverage annotations and metadata information in job execution. Examples of these systems are the high-level languages of Hadoop including Apache Pig [39], Hive [69], and Jaql [14]. These query languages offer some optimizations while compiling the high-level scripts into map-reduce jobs. Yet, they do not have sophisticated query optimizer, and thus they rely on users' annotations of the query script and use that as hints for the query optimizer. For example, a join statement can be annotated with keyword *"replicated"* in Pig to indicate that one of the two joined datasets is small, and it should be sent out (broadcast) to every mapper reading a portion of the big dataset. In this case, the join operation can take place as a map-only job instead of an expensive map-reduce job.

Other systems that leverage annotations during execution are the Nova [61] and Stubby [55] systems. Nova is a workflow management system on top the Pig/Hadoop infrastructure. It uses process-related and system-generated annotations to provide execution hints to the system such as the transfer mode of the data between processes, the output format and schema of each task, and the behavior of each task. On the other hand, Stubby uses the annotations to collect execution statistics, profiling jobs, and providing execution hints. Annotations can be related to a dataset, a specific operator in a workflow, or execution statistics. Examples of the dataset-related annotations can the physical layout of the dataset or any special ordering or partitioning properties about the data. Stubby uses the annotations as a mean to communicate information needed for the workflow optimization.

4.3   Provenance-Centric Metadata Approach

Lineage or provenance tracking means tracking the source of a given piece of data, e.g., from where it comes, which input records have contributed to a given output record, which derivations and transformations have been applied to get such output. Provenance tracking is very important in many applications because in some cases the trustworthy of the data cannot be assessed without knowing the source of the data. In some other cases it can be important for applications to go back to the source data and analyze why the output includes such values.

Lineage tracing has been recently studied in the context of Hadoop [6, 9, 45, 62]. Unlike relational database systems where complex transformations are possible though complex SQL queries, in Hadoop the transformation is only possible though the map and reduce phases. However, database execution is an *open box* in the sense that the system knows exactly the semantics of each operator and the type of transformation being applied to each tuple. In contrast, Hadoop execution is *black box* and the system may not know what type of transformation is being applied inside the map and reduce functions. This is especially true if the job is expressed in java.

Ramp system [45, 63] tracks the lineage of the data generated from map-reduce jobs by assigning artificial object identifiers (OIDs) for each input data record, and then producing each output record along with its provenance information (the contributing input OIDs). The OIDs are assigned to the <key, value> pairs generated from the input formats, and this OID reflects the offset of each input record in its HDFS file. And then, Ramp uses system-generated wrappers around the mapper and reducer functions as well as the RecordReader and RecordWriter functions within the InputFormat and OutputFormat, respectively (See Fig. 10). These wrappers will transparently carry the input OIDs to the output side while bypassing the user-defined map and reduce functions. For example, referring to Fig. 10a, the wrapper around the InputFormat RecordReader automatically adds a unique OID (p in the example) to each generated record, and then the wrapper around the user's mapper function extracts this provenance information and bypass it to the output record. Similar extensions have been designed for the reduce-side functions (Fig. 10b).

Since the map and reduce functions are black boxes with no known semantics to the system, Ramp puts some restrictions to ensure correct tracking of provenance information. For example, Ramp supports one-to-one and one-to-many input-to-output granularity in the mapper function. This means that Ramp can track the provenance if each input record to the map function generates zero, one or many outputs. However, many-to-one input-to-output can not be tracked. For example, if the map function internally performs some buffering, and then based on some internal semantics the function periodically generates output records, then the provenance cannot be tracked under such *hidden* behavior. Similar restrictions apply to the reduce side. For example, Ramp assumes that all inputs values corresponding to a key contribute to the output record. If the function internally does not follow this logic, then the provenance information will not be correctly tracked.

Several other provenance tracking systems have been proposed over Hadoop. This includes the *Kepler+Hadoop* system [25] that tracks the provenance within the

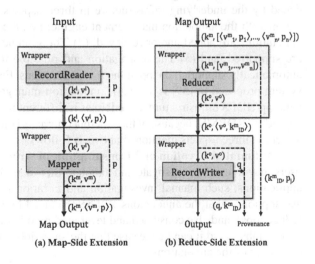

Fig. 10 Ramp extensions for provenance tracking [63]

(a) Map-Side Extension   **(b) Reduce-Side Extension**

scientific workflows, and Newt [57] that uses fine-grained lineage information for debugging and error tracking. Most of the aforementioned techniques encounter high overheads due to provenance tracking. This is because the size of the provenance information can be very large, and it may need to be carried out though the shuffling and sorting phases of MapReduce. The work proposed in HadoopProv [6] and MrLazy [7] overcomes this drawback by separating the provenance tracking of the map and reduce phases, where each phase writes its provenance information to disk separately, and no provenance information is shuffled in between these two phases. And then, only when needed, another job can join the separate results to construct the final output-to-input lineage information. This join operation is regarded as an offline task, and thus the involved overhead is no longer carried on the users' jobs.

All of the above mentioned techniques are coarse-grained techniques in the sense that they track the provenance at the record level, where a record is the object created from the underlying InputFormat function. These techniques cannot track the provenance at the attribute level since the map and reduce functions are assumed to be black boxes. The Lipstick technique [9] is distinct from these systems in that it is a fine-grained attribute-level provenance tracking system. However, Lipstick uses Apache Pig as its query interface, which enables the system to understand and track the logic and the semantics of the queries.

4.4 Data-Centric Metadata Approach

Unlike the other two approaches, the data-centric techniques are generic annotation management frameworks that do not bind the annotations to specific semantics. Instead, they enable applications to annotate and curate their data freely [59]. Conceptually, these generic frameworks can be leveraged in implementing the other types including the *execution-centric* and *provenance-centric* techniques.

In general, designing generic annotation management engines is greatly influenced by the underlying infrastructure in three aspects, which are: (1) The interaction with the annotation management engine, i.e., the mechanisms by which the annotations are added and/or retrieved, (2) The granularities at which annotations are supported, and (3) The propagation and possible transformations that can be automatically supported on top of the raw annotations. For example, the CloudNotes system proposed in [59] is a generic annotation management engine on top of the Hadoop/HDFS infrastructure. CloudNotes is different from those generic annotation engines in RDBMSs because of the inherent characteristics of Hadoop/HDFS, which affect the three aspects mentioned above as follows:

• **Automated Creation and Consumption of Annotations**: In RDBMSs, end-users may manually investigate and annotate their data. However, in Hadoop-based applications, such manual investigation and curation is not practical. Instead, the assumption is that the annotations will be produced by automated processes (map-reduce jobs), and also consumed and leveraged by other automated processes (map-reduce jobs). And it can be the case that the same job acts as both a producer and a consumer of the annotations.

• **Single-Granularity Annotations**: Annotation management engines typically support annotating the data at the finest granularity provided by the underlying data model. For example, in RDBMSs, annotations can be at the granularity of table cells, rows, columns, etc. [15, 42], and in array-based systems, annotations are defined at the granularity of array cells [72]. Supporting annotations at a smaller granularity, e.g., a sub-value within a table cell, becomes an application-specific task and encoded by the application. CloudNotes inherits the same principle. Since HDFS has a single unit of granularity, which is the object formed from the InputFormat layer and passed to the mapper layer, the system supports annotating the data at this granularity.

• **Blackbox Annotation Propagation**: Hadoop uses a blackbox map-reduce computing paradigm, where the actual computations and transformations applied to the data are unknown. As a result, CloudNotes does not provide automated transformation rules for the annotations. Instead, it provides the application developers with interfaces to integrate the annotations into the processing cycle as fits each application's semantics.

CloudNotes works by extending the internals of the Hadoop infrastructure to accept, store, and propagate the annotations attached to each data object (See Fig. 11). It provides interfaces in the map functions to be able to either add new annotations to its input records, add new annotations to its output records, or retrieve the annotations on the input records. Similarly, it provides interfaces for the reduce functions to be able to either retrieve their inputs' annotations (which are passed from mappers to reducers), or add new annotations to their output records. As illustrated in Fig. 11, reducers cannot annotate their inputs because this input is intermediate and gets automatically purged after the job completion.

Introducing the annotations into Hadoop creates several challenges including: (1) The data is no longer assumed to be read-only because the annotations associated with the data may change over time. Therefore, a concurrency control module needs to be developed to ensure correct execution among multiple jobs. (2) Annotations should automatically and transparently propagate with the data. And hence, effective buffering and proactive prefetching techniques need to be developed. It is worth mentioning that the techniques in the *execution-centric* and *provenance-centric* categories focus on creating the annotations, but they do not address how these annotations propagate in future queries whenever the data is accessed. And (3) Annotation jobs, i.e., the map-reduce jobs that only targeted to add more annotations, can be lazily evaluated as long as no other job is asking for these annotations. Therefore, possible optimizations and batching of jobs become feasible. CloudNotes system proposes different techniques to address these challenges.

CloudNotes proposes different storage schemes for the annotations, where the annotations can be either stored in HBase or HDFS. In both schemes, the storage is transparent from the end-users. Whenever a map-reduce job access a specific HDFS file, the annotations attached to each record in this file automatically propagate to the map functions and get cached locally (either in main memory or local disk) for fast retrieval by upon request.

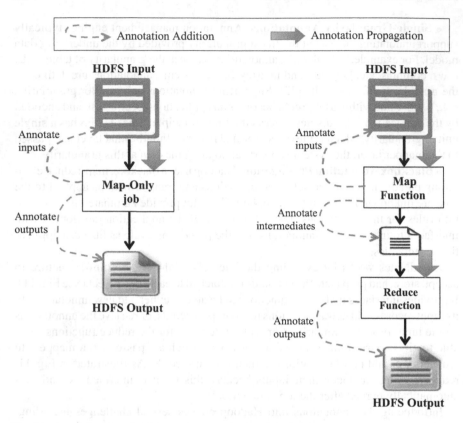

Fig. 11 Annotation flow (addition and propagation) in CloudNotes system

5 Conclusion

This chapter covered several advanced techniques related to data indexing, organization, and curation in the context of the emerging Hadoop MapReduce infrastructure. Several data indexing techniques have been presented covering the spectrum of record-level versus split-level indexing, static versus adaptive indexing, and Hadoop-centric versus hybrid (Hadoop plus relational DBMS) indexing. In addition to data indexing, the layout on disk also plays an important role in query processing and optimization. The chapter covered several advanced techniques for special data organization and layouts including data colocation, result materialization and caching, prepartitioning, and different types of layouts such as row-oriented, column-oriented, or PAX layouts. Both the indexing and special organization techniques have shown in literature to speedup query processing by orders of magnitudes. Yet, under non-traditional workloads, additional optimizations can be applied to further enhance the performance. The chapter covered several of these non-traditional workloads and their optimizations including recurring workloads and online analytics. Finally, the chapter covered the state-of-art techniques in big data curation and their leverage either in query and workload optimizations (the execution-centric approaches), or capturing higher-level semantics and business logic (the provenance-centric and data-centric approaches).

References

1. D.J. Abadi, Tradeoffs between parallel database systems, hadoop, and hadoopdb as platforms for petabyte-scale analysis, in *SSDBM* (2010), pp. 1–3
2. D.J. Abadi, D. Carney, U. Çetintemel, M. Cherniack, C. Convey, C. Erwin, E.F. Galvez, M. Hatoun, A. Maskey, A. Rasin, A. Singer, M. Stonebraker, N. Tatbul, Y. Xing, R. Yan, S.B. Zdonik, Aurora: a data stream management system, in *SIGMOD Conference* (2003), p. 666
3. D.J. Abadi, Y. Ahmad, M. Balazinska, U. Çetintemel, M. Cherniack, J.-H. Hwang, W. Lindner, A. Maskey, A. Rasin, E. Ryvkina, N. Tatbul, Y. Xing, S.B. Zdonik, The design of the borealis stream processing engine, in *CIDR* (2005), pp. 277–289
4. A. Abouzeid, K. Bajda-Pawlikowski, A.R. Daniel Abadi, A. Silberschatz, HadoopDB: an architectural hybrid of MapReduce and DBMS technologies for analytical workloads, in *VLDB* (2009), pp. 922–933
5. A. Abouzied, K. Bajda-Pawlikowski, J. Huang, D.J. Abadi, A. Silberschatz, Hadoopdb in action: building real world applications, in *SIGMOD Conference* (2010), pp. 1111–1114
6. S. Akoush, R. Sohan, A. Hopper, HadoopProv: towards provenance as a first class citizen in MapReduce, in *USENIX Workshop on the Theory and Practice of Provenance* (2013)
7. S. Akoush, L. Carata, R. Sohan, A. Hopper, MrLazy: lazy runtime label propagation for MapReduce, in *HotCloud* (2014)
8. A.M. Aly, A. Sallam, B.M. Gnanasekaran et al., M3: stream processing on main-memory mapreduce, in *ICDE* (2012), pp. 1253–1256
9. Y. Amsterdamer, S.B. Davidson, D. Deutch, T. Milo, J. Stoyanovich, V. Tannen, Putting lipstick on pig: enabling database-style workflow provenance, in *PVLDB* (2011), pp. 346–357
10. Apache. Oozie: hadoop workflow system. http://yahoo.github.com/oozie/
11. N. Backman, K. Pattabiraman, R. Fonseca et al., C-mr: continuously mapreduce workflows on multi-core processors, in *Proceedings of 3rd International Workshop on MapReduce and Its Applications Date* (2012), pp. 1–8
12. A. Balmin, T. Kaldewey, S. Tata, Clydesdale: structured data processing on hadoop, in *Proceedings of the ACM SIGMOD International Conference on Management of Data, SIGMOD 2012, Scottsdale, AZ, USA, May 20–24* (2012), pp. 705–708
13. A. Balmin, K.S. Beyer, V. Ercegovac, J. McPherson, F. Özcan, H. Pirahesh, E.J. Shekita, Y. Sismanis, S. Tata, Y. Tian, A platform for extreme analytics. IBM J. Res. Develop. **57**(3/4), 4 (2013)
14. K. Beyer, V. Ercegovac, R. Gemulla, A. Balmin, M.Y. Eltabakh, C.-C. Kanne, F. Ozcan, E. Shekita, Jaql: a scripting language for large scale semi-structured data analysis, in *PVLDB*, vol. 4 (2011)
15. D. Bhagwat, L. Chiticariu, W. Tan, An annotation management system for relational databases, in *VLDB* (2004), pp. 900–911
16. Y. Bu, B. Howe, M. Balazinska, M.D. Ernst, Haloop: efficient iterative data processing on large clusters. Proc. VLDB Endow. **3**(1–2), 285–296 (2010)
17. P. Buneman et al., On propagation of deletions and annotations through views, in *PODS* (2002), pp. 150–158
18. P. Buneman, A. Chapman, J. Cheney, Provenance management in curated databases, in *SIGMOD* (2006), pp. 539–550
19. P. Buneman, J. Cheney, W.-C. Tan, S. Vansummeren, Curated databases, in *Proceedings of the 27th ACM symposium on Principles of database systems (PODS)* (2008), pp. 1–12

20. P. Buneman, S. Khanna, W. Tan, Why and where: a characterization of data provenance. Lect. Notes Comput. Sci. **316–333**, 2001 (1973)
21. S. Chen, Cheetah: a high performance, custom data warehouse on top of mapreduce. PVLDB **3**(2), 1459–1468 (2010)
22. L. Chiticariu, W.-C. Tan, G. Vijayvargiya, DBNotes: a post-it system for relational databases based on provenance, in *SIGMOD* (2005), pp. 942–944
23. T. Condie, N. Conway, P. Alvaro, J.M. Hellerstein, K. Elmeleegy, R. Sears, Mapreduce online, in *NSDI* (2010), pp. 313–328
24. T. Condie, N. Conway, P. Alvaro, J. M. Hellerstein, J. Gerth, J. Talbot, K. Elmeleegy, R. Sears, Online aggregation and continuous query support in mapreduce, in *SIGMOD* (2010), pp. 1115–1118
25. D. Crawl, J. Wang, I. Altintas, Provenance for MapReduce-based data-intensive workflows, in *WORKS Workshop* (2011), pp. 21–30
26. Y. Cui, J. Widom, Lineage tracing for general data warehouse transformations, in *VLDB* (2001), pp. 471–480
27. J. Dittrich, J.-A. Quiané-Ruiz, A. Jindal, Y. Kargin, V. Setty, J. Schad, Hadoop++: making a yellow elephant run like a cheetah (without it even noticing). VLDB **3**, 518–529 (2010)
28. J. Dittrich, J. Quiané-Ruiz, S. Richter, S. Schuh, A. Jindal, J. Schad, Only aggressive elephants are fast elephants. PVLDB **5**(11), 1591–1602 (2012)
29. T. Donnelly, 9 Brilliant Inventions Made by Mistake. Inc. Accessed 24 Aug 2012
30. A. Eldawy, M.F. Mokbel, Spatialhadoop: a mapreduce framework for spatial data, in *31st IEEE International Conference on Data Engineering, ICDE 2015, Seoul, South Korea, April 13–17* (2015), pp. 1352–1363
31. I. Elghandour, A. Aboulnaga, Restore: reusing results of mapreduce jobs. Proc. VLDB Endow. **5**(6), 586–597 (2012)
32. M.Y. Eltabakh, W.G. Aref, A.K. Elmagarmid, M. Ouzzani, Y.N. Silva, Supporting annotations on relations, in *EDBT* (2009), pp. 379–390
33. M.Y. Eltabakh, Y. Tian, F. Özcan, R. Gemulla, A. Krettek, J. McPherson, Cohadoop: flexible data placement and its exploitation in hadoop. PVLDB **4**(9), 575–585 (2011)
34. M.Y. Eltabakh, F. Özcan, Y. Sismanis, P. Haas, H. Pirahesh, J. Vondrak, Eagle-eyed elephant: split-oriented indexing in Hadoop, in *Proceedings of the 16th International Conference on Extending Database Technology (EDBT)* (2013), pp. 89–100
35. A. Floratou, J.M. Patel, E.J. Shekita, S. Tata, Column-oriented storage techniques for mapreduce. PVLDB **4**(7), 419–429 (2011)
36. A. Floratou, U.F. Minhas, F. Özcan, Sql-on-hadoop: full circle back to shared-nothing database architectures. PVLDB **7**(12), 1295–1306 (2014)
37. A. Floratou, F. Özcan, B. Schiefer, Benchmarking sql-on-hadoop systems: TPC or not tpc? in Big Data Benchmarking - 5th International Workshop, WBDB, Potsdam, Germany, August 5–6, 2014. Revised Selected Papers **2014**, 63–72 (2014)
38. V.R. Gankidi, N. Teletia, J.M. Patel, A. Halverson, D.J. DeWitt, Indexing HDFS data in PDW: splitting the data from the index. PVLDB **7**(13), 1520–1528 (2014)
39. A.F. Gates, O. Natkovich, S. Chopra, P. Kamath, S.M. Narayanamurthy, C. Olston, B. Reed, S. Srinivasan, U. Srivastava, Building a high-level dataflow system on top of map-reduce: the pig experience. Proc. VLDB Endow. 1414–1425 (2009)
40. W. Gatterbauer, M. Balazinska, N. Khoussainova, D. Suciu, Believe it or not: adding belief annotations to databases. Proc. VLDB Endow. **2**(1), 1–12 (2009)
41. F. Geerts, J. Van Den Bussche, Relational completeness of query languages for annotated databases, in *DBPL* (2007), pp. 127–137
42. F. Geerts et al., Mondrian: annotating and querying databases through colors and blocks, in *ICDE* (2006), p. 82
43. F. Geerts, A. Kementsietsidis, D. Milano, MONDRIAN: annotating and querying databases through colors and blocks, *Proceedings of the 22nd International Conference on Data Engineering, ICDE 2006, 3–8 April 2006* (GA, USA, Atlanta, 2006), p. 82

44. K. Ibrahim, D. Xiao, M.Y. Eltabakh, Elevating annotation summaries to first-class citizens in insightnotes, in *Proceedings of the 18th International Conference on Extending Database Technology, EDBT 2015, Brussels, Belgium, March 23–27* (2015), pp. 49–60
45. R. Ikeda, H. Park, J. Widom, Provenance for generalized map and reduce workflows, in *CIDR* (2011), pp. 273–283
46. D. Jiang, B. C. Ooi, L. Shi, S. Wu, The performance of mapreduce: an in-depth study. Proc. VLDB Endow. 472–483 (2010)
47. A. Jindal, J. Quiané-Ruiz, J. Dittrich, Trojan data layouts: right shoes for a running elephant, in *ACM Symposium on Cloud Computing in conjunction with SOSP 2011, SOCC '11, Cascais, Portugal, October 26–28* (2011), p. 21
48. T. Kaldewey, E.J. Shekita, S. Tata, Clydesdale: structured data processing on mapreduce, in *15th International Conference on Extending Database Technology, EDBT '12, Berlin, Germany, March 27–30, 2012, Proceedings* (2012), pp. 15–25
49. G. Karvounarakis, T.J. Green, Semiring-annotated data: queries and provenance. SIGMOD Rec. **41**(3), 5–14 (2012)
50. P. Larson, J. Zhou, View matching for outer-join views. VLDB J. **16**(1), 29–53 (2007)
51. C. Lei, E. Rundensteiner, M.Y. Eltabakh, Redoop: supporting recurring queries in Hadoop, in *Proceedings of the 16th International Conference on Extending Database Technology (EDBT)* (2013)
52. C. Lei, Z. Zhuang, E.A. Rundensteiner, M.Y. Eltabakh, Shared execution of recurring workloads in mapreduce. PVLDB **8**(7), 714–725 (2015)
53. B. Li, E. Mazur et al. A platform for scalable one-pass analytics using mapreduce, in *SIGMOD* (2011), pp. 985–996
54. Q. Li, A. Labrinidis, P.K. Chrysanthis, ViP: a user-centric view-based annotation framework for scientific data, in *Proceedings of the 20th international conference on Scientific and Statistical Database Management (SSDBM)* (2008), pp. 295–312
55. H. Lim, H. Herodotou, S. Babu, Stubby: a transformation-based optimizer for MapReduce workflows. PVLDB **5**(11), 1196–1207 (2012)
56. Y. Liu, S. Hu, T. Rabl, W. Liu, H. Jacobsen, K. Wu, J. Chen, J. Li, Dgfindex for smart grid: enhancing hive with a cost-effective multidimensional range index. PVLDB **7**(13), 1496–1507 (2014)
57. D. Logothetis, S. De, K. Yocum, Scalable lineage capture for debugging DISC analytics, in *SOCC* (2013), pp. 17:1–17:15
58. P. Lu, G. Chen, B.C. Ooi, H.T. Vo, S. Wu, Scalagist: scalable generalized search trees for mapreduce systems [innovative systems paper]. PVLDB **7**(14), 1797–1808 (2014)
59. Y. Lu, Y. Li, M.Y. Eltabakh, Decorating the cloud: enabling annotation management in MapReduce. PVLDB **5**(11), 1–26 (2016)
60. T. Nykiel, M. Potamias, C. Mishra, G. Kollios, N. Koudas, Mrshare: sharing across multiple queries in mapreduce. Proc. VLDB Endow. 494–505 (2010)
61. C. Olston, G. Chiou, L. Chitnis, F. Liu, Y. Han, M. Larsson, A. Neumann, V.B. N. Rao, V. Sankarasubramanian, S. Seth, C. Tian, T. ZiCornell, X. Wang, Nova: continuous pig/hadoop workflows, in *SIGMOD Conference* (2011), pp. 1081–1090
62. H. Park, R. Ikeda, J. Widom, Ramp: a system for capturing and tracing provenance in mapreduce workflows. PVLDB **4**(12), 1351–1354 (2011)
63. H. Park, R. Ikeda, J. Widom, Ramp: a system for capturing and tracing provenance in mapreduce workflows, in *VLDB*. Stanford InfoLab (2011)
64. M. Ray, E.A. Rundensteiner, M. Liu, C. Gupta, S. Wang, I. Ari. High-performance complex event processing using continuous sliding views, in *EDBT* (2013), pp. 525–536
65. S. Richter, J. Quiané-Ruiz, S. Schuh, J. Dittrich, Towards zero-overhead adaptive indexing in hadoop, in *CoRR* (2012). arXiv:abs/1212.3480
66. A. Shinnar, D. Cunningham, B. Herta et al., M3r: increased performance for in-memory hadoop jobs. PVLDB 1736–1747 (2012)
67. M. Stonebraker et al., Mapreduce and parallel dbmss: friends or foes? Commun. ACM **53**(1), 64–71 (2010)

68. W.-C. Tan, Containment of relational queries with annotation propagation, in *DBPL* (2003)
69. A. Thusoo, J.S. Sarma, N. Jain, Z. Shao, P. Chakka, S. Anthony, H. Liu, P. Wyckoff, R. Murthy, Hive - a warehousing solution over a map-reduce framework. PVLDB, 1626–1629 (2009)
70. G. Wang, C.-Y. Chan, Multi-query optimization in mapreduce framework. PVLDB **7**(3), 145–156 (2013)
71. A. Woodruff, M. Stonebraker, Supporting fine-grained data lineage in a database visualization environment, in *ICDE* (1997), pp. 91–102
72. E. Wu, S. Madden, M. Stonebraker, SubZero: a fine-grained lineage system for scientific databases, in *ICDE* (2013), pp. 865–876
73. D. Xiao, M.Y. Eltabakh, InsightNotes: summary-based annotation management in relational databases, in *SIGMOD Conference* (2014), pp. 661–672
74. D. Zhang, M. Ray, M. Liu, D. Dougherty, E.A. Rundensteiner, Nested complex event processing: predicate specification and evaluation, in *Transactions on Large-Scale Data- and Knowledge-Centered Systems V*. Special Issue on Advanced Data Stream Management and Processing of Continuous Queries (Springer, Berlin, 2013)
75. J. Zhou, P. Larson, H.G. Elmongui, Lazy maintenance of materialized views, in *Proceedings of the 33rd International Conference on Very Large Data Bases, University of Vienna, Austria, September 23–27, 2007* (2007), pp. 231–242
76. J. Zhou, P. Larson, J. Goldstein, L. Ding, Dynamic materialized views, in *Proceedings of the 23rd International Conference on Data Engineering, ICDE 2007, The Marmara Hotel, Istanbul, Turkey, April 15–20* (2007), pp. 526–535

Big Data Query Engines

Mohamed A. Soliman

Abstract Big data analytics are techniques that are used to analyze large datasets in order to extract patterns, trends, correlations and summaries. Analytics are used in several big data applications ranging from the generation of simple reports to running deep and complex query workloads. The insights drawn by running big data analytics depend primarily on the capabilities of the underlying query engine, which is responsible for translating user queries into efficient data retrieval and processing operations, as well as executing these operations on one or multiple nodes in order to find query answers. Classically, parallel database systems have been adopted in various domains, particularly enterprise data warehouses, as the data processing platform for running big data analytics. An SQL-based query engine, running on a shared-nothing cluster, is typically used by these platforms. Scalability is realized by partitioning data across multiple machines that communicate via a high speed interconnect layer. These systems often rely on dedicated expensive hardware resources in order to scale-out query processing and provide fault tolerance. With the emergence of Hadoop, it became possible to use cheap commodity hardware for achieving linear scalability and fault tolerance. A typical Hadoop environment involves a software stack running in one ecosystem, while sharing hardware resources across different systems, called tenants. Earlier Hadoop query engines leveraged programming frameworks such as MapReduce to run analytics using programs executed on a distributed file system. The Hadoop Distributed File System (HDFS) has been effectively used for batch processing of simple analytics. The need for coding and manual optimization of analytics, the lack of support to complex queries and the limited interactive processing capabilities, have triggered the need for adopting new technologies with more expressive query languages and advanced query processing techniques. Integrating parallel database systems into Hadoop ecosystem is an obvious approach to combine the advantages of both worlds. In this respect, multiple challenges needed to be addressed to fit a parallel database query engine in Hadoop software stack. Data placement, query optimization, query execution and resource management are some of the technical problems that are actively studied in this area. In this chapter, we discuss the state-of-the-art of query engines in parallel database systems,

M.A. Soliman (✉)
Datometry Inc., San Francisco, CA, USA
e-mail: mohamed.soliman@datometry.com; mohamed.fathi@gmail.com

© Springer International Publishing AG 2017   179
A.Y. Zomaya and S. Sakr (eds.), *Handbook of Big Data Technologies*,
DOI 10.1007/978-3-319-49340-4_6

Hadoop-based systems, as well as the hybrid systems that integrate parallel databases and Hadoop technologies. We present the architectures of multiple example systems and highlight their similarity and differences. We also give an overview of the research problems and proposed techniques in the areas of query optimization and execution.

1 Introduction

The architecture of Hadoop-based data processing systems and Massively Parallel Processing (MPP) databases are similar in many aspects. Both architectures store big data by slicing it across a large number of shared-nothing independent nodes. Scalability is achieved by parallelizing query evaluation over these independent nodes. Fault tolerance is realized by replicating the same data blocks on multiple nodes.

Earlier technologies have integrated MPP databases and Hadoop-based data processing systems by building connectors that port data from one platform to the other. More recent technologies have achieved tighter integration of both worlds by adopting different approaches. Adapting MPP database systems to run in Hadoop ecosystems, strengthening the capabilities of Hadoop-based query engines to match parallel databases, and building parallel database systems designed specifically for Hadoop are some of the approaches that have been adopted in this respect.

1.1 MPP Query Engines

MPP database systems are primarily based on relational database technologies. A major requirement of these technologies is the existence of a relational schema that describes the structure of different data entities. A schema is defined as a set of relations, where each relation represents a logical data entity. Physically, each relation is stored as a table, which maps to one or more files on disk. The rows in each table capture relation instances, while the columns capture relation attributes. Normalization methods are used to eliminate redundancy and maintain data integrity by enforcing dependencies among relations.

In MPP systems, a database is partitioned into smaller databases stored at different physical machines (nodes) in an underlying cluster. Data partitioning in MPP systems is typically based on horizontal partitioning of data across nodes. Data storage and query evaluation are dual features that need to be provided by each node in the MPP cluster.

The query language in MPP databases is typically the Structured Query Language (SQL), which is a declarative language widely used in various domains for exploring large datasets and implementing business logic. SQL has a number of key strengths including well-defined semantics, industry-grade standards, and expressive power that allows specifying complex query constructs concisely.

Fault tolerance is the ability of the system to recover form disasters or unexpected errors that result in data loss such as disk failures. Fault tolerance in MPP systems is typically achieved by using replicas, which are nodes that maintain a copy of the original data. Consistency between the original nodes and their replicas is achieved by sharing the query logs, or by using publisher/subscriber mechanisms to synchronize data updates.

A given query is transformed by the query optimizer into an execution plan that defines the sequence of steps that need to be performed to compute query results. The optimizer-generated plan includes explicit data movement operations. The cost of moving data is taken into account during optimization.

A query executing in an MPP database can include several pipelined execution stages, with explicit communication between nodes at each stage. Query evaluation is parallelized by running data processing operations in parallel on different nodes. During query evaluation, nodes need to communicate with each other by sending and receiving partial query results.

1.2 Hadoop Query Engines

Hadoop data processing environments adopt a more flexible data model. Input data, both structured and unstructured, following different formats are consumed by the data processing engines in the Hadoop ecosystem. Relationships among data entities and integrity constraints are not strictly enforced. Users are typically required to implement consistency check logic to maintain the correct behavior of their data processing applications.

Query languages in Hadoop environments require writing imperative programs that utilize programming frameworks to implement the data processing task at hand. The earlier programming frameworks provided basic APIs for manipulating key/value pairs. The more recent frameworks introduced advanced programming techniques including functional programming and high-order functions to allow expressing more complex operations in a concise form.

Data partitioning in Hadoop environments is managed by a scalable distributed file system, the Hadoop Distributed File System (HDFS). Data is partitioned into small blocks stored at different nodes in the cluster. In order to achieve fault tolerance, new nodes are automatically added to store data blocks that were stored at failed nodes, while updating the HDFS data location tracking service.

Query execution in Hadoop environments requires building a data processing pipeline that divides the required data processing into multiple stages, and may require materializing the intermediate results between stages. The first widely adopted query execution framework of this nature is MapReduce [5].

The flexibility of Hadoop allows separating data storage and computation across different cluster nodes. In this environment, data could be stored on a subset of cluster nodes, and processed by a different subset of nodes.

1.3   Chapter Organization

We present the state-of-the-art of big data query engines in different environments, and discuss a number of research challenges related to big data query processing. We primarily focus on system architectures and the research challenges in the query optimization and query execution areas.

The remainder of this chapter is organized as follows. In Sect. 2, we present the architectures of some examples of MPP query engines. In Sect. 3, we present the architectures of some examples of Hadoop query engines. In Sect. 4, we present the architectures of some examples of hybrid systems that integrate MPP and Hadoop technologies. We then discuss in Sect. 5 the technical challenges involved in building query optimizers in big data systems. We describe the techniques used in building query executors for big data systems in Sect. 6. We summarize this chapter in Sect. 7 with final remarks.

2   Massively Parallel Query Engines

In this section, we present the system architectures of some examples of MPP query engines. The presented systems are only a small subset of the available solutions. Many of the design principles adopted by the presented systems are common in many query engines.

2.1   Teradata

Teradata [29] is probably the earliest MPP system that was commercially available and widely adopted by many enterprise customers. Teradata database is based on Symmetric Multiprocessing (SMP) technology, combined with a communications network that connects the SMP nodes to provide a Massively Parallel Processing system architecture. A node in Teradata architecture is a combination of hardware and software components running multiple symmetric CPUs. Each node has one or more disk arrays. An MPP configuration is two or more loosely coupled nodes. An interconnect layer, called BYNET, is used to link nodes on the MPP system. BYNET provides the communication means that is needed to provide message exchange among the nodes.

The Teradata MPP architecture is based on the concept of virtualized processors (vprocs) that abstract the details of the underlying physical node. Vprocs are software processes that run on Teradata node. There are multiple types of vprocs including the following:

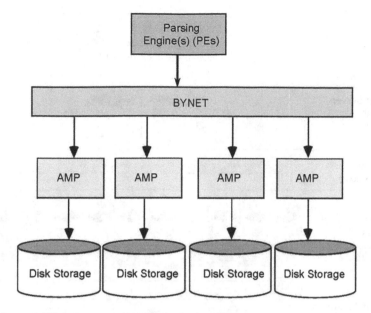

Fig. 1 Access Module Processors (AMPs) in teradata [29]

- Access Module Processor (AMP): A vproc that performs query execution. Each AMP owns a portion of the overall database storage. An AMP exclusively manages a disk space in its underlying node. During query processing, each AMP handles the tasks of sorting, joining and aggregation.
- Parsing Engine (PE): A vproc that performs query parsing and query optimization.

Teradata architecture is shared-nothing. The AMP and PE vprocs share neither memory nor disk across processing nodes. Figure 1 shows the architecture of AMPs in Teradata system.

Each table row is owned by exactly one AMP. That AMP is the only subsystem that can create, read, update, or lock its data. The local control on logging and locking in each AMP enhances system parallelism and reduces BYNET traffic.

Load balance in Teradata database is achieved by distributing table rows evenly across its AMPs and by giving the AMPs the responsibility for the data they own. Teradata is a self-organizing parallel database system. Tables do not have explicit distribution keys. Instead, table rows are hashed across the AMPs of a system using the row hash value of their primary index. When a new row is inserted, the value of its primary index column(s) are used to compute a hash value that maps to one AMP. The row is sent to the selected AMP, where it gets stored in the disk space managed by the AMP.

Users need to carefully choose the primary index for each table, so that rows that are frequently joined hash to the same AMP. This is crucial for performance as it allows eliminating the need to send rows across the BYNET in order to join them.

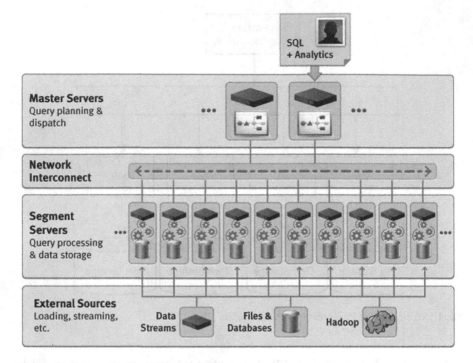

Fig. 2 Greenplum database architecture [23]

2.2 Greenplum

Pivotal Greenplum Database (GPDB) [23] is a massively parallel processing analytics database. GPDB adopts a shared-nothing computing architecture with two or more cooperating processors. Each processor has its own memory, operating system and disks. GPDB leverages this high-performance system architecture to distribute the load of petabyte data warehouses, and use system resources in parallel to process a given query. Figure 2 shows a high level architecture of GPDB.

Storage and processing of large amounts of data are handled by distributing the load across several servers or hosts to create an array of individual databases, all working together to present a single database image. The master is the entry point to GPDB, where clients connect and submit SQL statements. The master coordinates work with other database instances, called segments, to handle data processing and storage. When a query is submitted to the master, it is optimized and broken into smaller components dispatched to segments to work together on delivering the final results. The interconnect is the networking layer responsible for inter-process communication between the segments.

When creating a new table in GPDB, the table can be associated with a distribution method. The supported methods include hashed distribution and random distribution.

The goal of data distribution is to spread data across many nodes and disks in order to scale out query processing.

During query execution, data can be distributed to segments in multiple ways including hashed distribution, where tuples are distributed to segments based on some hash function, replicated distribution, where a full copy of a table is stored at each segment and singleton distribution, where the whole distributed table is gathered from multiple segments to a single host (usually the master).

A query is submitted to the master node, where it gets parsed and optimized into a query plan. The master node obtains catalog information required to optimize the given query by looking up data objects in the system catalog. After query plan is created, a copy of the plan is dispatched to each node.

The query plan is an operator tree that captures the order of query execution. During query execution, a leaf operator in the query plan, running on a given node, reads data from its local node storage and returns data to upper level operators. Upper level operators carry on query execution while communicating with query operators running on other nodes. This communication is enabled through the interconnect layer.

2.3 Vertica

Vertica [16] is the commercialized analytical database system resulting from the C-Store research project [28]. The architecture of Vertica is based on the concept of column projections, which are restricted forms of materialized views. A column projection is a sorted subset of the attributes of a table. The physical data organization in Vertica stores tables as column projections, rather than self-contained units. Each projection is possibly stored in a separate data file that could be compressed. A super projection that contains all table columns is also maintained to allow for reconstructing the full table when needed.

The main insight behind using column projections is that in many analytical queries, users are only interested in a subset of table columns. Reading data from full tables is not often required. Hence, by materializing and sorting vertical fragments of each table, there is a good chance that a small subset of these fragments is sufficient to answer a given query, without the need to scan and read the full tables.

Vertica supports join operations by materializing prejoin projections. By physically materializing the join results between a fact table and multiple dimension tables, many join queries have an improved performance. However, this comes with the expense of maintaining the prejoins in the physical storage and keeping them in sync with other projections in the presence of data updates. This cost can be lessened by employing compression and encoding techniques.

Figure 3 illustrates the high level design of column projections in Vertica using a sample sales table. Two projections are maintained for sales table. The first projection is a super projection, while the second projection contains two columns. Both projections can be horizontally partitioned and stored at multiple nodes to allow for parallel query processing.

Fig. 3 Column projections
in Vertica [16]

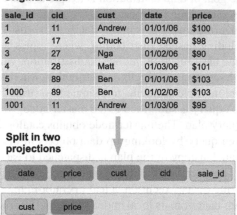

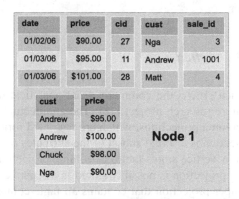

(a) Partitioning a table into projections

(b) Maintaining projection fragments on several
nodes

3 Hadoop Query Engines

In this section, we describe the architectures of a number of Hadoop query engines. Earlier Hadoop query engines required implementing elaborate imperative programs for performing data processing tasks. The underlying framework provided simple APIs as the building blocks of these programs. More recent solutions extended Hadoop programming frameworks with more advanced techniques such as functional programming. A number of Hadoop query engines have also considered adding a declarative SQL-like interface to be used as the main query language for writing big data applications.

3.1 MapReduce

MapReduce [5] is a programming framework that can be used for processing large data sets. The input to a MapReduce program is stored in a distributed file system as a set of partitions stored on different nodes in the computing cluster. The framework requires users to specify two basic functions:

- Map: A function that transforms a key/value pair into a set of intermediate key/value pairs. A typical map function processes input records by applying filtering and transformation operations, and generates as output a set of intermediate key/value pairs. Before the map function produces its output, a hash function is used to split output records into a set of disjoint partitions. The map function eventually generates a file for each partition and stores these files to the disk of local nodes.
- Reduce: A function that merges intermediate values that are associated with the same key, producing potentially a smaller set of values compared to the input. Each instance of the reduce function reads a number of partition files from a number of cluster nodes. Each partition file is consumed by one instance of the reduce function. The output of the reduce function is also stored to a file in the distributed file system.

Programs written using the previous framework can be easily parallelized and executed on a large cluster of commodity machines.

The structure of input files in MapReduce programs must be handled by the user program. While support for simple data types of key/value pairs is readily available, parsing and processing more complex structures requires adding logic to user program to perform these operations. When input data is shared by multiple applications, the structure of the input files must be enforced and maintained by some external service [21]. This problem does not usually exist in database systems, where catalog services are used to abstract the definition and maintenance of metadata (description of how data is structured) from the logic of the underlying applications.

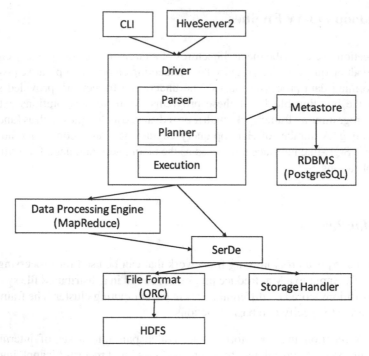

Fig. 4 The architecture of Hive [12]

3.2 Hive

Hive [12, 30] is a query translation layer on top of MapReduce. The design of Hive exposes a dialect of SQL to Hadoop users. The expressed queries are translated to graphs of MapReduce jobs. The expressiveness power of SQL allows users to formulate complex queries in a declarative manner, without the need to write MapReduce programs that specify the exact operations to be performed.

While abstracting the query language interface is key to improve the usability of Hadoop query engines, two additional aspects are also addressed by Hive to improve the productivity of Hadoop query engines:

- Storage: storing datasets in an efficient format that allows fast data access and better use of storage resources.
- Query optimization and execution: A sophisticated query optimizer and an advanced query executor are highly needed to improve query response times and optimize utilization of computation resources.

Figure 4 shows the architecture of Hive [12]. Two interfaces are exposed to clients to submit queries: (1) Command Line Interface (CLI), and (2) HiveServer2. The client facing interface is used to submit queries to the Driver component. The Driver component transforms the incoming query into a parse tree representation. The Planner component receives the query parse tree and then, based on the type of submitted query, a specific planner implementation is chosen to analyze the given query parse tree. During query analysis, catalog information regarding metadata objects (e.g., table definitions) may be needed. This information is obtained by contacting the Metastore, which is implemented using a Relational Database Management System (RDBMS), typically PostgreSQL [25].

The Planner component in Hive generates a directed acyclic graph of MapReduce tasks that represents the required data processing operations. The graph is submitted to the Data Processing Engine for execution.

To read/write a table with a specific file format, Hive assigns the corresponding file reader/writer to MapReduce tasks reading/writing this table. For a given file format, a serialization-deserialization library (SerDe) is used to serialize and deserialize data.

After all MapReduce jobs have finished, the Driver fetches the results of the query, and sends the results back to the query client. Besides processing data directly stored in HDFS, Hive can also process data stored in other storage systems, e.g. HBase [11]. For those systems, a corresponding Storage Handler is needed. For example, the HBase storage handler is used when a query needs to read or write data from or to HBase.

3.3 Spark

The key advantage of the MapReduce framework is the seamless scalability, fault tolerance and parallelization capabilities it provides for big data applications. However, managing distributed memory in the computing cluster is missing in MapReduce. Intermediate results are typically stored to temporary files on disks and may not be shared across applications. This means that it is not possible to efficiently reuse previous computations in new data processing tasks.

A number of classes of big data applications can largely benefit from reusing previous computations. For example, in machine learning and graph algorithms, iterative computation is used to analyze and process a large dataset. In each iteration, the previously computed values are used to generate new values. The intermediate computation used by one application may also be usable in other applications. Data reuse is highly needed to improve the efficiency of this type of applications.

Apache Spark [33] is a Hadoop-based computational framework that mainly targets iterative applications. The framework provides a general-purpose query engine, a set of functional programming APIs, and a number of libraries for streaming, graph processing and machine learning.

The functional programming APIs provided by Spark allow users to manipulate distributed data collections called Resilient Distributed Datasets (RDDs). An RDD is a collection of data objects that are partitioned across a cluster. A number of operations can be applied to RDDs including map, filter, and reduce operations.

RDDs maintain their lineage information in a reliable way. The lineage is the sequence of operations that were performed to construct an RDD. Maintaining lineage information is the key for fault tolerance in Spark, which allows recovering lost data using the lineage graph of the RDDs. Replaying the sequence of operations encoded in the RDD lineage allows reconstructing an RDD in the event of data loss.

The query evaluation on RDD is performed lazily. All data processing operations that do not require producing output are implicitly encoded in the RDD lineage, but not actually performed. When an output operation, for example producing some aggregated value, is requested, the sequence of data processing operations (encoded in the lineage) is triggered.

A first effort towards building a relational query interface on top of Spark was a system called Shark [31], which is a shorthand for Spark on Hive. Shark modified the Apache Hive system to run on Spark as the underlying computation framework that is used instead of MapReduce. Traditional query optimizations, such as columnar processing, are implemented on top of the Spark engine. However, Shark could only be used to query external data stored in the Hive catalog. This limitation means that it was not possible to use Shark for processing data inside a Spark user program.

A recent solution has addressed this limitation by introducing a native SQL layer on top of Spark. The resulting system is called SparkSQL [2]. We discuss the details of query optimization in SparkSQL in Sect. 5.3.

4 SQL on Hadoop

While Hadoop's scalability and fault-tolerance match the requirements of running analytics on big data, the need to formulate analytics as complex MapReduce programs as well as the lack of support to interactive data exploration (with short response times) were considerable limitations.

To address these limitations, SQL-like declarative languages, such as HiveQL in Hive [30], were first developed in a query language layer on top of Hadoop. HiveQL queries are compiled into MapReduce jobs. While HiveQL improved the usability of Hadoop ecosystem for running complex big data analytics, the generated MapReduce jobs had typically poor performance. It became obvious that more advanced query optimization and execution techniques are still lacking in Hadoop environments.

A number of systems (e.g., HAWQ [24], Impala [14] and Presto [3]) have implemented from scratch (or ported) full-fledged relational query engines to work on top of HDFS. There are two main advantages of these systems, compared to the native Hadoop-based query engines:

- Query optimizers are sophisticated enough to generate more efficient query plans. This means that the probability of running analytics based on bad execution plans, which often translates to long execution times, is relatively small.
- Query execution is not following the MapReduce execution style. This means that using these query engines for data exploration in Hadoop is possible. A key advantage of these engines is pipelining intermediate query results without the need to materialize them, in contrast to MapReduce.

In this section, we describe the system architectures of some examples of SQL-on-Hadoop systems, and highlight their main similarities and differences.

4.1   HAWQ

HAWQ [24] is a massively parallel processing SQL engine on top of HDFS. HAWQ originated as a redesign of Greenplum database into a hybrid of MPP database and Hadoop technologies. The layered architecture of Greenplum database (Fig. 2) is adopted and reused to build an SQL engine that relies on the Hadoop distributed file system for data replication and fault tolerance.

By building on the extensive query language and optimization capabilities of Greenplum database, HAWQ has a high degree of SQL standard compliance as well as extensive query optimization capabilities. In particular, HAWQ employs Orca [27], an industry-grade optimizer, at its core, to devise efficient query plans minimizing the cost of accessing data in Hadoop clusters. The architecture of HAWQ combines the benefits of using a state-of-the-art cost-based optimizer with the scalability and fault-tolerance of Hadoop to enable interactive processing on Big Data in Hadoop environments. We describe the architecture of Orca query optimizer in Sect. 5.2.

The main difference between HAWQ and Greenplum Database is the underlying data storage characteristics. Greenplum database assumes dedicated data servers that handles data storage, while HAWQ runs on a cluster of commodity machines. HAWQ achieves its scalability and fault tolerance by relying on Hadoop's ability to seamlessly scale the underlying cluster, and its built-in replication mechanisms.

The first design of HAWQ [4], shown in Fig. 5, re-architected various components of Greenplum database including distributed transaction processing, fault tolerance, unified catalog service and metadata dispatch. HAWQ's architecture is based on three type of nodes: master, HDFS name node, and segment nodes. The segment nodes run both HAWQ's compute units (each compute unit manages a physical partition of the database) as well as the typical HDFS data nodes.

The master node is the main entry point of the system, where user queries are submitted. The master node is responsible for authenticating users based on given credentials, parsing incoming queries, invoking query optimizer to produce execution plan, and finally dispatching query plans to segment nodes to initiate query execution. The master node has an accompanying standby master, which is kept in-sync with the primary master by sharing log files.

Each physical node in HAWQ's cluster runs an HDFS data node as well as multiple compute units. The compute units leverage multi-core architecture during query execution. In a typical configuration, each compute unit is assigned to a processor core in the physical node that the unit runs on. This allows operations like scanning input data to be conducted in parallel by initiating multiple compute units on the same physical node, each utilizing a dedicated processing core.

HDFS data nodes are collocated with compute units on the same physical machines to leverage data locality. Nodes access HDFS storage layer through lib-

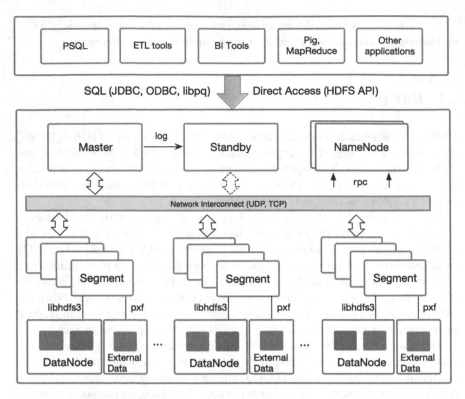

Fig. 5 Initial HAWQ architecture [4]

hdfs3, an HDFS library. Pivotal extension framework (PXF) is an extensible framework, which enables SQL access to external data sources such as HBase [11] and Hive [12].

A major requirement that was not addressed in the first design of HAWQ is elasticity within Hadoop ecosystem. This is important to allow HAWQ to share resources with other Hadoop tenants. Computing and storage resources should be dynamically allocated during query execution by negotiating with Hadoop resource managers.

The redesign of HAWQ is called Apache HAWQ [22] and its architecture is shown in Fig. 6. In this architecture, on each physical host, HAWQ runs a segment node, an HDFS data node and a node manager. The master nodes of Apache HAWQ, HDFS and YARN [32] (Hadoop utility for resource management) run on separate nodes. Apache HAWQ is integrated with YARN for resource management.

When a query is submitted to Apache HAWQ, a set of virtual segments are allocated according to the estimated cost of a query, as given by HAWQ's query optimizer, and the current usage of resources in the Hadoop cluster, as given by YARN. After virtual segments are determined, the query is dispatched to the corresponding physical hosts. The set of physical hosts that will execute a query can be a subset

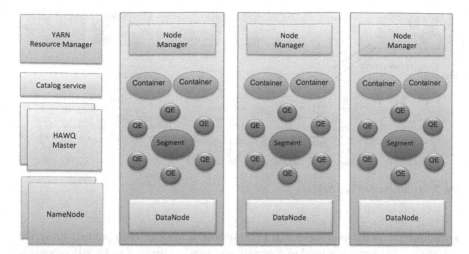

Fig. 6 Apache HAWQ architecture [22]

of physical hosts of the whole Hadoop cluster. The HAWQ resource enforcer on each node monitors and controls the real time resources used by the query to avoid resource usage violations. Nodes can be added dynamically without data redistribution. When a new node is added, the HAWQ master is notified to make the additional resources given by the new node visible for future queries.

4.2 Impala

Impala [14] is a massively-parallel query execution engine, developed specifically to run on Hadoop clusters. The design of Impala decouples the query engine from the underlying storage engine. The architecture of Impala is based on multiple components that interact together to provide query processing and data storage functionalities. The architecture is shown in Fig. 7. The architecture is based on three main services:

- Impalad: A daemon service that is responsible for accepting queries from client processes and managing query execution across the cluster. This daemon service is also responsible for executing individual query fragments on behalf of other Impalad's. One Impalad is deployed on every machine in the cluster. When an Impalad operates in the first role (managing query execution), it is said to be the coordinator for that query. All Impalad's may operate in all roles. This property is important for fault-tolerance and load-balancing. Each node also runs a datanode process, which is responsible for accessing data in the underlying HDFS. This allows taking advantage of data locality by reading from the filesystem without going through the network.

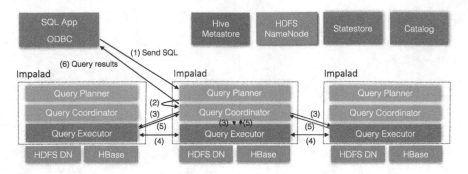

Fig. 7 Architecture of Impala [14]

- Statestored: A daemon service that is responsible for metadata publish-subscribe service. This service is used to disseminate metadata updates to Impala processes in the cluster. There is a single statestored instance.
- Catalog: A daemon service that serves as the catalog repository. Using catalog, Impala daemons may execute data definition language (DDL) commands (e.g., creating data objects such as relational tables). The object definitions are reflected in external catalog stores such as the Hive Metastore. Changes to the system catalog are broadcast via the statestore.

Impala synchronizes cluster-wide metadata by leveraging its symmetric-node architecture, where all nodes are able to accept and execute queries. Therefore, a fresh version of system-wide catalog must be obtainable by all nodes at any point of time. A holistic view of resource utilization in the system must also be available at all nodes, so that queries can be optimized and scheduled properly.

These capabilities are provided by pushing metadata updates to interested nodes. This is implemented by the satestore, which is a publish-subscribe service that pushes metadata updates to a set of subscribers. The statestore maintains a table that stores a set of topics. Processes that are interested in receiving updates on a particular topic subscribe with the statestore service.

After registration, the statestore periodically sends two message types to each subscriber:

- Topic update message: A message that has all changes to a topic since the last update. In response to a topic update, each subscriber sends a list of changes in its subscribed topics. Those changes are guaranteed to be applied by the time the next update is received.
- Keepalive message: The statestore uses keepalive messages to maintain the connection with subscribers. When keepalive message is not received from a subscriber after some time threshold, the subscription times-out, and the subscriber needs to re-register with statestore.

We discuss query optimization and query execution techniques in Impala in Sects. 5.5 and 6.4, respectively.

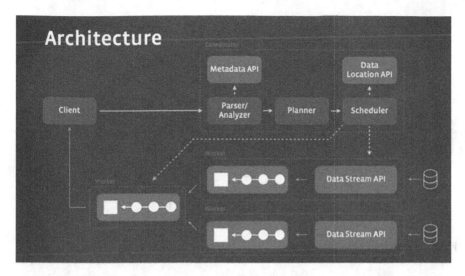

Fig. 8 Architecture of Presto [3]

4.3 Presto

Presto [3] is a distributed SQL query engine optimized for interactive processing in Hadoop clusters. Unlike MapReduce framework, which is a computation framework primarily designed for batch processing, Presto is designed to support data exploration tasks that typically involve running analytics with low latency requirements.

Figure 8 shows the architecture of Presto. The query client submits SQL statements to the coordinator component. The submitted queries are parsed, optimized and turned into query execution plans. While optimizing a query, a Metadata interface to Hive Metastore is used to obtain definitions of data and query objects such as tables and functions.

The query execution plan is turned by the scheduler component into an execution pipeline. The pipeline assigns work to nodes closest to the data. The query execution progress is monitored by the scheduler. The pipeline workers produce output by pulling data from the previous workers, apply corresponding data processing tasks, and sending the results upon request to later stages in the pipeline. An installation of Presto's components is shown in Fig. 9.

Presto query execution engine does not spawn MapReduce jobs. The query execution model employs a custom engine with special query operators designed to support SQL operations. Query processing is done in memory and intermediate results are pipelined across the network between stages. This avoids unnecessary I/O where intermediate results need to be written to disk between query processing stages. The pipelined execution model in Presto can run multiple stages at once, and streams data from one stage to the next as it becomes available. This reduces end-to-end latency of queries. However, in-memory query execution in Presto has its pitfalls

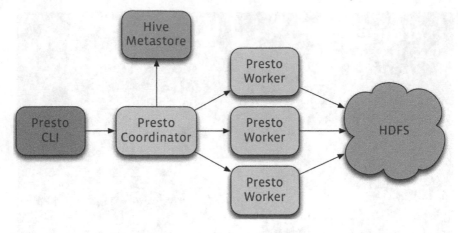

Fig. 9 Components of Presto [3]

when processing a large data set that exceeds the total available memory. In this case, query execution fails.

Presto has a storage abstraction that allows using disparate data sources. Storage connectors, to different types of data sources are designed by writing interfaces for fetching metadata getting data locations, and accessing the contents. The storage connectors support ingesting data from multiple sources including Hive and HBase.

Presto is offered as a cloud service by Qubole [26]. By using Presto cloud service, analytics can run on cloud data storage (e.g., Amazon storage service). This can be useful for running interactive workloads for data exploration, or integrating data from multiple sources by using storage connectors. The cloud elasticity is leveraged so that resources needed for data storage and query execution dynamically shrink or expand depending on the requirements of each query workload.

5 Query Optimization

The job of a query optimizer is to turn a user query into an efficient query execution plan. The optimizer typically generates the execution plan by considering a large space of possible alternative plans and assigning an estimated cost to each alternative in order to pick the cheapest one. The art of query optimization is a combination of technically challenging problems (e.g., plan enumeration, cost estimation, statistics derivation, property enforcement and join ordering), some of them are known to be NP-Hard.

Query optimizer is one of the most performance-sensitive components in a database system. Differences in query plans may result in several orders of magnitude of difference in query performance, significantly more than any other contributing factor. Big data has triggered a renewed interest in query optimization techniques as a set of principled methods that could, when combined with scalable hardware resources and robust execution engines, make analyzing and processing petabytes of data finally possible. The increased amounts of data that need to be processed in today's world stress on the importance of building an intelligent query optimizer at the core of any query engine.

The impact of having a good optimizer on query performance is known to be substantial. The extensive research done by the database community in query optimization over the past decades provides a plethora of techniques that can be leveraged and adopted in new data processing environments such as Hadoop.

In this section, we discuss query optimization in big data systems. We start by presenting a number of query optimization research problems. We then describe the architectures of query optimizers of some examples of big data query engines. We highlight the main similarities and differences of discussed optimizers, and present a number of technical problems tackled by these optimizers.

5.1   Research Problems

Building a query optimizer is not an easy undertaking. It takes a lot of design and development efforts to get the query optimizer in the required shape. There are many challenges involved in the design of query optimizers in big data query engines. In this section, we give an overview of some of these challenges.

Modularity. Building query optimizer as a complex monolithic software component does not allow adapting the optimizer design to the changing processing environments of big data. New data formats and query execution engines are constantly adopted in big data domains. The optimizer design needs to be modular to allow plugging new components that consume data in new formats, and exploit the query processing capabilities of new query execution engines.

An crucial interface to query optimizer is metadata acquisition. The optimizer makes extensive use of metadata (e.g., table/index definitions and data statistics) during plan enumeration, transformation rules, statistics derivation and cost computation. Using a highly extensible abstraction of metadata and system capabilities is important to allow query optimizer to deal with new data types and processing environments. Designing the optimizer's statistical and cost models as pluggable components is an important challenge that needs to be addressed in this respect.

Extensibility. Having all elements of a query and its optimizations as first-class citizens in optimizer's architecture is an important challenge that impacts optimizer design. When an optimizer is designed based on a strict set of optimizations and query elements, adding new optimizations becomes technically hard. This often leads to resorting to multiple optimization phases, where the decisions made in earlier phases are revisited in later phases.

One example of new requirements that were triggered by big data processing environments is data distribution. In big data environments, scalability is realized by distributing data load across machines in the cluster. Such requirement did not exist in earlier database systems. Optimizers ported from older systems, or the ones that adopted older designs handle data distribution by adding a plan parallelization phase, where data distribution is considered after some optimization decisions have been already made.

Multi-phase optimizer design impacts the quality of final query plan since the available information does not impact all optimizer's decisions. Multi-phase optimizers are notoriously difficult to extend as new optimizations or query constructs often do not match the predefined phase boundaries.

Addressing this challenge by abstracting query elements, data properties and query optimizations as first-class constructs that optimizer treats similarly can avoid the problems of multi-phase optimization where certain constructs are dealt with as an afterthought.

Exploiting multi-core architectures. Query optimization is probably the most CPU-intensive operation that a query engine performs. The space of plan alternatives is combinatorial by definition. There are pressing needs for efficient exploration and pruning of the plan space in order to produce high-quality execution plans.

Exploiting multi-core architectures is an important tool that allows optimizers to scale. By dividing the optimization tasks into a set of work units, using multiple CPU cores for query optimization becomes possible.

A crucial challenge in this respect is formulating and capturing dependencies among optimizer's work units, and designing a work unit scheduler that assigns optimization tasks to processing cores and efficiently gathers the computation results of these units to generate the optimizer's plan space.

Testing optimizer accuracy. Evaluating the accuracy of query optimizers objectively is a difficult problem. Benchmarks developed for assessing the query performance test the system as a whole end-to-end. However, no benchmarks are currently available to test a query optimizer in isolation [10]. Being able to compare the accuracy of optimizers across different products independently is highly desirable.

The most performance-critical element in a cost-based optimizer is probably the accuracy of its cost model as it determines how to prune inferior plan alternatives. There is no standard way to test an optimizer's accuracy. The cost units used in the cost model and displayed with the plan do not reflect anticipated wall clock time but are used only for comparison of alternative plans pertaining to the same input query. Comparing this cost value with the actual execution time does not permit conclusions about the accuracy of the cost model.

Moreover, the optimization results are highly system-specific and therefore defy the standard testing approach where results are compared to a reference or baseline to check if the optimizer finds the 'correct' solution: the optimal query plan for System A may widely differ from that for System B because of implementation differences in the query executors and the optimizers. These differences can lead to choosing radically different plans. Building testing infrastructure for evaluating query optimizers is an important research problem.

Fig. 10 Interaction between
Orca and database
system [27]

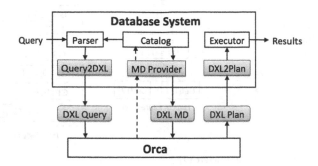

5.2   Orca

Orca [1, 6, 20, 27] is the query optimizer of Pivotal's data management products, including Greenplum database [23] and HAWQ [24]. While many query optimizers are tightly-coupled with their host systems, a unique feature of Orca is its ability to run outside the database system as a stand-alone optimizer. This ability is crucial for supporting query engines with different computing architectures (e.g., MPP and Hadoop) using one optimizer. It also allows leveraging the extensive legacy of relational optimization in new query processing paradigms. Furthermore, running the optimizer as a stand-alone product enables elaborate testing without going through the monolithic structure of a database system.

Figure 10 shows the interaction between Orca and an external database system. The database system needs to include translators that consume/emit data using Data Exchange Language (DXL), which is an XML-based language that is used to define an interface for accessing Orca. The Query2DXL translator converts a query parse tree into a DXL query, while the DXL2Plan translator converts a DXL plan into an executable plan. The implementation of such translators is done completely outside Orca, which allows multiple systems to interact with Orca by providing the appropriate translators.

The input to Orca is a DXL query. The output of Orca is a DXL plan. During optimization, the database system can be queried for metadata (e.g., table definitions). Orca abstracts metadata access details by allowing database system to register a metadata provider (MD Provider) that is responsible for serializing metadata into DXL before being sent to Orca. Metadata can also be consumed from regular files containing metadata objects serialized in DXL format.

The design of Orca is based on the Cascades optimization framework [9]. The space of plan alternatives generated by the optimizer is encoded in a compact in-memory data structure called the Memo. The Memo structure consists of a set of containers called *groups*, where each group contains logically equivalent expressions. Memo groups capture the different sub-goals of a query (e.g., a filter on a table, or a join of two tables). Group members, called *group expressions*, achieve the group goal in different logical ways (e.g., different join orders). Each group expression is

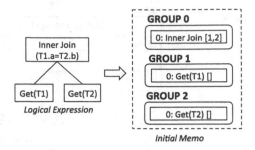

Fig. 11 Memo structure [9]

an operator that has other groups as its children. This recursive structure of the Memo allows compact encoding of a huge space of possible plans.

A DXL query message is shipped to Orca, where it is parsed and transformed to an in-memory logical expression tree that is copied-in to the Memo. For example, Fig. 11 shows the logical expression of a simple inner join query between two tables T1 and T2. The logical expression creates three groups for the two tables and the Inner-Join operation. Group 0 is called the *root group* since it corresponds to the root of the logical expression. The dependencies between operators in the logical expression are captured as references between groups. For example, Inner-Join [1, 2] refers to Group 1 and Group 2 as children.

Orca optimizes queries in top-down fashion by computing optimization requests in the Memo. Each optimization request specifies physical properties (e.g., data distribution and sort order) that need to be satisfied. A Memo group is optimized under a given optimization request by finding the best plan, rooted at the group, that satisfies the required properties at the least cost.

Figure 12 shows how Orca processes optimization requests in the Memo for a simple join query between two tables T1 and T2 based on the join condition (T1.a = T2.b). The query results are required to be sorted on column T1.a. Assume

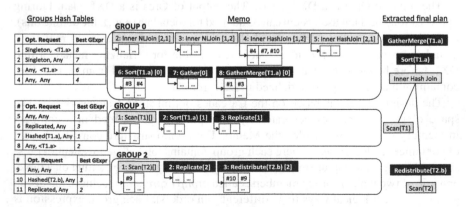

Fig. 12 Processing optimization requests in Orca [27]

that relation T1 is hash-distributed on column T1.a, and relation T2 is hash-distributed on column T2.a.

The initial optimization request is req. #1: {Singleton, $< T1.a >$}, which specifies that query results are required to be gathered to a single node (typically the master) based on the order given by column T1.a. The group hash tables are shown, where each request is associated with the best group expression (GExpr) that satisfies it at the least estimated cost. The black boxes indicate enforcer operators that are plugged in the Memo to deliver sort order and data distribution:

- Gather operator gathers tuples from all segments to the master.
- GatherMerge operator gathers sorted data from all segments to the master, while keeping the sort order.
- Redistribute operator distributes tuples across segments based on the hash value of given argument.
- Sort operator enforces a given sort order to the partition of tuples residing on each node.

Figure 13 shows the detailed optimization of req. #1 by InnerHashJoin[1,2], which is a hash-based algorithm for joining two relational inputs given by groups 1 and 2. For the join condition (T1.a = T2.b), one of the alternative plans is aligning child distributions based on join condition, so that tuples to be joined are co-located. This is achieved by requesting Hashed(T1.a) distribution from group 1 and Hashed(T2.b)

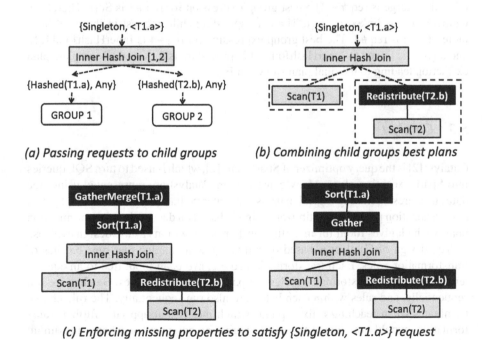

(a) Passing requests to child groups   (b) Combining child groups best plans

(c) Enforcing missing properties to satisfy {Singleton, <T1.a>} request

Fig. 13 Plan generation in Orca [27]

distribution from group 2. Both groups are requested to deliver ANY sort order. After child best plans are found, InnerHashJoin combines child properties to determine the delivered distribution and sort order. Note that the best plan for group 2 needs to hash-distribute T2 on T2.b, since T2 is originally hash-distributed on T2.a, while the best plan for group 1 is a simple Scan, since T1 is already hash-distributed on T1.a.

When it is determined that delivered properties do not satisfy the initial requirements, unsatisfied properties have to be *enforced*. Property enforcement in Orca in a flexible framework that allows each operator to define the behavior of enforcing required properties based on the properties delivered by child plans and operator local behavior.

Enforcers are added to the group containing the group expression being optimized. Figure 13 shows two possible plans that satisfy req. #1 through property enforcement. The left plan sorts join results on segments, and then gather-merges sorted results at the master. The right plan gathers join results from segments to the master, and then sorts them. These different alternatives are encoded in the Memo and it is up to the cost model to differentiate their costs.

Finally, the best plan is extracted from the Memo based on the linkage structure given by optimization requests. Figure 12 illustrates plan extraction in Orca. The local hash tables of relevant group expressions are illustrated. Each local hash table maps incoming optimization request to corresponding child optimization requests. The best group expression of req. #1 in the root group is first looked-up, which leads to GatherMerge operator. The corresponding child request in the local hash table of GatherMerge is req #3. The best group expression for req #3 is Sort. Therefore, GatherMerge is linked to Sort. The corresponding child request in the local hash table of Sort is req #4. The best group expression for req #4 is InnerHashJoin[1,2], and so Sort is linked to InnerHashJoin. This procedure is followed to complete plan extraction leading to the final plan shown in Fig. 12.

5.3 Catalyst

Catalyst [2] is the query optimizer of SparkSQL [2], which is used to turn SQL queries into Spark execution plans. Many constructs in Catalyst are represented using tree data structures, including logical expressions, intermediate expressions and physical query execution plans. The optimizer design is based on decoupling the optimization phases which gives room for including further optimizations as the design evolves.

The design of Catalyst is based on a query rewrite engine that runs batches of transformation rules. Each transformation rule converts an input tree to an equivalent output tree. Transformation rules are partitioned into batches. Batches are run sequentially, and rules within each batch are also run sequentially. The rule engine terminates when reaching a fixed point, which means that applying further transformations would not change the input expression anymore, or when a maximum number of iterations is reached.

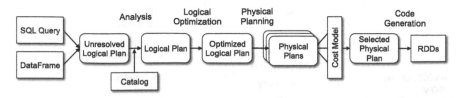

Fig. 14 Query optimization phases in SparkSQL [2]

Query optimization in Catalyst takes place in a number of sequential phases, as depicted in Fig. 14. We describe in the following the different phases that Catalyst optimizer goes through to produce the final query execution plan. We give a step-by-step illustration using the example shown in Fig. 15.

- Query Parsing: A HiveQL-based parser is used to generate an abstract syntax tree for an incoming SQL query. Figure 15a shows the tree representation of an example SQL query.
- Metadata Lookup: The abstract syntax tree is preprocessed by resolving table names referenced in the query by looking up the MetaStore to obtain table definition. Figure 15b shows a MetaStoreRelation created for an unresolved relation in the query after looking up the MetaStore.
- Resolving References: A globally unique identifier is assigned to each attribute (column) referenced in the query tree. Then, derivation of data types of every data item in the query tree takes place. Figure 15c shows unique numerical identifiers assigned to columns in the query tree.
- Logical Planning: Multiple logical rewrites are conducted to simplify the input query tree. These rewrite operations include the following:

  - Removing redundant query operators.
  - Constant folding, where inlined constant expressions are reduced to simple atomic values.
  - Filter simplification, where trivial filters are removed, and subtrees that are known to return no results are pruned.
  - Filter push down, so that filtering operations are applied as early as possible.
  - Eliminate unused column references that are not needed to compute the final query answers.

  Figure 15d shows the transformed query tree after applying the previous logical rewrites. For example, the Filter operator has been pushed down, while the Subquery operator was eliminated because of being a redundant operator.
- Physical Planning: In this phase, the optimized logical plan is transformed into a number of possible physical plans. This is done by recognizing operators in the logical plan that can be directly transformed into physical counterparts. For example, a filter on a MetaStore relation can be transformed into HiveTableScan. In addition, a list of strategy objects, each of which can return a list of physical plan options, is used to transform logical operators to equivalent physical ones.

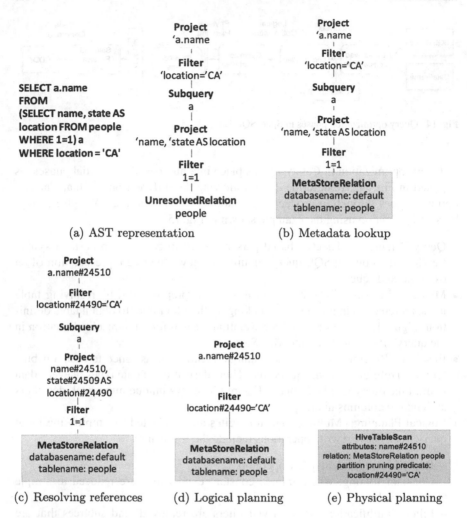

Fig. 15 Query optimization example in SparkSQL

Examples of strategy objects include HashJoin and NestedLoopJoin strategies. If
a given strategy object is unable to plan all of the remaining operators in the tree,
it creates a placeholder to be filled by other strategy objects. Figure 15e shows the
final physical plan, which is reduced to a simple HiveTableScan with an embedded
filter.

- Plan Parallelization: In this phase, data exchange operators are plugged into the
  generated physical plans to establish the required data distributions. Distribution
  requirements could arise from operator local requirements, or from query-specific
  requirements. Physical operators may not be able to establish data distributions on
  their own. In this case distribution enforcers are used to guarantee delivering the
  required data distribution. The enforced distributions include the following:

- Clustered distribution: Tuples that share the same values of clustering expressions are co-located.
- Ordered distribution: Tuples are sorted according to ordering expressions.
- All Tuples distribution: A single partition is created where all tuples are co-located.

Similar to many other operations, distribution enforcement is a transformation rule in SparkSQL. A transformation rule is used to ensure that distribution of input data meets the distribution requirements of each operator by inserting Exchange (data movement operators) when required. Given a physical plan, the Add-Exchange rule performs the following operations:

- Check if every child distribution satisfies the distribution required from that child
- Check if children distributions are compatible with each other. For example, in order for a join operation to be conducted correctly, the data of joined inputs must be either co-located or hashed to the same nodes, or, alternatively, one of the two inputs needs to be replicated to all nodes.
- If a child does not meet required distribution, or is not compatible with other children, an exchange operator is added.

5.4 V2Opt

V2Opt is the latest query optimizer of Vertica MPP database system [16]. The goal of building a new optimizer in Vertica database is to integrate the awareness of data distribution, sort orders, and non-star schemas into all optimizer decisions.

The design of V2Opt is extensible and allows adding new optimizations with small modifications to the codebase. In V2Opt, the physical properties of a given query are first identified. These properties includes column selectivity, sort order and integrity constraints. The identified physical properties are used by a cost-based pruning strategy. The underlying cost model, used in pruning, combines different factors including data compression and CPU/Network transfer costs. The objective of pruning is to retain the most important properties that need to influence optimizer decisions. The collected properties are used to reduce optimization search space, and integrate data distribution into join order enumeration phase. By having optimizer extensibility mechanisms in place, new physical properties could be added to the optimizer without changing the entire architecture.

5.5 Impala Query Optimizer

The query optimizer of Impala [14] generates execution plans by following a two-phase approach:

- Phase 1 (Single node planning): In this phase, a query plan that runs on a single node (i.e., without considering data partitioning in the Hadoop cluster) is generated.
- Phase 2 (Plan parallelization): In this phase, a parallelized query plan that processes distributed data in the cluster is generated.

In the first optimization phase, the query parse tree is translated into a non-executable single-node plan tree. The single node plan consists of different types of operators including table scan, join, aggregation, sort and analytic evaluation functions. In this phase, filters specified in the given query are pushed down to be applied as close as possible to data sources. This is important to early-prune parts of the data that are not needed to compute the final query answer. New filters could also be inferred from existing filters to prune data more aggressively. Another important optimization that takes place in this phase is join ordering, where an efficient evaluation order of relational joins is identified. Heuristics are typically used to avoid exhaustive enumeration of the join orderings space.

In the second optimization phase, the single-node plan is turned into a distributed (parallel) execution plan. The optimization objective of this phase is to minimize data movement and maximize scan locality.

Data movement is controlled by adding Exchange operators between plan nodes, and by adding extra non-exchange plan nodes to minimize data movement across the network (e.g., local aggregation nodes). During this second phase, the join strategy for every join node is decided, including broadcast (one input of the join is replicated on all nodes) and partitioned/hashed (the two join inputs are partitioned across nodes using the join expression, so that tuples that would join together reside on the same node).

All aggregations are executed as a local aggregation followed by a merge aggregation operation. The local aggregation output is partitioned on the grouping expressions and the merge aggregation is done in parallel on all nodes. Sort is parallelized similarly using a local sort followed by a single-node merge operation.

At the end of the second phase, the distributed plan tree is split into a set of plan fragments at the boundaries given by the exchange operators. Each plan fragment constitutes a unit of execution encapsulating a portion of the plan tree that operates on the same data partition on a single machine.

Figure 16 gives an example for a query joining two HDFS tables (t1, t2) and one HBase table (t3) followed by an aggregation and sort with a limit, where only the top-n sorted answers are required. The single-node plan is shown on the left, while the distributed and fragmented plan is shown on the right. The colored rounded rectangles indicate plan fragments and arrows indicate data exchange operations.

The table data is randomly partitioned. Tables t1 and t2 are joined using the partitioned strategy, while their join results are joined with t3 using the broadcast strategy. Each scan node is placed in its own fragment since scan results are exchanged to the join node, which requires a hash-based partition of the data. The following join with t3 is a broadcast join placed in the same fragment as the join between t1 and t2 because a broadcast join preserves the existing data partition.

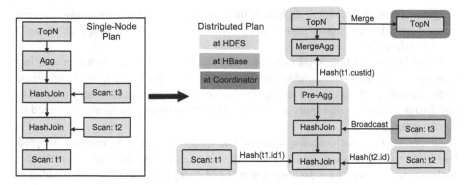

Fig. 16 Two-phase query optimization in Impala [14]

A two-phase distributed aggregation is performed after the second join. The local aggregation is computed in the same fragment as the second join. The local aggregation results are hash-exchanged based on the grouping keys, and then merge-aggregated again to generate the final aggregation result. The same two-phased approach is used to compute the top-n query results. The final top-n step is performed at the coordinator, which returns the results to the user.

6 Query Execution

Query executor is responsible for executing the data processing operations, as given by the plan generated by the optimizer, to the underlying data. The end results of query execution is producing the required query answers.

A key feature of many query executors is the ability to pipeline query results. This means that intermediate query results are progressively staged in a data processing pipeline, where each node in the pipeline performs a single operation, until the final query answer is obtained. Each operator in the pipeline is able to produce intermediate output results without reading the full input data. Pipelining may not be possible for certain types of operations that require consuming the full input before producing the output. These operations are called 'blocking'. One example is the Sort operation.

Pipelining in many query executors is based on the pull-based iterator model [8]. This model abstracts the details of query execution using three main APIs:

- Open: This API is used to start pulling data from a child operator by initializing operator's internal state.
- GetNext: This API is used by an operator to pull the next result from its child.
- Close: This API is used to signal end of processing, and triggers the cleanup of used resources.

Query plans are modeled as binary trees of operators communicating through iterator calls. Each operator pulls data incrementally from its child operators, until input is exhausted.

Consider for example a join operator that has two child operators representing scan operations on two input relations. The join operator pulls data from the two scan nodes, computes the joins and returns the join results to the client. Query execution terminates when the two inputs are exhausted, and no more join results still need to be computed.

In this section, we describe some of the query execution techniques adopted by big data query engines and highlight the impact of these techniques on system performance. We start by presenting a number of query execution research problems. We then describe how some of these problems are tackled by current proposals.

6.1 Research Problems

We give an overview of some of the research problems involved in the design and development of query executors in big data query engines.

Memory Management. The efficiency of query executor is largely determined by its ability to manage memory intelligently. Query processing on big data often leads to huge intermediate query results that cannot fit in memory. When a query executor has the ability to process data beyond memory limits, the query engine becomes much more useful at supporting complex data analytics.

One of the techniques that many executors use to handle data beyond memory limits is the management of spill files. In this technique, an operator state is allowed to overflow the available memory by writing data to disk. For example, a hash-join operator can store part of the hash table to disk files when it does not fit in available memory.

A crucial challenge in these techniques is how to determine the memory requirements of different query operators before query execution starts. The determination of memory requirements is essential in Hadoop multi-tenant ecosystem where applications need to reserve resources a priori by communicating with resource manager. Data statistics, optimizer's cost estimates, and state of query execution are important factors that need to be considered to determine memory requirements. Building predictive models that determine memory requirements is an important research problem.

Adaptive Execution. During query execution, the state of available resources change. For example, the available memory and/or network sources could become more scarce or abundant based on the current system state.

By designing query execution engine to be adaptive, the query engine becomes able to adapt its needs according to current system state. For example, a particular join algorithm might be efficient in the presence of abundant memory, and much less efficient if intermediate results need to be spilled to disk. In this case, changing the join algorithm in the midst of query execution could be a solution to work around the availability of system resources. Designing operators with such algorithmic flexibility is an interesting research problem.

Handling Different Data Models. Big data imposes the need to handle data following different models. While relational data is still the main data model adopted by many query engines, other data models are increasingly used. For example, nested data models (e.g., JSON [13]) are increasingly used by web services to exchange data.

Building a query executor that is able to process data following different models is crucial for handling the changing environments of big data. While some of the proposed techniques handle new data models by building transformers that convert data from one model to another, integrating native support of different data models within the execution engine is an important research problem.

6.2   Hadoop-Based Execution Engines

Hadoop-based execution engines rely on the distributed file system to achieve scalability and fault-tolerance. The execution engines provide libraries that integrate with the distributed file system to allow users to formulate analytics as imperative programs.

The first Hadoop-based query execution engine was the MapReduce framework [5] (cf. Sect. 3.1). In this framework, a library provides users with the APIs required for distributed query execution. Users need to program analytics using the APIs provided by the MapReduce library. The framework takes care of automatic distribution of query execution. The invocations of the map function are distributed across multiple machines by partitioning the input data into a set of splits. Each split can be processed on a different machine. The invocation of the reduce function are distributed by hashing the intermediate key space into disjoint partitions.

When a MapReduce program is executed, the following sequence of actions occurs, as illustrated in Fig. 17:

- A copy of user program is started on each machine used in query execution. A special copy of the user program is called the master. The rest of the copies are called workers, which are assigned work by the master. The master initially picks idle workers and assigns each one a map task or a reduce task.
- When a worker is assigned a map task, it reads the contents of the corresponding input split, and maps each key/value pair to an intermediate key/value pair buffered in memory.
- Periodically, the buffered pairs are hashed into partitions stored on local disk. The locations of these buffered pairs on the local disk are passed back to the master, which is responsible for forwarding these locations to the reduce workers.
- When the master notifies a reduce worker about the location of a hash partition of intermediate data, the worker reads the partition by issuing remote procedure

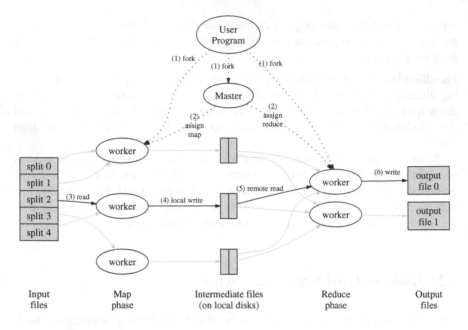

Fig. 17 Execution overview of MapReduce [5]

calls. The intermediate data are then sorted by the intermediate keys, so that all occurrences of the same key are grouped together. The sorting is needed because typically many different keys map to the same reduce task. If the amount of inter-mediate data is too large to fit in memory, an external sort is used.

- The reduce worker iterates over the sorted intermediate data and for each unique intermediate key encountered, it passes the key and the corresponding set of intermediate values to the reduce function. The output of the reduce function is appended to a final output file for this reduce partition.
- When all map tasks and reduce tasks finish, the master wakes up the user program. At this point, the MapReduce call in the user program returns back to the user code.

Hive [12] translates the incoming queries to executable jobs formulated using the previous MapReduce framework. The translation is performed by the Planner component (Fig. 4), which converts the query tree into an operator tree that repre-sents the required data processing operations. To allow for parallel query evaluation, the Planner determines for each operator whether it requires its input records to be partitioned based on some criteria or not. If an operator requires partitioned inputs, a ReduceSink operator is inserted before the operator to indicate the need to parti-tion the input data in a particular way before processing. For example, A GroupBy operator requires a ReduceSink operator to be inserted in the data pipeline before the GroupBy to inform the MapReduce execution engine that rows need to be partitioned based on the grouping key to allow for parallel execution of GroupBy operator.

The Hive operator tree is then passed to the MapReduce task compiler, where operations are broken to multiple stages represented by executable tasks. The MapReduce task compiler generates a directed acyclic graph of Map/Reduce tasks assembled in MapReduce jobs based on an operator tree. In the execution of a Map/Reduce task, operators inside a task are first initialized and then they process rows fetched by the MapReduce engine in a pipelined fashion.

6.3   Parallel Databases Execution Engines

We discuss the query execution techniques adopted by a number of parallel database systems. The main objective of parallel execution is achieving load balance by utilizing available resources in the compute cluster to provide the best possible query throughput. In these systems, different parts of the query plan can execute in different processes, both within a single host, and across different hosts. In a shared-nothing MPP architecture, processes within the same host share a common filesystem, while processes in different hosts communicate through a network.

Greenplum Database. In Greenplum database [23], the query optimizer produces parallel query plan by utilizing special query operators, called Motion operators, which implement the Volcano's Exchange operator [7]. Each Motion operator acts as both sender and receiver of data. It also defines the boundaries of what is known as a 'slice' in query execution plan. A query plan slice is a process that runs on a particular node and exchanges data with other processes running on other nodes.

The goal of Motion operators is to establish a given data distribution. For example, to establish a hashed distribution on column x, an instance of Redistribute(x) Motion operator, running on node n, sends tuples on n to other nodes based on the hash value of x, and also receives tuples from other Redistribute(x) operator instances running, in parallel, on other nodes. Similarly, a Broadcast Motion and Gather Motion operators are used to establish replicated and singleton distributions, respectively.

Figure 18 shows how query plan is sliced in Greenplum database for an example query. In this example, the plan is split into three slices using the Motion nodes as the slice boundaries. On a given node, three processes are active in parallel. The first process scans table customer (slice 3), the second process performs the join with table orders (slice 2), while the third process gathers the final results to the master node, where query results need to be returned to client (slice 1).

Figure 19 shows the interaction of plan slices during query execution. Slice 3 hashes input data based on the column cust_id, and sends the hashed data to other nodes based on the computed hash value. Slice 2 receives data from scan processes running on other nodes, and also sends the join results to slice 1, where the final query results are gathered.

Vertica Database. The Vertica execution engine [16] executes parallel query plans on MPP cluster. An example of Vertica query plan is shown in Fig. 20. Query execution in Vertica is multi-threaded and pipelined. At a given time, multiple operators can be

```
SELECT c.name, o.price
FROM Customer c, Orders o
WHERE o.cust_id = c.cust_id
AND c.city = 'San Francisco'
```

Customer is distributed randomly
Orders is distributed by cust_id

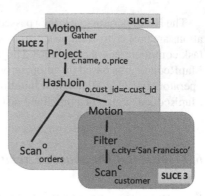

Fig. 18 Sliced query plan in Greenplum database [23]

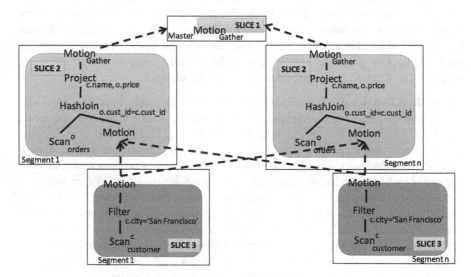

Fig. 19 Parallel query execution in Greenplum database [23]

running in parallel on different nodes. Within each operator, multiple threads could also be running in parallel.

The execution engine of Vertica is vectorized, where each block of row is requested and processed as a unit. This execution model is different from the traditional iterator model, where single rows are requested and processed one by one. Each query operator can use one of several possible physical implementations. The choice of which implementation is suitable at each query plan operator is controlled by the query optimizer.

In a Vertica query plan, the Send and Recv operators are used to move data across the nodes in the cluster. Send operator redistributes tuples from one node to other nodes, while Recv operator receives and processes the incoming data. Two types of

Fig. 20 Parallel query
execution in Vertica [16]

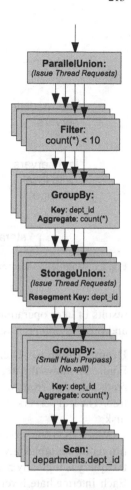

tuple redistribution are supported: broadcast, where tuple is sent to all nodes, and
hashed, where tuple is sent to a node based on some hash expression.

Each Send and Recv pair retains the order of tuples in the input stream. The send
operator hashes its outgoing tuples, such that identical values end up at the same
node in the computing cluster. This allows the operator running on each node to
produce the full results independently of other nodes. Data on the same node can
also be partitioned and processed by multiple threads utilizing separate cores to keep
all cores fully utilized. Figure 20 shows multiple GroupBy operators that are running
in parallel requesting data from the Storage Union operator which repartitions the
data such that the GroupBy is able to produce complete results.

The execution engine of Vertica also leverages sideways information passing
techniques to improve the performance of query execution. In these techniques,
a scan operator is extended with auxiliary filters that are used to improve query
performance. These auxiliary filters early-prune tuples that will not contribute to the

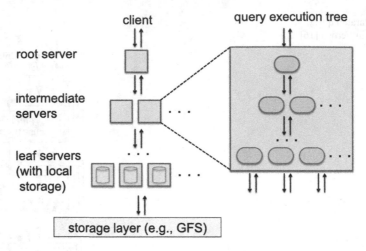

Fig. 21 Multi-level query execution in Dremel [18]

results of later operations like join. This allows avoiding the overhead of scanning and processing unnecessary data.

Dremel Database. Dremel [18] uses a multi-level serving tree to execute queries. Figure 21 shows the tree execution model. A root server receives incoming queries, reads metadata from the tables, and routes the queries to the next level in the serving tree. The leaf servers communicate with the storage layer or access the data on local disk.

When the root server receives a query request, it determines all horizontal partitions of the tables involved in the query. At the root level of the processing tree, the query is logically rewritten as a union of queries that operate on table partitions. Each intermediate level in the processing tree performs a similar rewriting. When the queries reach the tree leaves, the involved tables are scanned in parallel.

When query results are propagated upward in the processing tree, the intermediate servers perform parallel aggregation of partial results. This execution model is suitable for aggregation queries returning small and medium-sized results, which are a very common class of interactive queries. Other classes of queries, such as joins or large aggregations, may need to exchange data between servers at the same level before producing the intermediate results. Other mechanisms in of parallel query execution need to be leveraged in this case.

6.4 Code Generation

Query execution can largely benefit from reducing the number of CPU instructions during query evaluation. Code generation is a technique used by query execution engines [15, 19] to convert some intermediate representation of query plan elements

```
IntVal my_func(const IntVal& v1, const IntVal& v2) {
  return IntVal(v1.val * 7 / v2.val);
}
```

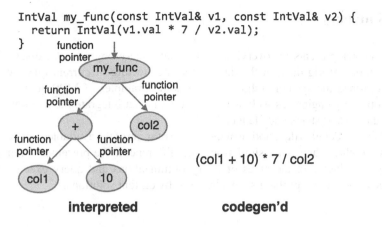

(col1 + 10) * 7 / col2

interpreted   **codegen'd**

Fig. 22 Code generation in Impala [14]

into machine code. Code generation can boost query performance by directly performing low-level operations using native machine code, rather than high level operations that still need to be compiled into machine instructions. For example, when a loop contains a function call, the function could be executed a large number of times (e.g., for every tuple). Optimizing this function by generating code that avoids some instructions in the function body can have a great impact on query performance.

One of the code generation tools that have been studied in the context of query engines is LLVM [17]. LLVM is a library and a compiler that allows users to write applications using a modular design, and have just-in-time (JIT) compilation within a running process. LLVM abstracts the details of CPU registers to simplify programming. The generated machine instructions are portable across different architectures.

In impala [14], query specific versions of functions are generated using LLVM to improve query performance. For example, when the type of an object instance is known at runtime, code generation can be used to replace virtual function calls with direct calls to the correct function, which can be inlined. This is useful when evaluating expression trees, as illustrated in Fig. 22. Each expression type is implemented by overriding a virtual function in the expression base class. Many of these expression functions are quite simple, e.g., adding two numbers. By resolving the virtual function calls with code generation and then inlining the resulting function calls, the expression tree can be evaluated directly with no function call overhead.

Presto [3] is written in Java. The Presto engine dynamically compiles certain portions of the query plan down to byte code which allows the Java virtual machine to optimize and generate native machine code. Through careful use of memory and data structures, Presto avoids typical issues of Java code related to memory allocation and garbage collection.

7 Summary

This chapter presents an overview of big data query engines. We discussed the implications of big data on the design and architecture of current query engines. We described the system architectures of some examples of MPP query engines, Hadoop query engines as well as hybrid systems that integrate the technologies of MPP database systems and Hadoop.

We gave a detailed description of the design of some of the available query engines, and highlighted their main similarities and differences. We presented a number of research problems in the areas of query optimizations and query execution, and discussed how these problems are addressed by current solutions.

References

1. L. Antova, A., El-Helw, M.A., Soliman, Z., Gu, M. Petropoulos, Waas, F. Optimizing queries over partitioned tables in MPP systems, in *Proceedings of the 2014 ACM SIGMOD International Conference on Management of Data* (2014)
2. M. Armbrust, R.S. Xin, C. Lian, Y. Huai, D. Liu, J.K. Bradley, X. Meng, T. Kaftan, M.J. Franklin, A. Ghodsi, M. Zaharia, Spark SQL: relational data processing in spark, in *Proceedings of the 2015 ACM SIGMOD International Conference on Management of Data* (2015)
3. L. Chan, Presto: Interacting with petabytes of data at Facebook (2016). http://prestodb.io
4. L. Chang, Z. Wang, T. Ma, L. Jian, L. Ma, A. Goldshuv, L. Lonergan, J. Cohen, C. Welton, G. Sherry, M. Bhandarkar, Hawq: a massively parallel processing SQL engine in hadoop, in *Proceedings of the 2014 ACM SIGMOD International Conference on Management of Data* (2014)
5. J., Dean, S. Ghemawat, MapReduce: simplified data processing on large clusters, in *OSDI* (2004), pp. 10–10
6. A. El-Helw, V. Raghavan, M.A. Soliman, G. Caragea, Z. Gu, M. Petropoulos, Optimization of common table expressions in MPP database systems, in *Proceedings of the VLDB endowment* (2015)
7. G. Graefe, Encapsulation of parallelism in the volcano query processing system, in *SIGMOD* (1990)
8. G. Graefe, Query evaluation techniques for large databases. ACM Comput. Surv. **25**(2), 73–169 (1993)
9. G. Graefe, The cascades framework for query optimization. IEEE Data Eng. Bull. **18**(3), 19–29 (1995)
10. Z. Gu, M.A. Soliman, F.M. Waas, Testing the accuracy of query optimizers, in *DBTest* (2012)
11. HBase: Apache HBase (2016). https://hbase.apache.org
12. Huai, Y., Chauhan, A., Gates, A., Hagleitner, G., Hanson, E.N., O?Malley, O., Pandey, J., Yuan, Y., Lee, R., Zhang, X.: Major technical advancements in apache hive, in *SIGMOD* (2014)
13. JSON: JSON (2016). http://www.json.org/
14. M. Kornacker, J. Erickson, Cloudera Impala: Real-Time Queries in Apache Hadoop, for Real (2012). http://www.cloudera.com/content/cloudera/en/products-and-services/cdh/impala.html
15. K. Krikellas, S. Viglas, M. Cintra, in *ICDE* (2010)
16. A. Lamb, M. Fuller, R. Varadarajan, N. Tran, B. Vandiver, L. Doshi, C. Bear, The vertica analytic database C-store 7 years later. VLDB Endow **5**(12), 1790–1801 (2012)

17. C. Lattner, V. Adve, Llvm: a compilation framework for lifelong program analysis and transformation, in *Proceedings of the International Symposium on Code Generation and Optimization: Feedback-directed and Runtime Optimization* (2004)
18. S. Melnik, A. Gubarev, J.J. Long, G. Romer, S. Shivakumar, M. Tolton, T. Vassilakis, Dremel: interactive analysis of web-scale datasets. PVLDB **3**(1), 330–339 (2010)
19. Neumann, T.: Efficiently compiling efficient query plans for modern hardware, in *Proceedings of the VLDB Endow*
20. Orca Open Source (2016). https://github.com/greenplum-db/gporca
21. A. Pavlo, E. Paulson, A. Rasin, D.J. Abadi, D.J. DeWitt, S. Madden, M. Stonebraker, A comparison of approaches to large-scale data analysis, in *SIGMOD 2009* (2009)
22. Pivotal: Apache HAWQ (2016). https://blog.pivotal.io/big-data-pivotal/products/introducing-the-newly-redesigned-apache-hawq
23. Pivotal: Greenplum Database (2016). http://greenplum.org/
24. Pivotal: HAWQ (2016). http://hawq.incubator.apache.org/
25. PostgreSQL: PostgreSQL (2016). http://www.postgresql.org/
26. Qubole: Presto as a service (2016). https://www.qubole.com/
27. M.A. Soliman, L. Antova, V. Raghavan, A. El-Helw, Z. Gu, E. Shen, G.C. Caragea, C. Garcia-Alvarado, F. Rahman, M. Petropoulos, F. Waas, S., Narayanan, K. Krikellas, R. Baldwin, Orca: a modular query optimizer architecture for big data, in *Proceedings of the 2014 ACM SIGMOD International Conference on Management of Data* (2014)
28. M. Stonebraker, D.J. Abadi, A. Batkin, X. Chen, M. Cherniack, M. Ferreira, E. Lau, A. Lin, S. Madden, E.J., O'Neil, P.E., O'Neil, A. Rasin, N. Tran, S.B. Zdonik, C-Store: a column-oriented DBMS, in *VLDB* (2005)
29. Teradata (2013). http://www.teradata.com/
30. A. Thusoo, J.S. Sarma, N. Jain, Z. Shao, P. Chakka, N. Zhang, S. Anthony, H. Liu, R. Murthy, Hive - a petabyte scale data warehouse using hadoop, in *ICDE* (2010)
31. R.S. Xin, J. Rosen, M. Zaharia, M.J. Franklin, S. Shenker, I. Stoica, Shark: SQL and rich analytics at scale, in *SIGMOD* (2013)
32. Yarn: Yarn (2016). http://hortonworks.com/hadoop/yarn/
33. M. Zaharia, M. Chowdhury, T. Das, A. Dave, J. Ma, M. McCauley, M.J. Franklin, S. Shenker, I. Stoica, Resilient distributed datasets: a fault-tolerant abstraction for in-memory cluster computing, in *NSDI 2012* (2012)

Large-Scale Data Stream Processing Systems

Paris Carbone, Gábor E. Gévay, Gábor Hermann, Asterios Katsifodimos, Juan Soto, Volker Markl and Seif Haridi

Abstract In our data-centric society, online services, decision making, and other aspects are increasingly becoming heavily dependent on trends and patterns extracted from data. A broad class of societal-scale data management problems requires system support for processing unbounded data with low latency and high throughput. Large-scale data stream processing systems perceive data as infinite streams and are designed to satisfy such requirements. They have further evolved substantially both in terms of expressive programming model support and also efficient and durable runtime execution on commodity clusters. Expressive programming models offer convenient ways to declare continuous data properties and applied computations, while hiding details on how these data streams are physically processed and orchestrated in a distributed environment. Execution engines provide a runtime for such models further allowing for scalable yet durable execution of any declared computation. In this chapter we introduce the major design aspects of large scale data stream processing systems, covering programming model abstraction levels and runtime concerns. We then present a detailed case study on stateful stream processing with

P. Carbone (✉) · S. Haridi
KTH Royal Institute of Technology, Stockholm, Sweden
e-mail: parisc@kth.se

S. Haridi
e-mail: haridi@kth.se

G.E. Gévay · G. Hermann · A. Katsifodimos · J. Soto · V. Markl
TU Berlin, Berlin, Germany
e-mail: gevay@tu-berlin.de

G. Hermann
e-mail: mail@gaborhermann.com

A. Katsifodimos
e-mail: asterios.katsifodimos@tu-berlin.de

J. Soto
e-mail: juan.soto@tu-berlin.de

V. Markl
e-mail: volker.markl@tu-berlin.de

© Springer International Publishing AG 2017
A.Y. Zomaya and S. Sakr (eds.), *Handbook of Big Data Technologies*,
DOI 10.1007/978-3-319-49340-4_7

Apache Flink, an open-source stream processor that is used for a wide variety of processing tasks. Finally, we address the main challenges of disruptive applications that large-scale data streaming enables from a systemic point of view.

1   Introduction

Today's modern societies are increasingly employing big data analytics systems (BDAS) to analyze data and enable sound judgment. Among them are the data stream processing systems (DSPS). DSPS are specially designed systems that are able to manage infinite streams of data with low latency and high throughput. They are widely employed in active database, complex event processing, and publish-subscribe system applications and rooted in database systems, data warehouses, and information flow programming systems. Gradually, they have incorporated additional technological capabilities, including (Map-Reduce-like) scalability, declarativity, and expressivity (akin to relational programming models, such as SQL), and the efficiency of data warehousing technologies. In this chapter, we analyze the state of the art of this ecosystem from two distinct perspectives: (i) *programming models* for scalable data stream processing, (ii) *systems* and *runtimes* that can execute applications expressed in these models. Furthermore, we will reason behind architectural choices and semantics, and finally provide an analysis of emerging problems and solutions within this domain (Fig. 1).

In this section, we first categorize in Sect. 1.1 the main system precursors and ideas that have influenced the current state of the art in data stream processing. In

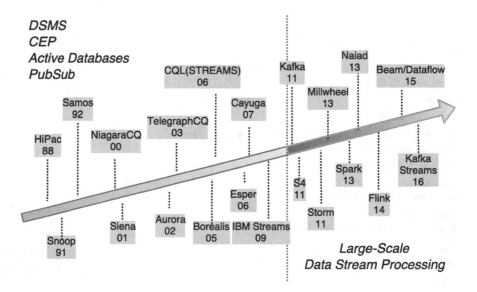

Fig. 1 Evolution of data stream processing systems

Sect. 1.2 we provide a synopsis of the ecosystem of data stream processing, followed by the distinctive features of these systems from both the programming model and execution perspectives, in Sect. 1.3. We should highlight that the main focus of this chapter is *systems* and *programming models*, thus, we will omit algorithmic aspects or domain-specific use cases in data streaming.

1.1 Stream Processing and Its Precursors

Modern data-stream processing systems, which we will call *stream processors* (or SPs for short), are rooted in several well established topic areas of data management such as active databases, complex event processing, publish-subscribe systems, and data-stream management systems. To put stream processing into the greater context of data management, next we present a brief description of each of these domains.

Active Database Systems (ADS) are an explicit form of event-based database systems, implemented originally as extensions to existing DBMSs. HiPac [41], Snoop [71], and Samos [48] enriched traditional relational query support with event-based actions, conditionally applied to data upon ingestion time. The usage of ADS (and the heavy use of database triggers) ranged from basic *Extract-Transform-Load* operations (ETL) to more sophisticated event-based detection tasks. The latter category of queries is also known as *standing* or *continuous queries*, and they were typically queries that were used to monitor database entries for specific conditions. NiagaraCQ [37] and OpenCQ [65] are two noteworthy examples of systems supporting continuous queries. One of the major downsides of ADS was their limitation to centralized, non-scalable execution engines, an inherent property of the monolithic DBMSs that hosted this technology. Furthermore, they were tightly coupled to specific DBMSs, thus, limiting broader adoption. Another downside of ADS was their limited expressivity when it comes to defining complex patterns for event processing, an ability that Complex Event Processing (CEP) systems were created to cover. Nevertheless, the impact of ADS in the development of SPs is undoubtedly noteworthy. We often see modern, large-scale stream processing systems committing to a similar architecture such as S-Store [31], an in-memory scalable stream processing system that is built on top of H-Store [58], a shared-nothing distributed relational database with strong transactional properties.

Complex Event Processing (CEP) systems [40] are among some of the most related technologies to modern SPs and are in use throughout several enterprise applications such as sensor-network monitoring, intrusion detection systems, and high-performance financial systems. Complex rule-based pattern matching is a first-class citizen in CEP systems, with a plethora of open-source and proprietary solutions, each exposing a distinct rich, declarative event-based language. One of the major contributions of the CEP ecosystem (e.g., Cayuga [24], Esper [20], and STREAMS [13]) is the introduction of sliding window semantics, among others. Windows are used to group event sequences into evolving views or sets, upon which aggregations and patterns can be evaluated. Streaming windows are now an integral part of many SPs [10, 64, 75]. Despite their rich semantics, the architecture of most CEP systems remains monolithic and centralized. Thus unnameable for large-scale, highly available deployments in clusters environments [25].

Publish-subscribe systems were mainly developed and used as distributed messaging middleware, which routes messages from data producers to data consumers [66] i.e., subscribers. In addition to trivial topic-based routing, subscriptions in several publish-subscribe systems can be expressed through complex rules such as event sequences and repetitions (e.g., Siena [29]) or even through arbitrarily complex queries [59]. Publish-subscribe systems served as a basis for the development of large-scale event-processing systems that run continuous queries. Moreover, they heavily influenced the development of modern messages queues and distributed log systems such as Apache Kafka [60] which adopt a similar subscription-based pattern for distributed data stream ingestion.

Data Stream Management Systems (DSMS) introduced many architectural primitives that reside at the core of many SPs today. Systems such as Aurora* [38], Borealis [7], and StreamBase [80] allowed for distributed, managed deployments of stream processing pipelines, consisting of tasks and data dependencies in a dataflow graph defined by the user through a graphical interface. A scheduler managed the deployment of parallel tasks which could be later re-optimised for improved latency or throughput, depending on dynamic properties such as the current workload. At the same time, other DSMSs focused on expressivity and integration with DBMSs by enriching SQL with windowing semantics, CQL [15] being a noteworthy example of such an effort. In general, DSMSs were mainly developed as research prototypes, yet, they went on to influence the emerging data stream processing field.

1.2 Large-Scale Data Stream Processing on Commodity Clusters

MapReduce [43] and the development of open source software stacks for distributed data processing on commodity clusters (e.g., Apache Hadoop [1], Apache Spark [89]) initially covered a major need for *batch* or *offline* data processing. However, low-latency and high-throughput computing emerged as an open problem. To this end, a new class of SPs targeting low-latency workloads started to evolve.

Some of the first open source SPs for commodity clusters were Yahoo S4 [74] and Twitter Storm [5]. S4 and Storm integrated seamlessly with the rest of the open source ecosystem, such as distributed logs (e.g., Apache Kafka [60]), message queues, and distributed file systems (e.g., HDFS). In addition, S4 and Storm offer low-level distributed dataflow programming models, suitable for engineering arbitrary task pipelining while hiding basic physical implementation concerns (e.g., distributed data exchange). More SPs gradually emerged in the same ecosystem and provided richer semantics and higher-level programming abstractions for data streams in order to simplify the writing of data stream analysis applications. Examples of such systems are Apache Flink [26], Beam [10], Samza [3], Spark Streaming [90], APEX [82], and more recently Kafka Streams [56]. The development of this latter class of SPs provided more declarative programming semantics such as custom windowing, functional programming primitives and stream SQL queries being heavily inspired by research projects such as Stratosphere [11], Naiad [73], and FlumeJava [32]. At the same time they provided stronger processing guarantees such as exactly-once operator state access and high-availability, thus, making them suitable for critical production deployments.

1.3 Distinctive Features of Data Stream Processing Systems

In order to facilitate the description and categorization of various stream-processing systems throughout this chapter, we first list a number of distinctive features that SPs exhibit.

– **Continuous Uninterrupted Execution**. The ability of a system to execute a continuous query or stream topology without inducing additional delays such as queuing, re-scheduling, halting, etc. This is the norm in shared-nothing stream processors, as we show later in this chapter.

– **Durability**. As with every distributed system, process or network failures are bound to happen, especially often in a long-running operators. These failures need to be handled transparently to system users. Systems differ in terms of their assumptions (e.g., repeatable persistent sources) and in the guarantees they can offer to applications under failures. A plethora of techniques have been studied in the past [9, 17, 27, 73] in order to deal with failures in continuous processing, each of them imposing additional computational and resource overhead to offer certain properties.

– **Low Latency**. Data-stream processors have the ability to process data and make incremental computations at ingestion rate. This drops end-to-end latency by orders of magnitude compared to batch executions or typical daily database integration jobs. Low latency processing can serve critical applications where timely and thus actionable knowledge is paramount. For that reason it serves as a major incentive for using SPs today. Moreover, low latency or response time is considered as a common benchmarking measurement [14, 19], used to stress and compare the performance of different SPs.

– **Explicit State Management**. State summarizes results of computation that results ran on a stream processor (e.g. a counter). From a systems perspective, state needs to be declared and managed explicitly in order to enable fault tolerance and repartitioning capabilities in a transparent manner [47]. Alternatively, when state is managed in the application layer (e.g. when programming in Apache Storm v0.10), there is a need for custom synchronization with external storage systems. However, a state-agnostic approach disables prospects for state reconfiguration and efficient native fault tolerance mechanisms [9, 27, 30].

– **Programming Primitives**. Systems provide primitives in their APIs that aid programmers with built-in implementations for common tasks. For example, aggregations [6], like summations could be implemented with low-level dataflow API by using custom managed state. However, stream processing applications typically contain parts where values must be aggregated. Thus, some systems provide an aggregation primitive to aid the development process. There are similar primitives [90] in many system APIs for other frequent cases, such as filtering or grouping by a key. Most frameworks also provide windowing primitives [6, 10, 28, 64] that group data points based on time. Frameworks are introducing more and more higher-level abstractions in their APIs for advanced expressivity [12].

– **Efficient Plan/Topology Execution**. As with database queries, executing a continuous stream processing query in a SP as declared by the user could lead to potentially redundant resource usage and inefficient pipelining of operators. Thus, several SPs optimize the execution of certain queries in order to eliminate redundancy, increase the efficiency of resource utilization and throughput, while also reducing latency [55]. As with programming languages, most optimisations are applied transparently upon query translation to physical dataflow operators or some other intermediate-level representation of the computation at hand. Typical optimisations that are adopted in several SPs are: *operator fusion* (also known as superbox scheduling), *operator reordering*, *state sharing*, and *batching* [55].

– **Elasticity and Dynamic Reconfiguration**. A stream processor can potentially execute a job for weeks or years continuously. The workloads of such long-running jobs can change considerably over time. Thus, it is handy for such a system to be able to adapt resource consumption (e.g., workers, memory) dynamically, according to changing demands. This requires general support for reconfiguration and additional monitoring of several runtime parameters. Furthermore, physical operators should allow for partionable state (when that is applicable) in order to scale in or out [30] according to the workload.

– **Sustainable Flow Control**. Network channels and in-memory buffers that reside within stream processors have a finite capacity, which places certain constraints on the ingestion throughput. In the most general case, data is being pulled at ingestion points in a pipeline (e.g., stream sources) and pushed throughout different computation tasks. In order to sustain an overall continuous ingestion, despite throughput imbalance among operators, an SP can either discard events [6] or trigger a flow control mechanism such as back-pressure [61]. Back-pressure allows SPs to "keep up" with processing at sustainable rates. This is a highly important property, which is missing in several existing stream processing solutions (e.g., Apache Storm [5]).

1.4   Chapter Overview

In order to provide a clear picture of the current state of the art of open source SP technology we present a snapshot of the most distinct, and perhaps the most sophisticated modern stream-processing systems, alongside their core features. In this analysis, we exclude academic projects (e.g. STREAM [13], Amos II [45], Aurora [6], TelegraphCQ [34], Cayuga [24], FLUX [77], S-Store [31]) and several proprietary systems (e.g., IBM SPL [54], Microsoft Trill [33] and FlumeJava [32]), either due to the lack of sufficient information about their internals or because they do not operate in a distributed scale-out setting, which is the main interest in our study. It is important to state that many of these systems have a pioneering role in establishing the fundamentals of modern data stream processing.

The rest of the chapter is structured as follows: in Sect. 2 we introduce the common denominators of programming primitives used throughout different stream processing frameworks categorized by their abstraction level (i.e., low-level dataflow programming primitives and higher-level functional APIs). We further present windowing semantics and their usage, along with different notions of time. In Sect. 3 we analyze the characteristics of the most common large-scale stream processing engines. We further provide a clear description of processing guarantees, flow control and other important problems that arise in a sustainable execution of stream computations along with solutions. Next, in Sect. 4 we present the Apache Flink stack, a case study of a fault-tolerant, large-scale stream processing platform, alongside a detailed overview of its asynchronous snapshotting mechanism. Finally, in Sect. 5 we enumerate several application domains where advances in data stream processing technology have offered tremendous impact, followed by Sect. 6 where we conclude this chapter with an outlook of the data stream processing field.

2   Programming Models

In this section, we discuss different models for developing data-stream processing applications. Currently, there is a wealth of frameworks and systems that enable data-stream processing on commodity clusters, offering various capabilities and programming interfaces. As stated in the introduction (Sect. 1), we do not discuss research projects or proprietary systems. Instead, we concentrate on large-scale, open-source stream processing frameworks, namely, Storm [5], Heron [61], APEX [82], Flink [26], and Spark Streaming [4, 90].

The goal of this section is to provide a brief overview of these programming abstractions, and show how a set of seemingly different programming concepts can share a set of common fundamentals. The structure of this section is as follows: first, we describe different levels of programming abstractions for stream processing (Sect. 2.1), then we discuss these levels in more detail (Sects. 2.2 and 2.3), finally we discuss windowing (Sect. 2.4), a programming abstraction that we can find in most stream processing APIs, regardless of how declarative they are.

Fig. 2 Abstraction levels of
streaming applications

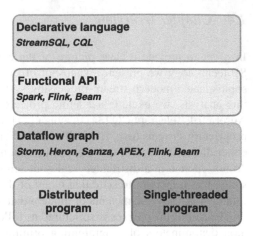

2.1 *Programming with Streams*

There are different abstraction levels that a programmer can use to express streaming computations. Figure 2 depicts how these abstractions build upon each other. Behind these abstractions, stream processing frameworks hide execution details from the programmers, and manage them in the background.

Low-Level Dataflow Programming. Historically, stream-processing engines such as Aurora [6], implemented the execution model of a *dataflow* program. In short, a dataflow program is represented as a directed graph, whose nodes (or, operators) represent a computation and whose edges represent connections among dataflow nodes. The programmers can, in principle, construct arbitrarily complex dataflow programs by implementing operators and by connecting them via dataflow graph edges. Modern stream-processing systems, such as Storm [5] and APEX [82], provide a dataflow programming abstraction, where a stream-processing application can be expressed using a dataflow graph and whose operators are provided by a programmer in the form of imperative user-defined functions and thus may contain arbitrarily complex business logic. A stream-processing system distributes a dataflow graph across multiple machines and is responsible for managing the partitioning of data, the network communication, as well as program recovery in case of machine failure.

Dataflow programming offers programmers complete freedom to programmers to implement their business logic, but require them to have good knowledge of the execution internals. This is because dataflow programming reveals the internals of the underlying execution engine, which handles the execution of the dataflow graph. As a result, programmers need to manually specify execution strategies in their programs and have to change their implementation whenever data statistics (e.g., distribution of values or data rate) or the size of the deployment changes (e.g., running the same program in a larger cluster). Moreover, common use cases, such as aggregation and stream-windowing (Sect. 2.4) have to be manually implemented by the programmer. More details on dataflow programming can be found in Sect. 2.2.

Functional APIs. Instead of forcing programmers to manually specify low-level dataflow graphs, stream-processing frameworks such as Spark [4] or Flink [26] offer higher-level *functional APIs*. These APIs are more declarative than low-level dataflow programming by giving programmers the ability to specify data-stream programs as transformations on data-streams. In functional APIs, common tasks such as aggregating values (e.g., calculating the sum of numbers in a stream) can be specified very concisely, relieving programmers from having to write large amounts of boilerplate code found in low-level dataflow programs.

High-Level Declarative Languages. In the past, several research projects in stream processing, such as CQL [15] and TelegraphCQ [34] have proposed a declarative SQL-like language for data stream processing. At the time of writing, there are ongoing efforts for providing a similar declarative language for large-scale stream processing systems. Such a declarative language offers less control to programmers for low-level execution strategies. Declarativity has the disadvantage of limiting the opportunities for fine-tuning the performance of applications. However, a declarative language allows for automatic optimization and shifts the responsibility of optimization from the programmer to the system. It also allows inexperienced users to write streaming applications without knowledge of the system's internals. Since there is no high-level language implementation on top of the systems we discuss here, we will omit discussing high-level languages.

In the sequel, we describe dataflow programming and functional APIs.

2.2 Lower-Level Dataflow Programming

Dataflow programming has been used by research projects such as Aurora [6] to express streaming programs. Later on, it turned out that dataflow programming is useful for parallelizing and executing programs in a distributed setting [38, 53]. More recent systems, such as Storm [5, 86], Samza [3], and APEX [82], apply ideas from past research and provide similar dataflow abstractions. In this section, we show a logical dataflow programming model (with pseudocode examples) resembling the APIs of these three frameworks and we explain what details should be specified by the programmer concerning distributed execution.

Logical Dataflow. A dataflow program is represented as a directed graph of *operators*, and a set of edges connecting those operators. The resulting graph is typically referred to as a *dataflow graph*. Operators are independent processing units defined by the programmer, which take input and produce output. The programmer has to implement an operator by defining a computational routine for each input record. The dataflow graph that connects the operators represents the data flowing through the program; edges between operators in the graph represent unidirectional data

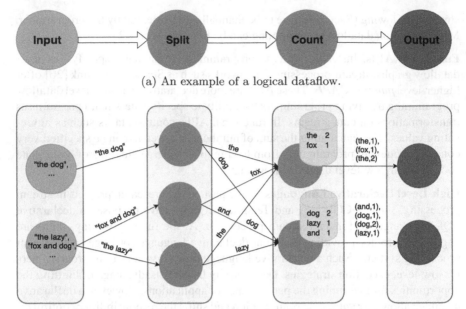

(a) An example of a logical dataflow.

(b) The physical dataflow associated with the above logical one.

Fig. 3 An example logical dataflow of the word count program (*top*) and its physical execution graph (*below*)

exchange and define how the outputs of some operators become the inputs of other operators.

Figure 3a shows a simple example of a logical dataflow program. The task here is to count the occurrences of words received on the input. This is the widely known word count example popularized by the paper on MapReduce [43]. Besides the input and output, there are two operators, namely *Split* and *Count*, which are defined by the programmer. *Split* takes a text line and splits it into words, and *Count* counts the occurrences. The operators are independent: they can only communicate with each other by their input and output connections.

In Fig. 4 we show how the *Split* operator is written using Java-like pseudocode. The function `onArrivingDataPoint` is an event handler. Each input record is passed to this handler via a distinct invocation, whereas the output of each call is a collection of records the operator emits. This example splits a string line into words using a space delimiter and puts the words into an array. Each word in the array is then emitted, one by one, to the operator's output via a "Collector" interface. Bear in mind that the output elements i.e., the words might not be in the return value of the method. Instead, dataflow-level APIs typically provide a `Collector` to the event handler, which can be used to emit output records one by one, via its `emit` method.

Physical Dataflow A logical dataflow graph is deployed in a distributed environment, in the form a *physical dataflow graph*. Before execution, systems typically create several parallel instances of the same operator, which we refer to as *tasks*. A system

```
//Input:   "the fox jumps", "lazy dog", ...
//Output:  "the", "fox", "jumps", "lazy", "dog", ...

Collector collector;

onArrivingDataPoint(String line) {
  String[] words = line.split(" ");
  for (String word : words) {
    collector.emit(word);
  }
}
```

Fig. 4 Pseudocode of a dataflow operator that splits its input lines into words

is able to scale out by distributing these tasks across many machines, akin to a MapReduce execution. In Fig. 3b we can see a physical dataflow of the word count example. The circles represent task instances of the operators, which may be located on different machines. We can reason about execution similarly to a logical dataflow: the data flows through these tasks, every task receives input elements from tasks corresponding to the previous operator and produces outputs to the tasks forwarding them to the next operator.

In low-level dataflow programming, the programmers can control the physical dataflow execution such as the *degree of parallelism* (e.g. in Fig. 3b, three tasks are assigned to the *Split* operator and two to the *Count* operator).

Stateful Operators. Unlike a simple operator such as *Split*, certain operators need to keep mutable *state*. For instance, in the word counting example, counting the word occurrences received by an operator, requires storing the words received thus far along with their respective counts. Thus, the *Count* operator must keep a state of the current counts. In contrast, the *Split* operator is stateless because a line can be split without any prior knowledge (i.e., independent of any previously received lines). In Fig. 5, we show how the *Count* operator can be implemented via a special

```
//Input:  the, dog, the, lazy, dog, ...
//Output: (the,1), (dog,1), (the,2), (lazy,1), (dog,2), ...

KeyValueState<String, Integer> wordCounts;
Collector collector;

onArrivingDataPoint(String word) {
  Integer count = wordCounts.get(word, 0);
  count++;
  wordCounts.put(word, count);
  collector.emit(new Pair(word, count));
}
```

Fig. 5 Pseudocode of a stateful dataflow operator that counts occurrences of words

class provided by the framework, `KeyValueState`, which can store key-value pairs. The state is read (`get` method) and updated (`put` method) in every call of the `onArrivingDataPoint` event handler. Explicit mutable state [47] is often a requirement, thus, several APIs with higher-level abstractions (e.g., Flink and Spark) provide an interface for that purpose.

Although it is often feasible to specify state without making the system aware of it, it is not recommended to do so. Since the systems we discuss are built on a JVM (Java Virtual Machine), operator logic is usually implemented as a custom event handler that implements a Java interface. This class could have a member variable for storing the state. For instance, we could store the word counts in a simple Java `HashMap`. However, the system would not know that we are maintaining a state in our operator and this would raise problems. First, if we store a bigger state (e.g., a data structure like a Java `HashMap`), then we have to manually take care of the situation when this data structure grows too large. If we assume that this will never happen, but our application becomes more popular than expected, then the system might run out of memory. Second, checkpointing a large state that the system knows nothing about can be problematic (see Sect. 4.3). Using the state abstractions of the APIs enables the runtime systems to deal with larger states and take care of making them fault tolerant.

Partitioning Strategies in Physical Dataflows. *Partitioning* strategies determine the allocation of records between the parallel tasks of two connected logical operators. Thus, they give control over data exchange patterns that fundamentally occur in physical dataflow. There are several standard strategies [49] tailored to data streaming in particular, each of them offering certain benefits. With *random* partitioning each output record of a task is shipped to a uniformly random assigned task of a receiving operator, distributing the workload evenly among tasks of the same operator. *Broadcast* is another typical strategy, which can be used to send records to every parallel task of the next operator. Furthermore, partitioning *by key* guarantees that records with the same key (e.g., declared by the user) are sent to the same parallel task of consuming operators, similar to the shuffle phase of the well-known MapReduce model [43]. We should keep in mind that most system APIs allow for user-defined partitioning functions as well, which can dynamically select the partition (e.g., parallel task) for each output record. Custom partitioning functions can be used to implement application-specific partitioning functions (e.g., geo-partitioning [57] or machine learning model selection [42]).

Since the operator state is also distributed among parallel stateful tasks and no global state is maintained or accessed, it is particularly crucial to have control of the partitioning at stateful operators. An example of this need can be observed in Fig. 3b, where it is necessary to ship records of the same word to the same task between the *Split* and the *Count* operators. This is necessary to compute the correct state (e.g., complete count per word). In that case, partitioning by key is the preferred partitioning strategy, where the key is the word (e.g., first element of each tuple). As in MapReduce, the same words end up at the same task, hence on the same machine, so that counting can be done locally.

Table 1 Dataflow abstraction terminology

	Dataflow graph	Operator	Task
Storm 0.10.x [5]	Topology	Bolt	Task
Samza 0.9.x [3]	Dataflow graph	Job	Task
APEX 3.2.x [82]	DAG	(Logical) operator	Physical operator

Differences Between APIs. Although all systems with a dataflow API discussed here use this same programming model of a logical dataflow with configurations for the physical dataflow (like parallelism hints and partitioning) there are slight differences. Each system has a distinct terminology, so we summarize the naming of the main concepts in Table 1. Differences also arise in various other aspects apart from the terminology (e.g., state management interface). However, we omit them here for practical reasons and leave the discussion of specific features to the documentation of the respective systems.

2.3 Functional APIs

Several frameworks provide more declarative APIs than the ones discussed in the previous section. This means that certain details of *how* to execute the computations can be omitted, and programmers need only specify *what* should be computed.

Collection Abstractions for Streams. In order to make APIs comparable to regular collection APIs (e.g., List in Java or Scala), some stream-processing frameworks (e.g., Spark [4], Flink [26], and Trident [85]) introduce a collection type representing data streams, which supports operations that resemble those of regular collections.

For example, consider the following map operation on a Scala list:

```
val list1 = List(1, 2, 3)
val list2 = list1.map(x => x + 1)
```

map applies the given function to each element of the given list, and returns a new list containing the results. Here, list2 will contain (2, 3, 4). Functions like map, are *higher-order functions*: in addition to taking data as their input, they also take other functions.[1] Functions that are passed as arguments to higher-order functions by programmers are called *User Defined Functions* (or UDFs in short). For example, the "x => x + 1" is a User Defined Function, which adds 1 to its argument.

The APIs of stream processing frameworks borrowed many higher-order functions from the standard libraries of functional programming languages. For example, the flatMap operation takes a UDF that maps each element of the input stream to a collection of (zero or more) elements per invocation. A typical example is word

---

[1]Generally in functional programming, higher-order functions might also produce functions as their outputs, but this does not appear in stream processing.

```
// A Context is used to specify the input, and for configuration
SystemContext ctx

// Read text from a TCP connection line by line
DataStream text = ctx.socketTextStream(host, port, '\n')

DataStream wordCounts = text
      // split the lines to words
      .flatMap { line => line.split(" ") }
      // add the initial count (1) to every word in a new field
      .map { word => (word, 1) }
      // group by the word (0th field)
      .partitionByKey(0)
      // sum up the counts by words (1th field)
      .sum(1)

// print the results
wordCounts.print()
// start the execution of the dataflow graph
ctx.execute
```

Fig. 6 Word count example in pseudocode using a functional data stream processing API

count (Figs. 4 and 5) when the input elements are lines of text, but we want to process individual words instead. We can see how to implement the same program in a functional collection API in Fig. 6. The program takes input from a TCP socket and simply prints the counts to standard output. We use a flatMap to split lines and a *map* to create (word, count) pairs (we later describe how keyBy and sum work). Note that the word "flat" is meant to say that the output is not a nested structure (e.g., a stream of collections), but the output collections of the UDF are "flattened" into a single stream.

The functions map and filter are special cases of flatMap: map always emits exactly one output element for each input element and filter always emits either zero or one output element for each input element. This allows making their UDFs slightly simpler: the output type need not be a collection, but only a single element in the case of map and a Boolean value in the case of filter (indicating whether the current input element should be "filtered out").

Notice that in the low-level dataflow model introduced in Sect. 2.2, all operators act like flatMaps: for every input element, they can emit any number of output elements.

Aggregations. Aggregations are one of the most common operations on data streams. For example, we might need to compute a rolling sum or maximum aggregate value from all the input elements that have arrived so far (or count them like in the word count example in Sect. 2.2). A *rolling aggregation* means that the current aggregate is emitted after every input element, thereby creating a stream of intermediate aggregates. For example, if the input elements are $1, 2, 3, 4, \ldots$ then a rolling sum emits $1, 3, 6, 10, \ldots$. This is in contrast to aggregations that we can see in batch processing, where the entire input needs to be ingested to produce a final aggregate value.

Stream records often contain a *key field* upon which users intend perform a rolling aggregation separately for each distinct key that occurs, similar to an aggregate over a `groupBy` in SQL. This type of aggregation is also known as *keyed aggregation*. For example, imagine that we are operating some online game, where a player can play a game at any time and achieve some separate score in each game. We would like to record the highest score achieved by each of our players separately and produce a high score table. Let us assume that stream records in that application are *(playerID, achieved score)* pairs, referring to occurrences of players with unique IDs playing a game and achieving a certain score. A rolling `max` aggregation per player ID would produce an output stream that has elements of the form *(playerID, highest score ever achieved by this player)*. Then to produce the high score table, we need a further (non-keyed) aggregation operation that maintains the top-K input elements, which is an example of a more complicated application logic occurring as an aggregation function.

Figure 6 shows a version of the word count application, written in a functional data stream processing API, which uses a *keyed aggregation*. Contrary to the low-level dataflow version of word count presented in Figs. 4 and 5, only two higher level operations are required to achieve the same result, namely declaring the `key` used to partition the stateful computation (via `partitionByKey`) and applying an `aggregation function` (`sum` on the second field).[2] In contrast, the low-level dataflow version of the same application requires the programmer to provide manual task-level behavior and state handling, which can often be avoided by using high level primitives.

2.4 Stream Windows

Stateless *maps, flatMaps,* and *filters* deal with one element at a time, whereas aggregations take the entire stream into account. Thus, none of these options satisfy our needs if we are interested in hourly, daily, or maybe yearly statistics. This can be the case, for example, if we are counting the clicks that a certain webpage received: what we are probably interested in is the number of clicks during some custom time periods of a certain granularity and not since our data processing application first started to operate. To satisfy this need, we can use *windowing* techniques to logically group records of an infinite stream into finite sets, upon which we can perform aggregations or other custom operations. Windows are typically declared in terms of predefined templates (e.g., time and count [15, 90]). However, there are directions

---

[2]Mind that `sum` in this example is a pre-defined aggregation function, however, a UDF can also be typically provided to declare an incremental computation.

towards more complex compositional window definitions such as delta-based [53], session [10] and user-defined windows [28], among others [44].

Notions of Time. One of the common ways of specifying windowing is through time intervals. However, time can often be an ambiguous measure in data stream processing. We first introduce the different notions of time that are typically considered in streaming applications. The input events of an application are usually created by an external source prior to their processing. Therefore, the following two notions of time can be identified [23, 79]:

- **Event time**: The time corresponding to when an event was generated externally. It is commonly provided by a *timestamp* field using the local clock of its source (e.g., a sensor).
- **Processing time**: The time corresponding to when a system processes the event (measured by the system local clock).

Notice that the event time of the records are a property of the *input*, whereas the exact processing time of events depends on the actual *execution* of our application, which can be affected by arbitrary network delays, system workload changes, and other factors. It is evident that processing time is not a consistent metric for progress in distributed stream processing due to an inherent absence of a global synchronized clock. Thus, event time is typically a preferable choice to specify time-based operations in an application (e.g., parallel time windows) given that there is a known clock skew among the sources of the data streams.

Tumbling and Sliding Windows. The simplest type of window is the *tumbling time window*, where we can specify the window size in some time units and the stream is split into non-overlapping, adjacent time intervals that have a given size. We can see an illustration of 3 second tumbling windows in Fig. 7a.

For example, imagine that we run an e-shop and we would like to know the sum of the prices of sales and we create a stream processing application to calculate it. One way to implement this is to use a higher-level API discussed before by applying an aggregation on a stream of sales containing their price. However, what we need is probably not the sum of every price since the start of our analytics program: we would not like to sum prices of items ordered two years ago because that data is not so important anymore. Instead we might be interested in the sum of the prices of orders *every three hours* (or days, weeks, etc.), which we can obtain by specifying a tumbling window and then applying the *sum* aggregation on it.

We often want more frequent updates than the window size, in which case we can use a *sliding window*: in addition to a window size, we also specify a *slide size*, which declares the frequency of our window computation (see Fig. 7b). Note that here the windows might overlap if the slide size is smaller than the window size or there might be "gaps" between the windows, if it is bigger. In fact, tumbling windows are a special case of sliding windows, where the slide size is the same as the window size. These basic types of windows are supported by almost every stream processing API, even lower-level ones, such as APEX.

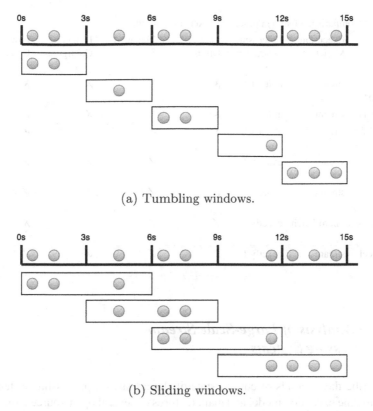

(a) Tumbling windows.

(b) Sliding windows.

Fig. 7 a Tumbling windows. **b** Sliding windows

3 System Support for Distributed Data Streaming

In the previous section we went through different programming models and unique high level concepts involved in distributed data stream processing. The focus of this section is the fundamentals of systems that implement such concepts, which we will call *stream processors* or SPs for brevity. The main purpose is to give an anatomy of existing systems, offer insights, and reason about different design considerations.

Perhaps the main property that differentiates a system architecture tailored for data streams is its ability for continuous, non-interrupted execution. An inherent benefit of a system that is built for continuous processing is the ability to subsume all other known types of computation (e.g., data-intensive batch processing or database query processing) in a universal way [84]. We will cover some of these concepts in this chapter, but before that, we describe all components and mechanisms that currently reside within open source stream processing systems today.

Table 2 A comparison of stream processing execution engines

	Execution Model	Processing Guarantees	Plan Optimiser	Managed State	Dynamic Configura-tion	Flow Control
Storm 0.10.x [5]	stream	at least 1	✗	✗	✗	✗
Heron [61]	stream	at least 1	✗	✗	✗	✓
Samza 0.9.x [3]	stream	at least 1	✗	✓	✗	✓
APEX 3.2.x [82]	stream	exactly 1	✓	✓	✓	✓
Flink 1.0 [26]	stream	exactly 1	✓	✓	✗	✓
Spark 1.6.x [90]	micro-batch	exactly 1	✓	✓	✗	✗
MillWheel [9]	stream	exactly 1	✗	✓	✓	✓

3.1 An Analysis of Large-Scale Stream Processing Systems

We describe the internals of current state of the art stream processing systems by discussing the execution model and main features of several open-source large-scale SPs. As already mentioned in the introduction (Sect. 1), we exclude academic projects and several proprietary systems from our discussion. In Sect. 2 we summarized the main systems of interest alongside their adoption of distinctive features. We will cover most of the features illustrated in the table above in the following sections in greater detail.

The execution model operates at the heart of a stream processor and encapsulates the coordination of any streaming computation. In fact, SPs exhibit alternative design considerations that enable continuous processing. However, any system can be categorized under two principal architectures: I) Stream-Dataflow and II) Micro-batching. In Fig. 8 we visualize these two main architectures in an abstract way. The *stream-dataflow* approach naturally distributes stream processing among pre-allocated processes (also known as long-running tasks) throughout a cluster that typically map to event-handling entities in a respective programming model along with mutable local state. On the other hand, micro-batching emulates stream processing on top of a batch-centric execution, orchestrated through an infinite sequence of input batches. Consequently, applications which are executed on a stream-dataflow architecture can react to every incoming record at the time of its ingestion on a parallel task whereas in the micro-batch model the application will have to react to sets of events when a batch eventually gets scheduled and processed. Both designs are used widely for different reasons, which we will analyze in more detail further below.

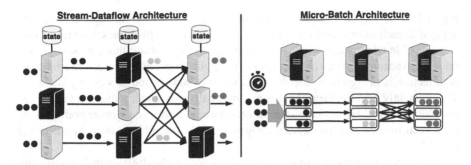

Fig. 8 Distributed data stream processing architectures

3.2 Execution Models

The Stream-Dataflow Approach. The stream-dataflow approach provides a more direct mapping of semantics and execution entities, as already mentioned in Sect. 2.1. According to that architecture, an application is specified and given to the system for execution in the form of a dataflow graph. A dataflow graph, is in most cases a Directed Acyclic Graph (DAG), which consists of stateful tasks and data dependencies i.e., between them. Tasks are independent processing entities and dependencies represent data streams of records. A task encapsulates the logic of a predefined operator (e.g., filter, map, window, aggregate, join) or a routine with user-specified logic. A data stream between two operators simply represents an infinite sequence of data elements/events produced by a task, which are available for further consumption.

Typically, a driver or master node receives a dataflow graph, and schedules tasks among the available cluster resources, which will be invoked once, and executed continuously throughout the possibly infinite lifetime of an application. Each task maintains and accumulates state independently (managed locally or externally) while processing data records from subscribed streams on-the-fly. This practically means that encapsulated operators in a physical task are repeatedly invoked for every respective record that is delivered for consumption. Depending on the dataflow system design, the dataflow can also contain control events (e.g., synchronization barriers) that trigger system specific actions, and are thus hidden from the operator logic. Tasks are the finest grain elements of computation, however, they might also encapsulate a chain of multiple operators as we describe further in Sect. 3.5.

The stream-dataflow architecture has been adopted by a plethora of systems throughout the large-scale computing open source ecosystem such as Apache Storm [5], Samza [3], Flink [26], APEX [82], Heron [61] and proprietary SPs alike, such as Google's MillWheel [9] (a precursor of Google Cloud Dataflow [50], the cloud runner of Apache Beam [10, 83]).

Discussion: There are certain fundamental benefits that the stream-dataflow approach offers. First and foremost, it supports event-based granularity for any custom processing task. This means that users can define arbitrary processing logic that can be applied in each individual event of an infinite sequence. This imposes no restriction (e.g., flexible windowing) whatsoever in the type of operations that can be declared, thus, it can accommodate a rich ecosystem of applications. Furthermore, the dataflow architecture is proven to be capable of delivering sub-second processing latencies. This is mainly because everything is automatically pipelined, without intermediate scheduling and additional communication between master and worker nodes, as we have seen in the batch processors designs (e.g., Hadoop [1] master, Spark driver [4, 89]).

The main arguments against stream-dataflow approaches [90] are their complexity when it comes to flow control mechanisms and fault tolerance but also the increased complexity of applying flexible runtime reconfiguration (e.g., auto-scaling or state sharding), all due to the nature of long-running tasks. Events are delivered and consumed in arbitrary order across parallel tasks, and as a result, there is a lack of a coarse-grain unit for transactional processing. Initial efforts to provide fault tolerance in the stream-dataflow design [17] enforce costly decentralized communication and state management mechanisms that can impose heavy overhead to the execution of the stream application, such as active replication [77] or passive backup with acknowledgments. However, recent advances in this domain circumvent a large portion of such overhead by lifting assumptions to repeatable, durable data sources [2, 3] with snapshotting [27] and lightweight transactional processing mechanisms that can guarantee consistent, yet non-disruptive, processing.

Flow control can also be challenging in the stream-dataflow architecture, due to the skew between the processing and ingestion rates [61]. This is typically handled via combining network channel and data source back-pressure. Finally, runtime reconfiguration is pending downside of stream-dataflow systems. First, because a reconfiguration process can violate important continuous processing and latency constraints, and secondly, due to the difficulty of repartitioning local state in operators at runtime. Several recent approaches in tackling these problems in dataflow-based SPs are covered further in this chapter (Sects. 3.3 and 4.3).

The Micro-Batch Approach. Micro-batching is a solution that enables processing data streams on batch processing systems. It is a simple concept that became mainstream through the widespread use of the Apache Spark stack [90]. With micro-batching we can treat a "stream" computation as a sequence of transformations on bounded sets by discretizing a distributed data stream into batches, and then scheduling these batches sequentially in a cluster of worker nodes. Discretization typically follows a user-specified time-interval [90], which reflects the granularity at which the input data will be partitioned in terms of processing time. When such a timeout occurs all input records collected during that interval are replicated (for durability), and then scheduled together as a single unit (e.g., RDDs on Spark [4]) for batch transformations.

The idea of re-using batch processing apparatus for processing data streams is very attractive due to the existing ecosystem of batch processing frameworks and systems, and has been approached before with Comet [52], Hadoop Online [39] and

CBP [67]. However, Spark's approach is a noteworthy contribution in fault-tolerant data streaming, regardless of its limitations.

Discussion: There are several benefits to micro-batching, some of which are not yet highlighted or tackled by other architectures. First and foremost, it exploits native fault tolerance in batch computing out-of-the-box and thus eliminates the need for any additional control apart from data replication at the ingestion phase [90]. Secondly, it can gracefully handle stragglers, thus, eliminating the need for sophisticated flow control and simply rely on the ability to execute redundant processing. Straggler elimination is, again, an ability inherited from the batch-processing runtime (e.g., Hadoop [1] and Spark [89]). Finally, micro-batching blends by definition batch and stream processing, thus, allowing intermediate results from batch applications to be joined with micro-batch transformations trivially to aid pipelining.

Perhaps the most critical performance-related downside of micro-batching is the discretization latency. Discretized data ingestion imposes a fixed lower bound on the processing overhead, which reaches an order of several seconds at a minimum [90], excluding the additional replication and scheduling times. This makes micro-batching rather less attractive for use-cases that rely on timely computations. Furthermore, during ingestion across consecutive batches, computational resources are not actively used, which might result in under-utilization of a certain degree. Finally, micro-batching restricts the granularity and thus expressibility of programming models that are built on top, to a batch level in contrast to dataflow systems that allow for record-level operations.

3.3 Processing Guarantees Upon Failure

The general terms people refer to when talking about processing guarantees and fault tolerance for different stream processors are *at-least-once*, *at-most-once*, and *exactly-once* processing. Unfortunately, even these terms lack clarity on what they actually mean in practice which often results in some general confusion. Since this is a rather important concept (but not as complicated as it might seem) we will attempt to elaborate on their actual meaning and hopefully offer some fundamental insights and reasons behind mainstream state management mechanisms.

First and foremost, processing guarantees refer to the state of an application. For example, when a system ensures *exactly-once* processing guarantees, it verifies that any application it runs will consume its input without record losses whereas all declared *internal states* will be updated once per record, even in the presence of failures. Bear in mind that this does not mean that the output of the application will be consistent under failures. Output guarantees are beyond the focus of this study and typically require strict transactional processing and version control between processing and storage systems to eventually be achieved, in addition to other assumptions such as deterministic processing and idempotency [9].

Fig. 9 Task actions and
guarantees

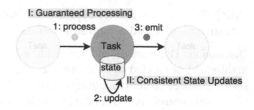

Preliminaries. We approach the general problem from the perspective of a physical
dataflow graph. As mentioned earlier, from a macroscopic point of view an execution
is modeled in the general case as a DAG of tens or hundreds of tasks interconnected
by streams. The main problem here is that any task can fail at any time, thus, the
system should deal with such a failure in order to satisfy the promised guarantees.
Before going further into guarantees, we will first go back to a microscopic level and
observe task behavior.

As seen in Fig. 9, a physical task is an independent entity (e.g., a process or a JVM
thread) that executes the following steps per input event:

1. Receives an input record,
2. Updates its state,
3. Optionally emits output records.

With those steps in mind, followed by every task managed by an SP, we propose
the composition of processing guarantees simply by using the following two distinct
properties, each task should eventually satisfy:

- **Property I: Guaranteed Processing**. *All records in a task's input dependencies
  should be eventually delivered to the task and fully processed (i.e., steps 1–3)*
  at-least-once.
- **Property II: Consistent State Updates**. *Each input record should lead to exactly
  one state update.*

Network Assumptions. Throughout this overview we will make the following
assumptions when it comes to the network links between every two tasks: no duplicate records are delivered, and all events sent are eventually delivered through a
channel in the same order that they were sent. This is typically satisfied by perfect
FIFO links, which network protocols such as TCP can guarantee.

Process and Failure Assumptions. We assume that every task that has not failed
follows a correct behavior, i.e., it executes steps 1–3 described in Fig. 9 for every
input record that it receives. Failures can normally happen before record delivery
or upon executing any of these steps. In any of these cases we consider the task as
having failed and the record that triggers the task computation as not having been
fully processed.

Now that we have seen the basic properties let us go deeper into processing
guarantees.

At-Most-Once Processing. This means that neither property **I** or **II** are guaranteed. A system that is turned off can still guarantee this, since zero records delivered and processed satisfy *at-most-once* processing. It is often fine to go with no guarantees. For example, a system can discard tuples when the input rate is too much to handle (e.g., an approach adopted by Aurora [6]). This is also the case for any existing system that runs with state management mechanisms turned off. For example, tasks in Apache Storm can re-start from scratch with a null state, which is fine in certain use cases such as approximate streaming. One can argue that this approach eliminates the reason for using an SP, however, several applications can still operate with no processing guarantees such as approximate stream analytics and best-effort complex event processing. The downside of "no guarantees" is countered by crucial benefits such as low latency and maximum throughput, since no mechanism needs to alter the execution of the application to enforce "correct" results so data flows through the system with no disruptions.

At-Least-Once. When a system offers *at-least-once* processing guarantees it means that property **I** is satisfied (no input loss) given the aforementioned network and process guarantees. Duplicate processing of a task's input can still occur. This can be fine in many situations where we are just interested in processing everything and do not care whether we process a few records (and their productions in the dataflow graph) more than once such as searching for a specific tweet during the day or generally executing any idempotent operator. It is also a fine compromise between high throughput, low latency, and correctness. We have seen *at-least-once* guarantees in two different flavors in the existing ecosystem today: relying on a repeatable logging system or natively through record acknowledgments and source-level replication. The two approaches work as follows:

Repeatable stream logs. A repeatable stream log, such as Apache Kafka [60], completely persists distributed streams from its producers and additionally allows for "repeating" stream consumption from a given offset. Repeatability unlocks the ability to replay any distributed sequence of records and thus guarantee processing (property **I**). Several systems, such as Apache Samza [3] and Kafka Streams [56] build natively on-top of a repeatable log to satisfy *at-least-once* processing. This way they redirect the materialization of every intermediate output stream to the log system. In the case of a task failure, an input is replayed from a previous offset and potentially re-process records (so property **II** is not guaranteed). This approach is modular, elegant but unfortunately incurs heavy overhead in order to persist intermediate streams or state into a persistent log (that in turn replicates data for durability).

Record Acknowledgments This approach makes no assumptions about persistent logs and thus implements input persistence and execution monitoring natively. Apache Storm [5] employed this strategy from its very early development stages. The mechanism of Storm relies on creating a bookkeeping entry per record r seen at the sources and eventually removing it when r and all its productions (subsequent output consumptions) in the dataflow graph have been completed. A hand-tuned timer, which is set by the user, alerts the sources in case a record bookkeeping entry takes

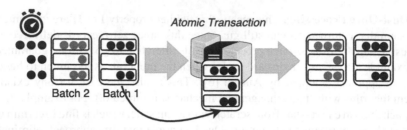

Fig. 10 Exactly-once processing in micro-batching systems

too much to be using an alarm. That can happen due to failed tasks that "break" the dataflow graph or often due to heavy load incurred by the application. In any case, when the alarm triggers for a record it is being resubmitted to the dataflow graph for execution. The benefit of using record acknowledgments is that upstream backup at the sources eliminates the need for external dependencies such as a repeatable log. However, bookkeeping incurs heavy computational and memory overhead while also enforcing explicit association between input and output records on each task from the programming model.

Exactly-Once Processing. This brings us to one of the most challenging concepts in distributed stream processing: *exactly-once* processing guarantees. To achieve this, each task has to satisfy both property **I** (guaranteed processing) and **II** (consistent state updates). Before we dig into how stream-dataflow systems can achieve that, we will recap the approach of micro-batch systems regarding fault tolerance.

Peeking into Guarantees in Micro-batching. Micro-batch stream processing systems (e.g., D-Streams [90]) provide exactly-once processing guarantees as an inherent property of the host system. To illustrate this, we depict how an execution of a discretized stream works in Fig. 10. As mentioned earlier, an input data stream is pre-partitioned into batches. Each batch first gets replicated and then scheduled for transformation as an atomic transaction that either completes or not. If a worker fails, then the batch that is currently processed (or most specifically that respected partition) is rescheduled and eventually the computation completes.

This guarantees exactly-once processing and satisfies the two core properties. Property **I** (no loss) is satisfied, since each input batch is durably replicated and thus no input in a batch is lost. Property **II** (consistent state updates) is harder to notice since we have not talked about state in the micro-batch model. In micro-batch systems such as the case of Apache Spark [90] all state updates are encoded as immutable transformations between two consecutive batches, thus, they can either succeed completely or not (in the latter case the full computation is repeated). That automatically satisfies property **II** and makes micro-batching inherently capable of exactly-once processing.

Among the major benefits of fault-tolerance with micro-batching we should highlight that the processing rate of every batch is constant and periodic, thus, durability mechanisms yield a constant overhead in the overall execution. On top of this, recov-

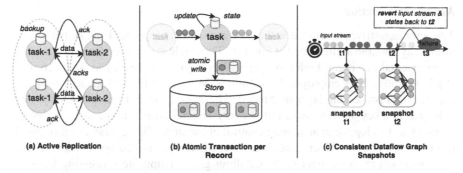

Fig. 11 Exactly-once processing for distributed dataflows

ery can be executed in parallel in a cluster environment since all worker tasks are stateless and available resources can be harnessed maximally.

Exactly-once Stream-Dataflow Systems

In the area of stream-dataflow systems, exactly-once processing techniques have evolved from very strict transactional-processing approaches [17] such as two-phase commit alternatives (e.g., in FLUX [78]) to lightweight snapshots [27].

Active Replication

Active replication [17] is a costly approach of handling failures in distributed dataflow processing, with heavy resource utilization and associated protocol complexity. This technique was incorporated by FLUX [77], the DPC Borealis protocol [16], and other approaches [25] where non-interruptive, highly available execution is the main concern.

As illustrated in Fig. 11a the basic idea is that all computation is duplicated into pairs of identical tasks. This yields a fault tolerant execution where at least one replica of each task is considered to be running at any time. Along with twice the computational resources, network traffic can even reach up to quadruple amounts of records in transit [25]. Furthermore, the need for message-intensive two-phase commit protocols and acknowledgments during the execution (for state synchronization) can negatively impact throughput [90].

With active replication, property **I** is satisfied, since all input is buffered and eventually processed on at least one of the replicas (garbage collection handles unneeded input records). Property **II** is also granted since at least one replica processes all inputs streams (*exactly-once*) and updates its state at any given time.

The main benefit of this approach is that recovery is trivial and high availability is guaranteed from the fact that at least one replica is running at a time among each task pair. Finally, tolerance on the number of failed tasks can be further increased by increasing the number of active replicas accordingly.

Atomic Transactions

This approach strives to achieve, at the task granularity, properties **I** and **II** in the most direct way possible as illustrated in Fig. 11b. The single assumption is the availability of an external persistent, high throughput, key-value store (e.g., Cassandra [63] or BigTable [36]). It is employed by Google's MillWheel [9] and Trident [85] (an experimental processing layer on Storm). The main concept is simple. Everything is logged in a key-value store or rather every record consumption and state change on every task is logged in an atomic commit at the store. This guarantees property **I** (no input loss) but not **II** since duplicate records can be emitted upon task recovery and processed by consumers in the dataflow graph. Duplicate processing leads to inconsistent state updates which violates property **II**. There are several ways of dealing with duplicate elimination and thus, satisfy exactly-once processing. For example, as we have seen in Google's MillWheel system, bloom filters can be used to mark duplicate input records effectively.

Furthermore, several optimisations are still possible with atomic transactions such as batching updates together in a single transaction, acknowledging record senders for garbage collection and better flow control and using sophisticated key value stores such as BigTable [36] with blind write support and high append-throughput [9]. This technique also allows for task-level recovery which is considerably more efficient than restarting for example the whole dataflow graph, especially when the number of tasks is too large to reconfigure. However, this approach is often rather complicated and hard to achieve in an average setup using lower-end commodity cluster infrastructure with limited memory and storage capacity for all required logs alongside a high-throughput transactional key-value store.

Consistent Snapshots

The concept of distributed snapshots is not a new one. Chandy and Lamport [35] originally conceived the idea of encapsulating the complete "picture" of a distributed system and be able to use it in order to potentially re-execute it from there or simply use it to find out more about its previous execution (i.e., safety and liveness properties). Snapshots in distributed systems can be used to revert the whole state of an arbitrary execution back to a savepoint and that is typically enough to guarantee *exactly-once* semantics. As illustrated in Fig. 11c, imagine taking periodic snapshots of a distributed dataflow graph (think of auto-saving in video games). For any task failure that can possibly happen we can eagerly roll back the whole dataflow graph to the last saved savepoint by resetting task states back to their saved values. Additionally, it is also required to be able to replay the input stream from the snapshot upon recovery, which is trivial given a durable repeatable log such as Apache Kafka [2, 60].

One of the recently adopted approaches in the domain of stream processors is to asynchronously execute a snapshott phase during the regular operation of the system and thus avoid disrupting continuous processing. We provide a comprehensive analysis of the original algorithm (ABS) [27] that was implemented in Apache Flink (more in Sect. 4.3). Apache Storm's v.1.0 is planned to incorporate ABS, thus, making it the most popular technique for *exactly once* processing. Additionally, the Apache

APEX system [82] takes a very similar route when it comes to snapshots, however, it uses the same mechanism to draw snapshots and apply windowing and thus enforces a batch-like execution that could otherwise be avoided. The main known downside for employing dataflow graph snapshots as a fault tolerance mechanism is the recovery granularity. That is because the whole dataflow graph has to reset back to a savepoint, however, partial recovery is still possible and considered as an improvement [27].

3.4 Flow Control

Any system has limits in the data rates that it can sustain, and stream processors are no exceptions. Thus, it is often required to regulate the data ingestion rate due to the imbalance between the data source rate and the processing rate that the system can handle at any time. Unfortunately, this is a highly dynamic problem, since both input and processing rates might vary considerably during a live execution.

Data spikes are common for several data sources such as mobile devices that suddenly get access to a wireless network and transmit events that they have buffered during the day. The same applies to logging systems in big data centers that transmit historical logs periodically in spikes when their cache is full or a file has finished.

Stream processors can also exhibit variable processing capacities during their execution. For example, heavy-load routines such as garbage collection or snapshotting to disk can introduce staleness periods, which can in turn affect continuous processing. Finally, workload imbalance across different tasks within a dataflow execution itself makes flow control mechanisms crucial for sustainable processing in order to avoid situations where network buffers are overflown.

There are two main strategies for enforcing flow control in such situations [61]: dropping data or employing back-pressure.

Dropping records. This is a common and effective strategy used by several systems that do not have strong processing guarantees such as Apache Storm. Simply put, when an input network buffer in a task gets full (no available resources to deserialize incoming records from the network) the task discards new records that might arrive. The discarded records either get lost forever (which works for *at-most-once* guarantees) or resent (combined with *record acknowledgments* and upstream backup in an *at-least-once* approach). As expected, this strategy does not work with *exactly-once* processing guarantees.

Backpressure Mechanisms. There are several types of backpressure used in production systems today, often employed in combination. An evident approach is to utilize *TCP backpressure* which was used in Twitter's Heron [61] among others. With TCP windowing, sender and receiver nodes adjust their processing rates depending on the size of the send/receive network buffers. Gradually, backpressure is realized from the congestion point back to the sources of the dataflow graph and the overall execution graph rate is calibrated. In practice, this mechanism can be too aggressive and potentially slow down *additional* tasks in the topology than the ones that are actually needed such as the successors of the congested tasks in the dataflow graph [61]. This is mainly due to the fact that network buffers are multiplexed and shared among arbitrary tasks in the topology. Thus, the flow control in this case is non-selective and is applied collectively with potentially high impact in the general throughput.

Another strategy that has proven to be more effective is *source backpressure*. The main idea is that we selectively inform the dataflow sources to go into "backpressure mode". In this mode the sources are halted and thus data injection stops until back-pressure is off. The cost of this approach is the introduction of additional control messages and protocols which is generally not recommended. However, this is a more direct way and happens to work well in harmony with repeatable and durable logging systems such as Apache Kafka. An alternative approach involves the iden-tification of different stages in an execution graph and employ this strategy at the beginning of each stage.

Discussion: Backpressure mechanisms are crucial in data stream processing. A gen-erally good practice is to use a combination of TCP and source-based backpressure to have a well regulated and adaptive traffic flow in the system [61].

When it comes to micro-batch systems, the problem resides solely at the driver node that discretizes a stream. Under time periods with high input spikes, the system might be out of resource capacity to replicate batches and then schedule them for further processing in the same constant rate. However, a form of source backpressure strategy can be possibly employed at the ingestion point of the input stream and coordinated through the driver that configures the execution.

3.5 Execution Plan Optimisations

There is a long list of possible optimisations that can be employed by SPs [55]. Most of such optimisations occur during an intermediate compilation phase from a logical to a physical plan. Many types of such optimisations require explicit declaration of the data types and processing logic that is being executed in each step. However, there are certain optimization strategies that can be applied without any semantical knowledge. In the context of general dataflow graphs, operators can be collocated, when applicable, in order to utilize less resources while achieving better throughput. This class of optimisations is used in several dataflow processing systems such as Apache Flink and Apache APEX and is known as "operator fusion".

Operator Fusion. Fusion, also known as *chaining* is inspired by "Loop Fusion", a well known compiler optimisation technique that replaces multiple loops that iterate over the same sequence with a single loop and thus potentially improve runtime performance due to data locality [72]. In the context of distributed dataflow graphs, this translates into merging multiple operators into a single execution task instead of allocating several parallel tasks. In addition to resource savings this has proven to improve performance considerably mainly due to reusing input data and avoiding network channels. In addition to open source system adaptation, Google's managed

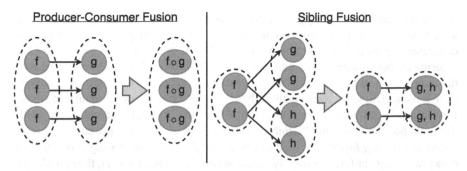

Fig. 12 Examples of fusion in dataflow execution graphs

cloud executor system for the Dataflow Model [50] is utilizing operator fusion. This is an inherent optimisation technique adopted from Google's FlumeJava [32], an internal dataflow pipeline processing system and a precursor of the Dataflow model. This has proven to result in cost savings for the users of the service while increasing throughput.

In brief, there are two common scenarios where operator fusion can be applied de facto [32] during a physical plan translation: *producer-consumer* and *sibling fusion*. We depict both of these optimisation strategies in Fig. 12. As illustrated, *producer-consumer* fusion is applicable when there is a one-to-one parallelism and data dependency between two subsequent tasks in the execution graph, for example f and g. In that case, only a single fused task $f \circ g$ is created instead. This way, in practice, we avoid redundant record serializations, deserializations and network latencies that we would otherwise have between tasks f and g. In the case of *sibling fusion*, we can merge two or more tasks, for example g and h, into a composite task, since g and h are consumers of the same data. Again, this way we transmit and serialize records once for each partition, which gives a significant benefit to execution efficiency.

4 Case Study: Stream Processing with Apache Flink

The domain of big data processing is undergoing disruptive paradigm shifts and trends. We have witnessed an unavoidable shift from an era where database management systems were the dominant means of storing and analyzing data at small scale to the MapReduce paradigm, where large scale processing replaced centrally indexed data with a partitioned view.

Distributed file systems such as the Hadoop Distributed File System (HDFS) became the norm for storing data in a scalable manner and a proliferation of several batch-based data processing frameworks, such as Apache Spark [89] served as dedicated endpoints to distributed data management and analysis. The recent adoption of resilient, scalable logging systems, such as Apache Kafka [2] increased the awareness and need for continuous and incremental processing of data, with several dedicated stream processors such as Apache Storm [5], Samza [3] and Flink [26] entering the big picture. This eventually led to a "split-brain" problem in hundreds of data centers, where people had to manage and maintain data in both static and stream-centric manner. Several design patterns had been put forward towards managing these fundamentally different workloads under a structurally coherent architecture, with the most popular being the "Lambda Architecture" [70]. Concepts such as *batch*, *speed* and *serving layer* were used to categorize the overly increasing ecosystem of data processing platforms primarily based on timely needs. However, the complexity of using different batch and streaming runtime architectures together increased the complexity of writing and maintaining data management pipelines.

The Kappa architectural pattern proposal [84] strived to provide an answer to this problem by fusing the "batch" and "stream" layers together and using streams as a first class citizen. The basic idea is that streams can be used to generate sets, tables or other static data representations and subsume batch processing by pipelining computation seamlessly.

Apache Flink [26] is a general unified data processing framework and a materialization of the ideas behind the Kappa architecture. In this section we offer a brief overview of the Apache Flink stack, the main entities that implement the core system properties and some further insights behind its lightweight snapshot-based fault tolerance mechanism.

4.1 The Apache Flink Stack

The Apache Flink system [26] offers a complete software stack of libraries for programming and executing distributed dataflow applications as depicted in Fig. 13a. At the core of the programming model, there are two main building blocks, exposed as abstract data types, one for streams (DataStream API) and one for finite sets

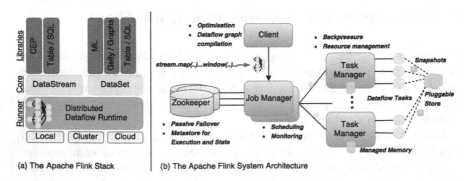

(a) The Apache Flink Stack   (b) The Apache Flink System Architecture

Fig. 13 An overview of the Apache Flink stack and architecture

(DataSet API). In Sect. 2, we offered a brief overview of the capabilities of functional APIs like the DataStream API, alongside other known programming models; whereas here we will focus on the big picture first and then explain how such a system is architectured by analyzing in more detail its snapshotting capabilities.

The users of Apache Flink can either specify data transformations using the core APIs or declare computation logic using one of the higher level libraries. The high level programming libraries of Flink are an example of the capabilities of a dataflow processing system, by exposing data representations such as relational, graph, event-based (complex event processing) and machine learning constructs over a single runtime. Programs written in any of these domain specific libraries translate into stream or set transformations. From there, all transformations are analyzed and translated further into logical dataflow operators and finally into an optimized, physical graph of tasks.

4.2 The Apache Flink System Architecture

There are three main runtime components involved in the compilation and execution of a dataflow graph, as sdepicted in Fig. 13b. The general architecture resembles characteristics from other principal system designs such assssss Hadoop [1]. First, a *client* component is responsible for compiling and optimizing operators (e.g., map, reduce, window) into dataflow tasks and then submitting an execution graph for execution at the *Job Manager*. The *Job Manager* has the role of the master node, as in most typical distributed computing systems and handles job execution and monitoring. It schedules individual tasks to *Task Managers* and collects meta-information about the execution. The *Task Managers* are the workers and are responsible for managing local resources such as slots and memory allocation throughout the continuous execution of a dataflow job. Tasks can snapshot their internal state as we will see next in Sect. 4.3, and Flink provides a modular pluggable interface for many types of storage back-ends that can be used for that purpose (e.g., Native RocksDB, HDFS, in-memory). Finally, Flink guarantees high availability at the *Job Manager* level (which suffices for maintaining a continuous uninterrupted execution). That means that the *Job Manager* is not a single point of failure, and standby processes can take over its role at any time whenever it fails. The distributed coordination and crucial state needed for recovering all job executions are both handled by Zookeeper, a distributed open-source system that is used for coordinating distributed operations that involve locking and leader election.

4.3 Lightweight Asynchronous Snapshots

As we have seen in Sect. 3, there are different solutions for fault tolerance in streaming dataflow systems. As a fault tolerance example, we show how state snapshotting

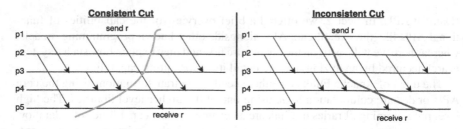

Fig. 14 Distributed snapshots as cuts

works in Flink. We briefly describe preliminary concepts (e.g., snapshots, cuts and halting snapshots), then the asynchronous snapshotting algorithm (ABS) applied by Flink.

Snapshots as Cuts. Chandy and Lamport first defined distributed snapshots [35] in the most general case as *cuts* in the timeline of events exchanged between independent processes. We depict some example cuts in Fig. 14. Each line depicts the timeline of a process, and arrows represent messages sent and delivered across processes. A cut can, therefore, be depicted as a line that crosses all process timelines. The content of a snapshot is the current internal state of each process at the point in its timeline that the cut crosses.

If a distributed snapshot is taken "correctly," one can simply reset a full distributed system from that exact snapshot seamlessly. A correct snapshot, however, needs a more precise definition. Intuitively, the state we persist within a cut should reflect a valid execution for a correct snapshot. According to Chandy and Lamport [35] that can be only true when all prior actions that resolve into a process state are also reflected in the cut. These prior actions are also known as "causal dependencies." Figure 14 illustrates what a consistent cut is, in a simplistic way. The first (green) cut in Fig. 14 is correct, since all causal dependencies are reflected in the state, i.e.,s every event that was received had also been sent according to that cut. The second cut is incorrect, since the snapshotted state of process $p5$ depends on receiving an event that was never sent by $p1$ according to the same cut.

Halting Snapshots. This notion of distributed snapshots is fundamentally suitable for dataflow graphs. Microsoft's Naiad [73] was one of the first known distributed dataflow systems that applied snapshotting to implement fault tolerance. The idea was to *halt* the whole graph execution, then persist all process states and events in transit, and finally continue back to normal execution. The problem with that approach is, evidently, that it violates the most important property of streaming systems: continuous, uninterrupted processing.

Asynchronous Snapshots. The problem of execution halting motivated the idea of Asynchronous Barrier Snapshotting (ABS) [27], the algorithm behind Apache Flink's fault tolerance mechanism. As the name implies, the goal is to take a consistent snapshot *asynchronously*, without halting the system execution. To do this under regular data ingestion, one has to superimpose the snapshotting process while tasks

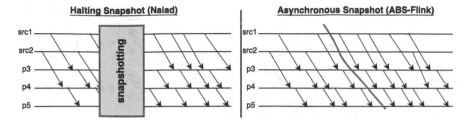

Fig. 15 Halting versus asynchronous snapshots

consume and process records. The ABS algorithm achieves that while minimizing the size of the required state. The basic idea is to intermix data records with barriers that are injected at the sources of the dataflow graph, and then subsequently visit the rest of the tasks in topological order while triggering state writes to external persistent storage.

The difference between Naiad's *halting* approach and the ABS approach can be seen in Fig. 15. One important property visualized in the figure is that no records in transit are part of the cut, i.e. no arrow is being overwritten by the cut line. That is a pleasant side effect of the ABS algorithm, and occurs because the snapshotting process is pipelined topologically, thus there is no need for buffering pending records while waiting for any termination condition (see Chandy and Lamport's approach [35]).

The asynchronous snapshotting algorithm was first implemented on Apache Flink and is managed by the *Job Manager* node. Periodically, the *Job Manager* initiates a snapshot phase for each running execution graph. The special barriers mentioned before are injected into each respective partitioned data stream starting from the sources and all tasks independently snapshot their state and acknowledge the completion of their local snapshot to the *Job Manager*, along with a reference to their replicated state (which can be datastore specific). Once all tasks have acknowledged their success, the *Job Manager* marks the snapshot as complete, and can consider it as a valid savepoint for recovery.

We depict the whole snapshotting process in detail in Fig. 16. In that figure we visualize all steps and state that is involved during checkpointing. On the left we can see a distributed data stream of records enriched with checkpoint barriers (in that case the barriers of the $n + 1$ snapshot). The current tasks in the DAG that have already checkpointed their internal state in the current snapshot (n) have a red color while the pending tasks are yellow.

One important detail in the algorithm is its *aligning* phase. When tasks have multiple inputs, they "align" their input consumption, by blocking channels where they have already received barriers. This way they ensure that all causal dependencies are consumed before proceeding to snapshotting and propagating the barriers further. In the example of Fig. 16 the two last tasks of the DAG (known as sinks) have entered their aligning phase. When a global snapshot is complete it is being persisted and available at the system runtime's disposal in case a rollback is needed. On the right

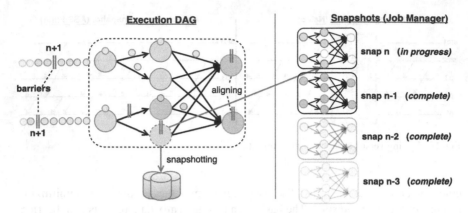

Fig. 16 An illustration of the ABS algorithm in Apache Flink

side of Fig. 16 we can observe a number of global snapshots that are being maintained at the *Job Manager* (in the form of metadata).

Recovering from Snapshots The recovery process from a snapshot is straightforward. The *Job Manager* first selects the latest, complete snapshot and then injects the state handles to the newly scheduled tasks. Every task retrieves its old state before starting its regular execution, which it sets as its initial state. Finally, all dataflow sources have to start generating records from where they were left during their snapshot. A typical approach to solve this problem is to use a persistent log such as Kafka and store the current offset read so far to the task state.

For example, consider again Fig. 16. If a task failure occurs while executing snapshot n, the system will instantly restart the execution graph from the states pinpointed by snapshot $n - 1$, since it is the last complete savepoint the system has at that point in time. In addition, it is required by the sources of the execution graph to rollback their ingestion to the input stream offset that they once had saved upon persisting snapshot $n - 1$. Background tasks of the system execute any appropriate garbage collection by removing older snapshots while also asynchronously persisting in Zookeeper all the meta-information required to fully recover the Apache Flink master with a valid view of the pipelines executed.

5 Applications, Trends and Open Challenges

Data stream processing is still a rapidly evolving domain in data management. At one hand, systems incorporate new abilities that enable new types of applications and services to build on top. On the other hand, new types of applications and needs inspire the development of more sophisticated features both related to programming models and system abilities and guarantees.

Perhaps the most well-known trends in data stream processing systems are the unification of batch and stream processors, efficient pipelining, and stream state exposure. However, in this section, we will focus on a few unconventional, yet emergent challenges in data stream processing hidden behind the spotlight, and offer further insights and directions on how to achieve them.

5.1 Graph Stream Processing

The domain of graph analytics efforts are currently split between dynamic graph database management systems (e.g., Neo4j [87]) and static graph processing frameworks (e.g., GraphX [88], Pregel [69] and GraphChi [62]). The most significant downside of the current state of the art is the focus on static graph snapshots, disregarding the fundamentally dynamic nature of graph-structured data. One can easily spot the potential for a direct, continuous ingestion of a graph (e.g., as a stream of edges)ssssss. Users in social networks relate to other users via actions such as *likes, reposts, tweet mentions, friend requests etc.* Such actions can be processed and pipelined within a composite and complex graph processing task.

The idea of processing graph data in a streaming fashion has been studied extensively in the past under different contexts, the semi-streaming model [18, 46, 81] being one of the most studied ones. According to that model, several common graph properties can be derived in a single pass. Some examples of common properties and aggregations where this applies are: bipartiteness checks, shortest paths estimation, degrees, triangle and triangle counts, among others. The best-effort, low-latency computation of such complex properties can offer many benefits in the future. For example, we can achieve efficient adaptive routing in software defined networks (SDNs), hotspot and fraud detection in network security, and influence or epidemics estimation in very large networks of human-generated data.

This paradigm shift in graph analytics can be achieved through two main directions: efficient summary data structures for graph data, and effective integration and support in existing data stream processing systems. For summaries and data structures, there is a plethora of existing approaches to be leveraged and extended, each targeting different graph approximations such as spanners and sparsifiers [8, 76]. The general idea is that by using such internal representations we can compact and continuously update compressed graph state while also maintaining some specific properties. For example, graph spanners [76] are sparse subgraphs that preserve approximate distance information between each pair of vertices that have appeared in the graph thus far.

As always, for implementing and extending such novel graph processing techniques, there should be proper system support. Current stream processors lack even a basic programming model for incorporating graph stream semantics. Existing graph-centered programming models and libraries such as Pregel [69] and GraphX [88] are architected around iteratively processing finite sets of data such as adjacency lists, which makes the adaptation of a stream-centered model hard to achieve. A few initial approaches such as CellIQ [57] and Chronos [51] present an intermediate step for incremental graph processing, that is based on continuous snapshots under coarse window processing constraints. Nevertheless, there is as of yet no programming model that operates at the granularity of individual events that allows for flexible graph state representations and aggregations. We strongly believe that such a direction can impact the graph processing domain and change the way we think about complex data.

5.2 Online Learning

Machine learning (ML) techniques are becoming increasingly mainstream and an integral part of modern analytics pipelines. The streaming execution poses several attractive benefits in ML such as low latency in model building, however, it also brings new challenges. First and foremost, the majority of existing ML techniques that originally operate on static data sets are inapplicable. That is mainly due to their bounded data assumptions (e.g., ss knowing all possible data points for clustering, or the inability to incorporate concept drift adaptation when dynamic changes occur in the data)sss.

Among systems for online learning, Weka Online, MOA [22] and Apache SAMOA [42] serve as specialized programming libraries and offer a collection of known algorithm implementations such as Vertical Hoeffding Trees and stream clustering. Unfortunately, despite their unique features, the limited scope of these frameworks, combined with exposure of non-declarative low-level system constructs make their general adoption and integration with wider analytical pipelines a challenging task. Apache SAMOA is the only system currently that achieves large scale deployments for online ML, and it does so by using known dataflow-based stream processors as runners (e.g.,s Apache Storm, Flink and Samza).

Finally, upcoming programming models for online ML should incorporate *concept drift* as a first class citizen, that is, the gradual change in the distribution of values exhibited as a stream. That is a vital property for ML system pipelines that can potentially operate continuously. Many classes of online learning algorithms would need to be revisited and designed with that concept in mind. There are a few noteworthy examples towards that direction such as adaptive stream windows [21], which consider concept drift in discretizing evolving streams based on data trends.

5.3 Complex Event Processing

To derive useful information from data streams, we often have to detect patterns in events. A mobile phone network operator can benefit from building a custom stream processing application that analyzes the operation of its systems. Cellular networks consist of lots of different types of network nodes, which all produce network logs in different formats. Manually making sense of all of that information when debugging a problem can be hard.

Complex event processing (CEP) aids the development of such applications [68]. It focuses on identifying important events, such as failures and business opportunities, and reacting to them as soon as possible. At a network operator, for instance, it could be important to detect the situation when there are lots of call drops in a certain geographical area. A CEP application might generate an alert to the engineers working in the network operations center, and it can even supply them with additional information. For example, it can take all log entries that contain information about call drops, group them by a reason code field, and perform a counting aggregation to help with identifying the root cause of the problem.

Distributed data stream processing systems could help managing a large amount of such event data. Future work in this area includes providing a rich set of features for CEP for the open source systems discussed here. However, there are efforts in that direction. For example, Apache Flink now provides a CEP API in their 1.0 version.

6 Conclusions and Outlook

In this chapter we focused on data stream processing, an increasingly popular paradigm in the general field of data management. The applications of stream processing are vast and vary from computing rolling aggregations to building extremely complex data pipelines consisting of asynchronous microservices. The emergence of distributed, durable logs and their universal adoption in modern data processing backends inspired the creation of sophisticated stream processors with strong processing guarantees and rich programming models. Modern stream processing systems are able to manage application state fault tolerance as well as efficient partitioning transparently without human intervention. Furthermore, many data stream processing systems today can support different notions of time and serialize operations consistently under such time assumptions.

We have shown programming abstractions for implementing distributed stream processing applications: a programming model for lower-level *dataflows*, a model for higher-level *functional APIs*, and windowing abstractions, which help processing infinite datasets by partitioning them into finite subsets.

We have also seen in detail a universal execution model for data stream processing, the *dataflow graph*, which models stateful operators and data dependencies between them. However, we have seen that continuous processing can also be emulated on widespread batch processing systems using time-discretized micro-batching techniques. Existing runtime engines for data stream processing can effectively deal with the need for flow control using backpressure mechanisms and often can optimize the physical execution of dataflow graph by applying several optimisations such as operator sharing and task fusion.

We made a deep dive into Apache Flink, one of the most popular stream processing systems today, and reasoned about its architecture. Flink is an example of using a stream processing architecture as the basis for any distributed computation, including the execution of batch processing tasks efficiently. We further motivated the need for lightweight state management mechanisms and offered some deep intuition behind ABS, Apache Flink's snapshotting mechanism. The combination of durable logs with repeatable computations in stream processing systems suffices to achieve exactly once processing guarantees. Recovery from distributed state snapshots achieves that in a lightweight way.

Finally, we went through different future directions in the field and analyzed a few special cases with eminent challenges. We envisioned on-line graph processing capabilities that can currently be implemented in several modern systems for stream processing. A combination of efficient approximate data structures for graph processing with rich expressive capabilities for streaming computation by modern stream processors can offer disruptive solutions in the field of graph and complex data analytics. Online machine learning (ML) and stream mining can also benefit from several recent developments in the field. A need for declarative ML models and proper integration with high level stream processing semantics are needed in order to achieve a broader integration with analytical pipelines. Furthermore, concept drift is a crucial aspect of stream mining that needs to be an integral part in new potential programming model.

The evolution of stream processors is far from done. In the following years we are going to see more standardization and broader integration of such systems for general use in analytics pipelines. Furthermore, dedicated programming libraries for graph processing, complex event processing and other domain-specific usages will proliferate, thus, shifting many processing tasks from bulk to low latency streaming. We foresee a great interdisciplinary benefit of adopting stream processors in the industry. We also expect a potential, yet gradual replacement of database and batch processing technologies with unified data stream processing systems. The generality and simplicity of distributed dataflow systems can be a main driver for adoption and a good solution to the highly diverse and complex data processing ecosystem nowadays.

References

1. Apache Hadoop project, https://hadoop.apache.org/
2. Apache Kafka project, http://kafka.apache.org/
3. Apache Samza project, http://samza.apache.org/
4. Apache Spark project, http://spark.apache.org/
5. Apache Storm project, http://storm.apache.org/
6. D.J. Abadi, D. Carney, U. Çetintemel, M. Cherniack, C. Convey, S. Lee, M. Stonebraker, N. Tatbul, S. Zdonik, Aurora: a new model and architecture for data stream management, in *VLDBJ* (2003)

7. D.J. Abadi, Y. Ahmad, M. Balazinska, U. Cetintemel, M. Cherniack, J.H. Hwang, W. Lindner, A. Maskey, A. Rasin, E. Ryvkina et al., The design of the Borealis stream processing engine, in *CIDR* (2005)
8. K.J. Ahn, S. Guha, A. McGregor, Graph sketches: sparsification, spanners, and subgraphs, in *Proceedings of the 31st symposium on Principles of Database Systems*. ACM (2012), pp. 5–14
9. T. Akidau, A. Balikov, K. Bekiroglu, S. Chernyak, J. Haberman, R. Lax, S. McVeety, D. Mills, P. Nordstrom, S. Whittle, MillWheel: Fault-tolerant stream processing at internet scale, in *VLDB* (2013)
10. T. Akidau, R. Bradshaw, C. Chambers, S. Chernyak, R.J. Fernández-Moctezuma, R. Lax, S. McVeety, D. Mills, F. Perry, E. Schmidt et al, The dataflow model: a practical approach to balancing correctness, latency, and cost in massive-scale, unbounded, out-of-order data processing, in *VLDB* (2015)
11. A. Alexandrov, R. Bergmann, S. Ewen, J.C. Freytag, F. Hueske, A. Heise, O. Kao, M. Leich, U. Leser, V. Markl et al., The Stratosphere platform for big data analytics. VLDB J. - Int. J. Very Large Data Bases **23**(6), 939–964 (2014)
12. A. Alexandrov, A. Kunft, A. Katsifodimos, F. Schüler, L. Thamsen, O. Kao, T. Herb, V. Markl, Implicit parallelism through deep language embedding, in *ACM SIGMOD* (2015), pp. 47–61
13. A. Arasu, B. Babcock, S. Babu, J. Cieslewicz, M. Datar, K. Ito, R. Motwani, U. Srivastava, J. Widom, Stream: The stanford data stream management system, Book chapter (2004)
14. A. Arasu, M. Cherniack, E. Galvez, D. Maier, A.S. Maskey, E. Ryvkina, M. Stonebraker, R. Tibbetts, Linear road: a stream data management benchmark. in *Proceedings of the Thirtieth International Conference on Very Large Data Bases, VLDB Endowment*, vol. 30 (2004), pp. 480–491
15. A. Arasu, S. Babu, J. Widom, The CQL continuous query language: semantic foundations and query execution, in *VLDBJ* (2006)
16. M. Balazinska, H. Balakrishnan, S.R. Madden, M. Stonebraker, Fault-tolerance in the Borealis distributed stream processing system. ACM Trans. Database Syst. (TODS) **33**(1), 3 (2008)
17. M. Balazinska, J.H. Hwang, M.A. Shah, Fault-tolerance and high availability in data stream management systems., in Encyclopedia of Database Systems (Springer, 2009), pp. 1109–1115
18. L. Becchetti, P. Boldi, C. Castillo, A. Gionis, Efficient semi-streaming algorithms for local triangle counting in massive graphs, in *Proceedings of the 14th ACM SIGKDD International Conference on Knowledge Discovery and Data Mining* (ACM, 2008), pp. 16–24
19. Benchmarking streaming computation engines at Yahoo! https://yahooeng.tumblr.com/post/135321837876/benchmarking-streaming-computation-engines-at
20. T. Bernhardt, A. Vasseur, *Esper: Event Stream Processing and Correlation*. ON-Java (O'Reilly, Springfield, 2007)
21. A. Bifet, R. Gavaldà, Adaptive learning from evolving data streams, in *Advances in Intelligent Data Analysis VIII* (Springer, Berlin, 2009), pp. 249–260
22. A. Bifet, G. Holmes, R. Kirkby, B. Pfahringer, Moa: Massive online analysis. J. Mach. Learn. Res. **11**, 1601–1604 (2010)
23. I. Botan, R. Derakhshan, N. Dindar, L. Haas, R.J. Miller, N. Tatbul, Secret: A model for analysis of the execution semantics of stream processing systems, in *VLDB* (2010)
24. L. Brenna, A. Demers, J. Gehrke, M. Hong, J. Ossher, B. Panda, M. Riedewald, M. Thatte, W. White, Cayuga: a high-performance event processing engine, in Proceedings of the 2007 ACM SIGMOD International Conference on Management of Data (ACM, 2007), pp. 1100–1102
25. P. Carbone, K. Vandikas, F. Zaloshnja, Towards highly available complex event processing deployments in the cloud, in *Seventh International Conference on Next Generation Mobile Apps, Services and Technologies (NGMAST)* (IEEE, 2013), pp. 153–158
26. P. Carbone, S. Ewen, S. Haridi, A. Katsifodimos, V. Markl, K. Tzoumas, Apache Flink: Stream and batch processing in a single engine. IEEE Data Engineering Bulletin (2015)
27. P. Carbone, G. Fóra, S. Ewen, S. Haridi, K. Tzoumas, Lightweight asynchronous snapshots for distributed dataflows (2015). arXiv preprint arXiv:1506.08603
28. P. Carbone, J. Traub, A. Katsifodimos, S. Haridi, V. Markl, Cutty: Aggregate sharing for user-defined windows, in *Proceedings of the 25th ACM International on Conference on Information and Knowledge Management* (ACM, 2016)

29. A. Carzaniga, D.S. Rosenblum, A.L. Wolf, Design and evaluation of a wide-area event notification service. ACM Trans. Comput. Syst. (TOCS) **19**(3), 332–383 (2001)
30. R. Castro Fernandez, M. Migliavacca, E. Kalyvianaki, P. Pietzuch, Integrating scale out and fault tolerance in stream processing using operator state management, in *Proceedings of the 2013 ACM SIGMOD international conference on Management of data* (ACM, 2013), pp. 725–736
31. U. Cetintemel, J. Du, T. Kraska, S. Madden, D. Maier, J. Meehan, A. Pavlo, M. Stonebraker, E. Sutherland, N. Tatbul et al., S-store: A streaming newSQL system for big velocity applications. Proc. VLDB Endow. **7**(13), 1633–1636 (2014)
32. C. Chambers, A. Raniwala, F. Perry, S. Adams, R.R. Henry, R. Bradshaw, N. Weizenbaum, FlumeJava: easy, efficient data-parallel pipelines, in *ACM Sigplan Notices*, vol. 45 (ACM, 2010), pp. 363–375
33. B. Chandramouli, J. Goldstein, M. Barnett, R. DeLine, D. Fisher, J.C. Platt, J.F. Terwilliger, J. Wernsing, Trill: A high-performance incremental query processor for diverse analytics. Proc. VLDB Endow. **8**(4), 401–412 (2014)
34. S. Chandrasekaran, O. Cooper, A. Deshpande, M.J. Franklin, J.M. Hellerstein, W. Hong, S. Krishnamurthy, S.R. Madden, F. Reiss, M.A. Shah, TelegraphCQ: continuous dataflow processing, in *Proceedings of the 2003 ACM SIGMOD International Conference on Management of Data* (ACM, 2003), pp. 668–668
35. K.M. Chandy, L. Lamport, Distributed snapshots: determining global states of distributed systems. ACM Trans. Comput. Syst. (TOCS) **3**(1), 63–75 (1985)
36. F. Chang, J. Dean, S. Ghemawat, W.C. Hsieh, D.A. Wallach, M. Burrows, T. Chandra, A. Fikes, R.E. Gruber, Bigtable: A distributed storage system for structured data. ACM Trans. Comput. Syst. (TOCS) **26**(2), 4 (2008)
37. J. Chen, D.J. DeWitt, F. Tian, Y. Wang, Niagaracq: A scalable continuous query system for internet databases, in *SIGMOD Record* (ACM, 2000)
38. M. Cherniack, H. Balakrishnan, M. Balazinska, D. Carney, U. Cetintemel, Y. Xing, S.B. Zdonik, Scalable distributed stream processing. CIDR. **3**, 257–268 (2003)
39. T. Condie, N. Conway, P. Alvaro, J.M. Hellerstein, K. Elmeleegy, R. Sears, Mapreduce online. NSDI. **10**, 20 (2010)
40. G. Cugola, A. Margara, Processing flows of information: From data stream to complex event processing. ACM Comput. Surv. (CSUR) **44**(3), 15 (2012)
41. U. Dayal, B. Blaustein, A. Buchmann, U. Chakravarthy, M. Hsu, R. Ledin, D. McCarthy, A. Rosenthal, S. Sarin, M.J. Carey et al., The HiPAC project: Combining active databases and timing constraints. ACM Sigmod Rec. **17**(1), 51–70 (1988)
42. G. De Francisci Morales, A. Bifet, Samoa: Scalable advanced massive online analysis. J. Mach. Learn. Res. **16**(1), 149–153 (2015)
43. J. Dean, S. Ghemawat, Mapreduce: simplified data processing on large clusters. Commun. ACM **51**(1), 107–113 (2008)
44. N. Dindar, N. Tatbul, R.J. Miller, L.M. Haas, I. Botan, Modeling the execution semantics of stream processing engines with secret. VLDB J. **22**(4), 421–446 (2013)
45. D. Elin, T. Risch, Amos II java interfaces. Uppsala University report (2000)
46. J. Feigenbaum, S. Kannan, A. McGregor, S. Suri, J. Zhang, On graph problems in a semi-streaming model. Theor. Comput. Sci. **348**(2), 207–216 (2005)
47. R.C. Fernandez, M. Migliavacca, E. Kalyvianaki, P. Pietzuch, Making state explicit for imperative big data processing, in *Proceedings of the 2014 USENIX Annual Technical Conference (USENIX ATC 14)* (2014), pp. 49–60
48. S. Gatziu, K.R. Dittrich, Samos: An active object-oriented database system. IEEE Data Eng. Bull. **15**(1–4), 23–26 (1992)
49. B. Gedik, Partitioning functions for stateful data parallelism in stream processing. VLDB J. **23**(4), 517–539 (2014)
50. Google Cloud Dataflow, https://cloud.google.com/dataflow/
51. W. Han, Y. Miao, K. Li, M. Wu, F. Yang, L. Zhou, V. Prabhakaran, W. Chen, E. Chen, Chronos: a graph engine for temporal graph analysis, in *Proceedings of the Ninth European Conference on Computer Systems* (ACM, 2014), p. 1

52. B. He, M. Yang, Z. Guo, R. Chen, B. Su, W. Lin, L. Zhou, Comet: batched stream processing for data intensive distributed computing, in *Proceedings of the 1st ACM Symposium on Cloud Computing* (ACM, 2010), pp. 63–74

53. M. Hirzel, H. Andrade, B. Gedik, V. Kumar, G. Losa, M. Nasgaard, R. Soule, K. Wu, SPL stream processing language specification. NewYork: IBMResearchDivisionTJ. WatsonResearchCenter, IBM ResearchReport: RC24897 (W0911–044) (2009)

54. M. Hirzel, H. Andrade, B. Gedik, G. Jacques-Silva, R. Khandekar, V. Kumar, M. Mendell, H. Nasgaard, S. Schneider, R. Soulé et al., IBM streams processing language: analyzing big data in motion. IBM J. Res. Develop. **57**(3/4), 7–1 (2013)

55. M. Hirzel, R. Soulé, S. Schneider, B. Gedik, R. Grimm, A catalog of stream processing optimizations. ACM Comput. Surv. (CSUR) **46**(4), 46 (2014)

56. Introduction to Kafka Streams, http://www.confluent.io/blog/introducing-kafka-streams-stream-processing-made-simple

57. A. Iyer, L.E. Li, I. Stoica, CellIQ: real-time cellular network analytics at scale, in *12th USENIX Symposium on Networked Systems Design and Implementation* (NSDI 15) (2015), pp. 309–322

58. R. Kallman, H. Kimura, J. Natkins, A. Pavlo, A. Rasin, S. Zdonik, E.P. Jones, S. Madden, M. Stonebraker, Y. Zhang et al., H-store: a high-performance, distributed main memory transaction processing system. Proc. VLDB Endow. **1**(2), 1496–1499 (2008)

59. K. Karanasos, A. Katsifodimos, I. Manolescu, Delta: Scalable data dissemination under capacity constraints. Proc. VLDB Endow. **7**(4), 217–228 (2013)

60. J. Kreps, N. Narkhede, J. Rao et al, Kafka: A distributed messaging system for log processing. NetDB (2011)

61. S. Kulkarni, N. Bhagat, M. Fu, V. Kedigehalli, C. Kellogg, S. Mittal, J.M. Patel, K. Ramasamy, S. Taneja, Twitter Heron: Stream processing at scale, in *ACM SIGMOD* (2015)

62. A. Kyrola, G. Blelloch, C. Guestrin, Graphchi: Large-scale graph computation on just a pc, in *Presented as part of the 10th USENIX Symposium on Operating Systems Design and Implementation (OSDI 12)* (2012), pp. 31–46

63. A. Lakshman, P. Malik, Cassandra: a decentralized structured storage system. ACM SIGOPS Oper. Syst. Rev. **44**(2), 35–40 (2010)

64. J. Li, D. Maier, K. Tufte, V. Papadimos, P.A. Tucker, Semantics and evaluation techniques for window aggregates in data streams, in *ACM SIGMOD* (2005)

65. L. Liu, C. Pu, W. Tang, Continual queries for internet scale event-driven information delivery. IEEE Trans. Knowl. Data Eng. **11**(4), 610–628 (1999)

66. Y. Liu, B. Plale et al., Survey of publish subscribe event systems. Computer Science Dept, Indian University **16** (2003)

67. D. Logothetis, C. Olston, B. Reed, K.C. Webb, K. Yocum, Stateful bulk processing for incremental analytics, in *Proceedings of the 1st ACM Symposium on Cloud Computing* (ACM, 2010), pp. 51–62

68. D. Luckham, *The power of events*, vol. 204 (Addison-Wesley Reading, Boston, 2002)

69. G. Malewicz, M.H. Austern, A.J. Bik, J.C. Dehnert, I. Horn, N. Leiser, G. Czajkowski, Pregel: a system for large-scale graph processing, in *Proceedings of the 2010 ACM SIGMOD International Conference on Management of Data* (ACM, 2010), pp. 135–146

70. N. Marz, J. Warren, *Big Data: Principles and Best Practices of Scalable Realtime Data Systems* (Manning Publications Co., Greenwich, 2015)

71. D. Mishra, SNOOP: an event specification language for active database systems. Ph.D. thesis, University of Florida (1991)

72. S.S. Muchnick, *Advanced Compiler Design Implementation* (Morgan Kaufmann, Burlington, 1997)

73. D.G. Murray, F. McSherry, R. Isaacs, M. Isard, P. Barham, M. Abadi, Naiad: a timely dataflow system, in *ACM SOSP* (2013)

74. L. Neumeyer, B. Robbins, A. Nair, A. Kesari, S4: Distributed stream computing platform, in *Proceedings of the 2010 IEEE International Conference on Data Mining Workshops* (IEEE, 2010), pp. 170–177

75. K. Patroumpas, T. Sellis, Window specification over data streams, in *Current Trends in Database Technology–EDBT 2006* (Springer, Berlin, 2006), pp. 445–464
76. D. Peleg, A.A. Schäffer, Graph spanners. J. Graph Theory **13**(1), 99–116 (1989)
77. M.A. Shah, J.M. Hellerstein, S. Chandrasekaran, M.J. Franklin, Flux: An adaptive partitioning operator for continuous query systems, in *Proceedings of the 19th International Conference on Data Engineering* (IEEE, 2003), pp. 25–36
78. M.A. Shah, J.M. Hellerstein, E. Brewer, Highly available, fault-tolerant, parallel dataflows, in *Proceedings of the 2004 ACM SIGMOD International Conference on Management of Data* (ACM, 2004), pp. 827–838
79. U. Srivastava, J. Widom, Flexible time management in data stream systems. in *Proceedings of the Twenty-Third ACM SIGMOD-SIGACT-SIGART Symposium on Principles of Database Systems* (ACM, 2004), pp. 263–274
80. StreamBase I: Streambase: Real-time, low latency data processing with a stream processing engine (2006)
81. J. Thaler, Semi-streaming algorithms for annotated graph streams (2014). arXiv preprint arXiv:1407.3462
82. The Apache APEX project, https://www.datatorrent.com/apex/
83. The Apache Beam System, https://wiki.apache.org/incubator/BeamProposal
84. The Kappa Architecture by Jay Kreps, http://milinda.pathirage.org/kappa-architecture.com/
85. The Trident Stream Processing Programming Model, http://storm.apache.org/releases/0.10.0/Trident-tutorial.html
86. A. Toshniwal, S. Taneja, A. Shukla, K. Ramasamy, J.M. Patel, S. Kulkarni, J. Jackson, K. Gade, M. Fu, J. Donham et al, Storm @ Twitter, in *Proceedings of the 2014 ACM SIGMOD International Conference on Management of Data* (ACM, 2014), pp. 147–156
87. J. Webber, A programmatic introduction to Neo4j, in *Proceedings of the 3rd Annual Conference on Systems, Programming, and Applications: Software For Humanity* (ACM, 2012), pp. 217–218
88. R.S. Xin, J.E. Gonzalez, M.J. Franklin, I. Stoica, GraphX: A resilient distributed graph system on Spark, in *First International Workshop on Graph Data Management Experiences and Systems* (ACM, 2013), p. 2
89. M. Zaharia, M. Chowdhury, M.J. Franklin, S. Shenker, I. Stoica, Spark: Cluster computing with working sets. HotCloud **10**, 10–10 (2010)
90. M. Zaharia, T. Das, H. Li, S. Shenker, I. Stoica, Discretized streams: an efficient and fault-tolerant model for stream processing on large clusters, in *Proceedings of the 4th USENIX Conference on Hot Topics in Cloud Ccomputing* (USENIX Association, 2012), pp. 10–10

Part II
Semantic Big Data Management

Semantic Data Integration

Michelle Cheatham and Catia Pesquita

Abstract The growing volume, variety and complexity of data being collected for scientific purposes presents challenges for data integration. For data to be truly useful, scientists need not only to be able to access it, but also be able to interpret and use it. Doing this requires semantic context. Semantic Data Integration is an active field of research, and this chapter describes the current challenges and how existing approaches are addressing them. The chapter then provides an overview of several active research areas within the semantic data integration field, including interactive and collaborative schema matching, integration of geospatial and biomedical data, and visualization of the data integration process. Finally, the need to move beyond the discovery of simple 1-to-1 equivalence matches to the identification of more complex relationships across datasets is presented and possible first steps in this direction are discussed.

1 An Important Challenge

The world around us is an incredibly complex and interconnected system – one filled with phenomena that cannot be understood in isolation. At the same time, the volume and complexity of the data, theory, and models established to explain these phenomena have led scientists to specialize further and further, to the point where many researchers now spend their entire careers on extremely narrow topics, such as the characteristics of one particular class of star, or the habits of a single species of fish. While such specialization is important to increase humanity's depth of knowledge about many subjects, some of the greatest leaps forward in our understanding come at the intersection of traditional scientific disciplines. These advances require the

---

M. Cheatham (✉)
DaSe Laboratory, Wright State University, Dayton, OH, USA
e-mail: michelle.cheatham@wright.edu

C. Pesquita
Universidade de Lisboa, Lisbon, Portugal
e-mail: cpesquita@di.fc.ul.pt

© Springer International Publishing AG 2017   263
A.Y. Zomaya and S. Sakr (eds.), *Handbook of Big Data Technologies*,
DOI 10.1007/978-3-319-49340-4_8

integration of data from many different scientific domains, and this integration must be done in a way that preserves the detail, uncertainty, and above all the *context* of the data involved.

Preserving these properties can be achieved through semantic data integration, a process through which semantically heterogeneous data can be integrated with minimal loss of information. This type of data integration is particularly relevant in domains where data models are diverse and entity properties are heterogeneous. For instance, health information systems, and in particular medical records employ a diversity of vocabularies to describe relevant entities. Health care facilities routinely use different software providers for different aspects of their functioning (outpatient, emergency, surgery, laboratory, billing, etc), each with their own set of vocabularies that many times employ different labels and assign different properties to the same entities. Moreover, the controlled vocabularies many times lack the information necessary to understand the data they describe. For instance, if during an emergency room visit the patient is assigned a primary diagnostic of "Acute upper respiratory infection" using ICD-10, how can we understand that the results of the lab test "Virus identified in Nose by Culture" coded using LOINC, are relevant for the diagnosis? Semantic data integration can provide the means to achieve the meaningful integration of data necessary to support more complex analysis and conclusions.

Unfortunately, semantic data integration is a challenging proposition, particularly for scientific data. Many obstacles stand in the way of synthesizing all of the data about an entity. One of the most obvious of these is accessing the data in the first place. Much of the data underpinning past and present scientific publications is not readily accessible – it exists only in isolated databases, as files on a grad student's computer, or in tables within PDF documents. Moreover, there can be various obstacles to retrieving this data, particularly due to a lack of consistency. For instance, some repositories might be accessible via websites or structured query mechanisms while others require a login and use of secure file transfer or copy protocols. Financial and legal concerns also inhibit data integration. Some data might be stored in proprietary databases or file formats that require expensive software licenses to read, and licenses indicating what users are allowed to do with the data can be missing or restrictive, resulting in legal uncertainty. These types of concerns led to a push towards Linked Open Data, which is described in the next section.

Of course, accessibility is only the first step to semantic data integration. For data to be truly useful, scientists need to be able to interpret and use it after they acquire it. Doing this requires semantic context. By semantic context, we mean the situation in which a term or entity appears. As a simple example, 'chair' would be considered a piece of furniture if it was seen in close proximity to 'couch' and 'table', but as a person if used in conjunction with 'dean' and 'provost'. Similarly, if temperature is included in a dataset that contains entirely Imperial units, it might be assumed to be in Fahrenheit rather than Celsius, particularly if the values correspond to what might be expected (e.g. values near 98° for body temperatures). In relational databases and spreadsheets, semantic context is sometimes lacking because important information about what the various data fields mean and how they relate to one another is often implicit in the names of database tables and column headers, some of which are incomprehensible to anyone other than the dataset's creator. What is needed is a way to express the semantic connections between different pieces of data in a way that is expressive enough to capture nuanced relationships while at the same time formalized and restrictive enough to allow software as well as humans to make inferences based on the links. Ontologies, described in Sect. 1.2, have been proposed for this purpose.

Even when data is made accessible by following the Linked Open Data principles and is organized according to a machine-readable ontology, challenges *still* remain. An ontology imposes order on a domain of interest, but order is in the eye of the beholder: if five different publishers of the same type of data were tasked to develop an ontology with which to structure their data, the result would very likely be five different ontologies. One obvious approach is to try to get all data publishers from a domain to agree on a single ontology. This tends to be unfeasible in many instances, for example due to a provider's data causing a logical inconsistency when it is shoe-horned into the agreed upon ontology. A "one ontology to structure them all" approach also conflicts with the inherently distributed paradigm championed by the Semantic Web. An alternative to this strategy is to allow data providers to create or choose whatever ontology best suits their data, and then to establish links that encode how elements of this ontology relate to those within other ontologies.

Establishing semantic links between ontologies and the data sets that they organize can be very difficult, particularly if the datasets are large and complex, as is routinely the case in scientific domains. The fields of ontology alignment and co-reference resolution seek to develop tools and techniques to facilitate the identification of links between datasets. Scientific datasets are particularly challenging to align for several reasons. Perhaps most obviously, such datasets can be extremely large, often over a petabyte of data, which is more than enough to swamp most existing data integration techniques. Additionally, scientific datasets generally have a spatiotemporal aspect, but current alignment algorithms struggle with finding relationships across this type of data because of the variety of ways to express it. For example, spatial regions can be represented by geopolitical entities (whose borders change over time), by the names of nearby points of interest, or by polygons whose points are given via latitude and longitude. Similarly, issues pertaining to measurement resolution, time zones, the international dateline, etc. can confuse the comparison of timestamps of data observations. Furthermore, scientific datasets frequently involve data of very different modalities, from audio recordings of dolphin calls to radar images of storms, to spectroscopy of cellular organisms. Such data is also collected at widely differing scales, from micrometers to kilometers. And oftentimes the data that needs to be integrated is from domains with only a small degree of semantic overlap, as is the case with, for example, one dataset containing information about NSF project awards and another with the salinity values for ocean water collected during oceanography cruises, several of which were funded by NSF.

We have identified a number of challenges in semantic data integration, namely: the accessibility of the data; providing data with semantic context to support its interpretation; and the establishment of meaningful links between data. These challenges are expanded in the following subsections. Section 2 addresses several state of the art topics in semantic data integration, while Sect. 3 lays out the path forwards in this area.

1.1 Linked Data

Tim Berners-Lee originally envisioned a world wide web that is equally accessible to both humans and computers [5]. Unfortunately, even after several decades we have yet to make this vision a reality. When we look at a webpage today, say, one that presents data about the publications of a group of researchers, we are likely to find that data within an HTML table with columns containing headers such as "Researcher", "Title", "Journal", "Publication Year", etc. If we additionally want to know which researchers are publishing in journals with a high impact factor, we would need to look for the journal's title in the appropriate column of the table, search for the journal's website using a search engine, and find the impact factor on the journal's website by looking for it (hopefully) on the journal's homepage. This is tedious for humans, but extremely difficult for computers. For example, recognizing that the table contains information about researchers' publications and identifying the meaning of each of the columns requires background knowledge and natural language processing, as does realizing that a journal's impact factor is not in the table. Pulling the journal's title out of the HTML table and submitting it to a search engine requires writing code that depends on the format of the table and the API of the search engine, both of which are likely to break if the website or search engine provider makes any changes to those resources. After the query has been made, determining if a particular query result actually contains the impact factor for the journal in question again requires natural language processing. Furthermore, the provider of the data concerning these researchers' publications may not consent to its use for the type of analysis we seek to perform.

Publishing information as linked data alleviates many of these challenges. Linked data builds upon existing web standards such as HTTP, RDF, and URIs to create web pages that are machine-readable and, ideally, machine-understandable. According to Berners-Lee,[1] the four rules of linked data are:

1. Use URIs to denote things.
2. Use HTTP URIs so that these things can be referred to and looked up ("dereferenced") by people and user agents.
3. Provide useful information about the thing when its URI is dereferenced, leveraging standards such as RDF and SPARQL.
4. Include links to other related things (using their URIs) when publishing data on the Web.

---

[1] http://www.w3.org/DesignIssues/LinkedData.html.

Linked data is generally published as RDF subject-predicate-object triples. For instance, the following triple indicates an article with the URI cspublications.org/TheSemanticWeb was written by Tim Berners-Lee.

```
<www.w3.org/People/Berners-Lee>
swrc:author
<cspublications.org/TheSemanticWeb> .
```

Similarly, the triples below specify that the article is titled "The Semantic Web", that it was published in 2001, and that the journal it was published in has the URI cspublications.org/ScientificAmerican.

```
<cspublications.org/TheSemanticWeb>
swrc:title
"The Semantic Web"@en .

<cspublications.org/TheSemanticWeb>
swrc:year
"2001"^^xsd:date .

<cspublications.org/TheSemanticWeb>
swrc:journal
<scientificamerican.com> .
```

The expectation is that following the URI scientificamerican.com allows us to learn more information about the journal in which this article was published *even if that information comes from an entirely different data source.*

Publishing data as RDF rather than HTML separates information about data's meaning and context from information about how to format it for presentation. This enables software applications to easily access the data. Additionally, it is possible to express the terms of use for a data set as linked data as well, thus allowing software agents to read and respect these constraints. While this detail is often overlooked, legal issues are often as big of a hindrance to data re-use as technical concerns. Fortunately, addressing this issue is not difficult. Many commonly used licenses have already been encoded in RDF, and datasets can simply add the appropriate triple to refer to them. For example, the triple below indicates that this dataset is available according to the conditions of version 3.0 of the Creative Commons "Share-Alike" license.

```
<cspublications.org/publications.rdf>
cc:license
<http://creativecommons.org/licenses/by-sa/3.0/> .
```

A very large amount of data has already been published as linked open data: according to the most recent survey, there are hundreds of linked datasets, which contain billions of facts about a wide variety of subjects, from music, to biology, to social networks [93]. The website www.linkeddata.org contains pointers to many such datasets. As the linked open data cloud continues to grow, the ability of data providers to contextualize their data by linking it to already-existing data will encourage the creation of more linked data, creating a virtuous feedback loop.

Keeping with our example of medical records, recent work has transformed a clinical datawarehouse into a semantic clinical datawarehouse by applying the Linked Data principles [78]. This enabled clinical data to be integrated with publicly available biomedical repositories, enabling for instance the identification of disease genes.

1.2 Ontologies

Tom Gruber, one of the early voices on knowledge representation (and the creator of Siri), defines an ontology as a "specification of a conceptualization." He elaborates that an ontology defines the concepts and relationships within a domain [35]. Figure 1 shows a snippet of the Semantic Web for Research Communities (SWRC) ontology [105]. The subset of entities shown represent key concepts within the publication domain. The entities shown in ovals, such as *Person* and *Publication* are called classes. A class represents a grouping of objects with similar characteristics. Classes are often arranged in a hierarchy using subclass relationships. For instance, in our

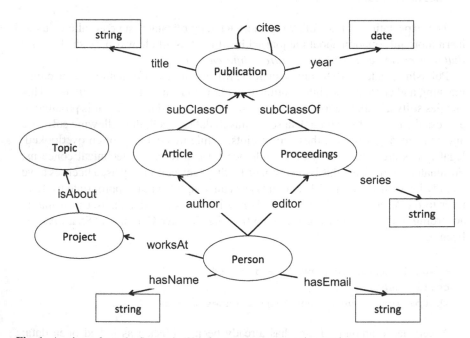

Fig. 1 A snippet from the Semantic Web for resource community ontology

example *Article* and *Proceedings* are both subclasses of *Publication* (i.e. every *Article* is a *Publication* but not every *Publication* is an *Article*). An instance (also sometimes called an individual) is a particular object. An instance has a type that is some class within the ontology. For example, an instance of type *Article* may be Weaving the Semantic Web and an instance of type *Person* may be Tim Berners-Lee. This is somewhat analogous to classes and instances of those classes in object-oriented programming languages. Relationships between instances, such as *hasName* and *author*, are called properties. All properties are directed binary relations that map an instance with a type from the domain to something in the range. These are represented as labeled arrows in Fig. 1, with the arrow pointing from the domain to the range. Properties that map an instance to another instance (e.g. *editor*, which maps an instance of type *Person* to an instance of type *Proceedings*) are object properties, whereas properties that map an instance to a literal value (e.g. *year*, which maps an instance of type *Publication* to a date value) are datatype properties. Common data types include integers, doubles, strings, and dates. Both object properties and data properties must involve an instance. A third type of property, called an annotation property, can be used to describe relationships between any types of entities (e.g. instances, classes or other properties).

Critically, an ontology should not require an agent, either human or computer, to understand the entity labels in order to leverage the ontology for data publication or consumption. Labels are human-centric, and the underlying goal of the Semantic Web is to put humans and machines on equal footing. Instead of relying on labels to convey meaning, the ontology designer should constrain the possible interpretation of entity labels through judicious use of logical axioms. For example, DBPedia, the linked data version of Wikipedia, contains a property called *hasGender*. The vast majority of uses of this property are to express a person's gender. However, because the domain and range of this property are very vague (i.e. any *Thing* can have a gender), some of the uses of *hasGender* are very different. For instance, DBPedia asserts that the name "Alexander" *hasGender* "Alexandria" and that a secondary school in England *hasGender* "unisex education." This can cause difficulty for software applications that are attempting to use the *hasGender* property. Misunderstandings can be avoided if the axioms are added to the ontology to constrain the possible meaning of the terms it uses. In this case, the domain of *hasGender* could be changed to be something like *LivingThing*, as shown below.

```
dbpedia:hasGender rdfs:domain dbpedia:LivingThing .
```

Constraints on ontology entities expressed through axioms, together with instance data published relative to those entities, enables a piece of software called a reasoner to infer additional facts that are not actually in the data. For example, if the dataset contained the fact that Tim Berners-Lee wrote "The Semantic Web" and the knowledge base contained an axiom stating that the domain of the property *wrote* is *Person*, a reasoner would be able to infer that Berners-Lee is a person, even if that fact was not explicitly in the knowledge base. A query to return all of the *Persons* in the knowledge base would then correctly include Tim Berners-Lee among the results. This is accomplished without any natural language processing, which can be error-prone in many situations.

Because constraints make the meaning of entity names and relationships more precise, they hold great potential to facilitate accurate data integration. Unfortunately, many existing ontologies do not contain significant numbers of axioms. However, as we will see in the next section, many existing data and schema integration systems are already capable of leveraging such axioms when they do exist.

There is a balance to be struck here: too few axioms can lead to many different interpretations of entities, making the ontology less useful; however, too many axioms can constrain the ontology so much that is only applicable in a narrow set of circumstances. For instance, it may seem reasonable to create an axiom that mandates that a *LivingThing* has exactly one gender, this is not the case for some beings. Ontologies are often encoded in the Web Ontology Language (OWL) [69]. Besides property domain and range and cardinality constraints, OWL allows one to state that two entities are equivalent or disjoint, that a property is reflexive, symmetric, transitive, or functional, or that one property is the inverse of another. All of this information: classes, properties, and axioms that restrict their interpretation, is called the schema, or T-box (for terminology), of the ontology. Conversely, the instance data, or A-box (for assertions), contains assertions about individuals using data from the T-box.

A more formal and extensive treatment of ontology design and representation can be found in [41]. Many ontologies exist today. Some of these, such as the Suggested Upper Merged Ontology (SUMO) [82] and the Descriptive Ontology for Linguistic and Cognitive Engineering (DOLCE) [32] begin modelling the world at the highest level of abstraction and working towards more detail. The top-level entities in DOLCE, for instance, are *Entity*, *Endurant*, *Perdurant*, and *Abstract*. There are also numerous domain-specific ontologies, such as the Gene Ontology, which models the structure and molecular processes of eukaryotic cells [2], and NASA's Semantic Web for Earth and Environmental Terminology (SWEET) ontology [86]. In the clinical domain, many providers have begun migrating from simple terminologies (such as ICD-10) to more complex ones that have an ontological foundation (such as SNOMED-CT) [14]. Lately many researchers have also begun to publish ontology "snippets," sometimes referred to as ontology design patterns, that model much more constrained topic areas. The website ontologydesignpatterns.org currently has dozens of ontology snippets, including models of a Hazardous Situation, a Species Habitat, and a Chess Game.

1.3   Ontology and Data Alignment

While the amount of linked data available on the Semantic Web has grown continually for more than a decade, the links between different datasets have not gown at the same rate. These links provide the context that makes the data more useful. The fields of ontology and data alignment attempt to discover links between datasets in an automatic or semi-automatic way. Ontology alignment systems tend to focus on finding relationships between schema-level entities, while co-reference resolution systems attempt to identify cases in which the same individual is referred to via different URIs.

1.3.1 Ontology Alignment

Engineering new ontologies is not a deterministic process – many design decisions must be made, and the designers' backgrounds and the application they are targeting will influence their decisions in different ways. The end result is that even two ontologies that represent the same domain will not be the same. They may use synonyms for the same concept or the same word for different concepts, they may be at different levels of abstraction, they may not include all of the same concepts, and they may not even be in the same language. And this is in the best case. In real-world datasets there are often problems with missing information, inconsistent use of the T-box when describing individuals, and logically inconsistent axioms. The goal of ontology alignment is to determine when an entity in one ontology is semantically related to an entity in another ontology (for a comprehensive discussion of ontology alignment, including a formal definition, see [23]).

An alignment algorithm takes as input two ontologies and produces a set of matches consisting of a URI specifying one entity from each ontology, a relationship, and an optional confidence value that is generally in the range of 0–1, inclusive. For example, Fig. 2 shows a second ontology describing publications. This ontology was

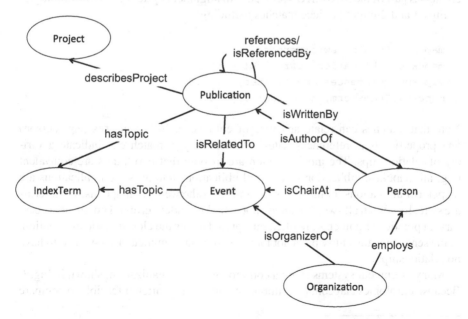

Fig. 2 A subset of the scientific publications ontology from the MAPEKUS project

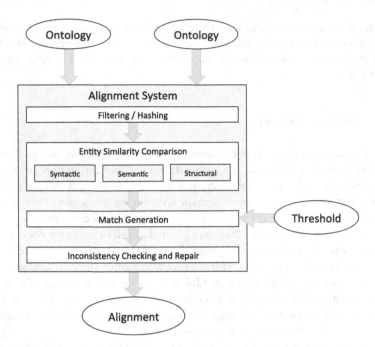

Fig. 3 General structure of an ontology alignment system

created as part of the MAPEKUS project.[2] An alignment system given the ontologies
in Figs. 1 and 2 might produce matches including:

```
mapekus:Person, swrc:Person, =, 1.0
mapekus:Publication, swrc:Publication, =, 0.9
mapekus:references, swrc:cites, =, 0.8
mapekus:IndexTerm, swrc:Topic, <, 0.6
```

Note that matches can relate any type of entities, including classes (e.g. Person)
and properties (e.g. references, cites). Additionally, a match can indicate a variety of relationships. The most common are to state that two things are equivalent
(e.g. all mapekus:Publications are swrc:Publications and all swrc:Publications are
mapekus:Publications) or that one subsumes the other (e.g. all mapekus:IndexTerms
are swrc:Topics but all swrc:Topics are not mapekus:IndexTerms). Though not necessary, in practice alignments are often interpreted under the closed world assumption,
in the sense that any entity pairs not mentioned in an alignment are assumed to have
no relationship.

Many alignment systems share a common general organization, shown in Fig. 3.
Because ontologies can contain millions of entities, it is often infeasible to compare

---

[2]http://mapekus.fiit.stuba.sk.

every entity in one ontology to every entity in the other. Therefore, alignment systems sometimes employ a filtering or hashing step to determine which entities to compare [20, 40]. Alignment systems typically use a combination of three different approaches to evaluate entity similarity: syntactic, semantic, and structural similarity metrics. Entity similarity is related to how much two entities have in common; it can be thought of as a measure of the degree one class, property, or individual could be used in place of another. Syntactic metrics compare entities from each of the ontologies to be aligned based on strings associated with the entities. The strings are generally the entity label, but can also include comments or other annotations of the entity. Semantic similarity metrics attempt to use the meanings of entity labels rather than their spellings. External resources such as thesauri, dictionaries, encyclopedias, and web search engines are often used to calculate semantic similarity [46, 107]. Structural techniques consider the neighborhoods of two entities when determining their similarity. For instance, two entities with the same superclass that share some common instances are considered more similar than entities that do not have these things in common. Graph matching techniques are often used for this [18, 31]. An alignment system may use zero or more of each type of similarity metric. The values from multiple approaches may be combined to form a single measure of similarity, or they may be used in a serial fashion to filter potential matches down to the most likely candidates. At some point, a final list of related entities is generated, frequently by including any matches with a confidence (similarity) value higher than some threshold. Additionally, alignment systems may use some form of inconsistency checking and repair after the matching process in order to ensure a merged ontology produced using the alignment is logically consistent [62, 83, 90].

Each year since 2005, the Ontology Alignment Evaluation Initiative has invited researchers to compare the performance of their alignment systems on a set of benchmark tasks. Current alignment systems have become very proficient at finding 1-to-1 equivalence relationships between classes and instances (the type of matches contained within the benchmarks). In fact, the top-performing systems now attain a 0.75 f-measure on one of the OAEI test sets that is designed to reflect real-world matching tasks [13]. This is nearing the level of consensus that humans familiar with ontology design have for alignment tasks involving this test set [11]. Unfortunately, the performance on finding relationships between properties is not nearly as good as that for classes and instances [12]. Additionally, there is some evidence that most of the accuracy of existing alignment systems is due to basic string similarity measures [10], which raises some concern that further gains may be more difficult to achieve.

1.3.2 Coreference Resolution

Coreference resolution algorithms attempt to determine when the same instance (i.e. individual) is referred to using different URIs. Note that because the term "ontology alignment" can either refer to aligning an entire ontology (the T-box and the A-box) or just the T-box, this section uses the term "schema alignment system" to refer to something that attempts to map only the T-box of an ontology.

Coreference resolution differs from schema alignment in several ways. One key difference is that the relationships sought by coreference resolution algorithms are only 1-to-1 equivalences: two individuals are either the same or distinct, whereas schema elements involve *sets* of individuals and can therefore have all of the traditional relationships that exist between sets, including subsumption, disjointness, and partial overlap. Another important contrast between coreference resolution and schema alignment is that the A-box of an ontology is often an order of magnitude larger than the T-box. This makes efficiency concerns even more important for coreference resolution algorithms than for schema alignment systems. Another distinction is that, while there is interplay in both directions between coreference resolution and schema alignment, it can be argued that coreferences generally place more constraints on schema alignments than the other way around. This is because many existing schema alignment systems employ some extensional comparators in the mapping process, i.e. they determine the likelihood that two schema elements are related by the degree of overlap between their instances. For example, if it is determined that *data1*: *Tim_Berners_Lee* in one dataset is equivalent to *data2*: *TimothyLee* in another dataset, and *data1*: *Tim_Berners_Lee* is a *data1*: *Scientist* while *data2*: *TimothyLee* is a *data2*: *ComputerProgrammer*, a schema alignment algorithm is more likely to conclude that the classes *data1*: *Scientist* and *data2*: *ComputerProgrammer* are related in some way (i.e. that they are not disjoint). This is done for classes in [22] and for properties in [36]. Because equality cannot be defined extensionally for individuals, questions about what it fundamentally means for two things to be identical tend to arise in coreference resolution research [37]. Coreference resolution can be thought of as data de-duplication, which has been an area of research for decades. For instance, there has been extensive research regarding recognizing the duplicate records, stretching back to at least 1969 [30]. Many of the approaches currently employed to resolve coreferences on the Semantic Web are adapted from techniques that were established decades ago in database integration systems. A good survey of such techniques can be found in [21].

Of course, there are obviously differences between databases and linked data published according to an ontology. The most obvious of these is that databases operate under the "closed world" assumption, meaning that if something is not present in the database, it is assumed not to exist. In contrast, the Semantic Web uses an open world assumption. Also, as Castano and his colleagues point out in [9], the structure of a linked dataset can differ greatly from a relational database that represents the same domain due to the expressiveness of ontology specification languages such as OWL compared to database table definition and column constraint capabilities. The more complex relationships expressible in ontologies may convey implicit knowledge that can be inferred by a reasoner. This additional information is not generally available when integrating two databases. In terms of focus specifically on integrating linked data, schema alignment has more of a research history than coreference resolution. For example, the annual Ontology Alignment Evaluation Initiative has existed since

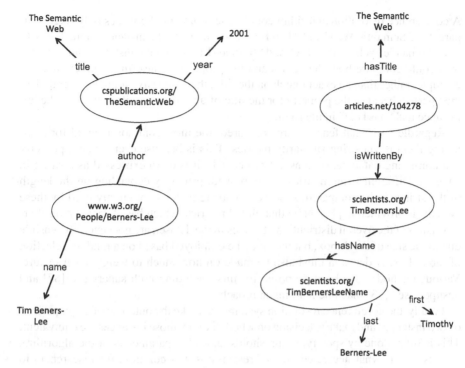

Fig. 4 A potential coreference

2005, but it has only had a track dedicated to evaluating performance on coreference resolution tasks since 2009.[3]

The general decisions made by the designer of a coreference resolution system are: what instances to compare, how to compare them, and how to determine if the result of that comparison implies that the two instances are equivalent.

As mentioned previously, there are generally many more instances in a dataset than schema entities. As a result, it is not considered feasible to compare every instance in one dataset to every instance in the other in order to determine if they are the same. Instead, some method of deciding whether two instances are "close enough" that they are worth comparing must be established. The choice of this method reflects a trade-off between recall and utility, i.e. an overly zealous filtering algorithm may miss some equivalences, while a conservative filtering approach may cause the system to take a long time to generate results.

If the filtering step has decided that two instances are close enough to warrant further scrutiny, the algorithm will compare them based on a selection of features. In most current coreference resolution systems, these features are either property values alone or property values together with property names. There is also a question of how deep within an instance's semantic neighborhood to go when extracting features. For an example, in Fig. 4 there are two instances, from different datasets.

---

[3]http://oaei.ontologymatching.org.

A coreference resolution algorithm could either compare the values only (e.g. compare "The Semantic Web" and "Tim Berners-Lee" from the instance on the left with "The Semantic Web", "Timothy" and "Berners-Lee" from the instance on the right), or it could compare both the values and the property names. In the latter case, for example, rather than an exact match on the title, the similarity would be slightly less than perfect because the property for the title of an article is called "title" in the left instance and "hasTitle" in the right.

Regardless of what features are compared, the most common method for comparing them is via string similarity metrics. This is because even when a property is a non-string datatype, such as a date or URL, it is often expressed as a string in datasets. Different string metrics are employed, primarily depending on the length of the strings to be compared. A survey of string metrics commonly used by these systems is provided in [10]. Note that global metrics, which based decision on characteristics of the overall distribution of values in the dataset, are not generally feasible due to the size of the A-box, but they may be employed based on a random selection of the A-box. A decision must also be made on how much to weight each feature. Various methods have been proposed for this, including both supervised [87] and unsupervised [73] machine learning approaches.

Finally, the coreference resolution system must take the outcome of a comparison of two instances and make a decision on whether or not those instances are equivalent. This is often done by specifying thresholds and other parameters of the algorithm. This is a somewhat neglected area of research – it is common for researchers to report that these values were "determined empirically" for the particular datasets being aligned. Among the small amount of work on this topic is an exploration by Paulheim and his colleagues of using interactive techniques to configure the threshold by asking a user targeted questions regarding the validity of potential matches with confidence values on either side of the current threshold and updating it accordingly [81].

2 Current State-of-the-Art

In the face of the performance plateau on current alignment benchmarks, many researchers have created innovative new alignment techniques that focus on various aspects or subproblems under the general umbrella of semantic data integration. This section explores a selection of this current work.

2.1 Interactive and Collaborative Approaches

While the performance of automated alignment systems is becoming quite good for certain types of mapping tasks, in practice no existing system generates alignments that are completely correct. Alignments tend to either lack some correct mappings, contain some incorrect mappings, or both. As a result, there is significant ongoing research on alignment systems that allow users to contribute their knowledge and expertise to the mapping process. These systems exist on a spectrum ranging from entirely manual approaches to semi-automated techniques methods that ask humans to chime in only when the automated system is unable to make a definitive decision. Because entirely manual alignment is feasible only for small datasets, most current research in this area focuses on semi-automated approaches. In contrast to fully manual approaches, semi-automated systems interact with the user(s) only intermittently, and then attempt to leverage this human-supplied knowledge to improve the scope and quality of the alignment. Interactive systems of this type differ in terms of what questions they ask users and how they make use of the responses. In addition to being judged on precision (how many of the mappings they generate are correct) and recall (how many of the correct mappings they generate), these systems are also generally gauged based on how much effort they require from the humans interacting with the system, in terms of the number of questions they must answer and the difficulty inherent in coming up with each answer.

An obvious approach to leveraging user input in an alignment system is to first use an automated approach to generate an alignment and then ask the user to verify (a subset of) the matches that were created. Invalid matches can then be pruned from the final alignment. ServOMBI implements this approach [52]. Clearly, this approach is capable of improving precision, particularly if the user is asked about matches that the automated system is most in doubt about, perhaps evidenced by confidence values near the threshold. However, since the user involvement comes at the end of the alignment process, this method cannot improve recall over what the automatic component achieves. On the other hand, this approach is suitable for adding an interactive component to *any* matching system, because it only require the end product of the tool.

A variation of this technique is to move the interactive questioning to within the matching process rather than conducting it at after the fact. This can have a very large impact on both precision and recall because most alignment system will only match an entity to one other entity, so any match that is incorrect may be doubly bad by causing the correct match to be missed. Furthermore, when a match is found, many algorithms use a technique called similarity flooding [63] to thoroughly explore the neighborhoods of both of the entities involved, sometimes with relaxed match criteria on the theory that things related to equivalent entities are more likely to also be equivalent. The general idea behind similarity flooding is that two entities that are connected to similar things are most likely similar themselves. For example, assume there is a class in one ontology called Man that is a subclass of Human and the domain of a property called hasAge, and there is class in a second ontology called Male that is a subclass of Person and the domain of a property called hasYears. If Human and Person and hasAge and hasYears have already been found to be highly similar, similarity flooding will increase the similarity value between Man and Male. When using similarity flooding, an incorrect decision during the matching process can cascade to cause a host of other incorrect decisions.

Several interactive systems attempt to ask the user for guidance at critical decision points (and only these points) during the mapping process in order to maximize their accuracy. One such system is LogMap 2, which arranges all potential mappings that it is unsure about in partial order based on the value of similarity metrics employed by the system. Starting at the beginning of this list, the system asks the user whether each potential mapping is valid, until the end of the list is reached or the user halts the process. When a user approves a match for an entity, any other potential matches for that entity are discarded. Any matches suggested by the algorithm that are logically inconsistent with the user-approved match are also discarded. Experimental results indicate that this interactive technique improves performance as long as the user responds accurately at least 70% of the time [50].

The AgreementMaker alignment system takes a different approach to integrating user feedback into the alignment process. Rather than choosing which mappings to ask the user about based on a single or aggregate similarity score, AgreementMaker asks about potential mappings on which its constituent matchers disagree. Specifically, the system uses four syntactic matchers, a structural matcher, and a semantic matcher (according to the terminology presented in Sect. 1.3). If the matchers are divided on a particular mapping, the user is asked to provide a decision. This decision is then used to update the certainty values on other potential mappings with the same pattern of matcher agreement/disagreement, and this update is considered when deciding what question to ask the user next. In this manner the system is able to significantly improve the alignment accuracy while asking relatively few questions overall [17].

The OAEI established an interactive matching track in 2013. Participating systems can make a programmatic call to an "oracle" that consists of a pair of URIs and a relation (currently limited to either equivalence or subsumption) and receive a true or false reply indicating whether or not the relation holds between the two entities.[4] Beginning in 2015, the track included tests with an imperfect oracle, i.e. the oracle was either correct all of the time, correct 90% of the time, 80% of the time, or 70% of the time. Also in 2015, the alignment tasks were expanded from ontologies related to conference organization to other tasks, including mapping larger biomedical ontologies. Four alignment systems have participated in this track each year (though not always the same four), and the results have improved annually. In 2013, the average performance of the system when interactions were possible was actually 3% *worse* in terms of f-measure than in a fully automatic setting. The best system performed 8% better, for an f-measure of 0.72. Two years later, the average performance was 20% better with interactions, and the best system performed 11% better, for an f-measure of 0.818. The number of requests to the oracle required to achieve these results has also decreased markedly since the first year [19, 34].

While the introduction of the interactive alignment track to the OAEI has clearly been productive in terms of encouraging research in this area and driving the improvement of interactive alignment systems, this is not a perfect approach to evaluating such systems. In particular, the type of queries that systems can pose to the oracle

---

[4]http://oaei.ontologymatching.org/2013/interactive/index.html.

is limited to asking yes or no questions regarding a particular match. One can easily think of many other types of questions that would be worthwhile, such as how certain a user is that a particular match is correct or how a user arrived at their decision on the correctness of a match. As Paulheim and his colleagues point out, the questions that are asked of a user and the way in which they are asked impact the size of the burden placed on the user. For instance, asking a user *what* relationship holds between two entities (or what other entity is equivalent to a given entity) is likely a more difficult question for a user to answer than *whether* a particular relationship holds [81]. Others have conducted more extensive evaluations of interactive matching systems for the bioinformatics domain, including usability, time requirements, and user satisfaction [55]. This type of user study is time consuming however, and it has not been performed in a standardized way for a large number of general matching systems.

Of course, the issue with the above methods is that ontology engineers and domain experts are generally very busy people, and they may not have much time to devote to manual or semi-automated data integration projects. As a result, some ontology alignment researchers have turned to generic large-scale crowdsourcing platforms, such as Amazon's Mechanical Turk.

Amazon publicly released Mechanical Turk in 2005. It is named for a famous chess-playing "automaton" from the 1700s. The automaton actually concealed a person inside who manipulated magnets to move the chess pieces. Similarly, Amazons Mechanical Turk is based on the idea that some tasks remain very difficult for computers but are easily solved by humans. Mechanical Turk therefore provides a way to submit these types of problems, either through a web interface or programmatically using a variety of programming languages, to Amazons servers, where anyone with an account can solve the problem. In general, this person is compensated with a small sum of money, often just a cent or two. The solution can then be easily retrieved for further processing, again either manually or programmatically. While there are few restrictions on the type of problems that can be submitted to Mechanical Turk, they tend towards relatively simple tasks such as identifying the subject of an image, retrieving the contents of receipts, business cards, old books, or other documents that are challenging for OCR software, transcribing the contents of audio recordings, etc. As of 2010, 47% of Mechanical Turk workers, called Turkers, were from the United States while 34% were from India. Most are relatively young (born after 1980), female, and have a Bachelors degree [44]. It is possible for individuals asking questions via Mechanical Turk (called Requesters) to impose qualifications on the Turkers who answer them. For instance, Requesters can specify that a person lives in a particular geographic area, has answered a given number of previous questions, has had a given percentage of their previous answers judged to be of high quality, or pass a test provided by the Requester. In addition, Requesters have the option to refuse to pay a Turker if they judge the Turkers answers to be of poor quality.

A group of researchers from Stanford University has recently published several papers on using Mechanical Turk to verify relationships within biomedical ontologies [66–68, 77]. Their results show that general purpose crowdsourcing platforms can be used to answer questions about the relationships between ontology entities, even if the domain modeled by the ontology is quite scientific. Mechanical Turk has also been used to validate existing alignments [11, 12]. Additionally, there is an interactive alignment system called CrowdMap that uses Mechanical Turk to generate alignments between two ontologies [92] (Fig. 5).

All of these systems reported good results, though some were hampered by scammers that answered questions randomly or with some other time-saving strategy in order to maximize their profit-to-effort ratio [77, 92]. Additionally, there is some indication that the way in which questions about potential mappings are asked may have a large impact on the utility of the general crowdsourcing approach. In the work described in [12], questions about potential equivalent properties were presented in the following form: "Does property label A mean the same thing as property label B?" Respondents were instructed to choose one of four options: they mean the same thing, one is a more general or more specific term than the other, they are related in some other way, or there is no relation. In order to provide some context, the

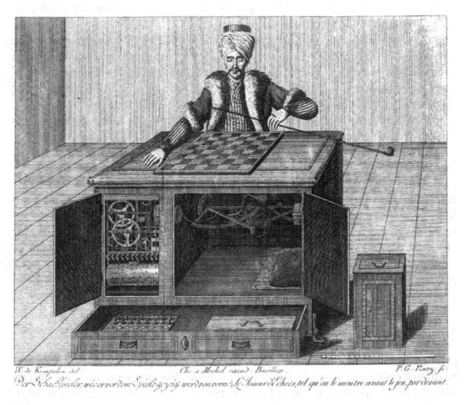

Fig. 5 An engraving of the original Mechanical Turk by Karl Gottlieb von Windisch. The Mechanical Turk was a famous chess-playing "automaton" from the 1700s that was actually operated by a human nestled inside the cabinet. It is the namesake of Amazon's Mechanical Turk platform, which allows developers to harness a plethora of human workers to solve tasks that remain difficult for computers

questions provided information about the domain and range of each property and up to five examples of instances with values for each property. The initial results showed that the general response on these nuanced verification questions were not very reliable. In all cases, the researchers responded by qualifying Turkers based on their performance on a small simple set of questions regarding possible mappings in order to gain access to the full range of tasks. This strategy proved reasonably effective.

Another strategy for dealing with scammers is to take money out of the equation. Instead of paying individuals to contribute to an alignment, the work can be packaged as a game that the user plays in order to earn a good score. This is the approach taken by the game SpotTheLink [109]. The game involves teaming up two random players, presents them both with an entity from the source ontology along with a description and image of the entity, if available, and asking them to collaboratively find an entity in the target ontology that is related and how it is related (equivalent, subclass, or superclass). Players only get points when they both agree. This game was built on top of OntoGame, which is a Java plug-in framework that provides services such as user login, randomly pairing users, and keeping score [99].

Another possible approach for avoiding scammers when crowdsourcing data alignments is to require people who wish to make sure of that data resource to first answer questions that contribute to its growth and quality (or to improve a separate, related data resource). This is the approach suggested by [60]. McCann and his colleagues point out that for this technique to work, the data resource must either not be available with no strings attached elsewhere, or it must be of a higher quality than alternative sources for the information. Additionally, the users must only be asked a limited number of questions, and they must have some control over when they will answer those questions.

2.2   Visualizing the Data Integration Process

Involving humans in the data integration process, as described in the previous section, requires some type of interface to enable those individuals to understand what questions are being asked of them, be aware of the context necessary to accurately answer those questions, and understand the implications of their answers. These needs lead to a set of requirements for data alignment interfaces. Several researchers have worked to enumerate these requirements. One of the first steps towards this was work by Falconer [25], which was then built upon by several others. There are several recurring points of emphasis in this work, which are addressed throughout the remainder of this section.

2.2.1   Presentation of Candidate Mappings

Users need a way to quickly see the mappings suggested by the automated mapping component, why they have been selected, and which mappings have been validated, refused, or remain to be considered. Because the number of candidate mappings may be quite large, there needs to be some way to manage them, for instance by clustering or by filter-based searching [33]. Logically organizing these potential mappings is key, because often validating one mapping enables a user to validate many additional mappings that have a similar underlying rationale [97].

VOAR is an application for working with ontology alignments that illustrates several of these concepts [96]. VOAR does not have a built-in automated alignment algorithm, but rather can call any such algorithm that implements a standard interface. Users can also load in multiple existing alignments and merge them or compute the intersection (i.e. only mappings that occur in all alignments are kept). When the user is validating and/or creating mappings, the class hierarchies are shown on either side of the interface, and a table of mappings, including the associated confidence value, is in the middle. Clicking on a mapping will highlight the relevant entities in the trees. VOAR also has another mode that allows users to visualize the alignment as a whole (Fig. 6). This shows which entities are involved in mappings, indicating areas within each ontology that are densely or sparsely related to the other. This view lists all entities from each ontology along the sides of the interface and connects related

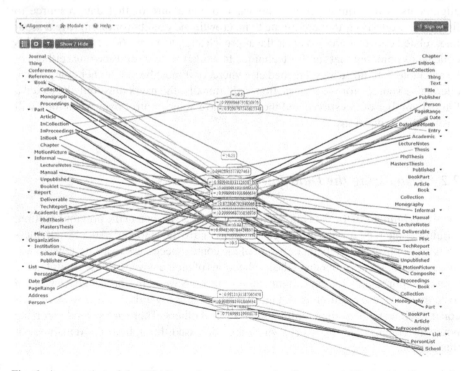

Fig. 6 A screenshot of the VOAR ontology alignment visualizaton tool. The entities from each ontology that are involved in mappings are listed along the edges, and the lines between them indicate equivalence relations

entities with a line. To assist the user in following these correspondences when there are many mappings, related entities and the line connecting them are color-coded.

BioMixer, a tool designed to visualize mappings among more than two biomedical ontologies at once, takes a different approach to showing an overview of all mappings within an alignment [111]. This tool provides several different ways to view mappings, including a matrix view in which the terms from each ontology are listed in alphabetical order along the top and left side of the matrix, and a colored square within the matrix indicates a mapping. This highlights clusters of mappings for similarly-named terms (which often serve as anchor points upon which to build more complete alignments). A different view enables the user to drill down into the part of the ontology surrounding a particular entity. This part of the tool uses displays the entity, its neighbors in the ontology, and its relationship to entities in the other ontology as a graph. The user is able to understand an entity's context and potentially identify missing mappings.

2.2.2 Presentation of the Ontologies

Presentation of the ontologies is also important. Most individuals will confirm the validity of a mapping based upon the neighbors surrounding the entities in both ontologies, and once they validate one mapping they are often able to confirm several others involving related entities [25]. For this reason, it is helpful to enable quick navigation between the list of potential mappings and the relevant entities in both ontologies. Further, by showing the entities in both ontologies that are involved in one or more potential mapping, the user can focus on these areas first and thereby improve their efficiency. Users also need to be able to add mappings that were missing from the list of suggestions. This necessitates the ability to navigate across both ontologies at varying levels of abstraction, including drilling down to view the details of any entity, and filtering on a wide range of criteria [33]. This is the area of visualization research that has received the most attention, as we will see later in this section.

A tree-based presentation of an ontology is only capable of displaying hierarchical information, such as the class hierarchy within an ontology. Other types of information contained in an ontology's axioms, such as property domain and range and cardinality constraints, are lost in a tree representation. This is particularly problematic for aligning the properties within ontologies [12]. To avoid this, many ontology visualization applications use a graph to display the ontologies. Kow and his colleagues take this approach in [53]. In this tool, candidate mappings are shown in a list that allows the user to accept or reject them. As with BioMixer, selecting a mapping in the list displays the relevant entities and their neighbors within the ontologies in a detailed graph view. While limiting the graph to the nearby neighbors of the entities in question keeps them from becoming cluttered, the overall context of the ontology is lost. A way to navigate across the ontologies at a high level of abstraction is needed. Kow's application enables this through a global view they call an "information landscape". This view shows all of the entities from both ontologies (color-coded red or green according to which ontology they belong), with similar concepts placed near one another. Clumps of entities are labeled with terms describing the group. The user can select areas of interest that seem likely to contain related entities, which automatically filters the mappings shown in the list. This method of filtering allows users to systematically explore an ontology at a high level of detail without losing track of the big picture.

2.2.3   Demonstration of Mapping Implications

The individual mappings that together make up an alignment are not independent of one another; there are some cases in which only one of two mappings can possibly be true. Sometimes mappings will result in a class being *unsatisfiable*, meaning that it is not possible for an instance to meet all of the requirements to be a member of that class. In other cases, one or more mappings, when taken together, may lead to an unintended and undesired inference. Visualization systems need to convey the implications of a potential mapping to users. Potential ways to achieve this include highlighting the relationships between mappings, allowing the user to temporarily add a mapping and observing its impact [45], and providing a mechanism for the user to indicate that a particular mapping is uncertain or subjective [25].

ContentMap is one application that attempts to provide details about the implications of mappings to users [49]. ContentMap uses several existing ontology alignment algorithms to generate a set of candidate mappings, which users can either accept or reject. The system then computes the *logical difference* between the entailments before and after the mappings are applied. Entailments that ContentMap suspects may be unintended (because they hold in the merged ontology but not in the individual ontologies) are presented to the user, who can indicate which ones are in fact undesired. The system then runs a mapping repair algorithm that attempts to remove the minimum number of axioms to alleviate the unintended entailments while preserving the entailments the user indicated were valid. Because computing the logical difference is quite difficult over expressive ontologies (there is no algorithm to do this for OWL 2 or OWLDL), ContentMap focuses only on alignments consisting entirely of subclass, equivalence, and disjointness relations.

Another data integration tool, MappingAssistant, takes a different approach to providing feedback to the user regarding the implications of a mapping [103]. MappingAssistant is based on the intuition that domain experts, who are not necessarily familiar with formal modeling constructs like ontologies or with logical entailments, nevertheless have a detailed understanding of the instance data. The system therefore conveys the implication of schema-level mappings to the user by selecting (using a clustering algorithm) a set of instances affected by the mapping rule and displaying them. Users can then indicate any instances that have been incorrectly classified, and the application will present a series of questions (expressed in natural language) to the user in order to determine which mapping has led to the incorrect classification.

2.2.4 Scalable to Large Ontologies

Assisting user in aligning ontologies with a large number of entities or many axioms constraining the relationships between entities can be particularly challenging for a visual interface. A myriad of issues come into play. For instance, many alignment systems display the class hierarchies as tree structures on either side of an interface, with lines between the trees indicating potential mappings. If the number of potential mappings is very large, users can quickly be overwhelmed by such a presentation [25]. Other interfaces display the ontologies and potential mappings in a graph, but again, the size and complexity of the graph grows with that of the ontologies, and the user can find this representation unwieldy [57]. Ivanova's work on requirements for large-scale ontology alignment make it clear that whatever strategy is used to represent the ontologies and potential mappings, it must not only "scale" visually, but also computationally – users cannot be made to wait after each interaction for the interface to update [45]. Users also cannot always be expected to align large ontologies by themselves in one setting. Consequently, tools should allow users to save their progress and to divide up the alignment task among multiple contributors [33].

One approach to dealing with the overwhelming complexity of a graph-based global view of an ontology is through employing a clustering algorithm to raise the level of abstraction at which the ontology is shown in the graph. Even though AlViz, developed in 2006, is older than many of the other visualization applications discussed in this section, we use it as an example of this approach since new work in the field still frequently cites it as a source of inspiration [57]. AlViz shows each ontology in its own graph and uses color coding of nodes to indicate areas of similarity and difference between the two. A slider on the side of each graph controls the level of abstraction. The size of a node gives an indication of the number of entities aggregated within it. Nodes are aggregated based on the similarity metric of an integrated alignment algorithm. Small world graphs such as those used by AlViz typically use a spring layout, which is known to have a cubic time complexity. Still, the original AlViz system was capable of displaying ontologies with 1000 entities and respond to user interactions without a lag.

The alignment system AML employs a different strategy to handle the complexity of graph-based views [85]. AML combines both ontologies and mappings in a single graph. However, instead of showing the full ontologies and alignment, it shows only a subgraph centered on a selected mapping, for which the neighborhood of classes and mappings between them can be shown up to five edges distance. This allows for a better understanding of neighboring mappings than typical tree-based visualizations, and is particularly relevant in the visualization of biomedical ontologies where multiple inheritance and the existence of different kinds of relations between classes is common. Users can then navigate the list of mappings to visualize the different subgraphs, mark mappings as correct or incorrect, and add new mappings.

The alignment system SAMBO implements several features to assist users in aligning large ontologies [56]. In particular, the tool allows users to cease calculations of mapping suggestions at any time and begin to approve or reject any of the mappings that have been suggested at that point. The user can also save their work and resume it later. Each saved session contains information about how many mappings have been validated, how many remain, and the last date the user worked on the alignment. Users also have the option to preprocess data in between sessions, to save time when they resume their work. The preprocessing step uses the class hierarchies of both ontologies to partition the ontologies into "mappable parts" such that the set of entities from the first ontology that are in a partition are highly likely to be mapped to an element from the second ontology that is also in the same partition. As a result, similarity metrics do not have to be applied between all pairs of entities, but only between those in the same partition. Furthermore, users can focus their attention on one partition rather than being overwhelmed by the entire ontologies. While the authors do not state this, these partitions might also be a way to divide the mapping validation task among multiple people.

Several researchers who have considered the requirements for a visualization system intended to facilitate data integration have also mentioned the need to allow users to annotate a particular mapping with its rationale and additional metadata as required for the particular use case, and a mechanism to debate or vote on mappings [33, 45]. Unfortunately, this information is not collected in as anything other than free text by most data integration tools. This issue is the subject of Sect. 3.3.

2.3 Integrating Geospatial Data

Many data sets, from user reviews of hotels and restaurants, to oceanographic measurements, to economic data, have a spatial component. Integrating data based on location can lead to important cross-domain insights relevant to a particular region. However, as mentioned in the introduction, spatial data is particularly difficult to align. There are many reasons for this. Of course, spatial data sets have all of the normal issues related to schema. For instance, one data set may refer to a building's location using the property "Address" while another one may use two properties: "City" and "State." There are also challenges specifically related to spatial data because of the many ways to express it. For example, location can be specified with an address, with latitude and longitude, or in reference to a nearby point of interest. There are also many ways to express a spatial region. For example, spatial regions can be represented by geopolitical entities (whose borders change over time), by polygons whose points are given via latitude and longitude, or by a point and a radius. Another issue is that spatial data is collected at widely different scales and with different resolution and coverage, which raise quality concerns when integrating several data resources. Furthermore, for both technical and social reasons, many spatial data sets are stored in relational databases, as images, or as vector data. These different representation formats necessitate different approaches to integration. This section surveys some of the current research related to integrating geospatial data.

2.3.1 Representing Geospatial Data

Many geospatial datasets have been published on the Semantic Web. Two of the largest and most well-known are GeoNames and LinkedGeoData. GeoNames has information about over 8 million geographic entities from around the world, including place name, coordinates, elevation and population. Much of the data was originally imported from official public sources, but it can now be edited by individuals. GeoNames is organized according to a relatively simple ontology involving nine features and 645 feature codes.[5] LinkedGeoData is based on the data from the Open-StreetMap project. OpenStreetMap's goal is to build a geographic knowledge base from the ground up, by allowing contributors to use aerial imagery and GPS devices to create and verify information.[6] GeoNames and LinkedGeoData are interlinked with one another and with DBPedia. Other geospatial datasets are region-specific. For instance, the UK Ordnance Survey, Great Britain's national mapping agency, has published gazetteer and administrative boundary information as linked data as part of the "Making Public Data Public" initiative within that country.[7] Publishing geospatial data according to the linked open data principles allows it to be integrated more easily. Consequently, useful applications that leverage linked geospatial data have begun to emerge, including for disaster management [80] and wildlife monitoring [54].

Much of the geospatial data that is currently available is represented using the Geography Markup Language (GML).[8] GML was created by the Open Geospatial Consortium (OGC) and has become an ISO standard. The schema is centered on the class *Feature*. A *Feature* can have a *Geometry*, such as point, line, polygon, curve or surface. GML also supports specifying a *Feature*'s location, using a coordinate reference system. Another way to represent geospatial data is using GeoSPARQL, which is an RDF vocabulary and a set of extensions to SPARQL to support spatial queries.[9] The GeoSPARQL vocabulary currently leverages many elements of GML, together with well-known text (WKT), to represent vector geometry objects on a map; simple feature, which contains spatial relationships such as intersects and within; region connected calculus (RCC8), to represent relationships between two regions such as tangential or partially overlapping; and DE-9IM, to represent topology.

Other OGC standards are closely related, including Keyhole Markup Language (KML) to specify how display geographic information on a map or other visualization.[10] Several ontologies have also been developed to represent higher-level concepts with a strong spatial element, such as a "Semantic Trajectory" to describe movement through space [43] and "Stimulus-Sensor-Observation" to model observations of phenomena collected at a particular time and place [47].

---

[5] www.geonames.org.

[6] http://linkedgeodata.org.

[7] http://data.ordnancesurvey.co.uk.

[8] http://www.opengeospatial.org/standards/gml.

[9] http://www.opengeospatial.org/standards/geosparql.

[10] http://www.opengeospatial.org/standards/kml.

2.3.2   Querying Geospatial Data

One way to query geographical data is by using the Web Feature Service (WFS).[11] This OGC standard can return results in GML or as shapefiles, a vector format dictated by the Environmental Systems Research Institute (Esri) and used by the popular GIS software platform ArcGIS. Examples of WFS queries are: "return the name of all towns that are along this line" and "return the name of all mountains within this bounding box." GeoSPARQL supports these types of queries as well, but over the full RDF vocabulary expressed in the GeoSPARQL standard.

Many datasets with a geospatial component are stored in relational databases rather than published as linked data. There are several reasons for this: databases are often an established part of a scientist's workflow, existing data analysis tools may require the data to be stored in a database, or collection systems may automatically publish to a database. However, there is still a need to incorporate semantics into queries of this data. One approach to achieving this is to allow users of the data to query it based on an ontology, and then to translate those queries into the language required by the database. This area of research is sometimes known as ontology-based data access (OBDA). An example of this is the work of Zhao and his colleagues, described in [114]. Their system uses an RDF ontology to enable semantic queries on a standard relational database containing geospatial data. Their approach is to translate queries based on their RDF ontology to WFS queries on the underlying database. There are some limitations to this approach. In particular, the ontology needs to be created manually for each dataset and application domain, and a table in the database can only map to one class in the ontology. Also, the ontology is specified in RDF rather than OWL in order to simplify the query rewriting. Later work supports more complex queries while extending this semantic querying capability to multiple geospatial datasets stored as GML [110].

2.3.3   Coreference Resolution and Alignment

Regardless of how it is stored, coreference resolution of geospatial data is particularly challenging due to noise and difference in coverage and resolution. Even extracting appropriate features on which to base the data integration task can be difficult, though recent work in that area by projects such as Brainwash show promise [1]. Once features have been collected, it has been common to use machine learning approaches to weight features of geospatial datasets such as location name, location type (e.g. mountain, desert, island) and coordinates, which are then used in standard classifiers such as SVMs and linear regression models [95]. More recent work takes a very similar approach. For instance, McKenzie and his colleagues integrate data from FourSquare and Yelp using a weighted combination of location name, location category (e.g. seafood restaurant, casual dining), geographic location, and an unstructured textual description. Their results were impressive, with 98% of places

---

[11] http://www.opengeospatial.org/standards/wfs.

of interests correctly aligned. An interesting element of their work was that a system based only on geographic coordinates was only 57% accurate. They indicate that this may be due to inaccuracies of mobile devices using GPS or wireless to calculate position [61]. Similarly, Li et al. merged point of interest data from Baidu (a search engine) and Sina (a social networking site) based on name, category and location. Their weighting method was based on the entropy of the various attributes. This method was chosen because the attribute values exhibited a non-linear similarity metric characteristic [58].

Aligning the schema of geospatial datasets can actually be somewhat easier than in the more general case. While geospatial datasets often have different labels for the same properties (e.g. "state" versus "administrative district"), labels of geospatial properties are selected from a smaller domain than are general properties. Furthermore, geospatial datasets typically have a large A-box, making extensional matching techniques useful when aligning the T-Box. For example, if one dataset has a property called "CensusCount" and another has a property called "Population," values for particular cities contained in both datasets allow an automated alignment system to conclude that these properties are likely equivalent.

2.3.4   Assessing Quality

When integrating data from multiple sources, quality becomes an important concern. This is particularly important for geospatial datasets. Whenever any continuous entity is measured, there will be inaccuracies inherent in the measured values due to limitations of coverage and resolution. Typical quality indicators include lineage, positional accuracy, attribute accuracy, logical consistency and completeness [7]. Additionally, interviews with consumers of geospatial data indicate the importance of metadata when assessing quality, such as the reputation of the data provider and the number of citations. Unfortunately, the majority of geospatial datasets do not have *any* quality information associated with them [59].

There have been some efforts to automatically derive quality information for geospatial datasets that lack it. For instance, work by Thakkar et al. is targeted toward situations in which many geographical datasets are being integrated, and only some of them have associated quality metrics. Quality is based upon completeness and positional accuracy. Completeness is the percentage of features that the source contains information on. Thakkar gives the following example: if there are 100 hospitals in an area and a source contains 25 of them, then the source is 25% complete. Positional accuracy is determined based on the number of features within a given bound, i.e. the location of 40% of the hospitals is accurate to within 10 m. They automatically assess the quality of an unknown data source by identifying a source with known quality that provides the same attribute and has at least some instances in common. The quality of the new source is then based upon comparison of a sample of the common subset. Once that source's quality has been evaluated, it can then be checked for overlap with any other sources within the system whose quality was previously unknown and used to assess their quality [108].

There has also been considerable research on assessing the quality of volunteered geographic information. For example, in 2010 Mooney and his colleagues evaluated OpenStreetMap data from 11 European countries based on sampling density and metadata tagging and their utility in correctly representing the shape of features such as lakes and forests (and found the quality to be quite low overall) [65]. Ballatore and Zipf take a higher level approach and consider the quality of the schemas used to organize the geospatial data. They argue that maintaining conceptual quality is straightforward when data producers and consumers are all colleagues, but that quality suffers when producers and consumers don't know one another, as is the case with volunteered geographic information. Their framework includes accuracy (existence of entities, categories and attributes necessary to accurately describe the geospatial features of interest), granularity (ability to describe the features at the desired level of abstraction), completeness (ability to describe all the features of interest), consistency (similar features are described with similar classes and properties), compliance (agreement of this schema with another one), and richness (number and variety of dimensions with which to describe a feature) [4].

2.4 Integrating Biomedical Data

Massive amounts of multimodal and diverse data are currently being generated by researchers, hospitals and mobile devices around the world, and their combined analysis presents unique opportunities for healthcare, science and society. The data can range from molecular to phenotypic, behavioral to clinical, individual to population, genetic to environmental. Maximizing the potential of this data through its meaningful integration can enable new directions for research, for instance discovering new drugs or determining the factors causing human disease.

Biomedical Big Data goes well beyond the recognized challenges in handling large volumes of data or large numbers of data sources, and presents specific challenges pertaining to the heterogeneity and complexity of data as well as to the complexity of its subsequent analysis. The availability of over 500 open biomedical ontologies in BioPortal [76] and dozens of biomedical datasets as Linked Open Data represents a unique opportunity to integrate clinical and biomedical data.

A first step in supporting the semantic integration of biomedical data is by making it available as Linked Data and having the entities and relationships referred to in the Linked Data defined according to ontologies. Exposing datasets as Linked Data enables the interconnection of distinct data items across providers, facilitating the integration of high volume and heterogeneous data sources (i.e. experimental data, libraries, databases) and also provides an aggregated view of biomedical data in a way that is machine interpretable and reusable, as well as semantically-enriched via links to ontologies. These links support the classification of data according to the concepts defined by a given ontology, which provides a perspective on the data.

The same data can be described under different ontologies, which provide different perspectives. For instance, patient data described under the Disease Ontology

[94] will provide a view of the diseases and disorders affecting the patient, while the same patient data described under the Symptom Ontology [3], will provide information about signs and symptoms but not the underlying causes. However, even when focusing on a single perspective, lets say diseases, the multitude of ontologies and controlled vocabularies currently in use to describe them impedes the seamless integration of data. Multiple ontologies for the same or closely related domains can and do exist, due to several reasons ranging from disconnected development, to development focusing on particular applications. This is especially true in the biomedical domain where there are for instance nine ontologies that describe neurological disease, ranging from highly specific ontologies covering a single disease (e.g. epilepsy, Alzheimers) to ontologies covering all kinds of human disease, such as the Disease Ontology. This results in several ontologies describing the same concepts under slightly different models.

These challenges are being addressed at several levels by the application of Semantic Web technologies.

2.4.1 Linked Biomedical Data

There have been several efforts to expose biomedical data as Linked Data, with the aim of providing structured and integrated access to the massive amounts of biomedical data distributed in numerous repositories [8, 64, 88, 89, 112]. This is a challenging endeavor since each biomedical dataset has a unique structure and vocabulary.

The Bio2RDF project [8] defines a set of simple conventions that allow the creation of a knowledge space of RDF documents as Linked Data. It uses a mashup approach that leverages normalized URIs and a common ontology, integrating publicly available data from some of the most popular databases in bioinformatics. However, few biomedical repositories expose their data as RDF, so the project built a toolbox to generate RDF files from locally stored databases or directly from HTML documents accessed via http requests. Although Bio2RDF facilitates the integration of heterogeneous datasets, achieving a complete syntactic and semantic normalization is not yet a reality. One of the reasons behind this, is that Linked Data serialized as RDF does not support the complex formal semantics that allow the inference of relationships between data items from heterogeneous datasets. This prevents a fuller integration, namely at the level of relations or types.

Another related project is Neurocommons [88], which is dedicated to creating an open source knowledge management platform for biological research and is specifically working on an open knowledge base of annotations to biomedical abstracts (in RDF) and the integration of major neuroscience databases into the annotation graph. Neurocommons is grounded in Semantic Web technologies, integrating OWL ontologies, RDF and SPARQL endpoints.

BioPortal also publishes their ontologies as RDF [89]. This dataset contains over 190 million triples, representing both metadata and content of the ontologies. It also publishes over 10 million mappings between ontologies, generated via both manual and automatic methods.

2.4.2 Cross-References and Mappings

To promote and facilitate integration, some biomedical ontologies already provide cross-references to equivalent or related concepts in other ontologies. These can be used not only for integrating ontologies, but also for the integration of data items described with the ontologies.

One notable effort in increasing the interoperability of biomedical ontologies has been the creation of logical definitions [71]. This is an initiative of the Open Biomedical Ontologies Foundry [101], a collective of ontology developers whose mission is to develop a family of interoperable ontologies that are both logically well-formed and scientifically accurate. One issue of biomedical ontologies is that although almost all classes have a textual definition, which can be interpreted by a human user, this is not accessible to a computer without sophisticated natural language processing. Therefore, efforts have been made to transform these definitions into a computable form as a set of logical definitions. Such logical definitions facilitate automated access to an ontology and complement text definitions. They could also potentially be used to reason over an ontology or to automatically derive relationships between classes, thus contributing to the integration of different ontologies. Developing and maintaining these computable definitions requires a lot of manual labor, leading to the development of strategies to partially automate the process [70]. More recently, the definition of composite relations as class expressions has also been explored through the alignment of classes in biomedical ontologies with foundational classes in a top-level ontology [42].

Another relevant resource is the UMLS [6], which provides a mapping structure among over 100 controlled vocabularies in the biomedical sciences, covering over 1 million biomedical concepts and 5 million concept names. UMLS is not originally available as RDF, but BioPortal through its UMLS2RDF project [89] has transformed the UMLS MySQL release into RDF triples. BioPortal also provides a set of 3.1 million mappings between the terms in UMLS vocabularies.

While these projects are contributing to the utility of biomedical data on the Semantic Web by establishing links between datasets, in many cases such mappings are unavailable, giving rise to the need to derive them automatically using ontology matching techniques.

2.4.3 Ontology Matching for Biomedical Ontologies

The specific characteristics of biomedical ontologies need to be taken into account when developing tools and techniques to explore them:

- large size: biomedical ontologies commonly have thousands of classes, which can represent both a computational and a visualization challenge. In Bioportal there are over fifty ontologies with more than ten thousand classes.
- complex vocabulary: biomedical ontologies typically encode several names for the same class, including one main label and several synonyms of different kinds (e.g., narrow synonym, broad synonym). This represents a challenge for lexical matchers, which need to be able to handle multiple labels and at different closeness degrees.
- multiple related domains with different points of view: it is fairly common to have the same biomedical domain being described according to different models. This can cause logical incoherences when two ontologies with different models are integrated. For instance, Fig. 7 illustrates a logical incoherence caused by two mappings between the National Cancer Institute Thesaurus Ontology and the Foundation Model of Anatomy Ontology. The logical incoherence arises because upon integration, *Fibrilar_Actin* becomes a subclass of both *Anatomic_Structure_System_or_Substance* and *Gene_Product*, which are disjoint classes. Solving these incoherences is far from trivial [83]
- rich axioms: biomedical ontologies have been evolving towards greater semantic richness establishing different kinds of relations between classes (e.g., regulates, adjacent to, participate in) and complex axioms (e.g., 'human patient and (has Age some float [>= 8]) participant in' WHO standard treatment for human brucellosis in adults and children eight years of age and older). Typically, ontology matching systems either just focus on taxonomic relations, or do not differentiate between different types of relations. This is especially relevant for structural matchers.

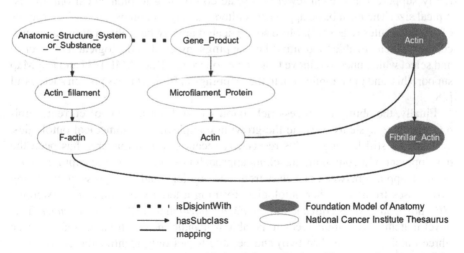

Fig. 7 Alignment between portions of the National Cancer Institute Thesaurus Ontology and the Foundation Model of Anatomy Ontology illustrating a logical incoherence

The relevance of ontology matching for biomedical ontologies has been recognized by the community, and the Ontology Alignment Evaluation Initiative [13] currently contains two tracks dedicated to biomedical ontologies: the anatomy track, and the large biomedical ontologies track. Both tracks illustrate the above mentioned challenges, and in the last few years some ontology matching systems have been quite successful in addressing the challenges of matching biomedical ontologies.

Regarding size and scalability, a well studied challenge [51], systems have had to evolve from the traditional encoding of the matching problem as a matrix of similarities to more efficient data structures. For instance, AgreementMaker [15], a system successfully used for matching biomedical ontologies [16] struggled with more than a few thousand classes, which inspired the development of AgreementMakerLight (AML) [27], based on more scalable data structures that can handle over 100 thousand classes in an ontology.

The complexity of the vocabulary in biomedical ontologies has also been addressed by several systems, which combine several matchers capable of exploiting different string and lexical similarities [15, 27, 72]. Some systems take this further by leveraging external resources as a source of synonyms. For instance, AML includes specifically tailored approaches to exploiting the rich synonyms of biomedical ontologies [84], that can use cross-references to extend synonyms and apply lexical techniques to derive new synonyms. It also contains high performing strategies to automatically select and utilize external ontologies as background knowledge [29]. Other systems use pre-defined external resources. For instance LogMap [48] makes use of external lexicons to derive spelling variants, and GOMMA [39] can explore external ontologies for synonyms.

Regarding the ability to handle logical incoherence, few ontology systems currently support it, and even fewer at a scale conforming to biomedical ontologies' typical size. The most basic approach is filters out any mappings that violate a series of semantic rules (e.g. [72]). More sophisticated approaches rely on automated procedures which are able to identify the mappings involved in the logical incoherence and select which ones to remove to achieve coherence. Both AML [91] and LogMap support this, and their application to the mappings in BioPortal has proven successful [28].

Finally, the ability to process rich axioms is still not a focus of current ontology matching systems. Despite the growing complexity of biomedical ontologies, systems are still lacking in this respect. A recent effort in this area has been the development of a compound matching approach [79], that is able to capture equivalence mappings between one class from one ontology and an expression involving two classes from two other ontologies, forming a ternary mapping. For example, *HP* : *aorticStenosis* is equivalent to an *FMA* : *aorta* that is *PATO* : *constricted* This novel matching paradigm needs to be able to handle the much larger search space (three ontologies instead of two) and be able to not only identify the equivalence mapping but compose the expression as well.

3   The Path Forward

Work such as that described in Sect. 2 has already begun to pay dividends. Techniques for semantic data integration have reached a level of maturity that has allowed them to be incorporated into commercial and open source tools from organizations such as Oracle, Apache and Microsoft. For example, Oracle 11g provides support for storing data as RDF, querying data from disparate sources seamlessly via SPARQL, and performing reasoning via SWRL-like rules [113]. Various aspects of the performance of these industry systems is being evaluated by the academic community [98] as well as utilized for domain-specific research [26, 100]. Additionally, these systems are being used by other commercial enterprises for applications ranging from entertainment media management to national intelligence [113]. However, many challenges clearly stand in the way of accurate and efficient data integration in the general case. This section considers some research threads that could potentially lead to future breakthroughs in semantic data integration.

3.1   *Moving Beyond 1-to-1 Equivalence Mappings*

Ideally, alignment systems should be able to uncover any entity relationships across two ontologies that can exist within a single ontology. Such relationships have a wide range of complexity, as shown in Fig. 8. The simplest type of relationship is 1-to-1 equivalence or disjointness of two entities (i.e. all instances of A are also instances of B or an instance of A is definitely not an instance of B). Assume that we have two ontologies, ont1 and ont2, that model a university. The relation ont1:Course = ont2:Class is an example of a 1-to-1 equivalence match, while ont1:registeredFor disjoint ont2:Teaching (i.e. someone cannot both register to take a course and teach it) is an example of a 1-to-1 disjointness relationship. The next complexity level

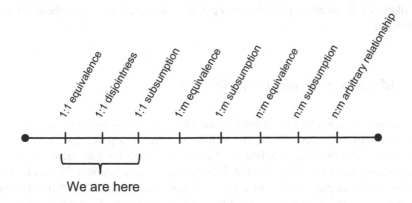

Fig. 8 Complexity range of entity relationships between ontologies

is subsumption relationships, i.e. that an entity in one ontology is a subclass or superclass of an entity in another ontology. ont1:Faculty ⊂ ont2:Employee is an example of this. Even harder to find are 1-to-many equivalence or subsumption relationships between entities, such as the union of ont2:AsstProf, ont2:AssocProf, and ont2:FullProf is equivalent to ont1:Professor. This causes a complexity problem. To find 1-to-1 relationships, an exhaustive search needs to compare every entity in the first ontology to every entity in the second ontology, which may be feasible for small ontologies. To find 1-to-m relationships an exhaustive approach would need to compare each entity in the first ontology to all possible combinations of m entities in the second ontology, which is not generally possible. Finding arbitrary n-to-m relationships is the most complex alignment task. By "arbitrary," we mean any type of relationship, not restricted to equivalence, disjointness, or subsumption. An example of this might be that a ont1:Professor with an ont1:hasRank value of "Assistant" is equivalent to an ont2:AsstProf. Such complex relationships would need to be expressed as logical rules or axioms.

Nearly all existing alignment systems fall at the simplest end of this scale. A few systems, including BLOOMS [46] and PARIS [104], are capable of finding subsumption relationships across ontologies. CSR [102] and TaxoMap [38] attempt to find 1-to-m equivalence and subsumption relationships. There has also been some preliminary explorations into identifying ternary compound mappings across biomedical ontologies [79]. In general though, most research activity in the field of ontology alignment remains focused on finding 1-to-1 equivalence relations between ontologies.

As mentioned previously, the performance of current alignment systems on tasks that focus on the identification of 1-to-1 equivalence relations has become quite good. However, alignment research may be in danger of becoming stuck in a "local maximum", and it might be time to make a concerted push towards discovering more complex semantic relationships. The computational complexity of this task makes it very unlikely that existing approaches to mapping discovery can be used to discover complex relationships. It is possible that existing algorithms from the fields of data mining and machine learning might be applied for this purpose, but significant effort will likely be required to identify appropriate techniques and tailor them for this application.

3.2   Advancing Alignment Evaluation

The Ontology Alignment Evaluation Initiative (OAEI) is now over a decade old, and it has been extremely successful by many different measures: participation, accuracy, and the variety of problems handled by alignment systems have all increased, while runtimes have decreased [24]. The OAEI benchmarks have become *the* standard for evaluating general-purpose (and in some cases domain-specific or problem-specific) alignment systems. In fact, you would be hard-pressed to find a publication on an ontology alignment system in the last ten years that did *not* use these benchmarks.

They allow researchers to measure their systems performance on different types of matching problems in a way that is considered valid by most reviewers for publication. They also enable comparison of a new systems performance to that of other alignment systems without the need to obtain and run the other systems. This is a huge boon for ontology alignment research.

Of course, benchmarks need to evolve over time in order to remain relevant. The OAEI suite of benchmarks contains eight tracks that test alignment systems in a range of contexts in which they might be used, but currently none of these tracks contain any complex relationships. In addition, the details of some of the test sets have led to the incorporation of behaviors in some alignment systems that may not be optimal. For instance, in several OAEI tracks an entity can be involved in at most one match, which may not be realistic for some real-world datasets. Similarly, entities are only matched to other entities of the same type in some tracks, e.g. classes to classes, instances to instances, etc. This is not realistic in all cases, particularly when the decision of when to represent something as an instance versus a class is not always clear cut.

As a specific example of the limitations of current alignment benchmarks, consider the case of property matching. Performance of current alignment systems on matching classes is on average three times better than on matching properties [12]. Researchers have suggested various reasons for this, including that the parts-of-speech used in property names differs from that used for class names [106], that taxonomies of properties are much less common that those of classes [74, 106], and that properties are reified in different cases than are classes [74]. Perhaps uncoincidentally, only one of the eight OAEI tracks involves any matches between properties, and those matches make up a small percentage of the total. This is a big cause for concern because many influential real-world linked datasets, such as DBPedia (the linked data version of Wikipedia) and YAGO, are strongly property-centric.

The OAEI is a community-driven effort, and its organizers are very willing to incorporate new benchmarks into the evaluation. Establishing new benchmarks is far from easy, however. Some of the existing OAEI testsets are synthetic, which means that the reference alignments are completely accurate. Synthetic benchmarks may not accurately reflect the type of challenges alignment systems face "in the wild" though. On the other hand, several of the OAEI testsets are based on real-world ontologies. The reference alignments were developed primarily by three graduate students, with feedback from "Consensus Workshops" held after each OAEI for several years. This method of benchmark creation is very resource-intensive and is therefore only feasible for small ontologies. In order to create the large reference alignments comprised of complex mappings needed to drive the field forward, more scalable methods of benchmark construction need to be explored.

As mentioned in Sect. 2.1, some researchers have turned to crowdsourcing platforms such as Amazon's Mechanical Turk to facilitate scalable ontology alignment. It may be possible to use such platforms to generate alignment benchmarks as well. However, there is some well-founded skepticism regarding the trustworthiness of crowdsourced alignment benchmarks. In particular, there is concern that the results may be very sensitive to how the question is asked. For instance, how much context from each ontology are users provided with? Are they able to rush through the work, or does some mechanism force or encourage them to give due consideration to each potential match? Does the best method for question presentation depend on the characteristics of the ontologies being aligned? How does the amount of monetary payment and bonuses affect performance? These are all very important questions, and if researchers in the ontology alignment field are going to accept work on complex alignments evaluated via crowdsourcing or a crowdsourced benchmark as valid, they must be addressed.

Another obstacle is that when creating an ontology alignment benchmark, one has to start from somewhere. It is too resource intensive to try to verify every potential relationship across all entities in both ontologies, even in the 1-to-1 equivalence case. This is a complete non-starter for complex alignments. The standard approach to this problem is to employ an ensemble of existing high-performing alignment systems to align the ontologies and then manually refine the results to create the reference alignment for the benchmark [92]. Unfortunately, this approach is not feasible for the creation of some types of benchmarks due to the lack of current alignment systems that attempt to the type of relationships required. It is something of a chicken-and-egg problem. For instance, it is very difficult to create a benchmark containing complex relationships when there are no alignment systems capable of identifying such relationships that can be used to create the benchmark. Solving this problem is an open area of research in this field.

3.3   Contextualizing Alignments

Data and schema integration is done for some *purpose*, and the mappings that should be included in a particular alignment are a function of that purpose. For example, alignments can be done to support distributed querying, or they can be used for logical reasoning. The characteristics for each type of alignment are different. For querying, recall (i.e. returning the relevant results) is generally an important aspect of the application using the alignment. This means that alignments to support query-centric applications arguably need to err on the side of expressing relationships that generally hold, even if some outliers lead those relationships to cause logical inconsistencies that confound reasoners. Conversely, applications that intend to employ a reasoner on the integrated data cannot generally make use of an alignment that contains any logical inconsistencies

Current alignment systems support these different use cases to some degree. For instance, AgreementMakerLight has the ability to detect and repair mapping inconsistencies, but it is also capable of leaving these in place [91]. However, there is currently no way to express in an alignment, after it has been created, which use case was targeted. It can be argued that the user of the alignment can simply check to see if it contains any inconsistencies, but this assumes that the user is employing the alignment, which may be only for the T-box of the ontologies, to the same A-box that was in place when the alignment was created, and that the A-box hasn't changed over time. Additionally, there are many applications that an alignment may have been created for beyond the simple query-versus-reasoning divide. A way to express the situations in which an alignment applies is needed.

In addition to a mechanism for expressing the applicability of an alignment, a standardized way to represent the rationale behind individual mappings within an alignment is also needed. For example, if an alignment asserts that Johnathan Smith who works at IBM is the same person as John Smith who organized the ABC Conference, it is helpful for the consumers of that alignment to know how this was determined. Perhaps in this case it is known that IBM was the primary sponsor of the ABC Conference and that John is a common nickname for Johnathan. Making this type of provenance available at the level of individual mappings is important for enabling consumers to make informed decisions about how to use the mappings within an alignment and how much confidence to place in them. While this need has been noted by both researchers and practitioners [75, 97], how best to represent this information is not currently clear.

By far the most common manner in which ontology alignment and coreference resolution systems represent their results is the Alignment API format. In this representation, each relationship between two ontologies (cell) is a "first-class citizen. In particular, each cell contains the URI of the entity that is the source of the relationship, the URI of the target entity, the relationship that holds between them (equality, subsumption, etc.), and the strength of that relationship (a decimal value between 0 and 1, inclusive). However, it also seems obvious that storing provenance information regarding who created a coreference and when would also be extremely useful. The creators of the Alignment API intended for such provenance information to be stored at the alignment level rather than at the level of individual cells. This is not well suited to projects in which coreferences may come from a variety of sources, including both people and automated algorithms, over a period of weeks, months, or years. Noy and her colleagues came to the same conclusion while collecting community-based mappings for the BioPortal ontology collection [75]. That work also reified coreferences, but it stored significantly more provenance information about the individual relations, including discussion and user comments, application context (conditions under which the relationship holds), mapping dependency (to express that this mapping holds if and only if some other mapping holds), mapping algorithm, creation date, creator (the person who uploaded the mapping), and external references (e.g. relevant publications). Unfortunately, this information is currently encoded largely as free-text, which violates the underlying Semantic Web principle that information about data and how it relates should be accessible to both humans and machines. Establishing an appropriate method for representing provenance and contextual information for alignments and individual mappings remains an important challenge for the field of semantic data integration.

Acknowledgements This work was supported in part by the National Science Foundation award 1440202 GeoLink - Leveraging Semantics and Linked Data for Data Sharing and Discovery in the Geosciences. It was also partially supported by Fundaç ão para a Ciência e Tecnologia (PTDC/EEI-ESS/4633/2014).

References

1. M.R. Anderson, D. Antenucci, V. Bittorf, M. Burgess, M.J. Cafarella, A. Kumar, F. Niu, Y. Park, C. Ré, C. Zhang, Brainwash: a data system for feature engineering, in *CIDR* (2013)
2. M. Ashburner, C.A. Ball, J.A. Blake, D. Botstein, H. Butler, J.M. Cherry, A.P. Davis, K. Dolinski, S.S. Dwight, J.T. Eppig et al., Gene ontology: tool for the unification of biology. Nat. Genet. **25**(1), 25–29 (2000)
3. K. Baclawski, C.J. Matheus, M.M. Kokar, J. Letkowski, P.A. Kogut, Towards a symptom ontology for semantic web applications, *The Semantic Web–ISWC 2004* (Springer, New York, 2004), pp. 650–667
4. A. Ballatore, A. Zipf, A conceptual quality framework for volunteered geographic information, *Spatial Information Theory* (Springer, New York, 2015), pp. 89–107
5. T. Berners-Lee, J. Hendler, O. Lassila et al., The semantic web. Sci. Am. **284**(5), 28–37 (2001)
6. O. Bodenreider, The unified medical language system (UMLS): integrating biomedical terminology. Nucleic Acids Res. **32**(suppl 1), D267–D270 (2004)
7. A.T. Boin, G.J. Hunter, Do spatial data consumers really understand data quality information, in *7th International Symposium on Spatial Accuracy Assessment in Natural Resources and Environmental Sciences* (Citeseer, 2006), pp. 215–224
8. A. Callahan, J. Cruz-Toledo, P. Ansell, M. Dumontier, Bio2RDF release 2: improved coverage, interoperability and provenance of life science linked data, *The Semantic Web: Semantics and Big Data* (Springer, New York, 2013), pp. 200–212
9. S. Castano, A. Ferrara, S. Montanelli, G. Varese, Ontology and instance matching, *Knowledge-Driven Multimedia Information Extraction and Ontology Evolution* (Springer, New York, 2011), pp. 167–195
10. M. Cheatham, P. Hitzler, String similarity metrics for ontology alignment, *The Semantic Web–ISWC 2013* (Springer, New York, 2013), pp. 294–309
11. M. Cheatham, P. Hitzler, Conference v2. 0: an uncertain version of the OAEI conference benchmark, *The Semantic Web–ISWC 2014* (Springer, New York, 2014), pp. 33–48
12. M. Cheatham, P. Hitzler, The properties of property alignment, in *Proceedings of the 9th International Conference on Ontology Matching-Volume 1317* (2014), pp. 13–24. http://CEUR-WS.org
13. M. Cheatham, Z. Dragisic, J. Euzenat, D. Faria, A. Ferrara, G. Flouris, I. Fundulaki, R. Granada, V. Ivanova, E. Jiménez-Ruiz et al., Results of the ontology alignment evaluation initiative 2015, in *10th ISWC Workshop on Ontology Matching (OM)* (2015), pp. 60–115 (No commercial editor)
14. R. Cornet, N. de Keizer, Forty years of SNOMED: a literature review. BMC Med. Inform. Decision Mak. **8**(Suppl 1), S2 (2008)
15. I.F. Cruz, F.P. Antonelli, C. Stroe, Agreementmaker: efficient matching for large real-world schemas and ontologies. Proc. VLDB Endow. **2**(2), 1586–1589 (2009)
16. I.F. Cruz, C. Stroe, C. Pesquita, F.M. Couto, V. Cross, Biomedical ontology matching using the agreementmaker system, in *ICBO* (2011)
17. I.F. Cruz, C. Stroe, M. Palmonari, Interactive user feedback in ontology matching using signature vectors, in *IEEE 28th International Conference on Data Engineering (ICDE)* (IEEE, 2012), pp. 1321–1324
18. B. Di Martino, Semantic web services discovery based on structural ontology matching. Int. J. Web Grid Serv. **5**(1), 46–65 (2009)
19. Z. Dragisic, K. Eckert, J. Euzenat, D. Faria, A. Ferrara, R. Granada, V. Ivanova, E. Jiménez-Ruiz, A.O. Kempf, P. Lambrix et al., Results of the ontology alignment evaluation initiative 2014, in *Proceedings of the 9th International Conference on Ontology Matching-Volume 1317* (2014), pp. 61–104. http://CEUR-WS.org
20. S. Duan, A. Fokoue, O. Hassanzadeh, A. Kementsietsidis, K. Srinivas, M.J. Ward, Instance-based matching of large ontologies using locality-sensitive hashing, *The Semantic Web–ISWC 2012* (Springer, New York, 2012), pp. 49–64

21. A.K. Elmagarmid, P.G. Ipeirotis, V.S. Verykios, Duplicate record detection: a survey. IEEE Trans. Knowl. Data Eng. **19**(1), 1–16 (2007)
22. J. Euzenat, Brief overview of t-tree: the tropes taxonomy building tool. Adv. Classif. Res. Online **4**(1), 69–88 (1993)
23. J. Euzenat, P. Shvaiko, *Ontology Matching*, vol. 18 (Springer, Heidelberg, 2007)
24. J. Euzenat, C. Meilicke, H. Stuckenschmidt, P. Shvaiko, C. Trojahn, Ontology alignment evaluation initiative: six years of experience, *Journal on Data Semantics XV* (Springer, New York, 2011), pp. 158–192
25. S.M. Falconer, M.-A. Storey, *A Cognitive Support Framework for Ontology Mapping* (Springer, New York, 2007)
26. Z. Fan, S. Zlatanova, Exploring ontologies for semantic interoperability of data in emergency response. Appl. Geomat. **3**(2), 109–122 (2011)
27. D. Faria, C. Pesquita, E. Santos, M. Palmonari, I.F. Cruz, F.M. Couto, The agreementmakerlight ontology matching system, *On the Move to Meaningful Internet Systems: OTM 2013 Conferences* (Springer, New York, 2013), pp. 527–541
28. D. Faria, E. Jiménez-Ruiz, C. Pesquita, E. Santos, F.M. Couto, Towards annotating potential incoherences in bioportal mappings, *The Semantic Web–ISWC 2014* (Springer, New York, 2014), pp. 17–32
29. D. Faria, C. Pesquita, E. Santos, I.F. Cruz, F.M. Couto, Automatic background knowledge selection for matching biomedical ontologies. PloS One **9**(11), e111226 (2014)
30. I.P. Fellegi, A.B. Sunter, A theory for record linkage. J. Am. Stat. Assoc. **64**(328), 1183–1210 (1969)
31. B. Gallagher, Matching structure and semantics: a survey on graph-based pattern matching. AAAI FS **6**, 45–53 (2006)
32. A. Gangemi, N. Guarino, C. Masolo, A. Oltramari, L. Schneider, Sweetening ontologies with DOLCE, *Knowledge Engineering and Knowledge Management: Ontologies and the Semantic Web* (Springer, New York, 2002), pp. 166–181
33. M. Granitzer, V. Sabol, K.W. Onn, D. Lukose, K. Tochtermann, Ontology alignment a survey with focus on visually supported semi-automatic techniques. Future Internet **2**(3), 238–258 (2010)
34. B.C. Grau, Z. Dragisic, K. Eckert, J. Euzenat, A. Ferrara, R. Granada, V. Ivanova, E. Jiménez-Ruiz, A.O. Kempf, P. Lambrix et al., Results of the ontology alignment evaluation initiative 2013, in *Proceedings of the 8th International Conference on Ontology Matching-Volume 1111* (2013), pp. 61–100. http://CEUR-WS.org
35. T.R. Gruber, A translation approach to portable ontology specifications. Knowl. Acquis. **5**(2), 199–220 (1993)
36. K. Gunaratna, K. Thirunarayan, P. Jain, A. Sheth, S. Wijeratne, A statistical and schema independent approach to identify equivalent properties on linked data, in *Proceedings of the 9th International Conference on Semantic Systems* (ACM, New York, 2013), pp. 33–40
37. H. Halpin, P.J. Hayes, When owl: sameas isn't the same: an analysis of identity links on the semantic web, in *LDOW* (2010)
38. F. Hamdi, B. Safar, N.B. Niraula, C. Reynaud, Taxomap alignment and refinement modules: results for OAEI 2010, in *Proceedings of the 5th International Workshop on Ontology Matching (OM-2010) Collocated with the 9th International Semantic Web Conference (ISWC-2010), CEUR-WS* (2010), pp. 212–220
39. M. Hartung, A. Gross, T. Kirsten, E. Rahm, Effective mapping composition for biomedical ontologies, in *Proceedings of Semantic Interoperability in Medical Informatics (SIMI-12), Workshop at ESWC*, vol. 12 (2012)
40. M. Hartung, L. Kolb, A. Groß, E. Rahm, Optimizing similarity computations for ontology matching-experiences from gomma, in *Data Integration in the Life Sciences* (Springer, New York, 2013), pp. 81–89
41. P. Hitzler, M. Krotzsch, S. Rudolph, *Foundations of Semantic Web Technologies* (CRC Press, Boca Raton, 2011)

42. R. Hoehndorf, M. Dumontier, A. Oellrich, D. Rebholz-Schuhmann, P.N. Schofield, G.V. Gkoutos, Interoperability between biomedical ontologies through relation expansion, upper-level ontologies and automatic reasoning. PloS One **6**(7), e22006 (2011)
43. Y. Hu, K. Janowicz, D. Carral, S. Scheider, W. Kuhn, G. Berg-Cross, P. Hitzler, M. Dean, D. Kolas, A geo-ontology design pattern for semantic trajectories, *Spatial Information Theory* (Springer, New York, 2013), pp. 438–456
44. P.G. Ipeirotis, Demographics of mechanical turk (2010)
45. V. Ivanova, P. Lambrix, J. Åberg, Requirements for and evaluation of user support for large-scale ontology alignment, *The Semantic Web. Latest Advances and New Domains* (Springer, New York, 2015), pp. 3–20
46. P. Jain, P. Hitzler, A.P. Sheth, K. Verma, P.Z. Yeh, Ontology alignment for linked open data, *The Semantic Web–ISWC 2010* (Springer, New York, 2010), pp. 402–417
47. K. Janowicz, M. Compton, The stimulus-sensor-observation ontology design pattern and its integration into the semantic sensor network ontology, in *Proceedings of the 3rd International Conference on Semantic Sensor Networks-Volume 668* (2010), pp. 64–78. http://CEUR-WS.org
48. E. Jiménez-Ruiz, B.C. Grau, Logmap: logic-based and scalable ontology matching, *The Semantic Web–ISWC 2011* (Springer, New York, 2011), pp. 273–288
49. E. Jiménez-Ruiz, B.C. Grau, I. Horrocks, R.B. Llavori, Logic-based ontology integration using contentmap, in *JISBD* (Citeseer, 2009), pp. 316–319
50. E. Jiménez-Ruiz, B.C. Grau, Y. Zhou, I. Horrocks, Large-scale interactive ontology matching: algorithms and implementation. ECAI **242**, 444–449 (2012)
51. E. Jiménez-Ruiz, C. Meilicke, B.C. Grau, I. Horrocks, Evaluating mapping repair systems with large biomedical ontologies
52. N. Kheder, G. Diallo, ServOMBI at OAEI (2015)
53. W.O. Kow, V. Sabol, M. Granitzer, W. Kienrich, D. Lukose, A visual SOA-based ontology alignment tool, in *Proceedings of the Sixth International Workshop on Ontology Matching (OM 2011)*, vol. 10 (2011)
54. K. Kyzirakos, M. Karpathiotakis, G. Garbis, C. Nikolaou, K. Bereta, I. Papoutsis, T. Herekakis, D. Michail, M. Koubarakis, C. Kontoes, Wildfire monitoring using satellite images, ontologies and linked geospatial data. Web Semant. Sci. Serv. Agents World Wide Web **24**, 18–26 (2014)
55. P. Lambrix, A. Edberg, Evaluation of ontology merging tools in bioinformatics. Pac. Symp. Biocomput. **8**, 589–600 (2003)
56. P. Lambrixa, R. Kaliyaperumalb, A session-based ontology alignment approach for aligning large ontologies
57. M. Lanzenberger, J. Sampson, AIViz-a tool for visual ontology alignment, in *Tenth International Conference on Information Visualization, IV 2006* (IEEE, 2006), pp. 430–440
58. L. Li, X. Xing, H. Xia, X. Huang, Entropy-weighted instance matching between different sourcing points of interest. Entropy **18**(2), 45 (2016)
59. V. Lush, L. Bastin, J. Lumsden, Geospatial data quality indicators (2012)
60. R. McCann, W. Shen, A. Doan, Matching schemas in online communities: a web 2.0 approach, in *IEEE 24th International Conference on Data Engineering, ICDE, 2008* (IEEE, 2008), pp. 110–119
61. G. McKenzie, K. Janowicz, B. Adams, A weighted multi-attribute method for matching user-generated points of interest. Cartogr. Geogr. Inf. Sci. **41**(2), 125–137 (2014)
62. C. Meilicke, Alignment incoherence in ontology matching. Ph.D. thesis, Universitätsbibliothek Mannheim (2011)
63. S. Melnik, H. Garcia-Molina, E. Rahm, Similarity flooding: a versatile graph matching algorithm and its application to schema matching. In *18th International Conference on Data Engineering, Proceedings* (IEEE, 2002), pp. 117–128
64. V. Momtchev, D. Peychev, T. Primov, G. Georgiev, Expanding the pathway and interaction knowledge in linked life data, in *Proceedings of International Semantic Web Challenge* (2009)
65. P. Mooney, P. Corcoran, A.C. Winstanley, Towards quality metrics for openstreetmap, in *Proceedings of the 18th SIGSPATIAL International Conference on Advances in Geographic Information Systems* (ACM, New York, 2010), pp. 514–517

66. J.M. Mortensen, Crowdsourcing ontology verification, *The Semantic Web–ISWC 2013* (Springer, New York, 2013), pp. 448–455
67. J.M. Mortensen, M.A. Musen, N.F. Noy, Crowdsourcing the verification of relationships in biomedical ontologies, in *AMIA Annual Symposium (Submitted, 2013)* (2013)
68. J.M. Mortensen, M.A. Musen, N.F. Noy, Ontology quality assurance with the crowd, in *First AAAI Conference on Human Computation and Crowdsourcing* (2013)
69. B. Motik, P.F. Patel-Schneider, B. Parsia, C. Bock, A. Fokoue, P. Haase, R. Hoekstra, I. Horrocks, A. Ruttenberg, U. Sattler et al., Owl 2 web ontology language: structural specification and functional-style syntax. W3C Recomm. **27**(65), 159 (2009)
70. C.J. Mungall, Obol: integrating language and meaning in bio-ontologies. Comp. Funct. Genomics **5**(6–7), 509–520 (2004)
71. C.J. Mungall, G.V. Gkoutos, C.L. Smith, M.A. Haendel, S.E. Lewis, M. Ashburner, Integrating phenotype ontologies across multiple species. Genome Biol. **11**(1), R2 (2010)
72. D. Ngo, Z. Bellahsene, Yam++: a multi-strategy based approach for ontology matching task, *Knowledge Engineering and Knowledge Management* (Springer, New York, 2012), pp. 421–425
73. A. Nikolov, M. dAquin, E. Motta, Unsupervised learning of link discovery configuration, *The Semantic Web: Research and Applications* (Springer, New York, 2012), pp. 119–133
74. N.F. Noy, C.D. Hafner, The state of the art in ontology design: a survey and comparative review. AI Mag. **18**(3), 53 (1997)
75. N.F. Noy, N. Griffith, M.A. Musen, *Collecting Community-Based Mappings in an Ontology Repository* (Springer, New York, 2008)
76. N.F. Noy, N.H. Shah, P.L. Whetzel, B. Dai, M. Dorf, N. Griffith, C. Jonquet, D.L. Rubin, M.-A. Storey, C.G. Chute et al., Bioportal: ontologies and integrated data resources at the click of a mouse. Nucleic Acids Res. gkp440 (2009)
77. N.F. Noy, J. Mortensen, M.A. Musen, P.R. Alexander, Mechanical turk as an ontology engineer?: using microtasks as a component of an ontology-engineering workflow, in *Proceedings of the 5th Annual ACM Web Science Conference* (ACM, New York, 2013), pp. 262–271
78. D.J. Odgers, M. Dumontier, Mining electronic health records using linked data. AMIA Summits Transl. Sci. Proc. **2015**, 217 (2015)
79. D. Oliveira, C. Pesquita, Compound matching of biomedical ontologies. Proc. Int. Conf. Biomed. Ontol. **2015**, 87–88 (2015)
80. J. Ortmann, M. Limbu, D. Wang, T. Kauppinen, Crowdsourcing linked open data for disaster management, in *Proceedings of the Terra Cognita Workshop on Foundations, Technologies and Applications of the Geospatial Web in conjunction with the ISWC* (Citeseer, 2011), pp. 11–22
81. H. Paulheim, S. Hertling, D. Ritze, Towards evaluating interactive ontology matching tools, *The Semantic Web: Semantics and Big Data* (Springer, New York, 2013), pp. 31–45
82. A. Pease, I. Niles, J. Li, The suggested upper merged ontology: a large ontology for the semantic web and its applications, in *Working Notes of the AAAI-2002 Workshop on Ontologies and the Semantic Web*, vol. 28 (2002)
83. C. Pesquita, D. Faria, E. Santos, F.M. Couto, To repair or not to repair: reconciling correctness and coherence in ontology reference alignments, in *OM* (2013), pp. 13–24
84. C. Pesquita, D. Faria, C. Stroe, E. Santos, I.F. Cruz, F.M. Couto, Whats in a nym? Synonyms in biomedical ontology matching, *The Semantic Web–ISWC 2013* (Springer, New York, 2013), pp. 526–541
85. C. Pesquita, D. Faria, E. Santos, J.-M. Neefs, F.M. Couto, Towards visualizing the alignment of large biomedical ontologies, *Data Integration in the Life Sciences* (Springer, New York, 2014), pp. 104–111
86. R.G. Raskin, M.J. Pan, Knowledge representation in the semantic web for earth and environmental terminology (sweet). Comput. Geosci. **31**(9), 1119–1125 (2005)
87. S. Rong, X. Niu, E.W. Xiang, H. Wang, Q. Yang, Y. Yu, A machine learning approach for instance matching based on similarity metrics, *The Semantic Web–ISWC 2012* (Springer, New York, 2012), pp. 460–475

88. A. Ruttenberg, J.A. Rees, M. Samwald, M.S. Marshall, Life sciences on the semantic web: the neurocommons and beyond. Brief. Bioinform. bbp004 (2009)
89. M. Salvadores, P.R. Alexander, M.A. Musen, N.F. Noy, Bioportal as a dataset of linked biomedical ontologies and terminologies in RDF. Semant. Web 4(3), 277–284 (2013)
90. E. Santos, D. Faria, C. Pesquita, F. Couto, Ontology alignment repair through modularization and confidence-based heuristics (2013). arXiv:1307.5322
91. E. Santos, D. Faria, C. Pesquita, F.M. Couto, Ontology alignment repair through modularization and confidence-based heuristics. PloS One 10(12) (2015)
92. C. Sarasua, E. Simperl, N.F. Noy, Crowdmap: crowdsourcing ontology alignment with microtasks, The Semantic Web–ISWC 2012 (Springer, New York, 2012), pp. 525–541
93. M. Schmachtenberg, C. Bizer, A. Jentzsch, R. Cyganiak, Linking open data cloud diagram (2014)
94. L.M. Schriml, C. Arze, S. Nadendla, Y.-W.W. Chang, M. Mazaitis, V. Felix, G. Feng, W.A. Kibbe, Disease ontology: a backbone for disease semantic integration. Nucleic Acids Res. 40(D1), D940–D946 (2012)
95. V. Sehgal, L. Getoor, P.D. Viechnicki, Entity resolution in geospatial data integration, in Proceedings of the 14th Annual ACM International Symposium on Advances in Geographic Information Systems (ACM, New York, 2006), pp. 83–90
96. B. Severo, C. Trojahn, R. Vieira, VOAR: a visual and integrated ontology alignment environment (2014)
97. A. Shepherd, C. Chandler, R. Arko, Y. Chen, A. Krisnadhi, P. Hitzler, T. Narock, R. Groman, S. Rauch, Semantic entity pairing for improved data validation and discovery, EGU General Assembly Conference Abstracts, vol. 16 (2014), p. 2476
98. H. Shi, K. Maly, S. Zeil, M. Zubair, Comparison of ontology reasoning systems using custom rules, in Proceedings of the International Conference on Web Intelligence, Mining and Semantics (ACM, 2011), p. 16
99. K. Siorpaes, M. Hepp, Ontogame: Weaving the Semantic Web by Online Games (Springer, New York, 2008)
100. S. Sizov, Geofolk: latent spatial semantics in web 2.0 social media, in Proceedings of the Third ACM International Conference on Web Search and Data Mining (ACM, 2010), pp. 281–290
101. B. Smith, M. Ashburner, C. Rosse, J. Bard, W. Bug, W. Ceusters, L.J. Goldberg, K. Eilbeck, A. Ireland, C.J. Mungall et al., The OBO foundry: coordinated evolution of ontologies to support biomedical data integration. Nat. Biotechnol. 25(11), 1251–1255 (2007)
102. V. Spiliopoulos, G.A. Vouros, V. Karkaletsis, On the discovery of subsumption relations for the alignment of ontologies. Web Semant. Sci. Serv. Agents World Wide Web 8(1), 69–88 (2010)
103. H. Stuckenschmidt, J. Noessner, F. Fallahi, A study in user-centric data integration. ICEIS 3, 5–14 (2012)
104. F.M. Suchanek, S. Abiteboul, P. Senellart, Paris: probabilistic alignment of relations, instances, and schema. Proc. VLDB Endow. 5(3), 157–168 (2011)
105. Y. Sure, S. Bloehdorn, P. Haase, J. Hartmann, D. Oberle, The SWRC ontology–semantic web for research communities, Progress in Artificial Intelligence (Springer, New York, 2005), pp. 218–231
106. V. Svátek, O. Šváb-Zamazal, V. Presutti, Ontology naming pattern sauce for (human and computer) gourmets, in Workshop on Ontology Patterns (2009), pp. 171–178
107. J.M. Taylor, D. Poliakov, L.J. Mazlack, Domain-specific ontology merging for the semantic web, Fuzzy Information Processing Society, 2005. NAFIPS 2005. Annual Meeting of the North American (IEEE, 2005), pp. 418–423
108. S. Thakkar, C.A. Knoblock, J.L. Ambite, Quality-driven geospatial data integration, in Proceedings of the 15th Annual ACM International Symposium on Advances in Geographic Information Systems (ACM, 2007), p. 16
109. S. Thaler, E.P.B. Simperl, K. Siorpaes, Spotthelink: a game for ontology alignment. Wissensmanagement 182, 246–253 (2011)

110. S. Tschirner, A. Scherp, S. Staab, Semantic access to inspire, in *Terra Cognita 2011 Workshop Foundations, Technologies and Applications of the Geospatial Web* (Citeseer, 2011), p. 75
111. E. Voyloshnikova, B. Fu, L. Grammel, M.-A.D. Storey, Biomixer: visualizing mappings of biomedical ontologies, in *ICBO* (2012)
112. A.J. Williams, L. Harland, P. Groth, S. Pettifer, C. Chichester, E.L. Willighagen, C.T. Evelo, N. Blomberg, G. Ecker, C. Goble et al., Open phacts: semantic interoperability for drug discovery. Drug Discov. Today **17**(21), 1188–1198 (2012)
113. A. Wu, X. Lopez, Building enterprise applications with oracle database 11g semantic technologies, *Presentation at Semantic Technologies Conference* (San Jose, 2009)
114. T. Zhao, C. Zhang, M. Wei, Z.-R. Peng, Ontology-based geospatial data query and integration, *Geographic Information Science* (Springer, New York, 2008), pp. 370–392

Linked Data Management

Manfred Hauswirth, Marcin Wylot, Martin Grund, Paul Groth
and Philippe Cudré-Mauroux

Abstract The size of Linked Data is growing exponentially, thus a Linked Data
management system has to be able to deal with increasing amounts of data. Additionally, in the Linked Data context, variety is especially important. In spite of its
seemingly simple data model, Linked Data actually encodes rich and complex graphs
mixing both instance and schema-level data. Since Linked Data is schema-free (i.e.,
the schema is not strict), standard databases techniques cannot be directly adopted
to manage it. Even though organizing Linked Data in a form of a table is possible,
querying a giant triple table becomes very costly due to the multiple nested joins
required typical queries. The heterogeneity of Linked Data poses also entirely new
challenges to database systems, where managing provenance information is becoming a requirement. Linked Data queries usually include multiple sources and results
can be produced in various ways for a specific scenario. Such heterogeneous data
can incorporate knowledge on provenance, which can be further leveraged to provide users with a reliable and understandable description of the way the query result
was derived, and improve the query execution performance due to high selectivity of
provenance information. In this chapter, we provide a detailed overview of current
approaches specifically designed for Linked Data management. We focus on storage
models, indexing techniques, and query execution strategies. Finally, we provide an
overview of provenance models, definitions, and serialization techniques for Linked
Data. We also survey the database management systems implementing techniques
to manage provenance information in the context of Linked Data.

1 Introduction

The nature of the World Wide Web has evolved from a web of linked documents to a
web including Linked Data [45]. Traditionally, we were able to publish documents on
the Web and create links between them. Those links however, allowed only to traverse
the document space without understanding the relationships between the documents

M. Hauswirth · M. Wylot (✉) · M. Grund · P. Groth · P. Cudré-Mauroux
Technical University of Berlin (TU Berlin), Berlin, Germany
e-mail: m.wylot@tu-berlin.de

© Springer International Publishing AG 2017
A.Y. Zomaya and S. Sakr (eds.), *Handbook of Big Data Technologies*,
DOI 10.1007/978-3-319-49340-4_9

and without linking to particular pieces of information. Linked Data allows to create meaningful links between pieces of data on the Web [7]. The adoption of Linked Data technologies has shifted the Web from a space connecting documents to a global space where pieces of data from different domains are semantically linked and integrated to create a global Web of Data [45]. Linked Data enables operations to deliver integrated results as new data is added to the global space. This opens new opportunities for applications such as search engines, data browsers, and various domain-specific applications. Web of Linked Data allows applications to operate on a machine processable unbound space of semi-structured data thus it enables them to deliver more complete answers as new data appears on the Web [45]. Moreover, this allows applications to join data from multiple independent and distributed data collections.

The Web of Linked Data is rapidly growing from a dozen data collections in 2007 to a space of hundreds data sources in April 2014 [5, 9, 65]. The number of linked datasets doubled between 2011 and 2014 [65], which shows an accelerating trend of data integration on the Web. The Web of Linked Data contains heterogeneous data coming from multiple sources, various contributors, produced using different methods, degrees of authoritativeness, and gathered automatically from independent and potentially unknown sources. Figure 1 shows the Linking Open Data cloud diagram created in April 2014; it depicts the scale and heterogeneity of Linked Data on the Web. Such data size and heterogeneity bring new challenges for Linked Data

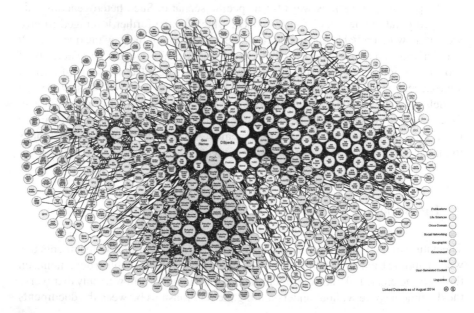

Fig. 1 The diagram shows the interconnectedness of datasets (nodes in the graph) that have been published by heterogeneous contributors to the linking open data community project. It is based on research conducted in April 2014

management systems (i.e., systems which allow to store and to query Linked Data). While small amounts of Linked Data can be handled in-memory or by standard relational database systems, big Linked Data graphs, which we nowadays have to deal with, are very hard to manage. Modern Linked Data management systems have to face large amounts of heterogeneous, inconsistent, and schema-free data.

2 Background Information

We now briefly introduce the basic concepts underpinning Linked Data technologies. We present a data model, vocabularies, and a data exchange format. Nevertheless, we expect the reader to be familiar with a number of basic techniques from the Database Systems, Linked Data, and Provenance areas. We refer the reader to the following books for an introduction to the fields related to this work:

- "Readings in database systems." Hellerstein, Joseph M., and Michael Stonebraker. MIT Press, 2005 [47].
- "Database systems: the complete book." Garcia-Molina, Hector. Pearson Education India, 2008 [31].
- "Linked data: Evolving the web into a global data space." Heath, Tom, and Christian Bizer. Synthesis lectures on the semantic web: theory and technology 1.1 (2011): 1–136 [46].
- "Provenance: an introduction to PROV." Moreau, Luc, and Paul Groth. Synthesis Lectures on the Semantic Web: Theory and Technology 3.4 (2013): 1–129 [54].

Linked Data extends the principles of the World Wide Web from linking documents to linking pieces of data and create a Web of Data; it specifies data relationships and provides machine-processable data to the Internet. It is based on standard Web techniques but extends them to provide data exchange and integration. The four main principles of the Web of Linked Data, as defined by Tim Berners-Lee [6], are:

1. Use URIs (Uniform Resource Identifier)[1] as names for things.
2. Use HTTP (Hypertext Transfer Protocol)[2] URIs so that people can look up those names.
3. When someone looks up a URI, provide useful information, using standards (Resource Description Framework,[3] SPARQL Query Language[4]).
4. Include links to other URIs, so that they can discover more things.

Linked Data uses RDF, the Resource Description Framework, as basic data model. RDF provides means to describe resources in a semi-structured manner. The information expressed using RDF can be exchanged and processed by applications. The

---

[1]http://www.w3.org/Addressing/.

[2]http://www.w3.org/Protocols/.

[3]http://www.w3.org/RDF/.

[4]http://www.w3.org/TR/sparql11-query/.

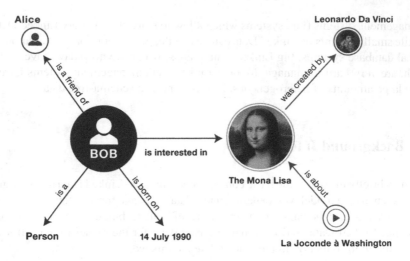

Fig. 2 An exemplary graph of triples [21]

ability to exchange and interlink data on the Web means that it can be used by applications other than those for which it was originally created, and that it can be linked to further pieces of information to enrich existing data. It is a graph-based format, optionally defining a data schema, to represent information about resources. RDF allows to create statements in the form of triples consisting of *Subject, Predicate, Object*. Subjects and Objects represent nodes on a graph while Predicates are labeled directed edges (Figs. 2 and 3). A statement expresses a relationship (defined by a predicate) between resources (subject and object). The relationship is always from subject to object (it is directional). The same resource can be used in multiple triples playing the same or different roles, e.g., it can be used as the subject in one triple and as the object in another. This ability enables to define multiple connections between the triples, hence creating a connected graph of data. The graph can be represented as nodes representing resources and edges representing relationships between the nodes. Figures 2 and 3 depict simple examples of RDF graphs.

Elements appearing in the triples (subjects, predicates, objects) can be of one of the following types:

IRI   (International Resource Identifier) identifies a resource. It provides a global identifier for a resource without implying its location or a way to access it. The identifier can be re-used by others to identify the same resource. IRI is a generalization of URI (Uniform Resource Identifier) allowing non-ASCII characters to be used. IRI can appear at all three positions in a triple (subject, predicate, object).

Literal   is a basic string value that is not an IRI. It can be associated with a datatype, thus can be parsed and correctly interpreted. It is allowed only as the object of a triple.

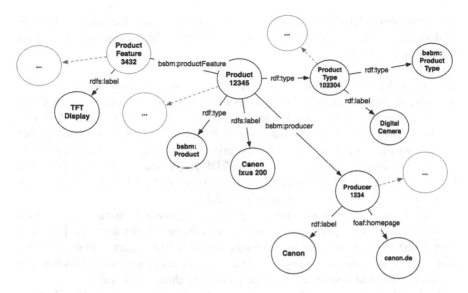

Fig. 3 Example showing an RDF sub-graph using the subject, predicate, and object relations given by the sample data

Blank node   is used to denote a resource without assigning a global identifier with an IRI, it is a local unique identifier used within a specific RDF dataset. It is allowed as the subject and the object in a triple.

The framework provides means to co-locate triples in a subset and to associate such subsets with an IRI [20]. A subset of triples constitutes an independent graph of data. In practice, it provides data managers with a mechanism to create a collection of triples. A dataset can consist of multiple named graphs and no more than one unnamed (default) graph.[5]

Even though RDF does not require any naming convention for IRIs and does not impose any schema on data, it can be used in combination with vocabularies provided by RDF Schema language [22]. RDFS is a semantic extension of RDF enabling to specify semantic characteristics of RDF data. It provides a data-modeling vocabulary for RDF data. It enables to state that an IRI is a property and that a subject and an object of the IRI have to be of a certain type. RDF schema allows to classify resources with categories, i.e. classes, types. Classes allow to regroup resources. Members of a class are called instances, while classes are also resources and can be described with triples. RDFS allows classes and properties to be hierarchical, as a class can be a sub-class of a more generic class. In the same way, properties can be defines as a specific property (sub-property) of a more generic one. RDFS enables also to specify a domain and a range of a predicate, i.e., types of resources allowed as subjects and objects. Properties are also resources that can be described by triples. An instance can be associated with several independent classes specifying different

---

[5]https://www.w3.org/TR/rdf11-concepts/#section-dataset.

sets of properties. RDFS defines also a set of utility properties allowing to link pieces of data, e.g., *seeAlso* to indicate a resource providing additional information about the resource of a subject.

Richer vocabularies (e.g., OWL) enable to express logical constraints on Web data. The OWL 2 Web Ontology Language [17] allows to define ontologies to give a semantic meaning to the data. An ontology models entities and interactions between them in a particular domain, it provides classes, properties, and data values. An ontology is exchanged along with the data as an RDF document, and defines vocabularies and relationships between terms, often covering a specific domain shared by a community. An ontology can also be seen as an RDF graph, where terms are represented by nodes and relationships between them are expressed by edges, thus they can be stored and indexed in the same way as RDF data.

Linked Data in general is a static snapshot of information, though it can express events and temporal aspects of entities with specific vocabulary terms [19]. A snapshot of the state can be seen as a separate (named) RDF graph containing a current state of the universe. Changes in data typically concern relationships between resources; IRIs and Literals are constant and rarely change their value.

Linked Data allows to combine and process data from many sources [6]. The basic triple representation of pieces of data combined together results in large RDF graphs. Such large amounts of data are made available as Linked Data where datasets are interlinked and published in the Web.

Linked Data can be serialized in a number of formats that are logically equivalent. The data can be stored in the following formats:

N-Triples   provides a simple, plain-text way to serialize Linked Data. Each line in a file represents a triple, the period at the end signals the end of a statement (triple). This format is often used to exchange large amount of Linked Data and for processing graphs with stream-oriented tools.

N-Quads   is a simple extension of N-Triples. It allows to add a fourth optional element in a line denoting a named graph IRI, which the triple belongs to.

Turtle   is an extension of N-Triples; it introduces a number of syntactic shortcuts, such as prefixes, lists, and shorthands for datatyped literals. It provides a trade-off between ease of writing, parsing, and readability. It does not support the notion of named graphs.

TriG   extends Turtle to support multiple named graphs.

JSON-LD   provides a JSON syntax for Linked Data. It can be used to transform JSON documents into Linked Data, and offers universal identifiers for JSON objects and a way in which a JSON document can link to an object in another document.

RDFa   is a syntax used to embed Linked Data in HTML and XML documents. This enables to aggregate data from web pages and use it to enrich search results or presentation.

RDF/XML   provides an XML syntax for Linked Data.

To facilitate querying and manipulating Linked Data on the Web, a semantic query language is needed. Such a language, named SPARQL Protocol and RDF Query

Language (shortly SPARQL), was introduced by the World Wide Web Consortium. SPARQL [18] can be used to formulate queries ranging from simple graph patterns to very complex analytic queries. Queries may include unions, optionals, filters, value aggregations, path expressions, subqueries, value assignment, etc. Apart from SELECT queries, the language also supports:

ASK   queries to retrieve binary "yes/no" answer to a query,
CONSTRUCT   queries to construct new RDF graphs from a query result.

All standards and Linked Data concepts are defined and explained in detail in documents published by the World Wide Web Consortium. We refer the reader to the following recommendations for further detail:

- RDF 1.1 Primer [21]
- RDF 1.1 Concepts and Abstract Syntax [19]
- RDF Schema 1.1 [22]
- RDF 1.1: On Semantics of RDF Datasets [20]
- OWL 2 Web Ontology Language [17]
- SPARQL 1.1 Overview [18]

This chapter is built on the PhD thesis "Efficient, Scalable, and Provenance-Aware Management of Linked Data" [70] and our previous papers on processing provenance in a native RDF store [73, 75].

3 Native Linked Data Stores

In this section, we first describe native RDF storage and query execution strategies. We define a native RDF systems as a system that was designed to store exclusively RDF data and thus that is fully optimized for persisting and querying this data. Naturally, traditional database design and architecture have had some impact on the design of native RDF stores. Moreover, a number of well-known techniques, (e.g., join processing) were adjusted for querying RDF data as we explain in the following.

In the area of native RDF stores, we distinguish three main systems trends: Quadruple stores (Sect. 3.1), Index-Permuted stores (Sect. 3.2) and Graph stores (Sect. 3.3). We first examine systems that store the data in table-like structures, storing additional information per triple, thus maintaining quadruples that keeps information related to the specific sub-graph the triples belong to. Then, we analyze index-permuted storage systems, which use a multiplicity of indexes to support high-performance query execution for arbitrary queries. Finally, we move on to graph stores, which represent one of the most natural ways to represent RDF data. Nodes represent, in this context, subjects and objects, while labeled edges represent predicates. In order to query a graph, an input query is translated into some graph pattern (see Fig. 16) that is matched against the full graph. Even for simple patterns, this can result in full traversals of the complete graph and thus may require significant processing time. Due to the fact that graph matching, especially against large graphs, is a very complex and time-consuming task, all existing approached trying to deal with the problem in fact partition graphs into subgraphs. Although here are few different techniques for doing that, we can distinguish two main trends. The first one uses classic graph partitioning algorithms like GGGP [52] used by [11] or METIS used by [49]. The second way to tackle graph partitioning is to try to discover recurring patterns/templates in the RDF graph to create subgraphs containing nodes describing certain topics within a defined scope, like what was proposed in [71, 72, 79]. All approaches proposes also a different way of indexing subgraphs, but the general trend is that there is one main index that allows to find certain subgraphs where the remaining part of the data can be found. More specific indices are also proposed for specific types of queries, like Dogma_ipd and Dogma_epd [10].

3.1 Quadruple Systems

The traditional way to persist RDF triples is to store triple statements directly in a table-like structure. By exploiting semantic information from the complete RDF graph, additional data can by annotated per triple and stored as a fourth element for each input triple. To improve query execution performance on top of this structure, various indexes can be built. In this section, we present systems and architectures that deal with persisting RDF triples in a most direct way.

Data Storage and Partitioning Virtuoso [28] by Erlin et al. stores data as RDF quads consisting of a graph element id, subject, predicate, and object. All the quads are persisted in one table. Each of the attributes can be indexed in different ways. From a high-level perspective, Virtuoso is comparable to a traditional relational database with enhanced RDF support. Virtuoso adds specific types (URIs, language and type-tagged strings) and indexes optimized for RDF. To partition the data in a clustered environment, Virtuoso uses hash-partitioning based on the subject of a triple. Since the number of resulting partitions is significantly higher than the number of worker nodes in the cluster environment, one node might receive multiple partitions. The distribution of the individual partitions can either be simply round-robin or follow more elaborate models to account for different hardware capacities of the nodes. The system allows moving partitions between nodes and insures data consistency during the process. To provide fault tolerance, Virtuoso allows each logical partition to be placed on multiple nodes.

In [40], Harris et al. propose a system called 4store. The system applies a simple storage model: it stores quads of (model, subject, predicate, object). In 4store, the model attribute is equivalent to Virtuoso's graph. Data is partitioned as non-overlapping sets of records among segments sharing a subject. To distribute segments across the cluster, round-robin is used allowing each node of the cluster to store one or more segments. To cover failing nodes in the cluster, 4store allows to increase the replication of the partitions. The number of replicas in the cluster corresponds to the number of nodes which can fail without causing any significant issue.

Indexing In Virtuoso, Erling et al. implement two indexes.[6] The default index (set as a primary key) corresponds to GSPO (graph, subject, predicate, object). In addition, it provides an auxiliary bitmap index (OPGS). The indexes are stored in compressed form. The GSPO index is used to deal with queries where the subject is known, e.g.:

select * **from** <lubm> **where** {<Professor1> ub:AdvisorOf ?y }

The auxiliary bitmap index is is applied to cases with known object and unknown subject, e.g.:

select * **from** <lubm> **where** { ?x ub:AdvisorOf <Student1> }

As strings are the most common values in the database, for example in URIs, Virtuoso compresses these strings by eliminating common prefixes. The system does not precalculate optimization statistics[7]; instead it samples data at query execution time. It also does not compute the exact statistics but just gets rough numbers of elements and estimates query cost to pick an optimal execution plan.

Harris et al. propose to store each of the quads in three indexes; in addition, they store literal values separately. 4store maintains a hash table of graphs where each entry points to lists of triples in the graph (M-Index in Fig. 4). Literals are indexed in a separate hash table (R Index in Fig. 4) and they are represented as (SPO). Finally, they consider two predicate-based indexes, referred to as P-Indices in Fig. 4. For each predicate, two radix tries are used where the key is either a subject or object, and respectively object or subject and graph are stored as entries. These indices can be used to select all quads having a given predicate and their subject/object. They can be seen as traditional predicate indices ($P \rightarrow OS$ and $P \rightarrow SO$).

Query Execution To build query execution plans Virtuoso divides query execution into multiple steps where each step takes as input the output from the previous step. Their query execution plan can hence be seen as a pipeline of steps. Most of the steps are individually executed. Sometimes, steps can be joined and executed as a unit. Most queries use predicate indices (P-Indices) in order to merge elements:

select * **from** <lubm> **where**
{?x a ub:Professor . ?x teacher_of <student> }

The query is executed as an intersection of elements from P-Indices (P:OS) a and $< teacher\_of >$; elements for $< Professor >$ are joined with elements related to $< student >$.

select * **from** <lubm> **where**
{ ?x a ub:Professor . ?x teaches_course ?c }

The second query is executed as a loop through Professor's (bitmap OPGS index); then, courses given by each professor are retrieved (from the GSPO index).

---

[6]In databases indexes are used to locate data without scanning the entire dataspace, thus to improve the speed of retrieval operations. For more details about database concepts we refer the reader to the positions in the literature introduced in Sect. 2.

[7]Optimization statistics are used in databases to choose the best execution plan for a query. They contain information describing distribution of various objects in a database.

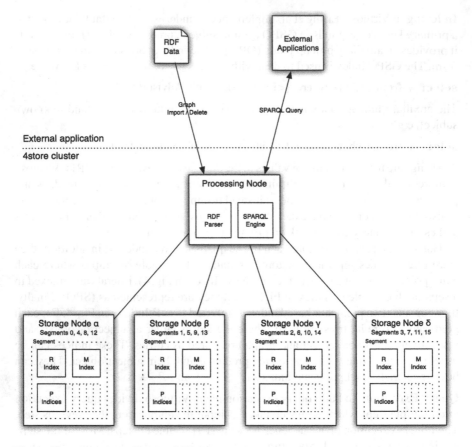

Fig. 4 4Store: system architecture [40]

select * **from** <lubm> **where**
{ ?x a ub:Professor . ?x ub:AdvisorOf ?y}

Such query is executed in four steps, one to translate URIs to IDs, a second one to get professors, a third one for students the professors advise. The last step translates results form IDs into URIs, since internally the system encode all strings as IDs to save space.

select * **from** <lubm> **where**
{ ?x a ub:Professor ; ub:advisorOf ?y ; ub:telephone ?tel }

The last query is also executed in four steps since the two properties are retrieved at the same tame, they are co-located because they have the same subject (GSPO partitioned on subject).

Subject	Predicate	Object
:the_matrix	:released_in	"1999"
:the_thirteenth_floor	:released_in	"1999"
:the_thirteenth_floor	:similar_plot_as	:the_matrix
:the_matrix	:is_a	:movie
:the_thirteenth_floor	:is_a	:movie

Distinct subjects: [:the_matrix, :the_thirteenth_floor]
Distinct predicates: [:released_in, :similar_plot_as, :is_a]
Distinct objects: [:the_matrix, "1999", :movie]

	:released_in	:similar_plot_as	:is_a
:the_matrix	0 1 0	0 0 0	0 0 1
:the_thirteenth_floor	0 1 0	1 0 0	0 0 1

Note: Each bit sequence represents sequence of objects (:the_matrix, "1999", :movie)

Fig. 5 BitMat: sample bit matrix [4]

3.2 Index Permuted Stores

The approach of index-permuted RDF storage exploits and optimizes traditional indexing techniques for storing RDF data. As most of the identifiers in RDF are URIs strings, one optimizations is to replace these arbitrary long strings with unique integers. As the data is sparse and many URIs are repetitive, this technique, allows to save memory. To increase the resulting performance, the indexes are built based on shorter encoded values rather than the uncompressed values.

Indexing and Data Storage One of the first approaches to exhaustive indexing was proposed by Harth et al. for a system called YARS [41]. The authors take into consideration quads of (Subject, Predicate, Object, Context[8]). Exhaustive indexing based on these attribute requires a total of 16 indexes. Harth et al. propose to use six indexes covering all major access patterns [SPOC, POC, OCS, CSP, CP, OS]. Their indexing approach leverages the property that B+tree indexes can be queried for range and prefix queries (see [31] Sect. 14.2 for more details on B-trees). If the key to such index is the full quad of subject, predicate, object and context, it becomes possible to query only a prefix of the key and use the remaining keys as values.

Atre et al. propose a system called BitMat [4] where they store data in compressed inverted index structures. They leverage the fact that RDF triples are fixed 3-dimensional entities. They propose a 3-dimensional bit-cube where each cell represents a unique triple and the cell value denotes the presence or absence of the triple. Figure 5 shows some sample RDF data and a corresponding bit matrix. The data is

---

[8]From the database perspective a *context* can be seen as a *graph*, thus there is no difference between those two concepts in terms of data management. Here, we keep the original terminology of the authors of each approach.

Fig. 6 Exhaustive indexing

then compressed using D-gap compression[9] on each row level. In this first approach, the authors only store S-O matrices, however in their next work [3] they introduce also a transposed matrix of O-P. Furthermore, they also slice out rows along the S and O dimension and store also P-S and P-O matrices.

Janik et al. introduce a system called BRAHMS [50], whose storage model evolves around permuted indexes. They store data in three hash tables (S-OP, O-SP, P-SO). The hash tables are organized in a logically contiguous memory block which can be dumped and loaded from disk during startup and shutdown, though the system itself works in-memory.

Several pieces of work like [59, 69] show that it is possible to use a new storage model that applies exhaustive indexing. The foundation for this approach is that any query on the stored data can be answered by a set of indices on the subject, predicates, and objects in different orders, namely all their permutations as shown in Fig. 6. In contrast to the concept of property tables where the table is only sorted by the subject [1], this allows fast access to all parts of the triples by sorted lists and fast merge-joins (see [31] Sect. 15.4.8 for more details on merge-join) on the elements.

The Hexastore index structure presented in [69] can be described as shown in Fig. 7. In this example, a spo index is described. The first level of the index is a sorted list of all subjects where each subject is associated to a list of sorted predicates. Each predicate links to a list of sorted objects. Queries that require many joins and unions in other storage systems can be answered directly by the index. In the case where the query requests a list of subjects that are related to two particular objects through any property, the answer can be computed by merging the subject lists of a osp index. Since the subject list of this osp index is sorted, this can be done in linear time.

The architectural drawback of this approach is the increase in memory consumption. Since every combination of possible query patterns is indexed, additional space is required due to the duplication of data. As the authors of [69] point out, less than a six-fold increase in memory consumption is required; the approach yields a worst-case five-fold increase since for the set of spo, sop, osp, ops, pso, pos indexes, one part can always be re-used: the initial sorted list of subjects, objects and predicates. Due to the replication of the data into the different index structures, updating and inserting into the index can become a second bottleneck.

Neumann et al. present RDF-3X [58] that relies on the same processing scheme with exhaustive indexing but further optimizes the data structures. As in Hexastore,

---

[9]http://bmagic.sourceforge.net/dGap.html.

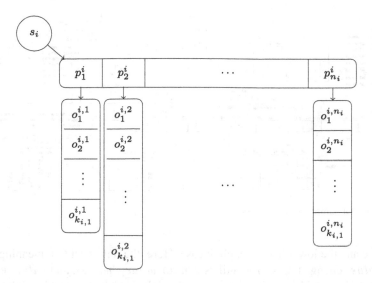

Fig. 7 Hexastore index structure, figure after [69]

they use dictionary encoding to replace variable-sized values by fixed integer IDs. In RDF-3X, the index data is stored in clustered B+ trees in lexicographic order.

The values inside the B+ tree are delta encoded (computed difference/delta between the ID attributed to the slot in the tree and the ID attributed to the previous slot) to further reduce the required amount of main memory to persist all data. Each triple (in one of the previously defined orders of spo, sop,...) is stored as a block of maximum 13 bytes. Since the triples are sorted lexicographically, the expected delta between two values is low, i.e., only a few bytes are consumed. Now the header of the value block contains two pieces of information: First a flag that identifies if $value_1$ and $value_2$ are unchanged and the delta of $value_3$ is small enough to fit in the header block; second, if this flag is not set, it then identifies a case number of how many bytes are needed to decode the delta to the previous block. In Fig. 8, we illustrate this example. The upper part of the illustration shows the general block

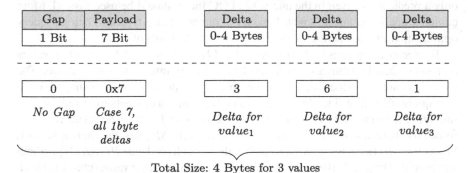

Fig. 8 RDF-3X compression example [58]

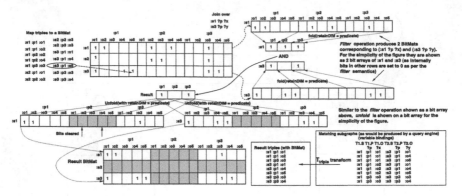

Fig. 9 BitMat: simple query execution [4]

structure and the lower half an explicit case. Here, the flag is set to 0 meaning more than $value_3$ changed. Case 7 identifies that for $value_1$, $value_2$, and $value_3$ exactly one byte changed. Using this information, the deltas can be extracted and the actual value of the triple can be decoded. In addition to the full index, RDF-3X stores additional aggregated indices to maintain information about how often a relation between two values occurs. This can be used to increase the query performance of those queries where unbound variables are used in the triple pattern, but are not projected and therefore can be used as multipliers for the output of result patterns. These count-aggregated index structure add another nine indexes to the previous indexes, six indexes for all pairs of two values, and three indexes for all single values.

In [62] Owens et al. propose a new storage model for Jena[10] called TDB. The approach stores data in three B+-tree indexes. They use SPO, POS, and OSP permutations. Each index contains all elements of all triples. The string values are encoded as 64bit identifiers.

Query Execution To execute simple queries, Harth et al. [41] evaluate which of the six indexes fits best to answer the query. Selecting the index depends on the access pattern; if a subject is specified in the query, the SPOC index should be used. If only a predicate is given in the query, the POC index should be used instead. More complex queries connected with logical operators require typical relational query optimization like reordering to efficiently execute all kinds of joins.

To execute the queries in BitMat, Atre et al. [4] use bitwise AND and OR operators on rows in their bit matrices, which results in a binary intersection of elements. The operations are performed directly on compressed data. To perform a simple single join query, they first filter by subject rows from matrices containing only triples satisfying a given query patterns (*S1* and *S2* rows in Fig. 9). The result rows are folded on objects, so that if any object is present for the *SP* pairs, the value is set to 1. Figure 9 shows two rows for *S1* and *S3* (3 cells for each predicate *P1, P2, P3*) without any specified object. In the example (Fig. 9) all cells have their value set to 1 for both

[10]https://jena.apache.org/documentation/tdb/.

subjects. Only the pair *S3* and *P2* are set to 0, because in the previously selected row for *S2*, for all objects related to *P2* we had value of 0, i.e., there is no triple (*S1, P2, X*). The following step performs AND operation on those two rows, which results in a row containing 1 s for predicates which are present for both rows ("Result" in Fig. 9, *P1* and *P3* are set to 1). Subsequently, the inverse operations are performed, i.e., between the initially selected rows for *S1* and *S3* and the result row from the previous step. An AND operation is applied on cells related to each predicate. This gives two rows for *S1* and *S3* containing the same values for predicates *P1* and *P3*, but cleared values (set to 0) for predicate *P2* and all objects related to *P2*. The two rows are combined it the same way as before with the initial matrix, which gives the final result. The rows for *S1* and *S3* are those which define the result, and since in the first step the *S2* row was not selected its values are cleared.

The authors also propose different algorithm to perform multiple-join operations, where first they create multi-join graphs capturing join variables. Then, for each join variable, they fold matrices associated to all possible triple patterns containing the variable. They perform bitwise AND on bitarrays. The final result is unfolded and in the end a result BitMat is generated by OR operations on all matrices associated with the triple pattern.

Janik et al. focus in their work [50] on the semantic association discovery problem. The problem itself refers to finding a semantic connection between two objects. Tackling this issue, they had to overcome the fact that, at that time, SPARQL did not fully support that kind of queries. It supported queries only with fixed distance, whereas to discover association one was interested in any association independently of a distance between objects (arbitrary transitive closures, which were not supported by SPARQL at the time). In BRAHMS, they mainly leveraged two graph algorithms to answer queries: depth-first search and breadth-first search.

TDB [62] divides a query into basic graph patterns [63], which are then matched onto the stored RDF data. Subsequently, the other operations are executed by replacing all known values for the variables. This is optimized by favoring elements that are expected to yield the fewest elements, based on statistics. Matching triple patterns is performed by choosing the most appropriate index. The system then performs a range scan of the index for finding particular elements.

3.3 Graph-Based Systems

RDF naturally forms graph structures, hence one way to store and process it is through graph-driven data structures and algorithms. Many graph algorithms are however known to be very computationally complex. In this section we present approaches trying to apply ideas from the graph processing world to efficiently handle RDF data.

Data Storage and Partitioning In TripleT [29], Fletcher et al. introduce the term of atom around which triples are co-located. A key k, regardless of its role in the triples, is selected and then all triples where k occurs are co-located together it improve data

Prefix: y= http://en.wikipedia.org/wiki/

vID	vLabel	adjList {(eLabel, nLabel)⁺}
001	y:Abraham_Lincoln	(hasName, "Abraham Lincoln") (BornOnDate, "1809-02-12"), (DiedOnDate, "1865-04-15") (DiedIn, y:Washington_D.C)
002	y:Washington_D.C	(hasName, "Washington D.C.") (FoundYear, "1790") (rdf:type, y:city)
003	y:United_States	(hasName, "United States") (hasCapital,y:Washington_D.C) (rdf:type, y:country)
004	y:Reese_Witherspoon	(hasName, "ReeseWitherspoon") (BornOnDate, "1976-03-22") (hasCapital, y:New_Orleans,_Louisiana) (rdf:type, y:Actor)
005	y:New_Orleans,_Louisiana	(FoundYear, "1718"), (locatedIn, y:United_States) (rdf:type, y:city)

Fig. 10 gStore: adjacency list table [79]

Fig. 11 gStore: signature graphs [79]

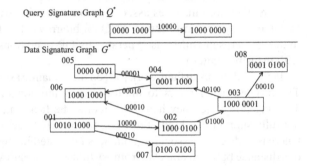

locality. For example, for k three buckets are created. One containing pairs (p,o) where k is subject, one containing pairs (s,o) where k is predicate, one containing pairs (s,p) where k is object. All those pairs are sorted. The storage model itself has an index with keys corresponding to subjects and objects.

Das et al. in their system called gStore [79], organize data in adjacency list tables. Each vertex is represented as an entry in the table with a list of outgoing edges and neighbors (Fig. 10). The entries take the following form [vID, vLabel, adjList], where vID is the ID of the vertex, vLabel is an URI and adjList is a list of outgoing edges and neighbors, which in fact results in shapes that are similar to the atoms proposed in TripleT. As a following step, a bitstring signature is assigned to each vertex (see Fig. 11).

Brocheler et al. [10] propose a system called DOGMA. Their approach is based on a balanced binary tree where each node is located on one disk page. Since the page size is fixed, the size of the subgraph located on a page is limited also. There can be many different indexes built for the same RDF database; the authors of DOGMA focus on minimizing potential cross edges between subgraphs located on different pages. Supposing we have two nodes N1 and N2, the fewer cross edges we have, the more independent the nodes become. When answering a query, we will most probably not have to open/read both of those subgraphs (see Fig. 12).

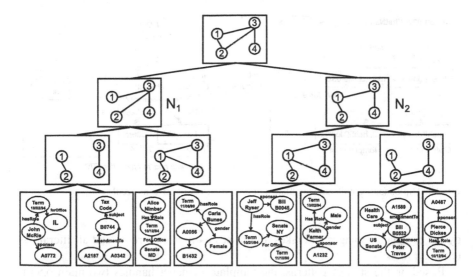

Fig. 12 DOGMA: index [10]

To partition graphs, the authors use the algorithm proposed in [52]. Their algorithm takes as input a weighted graph and partitions it in the way that the total weight of cross edges (between subgraphs) is minimized and the sum of the weights in each subgraph is approximately equal. They start by assigning a weight of 1 to each vertex and edge in an RDF graph and then coarsen the graph into a subgraph so that the latter contains about half of the vertices from the former, and so on with each of the subgraphs until each of the subgraphs has no more than the predefined number of vertices (i.e., to fit into a disk page). The coarsening algorithm randomly picks one vertex (v), then selects a maximally-weighted node (m). It merges the neighbors of the node m and m itself into one node, updates the weights and removes v. Edges from m to its neighbors are also removed. This process is rerun until the subgraph contains half or less vertices as the initial graph. Then, the index is built for all subgraphs.

One of the key innovations of Diplodocus [71, 72] revolves around the use of *declarative storage patterns* [24] to co-locate large collections of related values on disk and in main-memory. When setting-up a new database, the database administrator may give Diplodocus a few hints as to how to store the data on disk: the administrator can give a list of triple patterns to specify the *root nodes*, both for the template lists and the molecule clusters (see for instance Fig. 17, where "Student" is the root node of the molecule, and "StudentID" is the root node for the template list). Cluster roots are used to determine which clusters to create: a new cluster is created for each instance of a root node in the database. The clusters contain all triples departing from the root node when traversing the graph, until another instance of a root node is crossed (thus, one can join clusters based on the root nodes). Template roots are used to determine which literals to store in template lists. In case the administrator gives no hint about the root nodes, the system inspects the templates created

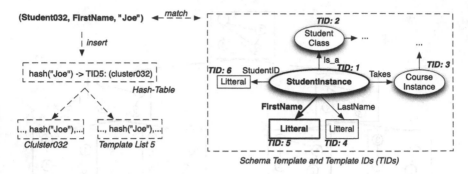

Fig. 13 Diplodocus: a molecule template along with one of its RDF molecules [72]

by the template manager and takes all classes as molecule roots and all literals as template roots.

Based on the storage patterns, the template manager handles two main operations in the system: (i) it maintains a schema of triple templates in main-memory and (ii) it manages template lists. Whenever a new triples enters the system, it is passed to the template manager, which associates template IDs corresponding to the triple by considering the type of the subject, the predicate, and the type of the object. Each distinct list of "(subject-type, predicate, object-type)" defines a new triple template. The triple templates play the role of an instance-based RDF schema in our system. In case a new template is detected (e.g., a new predicate is used), then the template manager updates its in-memory triple template schema and inserts new template IDs to reflect the new pattern it discovered. Figure 13 gives an example of a template. In case of very inhomogeneous data sets containing millions of different triple templates, wildcards can be used to regroup similar templates (e.g., "Student - likes - *").

The triple is then passed to the Cluster Manager, which inserts it in one or several molecules. If the triple's object corresponds to a root template list, the object is also inserted into the template list corresponding to its template ID. Templates lists store literal values along with the key of their corresponding cluster root. They are stored compactly and segmented in sublists, both on disk and in main-memory T.

Indexing Das et al. [79] build an S-tree for all vertices in their adjacency list table to reduce the search space (Fig. 14). The tree leaves correspond to vertices from the initial graph (G*), and each intermediate (parent) node is formed by performing bitwise OR operation on all children signatures. However, S-trees cannot support multi-way join processing; to solve this issue, the authors propose a VS-tree extension. Given an S-tree, leafs are linked according to the initial graph, and new edges are introduced depending on whether certain leafs are connected in G*. Specifically, two leafs in S-tree (001 and 002 in Fig. 14) are linked if there is an edge in G* between vertices corresponding to them. On the upper level in S-tree, super-edges are introduced between nodes if there is at least one connection between the children of those nodes. In other words, if there is a link between two leafs which does not

Fig. 14 gStore: S-tree [79]

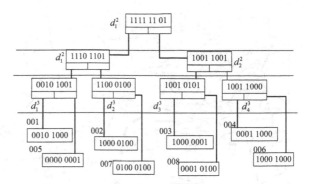

Fig. 15 gStore: VS-tree [79]

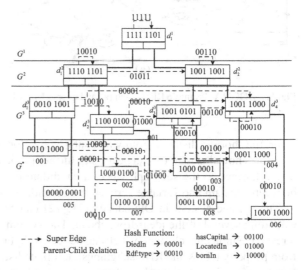

share a parent, a link between their parents is then created. Bitwise "O" operations over connecting edge labels of the children are performed to assign labels to such super-edges (see Fig. 15).

In the approach proposed by Brocheler et al. the storage model itself is an index mostly, though the authors also propose two additional indexes to help pruning the result candidates. The DOGMA internal partition distance (IPD) index stores, for each vertex v in node N, the distance to edge of the subgraph corresponding to N. During query execution, for two vertices (v, u) the algorithm looks for nodes for which the vertices belong (N!=M). N and M are at the same level of the tree and closest to the root. If such nodes do not exist, because the vertices are in the same leaf node of the tree, then the distance between them is set to 0, otherwise it is set to the maximal distance from each of them to the border of the subgraph the vertex belongs to (formally: $d(u, v) = max(ipd(v, N), ipd(u, M))$). The idea behind the DOGMA external partition distance (EPD) index is to maintain distances to other subgraphs. For each lowest-level subgraph, a color is assigned. For each vertex and

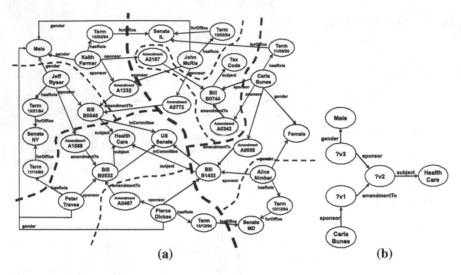

Fig. 16 DOGMA: Example RDF graph (**a**) and query (**b**) [10]

color, the shortest distance from v to a subgraph colored with c is stored. Figure 16 illustrates how this method can be used to further prune result candidates. Basically, if the distance to the subgraph where a candidate lies is bigger than the distance between constant and variable vertices, the candidate can be pruned.

The storage system in Diplodocus [71, 72] can be seen as a hybrid structure extending several of the ideas from above. The system is built on three main structures: RDF molecule clusters (which can be seen as hybrid structures borrowing both from property tables and RDF subgraphs), template lists (storing literals in compact lists as in a column-oriented database system) and an efficient hash-table indexing URIs and literals based on the clusters they belong to.

Figure 17 gives a simple example of a few molecule clusters—storing information about students—and of a template list—compactly storing lists of student IDs. Molecules can be seen as *horizontal* structures storing information about a given object instance in the database (like rows in relational systems). Template lists, on the other hand, store *vertical* lists of values corresponding to one *type* of object (like columns in a relational system).

Molecule clusters are used in two ways in the system: to logically group sets of related URIs and literals in the hash-table (thus, pre-computing joins), and to physically co-locate information relating to a given object on disk and in main-memory to reduce disk and CPU cache latencies. Template lists are mainly used for analytics and aggregate queries, as they allow to process long lists of literals.

Query Execution In TripleT, [29] Fletcher et al. try to minimize complex subject-object joins which in table-oriented systems involve many self-join operations. Thanks to their indexing scheme, they can perform a join as a single look-up on

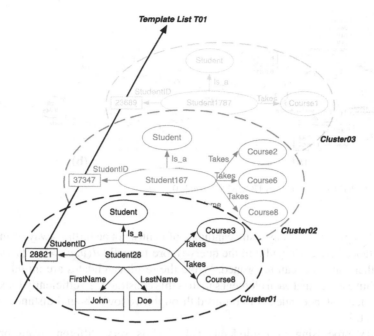

Fig. 17 Diplodocus: the two main data structures: molecule clusters, storing in this case RDF subgraphs about students, and a template list, storing a list of literal values corresponding to student IDs [72]

a common join variable (same value for subject and object) and then merge values related to the subjects and the objects.

To answer queries, gStore employs a top-down strategy over a VS-tree to find the match of a query (Q*) over a graph (G*). First, the system finds the top-matches of Q* in the VS-tree, and queues those matches. Then, it pops up one match from the queue and expands it to its children (all descendant and the node) and for each of them checks if it matches Q*. All valid matches are queued back again. This process is iterated until reaching the leaf entries of the VS-tree. Finally, the system finds matches of Q* over leaf entries in VS-tree, i.e., matches of Q* over G*.

To answer a query Brocheler et al. in DOGMA first retrieve for all variable vertices in Q* a set of result candidates w.r.t. the vertices (see Fig. 18). The sets are initialized with vertices that are connected to a defined vertex with a defined predicate. Then, for the vertex with the lowest cardinality of result candidates, each candidate is set as a value of the vertex, such that there is a new constant vertex. The algorithm can be rerun to prune result candidates for other vertices, and so on until the final result is found. The basic algorithm presented above is efficient enough for simple queries on neighboring vertices. Considering vertices located in different nodes, additional indices to help prune the result candidates would be needed. The authors propose a second algorithm, which verifies if two vertices are "in range". Let v be a variable vertex with set of result candidates, and c a constant vertex with a long range

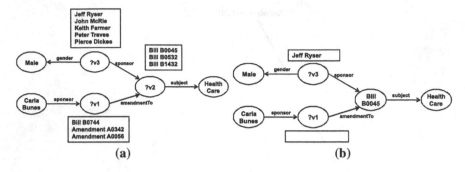

Fig. 18 DOGMA: execution of DOGMA_basic [10]

dependency on v. Then any result candidate of v must not be further away from c than the distance between c and v in the query (more formally $d(v, c) >= d(T(v), s)$). Any other candidate can be pruned. While the result candidates are initialized, the algorithm ensures that each element satisfies this constraint. To efficiently look up for a $d(v, c)$, a distance index is introduced through two lower-bound distance indexes (see Sect. 3.3).

Query processing in Diplodocus [71, 72] is very different from previous approaches to execute queries on RDF data, because of the three peculiar data structures in the system: a hash-table associating URIs and literals to template IDs and cluster lists, clusters storing RDF molecule clusters in a compact fashion, and template lists storing compact lists of literals. Triple patterns in Diplodocus are resolved by looking for a bound-variable (URI) in the hash-table, retrieving the corresponding cluster numbers, and finally retrieving values from the clusters when necessary. Since the RDF nodes are logically grouped by clusters in the hash-table, it is typically sufficient to read the corresponding list of clusters in the hash table (e.g., for "return all students following Course0"), or to intersect or take the union of several lists of clusters in the hash table (e.g., for "return all students following Course0 whose last names are Joe") to answer the queries. In most cases, no join operation is needed since joins are implicitly materialized in the hash-table and in the clusters. When more complex join occurs, Diplodocus resolves them using standard hash-join operators.

Diplodocus [71, 72], to deal with star-like queries invokes the hash-table to find the corresponding cluster, which contains then all the corresponding values. For bigger scopes, the system can join clusters based on the various root nodes they contain.

Many analytic queries can be solved in Diplodocus [72] by first intersecting lists of clusters in the hash-table, and then looking up values in the remaining molecule clusters. Large analytic or aggregate queries on literals can be resolved by taking advantage of template lists (containing compact and sorted lists of literal values for a given template ID), or by filtering template lists based on lists of cluster IDs retrieved from the hash-table.

4 Provenance for Linked Data

Understanding where and how a piece of data is produced (its provenance) has long been recognized as an important factor in determining the quality of a data item particularly in data integration systems [68]. Thus, it is no surprise that provenance has been of concern within the Linked Data community where a major use case is the integration of data sets published by multiple different actors [45].

The work on provenance and linked data builds upon prior work in the database, e-science and distributed systems communities. Moreau provides an extensive review of that literature contextualizing it with respect to the Web [55]. Instead of providing a comprehensive overview of the work, we provide a framework for situating provenance work with respect to linked data and then drill down into those works that have focused on provenance-aware linked data management systems.

Following [38], one can categorize work into three areas: content, management, and use. Work in the content area has focused on representations and models of provenance. In management, the work has focused on collecting provenance in software ranging from scientific databases [23] to operating systems or large scale workflow systems as well as mechanisms for querying it. Finally, provenance is used for a variety of applications including debugging systems, calculating trust and checking compliance.

With respect to provenance within linked data management systems, we first discuss the representations of provenance primarily for interoperable interchange (i.e. content), we then more heavily discuss provenance in linked data management systems (management). The usage of provenance, while important, is often application or domain specific.

4.1 Provenance Representations

Because data processing takes place across systems, there is a need to be able to interchange information about data was combined, recombined and processed. This has led to the development of a number of ontologies for the representation of that information including, Dublin Core,[11] the Proof Markup Language (PML) [25], Provenir [64], Provenance Authoring and Versioning [16], Provenance Vocabulary [42], and OPMV [76]. These ontologies shared many common capabilities to describe various types of information about how data was combined and processed including:

- how software executed to consume and produce data,
- that data was derived from other data,
- that data were composed of other data,
- who was involved in the manipulation or construction of data.

---

[11] http://dublincore.org/documents/dcmi-terms/.

These common characteristics were also evident to the broader provenance community through the development of the the the Open Provenance Model [56]. Given this, the World Wide Web Consortium formed a working group that developed a recommendation for the interchange of provenance on the web called PROV [39]. PROV incorporates the concepts from above and establishes a baseline for the interchange of provenance. We recommend the PROV Primer [33] as a good introduction to the concepts of PROV. A book length introduction is in given by Moreau and Groth [54]. For an in-depth description of how PROV came to be and the design decisions behind it, see [57]. Many of the ontologies described above that were pre-cursors to PROV have now been revised to extend PROV including PML[12] and PAV [15]. Furthermore, there is a mapping between PROV and Dublin Core.

The key realization here is that these ontologies provide the vocabulary to represent provenance information at differing level of details to enable interchange between systems. Thus, when developing a linked data management system, it's important to be aware that there is a difference between these sort of external, interchangeable representations and those used internally to manage data.

These representations all use the common Linked Data formats and importantly rely on the URL as identifiers to point to pieces of data and describe the provenance. At the dataset level, provenance is often attached to a dataset descriptor [35] often embedded in a Vocabulary of Interlinked Dataset file (VoID) [2]. VoID is an important part of Linked Data provenance as it allows one to define what constitutes a dataset and its associated metadata. The Dataset Descriptions: HCLS Community Profile[13] provides excellent guidance on what metadata (inclusive of provenance) should be provide by linked datasets.

Within linked data, provenance is attached using either reification [44] or named graphs [12]. Widely used datasets such as YAGO [48] reify their entire structures to facilitate provenance annotations. Indeed, provenance is one reason for the inclusion of named graphs in the current version of RDF 1.1 [77]. Other approaches, such as nanopublications [37], extensively use named graphs to enable subsets of linked data to be referred to and for their provenance to be described [37].

Tracking and generating these representations is the subject of the next section.

4.2 Provenance in Data Management Systems

Provenance within linked data management systems builds heavily on the work of the database systems community. See [14] for an extensive review. Here, we introduce some important concepts and relevant work.

Miles defined the concept of *provenance query* [53] in order to only select a relevant subset of all possible results when looking up the provenance of an entity. A good example of a classic database system that handles provenance is Perm [34], which can

---

[12]http://inference-web.org/wiki/PML_3.0.
[13]http://www.w3.org/TR/hcls-dataset/.

compute, store, and query relational. Provenance was computed by using standard relational query rewriting techniques. Perm supports the calculation of provenance both when queried for (the lazy approach) or when a new relation is created or data is inserted (the eager approach) depending on settings. Recently, Glavic showed that the provenance captured within such a standard relational system can be represented and interchanged using PROV [61].

An important point is that these traditional database approaches [8, 51] assume a strict relational schema, whereas RDF data is by definition schema free. To address these issues, a number of authors have adopted the notion of annotated RDF [30, 67]. This approach assigns annotations to each of the triples within a dataset and then tracks these annotations as they propagate through either the reasoning or query processing pipelines. Formally, these annotated relations can be represented by the algebraic structure of communicative semirings, which can take the form of polynomials with integer coefficients [36]. These polynomials represent how source tuples are combined through different relational algebra operators (e.g., UNION, JOINS). Theoharis et al. [66] provide a comprehensive theoretical foundations of tracing provenance in SPARQL workload. In terms of formalization, SPARQL poses, however, difficulties because of the OPTIONAL operator, which implies negation. The detailed explanation of theoretical basics goes beyond the scope of this chapter.

Zimmermann et al. [78] propose to annotate a triple with temporal data and a provenance value. Such provenance value refers to the source of a triple. The authors use a standard triple-oriented data model and include temporal and provenance annotation. A triple take the form of (Subject, Predicate, Object, Annotation), i.e., N-Quad (Sect. 2). Such statements can be stored in any triplestore supporting N-Quads. Zimmermann et al. proposes also a model to describe provenance of inferred triples with the logical operators $\vee$ and $\wedge$. Consider the following data:

(chadHurley; worksFor; youtube) : chad
(chadHurley; type; Person) : chad
(youtube; type; Company) : chad
(Person; sc; Agent) : foaf
(worksFor; dom; Person) : wrokkend
(worksFor; range; Company) : workont

It is possible to infer the following triple:

(chadHurley; type; Agent) : (chad $\wedge$ foaf $\wedge$ workont)
$\vee$ (chad $\wedge$ foaf)

Which logically is equivalent to:

(chadHurley; type; Agent) : chad $\wedge$ foaf

The proposed method to describe provenance of inferred triples could be possibly leveraged to trace provenance in query execution, however this avenue was not explored by the authors. Zimmermann et al. propose also a query language which allows to incorporate provenance information in the query execution. The basic idea to query over provenance values is similar to named graphs in SPARQL. The query incorporates information on the annotation, which is then taken into account during the execution.

A similar is described by Udrean et al. [67]. The authors extend the RDF schema for temporal, uncertainty, and provenance annotations. The main focus of this work is to develop a theoretical model to manage such metadata information. The authors propose also a query language which allows to query over such meta data. Contrary to the previous solution, here the authors annotate predicates with provenance information. In a similar way Nguyen et al. [60] propose to use a singleton property instead of RDF reification or named graphs to describe provenance. A triple take then a form of (Subject, Predicate: Annotation, Object). Such annotation added to a predicated could be later tracked to deliver a trace of the query execution, however it is not included in the work. Udrean et al. [67] in their work propose also a query language to include provenance information in the query execution process. A query then take a form similar to their annotated triple i.e., (Subject, Predicate : Annotation, Object). Query schema proposed in this work in not fully compatible with SPARQL. Queries can by expressed in SPARQL but the annotations are not taken into account in such case.

Flouris et al. [30] instead of annotating triples with URIs assign a color to each triple which allows to trace the provenance of the results. Their storage model is based on a statement table where they store all triples along with their colors. The statement table is extended for a fourth column which in result gives N-Quads (Sect. 2). During the query execution they trace all colors involved in the process and provide them as a provenance trace. The final lineage is delivered as a list of unique colors.

Ding et al. consider tracking provenance from the database perspective at a molecule level [27]. They define RDF molecule as the finest and lossless sub-graph resulting from the graph decomposition. Which boils down to triples if there are no blank nodes involved. In case the data set contains blank nodes, triples sharing the same blank node are placed in the same molecule. They consider provenance at document level which means that molecules are annotated with the URI they source from. Similarly to the previous approaches, in their implementation, they store data in a form of quads in a statement table. Likewise, the final output of their system consist of query results and a list of documents (URIs) which provided the triples used in the query execution.

The RDFProv [13] system allows to manage and query provenance that results from scientific workflows. RDFProv proposes a solution to manage scientific workflow provenance with the means of Semantic Web, i.e., represented as triples. Chebotko et al. [13] propose two algorithms to map a provenance ontology into relational database system. The first algorithm uses database views, the second one instead of using views uses tables, thus it replicates all data and result in more complex update operations since all independent relations have to be modified separately. To map an ontology, first of all, they store all data in a statement table, additionally for each $type$ (class of resources) they create three auxiliary views/relations co-locating triples to address different kind of workloads. The three views are as flows:

- a view for all instances of the *type*
- Subject(i,p,o) for triples whose subjects belong to the *type*
- Object(s,p,i) for triples whose objects belong to the *type*

To optimize the execution they create B+-tree indexes on columns (s,p,o), (s,o), and (p) of the statement table. Similar indexes are created on the auxiliary relations in case tables are employed.

As Damásio et al. have noted [26], many of the annotated RDF approaches do not expose how-provenance (i.e., how a query result was constructed). The most comprehensive implementations of these approaches are [67, 78]. However, they have only been applied to small datasets (around 10 million triples) and are not aimed at reporting provenance polynomials [36] (i.e., algebraic structures representing, using relational algebra operators, how data is combined) for SPARQL query results, focusing instead on inferred triples. Annotated approaches have also been used for propagating trust values [43]. Other recent work, e.g., [26, 32], has looked at expanding the theoretical aspects of applying such a semiring based approach to capturing SPARQL.

Contrary to the previous approaches, TripleProv [73] extends a native triple-store [72] to allow storing, tracing, and querying provenance information in processing RDF queries. TripleProv returns a description of the way the results of an RDF query were derived; specifically it gives an explanation which pieces of data and how were combined to produce the answer of a query. The system allows as well to tailor query execution with provenance information [75]. The user can input a provenance specification of the data he wants to use to derive the answer. For example, if he is interested with articles about "Obama", but he wants the answer to come only from sources attributed to "US News".

As an input to the system the user provides a query he wants to execute (workload query) and an RDF query describing provenance of the data he wants to be used in query processing (provenance query) (Fig. 19). The query execution process can vary depending of the strategy. Typically the system starts with executing the provenance query, then it optionally pre-materializes or co-locates data. Afterwards, TripleProv executes the workload queries, at the same time it collects information of entities used during the query execution and the way they are combined. The system returns:

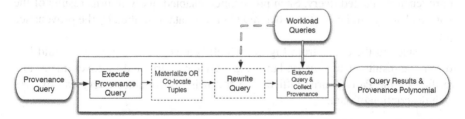

Fig. 19 TripleProv: executing provenance-enabled queries; both a workload and a provenance query are given as input to a triplestore, which produces results for both queries and then combine them to obtain the final results and the provenance polynomial [74]

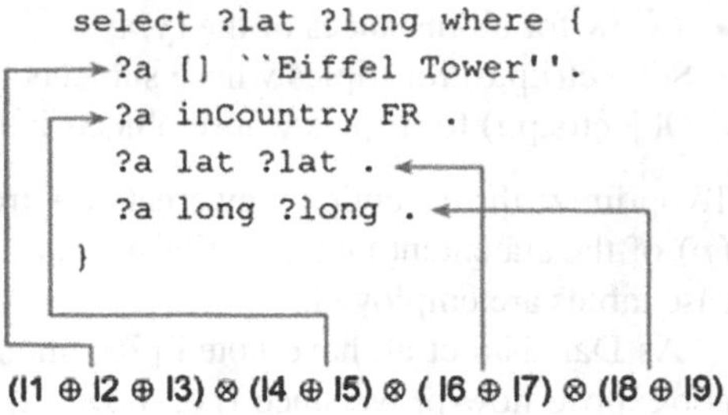

Fig. 20 TripleProv:
provenance polynomial
represents how the data is
combined to derive the query
answer using different
relational algebra operators
(e.g., UNION, JOINS) [74]

```
select ?lat ?long where {
   ?a [] ``Eiffel Tower''.
   ?a inCountry FR .
   ?a lat ?lat .
   ?a long ?long .
}
```

$$(I1 \oplus I2 \oplus I3) \otimes (I4 \oplus I5) \otimes (I6 \oplus I7) \otimes (I8 \oplus I9)$$

- results of the workload queries, restricted to those which are following the provenance specification;
- the provenance polynomial describing the way the results were derived.

TripleProv provides detailed information on each piece of data used to produce the
answer and the exact way it contributed to the results. To express this information the
system uses the notion of a provenance polynomial (Fig. 20), which is an algebraic
structure describing how the data was combined. A provenance polynomial provided
by TripleProv allows to pinpoint and trace back the exact pieces of data used to
produce the answer and the exact way how those pieces of data were combined. In
order to express the way the pieces of data were combined TripleProv uses two basic
algebraic operators. The first one ($\oplus$) to represent a union, and the second ($\otimes$) to
represent a join.

The Fig. 20 shows a simple star query (Basic Graph Pattern) and a provenance
polynomial pinpointing how each part of the query is tackled. In this example the
first triple pattern is satisfied with lineage I1, I2 or I3, while the second has been
satisfied with I4 or I5, third was processed with elements having a lineage of I6 or I7,
and the last one was processed with elements from I8 or I9. The triples were joined
on variable ?a, which is expressed by the join operation ($\otimes$) in the polynomial.

TripleProv allows to tailor RDF queries with provenance information [75]. The
user can provide to the system a description of the data which will be used in the
query processing. Such description (provenance query) is expressed in the same way
as the workload query. Together the workload query and the provenance query give
a provenance-enabled query. Such provenance-enabled query returns results of the
workload query, limited to those derived from the data described by the provenance
query.

Considering the query from Fig. 20, which is a workload query, we would like
to retrieve results of this query, but using only data attributed to government, and
verified by the Paris Tourist Office. The following provenance query can express
such description of data:

```
SELECT ? ctx WHERE {
         ? ctx prov: wasAttributedTo <government> .
         ? ctx prov: wasVeryfiedBy <PartisTouristOffice> .
    }
```

Sending those two queries to TripleProv will give to the user information about geolocation of Eiffel Tower in France, the information will be obtained from the data following the provenance description. Additionally, TripleProv will provide a trace of how particular pieces of data were combined to deliver the results.

Provenance is a central part of linked data management. Systems should be able to both maintain provenance but also interchange using common vocabularies and data formats.

References

1. D.J. Abadi, A. Marcus, S. Madden, K.J. Hollenbach, Scalable semantic web data management using vertical partitioning, in *Proceedings of the 33rd International Conference on Very Large Data Bases, University of Vienna, Austria, 23–27 September 2007* (ACM, 2007), pp. 411–422
2. K. Alexander, M. Hausenblas, Describing linked datasets — on the design and usage of void, the vocabulary of interlinked datasets, in *In Linked Data on the Web Workshop (LDOW 09), in Conjunction with 18th International World Wide Web Conference (WWW 09)* (2009). http://richard.cyganiak.de/2008/papers/void-ldow2009.pdf
3. M. Atre, V. Chaoji, M.J. Zaki, J.A. Hendler, Matrix "Bit" loaded: a scalable lightweight join query processor for RDF data, in *Proceedings of the 19th International Conference on World Wide Web, WWW 2010, Raleigh, North Carolina, USA, 26–30 April 2010* (ACM, 2010), pp. 41–50
4. M. Atre, J.A. Hendler, BitMat: a main memory bit-matrix of RDF triples, in *The 5th International Workshop on Scalable Semantic Web Knowledge Base Systems (SSWS2009)* (Citeseer, 2009), p. 33
5. S. Auer, J. Demter, M. Martin, J. Lehmann, Lodstats-an extensible framework for high-performance dataset analytics, in *Knowledge Engineering and Knowledge Management* (Springer, Berlin, 2012), pp. 353–362
6. T. Berners-Lee, Linked data-design issues (2006)
7. T. Berners-Lee, J. Hendler, O. Lassila et al., The semantic web. Sci. Am. **284**(5), 28–37 (2001)
8. O. Biton, S. Cohen-Boulakia, S.B. Davidson, Zoom*userviews: querying relevant provenance in workflow systems, in *Proceedings of the 33rd International Conference on Very Large Data Bases, VLDB '07* (VLDB Endowment, 2007), pp. 1366–1369
9. C. Bizer, A. Jentzsch, R. Cyganiak, State of the lod cloud. Version 0.3 (September 2011) 1803 (2011). http://lod-cloud.net/state/
10. M. Bröcheler, A. Pugliese, V. Subrahmanian, DOGMA: a disk-oriented graph matching algorithm for RDf databases, in *The Semantic Web-ISWC 2009* (Springer, Berlin, 2009), pp. 97–113
11. M. Bröcheler, A. Pugliese, V.S. Subrahmanian, DOGMA: a disk-oriented graph matching algorithm for RDF databases, in *Proceedings of the Semantic Web - ISWC 2009, 8th International Semantic Web Conference, ISWC 2009, Chantilly, VA, USA, October 25–29, 2009* (Springer, Berlin, 2009), pp. 97–113
12. J.J. Carroll, C. Bizer, P. Hayes, P. Stickler, Named graphs, provenance and trust, in *Proceedings of the 14th International Conference on World Wide Web* (ACM, 2005), pp. 613–622
13. A. Chebotko, S. Lu, X. Fei, F. Fotouhi, RDFProv: a relational RDF store for querying and managing scientific workflow provenance. Data Knowl. Eng. **69**(8), 836–865 (2010)
14. J. Cheney, L. Chiticariu, W.C. Tan, *Provenance in Databases: Why, How, and Where* (Now Publishers Inc., Breda, 2009)
15. P. Ciccarese, S. Soiland-Reyes, K. Belhajjame, A.J. Gray, C. Goble, T. Clark, Pav ontology: provenance, authoring and versioning. J. Biomed. Semant. **4**(1), 1–22 (2013). doi:10.1186/2041-1480-4-37

16. P. Ciccarese, E. Wu, G. Wong, M. Ocana, J. Kinoshita, A. Ruttenberg, T. Clark, The swan biomedical discourse ontology. J. Biomed. Inf. **41**(5), 739–751 (2008). doi:10.1016/j.jbi.2008. 04.010
17. Consortium WWW, OWL 2 Web Ontology Language (2012)
18. Consortium WWW, SPARQL 1.1 Overview (2013)
19. Consortium WWW, RDF 1.1 Concepts and Abstract Syntax (2014)
20. Consortium WWW, RDF 1.1: On Semantics of RDF Datasets (2014)
21. Consortium WWW, RDF 1.1 Primer (2014)
22. Consortium WWW, RDF Schema **1**, 1 (2014)
23. P. Cudré-Mauroux, H. Kimura, K.T. Lim, J. Rogers, R. Simakov, E. Soroush, P. Velikhov, D.L. Wang, M. Balazinska, J. Becla, D.J. DeWitt, B. Heath, D. Maier, S. Madden, J.M. Patel, M. Stonebraker, S.B. Zdonik, A demonstration of SciDB: a science-oriented DBMS. PVLDB **2**(2), 1534–1537 (2009)
24. P. Cudré-Mauroux, E. Wu, S. Madden, The case for rodentstore, an adaptive, declarative storage system, in *Biennial Conference on Innovative Data Systems Research (CIDR)* (2009)
25. P.P. da Silva, D.L. McGuinness, R. Fikes, A proof markup language for semantic web services. Inf. Syst. **31**(4), 381–395 (2006). doi:10.1016/j.is.2005.02.003
26. C.V. Damásio, A. Analyti, G. Antoniou, Provenance for sparql queries, in *Proceedings of the 11th International Conference on The Semantic Web - Volume Part I, ISWC'12* (Springer, Berlin, 2012), pp. 625–640. doi:10.1007/978-3-642-35176-1_39
27. L. Ding, Y. Peng, P.P. da Silva, D.L. McGuinness, Tracking RDF graph provenance using RDF molecules, in *International Semantic Web Conference* (2005)
28. O. Erling, I. Mikhailov, *Towards web scale RDF*, in *Proceedings of the SSWS* (2008)
29. G.H.L. Fletcher, P.W. Beck, Scalable indexing of RDF graphs for efficient join processing, in *Proceedings of the 18th ACM Conference on Information and Knowledge Management, CIKM 2009, Hong Kong, China, November 2–6, 2009* (ACM, 2009), pp. 1513–1516
30. G. Flouris, I. Fundulaki, P. Pediaditis, Y. Theoharis, V. Christophides, Coloring RDF triples to capture provenance, in *Proceedings of the 8th International Semantic Web Conference, ISWC '09* (Springer, Berlin, 2009), pp. 196–212. doi:10.1007/978-3-642-04930-9_13
31. H. Garcia-Molina, *Database Systems: The Complete Book* (Pearson Education, India, 2008)
32. F. Geerts, G. Karvounarakis, V. Christophides, I. Fundulaki, Algebraic structures for capturing the provenance of sparql queries, in *Proceedings of the 16th International Conference on Database Theory, ICDT '13* (ACM, New York, 2013), pp. 153–164. doi:10.1145/2448496. 2448516
33. Y. Gil, S. Miles, K. Belhajjame, H. Deus, D. Garijo, G. Klyne, P. Missier, S. Soiland-Reyes, S. Zednik (eds.), in *PROV model primer. W3C Working Group Note NOTE-prov-primer-20130430, World Wide Web Consortium* (2013). http://www.w3.org/TR/prov-primer/
34. B. Glavic, G. Alonso, The perm provenance management system in action, in *Proceedings of the 2009 ACM SIGMOD International Conference on Management of Data, SIGMOD '09* (ACM, New York, NY, USA, 2009), pp. 1055–1058
35. A.J. Gray, Dataset descriptions for linked data systems. IEEE Internet Comput. **18**(4), 66–69 (2014). doi:10.1109/MIC.2014.66
36. T.J. Green, G. Karvounarakis, V. Tannen, Provenance semirings, in *Proceedings of the Twenty-Sixth ACM SIGMOD-SIGACT-SIGART Symposium on Principles of Database Systems* (ACM, 2007), pp. 31–40
37. P. Groth, A. Gibson, J. Velterop, The anatomy of a nanopublication. Inf. Serv. Use **30**(1–2), 51–56 (2010). http://dl.acm.org/citation.cfm?id=1883685.1883690
38. P. Groth, Y. Gil, J. Cheney, S. Miles, Requirements for provenance on the web. Int. J. Digit. Curation **7**(1), 39–56 (2012). doi:10.2218/ijdc.v7i1.213
39. P. Groth, L. Moreau (eds.), PROV-overview. An overview of the PROV family of documents, in *W3C Working Group Note NOTE-Prov-Overview-20130430, World Wide Web Consortium* (2013). http://www.w3.org/TR/2013/NOTE-prov-overview-20130430/
40. S. Harris, N. Lamb, N. Shadbolt, 4store: the design and implementation of a clustered rdf store, in *5th International Workshop on Scalable Semantic Web Knowledge Base Systems (SSWS2009)* (2009), pp. 94–109

41. A. Harth, S. Decker, Optimized index structures for querying RDF from the web, in *IEEE LA-WEB* (2005), pp. 71–80
42. O. Hartig, Provenance information in the web of data, in *LDOW* (2009). http://ceur-ws.org/Vol-538/ldow2009_paper18.pdf
43. O. Hartig, Querying trust in RDF data with tSPARQL, in *Proceedings of the 6th European Semantic Web Conference on The Semantic Web: Research and Applications, ESWC 2009 Heraklion* (Springer, Berlin, 2009), pp. 5–20. doi:10.1007/978-3-642-02121-3_5
44. P. Hayes, B. McBride, RDF semantics, in *W3C Recommendation* (2004)
45. T. Heath, C. Bizer, Linked Data: Evolving the Web into a Global Data Space. Morgan and Claypool (Morgan & Claypool Publishers, 2011). doi:10.2200/S00334ED1V01Y201102WBE001
46. T. Heath, C. Bizer, Linked data: evolving the web into a global data space. Synth. Lectures Semant. Web: Theory technol. 1(1), 1–136 (2011)
47. J.M. Hellerstein, M. Stonebraker, *Readings in Database Systems* (MIT Press, Cambridge, 2005)
48. J. Hoffart, F.M. Suchanek, K. Berberich, G. Weikum, YAGO2: a spatially and temporally enhanced knowledge base from wikipedia. Artif. Intell. 194(0), 28–61 (2013). doi:10.1016/j.artint.2012.06.001, http://www.sciencedirect.com/science/article/pii/S0004370212000719 (Artificial Intelligence, Wikipedia and Semi-Structured Resources)
49. J. Huang, D.J. Abadi, K. Ren, Scalable SPARQL querying of large RDF graphs. PVLDB 4(11), 1123–1134 (2011)
50. M. Janik, K. Kochut, BRAHMS: a workbench RDF store and high performance memory system for semantic association discovery, in *Proceedings of the The Semantic Web - ISWC 2005, 4th International Semantic Web Conference, ISWC 2005, Galway, Ireland, November 6–10, 2005* (Springer, Berlin, 2005), pp. 431–445
51. G. Karvounarakis, Z.G. Ives, V. Tannen, Querying data provenance, in *Proceedings of the 2010 ACM SIGMOD International Conference on Management of Data* (ACM, 2010), pp. 951–962
52. G. Karypis, V. Kumar, A fast and high quality multilevel scheme for partitioning irregular graphs. SIAM J. Sci. Comput. 20(1), 359–392 (1998)
53. S. Miles, Electronically querying for the provenance of entities, in *Provenance and Annotation of Data*, vol. 4145, ed. by L. Moreau, I. Foster. Lecture Notes in Computer Science (Springer, Berlin, 2006), pp. 184–192. doi:10.1007/11890850_19
54. M. Luc, G. Paul, *Provenance: An Introduction to PROV* (Morgan and Claypool, 2013). http://eprints.soton.ac.uk/356858/
55. L. Moreau, The foundations for provenance on the web. Found. Trends Web Sci. 2(2–3), 99–241 (2010). doi:10.1561/1800000010, http://eprints.ecs.soton.ac.uk/21691/
56. L. Moreau, B. Clifford, J. Freire, J. Futrelle, Y. Gil, P. Groth, N. Kwasnikowska, S. Miles, P. Missier, J. Myers, B. Plale, Y. Simmhan, E. Stephan, J.V. den Bussche, The open provenance model core specification (v1.1). Future Gener. Comput. Syst. 27(6), 743–756 (2011). doi:10.1016/j.future.2010.07.005, http://www.sciencedirect.com/science/article/pii/S0167739X10001275
57. L. Moreau, P. Groth, J. Cheney, T. Lebo, S. Miles, The rationale of PROV. Web Semant.: Sci. Serv. Agents World Wide Web 35, Part 4, 235–257 (2015). http://dx.doi.org/10.1016/j.websem.2015.04.001, http://www.sciencedirect.com/science/article/pii/S1570826815000177
58. T. Neumann, G. Weikum, RDF-3X: a RISC-style engine for RDF. Proc. VLDB Endow. (PVLDB) 1(1), 647–659 (2008)
59. T. Neumann, G. Weikum, The RDF-3X engine for scalable management of RDF data. VLDB J. 19(1), 91–113 (2010)
60. V. Nguyen, O. Bodenreider, A. Sheth, Don't like RDF reification? Making statements about statements using singleton property, in *Proceedings of the 23rd International Conference on World Wide Web. International World Wide Web Conferences Steering Committee* (2014), pp. 759–770
61. X. Niu, R. Kapoor, B. Glavic, D. Gawlick, Z.H. Liu, V. Krishnaswamy, V. Radhakrishnan, Interoperability for provenance-aware databases using PROV and JSON, in *Proceedings of the 7th USENIX Conference on Theory and Practice of Provenance, TaPP'15* (USENIX Association, Berkeley, CA, USA, 2015), p. 6. http://dl.acm.org/citation.cfm?id=2814579.2814585

62. A. Owens, A. Seaborne, N. Gibbins, et al., Clustered TDB: a clustered triple store for jena (2008)
63. E. Prud'Hommeaux, A. Seaborne, et al., Sparql query language for RDF. W3C Recommendation (2008)
64. S.S. Sahoo, A. Sheth, Provenir ontology: towards a framework for escience provenance management, in *Microsoft eScience Workshop* (2009). http://knoesis.wright.edu/library/resource.php?id=741
65. M. Schmachtenberg, C. Bizer, H. Paulheim, Adoption of the linked data best practices in different topical domains, in *The Semantic Web–ISWC 2014* (Springer, 2014), pp. 245–260
66. Y. Theoharis, I. Fundulaki, G. Karvounarakis, V. Christophides, On provenance of queries on semantic web data. IEEE Internet Comput. **15**(1), 31–39 (2011). doi:10.1109/MIC.2010.127
67. O. Udrea, D.R. Recupero, V. Subrahmanian, Annotated RDF. ACM Trans. Comput. Log. (TOCL) **11**(2), 10 (2010)
68. Y.R. Wang, S.E. Madnick, A polygon model for heterogeneous database systems: the source tagging perspective, in *Proceedings of the Sixteenth International Conference on Very Large Databases* (Morgan Kaufmann Publishers Inc., San Francisco, CA, USA, 1990), pp. 519–533. http://dl.acm.org/citation.cfm?id=94362.94604
69. C. Weiss, P. Karras, A. Bernstein, Hexastore: sextuple indexing for semantic web data management. Proc. VLDB Endow. (PVLDB) **1**(1), 1008–1019 (2008). http://doi.acm.org/10.1145/1453856.1453965
70. M. Wylot, Efficient, scalable, and provenance-aware management of linked data. Ph.D. thesis, University of Fribourg (Switzerland) (2015)
71. M. Wylot, P.C. Mauroux, Diplocloud: Efficient and Scalable Management of RDF Data in the Cloud (2015)
72. M. Wylot, J. Pont, M. Wisniewski, P. Cudré-Mauroux, dipLODocus[RDF] - short and long-tail RDF analytics for massive webs of data, in *International Semantic Web Conference* (2011), pp. 778–793
73. M. Wylot, P. Cudre-Mauroux, P. Groth, TripleProv: efficient processing of lineage queries in a native RDF store, in *Proceedings of the 23rd International Conference on World Wide Web, WWW '14. International World Wide Web Conferences Steering Committee, Republic and Canton of Geneva, Switzerland* (2014), pp. 455–466
74. M. Wylot, P. Cudré-Mauroux, P. Groth, A demonstration of tripleprov: tracking and querying provenance over web data. Proc. VLDB Endow. **8**(12), 1992–1995 (2015)
75. M. Wylot, P. Cudré-Mauroux, P. Groth, Executing provenance-enabled queries over web data, in *Proceedings of the 24rd International Conference on World Wide Web, WWW '15. International World Wide Web Conferences Steering Committee, Republic and Canton of Geneva, Switzerland* (2015)
76. J. Zhao, Guide to the Open Provenance Model Vocabulary (2010). http://open-biomed.sourceforge.net/opmv/opmv-guide.html
77. J. Zhao, C. Bizer, Y. Gil, P. Missier, S. Sahoo, Provenance requirements for the next version of RDF, in *W3C Workshop RDF Next Steps* (2010)
78. A. Zimmermann, N. Lopes, A. Polleres, U. Straccia, A general framework for representing, reasoning and querying with annotated semantic web data. Web Semant. **11**, 72–95 (2012). doi:10.1016/j.websem.2011.08.006
79. L. Zou, J. Mo, L. Chen, M.T. Oezsu, D. Zhao, gStore: answering SPARQL queries via subgraph matching. PVLDB **4**(8), 482–493 (2011)

Non-native RDF Storage Engines

Manfred Hauwirth, Marcin Wylot, Martin Grund, Sherif Sakr and
Phillippe Cudré-Mauroux

Abstract The proliferation of heterogeneous Linked Data requires data management systems to constantly improve their scalability and efficiency. Linked Data can
be stored according to many different data storage models. Some of these attempt to
use general purpose database storage techniques to persist Linked Data, hence they
can leverage existing data processing environments (e.g., big Hadoop clusters). We
therefore look at the multiplicity of Linked Data storage systems which we categorize into the following classes: relational database-based systems, NoSQL-based
systems, massively parallel systems.

1 Introduction

RDF data can be stored in a multiplicity of different storage engines. Some of these
presented in a previous chapter are more tailored for storing RDF data while others
try using general purpose database storage engines to persist RDF data. We define
a *non-native RDF storage engines* as a specific engine that uses either traditional
relational storage concepts or builds on these concepts to integrate storing and query
execution on RDF data. The biggest differentiation to native RDF storage solutions

M. Hauwirth · M. Wylot (✉)
Open Distributed Systems, TU Berlin, Berlin, Germany
e-mail: m.wylot@tu-berlin.de

M. Hauwirth · M. Wylot
Open Distributed Systems, Fraunhofer FOKUS, Berlin, Germany
e-mail: manfred.hauswirth@fokus.fraunhofer.de

M. Grund · P. Cudré-Mauroux
eXascale Infolab, University of Fribourg, Fribourg, Switzerland
e-mail: pcm@unifr.ch

S. Sakr
University of New South Wales, Kensington, Australia
e-mail: ssakr@cse.unsw.edu.au

© Springer International Publishing AG 2017   339
A.Y. Zomaya and S. Sakr (eds.), *Handbook of Big Data Technologies*,
DOI 10.1007/978-3-319-49340-4_10

is hereby the translation of the RDF concepts into concepts that are native to the underlying engine instead of working directly with the RDF data. In this chapter, we therefore look at a multiplicity of RDF storage systems that leverage existing database systems to support the processing of RDF data. We categorize them into the following set of groups: relational database-based systems, NoSQL-based systems, massively parallel systems.

2 Storing Linked Data Using Relational Databases

Relational database management systems (RDBMSs) have been the backbone of almost all applications, ranging from traditional enterprise like applications to modern web applications. They have repeatedly shown that they are very efficient, scalable and successful in hosting types of data which have formerly not been anticipated to be stored inside relational databases. In addition, RDBMSs have shown their ability to handle vast amounts of data very efficiently using powerful indexing mechanisms. In this section, we present various relational-based approaches [39] for storing, indexing and querying RDF data.

2.1 Statement Table

The general data structure that is represented by a set of RDF triples is an edge-labeled directed graph. Figure 1 shows a subset of nodes from a sample dataset inspired by the data model from the Berlin SPARQL Benchmark [7]. In this representation, subjects and objects are stored as nodes with an edge and an associated edge property assigned: $[S] - P \rightarrow [O]$. As stated by [4, 8], subjects and objects can be interchanged. In addition, all triples are unordered [4].

Since RDF does not describe any specific schema for the graph, there is no easy way to determine a set of partitioning or clustering criteria to derive a set of tables to store the information. In addition, there is no definite notion of schema stability, meaning that at any time the data schema might change, for example when adding a new subject-object edge to the overall graph.

A trivial way for adopting a relational data structure to store RDF data is to store the input data as a linearized list of triples, storing them as ternary tuples [39]. In [4], this approach is called the "generic" approach. The RDF specification states that the objects in the graphs can be either URIs, literals, or blank nodes. Properties (predicates) always are URI references. Subject nodes can only be URIs or blank nodes. This allows to specify the underlaying data types for storing subject and predicate values. For storing object values this becomes a little more complex since the data type of the object literal is defined by the XML schema that is referenced by the property. A common way is to store the object values using a common string representation and perform some type conversion whenever necessary. An example table showing the same data set as in Fig. 1 is shown in Fig. 2.

An example on *Statement Table* approach is Jena1 [30]. Jena1 for relational databases stores data in a statement table. The URI and String are encoded in ID and two separate dictionaries are maintained for literals and resources/URIs. To distinguish literals from URI in the statement table there are two columns. In Jena2 [48] the schema is denormalized and URIs and simple literals are directly stored in the

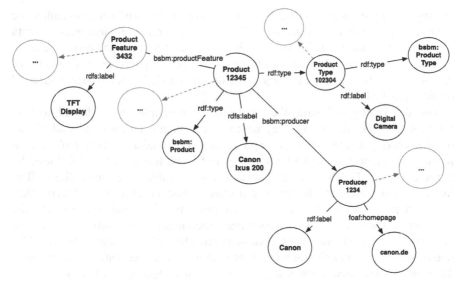

Fig. 1 Example showing an RDF sub-graph using the subject, predicate, and object relations given by the sample data

Subject	Predicate	Object
Product12345	rdf:type	bsbm:Product
Product12345	rdfs:label	Canon Ixus 2010
Product12345	bsbm:producer	bsbm-inst:Producer1234
...	...	...
Producer1234	rdf:label	Canon
Producer1234	foaf:homepage	http://www.canon.com
...	...	...

Fig. 2 A simple RDF storage scheme using a linearized triple representation. The illustration uses schema elements from the Berlin SPARQL benchmark [7]

statement table. The dictionary tables are used only to store long strings (exceeding a threshold). This allows to perform filters operation directly on the statement table, however it results also higher storage consumption, since string values are stored multiple times.

2.2 Optimizing Data Storage

Storing the triples as a linear list of triples is very simple and yet powerful, since it captures the complete essence of RDF data. However, the problem with this data structure is that additional information that is crucial for query processing needs to be analyzed at query run-time even though it is not likely to change. Examples of this issue are whether or not the object is a literal or a URI or if the edge is inferred or an original edge.

One disadvantage of storing the data inside a large triple table is that all fields in this table must be encoded as string with variable length. This generates additional overhead during data storage and data processing. The standard MySQL table storage format, for example, uses an 8 bit or if required 16 bit length identifier followed by the actual string. To process a set of fields it is not possible to perform a direct offset into the set of tuples, and the database storage engine has to interpret the complete row, or requires additional data structure for pointers into the variable length fields.

One possible way to optimize the storage structure is to apply dictionary encoding on the resources and literal values. Dictionary encoding allows to replace the variable-length string representation of a literal or resource by a fixed-length integer value. There are several possible ways to generate such an integer values. In order to avoid issues with duplicates, Harris et al. [22] proposed to use a hash function that allows hashing of all literal and resource values. The actual values are replaced with the hash value and the hash and the value is stored in an additional table for later reference. Figure 3 illustrates this scheme.

The downside of using a hash function to generate the encoded values lays within the properties of the hash function. Even though the probability of a hash collision can be low—depending on the actual hash implementation–they still can occur. As a consequence, the import system has to perform validity checks for all imported

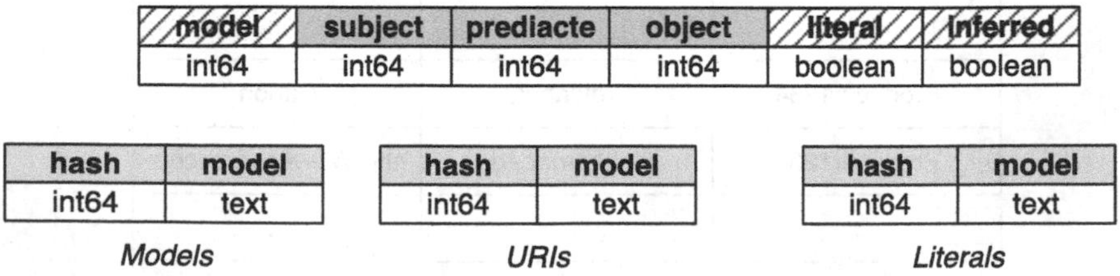

model	subject	prediacte	object	literal	inferred
int64	int64	int64	int64	boolean	boolean

hash	model		hash	model		hash	model
int64	text		int64	text		int64	text

Models   URIs   Literals

Fig. 3 Logical database design of the triple table in 3store. Illustration after [22]

literals and resources to evaluate if the generated hash value generates a collision. While it is easy to handle such collisions during insertion time as described by Harris et al. [22] it becomes more complicated to handle collisions at runtime when new triples are inferred based on the existing knowledge-base. If a collision happens, it is almost impossible to handle this without modifying the inferred value to generate a different hash value. In addition, Harris et al. [22] use a cryptographic hash function to calculate a hash key that has as few collisions as possible. The disadvantage of this approach is that computing a cryptographic hash consumes more CPU cycles compared to a simpler hash-function. In their example, the calculation of the MD5 hash takes about 1,000 CPU cycles. This limits a single CPU core on a modern 2.5 GHz CPU to calculating 2.5M hashes per second. Using such an expensive hash function can thus lead to CPU-bound behaviors even though the database does not operate at optimal speed.

To further reduce the probability of collisions, Harris et al. [22] use two different buckets to store hashes and values. One bucket for literals and one bucket for resources. An Entity-Relationship diagram showing the dependency for the two different triple types is shown in Fig. 4. Another advantage of using the MD5 hash function to represent resources or literals is that during query processing the actual

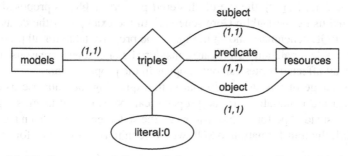

(a) ER diagram modeling a triple pattern with no literal object

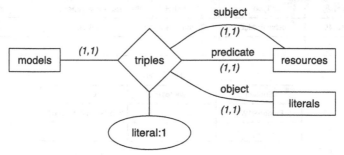

(b) ER diagram modeling a triple pattern having a literal object

Fig. 4 Dependency for the two different triple types [22]

lookup of a value is not performed by joining the resources hash table with the triple table as the query processor can directly use the built-in MD5 hash function.

2.3 Property Tables

Storing RDF triples in a single large statement table presents a number of disadvantages when it comes to query evaluation. In most cases, for each set of triple patterns that is evaluated in the query, a set of self-joins on the table is necessary to evaluate the graph traversal. Since the single statement table can become very large, this can have a negative effect on query execution. While the horizontal storage of semantic data has been first introduced by Agrawal et al. in [3], the authors of Jena and Sesame propose different ways to alleviate this problem by introducing the concept of property tables in [8, 48]. Instead of building one large table for all occurrences of all properties, they propose two different strategies that can be distinguished into two different concepts: clustered and normalized property tables.

Clustered Property Tables The goal of clustered property tables is to cluster commonly accessed nodes in the graph together in a single table to avoid expensive joins on the data. In [49], the use of clustered property tables is proposed for data that is stored using the Dublin Core schema.[1] In the example of the dataset of the Berlin SPARQL benchmark shown in Fig. 5, one property table for all products and a statement table for all other triples are considered. For efficiency reasons, a product record and all affected triples can only appear in the property table.

The advantage of this storage format is that querying the database using triple patterns that are materialized in the property can be evaluated using simple filter predicates instead of performing self-joins on the statement table. Given the query in Listing 10.1, the transformation to SQL would require two joins, one for each triple

Product Property Table

Subject	Type	Label	NumericProperty1	aaa
Product12345	bsbm:Product	Canon Ixus 2010	NULL	...
...	...	...	...	...

Left-Over Triples

Subject	Predicate	Object
Producer1234	foaf:homepage	http://www.canon.com
...	...	...

Fig. 5 Example illustrating clustered property tables. Frequently co-accessed attributes are stored together

---

[1]http://dublincore.org/.

pattern. However, if the metadata defines that all triples of the type bsbm: Product are stored in a property table, this can be translated into a simple predicate evaluation as shown in Listing 10.2.

Listing 10.1 Example SPARQL Query
```
SELECT ?a
WHERE (?a rdf:type bsbm:Product),
      (?a bsbm:NumericProperty1 10)
```

Listing 10.2 Translation of the SPARQL Query in the Listing 10.1 to SQL using the clustered property table approach
```
SELECT t.subject FROM clustered_products as t
WHERE t.NumericProperty1 = 10;
```

The consequences of this approach are that the schema must be known in advance. If the properties for a materialized type change during runtime, this requires table alternations that are costly and often require explicit table-level locking. In addition, multi-valued attributes cannot be easily represented using a clustered property table. If multi-valued attributes must be considered, designer has to choose either to not materialize the path of the attribute or, if the sequence of the attribute is bounded, to include all possible occurrences in the materialized clustered property table.

Properties tables were also implemented in Jena2 [48] together with a statement table. In that context, multiple-values properties are clustered in a separate table. The system also allows to specify the type of the column in the underlying database system for the property value. This can be further leveraged for range queries and filtering. For example, the property *age* can be implemented as an integer, which can then be efficiently filtered.

Normalized Property Table In this second approach of property tables, the database choses to store triples based on the occurrence of single properties. In RDF for each triple s p o a number of triples s srdf:stype rdf:Property can be inferred.[2] Based on this knowledge, the database can now select a subset of triples that will be materialized in these special normalized property tables. All other triples will be stored in a general statement table. Figure 6 shows an example for this pattern where all instances of srdf :type | and s | rdfs:textttlabel| are separated in distinct tables. Abadi et al. present in [1] an extension to this model where all distinct occurrences for a single property will be stored in a decomposed way in a property table. In particular, this approach, called as SW-Store [1], stores RDF data using a fully decomposed storage model (DSM) [14]. In this approach, the triples table is rewritten into *n* two-column tables where *n* is the number of unique properties in the data. In each of these tables, the first column contains the subjects that define that property and the second column contains the object values for those subjects while the subjects that do not define a particular property are simply omitted from the table

---

[2]https://www.w3.org/TR/2014/REC-rdf11-mt-20140225/.

<rdf:type>

Subject	Object
Product12345	bsbm:Product

<rdfs:label>

Subject	Object
Product12345	Canon Ixus 2010
Producer1234	Canon

Statement Table

Subject	Predicate	Object
Product12345	bsbm:producer	bsbm-inst:Producer1234
Producer1234	foaf:homepage	http://www.canon.com
...	...	...

Fig. 6 Example illustrating RDF property tables. For each existing predicate one subject-object table exists

for that property. Each table is sorted by subject, so that particular subjects can be located quickly, and that fast merge joins can be used to reconstruct information about multiple properties for subsets of subjects. For a multi-valued attribute, each distinct value is listed in a successive row in the table for that property. One advantage of this approach is that while property tables need to be carefully constructed so that they are wide enough but not too wide to independently answer queries, the algorithm for creating tables in the vertically partitioned approach is straightforward and need not change over time. Moreover, in the property-class schema approach, queries that do not restrict on class tend to have many union clauses while in the vertically partitioned approach, all data for a particular property is located in the same table and thus union clauses in queries are less common. The implementation of SW-Store relies on a column-oriented DBMS, C-store [44], to store tables as collections of columns rather than as collections of rows. In standard row-oriented databases (e.g., Oracle, DB2, SQLServer, Postgres, etc.) entire tuples are stored consecutively. The problem with this is that if only a few attributes are accessed per query, entire rows need to be read into memory from disk before the projection can occur. By storing data in columns rather than rows, the projection occurs for free only where those columns that are relevant to a query need to be read (Fig. 7).

2.4 Query Execution

Query execution in relational RDF engines pushes all computational logic of RDF query evaluation to the database to achieve the best performance and leverage available optimization strategies. In [13], Chong et al. present a system that builds on an Oracle relational database management system. Instead of supporting the complete syntax of an RDF query language like SPARQL focuses on the most important subset of these languages which is matching RDF triples. Therefore, they implement a table

Product Property Table

Subject	Type	Label	NumericProperty1	aaa
Product12345	bsbm:Product	Canon Ixus 2010	NULL	...
...	...	...	...	...

Left-Over Triples

Subject	Predicate	Object
Producer1234	foaf:homepage	http://www.canon.com
...	...	...

Fig. 7 Example illustrating clustered property tables. In this example, only commonly used predicates are clustered in property tables

function [13] that allows to rewrite a set of RDF triple filters to SQL. Figure 8 shows an example of a translation of a simple triple pattern query into the matching SQL query that is then issued against the database system.

The general processing schema using the s |RDF_MATCH()| function is based on self-joins on the statement table. Since the runtime of queries increases with the size of this statement table, Chong et al. propose using the built-in materialized join-view functionality of the underlaying database management system. Therefore, they allow defining a set of materialized views in the form of s |subject-subject|, s |subject-property|, s |ubject-object|, s |property-property, s |property-object|, and s |object-object| as long as the storage requirements are met. In addition to these generic materialized join-views, they propose defining an additional set of subject- property matrix materialized join views, which are basically a adaptation of the previously-described clustered property tables from section "Clustered Property Tables". While this allows to increase the query performance, since for each materialized join in the matrix table one less self-join has to be performed, selecting the optimal properties to build the matrix materialized join view is non-trivial and heavily workload and data dependent.

The advantage of using database inherent materialized views is that they are fully integrated to the tuple life-time process and can thus be automatically dropped and

```
SELECT  t.a age              SELECT  t.a age
FROM TABLE(RDF MATCH(        FROM (
    '(?r Age ?a)',              SELECT  u1.UriValue a, u1.Type,
    RDFModels('reviewer'),      FROM IdTriples t1, UriMap u1
    NULL)) t                    WHERE t1.PropertyID=29 AND
WHERE  t.a < 25;                t1.ModelId=1 AND
                                u1.UriID = t1.SubjectID) t
                             WHERE  t.a < 25;
```

Fig. 8 The above listing shows a translation of the triple definition using the RDF_MATCH() table function into SQL

rebuilt if required. Materialized views are as well independent of semantic schema changes, because they only have to be rebuilt in case the RDF model changes and can be considered as secondary storage. Jena1 [30] simply rewrites SPARQL query to a single SQL query which is then executed over the statement table. In Jena2 [48], it is often impossible to construct one SQL query to satisfy all triple patterns over multiple tables (conjunction of statement and property tables), thus the system first generates a group of SQL queries, one for each set of patterns that can be evaluated with a singe table, and the second containing patterns that span tables. The two groups of queries are then joined in nested loops.

3   No-SQL Stores

In this section we present different approaches regrouped under the NoSQL umbrella. The systems we present were not specifically tailored to handle RDF data in the first place, though they were adapted. We chose systems that represent a variety of NoSQL system types: document databases, key-value/column stores, and query compilation for Hadoop. The systems described in this section were are available online[3] and were benchmarked in [15].

Data Storage and Indexing The first system described in [17] is JenaHBase combines the Jena RDF query engine and the HBase data storage system. Apache HBase[4] is an open source, horizontally scalable, row consistent, low latency, random access data store inspired by Google's BigTable [10]. It relies on the Hadoop Filesystem (HDFS)[5] as a storage back-end and on Apache Zookeeper[6] to provide support for coordination tasks and fault tolerance. Its data model is a column oriented, sparse, multi-dimensional sorted map. The basic logical unit in HBase is a column. Several columns compose a row, which is uniquely addressed by a row key. A set of rows composes a table and all rows within a table are sorted lexicographically by the key. Columns are grouped into column families. An important distinction is that column families have to be specified at schema design time, while columns can be dynamically added. All the coordinates and data in HBase are an uninterpreted array of bytes.

The four basic operations in HBase are similar to most NoSQL stores: Put, Get, Scan, Delete. A *Put* operation inserts or updates a single row in a table. If the row key already exists, the cell value is updated with the new one. Similarly a *Get* operation retrieves a single row and a *Delete* operation removes a single row. The *Scan* operation provides an iterator over a set of rows, which can be retrieved in increments as the iterator progresses. The scope of each of these operations can be

---

[3]http://ribs.csres.utexas.edu/nosqlrdf/.
[4]http://hbase.apache.org/.
[5]http://hadoop.apache.org/hdfs.
[6]http://zookeeper.apache.org/.

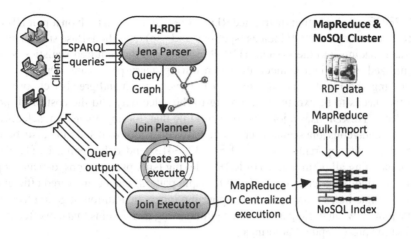

Fig. 9 The architecture of H2RDF system [34]

reduced by optionally specifying a *column family, column* or *timestamp*.[7] In practice, there are a number of benefits in using HBase for storing RDF. First, HBase has a proven track-record for scaling out to clusters containing roughly 1000 nodes.[8] Second, it provides considerable flexibility in schema design. This is important, because different applications have different access patterns. For example, one can choose a schema suited for read-writes when supporting RDF streaming or only for reads in a typical triple store environment. Finally, HBase is well integrated with Hadoop, a large scale MapReduce computational framework. This can be leveraged for efficiently bulk-loading data into the system and for running large-scale inference algorithms [46].

The HBase schema employed is based on the optimized index structure for quads presented by Harth et al. [23]. The authors use only triples so they build 3 index tables: *SPO, POS* and *OSP*. They map RDF URIs and most literals to 8-byte ids and use the same table structure for all indices: the row key is built from the concatenation of the 8-byte ids, while column qualifiers and cell values are left empty. This schema leverages lexicographical sorting of the row keys, covering multiple triple patterns with the same table. For example, the table *SPO* can be used to cover the two triple patterns: subject position bound i.e. *(s ? ?)*, subject and predicate positions bound i.e. *(s p ?)*. Additionally, this compact representation reduces network and disk I/O, so it has the potential for fast joins. As an optimization, they do not map numerical literals, instead they use a number's Java representation directly in the index. This can be leveraged by pushing down SPARQL filters and reading only the targeted information from the index. Two dictionary tables are used to keep the mappings to and from 8-byte ids (Fig. 9).

---

[7]In HBase *timestamp* adds an additional dimension to each cell besides *column family* and *column*.
[8]see e.g.,http://www.youtube.com/watch?v=byXGqhz2N5M.

H2RDF [33] is another distributed RDF store that has been built on top of HBase. H2RDF creates three RDF indices (spo, pos and osp) over the HBase store. During the data loading into the system, H2RDF collects all the required statistics which get utilized by the join planner algorithm during query processing. During query processing, the Join Planner iterates over the query graph and greedily chooses the join that needs to be executed, according to the selectivity and the cost of all possible joins. H2RDF uses a join executor module that, for any given join, selects the most advantageous join scenario, choosing between centralized and fully distributed, through the Hadoop framework. H2RDF+ [32, 34] extended the approach of H2RDF by maintaining all permutations of RDF indexing (spo, pso, pos, ops, osp and sop). Using this indexing scheme, all SPARQL triple patterns can be answered efficiently using a single index scan on the corresponding index. In addition, it guarantees that every join between triple patterns can be done using merge joins that can effectively exploit the pre-computed orderings.

Haque et. al in [21] proposes to use HBase with Apache Hive.[9] Hive is a SQL-like data warehousing tool that allows for querying using MapReduce. A property table is employed as the HBase schema. For each row, the RDF subject is compressed and used as the row key. Each column is a predicate and all columns reside in a single HBase column family. The RDF object value is stored in the matching row and column. Property tables are known to have several issues when storing RDF data [2]. However, these issues do not arise in the authors' HBase implementation. They distinguish multi-valued attributes from one another by their HBase timestamp. These multi-valued attributes are accessed via Hive's array data type. An advantage of property tables is the pre-computation of self-joins. In a traditional triple-table, a query with n triple patterns will require $n - 1$ self-joins. This holds true when each triple pattern contains the same subject. However, in a property table, a query with n triple patterns with the same subject will require a single row access, as opposed to $n - 1$ joins.

CumulusRDF[10] originally described and proposed in [26] by Ladwig and Harth is an RDF store which provides triple pattern lookups, a linked data server and proxy capabilities, bulk loading, and querying via SPARQL. The storage back-end of CumulusRDF is Apache Cassandra [27], a NoSQL database management system originally developed by Facebook [28]. Cassandra provides decentralized data storage and failure tolerance based on replication and failover. Cassandra's data model consists of nestable distributed hash tables. Each hash in the table is the hashed key of a row and every node in a Cassandra cluster is responsible for the storage of rows in a particular range of hash keys. The data model provides two more features used by CumulusRDF: super columns, which act as a layer between row keys and column keys, and secondary indices that provide value-key mappings for columns. The index schema of CumulusRDF consists of four indices [23] (SPO, PSO, OSP, CSPO) to support a complete index on triples and lookups on named graphs (contexts). The indices provide fast lockup for all variants of RDF triple patterns. The indices are

---

[9]http://hive.apache.org/query.

[10]http://code.google.com/p/cumulusrdf/.

stored in a "flat layout" utilizing the standard key-value model of Cassandra [26]. CumulusRDF does not use dictionaries to map RDF terms but instead stores the original data as column keys and values. Thereby, each index provides a hash based lookup of the row key, a sorted lookup on column keys and values, thus enabling prefix lookups.

The Rya system [35] has been implemented on top of Accumulo,[11] a distributed key-value and column-oriented NoSQL store that provides sorting of keys in lexicographical ascending order. Accumulo sorts and partitions all key-value pairs based on the Row ID part of the key. Rows with similar IDs are grouped into the same tablet/server for faster access. Rya stores the RDF triple (subject, predicate, and object) in the Row ID part of the Accumulo tables. In addition, it indexes the triples across three separate tables (SPO, POS, and OSP) that satisfy all the permutations of the triple pattern. These tables store the triple in the Accumulo Row ID and order the subject, predicate, object differently for each table. This solution utilizes the row-sorting scheme of Accumulo to efficiently store and query triples across multiple Accumulo tablets.

Couchbase[12] is a document-oriented, schema-less distributed NoSQL database system, with native support for JSON documents. Couchbase is intended to run in-memory mostly, and on as many nodes as needed to hold the whole dataset in RAM. It has a built-in object-managed cache to speed-up random reads and writes. Read and write operations first go to the in-memory object-managed cache, and defaults to fetching data from disk when the targeted document is not in cache. Updates to documents are first made in the in-memory cache, and are only later persisted to disk using the eventual consistency paradigm. The authors tried to follow the document-oriented philosophy of Couchbase when implementing our approach. To load RDF data into the system, they map RDF triples onto JSON documents. For the primary copy of the data, they put all triples sharing the same subject in one document (i.e., creating RDF molecules), and use the subject as the key of that document. The document consists of two JSON arrays containing the predicates and objects. To load RDF data, they parse the incoming triples one by one and create new documents or append triples to existing documents based on the triples' subject.

AMADA [5] has been presented as a platform for storing and querying RDF data which is full based on the Amazon Web Services (AWS) cloud infrastructure. AMADA operates in a Software as a Service (SaaS) approach, allowing users to upload, index, store, and query the RDF data. In particular, the RDF data is stored using Amazon Simple Storage Service (S3), the AWS store for large objects. AMADA builds its own data indexes using SimpleDB, a simple database system supporting SQL-style queries based on a key-value model that supports single-relation queries, i.e., no joins. In AMADA, The query execution is performed using virtual machines within the Amazon Elastic Compute Cloud (EC2). In practice, once a query submitted to the system, it get sent to a query processor module, running on an EC2 instance, which performs a look-up to the indexes in SimpleDB in order to find out

---

[11] https://accumulo.apache.org/.

[12] http://www.couchbase.com/couchbase-server/architecture.

the relevant indexes for answering the query, and evaluates the query against them. Results are written in a file stored in S3, whose URI is sent back to the user to retrieve the query answers.

Query Execution In JenaHBase, the Jena SPARQL engine executes a query over HBase. Jena represents a query plan through a tree of iterators. The iterators, corresponding to the tree's leafs, use the HBase data layer for resolving triple patterns e.g. *(s ? ?)*, which make up a Basic Graph Pattern (BGP). For joins, the authors use the strategy provided by Jena, which is indexed nested loop joins. As an optimizations, they pushed down simple numerical SPARQL filters, i.e., filters which compare a variable with a number, translating them into HBase prefix filters on the index tables. They used these filters, together with selectivity heuristics [45], to reorder subqueries within a BGP. In addition, they enabled joins based on ids, leaving the materialization of the ids after the evaluation of a BGP. Finally, they added a mini LRU cache in Jena's engine, to prevent the problem of redundantly resolving the same triple pattern against HBase.

For the second HBase approach (with Hive), at the query layer, the authors use Jena ARQ to parse and convert a SPARQL query into HiveQL. The process consists of four steps. Firstly, an initial pass of the SPARQL query identifies unique subjects in the query's BGP. Each unique subject is then mapped onto its requested predicates. For each unique subject, a Hive table is temporarily created. It is important to note that an additional Hive table does not duplicate the data on disk. It simply provides a mapping from Hive to HBase columns. Then, the join conditions are identified. A join condition is defined by two triple patterns in the SPARQL WHERE clause, $(s_1\ p_1\ s_2)$ and $(s_2\ p_2\ s_3)$, where $s_1 \neq s_2$. This requires two Hive tables to be joined. Finally, the SPARQL query is converted into a Hive query based on the subject-predicate mapping from the first step and executed using MapReduce.

CumulusRDF uses the Sesame query processor[13] to provide SPARQL query functionality. The Sesame [9] query processor translates SPARQL queries to index lookups on the distributed Cassandra indices; Sesame processes joins and filter operations on a dedicated query node.

The Couchbase implementation provides MapReduce views on top of the stored JSON documents. The JavaScript Map function runs for every stored document and produces 0, 1, or more key-value pairs, where the values can be null (if there is no need for further aggregation). The reduce function aggregates the values provided by the Map function to produce results. The query execution implementation is based on the Jena SPARQL engine to create triple indices similar to the HBase approach described above. The authors implement Jena's Graph interface to execute queries and hence provide methods to retrieve results based on triple patterns. They cover all triple pattern possibilities with only three Couchbase views, on $(?p?)$ $(??o)$ and $(?po)$. For every pattern that includes the subject, we retrieve the entire JSON document (equivalent to molecule [50, 51]), parse it, and provide results at the Java layer. For the query optimization, similar to the HBase approach above, selectivity heuristics are used. This approach is quite generic and relatively simple. They have not pushed

---

[13]http://www.openrdf.org/.

simple numerical filtering in the case of Couchbase. The authors also designed specific materialized views that are produced and updated using MapReduce functions. They produce a view for each type of query they execute. For example, they had a view on [$ProductType$, $propertyNumeric1$]. Since Couchbase allows range queries, with this index, they can obtain all products of a certain product type having a value greater than a given value. They can then retrieve all the documents that satisfy the condition and check if the other documents satisfy the remaining conditions of the query (e.g., for $ProductFeature1$ and $ProductFeature2$). This sits on a Java layer on top of their implementation.

4 Massively Parallel Processing for Linked Data

With ever larger data sets, distributing RDF data across multiple nodes becomes an important requirement. Instead of designing and implementing custom distributed RDF storage systems, one approach is to reuse existing infrastructure like Hadoop MapReduce and the Hadoop File System.

MapRedude is specifically designed to process large amounts of data. Processing RDF data with MapReduce based on a relational table-like storage model can be very demanding due to possibly high numbers of joins in RDF queries. If the joins produce large intermediate results, these must be distributed across the executor nodes requiring additional storage and network traffic. However, the advantage of Hadoop MapReduce and HDFS is that both systems are established on proven infrastructure systems being able to scale to thousands of nodes and almost arbitrary dataset sizes. As a consequence, optimizing data storage and query execution becomes a challenging and interesting aspect of native RDF database systems. The goal of this section is to present systems that leverage MapReduce and HDFS for large scale RDF storage and query execution.

4.1 Data Storage and Partitioning

Rohloff et al. in their work [38] propose a system called SHARD. While they do not introduce any novel storage model, they nevertheless expect data to be stored in a specific format (not ordinary triples). In the datafile, they expect each line to correspond to a star-like shape centering around a subject and all edges from this node. The files containing all the data is stored directly on HDFS without any specific partitioning scheme, by exploiting the replication factor of the underlying distributed file system.

Kurt owns car0 livesIn Cambridge
car0 a car madeBy Ford madeIn Detroit
Detroit a city Cambridge a city

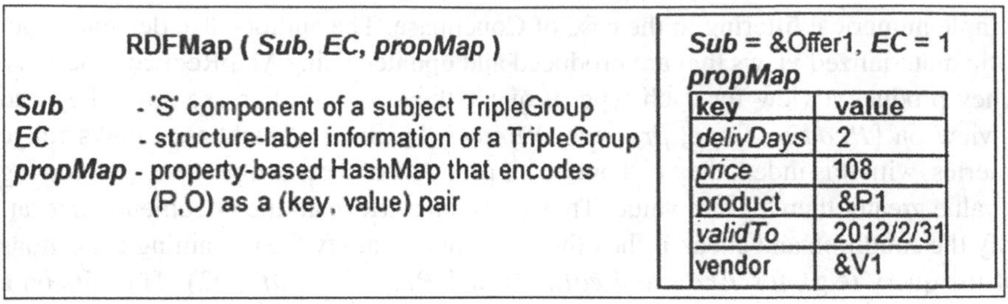

RDFMap (*Sub, EC, propMap*)		*Sub* = &Offer1, *EC* = 1	
		propMap	
Sub	- 'S' component of a subject TripleGroup	key	value
EC	- structure-label information of a TripleGroup	delivDays	2
propMap	- property-based HashMap that encodes	price	108
	(P,O) as a (key, value) pair	product	&P1
		validTo	2012/2/31
		vendor	&V1

Fig. 10 RAPID: RDFMap representing a TripleGroup [38]

The example above represents the following RDF triples:

<Kurt> <owns> <car0> .
<Kurt> <livesIn> <Cambridge> .
<car0> <a> <car> .
<car0> <madeBy> <Ford> .
<car0> <madeIn Detroit> .
<Detroit> <a> <city> .
<Cambridge> <a> <city> .

Ravindra et al. implement their system (RAPID+) [37] on top of Apache Pig.[14] They leverage a nested HashMap called RDFMap. Data is grouped in TripleGroup (implemented using a native bag data structure from Pig) around a subject which is a first-level key in the map, i.e., the data is co-located for a shared subject which is a hash value in the map. The nested element (i.e., the value from the previous map) (propMap) is a hash map with predicates as keys and objects as values. Figure 10 shows an example RDFMap. In fact, it forms a star-like substructures around subjects. They are in addition indexed on the first level by subject and on the second level by predicate.

The PigSPARQL [41] is another system that compiles SPARQL queries into the query language of Pig [31], a data analysis platform on top of the Hadoop framework. Pig language uses a fully nested data model and provides relational style operators (e.g., filters and joins). As illustrated in Fig. 11, in PigSPARQL, a SPARQL query is parsed to generate an abstract syntax tree which is then translated into a SPARQL algebra tree. Using this tree, PigSPARQL applies various optimizations on the algebra level such as the early execution of filters and the reordering of triple patterns by selectivity. Finally, PigSPARQL traverses the optimized algebra tree bottom up and generate for every SPARQL algebra operator an equivalent sequence of Pig Latin expressions. For the query execution, Pig automatically maps the resulting Pig Latin script into a sequence of Hdoop jobs. An advantage of PigSPARQL as intermediate layer that uses Pig between SPARQL and Hadoop is being independent of the actual Hadoop version or implementation details.

[14]https://pig.apache.org/.

Fig. 11 The architecture of
PigSPARQL system [41]

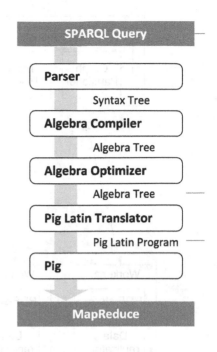

Huang et al. [24] propose a hybrid solution combining a single node RDF-store
(RDF-3X, see above) and Hadoop MapReduce to provide horizontal scalability. To
distribute triples across nodes, they leverage the METIS graph partitioning system.[15]
Hence, they co-locate triples forming a subgraph (star-like structure) on a particular
node. This enables to maximize the number of operations performed in parallel
on separate processing nodes avoiding expensive centralized cross-nodes joins. All
this allows reducing the amount of data that is transferred over the network for
intermediate results. Figure 12 shows the architecture of the system. Data is loaded
and partitioned on the master node while triples are distributed among workers. On
each node in the Hadoop cluster, there is an installation of the native RDF store
which receives and loads subsets of triples. The authors partition graph vertexes so
that each worked receives a subset of those vertexes that are close to each other in
the graph. Having all vertexes partitioned, the system assigns triples to worker in the
way that the triple is placed on the machine if its subject is among vertexes owned by
the worker. The process consist in two steps. First, the system divides vertices into
disjoint subsets. Then, it assigns triples to workers. Before partitioning vertices, the
system removes all triples where the predicate is `rdf:type`. Following this step, the
system prepares an input list of edges and vertices (an undirected graph) for METIS.
As an output from METIS, the system receives partitions of vertices that are disjoint.
Having all vertexes partitioned, the system starts placing triples on nodes in a cluster.
The basic idea is to place a triple on a partition if its subjectis among the vertices
assigned to the partition; this forms 1-hop star-like subgraph. This can be extended

---

[15]http://glaros.dtc.umn.edu/gkhome/views/metis.

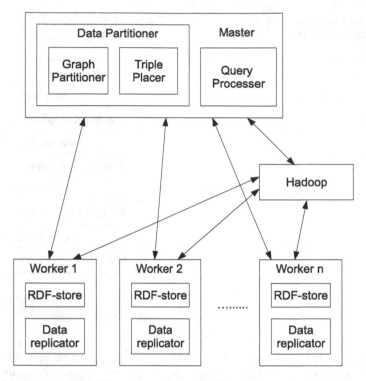

Fig. 12 MapReduce + RDF-3X: system architecture [24]

to further hops so that objects of triples are extended with triples considering them as subjects. The triple placement can also be performed on an undirected graph such that triples containing the vertex assigned to a partition as an object are also placed in it. Both of these extensions are trade-offs between duplicating data on worker nodes and query execution performance (the more extended the sub-graphs are, the less joins have to be performed in the final step).

CliqueSquare [16, 18] is another Hadoop-based RDF data management platform for storing and processing big RDF datasets. With the central aim of minimizing both the number of MapReduce jobs and the data transfer between nodes during query execution, CliqueSquare exploits the built-in data replication mechanism of the Hadoop Distributed File System (HDFS), each partition is has three replicas by default, to partition the RDF dataset in different ways. In particular, for the first replica, CliqueSquare partitions triples based on their subject, property, and object values. For the second replica, CliqueSquare stores all subject, property, and object partitions of the same value within the same node. Finally, for the third replica, CliqueSquare groups all the subject partitions within a node by the value of the property in their triples. Similarly, it groups all object partitions based on their property values. In addition, CliqueSquare implements a special treatment to triples where the property is *rdf:type* by translating them into an unwieldy large property partition. CliqueSquare then splits the property partition of *rdf:type* into several smaller partitions, according to their object value.

S2RDF[16] (*S*PARQL on *S*park for *RDF*) [42] introduced a relational partitioning schema for encoding RDF data called ExtVP (*Ext*ended *V*ertical *P*artitioning) that extends the Vertical Partitioning (VP) schema introduced by Abadi et al. [1] and uses a semi-join based preprocessing to efficiently minimize query input size by taking into account the possible join correlations between underlying encoding tables of the RDF data, join indices [47]. In particular, ExtVP precomputes the possible join relations between partitions (i.e. tables) of VP. The main goal of ExtVP is to reduce the unnecessary I/O, comparisons and memory consumption during executing join operations by avoiding the dangling tuples in the input tables of the join operations, i.e. tuples that do not find a join partner. Apparently, ExtVP comes at the cost of additional storage overhead in comparison to the basic VP encoding. Therefore, ExtVP does not use an exhaustive precomputations for all the possible join operations. Instead, an optional selectivity threshold for ExtVP can be specified to materialize only the tables where reduction of the original VP tables is large enough. This mechanism allows to control and reduce the size overhead while preserving most of its performance benefit. Therefore, during query execution, S2RDF can use the precomputed semi-join tables, if they exist, or alternatively uses the base encoding tables. S2RDF is built on top of Spark,[17] a general-purpose in-memory distributed data processing system, and execute SPARQL queries by translating them into SQL queries which are evaluated using Spark SQL [6], an SQL query processor based on Spark [52], over ExtVP encoding. S2RDF uses the Parquet[18] columnar storage format for storing the RDF data on the Hadoop Distributed File System (HDFS).

SparkRDF [11, 12] is another Spark-based RDF engine which splits the RDF graph into MESGs(Multi-layer Elastic SubGraph) based on relations (R) and classes (C) by creating 5 kinds of indexes(C,R,CR,RC,CRC) with different grains to support efficient evaluation for the diverse query triple patterns(TP). These index files are modeled as RDSG(Resilient Discreted SubGraph), a collection of in-memory semantic subgraph objects partitioned across machines. SPARQL queries are evaluated over these indices using a series of basic operators (e.g., filter, join). All intermediate results, which are also represented as the RDSG, are also maintained in the distributed memory to support further fast joins. SparkRDF uses a selectivity-based greedy algorithm to design a optimal execution order of query triple patterns (TPs) that aims of effectively reduce the size of intermediate results. In addition, it uses a location-free prepartitioning strategy that avoids the expensive shuffling cost for the distributed join operations. In particular, it ignores the partitioning information of index files, while repartitioning the data with the same join key to the same node. S2X (SPARQL on Spark with GraphX) engine [40] RDF engine has been implemented on top of GraphX [19], an abstraction for graph-parallel computation was added to Spark [52]. It combines graph-parallel abstraction of GraphX to implement the

---

[16]http://dbis.informatik.uni-freiburg.de/S2RDF.

[17]http://spark.apache.org/.

[18]https://parquet.apache.org/.

graph pattern matching part of SPARQL with data-parallel computation of Spark to build the results of other SPARQL operators. Similar approach has been followed by Goodman and Grunwald [20] for implementing an RDF engine on top the GraphLab framework, another graph-parallel computation platform [29].

Trinity.RDF [53] is a distributed in-memory RDF system which is based on Trinity [43], a distributed in-memory key-value store and a custom communication protocol based on the Message Passing Interface (MPI) standard. In particular, Trinity.RDF builds a graph interface on top of the key-value store. It randomly partitions an RDF graph across a cluster of commodity machines by hashing on the nodes. Thus, each machine holds a disjoint part of the graph. For any SPARQL query, a user submits his query to a proxy. Trinity.RDF performs parallel search on each machine where the machines may need to exchange data as a query pattern may span multiple partitions. In particular, Trinity.RDF decomposes a SPARQL query into a set of triple patterns and conduct a sequence of graph traversal to generate bindings for each of the triple pattern. The proxy generates a query plan and delivers the plan to all the Trinity machines which hold the RDF data. Each machine executes the query plan under the coordination of the proxy. When the bindings for all the variables are resolved, all Trinity machines return the bindings to the proxy where the final result is constructed and returned back to the user.

4.2 Query Execution

Ravindra et al. propose an intermediate algebra called Nested Triple Group Algebra (NTGA) to optimize their query execution process [37]. This approach minimizes the number of MapReduce cycles to answer the query. It also introduces algorithms to postpone the decompression of intermediate results so they can be kept in compact form, which in result reduces the number of I/O operations. The fundamental concept of NTGA is a TripleGroup [36], which is a group of triples sharing the same subject or object (star-like structure). Within one MapReduce operation, they pre-compute all possible star substructures, thus materializing all possible first-hop joins. Having computed all star-like structures, the system filters-out those stars that do not fulfill query constrains. In the next step, if necessary, the system joins stars. Figure 13 shows an example query and its execution plan. The first step (LOLoad) loads all data and at the same time also applies value-based filters on the data to avoid processing irrelevant triples. Then, during one Reduce operation, the LOCogroup operator groups triples and applies constrains on the groups, such that all irrelevant "stars" are filtered out. The last step in the flow is joining stars based on subjects or objects, which is achieved with the LORDFJoin operator.

Rohloff et al. introduce, in their SHARD system, a clause iteration algorithm [38] the main idea of which is to iterate over all clauses and incrementally bind variables and satisfy constrains (Fig. 14). During the first step, they identify all edges matching to a clause and remove duplicates. The output collection consists of keys (which are variable bindings from the clause) and NULL values. The following step identifies

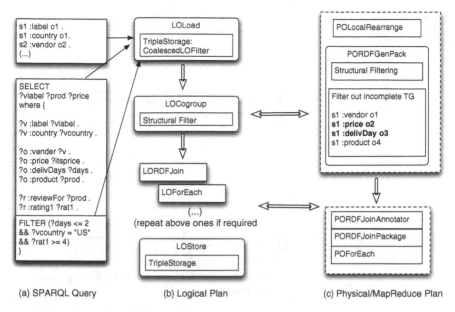

Fig. 13 RAPID+: query execution in [25]

Algorithm 1 SHARD: Iteration algorithm [38].

Require: triples, query
1: $mrOutput \longleftarrow NULL$
2: $mrInput \longleftarrow triples$
3: firstClauseMapReduce(mrInput, mrOutput, query.clause(0))
4: $boundVars \longleftarrow query.clause(0).getVars()$
5: **for** $i \leftarrow 1$ **to** $query.numClauses - 1$ **do**
6:     $mrInput \longleftarrow union(triples, mrOutput)$
7:     $curVars \longleftarrow query.clause(i).getVars()$
8:     $comVars \longleftarrow intersection(boundVars, curVars)$
9:     intermediateClauseMapReduce(mrInput,mrOutput, query.clause(i), comVars)
10: **end for**
11: $mrInput \longleftarrow mrOutput$
12: selectMapReduce(mrInput, mrOutput, query.select())
13: **return** $mrOutput$

---

edges matching to the remaining clauses (in the same way as previously). It also joins them with sub-graphs corresponding to the previously matched edges. The final step filters variables to obtain those requested in the SELECT clause. Algorithm 1 runs a `firstClauseMapReduce` MapReduce job to perform the first step. As an output, it returns sets of possible assignments to the variables of the first clause. `boundVars` tracks variables that were bound during this step. For the following example query:

```
SELECT ?person WHERE {
    ?person :owns ?car .
    ?car :a :car .
```

Fig. 14 SHARD: a schema
of the clause iteration
algorithm [38]

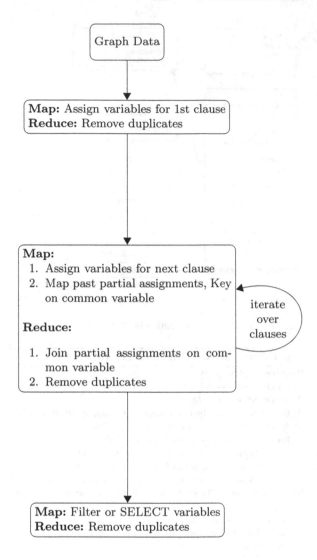

```
?car :madeIn :Detroit .}
```

variables `?person` and `?car` are bound and set to `boundVars`. The iterating step
runs the `intermediateClauseMapReduce` MapReduce job to perform the second step. It identifies triples matching to each clause (one by one) and then performs
joins over intermediate results of this step and all previous steps. For instance, after the
very first step, the system gets a set of bound variables `< (?car car0), null>`;
after the first iteration, it gets a map of variables bound during the second and
first steps `< (?car car0), (?person, Kurt)>`. The reduce phase combines
those two and returns `< (?car car0 ?person, Kurt)>`.

Huang et al. [24] take advantage of their partitioning scheme and of their back-end triplestore. Queries are decomposed into chunks executed in parallel and then reconstructed with MapReduce. They push as much of query processing as possible to the triplestore while the remaining part is consolidated by Hadoop. The system divides queries into two kinds. First, those that can be executed on a node, meaning that each node has sufficient data to generate complete result tuples. The second kind of query has to be decomposed into sub-queries executed on nodes, whose results are finally collected and joined at the master node.

The of query processing of CliqueSquare [16, 18] relies on a clique-based algorithm which produces query plans that minimize the number of MapReduce stages. The algorithm is based on the variable graph of a query and its decomposition into clique subgraphs. The algorithm works in an iterative way to identify cliques and collapse them by evaluating the joins on the common variables of each clique. The process ends when the variable graph consists of only one node. During the physical query execution, CliqueSquare exploits the different partitions RDF datasets to perform most common types of RDF queries locally at each node and minimize the data transfer through the network. In particular, it allows that most of the incoming queries to be processed in a single MapReduce job which enables a significant performance competitive advantage.

References

1. D.J. Abadi, A. Marcus, S. Madden, K.J. Hollenbach, Scalable semantic web data management using vertical partitioning, in *Proceedings of the 33rd International Conference on Very Large Data Bases, University of Vienna, Austria, September 23–27, 2007* (ACM, New York, 2007), pp. 411–422
2. D.J. Abadi, A. Marcus, S.R. Madden, K. Hollenbach, Scalable semantic web data management using vertical partitioning, in *Proceedings of the 33rd International Conference on Very Large Data Bases, VLDB '07* (2007), pp. 411–422
3. R. Agrawal, A. Somani, Y. Xu, Storage and querying of E-commerce data, in *VLDB 2001, Proceedings of 27th International Conference on Very Large Data Bases, September 11–14, 2001, Roma, Italy* (Morgan Kaufmann, Burlington, 2001), pp. 149–158
4. S. Alexaki, V. Christophides, G. Karvounarakis, D. Plexousakis, On Storing voluminous RDF descriptions: the case of web portal catalogs, in *WebDB* (2001), pp. 43–48
5. A. Aranda-Andújar, F. Bugiotti, J. Camacho-Rodríguez, D. Colazzo, F. Goasdoué, Z. Kaoudi, I. Manolescu, AMADA: web data repositories in the amazon cloud, in *21st ACM International Conference on Information and Knowledge Management, CIKM'12, Maui, HI, USA, October 29 - November 02, 2012* (2012), pp. 2749–2751. doi:10.1145/2396761.2398749
6. M. Armbrust, R.S. Xin, C. Lian, Y. Huai, D. Liu, J.K. Bradley, X. Meng, T. Kaftan, M.J. Franklin, A. Ghodsi, M. Zaharia, Spark SQL: relational data processing in spark, in *SIGMOD* (2015), pp. 1383–1394. doi:10.1145/2723372.2742797
7. C. Bizer, A. Schultz, The Berlin SPARQL benchmark. Int. J. Semant. Web Inf. Syst. **5**(2), 1–24 (2009)
8. J. Broekstra, A. Kampman, F. van Harmelen, Sesame: a generic architecture for storing and querying RDF and RDF schema, in *The Semantic Web - ISWC 2002, First International Semantic Web Conference, Sardinia, Italy, June 9-12, 2002, Proceedings* (Springer, Heidelberg, 2002), pp. 54–68

9. J. Broekstra, A. Kampman, F. Harmelen, Sesame: a generic architecture for storing and querying RDF and RDF schema, in *The Semantic Web ISWC 2002*, by eds. I. Horrocks, J. Hendler, Lecture Notes in Computer Science, vol. 2342 (Springer, Heidelberg, 2002), pp. 54–68. doi:10.1007/3-540-48005-6-7

10. F. Chang, J. Dean, S. Ghemawat, W.C. Hsieh, D.A. Wallach, M. Burrows, T. Chandra, A. Fikes, R.E. Gruber, Bigtable: a distributed storage system for structured data. ACM Trans. Comput. Syst. **26**(2), 4:1–4:26 (2008). doi:10.1145/1365815.1365816

11. X. Chen, H. Chen, N. Zhang, S. Zhang, SparkRDF: elastic discreted RDF graph processing engine with distributed memory, in *Proceedings of the ISWC 2014 Posters and Demonstrations Track a track within the 13th International Semantic Web Conference, ISWC 2014, Riva del Garda, Italy, October 21, 2014* (2014), pp. 261–264. http://ceur-ws.org/Vol-1272/paper_43.pdf

12. X. Chen, H. Chen, N. Zhang, S. Zhang, SparkRDF: elastic discreted RDF graph processing engine with distributed memory, in *IEEE/WIC/ACM International Conference on Web Intelligence and Intelligent Agent Technology, WI-IAT 2015, Singapore, December 6-9, 2015*, vol. I (2015), pp. 292–300. doi:10.1109/WI-IAT.2015.186

13. E.I. Chong, S. Das, G. Eadon, J. Srinivasan, An efficient SQL-based RDF querying scheme, in *Proceedings of the 31st International Conference on Very Large Data Bases, Trondheim, Norway, August 30 - September 2, 2005* (ACM, New York, 2005), pp. 1216–1227

14. G.P. Copeland, S. Khoshafian, A decomposition storage model, in *Proceedings of the ACM SIGMOD International Conference on Management of Data* (1985), pp. 268–279

15. P. Cudr–Mauroux, I. Enchev, S. Fundatureanu, P. Groth, A., Haque, A. Harth, F.L. Keppmann, D. Miranker, J. Sequeda, M. Wylot, NoSQL databases for RDF: an empirical evaluation, in *International Semantic Web Conference* (2013)

16. B. Djahandideh, F. Goasdoué, Z. Kaoudi, I. Manolescu, J. Quiané-Ruiz, S. Zampetakis, Cliquesquare in action: flat plans for massively parallel RDF queries, in *31st IEEE International Conference on Data Engineering, ICDE 2015, Seoul, South Korea, April 13-17, 2015* (2015), pp. 1432–1435. doi:10.1109/ICDE.2015.7113394

17. S. Fundatureanu, A scalable RDF store based on HBASE. Master's thesis, Vrije University (2012). http://archive.org/details/ScalableRDFStoreOverHBase

18. F. Goasdoué, Z. Kaoudi, I. Manolescu, J. Quiané-Ruiz, S. Zampetakis, Cliquesquare: flat plans for massively parallel RDF queries, in *31st IEEE International Conference on Data Engineering, ICDE 2015, Seoul, South Korea, April 13–17* (2015), pp. 771–782 (2015). doi:10.1109/ICDE.2015.7113332

19. J.E. Gonzalez, R.S. Xin, A. Dave, D. Crankshaw, M.J. Franklin, I. Stoica, GraphX: graph processing in a distributed dataflow framework, in *11th USENIX Symposium on Operating Systems Design and Implementation, OSDI '14, Broomfield, CO, USA, October 6–8, 2014* (2014), pp. 599–613. https://www.usenix.org/conference/osdi14/technical-sessions/presentation/gonzalez

20. E.L. Goodman, D. Grunwald, Using vertex-centric programming platforms to implement SPARQL queries on large graphs, in *Proceedings of the 4th Workshop on Irregular Applications: Architectures and Algorithms, IA3 '14* (IEEE Press, Piscataway, NJ, USA, 2014), pp. 25–32. doi:10.1109/IA3.2014.10

21. A. Haque, L. Perkins, *Distributed RDF triple store using HBase and Hive* (2012)

22. S. Harris, N. Gibbins, 3store: efficient bulk RDF storage, in *PSSS1 - Practical and Scalable Semantic Systems, Proceedings of the First International Workshop on Practical and Scalable Semantic Systems, Sanibel Island, Florida, USA, October 20, 2003* (CEUR-WS.org, 2003)

23. A. Harth, S. Decker, Optimized index structures for querying RDF from the Web, in *IEEE LA-WEB* (2005), pp. 71–80

24. J. Huang, D.J. Abadi, K. Ren, Scalable SPARQL querying of large RDF graphs. PVLDB **4**(11), 1123–1134 (2011)

25. H. Kim, P. Ravindra, K. Anyanwu, From sparql to mapreduce: the journey using a nested triplegroup algebra. PVLDB **4**(12), 1426–1429 (2011)

26. G. Ladwig, A. Harth, CumulusRDF: linked data management on nested key-value stores, in *The 7th International Workshop on Scalable Semantic Web Knowledge Base Systems (SSWS 2011)* (2011), p. 30
27. A. Lakshman, P. Malik, Cassandra: a decentralized structured storage system. SIGOPS Oper. Syst. Rev. **44**(2), 35–40 (2010). doi:10.1145/1773912.1773922
28. A. Lakshman, P. Malik, Cassandra: a decentralized structured storage system. SIGOPS Oper. Syst. Rev. **44**(2), 35–40 (2010). doi:10.1145/1773912.1773922
29. Y. Low, J. Gonzalez, A. Kyrola, D. Bickson, C. Guestrin, J.M. Hellerstein, Distributed GraphLab: a framework for machine learning in the cloud. PVLDB **5**(8), 716–727 (2012). http://vldb.org/pvldb/vol5/p716_yuchenglow_vldb2012.pdf
30. B. McBride, Jena: a semantic web toolkit. IEEE Int. Comput. **6**(6), 55–59 (2002)
31. C. Olston, B. Reed, U. Srivastava, R. Kumar, A. Tomkins, Pig Latin: a not-so-foreign language for data processing, in *Proceedings of the 2008 ACM SIGMOD International Conference on Management of Data* (ACM, New York, 2008), pp. 1099–1110
32. N. Papailiou, I. Konstantinou, D. Tsoumakos, P. Karras, N. Koziris, H2RDF+: high-performance distributed joins over large-scale RDF graphs, in *Proceedings of the 2013 IEEE International Conference on Big Data, 6-9 October 2013* (Santa Clara, CA, USA, 2013), pp. 255–263. doi:10.1109/BigData.2013.6691582
33. N. Papailiou, I. Konstantinou, D. Tsoumakos, N. Koziris, H2RDF: adaptive query processing on RDF data in the cloud, in *WWW (Companion Volume)*
34. N. Papailiou, D. Tsoumakos, I. Konstantinou, P. Karras, N. Koziris, H$_2$rdf+: an efficient data management system for big RDF graphs, in *International Conference on Management of Data, SIGMOD 2014, Snowbird, UT, USA, June 22-27, 2014* (2014), pp. 909–912. doi:10.1145/2588555.2594535
35. R. Punnoose, A. Crainiceanu, D. Rapp, SPARQL in the cloud using Rya. Inf. Syst. **48**, 181–195 (2015). doi:10.1016/j.is.2013.07.001
36. P. Ravindra, V.V. Deshpande, K. Anyanwu, Towards scalable RDF graph analytics on mapreduce, in *Proceedings of the 2010 Workshop on Massive Data Analytics on the Cloud* (ACM, New York, 2010), p. 5
37. P. Ravindra, H. Kim, K. Anyanwu, An intermediate algebra for optimizing RDF graph pattern matching on MapReduce, in *The Semantic Web: Research and Applications - 8th Extended Semantic Web Conference, ESWC 2011, Heraklion, Crete, Greece, May 29 - June 2, 2011, Proceedings, Part II* (Springer, Heidelberg, 2011), pp. 46–61
38. K. Rohloff, R.E. Schantz, Clause-iteration with mapreduce to scalably query datagraphs in the shard graph-store, in *Proceedings of the Fourth International Workshop on Data-intensive Distributed Computing* (ACM, New York, 2011), pp. 35–44
39. S. Sakr, G. Al-Naymat, Relational processing of RDF queries: a survey. SIGMOD Rec. **38**(4), 23–28 (2009). doi:10.1145/1815948.1815953
40. A. Schätzle, M. Przyjaciel-Zablocki, T. Berberich, G. Lausen, S2X: graph-parallel querying of RDF with GraphX, in *1st International Workshop on Big-Graphs Online Querying (Big-O(Q))* (2015)
41. A. Schätzle, M. Przyjaciel-Zablocki, T. Hornung, G. Lausen, Pigsparql: A SPARQL query processing baseline for big data, in *Proceedings of the ISWC 2013 Posters and Demonstrations Track, Sydney, Australia, October 23, 2013* (2013), pp. 241–244. http://ceur-ws.org/Vol-1035/iswc2013_poster_16.pdf
42. A. Schätzle, M. Przyjaciel-Zablocki, S. Skilevic, G. Lausen, S2RDF: RDF querying with SPARQL on spark. CoRR (2015). http://arxiv.org/abs/1512.07021
43. B. Shao, H. Wang, Y. Li, Trinity: a distributed graph engine on a memory cloud, in *Proceedings of the 2013 International Conference on Management of Data* (ACM, New York, 2013), pp. 505–516
44. M. Stonebraker, D.J. Abadi, A. Batkin, X. Chen, M. Cherniack, M. Ferreira, E. Lau, A. Lin, S. Madden, E.J. O'Neil, P.E. O'Neil, A. Rasin, N. Tran, S.B. Zdonik, C-Store: a column-oriented DBMS, in *Proceedings of the 31st International Conference on Very Large Data Bases (VLDB)* (2005), pp. 553–564

45. P. Tsialiamanis, L. Sidirourgos, I. Fundulaki, V. Christophides, P. Boncz, Heuristics-based query optimisation for SPARQL, in *Proceedings of the 15th International Conference on Extending Database Technology*

46. J. Urbani, S. Kotoulas, J. Maassen, N. Drost, F. Seinstra, F.V. Harmelen, H. Bal, Webpie: a web-scale parallel inference engine, in *Third IEEE International Scalable Computing Challenge (SCALE2010), held in conjunction with the 10th IEEE/ACM International Symposium on Cluster, Cloud and Grid Computing (CCGrid)* (2010)

47. P. Valduriez, Join indices. ACM Trans. Database Syst. **12**(2), 218–246 (1987). doi:10.1145/22952.22955

48. K. Wilkinson, C. Sayers, H.A. Kuno, D. Reynolds, Efficient RDF storage and retrieval in jena2, in *SWDB'03* (2003), pp. 131–150

49. K. Wilkinson, K. Wilkinson, Jena property table implementation, in *International Workshop on Scalable Semantic Web Knowledge Base Systems (SSWS)* (2006)

50. M. Wylot, P.C. Mauroux, Diplocloud: Efficient and Scalable Management of RDF Data in the Cloud (2015)

51. M. Wylot, J. Pont, M. Wisniewski, P. Cudré-Mauroux, dipLODocus[RDF] - short and long-tail RDF analytics for massive webs of data, in *International Semantic Web Conference* (2011), pp. 778–793

52. M. Zaharia, M. Chowdhury, M.J. Franklin, S. Shenker, I. Stoica, Spark: cluster computing with working sets, in *2nd USENIX Workshop on Hot Topics in Cloud Computing, HotCloud'10, Boston, MA, USA, June 22, 2010* (2010). https://www.usenix.org/conference/hotcloud-10/spark-cluster-computing-working-sets

53. K. Zeng, J. Yang, H. Wang, B. Shao, Z. Wang, A distributed graph engine for web scale RDF data. PVLDB **6**(4), 265–276 (2013). http://www.vldb.org/pvldb/vol6/p265-zeng.pdf

Exploratory Ad-Hoc Analytics
for Big Data

Julian Eberius, Maik Thiele and Wolfgang Lehner

Abstract In a traditional relational database management system, queries can only be defined over attributes defined in the schema, but are guaranteed to give single, definitive answer structured exactly as specified in the query. In contrast, an information retrieval system allows the user to pose queries without knowledge of a schema, but the result will be a top-k list of possible answers, with no guarantees about the structure or content of the retrieved documents. In this chapter, we present Drill Beyond, a novel IR/RDBMS hybrid system, in which the user seamlessly queries a relational database together with a large corpus of tables extracted from a web crawl. The system allows full SQL queries over a relational database, but additionally enables the user to use arbitrary additional attributes in the query that need not to be defined in the schema. The system then processes this semi-specified query by computing a top-k list of possible query evaluations, each based on different candidate web data sources, thus mixing properties of two worlds RDBMS and IR systems.

1 Exploratory Analytics for Big Data

While the term *Big Data* is most often associated with the challenges and opportunities of today's growth in data volume and velocity, the phenomenon is also characterized by the increasing *variety* of data [36]. In fact, data is collected in more and more different forms from increasingly heterogeneous sources. The spectrum of additional data sources ranges from large-scale sensor networks, over measurements from mobile clients or industrial machinery, to the log- and click-streams of ever

J. Eberius · M. Thiele (✉) · W. Lehner
Faculty of Computer Science, Database Technology Group,
Technische Universität Dresden, 01062 Dresden, Germany
e-mail: maik.thiele@tu-dresden.de

J. Eberius
e-mail: julian.eberius@tu-dresden.de

W. Lehner
e-mail: wolfgang.lehner@tu-dresden.de

© Springer International Publishing AG 2017
A.Y. Zomaya and S. Sakr (eds.), *Handbook of Big Data Technologies*,
DOI 10.1007/978-3-319-49340-4_11

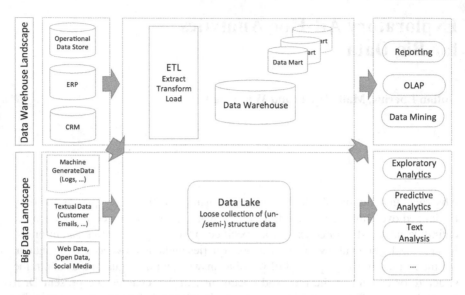

Fig. 1 The growing big data analytics landscape

more complex software architectures and applications. In addition, there is more publicly available data, such as social network data, as well as Web and Open Data. More and more organizations strive to efficiently harness all forms and sources of data in their analysis projects to gain new insights, or enable new features in their products.

However, conventional data warehouse infrastructures (upper part of Fig. 1) assume controlled ETL processes with well-defined input and target schemata, that define the data pipelines in the organization. The data sources typically are operational databases and common enterprise IT systems, such as Customer Relationship Management (CRM) and Enterprise Resource Planning (ERP) systems. Traditionally, the data sink in such an architecture has been the warehouse or data marts, whose schemata define what is immediately queryable for an analyst. If there is an ad-hoc information need that can not be satisfied with the current schema because external information has to be integrated, an intricate process has to be followed. Because the warehouse is a crucial piece of infrastructure, it is highly controlled: ad-hoc integration and analytics is not a feature that it is designed for. Still organizations today aim at generating value from all available data, which includes novel internal, but also increasingly external sources. Consider the lower part of Fig. 1, which depicts key changes to the traditional architecture. Beside the continued growth of data volume and its increasing heterogeneity, there are also changes at the data consumption side, where we can see a development towards agile and exploratory data analysis overcoming inflexible data warehouse infrastructures. Instead, new information management principles such as data lake [44, 47] MAD [12] arise, that aim to easily ingest, transform, and analyze data in an exploratory and agile manner. In addition, information needs are often ad-hoc or situational [42], or require the use of heterogeneous or unstructured data that are not integrated in a data warehouse. An instant integration of Big Data is not even desirable, as the future use cases of the data is not known. Implementing the data lake principle allows to store the mentioned variety of data sources. However, while a new wealth of data is available, the integration of a large variety of sources is still a complicated, laborious and mostly manual process that has to be performed by experts and that is required before queries on the combined data can be issued. This limits the ability of non-expert users to include more data into their analysis tasks in exploratory manner. Without additional tool support, the effort of data integration will most likely prevent those users from taking advantage of the wealth of Big Data today.

To illustrate the problems with today's data management tool let us consider the following scenario: Imagine you are a working in the marketing department of a company and you need to select customers that should be targeted by a specific campaign. To achieve this you have to group your customers according to different properties of the their home countries. While the customer master data is part of your enterprise data warehouse detailed country information, e.g. population, life expectancy, GDP, debt, etc., is missing. To get this data you need to identify relevant data sources manually, for example through a regular search engine. Then, the data has to be extracted and cleaned, i.e., converted into a form that is usable in a regular database. In a next step, it needs to be integrated into an existing database, which includes mapping of corresponding concepts, but may also include instance-level transformations. Only after this process is finished, the original query can be issued. Still, the result may not be what the user originally desired, or the user may want to see the query result when using a different Web data source. In this case, another iteration of the process is necessary. The overall process is extremely cumbersome and you will be likely miss a large fraction of all relevant data sources that are available on the Web [37].

To solve this problem we propose a novel method for *Top-k Entity Augmentation* (Sect. 2), that enables us to enrich a given set of entities with additional attributes based on a large corpus of heterogeneous data. This allows for ad-hoc data search and integration in Big Data environments with large collections of heterogeneous data such as data lakes or publicly available data. We extend this approach to an *Exploratory Ad-Hoc Analytics System* called DrillBeyond (Sect. 3), where we incorporate entity augmentation into traditional relational DBMS, to enable their efficient use in analytical query contexts. This enables users to issue analytical ad-hoc queries on a database while referencing data to be integrated as if it were already defined in the database, without providing specific sources or mappings. In case of our marketing scenario the country properties would be seamlessly integrated during query execution and instantly used to answer the user request.

In the remainder of this section, we will derive requirements (Sect. 1.1) for a novel data management architecture focused on ad-hoc integration of heterogeneous data sources with traditional data management systems. Given that, we will propose an architectural blueprint (Sect. 1.2) that also serves as the outline for the rest of the chapter.

Parts of the material presented in this chapter have already been published in [22] and [21].

1.1 Requirements

To put it succinctly, we identify two trends: first, we observe an ever increasing amount and diversity of available data sources, both inside organizations and outside. Organizations are stockpiling data in a quantity and of a variety not seen in the past. At the same time, developments such as the Open Data trend and sophisticated technologies for Web data extraction lead to higher availability of public data sources. As a consequence, there is an increasing demand to enhance and enrich data by integrating it with external data sources which introduces additional complexity. Second, there is a *growing demand for exploratory analytics and ad-hoc integration*, driven by the broadening spectrum of data users. This demands that the tools and processes of data integration become simpler, and able to cater to a larger audience. While data processing and analysis traditionally were the domain of IT departments and BI specialists, more and more users from other departments are becoming active data users and require easy access to all valuable data sources, both inside and outside organizational boundaries. In addition, the focus of data use has broadened: Traditional BI processes focused on defined data flows, typically from source systems to a data warehouse, and aimed at producing well-defined reports or dashboards. However, a new type of investigative data analysis, often associated with the role of the *data scientist*, increases the demand for ad-hoc integration of the variety of data sources mentioned above.

Still, even with all these new sources for data, the focus of most of the analysis tasks is on the respective core data of each organization. This most valuable data will still be stored in controlled data warehouse environments with defined schemata. Therefore, new ad-hoc data integration techniques need to take this analytics context into account. From these observations, we derive requirements for a novel data management architecture focused on fulfilling the need for exploratory analytics.

Exploratory Integration and Analytics The novel data sources discussed above do not lend themselves to classical data integration with centralized schemata and expert-defined mappings and transformations. Data from Web and Open sources, as well as from heterogeneous management platforms such as data lakes, should be integrated dynamically at the time it is needed by a user. The volume and variety of data sources in these scenarios makes single, centralized integration effort infeasible. Instead of consolidating all available sources as soon as they are available, sources should be retrieved and integrated based on a specific analytical need. In such a scenario, data search takes the place of exhaustive schemata covering all available information. We therefore propose methods and systems to enable exploratory analysis over large numbers of heterogeneous sources.

Minimal Up-front User Effort When developing methods for answering ad-hoc information needs, an important factor is the minimization of up-front costs for the user. In other words, our ambition is to minimize the user effort necessary before first results become available. Specifically, we want to reduce the reliance on IT personnel and integration experts, or the need to search data repositories or master integration tools before external sources can be included in a data analysis scenario. In the ideal case, a user should be able to declaratively specify the missing data and be presented with workable answers automatically. To facilitate this, we propose keyword queries that allow to specify the users' information need and which can be iteratively refined during an analytical session. While this approach minimizes the user effort it also introduces ambiguities that need to be resolved by the underlying retrieval system.

Trustworthy Automated Integration The two goals introduced above amount to reducing user effort and time-to-result in data search and integration. As a consequence, we propose novel methods for automating these highly involved processes. However, exact data integration has been called an "AI-complete" problem [30], and is generally considered not to be automatically solvable in the general case. In fact, all proposed methods return results with varying confidence values, instead of perfect answers, requiring human validation in many cases. We therefore introduce methods that automatize the data search and integration processes as much as possible, while also facilitating user understanding and verification of the produced results.

System Integration Finally, ad-hoc integration queries should not introduce an isolated, new class of systems into existing data management architectures. As discussed above, the wealth of Big Data collected by organizations or found on the Web is very promising. However, most analytical tasks will still focus on the core databases inside

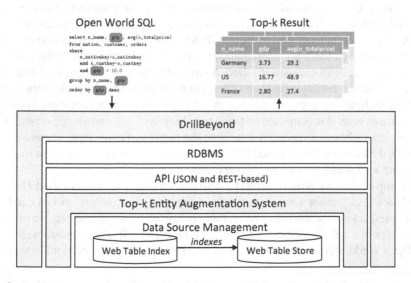

Fig. 2 Architecture overview of a combined database and information retrieval engine

organizations, with ad-hoc integrated sources supplementing, not replacing, them. Therefore, methods developed to support ad-hoc integration can not be deployed in a vacuum, but need to work hand in hand with systems managing core data. In detail, we propose an RDBMS/IR system hybrid system that allows querying and analyzing a regular relational database while using additional attributes for which the values could be found only in external data sources (Fig. 2).

1.2 Architecture Overview

In the following, we provide an architectural blueprint of a combined database and information retrieval engine in order to perform powerful queries over DBMS and heterogeneous data sources in an efficient, easy-to-use and seamless manner that fulfills the requirements outlined in Sect. 1.1. Our proposed architecture consists of a series of layers providing increasingly higher-level services. We assume a large collection of external tabular structured data, e.g. a corpus of Web tables, i.e., data-carrying tables extracted from the open Web, datasets published on an Open Data platform, or spreadsheets part of a corporate *data lake* [44]. Without limiting the generality of the proposed system we utilize Web tables for the remainder of this chapter, since they are freely available such as Dresden Web Table Corpus (DWTC) [19], that consists of 125M Web tables extracted from a public Web crawl. This corpus is indexed by an industry-standard document index server such as Solr[1] or Elastic-Search.[2]

On top of the inverted index we propose an *Entity Augmentation System* (EAS) that forms the basic building block for ad-hoc data integration. EAS's aim at extending a given set of entities with an additional, user-requested attribute that is not yet defined for them. In Sect. 2, we will present a novel approach to the problem which is especially suited for analytical uses cases. It is based on an extended version of the set cover problem, called *top-k consistent set covering* for which we introduce several algorithms. The EAS consists of a *Data Source Management System* providing storage and indexing facilities for Web tables, enabling the higher layers to retrieve raw Web tables based on keyword matches in data, schema or other metadata. Further it orchestrates several schema and instance matching systems, knowledge repositories and ranking schemes to create a candidate dataset D given an augmentation query. Finally, it includes a JSON-based REST API, which enables other systems to easily integrate with it and pose entity augmentation queries.

On top of that, we propose an RDBMS/IR system hybrid system called *DrillBeyond* (see Sect. 3), that allows querying and analyzing a regular relational database while using additional attributes not defined in said database. Therefore, we propose a new type of database queries, which we denote as *Open World Queries*. In short, in an Open World SQL, the user is allowed to reference arbitrary additional attributes

---

[1]http://lucene.apache.org/solr/.

[2]https://www.elastic.co/.

not defined in this database. An exemplary query is shown in Fig. 8. The goal is to enable users to specify information needs requiring external data declaratively, just as if only local data was used, without having to integrate data up-front. Leveraging the *Entity Augmentation System* a database can be augmented at query time with Web data sources providing those attributes. The system will not respond to Open World Queries with a single, perfect result, as it would be the case with normal database queries. Instead, it should produce a ranked list of possible answers, each based on different possible data sources and query interpretations. The users can then pick the result most suitable to their information need. To this end, our system tightly integrates regular relational processing with new data retrieval and integration operators that encapsulate our novel augmentation techniques. In Sect. 3 we describe the challenges in processing this new query type, such as *efficient processing of multi-result SQL queries*, and present how our novel DrillBeyond system solves them.

2 A Top-K Entity Augmentation System

In the previous section, we discussed the need to support ad-hoc information needs in analytical scenarios with tools for exploratory data search and lightweight integration. Specifically, we want to enable a user working on a dataset to effortlessly retrieve and integrate other relevant datasets from a large pool of heterogeneous sources. One compelling type of query in this context are so-called *entity augmentation queries*, or *EAQ*. These queries are defined by a set of entities, for example a set of companies, and an attribute that has been undefined so far for these entities, for example "revenue". The result of the query should then associate each of the entities with a value for the queried attribute, by automatically retrieving and integrating data sources that can provide them. We will call this attribute the *augmentation* attribute, and the system processing such queries Entity Augmentation System, or *EAS*. The user has to specify the augmented attribute just by a keyword, while the EAS decides on how to lookup data sources, how to match them to the existing data, and possibly how to merge multiple candidate sources into one result. This makes entity augmentation queries both powerful and user-friendly, and thus interesting for exploratory data analysis. Effectively, an EAS aims at automating the process of data search, as well as the following integration step.

In principle, any type of data source could be used for answering entity augmentation queries. For example, a large ontology such as YAGO [52] could answer some EAQ rather easily, though only if the augmentation attribute is defined in this knowledge base. In recent related work, several systems that process such queries on the basis of large Web table corpora have been proposed, for example *InfoGather* [56, 57], the *WWT* system [48, 50], and the *Mannheim Search Join Engine* [37]. An advantage of methods using tables extracted from the Web is that they offer more long tail information, and do not rely on the queried attribute being defined in a central knowledge base. Methods based on Web tables as their data source can, in principle, also be used with any other large collection of heterogeneous data sources. The techniques for automatic data search and integration introduced there could be applied to enterprise *data lakes* [44, 47] as well. In fact, many challenges, such as identifying relevant datasets in light of missing or generic attribute names, bridging semantic or structural differences, or eliminating spurious matches, have already been tackled by existing augmentation systems, which we review in Sect. 2.4.

Though solving these fundamental issues of data integration remains the most important factor for the success of an EAS, we argue that several other challenges are still unanswered in entity augmentation research. These challenges, are discussed in Sect. 2.1 and used to derive design requirements for a novel entity augmentation method in Sect. 2.2. Next, in Sect. 2.3, we will map the entity augmentation problem to an extended version of the *Set Cover* problem, which we call *Top-k Consistent Set Cover* and provide basic Greedy algorithm that solves this problem. Finally, we will survey related work in Sect. 2.4.

Parts of the material presented in this chapter have already been published in [22].

2.1 Motivation and Challenges

In this section, we will discuss an exemplary entity augmentation scenario based on a heterogeneous table corpus with partially overlapping sources. Our example scenario is depicted in Fig. 3, with the query represented as a table on the top, and the available candidate data sources below it. The query table consists of a set of five companies, and the augmentation attribute *"revenue"*. The candidate data sources depicted below the query table vary in coverage of the query domain, the exact attribute they provide, and their context. They are further annotated with their relevance with respect to the query, which is depicted as a numeric score on each source. In [22] we provide detailed scores and similarity measurements to compute the overall relevance score of the candidate data sources.

As an introductory example let us assume an algorithm that picks, for every queried entity, the respective value from the source with the highest relevance score. In our example, this naïve algorithm would pick values from the sources $S7$ for "Rogers", $S8$ for "AT & T", $S5$ for "Bank of China" and "China Mobile", and finally $S3$ for "Banco do Brasil", using the highest ranked available source for each queried entity. This means that the algorithm picks a large number of data sources, almost one distinct source for each entity. More sophisticated methods for pruning candidate tables and picking values from the remaining candidates can be used, such as correlating or clustering sources, mapping sources to knowledge bases, or quality-weighted majority voting (again, see Sect. 2.4). However, these methods do not fundamentally change the fact that the integration is performed on a by-entity basis, which leads to a large number of distinct sources being used to construct the result. It has been argued that in data integration, and especially selection of sources to integrate, "less is more" [18]. Choosing too many sources not only increases integration cost, but may even deteriorate result quality if low quality sources are added. For example,

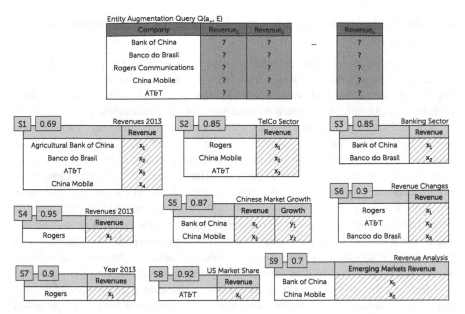

Fig. 3 Example augmentation scenario: query table and candidate data sources

adding S8 to the example result made the result inconsistent: In contrast to the other sources, this one is, in fact, concerned with US revenue only. This is just one intuitive example of a problem introduced by inadequate source selection and combination in the augmentation process. In the following, we will identify specific challenges that are insufficiently solved with existing by-entity fusion methods.

Trust and Lineage Our first argument is concerned with the usability of an entity augmentation result. In many application domains, the user can not blindly trust an automatic integration result, no matter how sophisticated the method used is. Rather, the result will serve as a starting point in a semi-automatic integration process, where the user will manually check the sources proposed by the system, correct errors made by the automated integration techniques, and even switch some of the sources for others. Choosing a large number of sources therefore increases the users' verification effort. We argue that existing fuse-by-entity models diminish *trust* and hinder understanding of data *lineage*, two properties that are important in the overall process of working with data, because the number of distinct sources they fuse is not considered in source selection. In this chapter, we therefore investigate methods that produce not only consistent augmentation results from several sources, but *minimal* augmentations, i.e., augmentations that use a minimal number of sources to facilitate the usage of the result. In the running example, such a result would be S2, S3, as it only uses two sources to augment all entities, even though the sources' average score is slightly worse than the score of the naïve solution introduced above. To summarize, when coercing a large number of data sources into one result, properties important for data analysis such as transparency, lineage and trustworthiness of the result are diminished.

Attribute Variations Another problem that we identified with related work is the underlying assumption of a single true value for any combination of entity and augmentation attribute. This single-truth idea, however, does not reflect the complex realities. For example, the augmentation attribute in our scenario was given simply as "revenue". However, the concept is actually more complex, with many variants such as "US revenue" or "emerging markets revenue" ($S8$ and $S9$) and derived attributes such as "revenue growth" ($S5$ and $S6$). Furthermore, many types of values come with temporal or spatial constraints, such as different years of validity, or may have been measured in different ways, for example using different currencies. Therefore, we argue that even when only sources of high relevance are picked, they may be correct and of high quality when considered on their own, but still do not form a consistent result for the overall query domain when combined. The differences in precise semantics can, in most cases, not be decided based on the extracted attribute values, but on the level of data sources, for example by considering the context of a table. Even though better methods for creating consistent augmentations for some important dimensions such as time and unit of measurement have been proposed [50, 57], source consistency is not considered as a general dimension in entity augmentation so far. To summarize the argument: due to the existence of subtle *attribute variations*, the notion of source consistency needs to be taken into account when combining several sources to a single augmentation result.

Unclear User Intent Extending our argument based on the intricacies of attribute variations, we will discuss an additional challenge: the problem of *unclear user intent*. In entity augmentation queries, the information need of the user is relatively underspecified, especially when compared to queries on a database with a defined schema. Even though entity augmentation also operates on structured data, for example on a large-scale Web table corpus, the user is still forced to pose queries on data whose extent or schema is unknown to him or her. Forming precise queries may therefore not always be possible, especially in the presence of attribute variations. In turn, the entity augmentation system may not always be able to pinpoint the exact attribute the user is interested in. To give one example, in the scenario in Fig. 3 it is unclear whether the user is interested in any specific year, or just in the highest ranked sources. In this example, a solution based on the sources $S1$, $S4$ may be more useful than the higher-ranked solution $S2$, $S3$ proposed above, because both sources explicitly state a certain year of validity. However, which solution the user would prefer can not be decided from the query alone.

Exploratory Search Even if the user can specify a precise query, in the case of situational or ad-hoc analysis, they may not even yet know which attribute variant is the most relevant. In those situations, the underspecified nature of entity augmentation queries may even be turned into an advantage. For example, a user may want to stumble over new aspects of his or her information need, similarly to the way it may happen with a Web search engine. In the example scenario, the user may have queried only for "revenue", but a solution showing "revenue growth" based on sources *S*5, *S*6 may, in some situations, give the ongoing analysis process a new perspective or direction. However, such serendipitous query answering is not supported with current augmentation systems. One partial exception is the Mannheim Search Engine [37], which allows so-called unconstrained queries, in which the system returns augmentations for all attributes it can retrieve from all available sources. However, this may leave the user with an unfocused, large and hard-to-comprehend result that is not connected to the information need at hand. In other words, the *exploratory* nature of entity augmentation queries is not done justice in current approaches.

Error Tolerance Finally, we note that all existing augmentation systems are based on techniques for automated schema matching, entity resolution and automated relevance estimation. All these components by themselves have a certain margin of error that, no matter how sophisticated the underlying methods become, can never be fully eliminated. In combining these systems to higher-level service such as entity augmentation, the individual errors will even multiply. Still, none of the existing systems, while of course striving to optimize their precision in various ways, offer explicit support for *tolerating* possible errors.

2.2 Requirements

Having introduced and discussed five significant open challenges in entity augmentation, we now want to discuss *requirements* for a novel entity augmentation method that alleviates these challenges. First, let us discuss the challenges *lineage* and *attribute variants*. We already discussed that using a large number of sources impedes the user's understanding. Further, we discussed that, to detect and correctly exploit the existing variants of the queried attribute, we need to take consistency between sources into account. Therefore, we investigate methods that produce both augmentation results that are both *consistent* and *minimal*, i.e., augmentations that use a minimal number of distinct sources, with each of them representing the same attribute variant. To determine this consistency of datasets, it is possible to measure the similarity between their attribute names, compare global properties of their value sets, or analyze the associated metadata such as dataset titles and context. We will discuss our notion of consistency in more detail and also give formal definitions in the following sections.

Now let us consider two further challenges: *unclear user intent* and *error tolerance*. Both of these challenges result from various forms of uncertainty: The first from uncertainty about the user intent, the second from uncertainty of the utilized basic methods such as schema matching and entity resolution, and the third from uncertainty in the sources. Information retrieval systems solve this problem by presenting not one exact, but a top-k list of possible results. For example, errors in relevance estimation are tolerable, as long as some relevant documents are included in the top-k result list. Unclear user intent or ambiguities in the query keywords can be resolved by returning documents based on multiple interpretations of the query, instead of focusing exclusively on the most likely interpretation [3]. Furthermore, the challenge of *exploratory search* can also be solved in a top-k setting by means of result list diversification, as is common practice in Web search [5] or recommender systems [58]. We argue that for entity augmentation a similar argument can be made: It is advantageous to provide not only one solution, but allow the user to choose from a ranked list of alternative solutions. In other words, we aim at extending entity augmentation to *diversified top-k entity augmentation*.

Let us reconsider our running example shown in Fig. 3. Instead of returning only one result based on $S2, S3$ as discussed above, one alternative would be a result based on $S1, S4$. It has a worse average score, but has clearly marked year in both sources, which may be more useful for the user on manual inspection, because of the clearly marked year information in the context.

Another aspect are attribute variations, which, due to the exploratory nature of entity augmentation queries, may also be of interest to the user. An example would be a third result based on $S6$ and the second column of $S5$ which represents changes in revenue instead of absolute revenue. Yet another exploratory result could be comprised of just $S9$, which does not cover all entities, but might give the users' analysis a new direction.

Note however, that we want to generate solutions that are real alternatives, such as the three examples above. Because of copying effects on the Web [13, 38], using only the most relevant sources for creating all k augmentation results however would lead to many answers being created from structurally and semantically similar sources. Furthermore, because we fuse results from several sources, naïve solutions would use the same, most relevant sources multiple times in various combinations, leading to *redundancy* in the result list. In the example, one such redundant result would be $S1, S7$. However, since $S7$ differs only superficially from source $S4$, the results $S1, S4$ and $S1, S7$ are very similar. We want to avoid slight variations of one solution as this would add little information to the top-k result. A meaningful top-k list needs to consider a *diverse* set of sources, exploring the space of available data sources while still creating correct and consistent answers. This has been recognized a long time ago in information retrieval [10], but has recently been explored for structured queries [14, 32], and even for Web table search [46], which is highly related to entity augmentation. We will define the notion of result diversity for our specific problem, as well as our means to achieve it, in the following sections.

In a nutshell, we can derive the following two goals: We aim at providing a *diversified top-k* list of alternative results, which are composed of *consistent and minimal* individual results.

2.3 Top-k Consistent Entity Augmentation

In this section we will describe our novel method of *Top-k Consistent Entity Augmentation*. Initially, we will formalize augmentation queries and the optimization objectives of our method in Sect. 2.3 and introduce top-k consistent *set covering* as an abstract framework for solving our problem in Sect. 2.3.

Entity Augmentation Queries We will now formalize our notion of Top-k Consistent Entity Augmentation, and introduce the optimization objectives that we aim at. First, consider a general entity augmentation query definition.

Definition 1 *(Entity Augmentation Query)* Let $E(a_1, \ldots_a n)$ denote a set of entities with attributes $a_1, \ldots, a_n$, and a_+ denote the augmentation attribute requested by the user. Then, an augmentation query is of the form:

$$Q_{EA}(a_+, E(a_1, \ldots, a_n)) = E(a_1, \ldots, a_n, a_+)$$

In other words, the result of such a query is exactly the set of input entities with the augmented attribute added. To create this new set of entities, the EAS has to retrieve values for attribute a_+ for each entity $e \in E$, which we will denote v_e. These values will be retrieved from a corpus of data sources $\mathcal{D}$ managed by the EAS, from which it selects a relevant subset D for each query Q_{EA}.

Sources $d \in D$ can provide a_+ values for some subset of E, denoted $\text{cov}(d)$, i.e., they cover E only partially in the general case. Individual values provided by a source d for an entity e are denoted $d[e]$ with $e \in \text{cov}(d)$. Given a heterogeneous corpus of overlapping data sources, an augmentation system will potentially retrieve multiple values for an entity e, the set of which we will denote by $V_e = \bigcup_{d_i \in D} d_i[e]$. Finally, the EAS assigns each data source a relevance score $rel : D \to [0, 1]$ with respect to the query. To determine this relevance score, various measures can be combined. Examples include the similarity between the queried attribute name and the respective attribute's name in the data source, or the quality of the match between the queried instances and those found in the data source. In addition, global measures for the quality, such as the PageRank of the source page, can be integrated.

As we described in Sect. 2.1, most systems from literature assume that they can reconcile the set of values V_e into a single correct augmentation value v_e for each entity. An example for such a fusion method would be majority voting, or clustering values and then picking a representative from the highest ranked cluster as in [37, 56].

As motivated in Sect. 2.2, our notion of an entity augmentation query differs in several aspects: First, instead of individual values, it picks subsets of sources that cover E, and second, it returns an ordered list of alternative solutions. In other words, its basic result is a list of top-k alternative *selections of sources*.

Definition 2 *(Top-k Source Selections)* Given a set D of relevant data sources, and a number k of augmentations to create, a top-k Source Selection is defined as:

$$Q_{EA}(a_+, E, k) = [c_1, \ldots, c_k \mid c_i \subset D \wedge \text{cov}(c_i) = E] \qquad (1)$$

We call one such set c_i a *cover* or an *augmentation*, and the list of these augmentations the query result.

Fig. 4 MaxRelevance

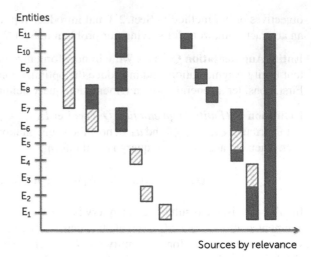

Fig. 5 MinSources

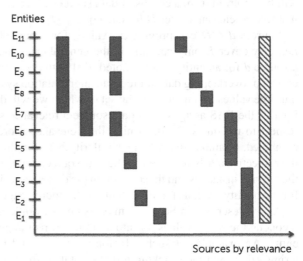

Definition 3 *(Cover/Augmentation)* A cover is an ordered subset of D that covers E, i.e., $c = [d_i, \ldots, d_x]$ with $\bigcup_{d \in c} \mathrm{cov}(d) = E$. If multiple data sources provide values for a distinct e, i.e., if $\exists_e (e \in \mathrm{cov}(d_i) \cap \mathrm{cov}(d_j))$, the augmented value for e is decided by the order of the sources in c, i.e., $v_e = d_i[e]$ with $i = min(\{i \mid e \in d_i \wedge d_i \in c\})$.

As discussed in Sect. 2.2, the aim is to enable the user to choose the most promising augmentation from the ranked list of alternatives. This leads to the question of how to construct these subsets $c_i \subset D$, in order to create a valuable list of alternatives for the user. We will now introduce the individual dimensions of this problem, *relevance*, *minimality*, *consistency* and *diversity*, discussing exemplary baseline strategies that optimize for one of each dimension.

Fig. 6 MaxConsistency

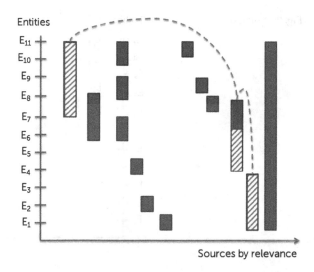

Entities

Sources by relevance

Relevance One naïve baseline strategy, which we call *MaxRelevance*, is depicted in Fig. 4. Starting from the highest ranked data source, it picks all the values it provides for entities that do not have a value yet, then continues to the next most relevant source, according to the relevance function introduced above. While this strategy obviously maximizes the average relevance of the created augmentation, a large number of distinct sources might be picked. This makes it harder for the user to understand the query result and assess its quality, and also has a high chance of producing inconsistencies between sources.

Minimality A naïve approach to solve the latter problem would be to prioritize data sources with large coverage of the queried entities. This strategy, called *MinSources*, is illustrated in Fig. 5. As is illustrated in this particular example, while solutions created this way use a minimal number of distinct sources, the other objectives, such as relevance, can be arbitrarily bad.

Consistency Next, consider the strategy *MaxConsistency*, in which sources are chosen based on a measure of source similarity, i.e., a function sim : $D \times D \rightarrow [0, 1]$, which is depicted using dashed arrows in Fig. 6. This function captures our notion of attribute variant consistency as discussed in Sect. 2.1. Utilizing such a function to guide source selection will increase the overall consistency of the created augmentations, but will create augmentations that are not necessarily minimal nor highly relevant. It is calculated from measures such as the similarity between the two data source's attribute names, their value sets, and by comparing the associated metadata such as dataset titles and context.

Diversity In addition, we will have to devise a method that is able to create multiple meaningful alternative solutions. A naïve solution would be to create one cover c from sources D, and then iteratively set $D' = D \setminus c$ and create the next cover from D'. This approach, called *NoReuse*, is illustrated in Fig. 7. It has two problems:

Fig. 7 NoReuse

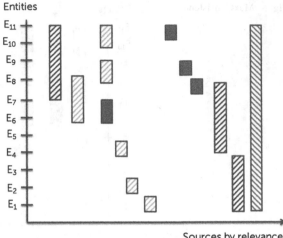

Firstly, each data source can be used in only one alternative, even though several combinations of good sources might be possible. Secondly, just prohibiting reuse of specific datasets does not necessarily lead to diversified solutions, as there may be data sources so that $\exists_{d_i,d_j} |d_i \neq d_j \wedge \text{sim}(d_i, d_j) \approx 1.0$. This occurs, for example, due to frequent copying-effects on the Web [13, 38]. Since we aim at minimizing the pairwise similarity of covers in the query result, we introduce a similarity function $\text{sim} : D \times D \rightarrow [0, 1]$ that compares covers instead of data sources. This lifts the similarity function to the domain of covers $\text{sim}_{\mathcal{A}} : C \times C \rightarrow [0, 1]$, whereas the aggregation function $\mathcal{A}$ can be an *average* or *max*.

We consider these four dimensions to be the decisive factors for a useful top-k entity augmentation result. What we therefore need, is a strategy that creates complete covers, while jointly optimizing all mentioned objectives.

Definition 4 *(Top-k Consistent Entity Augmentation)* A top-k Consistent Entity Augmentation query produces a top-k Source Selection (Definition 2) that is optimized with respect to the relevance, minimality, consistency, and diversity objectives.

In the next section, we will introduce our algorithmic approach to processing top-k Consistent Entity Augmentation queries.

Ranked Consistent Set Covering We propose a new approach for constructing entity set augmentations by modeling it as an extended form of the *Weighted Set Cover Problem*, one of Karp's original 21 NP complete problems [34].

Definition 5 *(Weighted Set Cover)* Given a universe of elements U and a family of subsets of this universe S, each associated with a weight w_i, the Weighted Set Cover problem is to find a subset $s \subset S$ with $\bigcup_s = U$, such that $\sum_{i \in s} w_i$ is minimized.

Intuitively speaking, the aim is to *cover* all elements in U using sets from S with minimal cost. In our problem domain, the algorithm input consists of a set of entities

E that are to be augmented, corresponding to U in the original problem, and a set of data sources $D = \{d_1, \ldots d_n\}$, as retrieved and matched by the underlying entity augmentation system, corresponding to S. The relevance score assigned to each datasource by rel(d) is used in place of the weights w.

So far, we could trivially map our problem to the well known Set Cover problem. Specifically, the *Relevance* and *Minimality* objectives defined in section "Entity Augmentation Queries" correspond closely to the objective $\sum_{i \in s} w_i$ in the set cover problem. Still, there are some crucial differences: In contrast to the original problem, where only a *single* minimal cover is required, the output we aim for is a ranked list of covers, denoted $C = [c_1, \ldots, c_n]$. Furthermore, as illustrated in section "Entity Augmentation Queries", the entity augmentation use case does not only require small covers with high individual relevance, but *consistent* covers, as defined in the *consistency* objective. And lastly, we also introduced the *diversity* objective, i.e., the covers should not consist of the same or similar datasets throughout the top-k list, but be complementary alternatives.

We will now incrementally develop our proposed algorithms for top-k consistent set covering. We start from the well known greedy algorithm for the Weighted Set Cover problem, which, given a universe U, a set of sets S with weights w, and a set of yet uncovered elements F, iteratively picks the set:

$$\underset{S_i \in S}{\operatorname{argmin}} \frac{w_i}{|S_i \cap F|} \qquad \textit{(Greedy Set Cover Algorithm Step)}$$

The algorithm chooses sets S_i until $F = \emptyset$, at which point a complete cover has been formed. Although the greedy algorithm does not create optimal covers, it is still the most frequently employed algorithm for the set covering problem. In fact, it has been shown that the algorithm, achieving an approximation ratio of $H(s') = \sum_{k=1}^{n} \frac{1}{k}$ is essentially the best possible polynomial-time approximation algorithm for the set cover problem.

Coverage and Relevance. We therefore also initially base our algorithm on the greedy set covering algorithm. With an initially empty cover c and a free entity set $F = E$, we can use the original greedy Set Cover algorithm to produce an ordered subset of D, by picking in each iteration the dataset d that maximizes:

$$\underset{d_i \in D}{\operatorname{argmax}} \operatorname{rel}(d_i) \cdot |\operatorname{cov}(d_i) \cap F| \qquad (2)$$

until $F = \emptyset$. Note that we maximize scores instead of minimizing weights as this is more intuitive for the problem domain.

An augmentation constructed in this way would roughly correspond to a middle ground strategy between the *MaxRelevance* and *MinSources* strategies discussed in section "Entity Augmentation Queries". This implies, however, that it can potentially create augmentations from very heterogeneous data sources.

Cover Consistency. To counteract this effect, we explicitly model consistency between the datasets that make up a cover. We utilize the similarity function between datasets sim : $D \times D \rightarrow [0, 1]$, as defined in section "Entity Augmentation Queries", which models the consistency between data sources. Given an initially empty cover c and an aggregation function $\mathcal{A}$ such as *average* or *max*, we can greedily generate covers using consistent datasets by picking in each iteration the dataset d that maximizes:

$$\underset{d \in D}{\text{argmax}} \, \text{rel}(d) \cdot |\text{cov}(d) \cap F| \cdot \text{sim}_{\mathcal{A}}(d, c) \tag{3}$$

This means we encourage picks of data sources that are similar to data sources that were already picked for the current cover. We assume as a special case that $\text{sim}_{\mathcal{A}}(d_i, \emptyset) = 1$, which implies that the first data source chosen will be the same as in regular set covering. Subsequent choices on the other hand will be influenced by already selected sources. This also implies that datasets with a low relevance or coverage, that are not picked initially, may still be chosen in a later iteration, if they fit well with those chosen so far. Since we require $|\text{cov}(d) \cap F|$ to be greater than zero, the algorithm will still make progress with every step, as only datasets that provide at least one new value can be selected.

Using objective function (3), the algorithm picks datasets to create covers that are not only highly relevant to the query, but also fit well together according to sim : $D \times D$.

However, using only this objective function, there is still no intuitive way of creating useful top-k augmentations. The naïve approach re-running the same algorithm with $D \setminus c$ as the set of candidate data sources would not lead to useful alternative solutions, as discussed in Sect. 2.3.

Top-k Results and Diversity. Let C denote the set of previously created covers, with $|C| \geq 1$. This set could be initialized with a single cover created, for example, using the greedy algorithm and objective function (3). Our core idea is to perform consecutive runs of the greedy algorithm using the same input datasets, with each run placing greater emphasis on datasets that are dissimilar to datasets picked in previous iterations, i.e., dissimilar to datasets in $\bigcup_C$. Implementing this idea naïvely however, for example by dividing function (3) by $\sum_{d_i \in \bigcup_C} \text{sim}(d, d_i)$ does not yield the expected results. While the second iteration might then choose datasets from a different part of the similarity space than the first iteration, the term becomes increasingly meaningless with more iterations as $\bigcup_C$ grows. This is because newly picked datasets are compared to a larger and larger subset of the candidate set D, leading to an increasingly uniform value for $\sum_{d_i \in \bigcup_C} \text{sim}(d, d_i)$.

Instead, we introduce a more complex dissimilarity metric based on individual entities in E and the datasets that were used to cover them in previous iterations. We define a function *coveredBy(e, C)* which yields the datasets that were used to augment the entity e in covers C created in previous iterations.

$$coveredBy(e, C) = \{d \mid \exists_c \in C : d \in c \wedge e \in \text{cov}(d)\} \tag{4}$$

We can then define our final scoring function as

Algorithm 1 Top-k consistent set covering: Greedy

---

function GREEDY- TOPK- COVERS(k, E, D)
   $C \leftarrow \emptyset$
   $U \leftarrow \begin{pmatrix} 0 \dots 0 \\ \vdots \ddots \vdots \\ 0 \dots 0 \end{pmatrix}_{|E| \times |D|}$ ▷ Usage matrix
   while $|C| < k$ **do**
      $c \leftarrow$ COVER(E, D, U)
      for all $(e \rightarrow d) \in c$ **do**   ▷ Update Usage Matrix
         $U[e, d] \leftarrow U[e, d] + 1$
      if $c \notin C$ **then**   ▷ Remove duplicates
         $C \leftarrow c$
   return C

function COVER(E, D, U)
   $c \leftarrow \emptyset$
   $F \leftarrow E$ ▷ Free set, uncovered entities
   while $|F| > 0$ **do**
      $d \leftarrow \text{argmax}_{d \in D} \frac{\text{rel}(d) \cdot |\text{cov}(d) \cap F| \cdot \text{sim}_A(d, c)}{\text{REDUNDANCY}(d, D, F, U)}$
      for all $e \in F \cap \text{cov}(d)$ **do**
         $F \leftarrow F \setminus e$ ▷ Update free set
         $c \leftarrow c \cup (e \rightarrow d)$ ▷ Update cover
   return c

function REDUNDANCY(d, D, F, U)
   $r, norm = 0, 0$
   for all $e \in F \cap \text{cov}(d)$ **do**   ▷ Coverable by d
      $u \leftarrow U[e]$ ▷ Sources used to cover e
      $r \leftarrow r + \sum_{i=0}^{|u|} u[i] * \text{sim}(d, D[i])$
      $norm \leftarrow norm + \sum u$
   return $\frac{r}{norm}$

$$\text{argmax}_{d \in D} \frac{\text{rel}(d) \cdot |\text{cov}(d) \cap F| \cdot \text{sim}_A(d, c)}{\text{redundancy}(d, F, C)} \qquad (5)$$

where

$$redundancy(d, F, C) = \sum_{e \in F \cap \text{cov}(d)} \text{sim}_A(d, \text{coveredBy}(e, C)) \qquad (6)$$

In other words, we penalize picks that would cover entities with data sources that are similar to datasets that were already used to cover these entities in previous iterations. By penalizing similarity to previous covers, we avoid using the same similar datasets repeatedly for all covers in the top-k list, but we also do not strictly disallow the re-use of data sources in new combinations. Objective function (5) forms the core of our proposed entity augmentation algorithms, which we will introduce in the next sections.

Basic Greedy Algorithm and Extensions With the scoring function in place, we can construct a greedy consistent set covering, shown in Algorithm 1, that produces consistent individual augmentations, as well as diversified solutions when run with $k > 1$. In Algorithm 1, the function *Greedy-TopK-Covers* produces k covers by calling the function *Cover* k times, while keeping an $|E| \times |D|$ matrix called U as state between the calls. While the *Cover* function performs the basic greedy set cover algorithm with the objective function defined above, the main function updates the matrix U after each iteration by increasing the entry for each entity/dataset combination that is part of the produced cover. The function *coveredBy* used in the *redundancy* term of objective function (5) is realized in the algorithm by summing up the matrix row values $U[e]$, which record the datasets used to cover e in previous covers. Note that the main function also discards duplicate solutions, which may occur if the influence of the *redundancy* function is not strong enough to steer the search away from an already existing solution. Still, the matrix U is updated even if a solution is rediscovered, so that further choices of the same data sources become more and more penalized, guiding the search into a different part of the solution space.

The greedy approach, while being easy to implement and fast to execute, will not necessarily construct the best possible list of solutions, as our evaluation in [22] shows. This is mainly due to exploring only a small part of the search space, i.e., considering only k different covers. Therefore, we developed two further algorithms as extensions of the basic framework: the first one is based on the observation that the first k solutions produced by Algorithm 1 may not necessarily be the best solutions. After the first solution has been produced the search is mainly guided by using different datasets for each solution, and thus new combinations of previously used data sets are often not considered in the basic greedy algorithm. One simple extension is called *Greedy**-algorithm, which uses the basic greedy algorithm to create more covers than requested, and then introduces a second phase to the query processing called *Select*, in which the k best solutions are selected from a pool of $s \times k$ possible solutions, with s being the scale factor. In comparison to the *Greedy* algorithm, the *Greedy** approach should find better solutions as it searches a larger portion of the search space, at the cost of a runtime that increases with the scale factor s, plus some overhead for the selection phase. However, the way it explores the solution space is relatively naïve. For this reason, we also developed a genetic approach which naturally fits to our problem as it intrinsically generates a pool of solutions from which k can be picked, and both consistency and diversity of the results can be modeled intuitively. Specifically, consistency can be modeled as part of the fitness function, and diversity can be introduced through a suitable population replacement strategy.

A detailed evaluation and comparison of all three algorithms is provided by [22].

2.4   Related Work

Entity Augmention An early publication on Web table-based entity augmentation is [8], which is concerned with automating the search for relevant Web tables. The paper does not aim at fully automated table extension, but proposes a set of operators that are to be used in a semi-automatic process, enabling the user to search for tables, extract their context, but also to extend a found table with information from a related table. This last operator, called *extend*, corresponds to our notion of entity augmentation. The paper proposes an algorithm called *MultiJoin*, that attempts to find matching Web tables for each queried entity independently, and then clusters the tables found to return the cluster with the largest coverage. However, it does not try to construct consistent solutions, but returns the set of possible values for each entity.

A strongly related work is the InfoGather system [56], and its extension Info-Gather+ [57]. The first system introduces Web table-based entity augmentation, as well as related operations such as attribute name-based table queries. InfoGather improved the state of the art especially by identifying more candidate tables than a naïve matching approach, while eliminating many spurious matches at the same time. They also introduce methods for efficiently computing the similarity graph between all indexed tables offline. InfoGather+ improves the system by tackling similar consistency issues as our work: it assigns labels for time and units of measurements to tables, and propagates these labels along the similarity graph described above to other tables where such labels can not be found directly. While InfoGather+ tackles the problem of producing more consistent results from various possible Web sources, it does not produce top-k results, or minimize the number of sources used.

The basic problem that there will be more than one correct answer for many augmentation queries, e.g., multiple *revenue* values for a single company because of different years of validity, is also explored in [50]. Specifically, the work targets quantity queries, i.e., queries for a numeric attribute of a certain entity. The earlier InfoGather+ already allows the user to specify a unit of measurement and a year-of-validity, and will only try to retrieve a single attribute value with these specific constraints. The QEWT system presented in [50], on the other hand, solves this problem by modeling the query answer as a probability distribution over the retrieved values, and then returning a ranked list of intervals as the final query answer. This work is similar in spirit to ours, in that it does not try to simplify complex real-world attributes into single values, but deals with the uncertainty of data explicitly.

In [37], a table augmentation system called *Mannheim SearchJoin Engine* is proposed that, in addition to entity augmentation given a specific attribute, also supports *unconstrained* queries, i.e., queries in which, given only a set of entities, all possible augmentation attributes are to be retrieved. Their method of dealing with multiple, possibly conflicting sources, by merging values using clustering and majority voting, is similar to [56].

In [45] a comprehensive system for so-called *transformation queries*, that largely correspond to entity augmentation queries, is envisioned. The paper's main contribution is that it proposes a system, named *DataXFormer*, that includes multiple transformation subsystems, based on Web tables, wrapped Web forms, as well as crowdsourcing, although it does not give specific methods of combining the subsystems. Furthermore, it also relies on returning a single value for each queried attribute.

Set Covering The set covering problem as one of Karp's original 21 NP complete problems [34] implies the need for heuristic solutions and led to many optimization techniques. Our methods for generating top-k covers are inspired by multi-start optimization methods [43], such as GRASP [26] and Meta-RaPS [15], which have been applied to the set cover problem among others in [7, 35]. On a high level, these methods combine multiple iterations of a randomized *construction* phase and a local *improvement* phase. In the first phase, a solution is created using some heuristic, e.g., the Greedy approach, but applying some form of randomization. The randomization allows the algorithm to create slightly different results in multiple runs. This is achieved for example by randomly making non-optimal decisions in the individual steps of the respective algorithm.

A second inspiration for our top-k approach is *tabu search* [27], in which an initial solution is the starting point for multiple iterations of a neighborhood search aimed at improving the solution. The distinguishing feature is the *tabu list*, which stores already visited parts of the solution space. It is used to prohibit the algorithm from returning to an already visited part of the solution space, unless a so-called aspiration criterion is met, or a certain amount of iterations has passed. In this way, the search is prevented from getting stuck in, or repeatedly revisiting, the same local optima. Our approach tracks which entities have already been covered by which datasets in previous solutions by using a *usage matrix* (see Algorithm 1 in Sect. 2.3). Similarly to tabu lists, this matrix is then used to discourage the algorithm from making those choices again, preventing future iterations from creating similar covers. This was partly inspired by the concept of tabu lists, except that our approach does not prohibit certain choices, but only adjusts their weights.

Fig. 8 Exemplary Open World SQL query, ad-hoc integrated attributes *gdp* and *creditRating* highlighted

```
select
    nation. creditRating ,
    avg(o_totalprice)
from nation, customer, orders
where
    n_nationkey=c_nationkey
    and c_custkey=o_custkey
    and nation. gdp  > 10.0
group by nation. creditRating
```

3 DrillBeyond – Processing Open World SQL

So far, we only studied top-k entity augmentation queries (see Sect. 2) in an isolated context, i.e. for a single table and simple attribute queries $Q_{EA}(E, a_+, k)$. However, it is natural to assume that ad-hoc data integration will be most useful in analytical Big Data scenarios, in which the user works with complex databases, and the augmentation query is only one step in a chain of analytical operations. We will exemplify this on a data analytics scenario illustrated in Fig. 8. There, we see a TPC-H query with two highlighted augmentation attributes "GDP" and "creditRating" that are not defined in the TPC-H schema, i.e. the query is not immediately processable in a traditional relational system with a closed schema. A possible way to approach the above query would be to export the "nation" relation, and feed it into a stand-alone augmentation system, such as the *REA* system introduced in the previous chapter. However, part of our solution to these challenges is to introduce the top-k entity augmentation paradigm, in which the system produces several alternative solutions from which the user has to choose from. This process would therefore result in a top-k list of possible augmentations for the exported table. Since a standard DBMS can not process the query based on a multi-valued augmentation, the user would have to choose one of the augmentations while in the independent context of the augmentation system, and then re-import the selected augmentation into the DBMS. Again, iterations of this process may be necessary if the initial result is not satisfactory.

In the next section, we will discuss the challenges that arise when entity augmentation is utilized in analytical scenarios involving traditional database management systems. From the identified challenges, we will derive the need for closer integration of top-k EA systems and RDBMS, and derive requirements for a hybrid system in Sect. 3.2. In Sect. 3.3, we will introduce the system architecture of our DrillBeyond system, and describe its core, the DrillBeyond plan operator. We will also discuss peculiarities of hybrid augmentation/relational query processing, and introduce Drillbeyond's ways of dealing with them. Finally, we will survey related work in Sect. 3.5.

Parts of the material presented in this chapter have already been published in [20, 21].

3.1 Motivation and Challenges

We already introduced how top-k entity augmentation can be applied in analytics scenarios to fulfill ad-hoc information needs. However, we also argued that in complex analytical scenarios, using a standalone entity augmentation system to answer ad-hoc information needs has several deficiencies that should be discussed in the following

Context-Switching First, there is a cost associated with context-switching, both with respect to *user effort*, but also with respect to *data locality*. The user would be required to move the data that is to be augmented into the specialized data search and integration system, such as REA presented in Sect. 2, inspect and verify the solution in this context, and then move the data back into the actual analytics system. On the one hand, this introduces a considerable overhead into the analysts workflow, requiring additional effort that may even discourage from performing certain ad-hoc exploratory queries at all. On the other hand, it also introduces physical overhead of moving the data between systems. This overhead may be negligible if a small dimension table, e.g. the "Nation" table in the example, is to be augmented. For larger tables, however, data transfer times can be significant, and further impede an interactive analytics workflow.

Incompatible Query Model A second challenge when introducing top-k entity augmentation into a traditional analytics workflow is the mismatch in query models. Augmentation systems such as REA produce top-k results, while other parts of the environment, such as DBMS, work with exact, single results. In this respect, top-k augmentation systems are more similar to information retrieval systems that handle the uncertainty of their results by producing a list of possible results. However, a traditional DBMS is not natively prepared to handle multi-variant data. This gap needs to be bridged in order to enable effective combination of the two system types.

Loss of Context Information Third, by using a separate system for augmentation queries, the broader context associated with the analytical task is not taken into account. A generic entity augmentation system uses only the set of entities and the augmentation attribute as input to guide its data search and integration process. However, the query context may contain valuable hints that can improve the augmentation system performance or precision, if the systems were able to exploit them. For example, in an SQL query, other tables that are joined with the augmented table may provide useful context for the data source retrieval and matching process in the augmentation system. Similarly, predicates used in the original query add semantics that can be used to improve the precision of downstream augmentation system. In an optimal combination of DBMS and entity augmentation system, such context information incurring in one system would be utilized in the other.

Unused Optimization Potential Finally, by using separate systems, query *optimization potential* is wasted. For example, DBMS use cardinalities and estimated selectivities to choose an optimal join order for a given query. If a manual entity augmentation has to be performed before the query can be executed, then the cardinality of the augmented relation, or the selectivity of predicates on the augmented attribute can not be exploited in this optimization process. Depending on the exact circumstances, it can be beneficial to intermingle the normal DBMS query processing and the augmentation query processing to achieve optimal performance.

3.2 Requirements

From the challenges we identified in Sect. 3.1, it is easily recognizable that a closer integration of the two system types DBMS and EAS (Entity Augmentation System) is necessary. In this section, we will introduce requirements for a hybrid system that is able to close this gap.

DBMS-integrated Entity Augmentation In the previous section, we discussed how the lack of integration between DBMS and EAS systems leads to an increased user effort for situational one-of analysis queries. This effort could be reduced if the DBMS would directly support looking up and integrating Web data sources as part of its query processing, and allow the specification of such queries declaratively in SQL. However, there are several differences to bridge. First, the two system types differ in the type of data they manage: DBMS deal with structured and cohesive databases, while EAS deal primarily with large heterogeneous corpora of Web data sources. Further, DBMS work with fully specified queries in a structured language, while EAS, lacking a defined schema, accept keyword queries. We therefore require a design that blurs the line between these classes of systems in all three aspects mentioned above: type of data managed, query language used, and nature of the query result. The resulting hybrid system should be able process mixed SQL/EA queries, which we will call *Open World SQL queries*. In these queries, the user may reference arbitrary additional attributes not defined in the schema. We will use Fig. 8 as our running example for such a query. The system will associate values of these additional attributes to instances at query processing time, avoiding an explicit data retrieval and integration step. This is achieved by executing top-k entity augmentation queries at runtime, which are integrated as a new type of subquery of regular relational queries. Since a top-k augmentation query will return multiple augmentations as described in Sect. 2.3, an Open World SQL query will, instead of returning a definitive query answer, return multiple alternative query results as well. Figure 9 gives an intuitive overview of our goal. It illustrates how an Open World query is processed by integrating entity augmentation into query processing, producing k alternative SQL query results.

In conclusion, the system should produce structured results of exactly the form specified by the user query, just as a regular DBMS would, but also presents several possible versions of the result, similarly to an information retrieval system. However, producing multiple alternative SQL results for a single query has performance implications, which we will discuss next.

Efficient Multi-result Query Processing A naïve approach to bridging the different query models of DBMS and top-k EA systems would be to process the EA query, and then process the SQL query multiple times. For example, if k possible augmentations are requested, the runtime of the SQL query is increased by this factor k. This is not acceptable, since in many Open World SQL queries, the majority of the processing time will still be spent processing local data, which does not change between runs of the query. For example, consider again the example Open World SQL depicted in Fig. 8. Here, a large part of the work consists of local joins between the relations

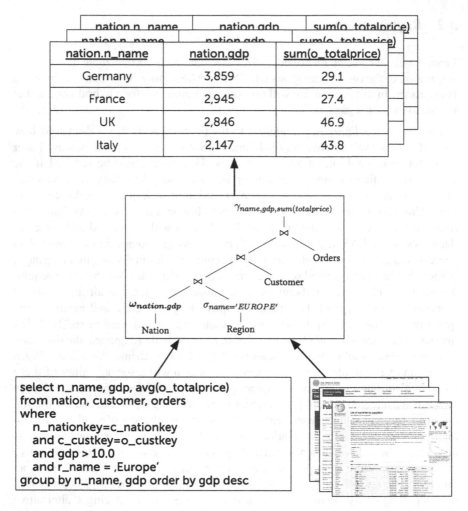

nation.n_name	nation.gdp	sum(o_totalprice)
Germany	3,859	29.1
France	2,945	27.4
UK	2,846	46.9
Italy	2,147	43.8

Fig. 9 RDBMS-integrated top-k augmentations using web tables

Customer and Order, and aggregation of the local attribute o_totalprice. When processing this query multiple times based on different augmentations, only the set of nations that pass the predicate on gdp, as well as the order of result tuples would change, but not the aggregates for the individual nations. Consequently, the hybrid system should process Open World SQL queries in a way that minimizes duplicate work between query variant executions.

Open World Query Planning The third requirement arises from the fact that properties of data sources used in augmentation are not fully known at plan-time. For instance, estimating the selectivity of a predicate over an augmentation attribute is not easy, as the set of data sources that will be used is not known at planning-time. The same holds for determining the open attributes' metadata, such as the data type, since we do not require the user to specify it in the query. Therefore, the system should be able to plan queries even if some attributes are only fully known at run-time.

With these requirements established, we will introduce the *DrillBeyond* system and its entity augmentation operator in the following section.

3.3 The DrillBeyond System

To solve the challenges identified in Sect. 3.1 and enable entity augmentation queries as part of relational query processing, we designed the *DrillBeyond* system. It is an RDBMS/EAS hybrid, that embeds entity augmentation sub-queries into standard RDBMS query processing. The next sections will detail the required changes to the RDBMS architecture to realize this mixed query processing.

System Architecture Processing open world SQL queries requires a top-k entity augmentation system, including a data source management system, as well as modifications to three core RDBMS components: the analyzer, the planner and the executor. Figure 10 gives an overview of the modified and the novel components, and further includes a high level description of the changes in control flow. The core augmentation functionality is introduced through the new *DrillBeyond plan operator*, which will be discussed in detail in Sect. drillbeyond:sec:DrillBeyondspssystem. In the following, we will first give a general overview of all the novel or modified DBMS components in DrillBeyond.

Data Source Management System A standard RDBMS is tailored to manage a relatively small set of relations that form a coherent schema. An EAS, on the other hand, manages a large corpus of heterogeneous individual Web data sources. We aim at enabling one system to process both kinds of data. The DrillBeyond system does not make assumptions regarding the nature of the data sources and system that they are managed by. A generic system that exposes an interface for keyword-based dataset search is sufficient. For example, when using our proposed EAS (see Sect. 2), an industry-standard document index server such as Solr[3] or ElasticSearch[4] is sufficient. The necessary source selection, matching and integration operations are performed in the integrated entity augmentation system, which we discuss next.

Entity Augmentation System This component implements the actual top-k entity augmentation processing inside the DBMS. It interfaces with the data source management system to retrieve Web data sources, and with the core RDBMS components to provide augmentation services to the executor and the planner. DrillBeyond extends the generic augmentation query definition (see Definition 1) using *query context hints H*, which are extracted from the respective outer SQL query. These hints are used to guide the search and the ranking of Web data sources. For example, if the

---

[3] http://lucene.apache.org/solr/.

[4] https://www.elastic.co/.

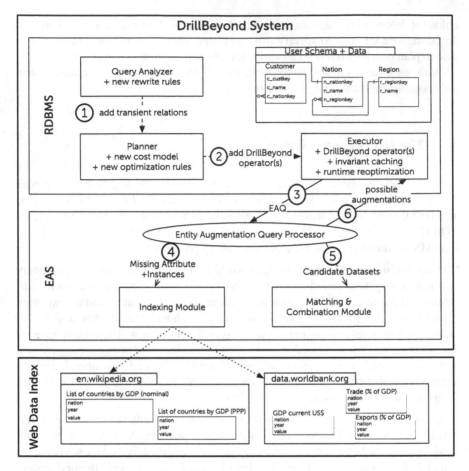

Fig. 10 System architecture and high level control flow

outer SQL query includes a numeric predicate on the augmentation attribute, this fact can be used as a query hint by instructing the augmentation system to only retrieve data sources that provide numeric values. We discuss the exact nature and usage of these context hints in section "Pushing SQL Query Context". The complete interface used by the executor is therefore $Q_{EA}(a_+, E, k, H)$. Having introduced the novel components necessary for entity augmentation in the DrillBeyond system, we will now give an overview of the modifications to existing RDBMS core components.

Query Analyzer The first step in DrillBeyond query processing is triggered by the query analyzer, which maps tokens in the SQL query string to objects in the database's metadata catalog. Unrecognized tokens, such as *gdp* lead to an error in a typical RDBMS. In the DrillBeyond system, we take a minimally invasive approach: we introduce transient metadata for the duration of the query, so that the regular analysis can continue. The query is then rewritten to include an additional join with a transient relation, effectively introducing a source for the missing attribute into the query processing, and also paving the way for the DrillBeyond operator to be placed by the regular join order planning mechanisms of the DBMS. The analyzer is also responsible for determining the type of all expressions in the query. In the case of augmentation attributes, which are not represented with a type in the database catalog, we include a type inference mechanism. It first tries to infer the data type syntactically, by considering filter and join predicates the attribute is used in, and comparing it with the types of regular attributes and constants in these expressions. If syntactic inference is not possible, DrillBeyond infers a type statistically. It uses the augmentation system to determine the most common data type occurring for attributes named a_+ in the dataset corpus using a fast probe query. Having created the necessary query metadata, the analyzed statement can be passed on to the modified query planner.

Query Planner In DrillBeyond, the query planer has several new tasks to perform compared to a regular RDBMS. First, it needs to place the DrillBeyond operator in the query plan. This placement is crucial to the execution time of the query but also influences the augmentation quality. Furthermore, while regular query plans are created to be optimal for a single execution, multi-solution processing requires plans that minimize the overhead of creating multiple result variants based on top-k augmentations. This is impeded by the lack of plan-time knowledge about the data sources, requiring plan adaption.

Executor The executor is modified to repeatedly execute the planned operator trees, creating the top-k query result. In each iteration, it orders the DrillBeyond operators to augment incoming tuples with values from a different augmentation. It further tags the SQL results produced with a distinct augmentation id, by adding a column carrying this id to each finished result. This allows external tools using the top-k SQL result to distinguish between the tuples belonging to alternative results. The majority of new functionality, however, is part of the DrillBeyond operator itself, which we will detail in the next section.

The DrillBeyond Operator In its basic form, the DrillBeyond plan operator, denoted ω, is designed to resemble a join operator, which facilitates integration with the existing system architecture. Specifically, it acts like an outer join: it adds new attributes to its input tuples based on join keys, but will not filter original tuples if no partner is found. Instead, it adds null values if the augmentation system can not produce a value. In this way, the part of the query operating on local data can still be processed. However, in contrast to a regular join, only one of the joined tables is known at plan-time, while the other table, as well as the join keys, are decided at query processing time. These run-time decisions are made by the entity augmentation system based on the input tuples of the operator. Specifically, the operator extracts distinct combinations of textual attributes from input tuples, as these are used to functionally determine the values of the augmented attribute.

Algorithm 2 shows the specifics of the state kept in the operator and the implementations of its iterator interface and helper functions. DrillBeyond uses a traditional

---

Algorithm 2 DrillBeyond operator

---

  function INIT
    $state \leftarrow$ '*collecting*'
    $tuplestore \leftarrow \emptyset$
    $augMap \leftarrow HashMap()$
    $n \leftarrow 0$ $\triangleright$ Current Iteration, runs from 0 to $k - 1$

  function NEXT
    if $state = $ '*collecting*' **then**
      COLLECT()
      AUGMENT()
      $state \leftarrow$ '*projecting*'
    return PROJECT()

  function COLLECT
    while *true* **do** $\triangleright$ Retrieve all tuples
      $t \leftarrow $ NEXT($childPlan$)
      if $t = NULL$ **then**
        break
      $tuplestore \leftarrow t$
      $augKey \leftarrow $ TEXTATTRS(t)
      if $augKey \notin augMap$ **then**
        $augMap[augKey] \leftarrow \emptyset$

  function AUGMENT
    $augReq \leftarrow (\forall k \in augMap \mid augMap[k] = \emptyset)$
    for all $augKey, [augValues...] \in $ SEND($augReq$) **do**
      $augMap[augKey] \leftarrow [augValues...]$

  function PROJECT
    $t \leftarrow $ NEXT($tuplestore$)
    if $t = NULL$ **then return** $NULL$
    $augKey \leftarrow $ TEXTATTRS(t)
    $t[a_+] = augMap[augKey][n]$
    return t

  function RESCAN
    $state \leftarrow$ '*collecting*'
    $tuplestore \leftarrow \emptyset$

  function NEXTVARIANT
    RESCAN($tuplestore$)
    $n \leftarrow n + 1$

---

row-based iterating executor. The conventional interface functions *Init()*, *Next()* and *ReScan()*, as well as the novel *NextVariant()* function, are called by the DBMS during regular query processing. The other functions shown in Algorithm 2 are used internally by the operator.

    The *Init()* function, which is called by the RDBMS executor before processing the query the first time, initializes operator state. This includes a tuple store for materializing the lower operator's output, a hash table mapping local textual attribute values to augmented values called *augMap*, and two variables *state* and *n*, determining the behavior of the operator when *Next()* is called.

The *Next()* function is called by the executor and produces augmented tuples. This is done in three phases: *Collect()*, *Augment()* and *Project()*. On the first call to *Next()*, since no augmented values are available, the first two phases are triggered. In the *Collect()* phase, the operator pulls and stores all tuples that the lower plan operator can produce, making DrillBeyond a blocking operator. The reasons for blocking are discussed in section "Augmentation Granularity". In this phase, the operator also stores the textual attributes originating from the augmented relation and its context in a hash table, to obtain all distinct combinations of textual values in the input tuples. In the *Augment()* phase, all entries in the augmentation map that do not yet have values associated with them are passed to augmentation system as one augmentation context. After successfully retrieving values for all collected tuples, the operator is put into the *projecting* state, and produces the first output tuple. Output tuples are produced in the *Project()* function by replaying the stored tuples and filling the augmentation attribute by looking up values in the hash table.

The *ReScan()* function is called by the DBMS executor when subtrees have to be re-executed, e.g., in dependent subqueries or below a nested loop join. Here, the operator empties its tuplestore and changes state to collect new input, but keeps its augmentation hash table, to prevent expensive re-augmentation for values that have already been seen. Finally, *NextVariant()* is an interface extension not seen in typical RDBMS operators, which is necessary for producing the multi-variant query results as discussed in Sect. 3.2. When called, the operator's tuplestore is prepared for another iteration over the stored tuples using *ReScan()*, and the iteration counter n is incremented. This makes sure that in a new execution of the query plan, operators below the DrillBeyond operator are not called again. Instead their materialized output will be replayed, and augmented with the next augmentation variant in the *Project()* function.

Entity Augmentation Query Q(a$_+$, E)

Country	GDP
Russia	?
UK	?
USA	?

ds$_1$ – American Countries

Country	GDP M USD
USA	X
Canada	Y

ds$_2$ – World GDPs

Country	GDP Mil. $
United Kingdom	X
Russian Federation	Z

ds$_3$ – European Economy

Country	GDP EUR
Russia	X
France	Y

ds$_4$ – EU Economy

Country	GDP EUR
UK	X

Fig. 11 Example augmentation problem

The functions *Init( )* and *Next( )* are part of the traditional iterator interface used in RDBMS and are called by the executor in regular query processing. The *NextVariant( )* function is different however, and requires novel functionality in the executor. Specifically, when the top operator of the plan returns null, the executor usually assumes that all data has been sent and stops the processing. DrillBeyond however, keeps a global iteration count n, running from 1 to k, to track the number of produced alternative results. In case the plan has finished, n is increased, *NextVariant( )* is called on the DrillBeyond operator, and then the whole plan is restarted. In case there is more than one augmentation attribute, and thus more than one DrillBeyond operator, the system produces a crossproduct of alternative results. This is achieved by systematically calling *NextVariant( )* only one operator in each iteration, to iteratively produce all combinations of possible results for all augmented attributes.

Having introduced the basic operator functionality, and the way it is called by the DBMS executor, we will discuss the operator characteristics in the following subsections.

Augmentation Granularity A naïve entity augmentation operator working *tuple-at-a-time* would be most compatible to the iterator-based query processing used in most traditional RDBMS operators. However, augmenting each tuple on its own implies looking up and matching Web data sources for each tuple individually. Consider the simplified augmentation example shown in Fig. 11, where the table to be augmented is on top, and the available Web data sources at the bottom. With the *tuple-at-a-time* style, the augmentation system may choose ds_1, ds_2 and ds_4 for the USA, Russia, and UK tuples respectively. These sources match each individual tuple best, and the individual tuples are what the augmentation system can process in this case.

Still, we can see that the augmentation system can not perform optimally with regard to the query as a whole, as it is not provided with the overall query context. While the chosen sources are the best fitting for each individual tuple, they do not form a consistent joint result, as the units of currency do not match. If the augmentation system is instead provided with the set the complete set of tuples as the input for one augmentation query, the more consistent solution comprised of ds_1 and ds_3 can be constructed, even though the individual entity matches are slightly worse. This is a similar argument to those we mentioned in Sect. 2 about result consistency. Extending single tuples individually would correspond to the *By-Entity Fusion* augmentation strategy as introduced in section "Entity Augmentation Queries" leading to results with similar deficiencies as discussed there. In [22], we showed that our approach to entity augmentation leads to higher quality results, but requires all entities to be queried in a group. Otherwise, the consistency of the source selection can not be ensured. We conclude that for reasons of result quality, the DrillBeyond operator needs to be a blocking operator, i.e. it consumes tuples from underlying operators until they are exhausted, then hands them over to the augmentation system, and produces the first result tuple only when it returns. Referring back to Algorithm 2, this blocking behavior is realized in the state "collecting" in function *Next( )*.

Context-Dependent Results Having established the DrillBeyond operator as blocking, we will now consider the question of where to place it in a query plan. Let us assume two queries. The first one performs the augmentation directly after the scan of the TPC-H `Nation` table, while in the second query ω is executed after `Nation` has been joined with table `Region` and was filtered for European countries only. In the first case, the augmentation system will retrieve, match, rank, and combine datasets for all countries in the local database. In the second case, the local join will remove tuples about non-European nations, so the results of the entity augmentation will be more likely to be based on data sources specifically about Europe. Furthermore, completing the join with the `Region` table does not only limit the scope of the query, it also adds context to each tuple by adding `Region`'s attributes. So even if the join would not act as a filter limiting the number of tuples, adding information about the region name will improve the accuracy of the augmentation system. For example, while augmenting a set of *City* tuples may be hard and error-prone because of the ambiguity of common city names such as "Springfield", augmenting after a join with a *State* will be a more realistic task for the augmentation system. We have identified the following property of the ω operator:

Definition 6 *(Selection Dependency)* The DrillBeyond operator is not associative with respect to selection in the general case. When augmenting relation R with attribute a_+ and selecting with predicate p on R, then $\omega_{R,a_+}(\sigma_p(R)) \neq \sigma_p(\omega_{R,a_+}(R))$.

Note that the augmentation system uses only the textual attributes $textAttr(R)$ for matching with Web data sources. Therefore, in the special case that the set of distinct textual attribute values is invariant under predicate p, the augmentation results are also invariant under selection with p. For example, if a selection $\sigma_p(Nation)$ returns at least one tuple for each `Nation` in its result set, then $\omega_{Nation,a_+}(\sigma_p(Nation)) = \sigma_p(\omega_{Nation,a_+}(Nation))$. This however, is not the general case, and can not be relied upon.

A similar dependency property holds for another form of query context, namely for the set of attributes of the input tuples of an operator ω. As mentioned, the key for searching matching Web datasets for R are its textual attributes $textAttr(R)$. For example, not projecting the attribute n_name of the TPC-H `Nation` relation, possibly because it is not part of the desired query result, will make augmentation impossible. No Web data source can be found if the natural key of the relation, the nation's name, is not part of ω's input.

Definition 7 *(Projection Dependency)* The DrillBeyond operator is associative with respect to projections only if it includes all textual attributes of R. Formally, when augmenting relation R with attribute a_+, and projecting to a set of attributes $\mathcal{A}$ with $\mathcal{A} \cap textAttr(R) \neq textAttr(R)$, then $\omega_{R,a_+}(\pi_{\mathcal{A}}(R)) \neq \pi_{\mathcal{A}}(\omega_{R,a_+}(R))$.

Given that the result of ω_{R,a_+} changes under projection and selection, we define the following placement rule that defines bounds on the placement of the operator.

Definition 8 *(Placement Bounds)* The DrillBeyond operator augmenting a relation R can only be placed with respect to the following conditions:

1. after all operators filtering R such as joins and selections
2. before any projection removing textual attributes of R

In other words, the DrillBeyond operator ω is always applied to the minimum number of distinct combinations of textual column values in the tuples of R, as these determine the matching process and its result.

Pushing SQL Query Context As mentioned in Sect. 3.1, context from the outer SQL query can be used as filter to improve the accuracy and runtime of the inner entity augmentation query, when compared to isolated augmentation. These filters work implicitly to improve the augmentation quality by narrowing the scope of the augmentation operation, and do not require any changes to the augmentation system or its API. However, we can further improve the augmentation by explicitly pushing additional query knowledge to augmentation system. Specifically, we push two types of information: *type information* and *predicates on augmented attributes*.

Type Information Though the user can specify a data type such as *text* or *double* for open attributes using SQL syntax, we do not expect those annotations to be provided. However, in many cases it is possible to infer the type of the open attributes from the surrounding query by applying methods of type inference to SQL. As mentioned in section "System Architecture", we integrated a type inference mechanism that determines an open attribute's data type both from the query expressions it takes part in, and from the Web data corpus used. We can pass type information to the augmentation system, which in turn uses this type information to restrict the set of candidate data sources to those matching the open attribute's type. Consider again the query shown in Fig. 8. From the constraint on the *gdp* attribute it can be inferred that it must be of a numeric type. This allows the augmentation system both to reduce its runtime and increase precision by pruning non-numeric candidate sources.

Predicates on augmented attributes In addition to the data type of open attributes, we can also push-down the predicates on open attributes themselves. This allows a similar, but more sophisticated, candidate pruning. We assume that users expect some filtering on the database instance level to happen when specifying a predicate, i.e., we assume that some domain-knowledge is encoded in the predicate. With the query shown in Fig. 8, the user clearly intends to filter low *GDP* countries, and has given a relatively large integer number as the specific condition. Though the user query is given only with the very general keyword *GDP*, by also considering the predicate, the augmentation system can improve its data source ranking. In the example, all candidate datasets that give the GDP as a percentage or rank can be ranked lower, because those datasets will not discriminate the entities with respect to the predicate. In other words, given the predicate above, using a dataset that provides *GDP* rank values will lead to all entities being filtered, which is clearly not the user-intent. To improve the augmentation systems ranking, we can therefore check how well the data source's values fit to the predicate that will be applied, i.e., how well the data source discriminates the entities with respect to the predicate. Additionally, we check whether the predicate value and the average of the data source's values are in the same order or magnitude. Again, the intent is measuring whether the data source is fit to evaluate the given predicate.

The improvements in precision and runtime of the EAS when considering predicate and type information are shown in [21].

Cost Model and Initial Placement Strategy The observations from the previous three sections might suggest that the DrillBeyond operator should be placed as late as possible in the query plan, to maximize the context knowledge available in the intermediate result. However, in addition to quality considerations, there are also performance considerations to be made. As mentioned, the operator is modeled to resemble a join from the perspective of the DBMS. We can therefore reuse the existing join optimization machinery to place the DrillBeyond operator. This however depends on a model for the operator runtime and the operator's output cardinality.

Output Cardinality In the most basic case, the operator produces exactly as many tuples as its input relation, as it just adds a single attribute to each tuple. However, we also must consider selectivity of possible predicates on augmented attributes, and the value of k, i.e., the number of alternative augmentations that are to be processed. Since at plan-time we do not assume any knowledge about which Web data sources will be used to augment an attribute, correctly estimating selectivity is almost impossible. A well-known solution to processing queries with unknown selectivities is run-time plan adaptation [33]. We therefore initially use the DBMS' default selectivities for different types of predicates, and employ run-time optimizations to compensate when more information is available. The variable k on the other hand is known at plan-time, and could be part of the cost model. However, with our execution strategy, the operator does not produce k tuples for each input tuple, but just a different one in each of k query executions. We therefore do not consider k at this point, but create optimal plans for single execution, and then later optimize their re-execution.

Cost Model The operator's runtime depends on three components: The first part is incoming tuple processing, the second entity augmentation, and the third is projecting tuples with the augmented attribute (see Algorithm 2). Since we designed the operator as blocking, the first part consists of reading all tuples from lower plan nodes, storing them, and computing all distinct combinations of all textual attributes, which are needed by the entity augmentation system for matching. In the second phase, the distinct combinations are then submitted to the augmentation system. The third phase then consists of iterating all stored tuples, and projecting the new attribute based on the combination of textual attributes found in each tuple.

We can estimate the cost of the operator using a similar model as for a hash join, as phases one and three, hashing the child relation, and then probing it against the augmented hash table correspond to the phases of a hash join. Additionally, there is the cost of phase two, the actual augmentation, which depends not on the number of tuples, but on the number of distinct entries in the augmentation hash table. The processing cost per entry depends on the augmentation algorithms, and is therefore not easy to model in the context of a generic DBMS cost model. However, we can assume that the cost per entry is in a different order of magnitude than the per-tuple cost of the relatively primitive database operations such as comparison or hashing. For our cost model we therefore assume an additional large constant factor $\mathcal{C}$ which is learned from previous executions of our augmentation system (see Sect. 2).

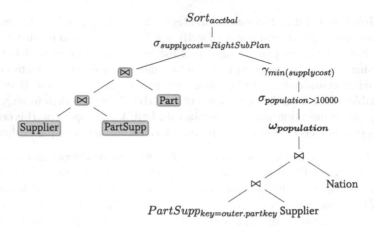

Fig. 12 Plan invariants

3.4 Processing Multi-result Queries

So far, we have considered the DrillBeyond operator in a single query result setting. However, as discussed in Sect. 3.2, we aim at translating the top-k result returned by the augmentation system into a top-k SQL result. The naïve method, given our operator, is to simply re-execute the query plan k times, and after each execution trigger the projection of a new augmentation result from the operator via the *NextVariant()* API depicted in Algorithm 2. This obviously leads to duplicated work, as only the output of the DrillBeyond operator changes between executions, while the other parts of the query plan operate the same way. Consider the query plan in Fig. 12. Here, only the values of the gdp attribute would change between query runs, while the other operations, notably the more expensive joins with the Customer and Orders relations, would not change. This means that simple re-execution would increase time to compute the query k-fold, with most of the effort being inefficient duplicate work. Our first approach to this problem is to identify and then maximize invariant parts of the multiple executions, preventing their re-execution by materializing intermediate results. As a first observation, note that all tuple flows below the DrillBeyond operator can actually never change between executions: in the example shown in Fig. 12, these operators are highlighted. The cost of those operations can be minimized by materializing the input to the DrillBeyond operator. This elementary optimization is already included in the basic operator implementation shown in Algorithm 2, in the form of the tuple store created by the operator. However, the changing augmentation output may influence the result of other operators further up the query plan. In the example, the aggregation of the gdp attribute will change its output in each iteration based on the augmentation values provided by the DrillBeyond operator. In addition, even the aggregation of the regular attribute totalprice is influenced by the changing gdp values: Since the selectivity of the predicate on gdp may vary using different augmentations, the set of Orders-tuples that has to be aggregated may vary in different executions as well.

In [21] we introduced a series of optimization strategies for these problems including invariant caching, augmentation operator splitting, selection pull-up, projection pull-up partial selection and run-time reoptimization. We also implemented the described system in PostgreSQL and evaluated it on modified TPC-H queries [21]. The evaluation shows the effectiveness of various optimizations in minimizing the runtime overhead of producing multiple SQL query results based on alternative augmentations. Finally, we showed that pushing SQL query context into the EA system can improve quality and performance of the EA processing compared to standalone processing.

The tight integration of augmentation and relational query processing and its various optimizations provided by DrillBeyond, enables the use of ad-hoc data search and integration in new contexts, and greatly increase the practicality of the entity augmentation methods.

3.5 Related Work

To the best of our knowledge, there is no previous work that allows the user to augment a relational database with external data at query-time by simply adding arbitrary new attributes to an SQL query. However, there are considerable amounts of works that can be related to the various individual aspects of DrillBeyond.

Self-service BI and Mashup Tools A work closely related in spirit to ours is [2]. In this paper, the authors discuss their vision for self-service BI, in which an existing data cube can be semi-automatically extended with so-called *situational data* in the course of an interactive OLAP session. However, in contrast to the work at hand, it is a vision paper, and does therefore not present a concrete system design. In [40], a vision for ad-hoc BI based on data extracted from the textual Web content is presented. They envision using cloud-computing and parallelized information extraction methods to answer OLAP queries over the general Web, e.g., by enabling ad-hoc extracting of product data, reviews and customer sentiments from the general Web. Further, there is a considerable amount of work aimed at bridging traditional OLAP and semantic Web technologies to enable novel, agile forms of business intelligence. [1] gives a comprehensive survey of these efforts to utilize the semantic Web to acquire and integrate relevant external data with internal warehouse data. For example, [25] propose a RDFS vocabulary for expressing multidimensional data cubes, so-called *Web Cubes* over semantic Web data, and show how traditional OLAP operations can be performed over combined internal and Web cubes.

One of the central properties of our proposed *Open World Queries* is their declarative nature: users specify their information need simply by an attribute name, i.e., a keyword query. While this approach is sufficient in many use cases, more complex Web data integration problems may require more sophisticated, programmatic specifications. One class of approaches to this problem, called *Mashup tools*, aims at allowing users with only basic programming knowledge to easily compose various data sources to higher level services [16]. Many of these tools have been proposed, which differ in their level of abstraction, community features, user interfaces, or support for discovery of data sources and mashups [28]. Similarly to our work, most research on mashups uses public Web data sources because of their general availability. However, mashups have also been recognized as a lightweight form of data integration for the enterprise context, which is exemplified by systems such as IBM's Damia [51]. In [54], a framework for mashup construction with strong focus on data integration is proposed. It adds more complex data transformation operators, such as a *fuse* operator for automatic object matching, and an aggregation operator to reduce multiple matches that result from a fuse operation to a concise representation. Furthermore, it introduced entity search strategies that minimize the number of queries that need to be issued to data sources such as entity search engines or Deep Web databases. These strategies where explored in more detail in [23], where different query generators are combined with query ranking and selection strategies to form an adaptive querying process, which also utilizes initial query results to optimize the choice of further queries. In [24], this feature set is further extended to include efficient, pipelined execution of such query processes, enabling stream-based processing and quick presentation of initial integration results. By combining approaches such as the ones outlined here, mashup tools can become a powerful alternative to traditional data integration processes. While they are more expressive than *Open World Queries*, they are only applicable in situations where the sources to be integrated are known, and a user with programming knowledge is available to define the mashup and necessary source adapters.

Hybrid DB/IR Systems A closer integration between the worlds of database and IR research has long been a goal of both communities [4, 11, 55]. There are several classes of DB/IR integration, and multiple system architectures for achieving them. First, there are works that aim at including IR capabilities, such as full text search into database engines, e.g., to support keyword queries on textual attributes of a database. Such systems have already been available in commercial RDBMS for some time [17, 31, 41].

Of course, the other direction of system integration is also possible: extending information retrieval systems with more database-like features. The authors of [49] proposed a method for processing semi-structured keyword queries over large, Web-extracted knowledge bases is presented. One novel aspect is the query language, which only adds a minimal amount of structure compared to pure keyword queries, but still allows users to formulate precise information needs without having to understand the large, heterogeneous schema of the underlying knowledge base. The other aspect is the disambiguation of the keywords in the query with respect to the concepts and relations in the knowledge base. It may be worthwhile to apply their method for graph-based query disambiguation to DrillBeyond.

Similar work has been done in the area of XML databases, where a large class of work is aimed at enabling ranked retrieval over collections of XML documents. A recent overview over this extensive field is given in [53]. The problems this research faces are related to those faced by DrillBeyond: underspecified, but structured queries have to be mapped to a large heterogeneous collection of datasets.

In [6], a hybrid DB/IR query type called context sensitive prefix search is introduced. Given a set of documents forming a *context*, and a word prefix, this query type aims at retrieving documents containing words with the given prefix, conditioned on the words also occurring in the context document set. The authors show how, by introducing further structure through special markup keywords, this abstract operator can be applied to implement many database and expressive information retrieval operations, including joins and aggregation.

Other works aim at processing structured queries based on Web information extraction, e.g., [9], or even at decomposing structured queries to keyword queries that can be posed to an information retrieval system [39]. The idea in this case is to enrich the structured query result with relevant documents.

In general, the large amount of related work in DB/IR hybrid technologies shows the necessity of fusing the two paradigms for many applications. By integrating entity augmentation directly into relational query processing, DrillBeyond also aims at supporting scenarios that require a combination of structured and semi-structured data.

4 Summary and Future Work

In the era of Big Data, the number and variety of data sources is increasing every day. However, not all of this new data is available in well-structured databases or warehouses. Rather, data is collected at a rate that often precludes traditional integration with ETL processes and global schemata. Instead, heterogeneous collections of individual datasets are becoming more prevalent, both inside enterprises in the form of *data lakes*, and in public spaces such as Web data sources. This new wealth of data, though not integrated, has enormous potential for generating value in exploratory or ad-hoc analysis processes, which are becoming more common with increasingly agile data management practices. However, in today's database management systems there is a lack of support for ad-hoc data integration of such heterogeneous data sources. Instead, integration of new sources into existing data management landscapes is a laborious process that has to be performed ahead-of-time, i.e., before queries on the combined data can be issued.

In this chapter, we introduced a combined database and information retrieval system that enables users to query a database as well as a heterogeneous data repository in a seamless and integrated way with standard SQL. Relevant sources are automatically retrieved and integrated at query processing-time, without further input from the user. The ambiguity resulting from the coarse query specification, as well as the uncertainty introduced by relying on automatically integrated data is compensated by returning a ranked list of possible results, instead of a single deterministic result as in a regular SQL query. This allows the user to choose the best alternative for the problem at hand.

To achieve that, we introduced a novel method for *Top-k Entity Augmentation* (Sect. 2) which is able to construct a top-k list of consistent integration results from a large corpus of heterogeneous data sources. This technique forms the basis for our *DrillBeyond* system (Sect. 3), which provides hybrid augmentation/relational query processing capabilities. This enables the use of ad-hoc data integration for exploratory data analysis queries, and improves both performance and quality when compared to using separate systems for the two tasks.

In conclusion, we introduced novel, automatic data augmentation methods that harness the large variety of data sources, while requiring minimal user effort and incorporated those methods into traditional relational DBMS, to enable their efficient use in analytical query contexts. To conclude this chapter, we will finally sketch opportunities to extend and build on our contributions in future work.

4.1 Future Work

In the current work, we applied the *Top-k Entity Augmentation* to Web tables only. However, we argue that underlying top-k consistent set covering is a general technique that can be applied to many different forms of data sources. For example, there we discussed related work on augmentation based on information extraction from general Web page text. Applying our approach to generate minimal but diverse covers based on Web pages instead of Web tables would be a promising approach to increase coverage. Furthermore, our approach does not yet consider correlations between sources as a factor of trust. Approaches from data fusion literature that detect and utilize such source correlations could be combined with our set covering approach to increase the precision of the generated covers.

With respect to the *DrillBeyond* system, a possible avenue for future work would be to investigate which parts of the concept could be adapted to modern analytical RDBMS architectures to increase efficiency. In our current work, we integrate entity augmentation with a classical, single-node row store DBMS. However, in many contemporary scenarios, analytical queries are executed on highly parallel, distributed column stores. Investigating how our proposed architecture and optimizations apply to these systems would increase our method's practical applicability. Furthermore, DrillBeyond so far only allows the usage of additional attributes, i.e., it allows only horizontal table augmentation. However, methods for vertical augmentation, or in other words, the ad-hoc integration of further tuples of an existing relation, have also been discussed in related work [29, 48]. These approaches could be integrated with relational query processing as well. Similarly, the materialization of completely new relations from Web Data using just a schema description has also been studied in isolation, but could be integrated with general query processing as well. Furthermore, although DrillBeyond does support joins over open attributes, we did not study the optimization of such joins. In summary, in future work the idea of Open World SQL queries could be generalized from additional attributes to all aspects of SQL and relation query processing.

References

1. A. Abello, O. Romero, T. Bach Pedersen, R. Berlanga, V. Nebot, M. Aramburu, A. Simitsis, Using semantic web technologies for exploratory olap: a survey. IEEE Trans. Knowl. Data Eng. **27**(2), 571–588 (2015)
2. A. Abelló, J. Darmont, L. Etcheverry, M. Golfarelli, J.N. Mazón, F. Naumann, T.B. Pedersen, S. Rizzi, J. Trujillo, P. Vassiliadis, G. Vossen, Fusion cubes: towards self-service business intelligence. Int. J. Data Wareh. Mining (IJDWM) (2012). (accepted)
3. R. Agrawal, S. Gollapudi, A. Halverson, S. Ieong, Diversifying search results. In: Proceedings of the Second ACM International Conference on Web Search and Data Mining, WSDM '09 (ACM, New York, 2009), pp. 5–14
4. S. Amer-Yahia, P. Case, T. Rölleke, J. Shanmugasundaram, G. Weikum, Report on the db/ir panel at sigmod 2005. ACM SIGMOD Rec. **34**(4), 71–74 (2005)
5. P. André, J. Teevan, S.T. Dumais, From x-rays to silly putty via Uranus: serendipity and its role in web search. In: Proceedings of the SIGCHI Conference on Human Factors in Computing Systems (ACM, New York, 2009), pp. 2033–2036
6. H. Bast, I. Weber, The complete search engine: Interactive, efficient, and towards IR and db integration, in *CIDR 2007: 3rd Biennial Conference on Innovative Data Systems Research*, ed. by G. Weikum (VLDB Endowment, Asilomar, CA, USA, 2007), pp. 88–95
7. J. Bautista, J. Pereira, A grasp algorithm to solve the unicost set covering problem. Comput. Oper. Res. **34**(10), 3162–3173 (2007)
8. M.J. Cafarella, J. Madhavan, A. Halevy, Web-scale extraction of structured data. SIGMOD Rec. **37**(4), 55–61 (2009)
9. M.J. Cafarella, C. Re, D. Suciu, O. Etzioni, M. Banko, Structured querying of web text, in *3rd Biennial Conference on Innovative Data Systems Research (CIDR)* (Asilomar, California, USA, 2007)
10. J. Carbonell, J. Goldstein, The use of MMR, diversity-based reranking for reordering documents and producing summaries, in *Proceedings of the 21st Annual International ACM SIGIR Conference on Research and Development in Information Retrieval, SIGIR '98* (ACM, New York, NY, USA, 1998), pp. 335–336
11. S. Chaudhuri, R. Ramakrishnan, G. Weikum, Integrating DB and IR technologies: what is the sound of one hand clapping, in *CIDR* (2005), pp. 1–12
12. J. Cohen, B. Dolan, M. Dunlap, J.M. Hellerstein, C. Welton, Mad skills: new analysis practices for big data. Proc. VLDB Endow. **2**, 1481–1492 (2009)
13. N. Dalvi, A. Machanavajjhala, B. Pang, An analysis of structured data on the web. Proc. VLDB Endow. **5**(7), 680–691 (2012)
14. E. Demidova, P. Fankhauser, X. Zhou, W. Nejdl, Divq: Diversification for keyword search over structured databases, in *Proceedings of the 33rd International ACM SIGIR Conference on Research and Development in Information Retrieval, SIGIR '10* (ACM, New York, NY, USA, 2010), pp. 331–338
15. G.W. DePuy, R.J. Moraga, G.E. Whitehouse, Meta-raps: a simple and effective approach for solving the traveling salesman problem. Transp. Res. Part E Logist. Transp. Rev. **41**(2), 115–130 (2005)
16. G. Di Lorenzo, H. Hacid, Hy Paik, B. Benatallah, Data integration in mashups. SIGMOD Rec. **38**(1), 59–66 (2009)
17. P. Dixon, Basics of oracle text retrieval. IEEE Data Eng. Bull. **24**(4), 11–14 (2001)
18. X.L. Dong, B. Saha, D. Srivastava, Less is more: selecting sources wisely for integration, in *Proceedings of the 39th international conference on Very Large Data Bases, PVLDB'13, VLDB Endowment* (2013), pp. 37–48
19. J. Eberius, K. Braunschweig, M. Hentsch, M. Thiele, A. Ahmadov, W. Lehner, Building the dresden web table corpus: a classification approach, in *2nd IEEE/ACM International Symposium on Big Data Computing, BDC* (2015)
20. J. Eberius, M. Thiele, K. Braunschweig, W. Lehner, DrillBeyond: enabling business analysts to explore the web of open data, in *PVLDB* (2012)

21. J. Eberius, M. Thiele, K. Braunschweig, W. Lehner, Drillbeyond: processing multi-result open world SQL queries, in *Proceedings of the 27th International Conference on Scientific and Statistical Database Management, SSDBM '15* (ACM, New York, NY, USA, 2015), pp. 16:1–16:12

22. J. Eberius, M. Thiele, K. Braunschweig, W. Lehner, Top-k entity augmentation using consistent set covering, in *Proceedings of the 27th International Conference on Scientific and Statistical Database Management, SSDBM '15* (ACM, New York, NY, USA, 2015), pp. 8:1–8:12

23. S. Endrullis, A. Thor, E. Rahm, Entity search strategies for mashup applications, in *2012 IEEE 28th International Conference on Data Engineering (ICDE)* (IEEE, New Jersey, 2012), pp. 66–77

24. S. Endrullis, A. Thor, E. Rahm, Wetsuit: an efficient mashup tool for searching and fusing web entities. Proceedings of the VLDB Endowment **5**(12), 1970–1973 (2012)

25. L. Etcheverry, A. Vaisman, Enhancing olap analysis with web cubes, in *The Semantic Web: Research and Applications*, vol. 7295, Lecture Notes in Computer Science, ed. by E. Simperl, P. Cimiano, A. Polleres, O. Corcho, V. Presutti (Springer, Berlin Heidelberg, 2012), pp. 469–483

26. T.A. Feo, M.G. Resende, Greedy randomized adaptive search procedures. J. Glob. Optim. **6**(2), 109–133 (1995)

27. F. Glover, Tabu search-part i. ORSA J. Comput. **1**(3), 190–206 (1989)

28. L. Grammel, M.A. Storey, A survey of mashup development environments, in *The Smart Internet, Lecture Notes in Computer Science* vol. 6400 (2010), pp. 137–151

29. R. Gupta, S. Sarawagi, Answering table augmentation queries from unstructured lists on the web. Proc. VLDB Endow. **2**(1), 289–300 (2009)

30. A. Halevy, A. Rajaraman, J. Ordille, Data integration: the teenage years, in *Proceedings of the 32nd International Conference on Very Large Data Bases, VLDB '06, VLDB Endowment* (2006), pp. 9–16

31. J.R. Hamilton, T.K. Nayak, Microsoft SQL server full-text search. IEEE Data Eng. Bull. **24**(4), 7–10 (2001)

32. M. Hasan, A. Mueen, V. Tsotras, E. Keogh, Diversifying query results on semi-structured data, in *Proceedings of the 21st ACM International Conference on Information and Knowledge Management, CIKM '12* (ACM, New York, NY, USA, 2012), pp. 2099–2103

33. Z.G. Ives, D. Florescu, M. Friedman, A. Levy, D.S. Weld, An adaptive query execution system for data integration, in *Proceedings of the 1999 ACM SIGMOD International Conference on Management of Data, SIGMOD '99* (ACM, New York, NY, USA, 1999), pp. 299–310

34. R.M. Karp, Reducibility among combinatorial problems, in *Complexity of Computer Computations* (1972)

35. G. Lan, G.W. DePuy, G.E. Whitehouse, An effective and simple heuristic for the set covering problem. Eur. J. Oper. Res. **176**(3), 1387–1403 (2007)

36. D. Laney, 3d data management: controlling data volume, velocity and variety. META Group Res. Note **6**, 70 (2001)

37. O. Lehmberg, D. Ritze, P. Ristoski, R. Meusel, H. Paulheim, C. Bizer, The mannheim search join engine, in *Web Semantics: Science, Services and Agents on the World Wide Web* (2015)

38. X. Li, X.L. Dong, K. Lyons, W. Meng, D. Srivastava, Truth finding on the deep web: is the problem solved? in *Proceedings of the 39th International Conference on Very Large Data Bases, PVLDB'13, VLDB Endowment* (2013), pp. 97–108

39. J. Liu, X. Dong, A.Y. Halevy, Answering structured queries on unstructured data, in *WebDB*, vol. 6 (Citeseer, 2006), pp. 25–30

40. A. Löser, F. Hueske, V. Markl, Situational business intelligence, in *Business Intelligence for the Real-Time Enterprise*, vol. 27, Lecture Notes in Business Information Processing, ed. by M. Castellanos, U. Dayal, T. Sellis (Springer, Berlin, 2009), pp. 1–11

41. A. Maier, D.E. Simmen, DB2 optimization in support of full text search. IEEE Data Eng. Bull. **24**(4), 3–6 (2001)

42. G. Marchionini, Exploratory search: From finding to understanding. Commun. ACM **49**(4), 41–46 (2006)

43. R. Martí, M.G. Resende, C.C. Ribeiro, Multi-start methods for combinatorial optimization. Eur. J. Oper. Res. **226**(1), 1–8 (2013)
44. H. Mohanty, P. Bhuyan, D. Chenthati, Big Data: A Primer (Springer, India, 2015)
45. J. Morcos, Z. Abedjan, I.F. Ilyas, M. Ouzzani, P. Papotti, M. Stonebraker, Dataxformer: an interactive data transformation tool, in *Proceedings of the 2015 ACM SIGMOD International Conference on Management of Data* (ACM, New Jersey, 2015), pp. 883–888
46. T.T. Nguyen, Q.V.H. Nguyen, M. Weidlich, K. Aberer, Result selection and summarization for web table search, in *2015 IEEE 31st International Conference on Data Engineering (ICDE)* (2015), pp. 231–242
47. D.E. O'Leary, Embedding ai and crowdsourcing in the big data lake. IEEE Intell. Syst. **29**(5), 70–73 (2014)
48. R. Pimplikar, S. Sarawagi, Answering table queries on the web using column keywords, in *Proceedings of the 36th Int'l Conference on Very Large Databases (VLDB)* (2012)
49. J. Pound, I.F. Ilyas, G. Weddell, Expressive and flexible access to web-extracted data: a keyword-based structured query language, in *Proceedings of the 2010 ACM SIGMOD International Conference on Management of Data, SIGMOD '10* (ACM, New York, NY, USA, 2010), pp. 423–434
50. S. Sarawagi, S. Chakrabarti, Open-domain quantity queries on web tables: Annotation, response, and consensus models, in *Proceedings of the 20th ACM SIGKDD International Conference on Knowledge Discovery and Data Mining, KDD '14* (ACM, New York, NY, USA, 2014), pp. 711–720
51. D.E. Simmen, M. Altinel, V. Markl, S. Padmanabhan, A. Singh, Damia: data mashups for intranet applications, in *Proceedings of the 2008 ACM SIGMOD international conference on Management of data, SIGMOD '08* (ACM, New York, NY, USA, 2008)
52. F.M. Suchanek, G. Kasneci, G. Weikum, Yago: a core of semantic knowledge, in *Proceedings of the 16th International Conference on World Wide Web, WWW '07* (ACM, New York, NY, USA, 2007), pp. 697–706
53. M.A. Tahraoui, K. Pinel-Sauvagnat, C. Laitang, M. Boughanem, H. Kheddouci, L. Ning, A survey on tree matching and XML retrieval. Comput. Sci. Rev. **8**, 1–23 (2013)
54. A. Thor, D. Aumueller, E. Rahm, Data integration support for mashups, in *Workshops at the Twenty-Second AAAI Conference on Artificial Intelligence* (2007)
55. G. Weikum, DB and IR: both sides now, in *Proceedings of the 2007 ACM SIGMOD International Conference on Management of Data* (ACM, New Jersey, 2007), pp. 25–30
56. M. Yakout, K. Ganjam, K. Chakrabarti, S. Chaudhuri, Infogather: entity augmentation and attribute discovery by holistic matching with web tables, in *Proceedings of the 2012 ACM SIGMOD International Conference on Management of Data, SIGMOD '12* (ACM, New York, NY, USA, 2012), pp. 97–108
57. M. Zhang, K. Chakrabarti, Infogather+: semantic matching and annotation of numeric and time-varying attributes in web tables, in *Proceedings of the 2013 International Conference on Management of data, SIGMOD '13* (ACM, New York, NY, USA, 2013), pp. 145–156
58. C.N. Ziegler, S.M. McNee, J.A. Konstan, G. Lausen, Improving recommendation lists through topic diversification, in *Proceedings of the 14th International Conference on World Wide Web, WWW '05* (ACM, New York, NY, USA, 2005), pp. 22–32

Pattern Matching Over Linked Data Streams

Yongrui Qin and Quan Z. Sheng

Abstract This chapter leverages semantic technologies, such as Linked Data, which can facilitate machine-to-machine (M2M) communications to build an efficient information dissemination system for semantic IoT. The system integrates Linked Data streams generated from various data collectors and disseminates matched data to relevant data consumers based on triple pattern queries registered in the system by the consumers. We also design two new data structures, *TP-automata* and *CTP-automata*, to meet the high performance needs of Linked Data dissemination. We evaluate our system using a real-world dataset generated from a Smart Building Project. With the new data structures, the proposed system can disseminate Linked Data faster than the existing approach with thousands of registered queries.

Keywords Internet of things · Linked data · Pattern matching

1 Overview

The Internet is a global system of networks interconnecting computers using the standard Internet Protocol (IP) suite. It has created a significant impact to the world by serving billions of users worldwide. Millions of private, public, academic, business, and government networks, of local to global scope, all contribute to the formation of the Internet. The traditional Internet has a focus on computers and can be called the Internet of Computers. In contrast, the Internet of Things (IoT) aims to connect everyday objects, such as coats, shoes, watches, ovens, washing machines, bikes, cars, even humans, plants, animals, and changing environments, to the Internet to enable communications/interactions between these objects [1]. The ultimate goal of

Y. Qin (✉)
School of Computing and Engineering, University of Huddersfield,
Huddersfield HD1 3DH, UK
e-mail: y.qin2@hud.ac.uk

Q.Z. Sheng
School of Computer Science, The University of Adelaide, Adelaide, SA 5005, Australia
e-mail: michael.sheng@adelaide.edu.au

© Springer International Publishing AG 2017   409
A.Y. Zomaya and S. Sakr (eds.), *Handbook of Big Data Technologies*,
DOI 10.1007/978-3-319-49340-4_12

IoT is to enable computers to see, hear and sense the real world. It is predicted by Ericsson that the number of Internet-connected things will reach 50 billion by 2020. In the era of IoT, it is envisioned that smart objects collect and share data at a global scale via the Internet.

As of 2012, 2.5 quintillion (2.5×10^{18}) bytes of data were being created daily.[1] In IoT, connecting all of the things that people care about in the world becomes possible. However, all these things will produce much more data than nowadays. The volumes of data are vast, the generation speed of data is fast and the data/information space is global. By exploiting such data in IoT, cities will become smarter and more efficient. Some promising IoT applications in future smart cities include resources management issues for modern cities [2], and effective urban street-parking management for reducing traffic congestion and fuel consumption [3]. Indeed, IoT is one of the major driving forces for *big data analytics*.

Given the scale of IoT, topics such as distributed processing, real-time data stream analytics, and event processing are all critical, and need to be revisited in order to improve upon existing technologies for applications of this scale [4, 5]. In this context, semantic technologies such as Linked Data (see http://linkeddata.org/), which aim to facilitate machine-to-machine (M2M) communications, play an increasingly important role [6]. Linked Data is part of a growing trend towards highly distributed systems, with thousands or potentially millions of independent sources providing structured data. Due to the large amount of data produced by various kinds of things, one challenging issue is how to *efficiently disseminate data* to relevant data consumers.

In this chapter, we focus on studying the Internet of Things (IoT) from a *data perspective*. As in IoT, data is processed differently compared with the processing in the traditional Internet environments (i.e., Internet of Computers). In the Internet of Computers, both the main data producers and consumers are human beings. However, in the Internet of Things, the main actors become *things*, which means things are the majority of data producers and consumers. Therefore, in IoT, addressable and interconnected things, instead of humans, act as the main data producers, as well as the main data consumers. Computers will be able to learn and gain information and knowledge to solve real world problems directly with the data fed from things. As an ultimate goal, computers enabled by IoT technologies will be able to sense and react to the real world for humans.

To deal with such challenge, it is imperative to efficiently retrieve the most relevant data from the big data generated in IoT and effectively extract useful information (e.g., in the process converting "data" into "information" or "knowledge"). We propose in this chapter an efficient data dissemination system for semantic IoT by leveraging semantic technologies, such as Linked Data. Our system will be very helpful and efficient in the retrieval of relevant data from the deluge of IoT data, which can then facilitate the extraction of required information. The system firstly integrates data generated from various data collectors. Then it transforms all the data into Linked Data streams in Resource Description Framework (RDF) format

---

[1]http://www-01.ibm.com/software/data/bigdata/.

(see www.w3.org/RDF). Meanwhile, data consumers can register their interests in the form of Basic Graph Patterns (BGPs) composed of simple triple patterns in the system. Based on these BGPs, the system disseminates matched Linked Data to relevant users. After receiving relevant data, these users can further make use of the data to extract information for their own purposes, such as environment monitoring, event detection, complex event processing, and so on. It should be noted that we will not discuss the data processing at the user side, instead we focus ourselves on how to efficiently match a large number of BGP queries against Linked Data streams in batch mode. We highlight our main contributions in the following.

- We identify new Linked Data dissemination needs in the context of the Internet of Things, which requires to process continuous data requests in batch mode efficiently.
- We develop two new data structures, Triple Pattern automata (TP-automata) and Conjunctive Triple Pattern automata (CTP-automata), for efficiently matching Linked Streams against a large number of single or conjunctive triple pattern queries based on automata techniques. We also develop novel techniques to evaluate these queries efficiently.
- We conduct extensive experiments using a real-world dataset generated in a Smart Building Project. The results show that: (1) when processing single triple patterns using TP-automata, the system can disseminate Linked Data at one million triples per second with 100,000 registered user queries, which is several orders of magnitude faster than existing techniques; (2) when processing conjunctive triple patterns using CTP-automata, the system can disseminate Linked Data at a speed of an order of magnitude faster than the existing approaches with thousands registered conjunctive queries.

The rest of this chapter is organized as follows. We present the framework and the technical details of our approach in Sect. 2. In Sect. 3, we report the results of an extensive experimental study. In Sect. 4, we review the related work. Finally, we present some concluding remarks in Sect. 5.

2 Linked Data Dissemination System

In order to disseminate high-quality information and provide high-performance matching services to data consumers (or subscribers), we aim to design a system that will not return false-negative match results. Therefore, we investigate pattern matching in this article. Pattern matching performs individual component matching between RDF triples and BGPs. It does not consider semantic relatedness between an RDF triple and a BGP. It may return false positive matching results but not false negative ones. Recent work on pattern matching includes Linked Data stream processing [7] and stream reasoning [8]. However, since these solutions are mainly designed for optimizations of individual query evaluations, they are not quite suitable for processing a large number of concurrent queries.

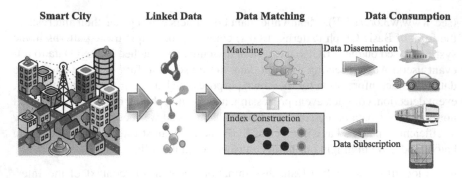

Fig. 1 System overview

An example of pattern matching is that pattern (?s, :is, :Student) will match triple (:James, :is, :Student) but will not match (:James, :is, :PhDStudent). Other types of matching include match estimation and semantic matching, both of which may return false-negative results. Again, take pattern (?s, :is, :Student) as an example. In match estimation, the main task is to estimate which dataset matches pattern (?s, :is, :Student) the best by using some summarization techniques among multiple datasets [9] to avoid querying all known datasets directly. In contrast, semantic matching will match semantically related triples compared to a specified pattern [10]. For example, pattern (?s, :is, :Student) may match (:James, :is, :PhDStudent) since the term :Student in the pattern is semantically related to :PhDStudent in the triple.

2.1 System Overview

Figure 1 shows an overview of our system in the smart city scenario. We assume that data generated by all kinds of things will be represented in the form of Linked Data streams using RDF (for the purpose of facilitating (M2M communications). Our system consists of two main components: the *matching component* and the *index construction component*. Data consumers (humans and/or smart things in the city) can register their interests as user queries in the system. The index construction component constructs an index for all user queries. The matching component evaluates the incoming Linked Data streams against the constructed index for efficiently matching triples to the user queries. Finally, the system disseminates the matched data to relevant data consumers for their further processing.

User Queries. Similar to [11, 12], triple patterns are adopted as the basic units of user queries in our system. A triple pattern is an expression of the form (s, p, o) where s and p are URIs or variables, and o is a URI, a literal or a variable. The eight possible triple patterns are: (1) (#s, #p, #o), (2) (?s, #p, #o), (3) (#s, ?p, #o), (4) (#s, #p, ?o), (5) (?s, ?p, #o), (6) (?s, #p, ?o), (7) (#s, ?p, ?o), and (8) (?s, ?p, ?o). Here, ? denotes a variable while # denotes a constant. Similar to data summaries in [9], we can also apply hash functions[2] to map these patterns into numerical values.

When a user query contains only one triple pattern, such queries are called single triple pattern queries. Mean while, a user query can also be expressed as a *conjunctive* triple pattern query composed of multiple triple patterns [11, 12]. Conjunctive queries can express data needs much more accurately compared with single triple pattern queries. A conjunctive query q has the form of:

$$?x_1, \ldots, ?x_n : (s_1, p_1, o_1) \wedge (s_2, p_2, o_2) \wedge \cdots \wedge (s_n, p_n, o_n)$$

where $?x_1, \ldots, ?x_n$ are variables, each (s_i, p_i, o_i) is a triple pattern, and each variable $?x_i$ appears in at least one triple pattern (s_i, p_i, o_i). Variables will always start with the '?' character. Variables $?x_1, \ldots, ?x_n$ are also called *answer variables* in order to distinguish them from other variables in the query.

Representations of Queries and Triples. In our Linked Data dissemination system, when the user queries (in the form of single or conjunctive triple pattern queries) are registered, all queries are transformed into numerical values. The reason for this is that the comparisons between numbers are faster than strings [9]. Note that, in such case, we will have three numbers for the three components in a query as described above. Then a suitable index can be constructed for efficient evaluation between Linked Data streams and user queries. Before a matching process starts, RDF triples in the data streams will be mapped into numerical values as well. Then, these numerical represented triples will be matched with conjunctive queries represented as numerical values in the constructed indexes.

2.2   TP-Automata for Single Triple Pattern Query Matching

Automata techniques have been adopted to process XML data streams [13]. They are mainly based on languages with SQL-like syntax, and relational database execution models adapted to process streaming data. In our system, to support *pattern matching*, we also apply automata to match each individual component of a triple with its counterparts of triple patterns in single triple pattern queries efficiently. We call this approach *Triple Pattern automata* (TP-automata).

As mentioned, operating on numbers is more efficient than operating on strings. Note that when we map triple patterns into numerical values, we treat variables in a triple pattern as a universal match indicator, e.g., represented by "?". This indicator will be mapped into a fixed and unique numerical value but not the whole range of a specific coordinate axis. This unique numerical value will be treated differently as well later in the triple evaluation process.

---

[2]There are many different hash functions that are suitable for this purpose. For more details, please refer to [9].

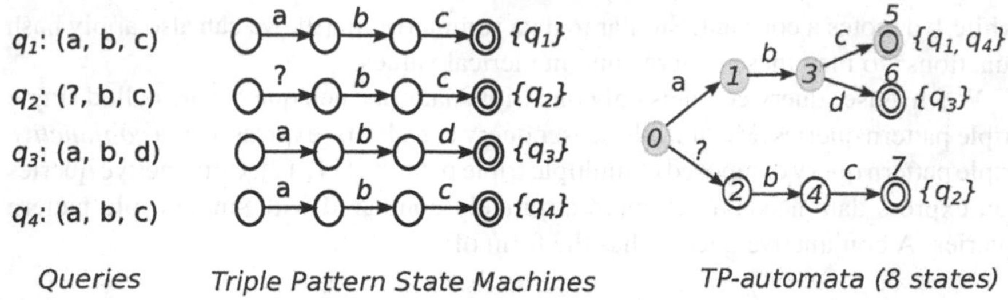

q_1: (a, b, c)

q_2: (?, b, c)

q_3: (a, b, d)

q_4: (a, b, c)

Queries   Triple Pattern State Machines   TP-automata (8 states)

Fig. 2 Structure of TP-automata

Figure 2 depicts the construction process of TP-automata. Firstly, user queries will be transformed into triple pattern state machines as shown in the middle of Fig. 2. As can be seen from the figure, each triple state machine contains an initial state, two internal states, one final state and three transitions. In the figure, the first circle of a state machine represents the initial state, the next two circles represent the two internal states and the doubled circle represents the final state. The three arrows associated with conditions are three transitions between different states. Similar to [13], these state machines can be combined into one machine by exploiting shared common states with same transitions. The combined machine, TP-automata, is shown on the right of Fig. 2. The shaded circles represent combined states.

To perform pattern matching over TP-automata, triples in the Linked Data stream will be firstly mapped into numerical values. For example, suppose a triple (s, p, o) is mapped into a 3D point (a, b, c). The system will match it against TP-automata in the following process. It firstly checks the initial state of TP-automata and looks for state transitions with condition a or condition ?. Following the state transitions, state 1 and state 2 become the current active states at the same time. It then looks for state transitions with condition b or ? from state 1 and state 2. Following the transitions, state 3 and state 4 become active states. Finally, following transitions with condition c or ? from state 3 and state 4, two final states, state 5 and state 7, are reached. By checking both final states, the system returns q_1, q_2, q_4 as the matching results. It should be noted that q_3: (a, b, d) will not match the input triple (a, b, c) as its object component's pattern is d, which does not match with c. The match process stops if and only if all current active states are final states or states with no satisfied transition.

2.3 CTP-Automata for Conjunctive Triple Pattern Query Matching

We also apply automata to match each individual component of a triple with its counterparts of triple patterns in conjunctive triple pattern query efficiently. Similarly, we call this approach *Conjunctive Triple Pattern automata* (CTP-automata).

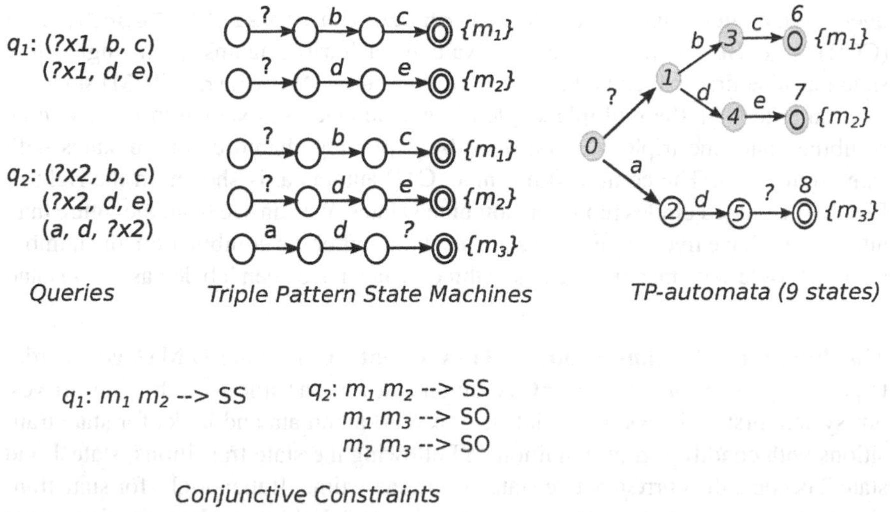

q_1: (?x1, b, c)
(?x1, d, e)

q_2: (?x2, b, c)
(?x2, d, e)
(a, d, ?x2)

Queries   Triple Pattern State Machines   TP-automata (9 states)

q_1: m_1 m_2 --> SS q_2: m_1 m_2 --> SS
m_1 m_3 --> SO
m_2 m_3 --> SO

Conjunctive Constraints

Fig. 3 Index structure and conjunctive constraints of CTP-automata

Construction of CTP-automata. Figure 3 depicts the construction process of CTP-automata. There are two conjunctive queries, q_1 : $(?x1, b, c) \wedge (?x1, d, e)$ and q_2 : $(?x2, b, c) \wedge (?x2, d, e) \wedge (a, d, ?x2)$. Accordingly, there are two triple patterns in q_1 and three triple patterns in q_2. Firstly, all the triple patterns in the conjunctive queries will be transformed into triple pattern state machines as shown in the middle of Fig. 3. As can be seen from the middle part of the figure, each triple state machine contains an initial state, two internal states, one final state, and three transitions. In the figure, the first circle of a state machine represents the initial state, the next two circles represent the two internal states and the doubled circle represents the final state. The three arrows associated with conditions represent three transitions between different states.

Suppose there are n conjunctive queries for the construction of CTP-automata and each query contains p patterns on average. Then the time complexity of the CTP-automata construction process will be $O(np)$. This is because we can add each pattern into the CTP-automata in an incremental manner and each pattern will require constant time to be inserted (we can adopt hashing based data structures to achieve constant time insertion of each pattern).

It is worth mentioning that we ignore variable names at this stage due to the fact that when processing triples in the Linked Data stream, at the first step, we have to evaluate these triples one by one and that variable naming does not have any relationships between different conjunctive queries. For example, $(?x1, b, c) \wedge (?x1, d, e)$ and $(?x2, b, c) \wedge (?x2, d, e)$ actually refer to the same conjunctive query. However, the variable naming does matter within the same conjunctive query. For example, in $(?x1, b, ?x2) \wedge (?x2, d, e)$, variables $?x1$ and $?x2$ refer to different triple components. We leave the resolution of different variable names within the same conjunctive query in the later stage, called `Conjunctive Constraints Resolution` (CCR) stage. Before that, we need to evaluate each triple against each single triple state machine first, which is the `Triple Pattern Matching` (TPM) stage.

Similar to [13], the multiple single triple state machines shown in Fig. 3 can be combined into one triple state machine by exploiting shared common states with same transitions. The combined machine, CTP-automata, is shown on the right of Fig. 3. The shaded circles represent combined states. We can see from the figure that, although we have five single triple state machines, after the combination, the number of single triple state machines drops to three, which have been labeled as m_1, m_2 and m_3, respectively.

Matching Triple Streams against CTP-automata. During the TPM stage, in order to perform *pattern matching* over CTP-automata, when a triple (a, b, c) arrives, our system firstly checks the initial state of CTP-automata and looks for state transitions with condition a or condition ?. Following the state transitions, state 1 and state 2 become the current active states at the same time. It then looks for state transitions with condition b or ? from state 1 and state 2. Following the transitions, only state 3 becomes active state and there is no transition triggered from state 2. Finally, following the transition with condition c or ? from state 3, one final state, state 6, is reached. By checking this final state, the system returns $\{m_1\}$ as the matching result. The matching process stops if and only if all current active states are final states or states with no satisfied transition.

At this TPM stage, the matching results are only intermediate results and the matched triples are just possible candidates which may satisfy some conjunctive queries. In order to determine which conjunctive queries have been satisfied, we need to further evaluate some `conjunctive constraints`, which will be detailed next.

Conjunctive Constraints Resolution (CCR) of CTP-automata. It should be noted that in order to match q_1 and q_2 in Fig. 3, all triple patterns they contain must be matched first. Take query $q_1 : (?x1, b, c) \land (?x1, d, e)$ as an example. Suppose that triples t_1 and t_2 match triple patterns $(?x1, b, c)$ and $(?x1, d, e)$, respectively. To ensure that query q_1 can be satisfied by t_1 and t_2, we need to check first that whether we have $t_1.s = t_2.s$. We call such conditions as *conjunctive constraints* of a conjunctive query. All conjunctive constraints must be satisfied before we can assure that a conjunctive query is satisfied. As mentioned before, the conjunctive constraints checking occurs at the CCR stage.

In this chapter, we have identified ten conjunctive constraints, including SS, PP, OO, SO, OS, SSPP, SSOO, PPOO, SOPP, OSPP. Constraint SS means that the subjects of two candidate triples must be matched. More details are shown in Table 1. These constraints can be used to determine whether a conjunctive query has been satisfied or not so far in the stream.

For example, in Fig. 3, query q_1's conjunctive constraint is $m_1m_2 \rightarrow$ SS and query q_2's conjunctive constraints are $m_1m_2 \rightarrow$ SS, $m_1m_3 \rightarrow$ SO and $m_2m_3 \rightarrow$ SO. Suppose that triples t_1, t_2, t_3 match triple pattern machines m_1, m_2, m_3, respectively.

Table 1 Conjunctive constraints

Conjunctive constraints	Description	Checking details
SS	The subjects of two candidate triples must be matched	$t_1.s = t_2.s$
PP	The predicates of two candidate triples must be matched	$t_1.p = t_2.p$
OO	The objects of two candidate triples must be matched	$t_1.o = t_2.o$
SO	The subject of a candidate triple in the first pattern machine matches the object of a candidate triple in the second pattern machine	$t_1.s = t_2.o$
OS	The object of a candidate triple in the first pattern machine matches the subject of a candidate triple in the second pattern machine	$t_1.o = t_2.s$
SSPP	The conjunction of both SS and PP constraints	$t_1.s = t_2.s$ and $t_1.p = t_2.p$
SSOO	The conjunction of both SS and OO constraints	$t_1.s = t_2.s$ and $t_1.o = t_2.o$
PPOO	The conjunction of both PP and OO constraints	$t_1.p = t_2.p$ and $t_1.o = t_2.o$
SOPP	The conjunction of both SO and PP constraints	$t_1.s = t_2.o$ and $t_1.p = t_2.p$
OSPP	The conjunction of both OS and PP constraints	$t_1.o = t_2.s$ and $t_1.p = t_2.p$

According to Table 1, for t_1, t_2 to satisfy q_1, we need to have $t_1.s = t_2.s$. Similarly, for t_1, t_2, t_3 to satisfy q_2, we need to have $t_1.s = t_2.s$, $t_1.s = t_3.o$ and $t_2.s = t_3.o$.

Dynamic Maintenance of the Matching Candidate List. In order to check conjunctive constraints, triples in the stream that match some triple pattern machines will be buffered for this purpose. Since the Linked Stream can be considered infinite, the buffered triple lists for triple pattern machines may grow all the time. To avoid this issue, we need to specify a sliding window to confine our matching scope. That is, we only consider matching within the sliding window.

Figure 4 shows two sliding windows with size T: w_1 and w_2, where only the most recent T triples will be considered for our matching. In order to evaluate conjunctive constraints, we need to update the buffered candidate triple list each time a triple arrives in or leaves the window. In Fig. 4, for w_1, we have got matching results for all three single triple pattern machines, $m1, m2, m3$, in Fig. 3. After receiving a new tripe, t_{i+T}, the oldest triple t_i will be removed from all candidate lists that contain t_i. In this example, only candidate list for $m1$ contains t_i and hence t_i will be removed from that candidate list. Further, suppose the new arriving triple t_{i+T} will be matched with machine $m3$. Then t_{i+T} will be added to the candidate list for $m3$. It is obvious that, each time when the sliding window moves forward by one triple, we should consider all the buffered lists affected by the leaving triple and the joining triple in the sliding window to verify conjunctive constraints.

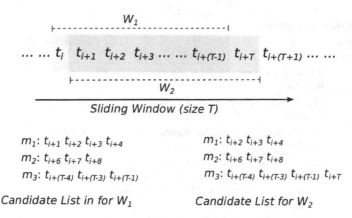

Fig. 4 Maintenance of candidate triple list

3 Experimental Evaluation

In this section, we report our experimental evaluation of the proposed approach. We will first describe the experimental settings, and then report the experimental results.

3.1 Experimental Setup

The dataset used in our experiments was generated in a Smart Building Energy Project [14]. The energy readings were collected from 4–19 August 2014. In total, there are around 6.2 million triples in the dataset. An event example is depicted in Fig. 5. This event is a power consumption event, showing the real-time power consumption in Room01 of building01. As shown in the event, the power consumption in Room01 at the moment of "2014 08 12T18:17:18" was 171.87 watts.

We evaluated the performance of our approach in terms of *average construction time* (in milliseconds) of the indexes and *average throughput* (in number of triples per second). We compared hash-based implementation (i.e., mapping triples and queries into numerical values, denoted as *HashMat* in the figures) with string-based implementation (i.e., using triples and queries as it is, denoted as *StringMat* in the figures). We also compared our methods with an existing approach, CQELS [7], which supports parallel query evaluation on Linked Data streams. Both TP-automata and CTP-automata engines, and CQELS,[3] were all implemented on Java Platform Standard Edition 7 running on Linux (Ubuntu 12.10, 64-bit Operating System), with

---

[3]The source code and documentation of CQELS can be obtained via http://code.google.com/p/cqels/.

```
@prefix do: <http://energy.deri.ie/ontology#>
@prefix dr: <http://../deri/deri rooms#>
:event1026fd7b0e5a    a                        events:PowerConsumptionEvent  .
:event1026fd7b0e5a    do:consumer              do:platform   .
:event1026fd7b0e5a    do:consumerType          dr:Room01   .
:event1026fd7b0e5a    do:consumerLocation      dr:building01   .
:event1026fd7b0e5a    do:powerUsage            :usage9739ccddc76d   .
:event1026fd7b0e5a    do:consumerDepartment    "facilities"   .
:event1026fd7b0e5a    do:atTime                :timedb2c06100b33   .
:usage9739ccddc76d    a                        dul:Amount   .
:usage9739ccddc76d    do:hasDataValue          171.87   .
:usage9739ccddc76d    do:isClassifiedBy        dr:watt   .
:timedb2c06100b33     a                        do:Instant   .
:timedb2c06100b33     do:inDDateTime           "2014-08-12T18:17:18"   .
```

Fig. 5 An event example

quad-core CPU@2.20GHz and 4 GB main memory. We ran each experiment 10 times in order to ensure consistency of results and reported the average experimental results.

3.2 Evaluation of TP-Automata

As an initial work, we used simple BGPs that contain only a single triple pattern in a query as queries in the experiment. We randomly generated BGPs using the seven patterns mentioned in Sect. 2 based on our dataset. We did not consider (?s, ?p, ?o) in our experiment as it requires every triple in the Linked Data stream. In such case, no query index is needed. We generated from 10,000 queries to 100,000 queries.

Firstly, average construction time is compared in Fig. 6. The construction times for both hash-based TP-automata and string-based TP-automata are similar to each other in most settings. For larger numbers of queries, such as 75 and 100 k queries, the construction of string-based indexes takes slightly longer time. Normally, the construction can be completed within a few hundred milliseconds. However, the construction time of CQELS takes much longer, which normally requires around ten thousand milliseconds.

Throughput performance of pattern matching is depicted in Fig. 7. It shows some large differences between CQELS and TP-automata based approaches (HashMat and StringMat). Generally, HashMat and StringMat can achieve throughput at the speed of nearly a million triples per second and are about four orders of magnitude faster than CQELS. The main reason for this is that CQELS is a much more comprehensive system focusing on optimizing evaluation of queries with complex operators and semantics but not on evaluation of a large set of concurrent and simple queries over Linked Data streams. In this regard, our approach can also be adapted to complement CQELS for dealing with our Linked Data dissemination scenario. Regarding HashMat and StringMat, in most cases, HashMat is about twice throughput speed compared with StringMat.

Fig. 6 Average construction time of TP-automata

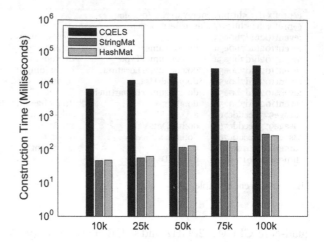

Fig. 7 Average throughput evaluation of TP-automata

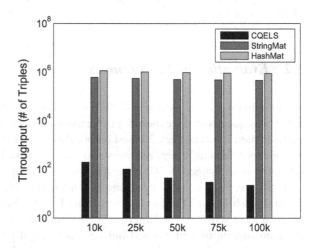

Finally, we also investigated the matching quality of hash-based TP-automata (HashMat) via Precision, Recall and F_1 Score [15]. This is because collisions are difficult to avoid in any hash-based approaches and false positives exist in hash-based TP-automata, which affects matching quality. Specifically, we look into Precision and F_1 Score when the Recall is 100%. As can be seen in Table 2, the Precision and F1 Score are 100% when the number of queries is 10 or 25 k. For larger numbers of queries (e.g., 50, 75 and 100 k), both Precision and F_1 Score are still greater than 99.99950%. This demonstrates that HashMat provides very high matching quality.

Table 2 Matching quality of hashmat (when the recall is 100%) in TP-automata

Queries (k)	Recall (%)	Precision (%)	F_1 Score (%)
10	100	100	100
25	100	100	100
50	100	99.99975	99.99987
75	100	99.99982	99.99991
100	100	99.99960	99.99980

Table 3 Workload parameters for the experiments of CTP-automata

Parameter	Range	Default	Description
Query number	100 to 2000	1000	The number of conjunctive triple pattern queries
Pattern number	1 to 5	3	The maximum number of triple patterns in a query
Window size	10 to 500	100	The window size, in terms of number of triples, for the evaluation of conjunctive triple pattern queries

3.3 Evaluation of CTP-Automata

Again, we used random walk method to generate conjunctive triple pattern queries in the experiments according to the data graph of the event data. The details of parameters we used for generating these queries are shown in Table 3. The parameters include query number, pattern number, and window size.

Construction Time. The average construction times of CTP-automata engines and CQELS engine is presented in Fig. 8. The construction times for both hash-based CTP-automata matching engine (HashMat) and string-based CTP-automata matching engine (StringMat) are close to each other in most settings and are always under 50 milliseconds. The construction of the string-based indexes takes slightly longer time. By contrast, the construction times of CQELS are much longer than CTP-automata engines. The main reason is that CQELS has to parse the conjunctive triple pattern queries using a SPARQL-like parser and then register the parsed queries in the processing engine. As shown from Fig. 8, the construction times of CQELS grow linearly with the number of conjunctive queries. When the query number is 100, the construction time is around 400 milliseconds. When the number of queries increases to 2000, the construction time reaches above 1610 milliseconds. This indicates that the construction of our CTP-automata engines is very fast.

Throughput. The throughput performance of *pattern matching* under varying query numbers is depicted in Fig. 9. It shows some similarities between HashMat and StringMat. In most cases, HashMat shows slightly better throughput speed compared

Fig. 8 Average construction
time of CTP-automata

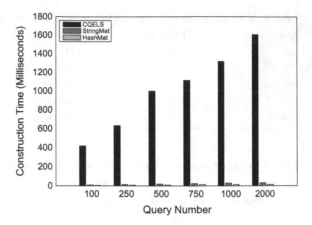

Fig. 9 Average throughput
evaluation of CTP-automata
(varying query number)

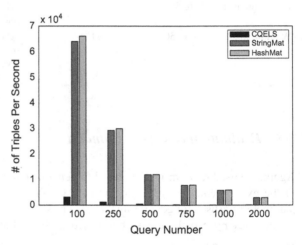

with StringMat. This indicates that although comparisons on strings are slower than
those on numbers, the differences betwen HashMat and StringMat are negligible.
The main reason for this is that the evaluation process of conjunctive queries spends
a large proportion of time to evaluate the conjunctive constraints on each query and
both HashMat and StringMat use the same strategy to evaluate all these conjunctive
constraints.

However, when compared with CQELS, both HashMat and StringMat outperform
CQELS significantly. To be specific, when the number of conjunctive queries is 100,
the throughput of HashMat and StringMat is more than 64,000 triples per second, and
for CQELS, just slightly more than 3,000. When the number of conjunctive queries
is 2,000, the throughput of HashMat and StringMat drops to slightly below 3,000
triples per second while CQELS has a throughput about just 50 triples per second.
From Fig. 9, we can observe that (1) HashMat and StringMat are normally 20 to 50
times faster than CQELS; (2) the throughput of HashMat, StringMat and CQELS all

Fig. 10 Average throughput evaluation of CTP-automata (varying pattern number)

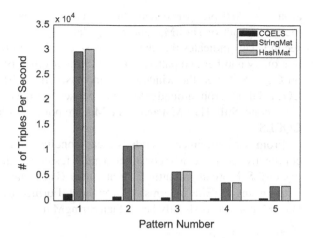

Fig. 11 Average throughput evaluation of CTP-automata (varying window size)

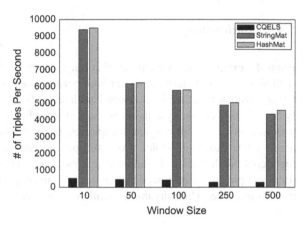

drops greatly when increasing the number of conjunctive queries. This also indicates that the evaluation of conjunctive constraints on each query takes a large amount of time and is difficult to share evaluation results between different conjunctive queries.

Figure 10 further demonstrates this finding. In the figure, we vary the maximum number of patterns of each conjunctive query. For the same amount of conjunctive queries, when the pattern number is only 1, the throughput of HashMat and StringMat is around 30,000 triples per second and for CQELS, it is around 1,200 triples per second, which is more than 20 times slower. When the pattern number is set to 5, the throughput of HashMat and StringMat drops to slightly lower than 3,000 triples per second and for CQELS, it drops to around 300 triples per second. This confirms that the evaluation of conjunctive constraints is time consuming. Similarly, HashMat and StringMat are both around an order of magnitude faster than CQELS.

Finally, Fig. 11 depicts the effect of window size, which is varied from 10 to 500. From the figure, we can observe that when the window size increases from 10 to 50, the throughput of HashMat and StringMat drops from 9,500 triples per second to around 6,200 triples per second. But when the window size increases from 50 to 500, the throughput of HashMat and StringMat only drops to around 4,500 triples per second. This indicates that the window size does not affect the throughput heavily like query number and pattern number. Similar effect of window size can be found on CQELS. When the window size increases from 10 to 500, the throughput of CQELS drops from around 500 triples per second to slightly lower than 300 triples per second. Still, HashMat and StringMat are both an order of magnitude faster than CQELS.

From our experimental study, we can conclude that CTP-automata indexes for conjunctive queries can be constructed much faster than the query registration process in CQELS. More importantly, CTP-automata (HashMat and StringMat) significantly outperforms CQELS in terms of throughput. Further, by using hashing techniques, HashMat performs slightly better than StringMat.

3.4 Limitations

From the experiments, we can see that our system can match Linked Data streams with single triple pattern queries at high performance, e.g., arriving at close to one million triples processed per second with 100,000 user queries registered in the system. However, our system cannot scale well when processing conjunctive triple pattern queries. As for conjunctive queries, the throughput of our system can only reach a few thousand triples processed per second with only 2,000 user queries registered. Such findings in our experiments suggest that there is an imperative need for developing more advanced techniques for handling a large number of conjunctive queries in Linked Data streams dissemination systems.

4 Related Work

Recent work on data summaries on Linked Data [9] transforms RDF triples into a numerical space. Then data summaries are built upon numerical data instead of strings as summarizing numbers is more efficient than summarizing strings. In order to transform triples into numbers, hash functions are applied on the individual components (s, p, o) of triples. Thus a derived triple of numbers can be considered as a 3D point. In this way, a set of RDF triples can be mapped into a set of points in a 3D space. To facilitate query processing over data summaries, a spatial index named QTree [9], which is evolved from standard R-tree [16], is adopted as the basic index. Data summaries are designed mainly for indexing various Linked Data sources and used for identifying relevant sources for a given query. However, data summaries are not suitable for our Linked Data dissemination system. Firstly, techniques on data summaries, such as QTree, do not consider variables in the BGPs but only RDF triples with concrete strings. Further, since data summaries are concise and imprecise representations of data sources [9], they just provide *match estimation*. Hence, query evaluation on them would return false negative results, which is not allowed in our system.

Pattern matching over streams has been studied in [17]. In order to represent each pattern query, a new query evaluation model is designed for processing pattern matching over RFID streams, by employing a new type of automaton that comprises a non-deterministic finite automaton (NFA) and a match buffer, named NFA^b. However, their techniques are not directly applicable to Linked Data stream processing as they do not specifically consider the characteristics of RDF data and conjunctive triple pattern matching.

In terms of triple pattern matching, a large body of work which focuses on optimizing individual query processing has also been put forward [12, 18–20]. Specifically, the problem of evaluating conjunctive triple pattern queries is studied in [12] in the context of Peer-to-Peer (P2P) networks. In [18], an indexing technique for efficient join processing on RDF graphs is proposed. The index is constructed upon RDF data directly but not join queries. Similarly, the work in [19] focuses on optimizing the processing of conjunctive triple pattern queries, especially star-shaped group based queries individually. Furthermore, optimization on RDF graph pattern matching on MapReduce is also studied in [20]. However, the common problem shared by these research efforts is that, they have not considered the scenarios of optimizing conjunctive triple pattern queries in batch mode, which is the focus of our work in this chapter.

Semantic matching has also been studied, which aims to match semantically related RDF triples against BGPs. It may provide false positive match results but not false negative. Both approximate event matching [10] and thematic event processing [21] apply semantic matching. Similarly, all these techniques will return false-negative matching results, which is not allowed in our system.

Some existing work on *pattern matching* of Linked Data, such as stream reasoning [8] and Linked Data stream processing [7], does not support large-scale query evaluation but focuses on the evaluation of a single query or a small number of parallel queries over the streaming Linked Data. Other existing work only studies pattern matching of multiple single triple patterns [22, 23], but not multiple conjunctive triple patterns. Therefore, the issue of supporting *pattern matching* over a large number of conjunctive triple patterns against Linked Data streams still remains open in these approaches.

5   Summary

In this chapter, we have leveraged semantic technologies, such as Linked Data, to build an efficient information dissemination system for semantic IoT. Firstly, in order to efficiently match a large number of user queries that contain only single triple patterns against Linked Data streams, we have proposed TP-automata, an automata based method designed for efficient pattern matching. In our evaluation, we show that

TP-automata can disseminate Linked Data at the speed of nearly one million triples per second with 100,000 registered user queries and is several orders of magnitude faster in terms of both index construction time and throughput compared with the state-of-the-art technique. Further, using hash-based TP-automata, the throughput is doubled compared with string-based TP-automata with high matching quality. Secondly, we have also investigated how to handle user queries with conjunctive triple patterns. In order to efficiently match a large number of conjunctive triple pattern queries against Linked Data streams in batch mode, similarly, we have proposed CTP-automata. In our evaluation, we show that CTP-automata can disseminate Linked Data an order of magnitude faster than the existing approaches. This confirms the efficiency and effectiveness of our proposed approach.

Acknowledgements We would like to thank the following researchers for their insightful feedback on our work: Edward Curry, Nickolas J.G. Falkner, and Ali Shemshadi.

References

1. Y. Qin, Q.Z. Sheng, N.J.G. Falkner, S. Dustdar, H. Wang, A.V. Vasilakos, When things matter: a survey on data-centric internet of things. J. Netw. Comput. Appl. (JNCA) (2016)
2. J. Gao, L.J. Guibas, N. Milosavljevic, D. Zhou, Distributed resource management and matching in sensor networks, in *Proceedings of the 8th International Conference on Information Processing in Sensor Networks (IPSN)* (IEEE, San Francisco, 2009), pp. 97–108
3. S. Mathur, T. Jin, N. Kasturirangan, J. Chandrasekaran, W. Xue, M. Gruteser, W. Trappe, Parknet: drive-by sensing of road-side parking statistics, in *Proceedings of the 8th International Conference on Mobile Systems, Applications, and Services (MobiSys)* (ACM, San Francisco, 2010), pp. 123–136
4. A.E. James, J. Cooper, K.G. Jeffery, G. Saake, Research directions in database architectures for the internet of things: a communication of the first international workshop on database architectures for the internet of things (DAIT 2009), in *Proceedings of the 26th British National Conference on Databases (BNCOD)* (Springer, Birmingham, 2009), pp. 225–233
5. P.M. Barnaghi, A.P. Sheth, C.A. Henson, From data to actionable knowledge: big data challenges in the web of things. IEEE Intell. Syst. **28**(6), 6–11 (2013)
6. P.M. Barnaghi, W. Wang, C.A. Henson, K. Taylor, Semantics for the internet of things: early progress and back to the future. Int. J. Semant. Web Inf. Syst. **8**(1), 1–21 (2012)
7. D.L. Phuoc, M. Dao-Tran, J.X. Parreira, M. Hauswirth, A native and adaptive approach for unified processing of linked streams and linked data, in *Proceedings of the 10th International Semantic Web Conference (ISWC)* (2011), pp. 370–388
8. D. Anicic, P. Fodor, S. Rudolph, N. Stojanovic, EP-SPARQL: a unified language for event processing and stream reasoning, in *Proceedings of the 20th International Conference on World Wide Web (WWW)* (2011), pp. 635–644
9. A. Harth, K. Hose, M. Karnstedt, A. Polleres, K.-U. Sattler, J. Umbrich, Data summaries for on-demand queries over linked data, in *WWW* (2010), pp. 411–420
10. S. Hasan, E. Curry, Approximate semantic matching of events for the internet of things. ACM Trans. Internet Techn. **14**(1), 1–23 (2014)
11. A. Seaborne, RDQL - a query language for RDF, in *W3C Member Submission* (2001)
12. E. Liarou, S. Idreos, M. Koubarakis, Evaluating conjunctive triple pattern queries over large structured overlay networks, in *ISWC* (2006), pp. 399–413
13. Y. Diao, M. Altinel, M.J. Franklin, H. Zhang, P.M. Fischer, Path sharing and predicate evaluation for high-performance XML filtering. ACM Trans. Database Syst. **28**(4), 467–516 (2003)

14. E. Curry, S. Hasan, S. O'Riain, Enterprise energy management using a linked dataspace for energy intelligence, in *SustainIT* (2012), pp. 1–6
15. D. Christopher, *Manning, Prabhakar Raghavan, and Hinrich Schütze*, Introduction to Information Retrieval (Cambridge, Cambridge University Press, 2008)
16. A. Guttman. R-trees: a dynamic index structure for spatial searching, in *SIGMOD* (1984), pp. 47–57
17. J. Agrawal, Y. Diao, D. Gyllstrom, N. Immerman, Efficient pattern matching over event streams, in *Proceedings of the ACM SIGMOD International Conference on Management of Data (SIGMOD'08)* (2008), pp. 147–160
18. G.H.L. Fletcher, P.W. Beck, Scalable indexing of RDF graphs for efficient join processing, in *CIKM* (2009), pp. 1513–1516
19. M.-E. Vidal, E. Ruckhaus, T. Lampo, A. Martínez, J. Sierra, A. Polleres, Efficiently joining group patterns in SPARQL queries. ESWC, Part **I**, 228–242 (2010)
20. P. Ravindra, H.S. Kim, K. Anyanwu, An intermediate algebra for optimizing RDF graph pattern matching on mapreduce. ESWC, Part **II**, 46–61 (2011)
21. S. Hasan, E. Curry, Thematic event processing, in *Proceedings of the 15th International Middleware Conference, Bordeaux, France, December 8-12, 2014* (2014), pp. 109–120
22. Y. Qin, Q.Z. Sheng, N.J.G. Falkner, A. Shemshadi, E. Curry, Towards efficient dissemination of linked data in the internet of things, in *Proceedings of the 23rd ACM International Conference on Conference on Information and Knowledge Management (CIKM)* (2014), pp. 1779–1782
23. Y. Qin, Q.Z. Sheng, E. Curry, Matching over linked data streams in the internet of things. IEEE Internet Comput. **19**(3), 21–27 (2015)

Searching the Big Data: Practices and Experiences in Efficiently Querying Knowledge Bases

Wei Emma Zhang and Quan Z. Sheng

Abstract Knowledge bases (KBs) are computer systems that store complex structured and unstructured facts, i.e., knowledge. KB are described as open shared database of the world's knowledge and typically use the entity-relational model. Most of the existing knowledge bases make their data in the RDF format. Tools including querying, inferencing and reasoning on facts are developed to consume the knowledge. In this chapter, we introduce a client-side caching framework aiming at accelerating the overall query response speed. In particular, we improve a suboptimal graph edit distance function to estimate the similarity of SPARQL queries and develop an approach to transform the SPARQL queries to feature vectors. Machine learning algorithms are leveraged using these feature vectors to identify similar queries that could potentially be the subsequent queries. We adapt multiple dimensional reduction algorithms to reduce the identification time. We then prefetch and cache the results of these queries aiming to improve the overall querying performance. We also develop a forecasting method, namely Modified Simple Exponential Smoothing, to implement the cache replacement. Our approach has been evaluated by using a very large set of real world queries. The empirical results show that our approach has great potential to enhance the cache hit rate and accelerate the querying speed on SPARQL endpoints.

1 Introduction

Knowledge Bases (KB) are widely used as one of the fundamental components in Semantic Web applications as they provide facts and relationships that can be automatically understood by machines (e.g., computer programs). Knowledge-based QA (KB-QA) systems are powered by knowledge bases and can automatically answer questions posed in natural languages. The knowledge bases usually use Resource

W.E. Zhang · Q.Z. Sheng (✉)
The University of Adelaide, North Terrace, Adelaide, SA 5005, Australia
e-mail: michael.sheng@adelaide.edu.au

W.E. Zhang
e-mail: wei.zhang01@adelaide.edu.au

© Springer International Publishing AG 2017   429
A.Y. Zomaya and S. Sakr (eds.), *Handbook of Big Data Technologies*,
DOI 10.1007/978-3-319-49340-4_13

Description Framework (RDF) as the data representation model. RDF allows the sharing and reuse of data across boundaries. SPARQL is the standard query language for RDF data [32]. In a KB-QA, a natural-language question is typically answered in two steps: (1) The question is transformed into a structured query (e.g., SPARQL query); and (2) The structured query is executed against a KB and the answers are returned [1, 2, 9, 41]. Many open KB provide interfaces to users. SPARQL endpoint is one of the widely used ways. SPARQL endpoints are interfaces that enable users to query these publicly accessible knowledge bases. As the SPARQL 1.1 specification introduces the SERVICE keyword, federated queries can be realized by using SERVICE to access data offered by other SPARQL endpoints. However, network instability and latency affect the query efficiency. Therefore, the most typical way for consumers who want to query public data is to download data dump and set up their own local SPRAQL endpoint. But data in a local endpoint is not up-to-date and hosting an endpoint requires expensive infrastructural support.

Many research efforts have been dedicated to circumvent this problem [24, 25, 36, 38, 40] and caching is one of the popular directions [34]. While most research efforts focus on providing a server-side caching mechanism, being embedded in triple stores, client-side caching has not been fully explored [25]. Our work provides a domain-independent client-side caching framework for SPARQL endpoints to facilitate the query answering process. Our approach is based on the observation that end users who consume RDF-modelled knowledge typically use programmatic query clients, e.g., software or services to retrieve information from SPARQL endpoints [24]. These queries usually have repetitive query patterns and only differ in specific elements of a triple pattern (a triple pattern includes three elements: subject, predicate and object), such as resources or literals. Moreover, they are usually issued subsequently. We illustrate two example queries in Fig. 1 to demonstrate two similar queries. Query 1 retrieves start year (i.e., the year their acting careers started) from the actors of the movie *Rain Man* and the year should be later than 1980. Query 2 requests the same information but for a different movie (*Eyes Wide Shut*). The differences between these two queries are the movie name (the underlined terms), which is the subject element of triple pattern "*movie* dbpedia-owl:starring ?actor" and the year in the Filter

Fig. 1 Example of similar queries. The queries only differ in the movie name and year

```
Query 1
SELECT ?actor ?year WHERE {
:Rain_Man dbpedia-owl:starring ?actor .
?actor dbpedia-owl:activeYearsStartYear ?year .
}
FILTER(?year>1980)

Query 2:
SELECT ?actor ?year WHERE {
:Eyes_Wide_Shut dbpedia-owl:starring ?actor .
?actor dbpedia-owl:activeYearsStartYear ?year .
}
FILTER(?year > 1960)
```

expression. We call the different element (movie name here) the *replacing element*. Thus, the similar queries are defined in our work as queries with same pattern and different *replacing elements* in its triple patterns.

By considering these observations, we propose a caching mechanism that is based on proactive fetching (i.e., prefetching) the query results of similar queries in advance. Since these similar queries are potentially subsequent queries, the cached results can accelerate the subsequent queries as the results are returned immediately rather than being retrieved from SPARQL endpoints. Thus, the average query response time will be reduced if the subsequent queries are already in the cache (cache hit). The key challenge to improve the hit rate centers on how to effectively generate queries that have a high possibility of being requested subsequently. We look into this issue and utilize machine learning techniques to suggest similar queries. We first define distance function to measure the distance between SPARQL queries by considering both Basic Graph Patterns (BGPs) and three keywords FILTER, BIND and VALUES. Then we develop an approach to transform SPARQL query to vector representation (referred to as Feature Modeling) using the distance function we defined. We train the *K Nearest Neighbour (KNN)* model with the feature vectors and suggest similar queries whose results need to be prefetched and cached. Three dimension reduction algorithms, Canonical Correlation Analysis (CCA) [16], Principal Component Analysis (PCA) [18] and Non-negative Matrix Factorization (NMF) [20] are adopted to accelerate the nearest neighbor calculations and achieve more accurate performance during KNN.

The suggestion process runs in a background thread to the query process. The training and mining process can be performed only once as a pre-computing step if there are records of historical queries. During the runtime, this background approach will give suggestions for similar queries based on the queries it has already processed. After generating similar queries, our algorithm prefetches the results of these queries and caches the (query, results) pairs. As the cache space is limited, less useful data should be removed from the cache. A cache replacement algorithm is introduced for this purpose. However, techniques for relational databases [7, 26, 29] cannot be directly applied into our client-side caching framework because our caching is record based, rather than traditional page-based caching algorithms. Moreover, our client-side application is not based on RDBMS and is not designed for server side as traditional caching algorithms do. In this work, we use a *time-aware frequency* based algorithm, which leverages the idea of a novel approach recently proposed for caching in main memory databases in *Online Transaction Processing (OLTP)* systems [23]. More specifically, we use Modified Simple Exponential Smoothing (MSES) to evaluate the frequencies of cached queries and remove the ones with the lowest scores from the cache.

The remainder of this chapter is structured as follows. We present some backgrounds in Sect. 2 and the framework in Sect. 3. We introduce how to identify and suggest similar queries in Sect. 4. Cache Replacement is discussed in Sect. 5. The evaluation and experimental results are reported in Sect. 6.

2 Background

Recent years, academics and industry researchers have put forth increasing efforts in knowledge bases, aiming at equipping the search engines with the understanding of users' natural language questions. With the aid of information provided by knowledge bases, useful facts and appropriate answers are extracted and selected automatically and more accurately from the large corpus of Web information. We will briefly introduce knowledge bases and its key components as well some industry involvements in this section.

2.1 Knowledge Base Preliminary

The knowledge base is originated from the expert system or called knowledge-based system which were first developed by Artificial Intelligence (AI) researchers. A knowledge-based system consists of two components: a knowledge base and an inference engine. The knowledge base represents facts about the world. The inference engine represents logical assertions and conditions about the world [33]. The first knowledge-based systems represented facts about the world as simple assertions in a flat database and used rules to reason these assertions. With the evolution of database systems, many kinds of database, such as graph database or object-oriented database are applied in maintaining KBs. Both structured and non-structure data can be managed by KBs. KBs in different field cover data from Government, publications, life sciences, media, geographic and social network etc.[1] There are also KBs that provide general information, such as DBpedia [22], Freebase [4] and YAGO [37]. Knowledge bases are recently adopted by many Semantic Web applications because it can provide facts and relationships as well as support inference. Semantic Web, also has name of Web 3.0, is the next major evolution in connecting information. It enables data to be linked between sources and to be understood by computers so that they can perform sophisticated tasks on the behalf of human beings.[2] Linked Data [3] is a project to link data on the Web. It actually connects the open knowledge bases.

Recent years, many industry giants also put their research efforts on knowledge bases. Google acquired Freebase in 2010 and applied the inference function in its search engine. In 2014, Freebase was replaced by WikiData project which is also a knowledge base. IBM's DeepQA project [10] also leverages knowledge bases to facilitate the natural language question answering.

---

[1] http://lod-cloud.net/.
[2] http://www.cambridgesemantics.com/semantic-university/.
[3] http://linkeddata.org/.

2.1.1 Knowledge Base Representing

RDF is widely used as data modeling language for the knowledge bases. RDF represents a relationship by a three-element tuple, i.e., triple (*subject, predicate, object*) in which *subject* and *object* are connected by *predicate*. A knowledge base can be represented by a set of triples which are connected to form a graph. So knowledge base modeled by RDF can also be called knowledge graph.

SPARQL Protocol and RDF Query Language (SPARQL) is the query language of RDF-modeled knowledge bases. The official syntax of SPARQL1.1 considers operators OPTIONAL, UNION, FILTER, SELECT and concatenation via a dot symbol (.) to group patterns. VALUES and BIND are to define sets of variable bindings. We consider these operators in this work as they appear most often in real world SPARQL queries [42, 43]. We use B, I, L, V for denoting the (infinite) sets of blank nodes, IRIs, literals, and variables. A SPARQL graph pattern expression is defined recursively as follows [31]:

- A valid triple pattern $T \in (I \cup L \cup V \cup B) \times (I \cup V) \times (I \cup V \cup L \cup B)$ is a graph pattern,
- If P_1 and P_2 are graph patterns, then expressions $(P_1 \text{ AND } P_2)$, $(P_1 \text{ UNION } P_2)$ and $(P_1 \text{ OPTIONAL } P_2)$ are graph patterns,
- If P is a graph pattern and R is a SPARQL build-in condition, then the expression $(P \text{ FILTER } R)$ is a graph pattern.

A *Basic Graph Pattern* (BGP) is a graph pattern represented by the conjunction of multiple triple patterns. It is the basic component of SPARQL queries. Although it cannot represent all features of SPARQL 1.1, it represent the core of the queries. Let $Q = (S_Q, P_Q)$ be the query where S_Q is the SELECT expression and $P_Q = P_1 \oplus ... \oplus P_n$ is the query pattern with $\oplus \in \{\text{AND, UNION, OPTIONAL, FILTER, BIND, VALUES, MINUS}\}$. When pattern feature $\oplus \in \{\text{AND, UNION, OPTIONAL, MINUS}\}$, graph pattern $P_i, i \in [1, n]$ can be recursively decomposed to sub-level graph patterns until the graph pattern is a BGP which can further be decomposed to triple patterns as $P_{bgp,i} = T_1 \oplus ... \oplus T_k$, where $\oplus = \text{AND}$. When pattern feature $\oplus \in \{\text{FILTER, BIND, VALUES}\}$, graph pattern P_i cannot be decomposed to BGPs and is represented as expressions. The decomposition of a SPARQL query is a recursive process with the result that can be regarded as a hierarchical tree. Further, it is easy to observe that query Q can also be represented as $Q = (S_Q, \{P_{bgp}, P_{filter}, P_{bind}, P_{value}\})$ where $P_{bgp}, P_{filter}, P_{bind}, P_{value}$ are BGP, FILTER, BIND and VALUE patterns in P_Q respectively. Note that each graph pattern can appear multiple times and in different depths in a query pattern which are not represented in our notation.

2.1.2 Querying Knowledge Base

Querying knowledge base remains a challenging topic among researchers and practitioners. The existing KB-QA systems can be categorized into curated KB-QA systems and open KB-QA systems. A curated KB-QA system is built upon a curated KB (e.g., Freebase and DBpedia), which is collaboratively and manually created [41]. The data schema is predefined. This type of KB-QA has long history with a number of implementations (e.g., DEANNA [39] and ParaSempre [3]). Open KB-QA system has been recently proposed [9]. An open KB is a large collection of n-tuple assertions that are automatically extracted from web corpora by means of open information extraction (open IE) techniques [2, 8, 11, 17]. The curated KB-QA is more precise and accurate while the open KB-QA does not require fixed schema and can extract up-to-date information automatically.

We focus on facilitating QA against curated KB in this chapter. We will describe a framework which works like a proxy layer between SPARQL endpoint, the knowledge bases' open interface, and the query issuer.

3 The Framework of Cache-Based Knowledge Base Querying

Figure 2 illustrates the cached-based client-side querying framework for SPARQL endpoint. The framework consists of two processes: Querying Process and Suggestion Process. The suggestion process is a background process that will not take the resources of querying process. In querying process, when a new query q is issued, the framework first checks if the query recording is enabled (①). The recorded historical queries will form the training queries in the suggestion process. Then the framework will check if an identical query (either cached as an issued query or a suggested query) has been cached (②). If the cache is not hit (i.e., the query is not in the cache), query result will be fetched from SPARQL Endpoint (③). The result will be returned to user (④) and the (query, result) pairs will be put into cache module (⑤). If the query q is not in the cache, its result will be fetched from cache directly (⑥) and then be returned to the user (⑦). The cache module maintains the cache and a cache replacement algorithm. Due to the limit space of the cache, cache replacement algorithm is applied to take care of which data should be kept in the cache and which should be removed from the cache when it is full. In the suggestion process, similar queries to q are suggested based on the training queries. The results of these similar queries are prefetched (⑧) and stored in the cache (⑨). Training queries are part of historical queries the user had issued and had been trained ahead of the querying and suggestion process.

4 Similar Queries Suggestion

Cache is designed on the purpose of facilitating the overall querying speed. We propose a framework to cache the similar queries in order to further improve the hit rate of cache. In this section, we discuss how to suggest the similar queries to current issued query. The three main steps in the suggestion process are as follows.

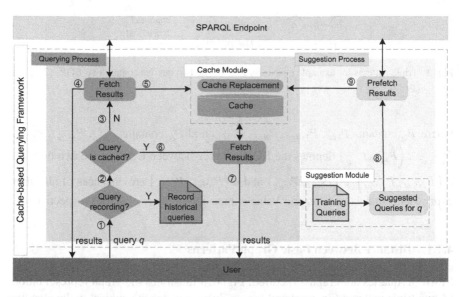

Fig. 2 Cached-based client-side querying framework for SPARQL endpoint

- Step 1: *Feature Modeling*. The key challenge to use machine learning algorithms is to model the SPARQL query to feature vector because many learning algorithms require vector representation of input data. The feature modeling for SPARQL query is to transform the query to a vector, where each attribute in the vector denotes a feature of the query. Two challenges of this step are how to define the distance function and how to transform the query to a vector.
- Step 2: *Training*. After obtaining the feature vectors, we need to train a suggestion model using historical queries as the training set. A trained model is the output.
- Step 3: *Suggestion*. When a new query q arrives, we first transform q to a feature vector using techniques introduced in Step 1. Then we put the vector to the trained suggestion model. Similar queries are suggested as output of the trained model.

We will focus on the discussion of Step 1 as it is the most challenging part of the framework and briefly discuss Step 2 and Step 3.

4.1 Query Distance Calculation

Before discussing the feature modeling approach, we first define the distance function which computes the distance between two given queries. The distance is defined by calculating the distance between patterns of the two queries:

$$d\left(P_Q, P_Q'\right) = d\left(P_{bgp}, P_{bgp}'\right) + d\left(P_{filter}, P_{filter}'\right) + d\left(P_{bind}, P_{bind}'\right) + d\left(P_{value}, P_{value}'\right)$$

$$(1)$$

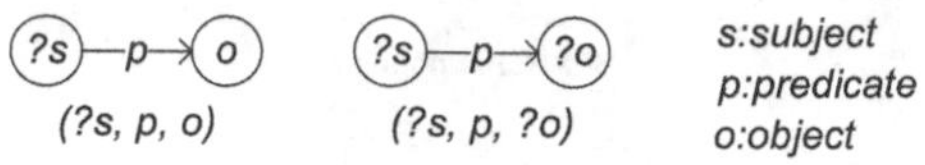

Fig. 3 Triple patterns for example queries in Fig. 1. The two queries contain two BGPs (?s, p, o) and (?s, p, ?o)

where P_Q contains $P_{bgp}, P_{filter}, P_{bind}, P_{value}$ and P'_Q contains $P'_{bgp}, P'_{filter}, P'_{bind},$ P'_{value}. $d\left(P_{bgp}, P'_{bgp}\right)$ denotes the BGP distance between BGPs of the two queries. $d\left(P_{filter}, P'_{filter}\right), d\left(P_{bind}, P'_{bind}\right)$ and $d\left(P_{value}, P'_{value}\right)$ are distances of all Filter expression, Bind expression and Value expression of the two queries respectively.

4.1.1 Distance Between Basic Graph Patterns

SPARQL queries are graph-structured. Figure 3 illustrates the graph representation of two triple patterns *(?s, p, o)* and *(?s, p, ?o)*. *s* denotes the *subject*, *p* denotes the *predicate* and *o* denotes the *object*, The question mark indicates that the corresponding component is a variable. However, it is hard to tell the differences of the two graphs, as they are structurally identical.

In our work, we formulate the problem of modeling BGPs with structurally different tripe patterns as follows:

Definition 1 (BGP Graph Modeling) Given $P_{bgp_i} = \{tp_1, tp_2, ..., tp_n\}$ denote the BGPs of an SPARQL query, tp_k, $k \in (1, n)$ is a triple pattern rooted at P_{bgp_i}. $ged(g_o, g_d)$ represents the Graph Edit Distance (GED) between graph g_o and graph g_d. BGP graph modeling is the task that models each tp_k to a graph g_{tp_k} satisfying $ged\left(g_{tp_k}, g_{tp_l}\right) > 0$ when $k \neq l$.

To address this problem, we propose to map the triple patterns to graphs that are able to uniquely represent each of the eight triple patterns. Figure 4 shows all the eight triple patterns and the corresponding graphs that are proposed by us. The black circles denotes conjunction nodes for clarity. They are not colored in graph modeling. To exemplify, we model the triple patterns of BGPs in the example queries in Fig. 1, to a graph, which is depicted in Fig. 5. We then use GED [35] as the distance between graphs which represent BGPs. GED between two graphs is the minimum amount of edit operations (i.e., deletion, insertion and substitutions of nodes and edges) needed to transform one graph to the other. We take the edit path from *(?s, p, o)* to *(?s, p, ?o)* in Fig. 4 as an example, and the steps are shown in Fig. 6. The GED between these two graphs is two. When $d\left(P_Q, P_{Q'}\right) = 0$, it indicates query Q and Q' are structurally the same.

The graphs of BGPs in the two example queries (Fig. 1) are the same. Their BGP distance (i.e., GED) equals to 0, namely $d\left(P_{bgp_1}, P_{bgp_2}\right) = 0$ in this case.

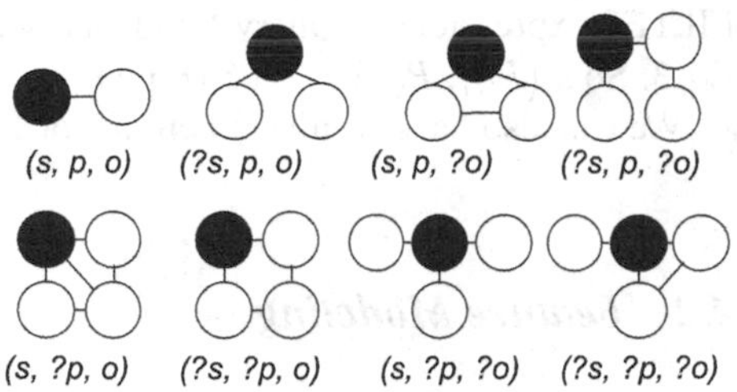

Fig. 4 Mapping triple patterns to graphs. 8 types of triple patterns are mapped to 8 structurally different graphs. Black nodes are conjunction nodes for clarity

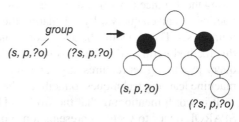

(s, p, o)   (?s, p, o)   (s, p, ?o)   (?s, p, ?o)

(s, ?p, o)   (?s, ?p, o)   (s, ?p, ?o)   (?s, ?p, ?o)

Fig. 5 Graph modelling for BGPs in example queries

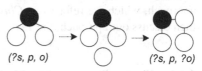

group

(s, p, ?o)   (?s, p, ?o)

(s, p, ?o)

(?s, p, ?o)

Fig. 6 Graph edit path from (?s, p, o) to (?s, p, ?o)

(?s, p, o)

(?s, p, ?o)

4.1.2 Distances Between Expressions

We calculate $d\left(P_{filter}, P'_{filter}\right)$, $d\left(P_{bind}, P'_{bind}\right)$ and $d\left(P_{value}, P'_{value}\right)$ only when $d\left(P_{bgp}, P'_{bgp}\right) = 0$. We define distance between two FILTER expressions as half of their *levenshtein* distance when the variables in these two expressions are identical, otherwise the distance is a fixed value 1. Thus the distance is in the range of $[0, 0.5]$ or equals to 1.

$$d\left(P_{filter,i}, P'_{filter,i}\right) = \begin{cases} \frac{levenshtein(E(i), E'(i))}{2max(length(E(i)), length(E'(i)))}, \\ \qquad\qquad if \ V(i) = V'(i) \\ 1, \qquad\quad else \end{cases} \tag{2}$$

where $E(i)$ and $E'(i)$ represent the FILTER expression for $P_{filter,i}$ and $P'_{filter,i}$. $V(i)$ and $V'(i)$ are variables in these two FILTER patterns respectively. When there are multiple FILTER expressions that can be compared, the total difference is defined as:

$$d\left(P_{filter}, P'_{filter}\right) = \sum_{i=1}^{m} d\left(P_{filter,i}, P'_{filter,i}\right) \tag{3}$$

FILTER expressions in Query 1 and Query 2 are similar as the distance is 0.05 using Eq. 3. So $d\left(P_{Q1}, P_{Q2}\right) = 0.05$ (Eq. 1).

We can also have similar functions for BIND and VALUE patterns.

4.2 Feature Modeling

Using the distance function introduced above (Eq. 1), it is intuitive to suggest similar queries to a given query q by calculating the distances between q and each query in training set. Then rank the distance scores and find the top K similar ones. However, this method is time consuming as the calculation of distances between q and each query in training set requires large amount of computation. Therefore, we leverage machine learning techniques to facilitate the suggestion process.

It is worth mentioning that the work in [15] proposes an approach to transform SPARQL query to vector representation. For comparison, we firstly introduce this approach, which we refer to as *Cluster-Based Feature Modeling* (Sect. 4.2.1) and then discuss our approach, the *Template-Based Feature Modeling* (Sect. 4.2.2).

4.2.1   Cluster-Based Feature Modeling

In cluster-based feature modeling, distances between each pair of queries in the training set are calculated using only BGP distance. Then *k-medoids* algorithm [19] is utilized to cluster the training queries by using distance scores that are calculated. The center queries of each cluster are selected and the distance scores between each center query and a query q is obtained to form a feature vector of q, where each score is regarded as an attribute of the feature of q. Thus the number of clusters equals to the number of dimensions (i.e., the number of feature attributes) of the feature vector of q. Figure 7 illustrates the cluster-based feature modeling. d_i is the distance between query q and the center query of the cluster C_i, and it represents a feature of q. The cluster-based modeling is based on the distances among each pair of all the training queries.

4.2.2   Template-Based Feature Modeling

The cluster-based feature modeling requires distances calculation between all queries. Moreover, the clustering process adds additional time consumption. To reduce the feature modeling time, we are motivated to develop a method that requires less time for computing distances and does not require the clustering process. So we propose to replace the center queries used in cluster-based feature modeling with representative queries that are generated by benchmark templates. Specifically, we generate queries from 18 out of 22 valid templates in the DBPSB benchmark [27]. We refer to these queries as *template queries*. By recording the distance scores between a query

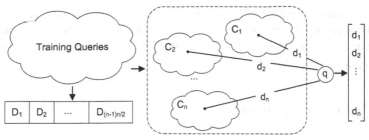

C_i : Cluster
d_i : Distance between q and center query of C_i
D_j : Distance between training queries

Fig. 7 The cluster-based feature modeling

Fig. 8 The template-based
feature modeling

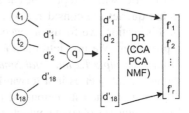

t_i : Template query
d'_i : Distance between q and t_i
f_j : Feature after dimensional reduction on d'_1 to d'_{18}

q with 18 template queries, we obtain a 18-dimension feature vector for q. The computation is then drastically reduced from $O(n^2)$ in cluster-based feature modeling to $O(n)$, where n is the number of queries. Therefore, our approach is feasible to apply to large size of training set. In addition, we further reduce the dimension of feature vectors leveraging three dimensional reduction algorithms, namely PCA, CCA and NMF. PCA aims to find a linear transformation to project the original data to a lower-dimensional space which is as informative as possible to capture as much of the variance of the original data in an unsupervised manner. CCA calculates the coefficient among all features and chooses the most uncorrelated features. NMF finds approximate decomposition of original data matrix and thus reduce the dimension by storing the two decomposed lower dimensional matrices. We implement them to reduce the dimension of query feature matrix.

Figure 8 depicts the template-based feature modeling. t_1 to t_{18} are template queries. d_1' to d_{18}' are distances between query q and 18 template queries. f_1' to f_r' are features that are obtained after applying dimensional reduction algorithm (i.e., CCA, PCA or NMF), where $r < 18$.

After the feature vectors are obtained from Step 1, we train a suggestion model with the feature vectors of training queries in Step 2. KNN is selected as the suggestion model because it outperforms Support Vector Machine Regression (SVR) [15]. Specifically, KNN is modified to a clustering algorithm and for each cluster, a KNN model is trained. In each KNN, *Euclidean Distance* is the distance measurement and a K Dimensional Tree (KD Tree) [12] is built to compute the nearest neighbors. When a new query is issued, we choose its K nearest neighbors obtained from Step 2 as suggested queries in Step 3.

5 Cache Replacement

Given the similar queries for a query q, we prefetch the results of these queries and put the (q_i, r_i) pairs into the cache during the caching process. As the cache has limit space, the obsolete data should be removed from the cache to give space to new data. In this work, we consider the recently most hit queries in the cache as hot queries. Hot queries will be kept in the cache, whereas queries in the cache that do not belong to hot queries are considered as obsolete, which will be removed from the cache. In this section, we focus on *cache replacement* and discuss how we realize cache replacement. We first introduce *Exponential Smoothing* (ES) and our modification *Modified Simple Exponential Smoothing* (MSES). Then we use MSES to process the historical queries in a forward way to compute hit frequencies for each query. According to the estimation, we rank the frequencies and consider those with the lowest values as obsolete, which will be removed from cache (Sect. 5.1). During the query processing, a new estimation will be made both for queries that are in the estimation record and for new queries. To update the cache, we develop two cache replacement strategies (Sect. 5.2).

5.1 Modified Simple Exponential Smoothing

The *Exponential Smoothing* (ES) is a technique to produce a smoothed data presentation, or to make forecasts for time series data, i.e., a sequence of observations [13]. It can be applied to any discrete set of repeated measurement and is currently widely used in smoothing or forecasting economic data in the financial markets. Equation 4 shows the simplest form of exponential smoothing. This equation is also regarded as *Simple Exponential Smoothing* (SES).

$$E_t = \alpha * x_t + (1 - \alpha) * E_{t-1} \tag{4}$$

where E_t stands for the smoothed observation of time t, x_t stands for the actual observation value at time t, and α is a smoothing constant with $\alpha \in (0, 1)$. From this equation, it is easy to observe that SES assigns exponentially decreasing weights as the observation becomes older, which meets the requirement of selecting the most frequently and recently issued queries. In our approach, we exploit SES to estimate access frequencies of queries. The reason behind our choice of SES is its simplicity and effectiveness [23]. Here x_t represents whether the query is hit at time t, thus it is either 1 if a cache hit; or 0 otherwise. Therefore we can modify the Eq. 4 to:

$$E_t = \alpha + E_{t_{prev}} * (1 - \alpha)^{t_{prev} - t} \tag{5}$$

where t_{prev} represents the time when the query is last hit and $E_{t_{prev}}$ denotes the previous frequency estimation for the query at t_{prev} [23]. The accuracy of ES can be measured by its standard error. We give the derivation of the standard error according to Eq. 5 and provide a theoretical proof that SES achieves better hit rates than the most used cache replacement algorithm LRU-2 in [43].

We perform cache replacement based on the estimation score calculated by MSES. Each time a new query is executed, we examine the frequency of cache hit of this query using MSES. If it is in the cache, we update the estimate for it. Otherwise, we just record the new estimate. We keep the estimate records for all processed queries if query recording is enabled. When the top H estimations are changed, the cache will be updated to reflect the new top H queries. Lower rank queries will be removed from the cache.

5.2 Replacement Algorithms

We use a forward algorithm to identify hot queries, which has been used to identify hot triples [43]. In this way, we regard the rest of the queries in cache are invalid. This algorithm works as follows. It process the historical queries from a beginning time to an ending time. A parameter H represents the number of queries to be classified as hot. The output is the H hot queries that will be cached. When encountering a hit to a query at time t, the algorithm updates this query's estimation using Eq. 5. When the scan is completed, the algorithm ranks each query by its estimated frequency and returns the H triples with the highest estimates as the hot set. There are three main advantages of the forward algorithm. Firstly, it is simple as we only need to choose a starting time and then calculate the new estimation when a query is hit again. Secondly, the algorithm enables us to update the estimation and the cache immediately after a new query is executed based on the previously recorded estimation. Thirdly, this algorithm implements an incremental approach that helps identify the *warm-up* stage and the *warmed* stage of the cache.

However, this algorithm also has several drawbacks. Specifically, it requires storing the whole estimation record which is a large overhead. Furthermore, the algorithm also consumes a significant amount of time when calculating and comparing the estimation values. To solve these issues, we consider improving the algorithm in two ways. One possible solution is that we do not keep the whole records. Instead we just keep a record after skipping certain ones. This is a naive sampling approach. We vary the sampling rate but it turns out that the performance of this sampling approach is not desirable. The other possible approach is that we maintain partial records by only keeping those within a specified range of time. Assume *last_hit_time* is the time a query is last hit. We find the earliest *last_hit_time*: $t_{earliest}$ in the hot triples and only keep the estimation records whose *last_hit_time* is later than $t_{earliest}$. Thus we only keep estimation records from $t_{earliest}$ to the hit time of currently processed query.

We provide two ways for cache replacement based on the two possible improved forward algorithms, namely the *Full-Records Based Replacement* and the *Improved Replacement*. In the full-records based replacement, each time when a new query is executed, we examine the frequency using MSES. If it is in the cache, we update the estimation for it. Otherwise, we record the new estimation. We keep the estimation records for all processed triples. When the top H estimations are changed, the cache will be updated to the new top H hot queries. In the improved replacement, we only

---

Algorithm 1: Algorithm for Improved Caching Replacement

Data: *Records, cachedQueryies, newAccQuery*
Result: Updated *cachedQueryies*
begin

1     $t_{latest} \longleftarrow max(last\_hit\_time, cachedQueryies)$
2     $t_{earliest} \longleftarrow min(last\_hit\_time, cachedQueryies)$
3     $est_{max} \longleftarrow max(est, cachedQueryies)$
4     $est_{min} \longleftarrow min(est, cachedQueryies)$
5     $Records \longleftarrow getPartialRecords(t_{latest}, t_{earliest})$
6   **if** *newAccQuery* **in** *cachedQueryies* **then**
7   *calculateNewEstimation()*
8   Update est, last_hit_time in *Records*
9        $t_{latest} \longleftarrow newAccQuery.last\_hit\_time$
10       $t_{earliest} \longleftarrow getEarliest(cachedQueryies)$
11   remove from *Records* the records with last_hit_time **less than** $t_{earliest}$

    else

12   **if** *newAccQuery* **in** *Records* **then**
13   *calculateNewEstimation()*
14   Update est, last_hit_time in *Records*
15   **if** *est* **between** est_{min} and est_{max} **then**
16   remove from *cachedQueryies* the records with est_{min} with minimum last_hit_time
17   *addToCached(newAccQuery)*
18           $t_{latest} \longleftarrow newAccQuery.last\_acc\_time$
19           $t_{earliest} \longleftarrow getEarliest(cachedQueryies)$
20   remove from *Records* the records with last_acc_time **less than** $t_{earliest}$

      else

21   *addToRecords(newAccQuery)*

---

keep estimation records from $t_{earliest}$ to the access time of the current processing query. Algorithm 1 describes the details on updating the cache by using part of the estimation records. The input of the algorithm is the whole estimation records, triples in cache and a new estimated triple. Line 1–4 initialize variables that represent the latest hit time, the earliest hit time, the minimal estimation and the maximum estimation in cached queries (i.e., the hot queries). Then the algorithm gets partial estimation records that are within the time range between $t_{earliest}$ and the hit time of the last processed query in the log (Line 5). If the new estimated query is in the cache, it shows a cache hit, and the algorithm updates the new estimation calculated by Eq. 5 and the *last_hit_time* of this triple in estimation records. It calculates the new $t_{earliest}$ and t_{latest} if the new estimated triple holds the previous $t_{earliest}$ and t_{latest}. If $t_{earliest}$ is changed, estimation records with *last hit time* earlier than $t_{earliest}$ will be removed (Line 6–11). If the new estimated query is not in the cache, which is a cache miss, the algorithm checks whether the new estimated query is in the estimation records. If so, it updates its estimation and *last_hit_time* in records. In addition, the cache needs to be updated if the estimation of the new estimated query is in the range of (est_{min}, est_{max}). This means it becomes a new hot query that should be placed into the cache. When the cache is updated, new $t_{earliest}$ and t_{latest} will be calculated, and the estimation records outside the time range will be removed (Line 15–20). If the new estimated query is not in the estimation records, it needs to be added to the records (Line 21).

6 Implementation and Experimental Evaluation

This section is devoted to the validation and performance study of our proposed approach. We first describe the setup of our evaluation environment (Sect. 6.1). Then the experimental results are discussed, including the evaluation of the effectiveness of our approach in terms of average hit rate, average query time and space usage (Sects. 6.2–6.4).

6.1 Setup

6.1.1 Datasets

We used real world queries gathered from USEWOD 2014 challenge. We analyzed the query logs from DBPedia's SPARQL endpoint[4] (DBpedia3.9) and Linked Geo Data's endpoint[5] (LinkedGeoData). The log files from these two datasets have the same format with 4 parts: anonymized IP address, time stamp, query, user ID. To extract queries, we processed the original query by decoding, extracting interesting values (IP, date, query string), identifying SPARQL queries from query strings and removing invalid queries (i.e., incomplete queries and queries with syntax errors according to SPARQL1.1 specification). We focused on SELECT queries in the experiments and retrieved 198,235 valid queries from DBpedia3.9 and 1,790,047 valid queries from LinkedGeoData. Within the SELECT queries, except for patterns which can be finally decomposed to BGPs (e.g., AND, UNION, OPTIONAL and MINUS), FILTER, VALUES and BIND are used, especially for FILTER, which occurs in 83.97% SELECT queries in DBpedia3.9 query logs and 50.72% SELECT queries in LinkedGeoData [42]. This actually provides a strong evidence that FILTER expressions should not be ignored when calculating similarity between queries.

---

[4]http://dbpedia.org/sparql/.

[5]http://linkedgeodata.org/sparql.

6.1.2 The System

We set up our own SPARQL Endpoint by installing local Virtuoso server and loading datasets into the Virtuoso. The server has the configuration of 64-bit Ubuntu 14.4 with 16GB RAM and 2.40GHz Intel Xeon E5-2630L v2 CPU. Our code runs on a PC with 64-bit Windows 7, 8GB RAM and 2.40GHZ Intel i7-3630QM CPU using Java SE7 and Apache Jena-2.11.2. We implemented GED using a suboptimal solution integrated in the Graph Matching Toolkit.[6]

6.2 Performance of Cache Replacement Algorithm

We firstly evaluated our cache replacement algorithm MSES, because it would be used in the following experiments. To evaluate the performance of MSES, we implemented various algorithms including the forward MSES, an Improved MSES, the Sampling MSES, and LRU-2. LRU-2 is a commonly used page replacement algorithm which we implemented based on record rather than page. All of LinkedGeoData valid queries obtained were processed in this experiment because the size of this query set is much larger than DBpedia3.9 query set. Thus we can observe the difference between Improved MSES and MSES algorithms. As the *Exponential Smoothing* has only one parameter α, the choice for α would affect the hit rate performance. We have verified that when $\alpha = 0.05$, the cache hit rate shows better performance [43].

Figure 9a shows the hit rates achieved by the four algorithms we implemented. It should be noted that in the experiment, the caching size was set to 20% of the total historical queries and α was set to 0.05 for MSES and its variants. We chose 20% as the caching size because it is neither too large (e.g., >50%) to narrow the performance differences among algorithms, nor too small (e.g., <10%) leading to inaccurate performance evaluation due to insufficient processed data. From the figure, we can see that the MSES and Improved MSES have the same hit rate until they have processed about 1.4 million RDF triples, after which MSES has a higher hit rate than Improved MSES. This is because MSES maintains the estimations for all processed records while the Improved MSES only keeps part of the estimations. The changing point denotes that from which, the Improved MSES maintains partial volume of estimation records. From the figure, we can also see that the Sampling MSES does not perform well. This figure only shows the hit rate of sampling MSES with the sampling rate of 50%, which is expected to have a high hit rate. The LRU-2 algorithm has the lowest hit rate of all the algorithms. The hit rates of all algorithms start from 0 and reach their first peak at certain points, and then go down and up. The direction to the first peak shows the warm-up stage and the rest of the lines are the warmed stage. This illustrates that we exploit an incremental approach, which includes a warm-up stage to calculate the hit rate.

---

[6]Graph Matching Toolkit: http://www.fhnw.ch/wirtschaft/iwi/gmt.

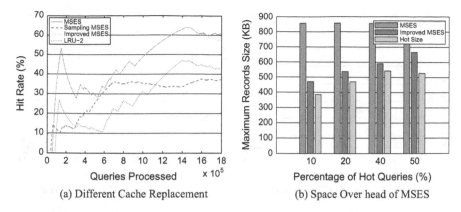

Fig. 9 Cache replacement performance (LinkedGeoData). Different cache replacement algorithms affect the hit rates largely (a). The improved MSES reduces the space overhead largely compared to the MSES (b)

Figure 9b gives the measurement of space usage by recording the estimations. As discussed before, MSES performs better than the Improved MSES. However, it consumes more storage space to maintain the estimation records for all processed triples. It also takes longer time to check the cache. Figure 9b shows the maximum space consumption for each algorithm. Note that we used all valid LinkedGeoData queries in this experiment. The columns are classified into four groups which represent the four different percentages of hot queries to all processed queries respectively. In each group, the left column represents the maximum space used by MSES, including the hot queries and the estimation records. The middle column represents the space usage of the Improved MSES that also includes the hot queries and the estimation records. The right column represents the size of the hot queries. From this figure, we can see that the Improved MSES consumes less space.

6.3   Comparison of Feature Modeling Approaches

In the experiments of this section, we compared our feature modeling approach (i.e., template-based feature modeling) with the state-of-the-art approach (i.e., cluster-based feature modeling), and evaluated the performance under the scenarios of applying and without applying suggestion/prefetching. We applied the dimensional reduction algorithms on both template-based feature modeling and cluster-based feature modeling.

Because the time consumption of cluster-based approach is tremendous, we did not use all valid queries as the training set. We randomly chose 21,600 training queries and 5,400 testing queries from the two query sets separately. The cache replacement algorithm we used in all testing cases is Improved MSES and we chose $\alpha = 0.05$. Because the larger size of cache, the higher hit rate would achieve, we only show experiment results when the number of queries in cache is set to 1,000.

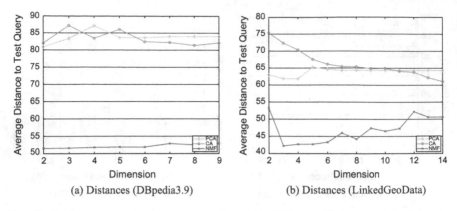

(a) Distances (DBpedia3.9)   (b) Distances (LinkedGeoData)

Fig. 10 Performance comparison among using CCA, PCA and NMF to reduce dimension (Template-based)

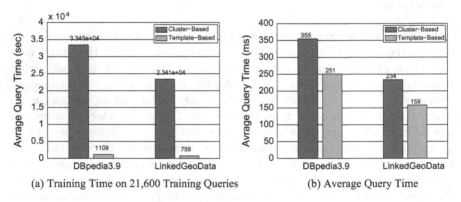

(a) Training Time on 21,600 Training Queries   (b) Average Query Time

Fig. 11 Time comparison on feature modeling approaches

6.3.1 Performance of Template-Based Feature Modeling

In cluster-based feature modeling, we also leveraged dimensional reduction algorithms. The performance of different algorithms is shown in Fig. 10. It is shown that NMF still outperforms other algorithms in extracting the most representative features. Figure 11 gives the impact of two feature modeling algorithm on time consumption. Figure 11a compares training time on 21,600 training queries. Cluster-based approach requires 33,446 s, which is more than 9 h for DBPedia3.9 queries, and 23,405 s (i.e., more than 6 h) for LinkedGeoData queries. Template-based approach largely reduces the time to 1,109 and 758 s, respectively. Figure 11b evaluates the average query time. Template-based approach also outperforms cluster-based approach.

Table 1 Performance comparison

	ASQC	No Cache	SECF
Hit	72.63%	NA	76.65%
AvgTime	264 ms	625 ms	251 ms
Space	7.15 MB	NA	7.15 MB/0.45 KB

Table 2 Server Performance

	ASQC	No Cache	SECF
AvgFreeMem	217.87 MB	224.30 MB	203.74 MB
AvgIO	11.49	7.72	21.84
AvgCPU	9.37 ms	10.09 ms	10.68 ms

6.4 Performance Comparison with the State-of-the-Art Work

We also compared our work with the Adaptive SPARQL Query Cache (ASQC) introduced in [25], as it is the first and complete work to cache SPARQL query in a client-side manner.

6.4.1 Systems Performance Comparison

In this experiment, we compared the average hit rate, average query time and space usage between our work SECF and ASQC. We also gave measurement when no cache was used. To compare our approach with ASQC, we modified the code of ASQC[7] to access our datasets. We performed the experiment on DBpedia3.9 dataset. We used Cluster-Based Feature Modeling, and Improved MSES with $\alpha = 0.05$. Table 1 presents the results. ASQC have slightly lower hit rate (72.63%) than SECF (76.26%). ASQC takes 264 ms in average for one query and SECF takes 251 ms. When no cache is implemented, the average query time increases to 625 ms. We did not include prefetching time as it is in separate thread. Space consumption evaluates how much memory the cache uses. In SECF, the total usage (before slash) for caching 1,000 queries, as shown in Table 1, includes cached queries and answers as well as the estimation records for cache replacement (after slash). We used the same implementation for cache in order to compare. The numbers indicate that most space are consumed by cached (query, result) pairs.

---

[7]http://wiki.aksw.org/Projects/QueryCache.

6.4.2 Server Overhead Comparison

In order to evaluate the impact of cache on the endpoint server, we monitored the memory and CPU usage as well as I/O on the server. We captured the usage every 20 s until the querying ends. Table 2 shows the average free memory (AvgFreeMem), average I/O (AvgFreeMem) and average CPU time (AvgCPU) including system CPU and user CPU time. We only present the result on querying DBpedia3.9 dataset due to limit space. From the result we find out that SECF and ASQC cause higher computation overhead (I/O and CPU) and memory usage on server compared to querying without cache and ASQC performs slightly better than SECF with more free memory (217.87 vs 203.74 MB), less I/O (11.49 vs 21.84) and less CPU time (9.37 vs 10.68 ms). It is because that SECF requires prefetching results for similar queries from server which leads to additional overhead.

6.5 Experimental Conclusion

We conclude some key observations from experiments in this section. When applying our suggestion and caching process, the overall hit rate improves which demonstrates the effectiveness of our caching framework. Compared to the state-of-the-art work ASQC, our work outperforms ASQC in terms of the average query time, but requires more system overhead on server. The new cache replacement algorithm we proposed in this work, namely MSES, achieves greater cache hit rate than the most used cache replacement algorithm LRU-2, thus it contributes to reduce the overall querying time. Improved MSES further reduces the space overhead by considering a part of estimation records without losing cache hit rate. The suggestion accuracy (evaluated by average distances in KNN) is improved when introducing dimensional reduction algorithms. NMF gives the best performance among the three most common algorithms. This section is devoted to the validation and performance study of our proposed approach. We first describe the setup of our evaluation environment (Sect. 6.1). Then the experimental results are discussed, including the evaluation of the effectiveness of our approach in terms of average hit rate, average query time and space usage (Sects. 6.2–6.4).

7 Related Work

Our work mainly addresses caching problems in two research areas, namely *Semantic Caching* and *Query Suggestion*. In this section, we review the recent representative works that are related to our work.

7.1 Semantic Caching

Semantic Caching involves techniques that keep previously fetched data for past queries. If subsequent queries use the same data, results can be returned immediately. It was originally developed for relational databases. Godfrey and Gryz [14] present a predicate-based caching schema in client server, which is a general logical foundation for semantic query caching. It is a comprehensive framework that addresses multiple issues regarding build and recover cache and the notions of semantic overlap and independence. Dar et al. [6] present a semantic schema of caching for client-server systems. They cache semantic regions rather than tuples or pages. They provide distance-based cache replacement policy, in which the distance that is farthest from the client's current location is discarded from cache. This approach is designed for SELECT queries.

In recent years, the semantic caching technique has been extended to triple stores that manage SPARQL queries. The work of Martin et al. [25] is the first step towards semantic caching for SPARQL queries, in which both the complete triple query result and the application object are cached. The work essentially builds a proxy layer between an application and a SPARQL endpoint. the cache layer considers only the identical queries issued afterwards and identical application object that could be potentially used. Shu et al. [36] improve this approach in a content-aware way by introducing query containment checking which evaluates whether a query can be answered by the result of a cached query. Thus not only the identical queries, but also the queries which succeed in the containment checking can utilize the result of the cached queries. But this method introduces large overhead on containment checking itself. Yang and Wu [40] develop an approach that caches intermediate result of basic graph patterns in SPARQL queries. It decomposes the query into BGPs and evaluates if the result of any BGP or join of BGPs is cached. The cached results which are hit will be returned and joined with the other parts of the query to form the final query result. This work does not address the impact of different join orders. Very recently, Papailiou et al. [30] introduce canonical labeling to identify isomorphic subgraphs in SPARQL query patterns, which are cached for subsequent querying. This solution implements a caching layer on top of the distributed partitions and dynamically updating the contents of the cache.

The Linked Data Fragments (LDF) approach [38] aiming at improving data availability can also be regarded as caching technique as it caches fragments of queryable data from servers that can be accessed by clients. So that each client is able to process SPARQL queries on replicated fragments cached from servers. This approach can be potentially used for federated queries. However, performance degradation issue needs to be addressed.

7.2   Query Suggestion

Query suggestion is an interactive approach used in search engines to better understand users' information needs. It plays an important role in improving the accuracy of searching. Query suggestions are usually made by mining query logs and session records of Web users' searching history [5]. They either aim to find similar queries in search logs and use those queries as suggestions, or identify pairs of queries which co-occur in the same query sessions.

Researchers recently have introduced these mining techniques into SPARQL processing. Lehmann et al. [21] propose a novel solution for querying knowledge bases. It leverages supervised machine learning framework to suggest SPARQL queries based on examples previously selected by users. This approach narrows the range of possible answers asked by users. With the learning techniques, no prior knowledge of the underlying schema or the SPARQL query language is required. More recently, Hasan [15] uses a machine learning method to predict the performance of SPARQL query performance. Specifically, a suggestion model is trained with previously issued queries and then based on this, the query time for new queries can be predicted. In our work, we extend this approach to suggest similarly-structured queries and prefetch and cache their answers.

Query relaxation (also called *query expansion* and *query augmentation*) can also be regarded as a kind of query suggestion. Its aim is to improve the recall of query. In this way, it accelerates the overall query processing. It discovers and suggests related information based on the expansion of original query. Features with similar meaning need to be identified and suggested to generate expanded query. In recent years, query expansion techniques have been used by several research efforts that focus on topics of SPARQL queries. Elbassuoni et al. [8] propose multiple types of relaxation to improve the recall of entity-relationship search. Lorey and Naumann [24] cluster similar SPARQL queries to different templates in order to detect recurring patterns in queries. These templates can be used to expand queries for query processing. Fokou et al. [11] investigate query relaxation over RDF data and focus on identifying parts of SPARQL query that are responsible of the failure of the query. It aims at providing users with alternative answers instead of an empty result.

8   Discussion and Conclusion

We discuss the issues from our experience in this work to give an idea to the prospective readers. We then finally give the conclusion.

- Training Size. In the training process, the larger the size of the training data, the better performance we can get. The reason is that more data variety is seen and the model will be less sensitive to unforeseen queries. However, in practice, it is time consuming to train queries of large size, although our approach has achieved great improvements in reducing the training and suggestion time.

- Space Overhead. As memory space is used for cache are mostly consumed by cached (query, result) pairs, directly caching the pairs in text can be improved by introducing data compression techniques to reduce the space consumption. However, how to effectively compress and decompress the data and provide effective indexing algorithms requires more research effort.
- Dynamic versus Static Mining. Our approach is based on periodically training of historical queries. Therefore, online training is left undiscussed in this work and will be an interesting topic to put efforts on.
- Sever Overhead. Compared to the state-of-the-art work, our work introduces larger overhead to server side. This is due to the fact that the prefetch process requests data in an additional thread to the querying process. We do not consider how to optimize the server performance in this work because we only focus on accelerating client-side querying. Thus this is out-of-scope of this work.

In summary, in this chapter, we introduce a client-side caching paradigm to improve the overall querying performance on the SPARQL endpoint. Our method to transform SPARQL queries to feature vectors greatly outperforms the state-of-the-art method. Based on the feature vectors, learning based approach is utilized to suggest queries whose results are prefetched and cached. We also design a distance measurement that is tailored to SPARQL queries and used by the learning algorithm to identify similar queries. Three dimensional reduction algorithms are introduced to the learning process and proved that they contribute to the reduction of overall query time. The proposed cache replacement algorithm MSES is evaluated effective and efficient. All the evaluations are performed on real-world queries. The results demonstrate the potential of our framework to speed up average querying process on SPARQL endpoints.

References

1. J. Bao, N. Duan, M. Zhou, T. Zhao, Knowledge-based question answering as machine translation, in *Proceedings of the 52nd Annual Meeting of the Association for Computational Linguistics (ACL 2014), Baltimore, USA* (2014), pp. 967–976
2. J. Berant, A. Chou, R. Frostig, P. Liang, Semantic parsing on freebase from question-answer pairs, in *Proceedings of the 2013 Conference on Empirical Methods in Natural Language Processing (EMNLP 2013), Seattle, USA* (2013), pp. 1533–1544
3. J. Berant, P. Liang, Semantic parsing via paraphrasing, in *Proceedings of the 52nd Annual Meeting of the Association for Computational Linguistics (ACL 2014), Baltimore, USA* (2014), pp. 1415–1425
4. K.D. Bollacker, C. Evans, P. Paritosh, T. Sturge, J. Taylor, Freebase: a collaboratively created graph database for structuring human knowledge, in *Proceedings of the ACM SIGMOD International Conference on Management of Data (SIGMOD 2008), Vancouver, Canada* (2008), pp. 1247–1250
5. H. Cao, D. Jiang, J. Pei, Q. He, Z. Liao, E. Chen, H. Li, Context-aware query suggestion by mining click-through and session data, in *Proceeding of the 14th ACM SIGKDD Conference on Knowledge Discovery and Data Mining (KDD 2008), Las Vegas, Nevada, USA* (2008), pp. 875–883

6. S. Dar, M.J. Franklin, B.T. Jónsson, D. Srivastava, M. Tan, Semantic data caching and replacement, in *Proceedings of the 22nd International Conference on Very Large Data Bases (VLDB1996), Mumbai (Bombay), India* (1996), pp. 330–341

7. P.J. Denning, The working set model for program behaviour. Commun. ACM **11**(5), 323–333 (1968)

8. S. Elbassuoni, M. Ramanath, G. Weikum, Query relaxation for entity-relationship search, in *Proceedings of the 8th Extended Semantic Web Conference (ESWC 2011), Heraklion, Crete, Greece* (2011), pp. 62–76

9. A. Fader, L. Zettlemoyer, O. Etzioni, Open question answering over curated and extracted knowledge bases, in *Proceedings of the 20th ACM SIGKDD International Conference on Knowledge Discovery and Data Mining (KDD 2014), New York, USA* (2014), pp. 1156–1165

10. D.A. Ferrucci, E.W. Brown, J. Chu-Carroll, J. Fan, D. Gondek, A. Kalyanpur, A. Lally, J.W. Murdock, E. Nyberg, J.M. Prager, N. Schlaefer, C.A. Welty, Building Watson: an overview of the DeepQA project. AI Magazine **31**(3), 59–79 (2010)

11. G. Fokou, S. Jean, A. Hadjali, M. Baron, Cooperative techniques for SPARQL query relaxation in RDF databases, in *Proceedings of the 12th Extended Semantic Web Conference (ESWC 2015), Portoroz, Slovenia* (2015), pp. 237–252

12. J.H. Friedman, J.L. Bentley, R.A. Finkel, An algorithm for finding best matches in logarithmic expected time. ACM Trans. Math. Softw. **3**(3), 209–226 (1977)

13. E.S. Gardner, Exponential smoothing: the state of the art-part II. Int. J. Forecast. **22**(4), 637–666 (2006)

14. P. Godfrey, J. Gryz, Answering queries by semantic caches, In *Proceedings of the 10th International Conference on Database and Expert Systems Applications (DEXA 1999), Florence, Italy* (1999), pp. 485–498

15. R. Hasan, Predicting SPARQL query performance and explaining linked data, in *Proceedings of the 11th Extended Semantic Web Conference (ESWC 2014), Anissaras, Crete, Greece* (2014), pp. 795–805

16. H. Hotelling, Relations between two sets of variates. Biometrika (1936), pp. 321–377

17. N.L. Johnson, A.W. Kemp, S. Kotz, *Univariate Discrete Distributions*, 2nd edn. (Wiley, New Jersey, 1993)

18. I. Jolliffe, *Principal Component Analysis*, Wiley Online Library (2002)

19. L. Kaufman, P. Rousseeuw, *Clustering by Means of Medoids*, (North-Holland, Amsterdam, 1987)

20. D.D. Lee, H.S. Seung, Learning the parts of objects by non-negative matrix factorization. Nature **401**(6755), 788–791 (1999)

21. J. Lehmann, L. Bühmann, AutoSPARQL: let users query your knowledge base, in *Proceedings of the 8th Extended Semantic Web Conference (ESWC 2011), Heraklion, Crete, Greece* (2011), pp. 63–79

22. J. Lehmann, R. Isele, M. Jakob, A. Jentzsch, D. Kontokostas, P.N. Mendes, S. Hellmann, M. Morsey, P. van Kleef, S. Auer, C. Bizer, DBpedia - a large-scale, multilingual knowledge base extracted from wikipedia. Semant. Web J. **6**(2), 167–195 (2015)

23. J.J. Levandoski, P. Larson, R. Stoica, Identifying hot and cold data in main-memory databases, in *Proceedings of 29th International Conference on Data Engineering (ICDE 2013), Brisbane, Australia* (2013), pp. 26–37

24. J. Lorey, F. Naumann, Detecting SPARQL query templates for data prefetching, in *Proceedings of the 10th Extended Semantic Web Conference (ESWC 2013), Montpellier, France* (2013), pp. 124–139

25. M. Martin, J. Unbehauen, S. Auer, Improving the performance of semantic web applications with SPARQL query caching, in *Proceedings of the 7th Extended Semantic Web Conference (ESWC 2010), Heraklion, Crete, Greece* (2010), pp. 304–318

26. N. Megiddo, D.S. Modha, ARC: a self-tuning, low overhead replacement cache, in *Proceedings of the Conference on File and Storage Technologies (FAST, San Francisco, California, USA* (2003)

27. M. Morsey, J. Lehmann, S. Auer, A.N. Ngomo, Usage-centric benchmarking of RDF triple stores, in *Proceedings of the 26th AAAI Conference on Artificial Intelligence (AAAI 2012), Toronto, Canada* (2012)

28. J.R. Movellan, A quickie on exponential smoothing. http://mplab.ucsd.edu/tutorials/ExpSmoothing.pdfa/

29. E.J. O'Neil, P.E. O'Neil, G. Weikum, The LRU-K page replacement algorithm for database disk buffering, in *Proceedings of the International Conference on Management of Data (SIGMOD 1993), Washington, D.C., USA* (1993), pp. 297–306

30. N. Papailiou, D. Tsoumakos, P. Karras, N. Koziris, Graph-aware, workload-adaptive SPARQL query caching, in *Proceedings of the International Conference on Management of Data (SIGMOD 2015), Melbourne, Australia* (2015), pp. 1777–1792

31. J. Pérez, M. Arenas, C. Gutierrez, Semantics and complexity of SPARQL. ACM Trans. Database Sys. 34(3) (2009)

32. R. Punnoose, A. Crainiceanu, D. Rapp, SPARQL in the cloud using Rya. Inf. Syst. **48**, 181–195 (2015)

33. S. Reid, Knowledge-based systems concepts, Techniques, Examples. http://www.reidgsmith.com/ (1985)

34. Q. Ren, M.H. Dunham, V. Kumar, Semantic caching and query processing. IEEE Trans. Knowl. Data Eng. **15**(1), 192–210 (2003)

35. A. Sanfeliu, K. Fu, A distance measure between attributed relational graphs for pattern recognition. IEEE Trans. Sys. Man Cybern. **13**(3), 353–362 (1983)

36. Y. Shu, M. Compton, H. Müller, K. Taylor, Towards content-aware SPARQL query caching for semantic web applications, in *Proceedings of the 14th International Conference on Web Information Systems Engineering (WISE 2013), Nanjing, China* (2013), pp. 320–329

37. F.M. Suchanek, G. Kasneci, G. Weikum. Yago: a core of semantic knowledge, in *Proceedings of the 16th International World Wide Web Conference (WWW 2007), Banff, Canada* (2007), pp. 697–706

38. R. Verborgh, O. Hartig, B.D. Meester, G. Haesendonck, L.D. Vocht, M.V. Sande, R. Cyganiak, P. Colpaert, E. Mannens, R.V. de Walle, Querying datasets on the web with high availability, in *Proceedings of the 13th International Semantic Web Conference (ISWC 2014), Riva del Garda, Italy* (2014), pp. 180–196

39. M. Yahya, K. Berberich, S. Elbassuoni, M. Ramanath, V. Tresp, G. Weikum, Natural language questions for the web of data, in *Proceedings of the 2012 Joint Conference on Empirical Methods in Natural Language Processing and Computational Natural Language Learning (EMNLP-CoNLL 2012), Jeju Island, Korea* (2012), pp. 379–390

40. M. Yang, G. Wu, Caching intermediate result of SPARQL queries, in *Proceedings of the 20th International World Wide Web Conference (WWW 2011), Hyderabad, India* (2011), pp. 159–160

41. P. Yin, N. Duan, B. Kao, J. Bao, M. Zhou, Answering questions with complex semantic constraints on open knowledge bases, in *Proceedings of the 24th ACM International Conference on Information and Knowledge Management (CIKM 2015), Melbourne, Australia* (2015), pp. 1301–1310

42. W.E. Zhang, Q.Z. Sheng, Y. Qin, K. Taylor, L. Yao, A. Shemshadi, SECF: improving SPARQL querying performance with proactive fetching and caching, in *Proceedings of the 31st ACM Symposium on Applied Computing(SAC 2016), Pisa, Italy* (2016), (To appear)

43. W.E. Zhang, Q.Z. Sheng, K. Taylor, Y. Qin, Identifying and caching hot triples for efficient RDF query processing, in *Proceedings of the 20th International Conference on Database Systems for Advanced Applications (DASFAA 2015), Hanoi, Vietnam* (2015), pp. 259–274

Part III
Big Graph Analytics

Management and Analysis of Big Graph Data: Current Systems and Open Challenges

Martin Junghanns, André Petermann, Martin Neumann
and Erhard Rahm

Abstract Many big data applications in business and science require the management and analysis of huge amounts of graph data. Suitable systems to manage and to analyze such graph data should meet a number of challenging requirements including support for an expressive graph data model with heterogeneous vertices and edges, powerful query and graph mining capabilities, ease of use as well as high performance and scalability. In this chapter, we survey current system approaches for management and analysis of "big graph data". We discuss graph database systems, distributed graph processing systems such as Google Pregel and its variations, and graph dataflow approaches based on Apache Spark and Flink. We further outline a recent research framework called GRADOOP that is build on the so-called Extended Property Graph Data Model with dedicated support for analyzing not only single graphs but also collections of graphs. Finally, we discuss current and future research challenges.

1 Introduction

Graphs are ubiquitous and the volume and diversity of graph data are strongly growing. The management and analysis of huge graphs with billions of entities and relationships such as the web and large social networks were a driving force behind the development of powerful and highly parallel big data systems. Many scientific and business applications also have to process and analyze highly

M. Junghanns (✉) · A. Petermann · E. Rahm
Database Research Group, Leipzig University, Leipzig, Germany
e-mail: junghanns@informatik.uni-leipzig.de

A. Petermann
e-mail: petermann@informatik.uni-leipzig.de

E. Rahm
e-mail: rahm@informatik.uni-leipzig.de

M. Neumann
Swedish Institute of Computer Science, Kista, Sweden
e-mail: mneumann@sics.se

© Springer International Publishing AG 2017
A.Y. Zomaya and S. Sakr (eds.), *Handbook of Big Data Technologies*,
DOI 10.1007/978-3-319-49340-4_14

457

interrelated data that can be naturally represented by graphs. Examples of graph data in such domains include bibliographic citation networks [29], biological networks [18, 109] or customer interactions with enterprises [83]. The ability of graphs to easily link different kinds of related information make them a promising data organization for data integration [85] as demonstrated by the so-called linked open data web[1] or the increasing importance of so-called *knowledge graphs* providing consolidated background knowledge [81], e.g., to improve search queries on the web or in social networks.

The flexible and efficient management and analysis of "big graph data" holds high promise. At the same time it poses a number of challenges for suitable implementations in order to meet the following requirements:

- *Powerful graph data model*: The graph data systems should not be limited to the processing of homogeneous graphs but should support graphs with heterogeneous vertices and edges of different types and with different attributes without requiring a fixed schema. This flexibility is necessary for many applications (e.g., in social networks, vertices may represent users or groups and relationships may express friendships or memberships) and is important to support the integration of different kinds of data within a single graph. Furthermore, the graph data model should be able to represent and process single graphs (e.g., the social network) as well as graph collections (e.g., identified communities within a social network). Finally, the graph data model should provide a set of powerful graph operators to process and analyze graph data, e.g., to find specific patterns or to aggregate and summarize graph data.
- *Powerful query and analysis capabilities*: Users should be enabled to retrieve and analyze graph data with a declarative query language. Furthermore, the systems should support the processing of complex graph analysis tasks requiring the iterative processing of the entire graph or large portions of it. Such heavy-weight analysis tasks include the evaluation of generic and application-specific graph metrics (e.g., pagerank, graph centrality, etc.) and graph mining tasks, e.g., to find frequent subgraphs or to detect communities in social networks. If a powerful graph data model is supported, the graph operators of the data model should be usable to simplify the implementation of analytical graph algorithms as well as to build entire analysis workflows including analytical algorithms as well as additional steps such as pre-processing the input graph data or post-processing of analysis results.
- *High performance and scalability*: Graph processing and analysis should be fast and scalable to very large graphs with billions of entities and relationships. This typically requires the utilization of distributed clusters and in-memory graph processing. Distributed graph processing demands an efficient implementation of graph operators and their distributed execution. Furthermore, the graph data needs to be partitioned among the nodes such that the amount of communication

---

[1]http://lod-cloud.net/.

and dynamic data redistribution is minimized and the computational load is evenly balanced.

- *Persistent graph storage and transaction support*: Despite the need for an in-memory processing of graphs, a persistent storage of the graph data and of analysis results is necessary. It is also desirable to provide OLTP (Online Transaction Processing) functionality with ACID transactions [46] for modifying graph data.
- *Ease of use/graph visualization*: Large graphs or a large number of smaller graphs are inherently complex and difficult to browse and understand for users. Hence, it is necessary to simplify the use and analysis of graph data as much as possible, e.g., by providing powerful graph operators and analysis capabilities. Furthermore, the users should be able to interactively query and analyze graph data similar to the use of OLAP (Online Analytical Processing) for business intelligence. The definition of graph workflows should be supported by a graphical editor. Furthermore, there should be support for visualization of graph data and analysis results which is powerful, customizable and able to handle big graph data.

Numerous systems have been developed to manage and analyze graph data, in particular graph database systems as well as different kinds of distributed graph data systems, e.g., for Hadoop-based cluster architectures. *Graph database systems* typically support semantically rich graph data models and provide a query language and OLTP functionality, but mostly do not support partitioned storage of graphs on distributed infrastructures as desirable for high scalability (Sect. 2). The latter aspects are addressed by distributed systems which we roughly separate into distributed graph processing systems and graph dataflow systems. *Distributed graph processing systems* include vertex-centric approaches such as Google Pregel [71] and its variations as well as extensions including Apache Giraph [10], GPS [97], GraphLab [69], Giraph++ [107] etc. (Sect. 3). On the other hand, *distributed graph dataflow systems* (Sect. 4) are graph-specific extensions (e.g., GraphX and Gelly) of general-purpose distributed dataflow systems such as Apache Spark [118] and Apache Flink [4]. These systems support a set of powerful operators (map, reduce, join, etc.) which are executed in parallel in a distributed system either separately or within analytical programs. The data between operators is streamed for a pipelined execution. The graph extensions add graph-specific operators and processing capabilities to simplify development of analytical programs including graph data.

Early work on distributed graph processing on Hadoop was based on the *MapReduce* programming paradigm [96, 99]. This simple model has been used for the development of different graph algorithms, e.g., [38, 65, 68]. However, MapReduce has a number of significant problems [14, 75] that are overcome with the newer programming frameworks such as Apache Giraph, Apache Spark and Apache Flink. In particular, MapReduce is not optimized for in-memory processing and tends to suffer from extensive overhead for disk I/O and data redistribution. This is especially a problem for iterative algorithms that are commonly necessary for graph analytics and can involve the execution of many expensive MapReduce jobs. For these reasons, we will not cover the MapReduce-based approaches for graph processing in this chapter.

In this chapter, we give an overview about the mentioned kinds of graph data systems, and evaluate them with respect to the introduced requirements. In particular we discuss graph database systems and their main graph data models, namely the resource description framework [63] and the property graph model [93] (Sect. 2). Furthermore we give a brief overview about distributed graph processing systems (Sect. 3) and graph dataflow systems with focus on Apache Flink (Sect. 4). In Sect. 5, we outline a new research prototype supporting distributed graph dataflows called GRADOOP (Graph analytics on Hadoop). GRADOOP implements the so-called Extended Property Graph Data Model (EPGM) with dedicated support for analyzing not only single graphs but also collections of graphs. In Sect. 6, we compare the introduced system categories w.r.t. introduced requirements in a summarizing way. Finally, we discuss current and future research challenges (Sect. 7) and conclude.

2 Graph Databases

Research on graph database models started in the nineteen-seventies, reached its peak popularity in the early nineties but lost attention in the two-thousands [7]. Then, there was a comeback of graph data models as part of the NoSQL movement [23] with several commercial *graph database systems* [6]. However, these new-generation graph data models arose with only few connections to early rather theoretical work on graph database models. In this section, we compare recent graph database systems to identify trends regarding used data models and their application scope as well as their analytical capabilities and suitability for "big graph data" analytics.

2.1 Recent Graph Database Systems

Graph database systems are based on a *graph data model* representing data by graph structures and providing graph-based operators such as neighborhood traversal and pattern matching [6]. Table 1 provides an overview about recent graph database systems including supported data models, their application scope and the used storage approaches. The selection claims no completeness but shows representatives from current research projects and commercial systems with diverse characteristics.

Supported data models: The majority of the considered systems supports one or both of two data models, in particular the property graph model (PGM) and the resource description framework (RDF). While RDF [63] and the related query language SPARQL [48] are standardized, for the PGM [93] there exists only the industry-driven de facto standard Apache TinkerPop.[2] TinkerPop also includes the query language Gremlin [92]. A more detailed discussion of both data models and their query languages follows in subsequent paragraphs.

---

[2]http://tinkerpop.apache.org/.

Table 1 Comparison of graph database systems

	Data model		Scope			Storage		
	RDF/SPARQL	PGM/TinkerPop	Generic	OLTP Queries	Analytics	Approach	Replication	Partitioning
Apache Jena TBD [11]	✓✓			✓		Native		
AllegroGraph [5]	✓✓			✓		Native	✓	
MarkLogic [72]	✓✓			✓		Native	✓	✓
Ontotext GraphDB [43]	✓✓			✓		Native	✓	
Oracle Spatial and Graph [82]	✓✓			✓		Native	✓	
Virtuoso [32]	✓✓			✓		Relational	✓	✓
TripleBit [117]	✓✓			✓		Native		
Blazegraph [105]	✓✓	✓/✓		✓	✓	Native RDF	✓	✓
IBM System G [21, 114]	✓✓	✓/✓	✓	✓	✓	Native PGM, Wide column store	✓	✓
Stardog [101]	✓✓	✓/✓		✓	○	Native RDF	✓	
SAP Active Info. Store [95]		✓/-		✓		Relational		
ArangoDB [60]		✓/✓		✓		Document store	✓	✓
InfiniteGraph [52]		✓/✓		✓		Native	✓	✓
Neo4j [77]		✓/✓		✓		Native	✓	
Oracle Big Data [17]		✓/✓			✓	Key value store	✓	✓
OrientDB [113]		✓/✓		✓		Document store	✓	✓
Sparksee [73]		✓/✓		✓		Native	✓	
SQLGraph [103]		✓/✓		✓		Relational		
Titan [108]		✓/✓		✓	○	Wide column store, Key value store	✓	✓
HypergraphDB [53]			✓	✓		Native		

A few systems are using generic graph models. We use the term *generic* to denote graph data models supporting arbitrary user-defined data structures (ranging from simple scalar values or tuples to nested documents) attached to vertices and edges. Such generic graph models are also used by most *graph processing systems* (see Sect. 3). The support for arbitrary data attached to vertices and edges is a distinctive feature of generic graph models and can be seen as a strength and a weakness at the same time. On the one hand, generic models give maximum flexibility and allow users to model other graph models like RDF or the PGM. On the other hand, such systems cannot provide built-in operators related to vertex or edge data as the existence of certain features like type labels or attributes are not part of the database model.

Application scope: Most graph databases focus on OLTP workload, i.e., CRUD operations (create, read, update, delete) for vertices and edges as well as transaction and query processing. Queries are typically focused on small portions of the graph, for example, to find all friends and interests of a certain user. Some of the considered graph databases already show built-in support for graph analytics, i.e., the execution of graph algorithms that may involve processing the whole graph, for example to calculate the pagerank of vertices [71] or to detect frequent substructures [104]. These systems thus try to include the typical functionality of graph processing systems by different strategies. IBM System G and Oracle Big Data provide built-in algorithms for graph analytics, for example pagerank, connected components or k-neighborhood [21]. The only system capable to run custom graph processing algorithms within the database is Blazegraph by its gather-apply-scatter (see Sect. 3) API.[3] Additionally, the current version of TinkerPop includes the virtual integration of graph processing systems in graph databases, i.e., from the user perspective graph processing is part of the database system but data is actually moved to an external system. However, indicated by a circle in the analytics column in Table 1, we could identify only two systems currently implementing this functionality.

Storage techniques: The majority of the considered graph databases is using a so-called *native* storage approach, i.e., the storage is tailored to characteristics of graph database models, for example, to enable efficient edge traversal. A typical technique of graph-optimized storage are adjacency lists, i.e., storing edges redundantly attached to their connected vertices [21]. By contrast, some systems implement the graph database on top of alternative data models such as relational or document stores. IBM System G and Titan are offering multiple storage options. The used storage approach is generally no hint for database performance [103]. Most systems can utilize computing clusters by replicating the entire database on each node to improve read performance. About half of the considered systems also has some support for partitioned graph storage and distributed query processing. Systems with non-native storage typically inherited data partitioning from the underlying storage technique but provide no graph-specific partitioning strategy. For example, OrientDB treats vertices as typed documents and implements partitioning by type-wise sharding.

---

[3]http://wiki.blazegraph.com/wiki/index.php/RDF_GAS_API.

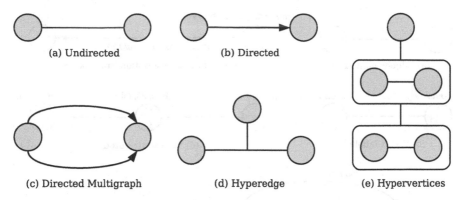

Fig. 1 Comparison of graph structures

2.2 Graph Data Models

A graph is typically represented by a pair $G = \langle V, E \rangle$ of vertices V and edges E. Many extensions have been made to this simple abstraction to define rich graph data models [6, 7]. In the following, we introduce varying characteristics of graph data models with regard to the represented graph structure and attached data. Based on that, we discuss RDF and the property graph model in more detail.

Graph structures: Figure 1 shows a comparison of different graph structures. Graph structures mainly differ regarding their edge characteristics. First, edges can be either undirected or directed. While edges of an *undirected* graph (Fig. 1a) are 2-element sets of vertices, the ones of a *directed* graph are ordered pairs. The order of vertices in these pairs indicates a direction from *source* to *target* vertex. In drawings and visualizations of directed graphs, arrowheads are used to express edge direction (Fig. 1b). In *simple* undirected or directed graphs, between any two vertices there may exist only one edge for undirected graphs and one edge in each direction for directed graphs. By contrast, *multigraphs* allow an arbitrary number of edges between any pair of vertices. Depending on the edge definition, multigraphs are directed or undirected. Most graph databases use directed multigraphs as shown by Fig. 1c.

The majority of applied graph data models support only binary edges. A graph supporting n-ary edges is called *hypergraph* [28]. In a hypergraph model edges are non-empty sets of vertices, denoted by *hyperedges*. Figure 1d shows a hypergraph with a ternary hyperedge. From the graph databases of Table 1 only HypergraphDB supports hypergraphs by default. A graph data model supporting edges not only between vertices but also between graphs is the *hypernode model* [86]. In this model we distinguish between primitive vertices and graphs in the role of vertices, the so-called *hypervertices*. Figure 1e shows a graph containing hypervertices. Except an early research prototype, there is no graph database system explicitly supporting this data model. However, using the concept of *n-quads*, it is possible to express hypervertices using RDF [22].

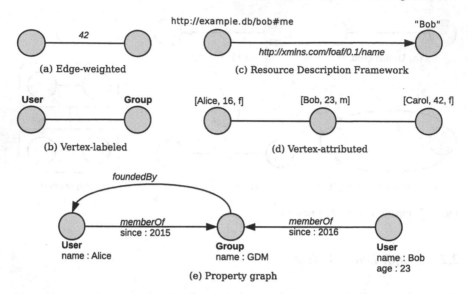

Fig. 2 Different variants of data attached to vertices and edges

Vertex- and edge-specific data: Another variation of graph data models relates to their support for data attached to the graph structure, i.e., their data content. Figure 2 illustrates different ways of attaching data to vertices and edges. The simplest form are *labeled graphs* where scalar values are attached to vertices or edges. For graph data management, labels are distinguished from identifiers, i.e., labels do not have to be distinct. An important special case of a labeled graph is a *weighted graph*, where edges show numeric labels (see Fig. 2a). Further on, labels are often used to add semantics to the graph structure, i.e., to give vertices and edges a type. Figure 2b shows a vertex-labeled graph where labels express different types of vertices. A popular semantic model using vertex and edge labels is the *Resource Description Framework* (RDF) [63], where labels may be identifiers, blank or literals. Figure 2c shows an example RDF graph.

Graph models supporting multiple values per vertex or edge are called *attributed*. Figure 2d shows an example vertex-attributed graph. The shown graph is *homogeneous* as all vertices represent the same type of entities and show a fixed schema (name, age, gender). A popular attributed model used by commercial graph databases is the *property graph model* (PGM) [93]. A property graph is a directed multigraph where an arbitrary set of key-value pairs, so-called *properties*, can be attached to any vertex or edge. The key of a property provides a meaning about its value, e.g., a property `name:Alice` represents a name attribute with value Alice. Property graphs additionally support labels to provide vertex and edge types.

Resource Description Framework: In its core, RDF is a machine-readable data exchange format consisting of *(subject, predicate, object)* triples. Considering subjects and objects as vertices and predicates as edges, a dataset consisting of such triples forms a directed labeled multigraph. Labels are either internationalized resource identifiers (IRIs), literals such as numbers and strings or so-called *blank nodes*. The latter is used to reflect vertices not representing an actual resource. There are domain constraints for labels depending on the triple position. Subjects are either IRIs or blank nodes, predicates must be IRIs and objects may be IRIs, literals or blank nodes. In contrast to other graph models, RDF also allows edges between edges and vertices, which can be used to add schema information to the graph. For example, the type of an edge :alice,:knows,:bob can be further qualified by another edge :knows,:isA,:Relationship. A schema describing an RDF database is a further RDF graph containing metadata and is often referred to as *ontology* [19]. RDF is most popular in the context of the semantic web where its major strengths are standardization, the availability of web knowledge bases to flexibly enrich user databases and the resulting reasoning capabilities over linked RDF data [111]. Kaoudi and Manolescu [58] comprehensively survey recent approaches to manage large RDF graphs and consider additional systems not listed in Table 1.

Property Graph Model: While RDF is heavily considered in research, the PGM and its de-facto standard Apache TinkerPop found lower interest so far. However, many commercial graph database products use TinkerPop and the approach appears to gain public interest, e.g., in popularity rankings of database engines.[4] With one exception, all of the considered PGM databases support TinkerPop. The TinkerPop property graph model describes a directed labeled multigraph with properties for vertices and edges. Basically, the PGM is schema-free, i.e., there is no dependency between a type label and the allowed property keys. However, some of the systems, for example Sparksee, use labels strictly to represent vertex and edges types and require a fixed schema for all of their instances. Other systems like ArangoDB manage schema-less graphs, i.e., labels may indicate types but can be coupled with arbitrary properties at the same time. In most of the databases upfront schema definition is optional.

Property graphs with a fixed schema can be represented using RDF. However, representing edge properties requires reification. In the standard way,[5] a logical relationship db:alice,schema:knows,db:bob is represented by a blank node _:bn and dedicated edges are used to express subject, object and predicate (e.g., _:bn,rdf:subject,db:alice). Properties are expressed analogously to vertices (e.g., _:bn,schema:since,2016). In consequence, every PGM edge is expressed by $3 + m$ triples, where m is the number of properties. Two of the graph databases of Table 1 store the PGM using RDF but both are using alternative, nonstandard ways of reification. Stardog is using n-quads [22] for PGM edge reification. N-quads are extended triples where the fourth position is an IRI to identify a graph. Used for edge reification, each of such graphs represents an PGM edge [27]. Blazegraph follows a further, non-standard approach to reification and implements custom RDF and SPARQL extensions [49].

---

[4]http://db-engines.com/en/ranking/graph+dbms.
[5]https://www.w3.org/TR/rdf-schema/#ch_reificationvocab.

2.3 Query Language Support

In [6], Angles named four operators specific to graph databases query languages: adjacency, reachability, pattern matching and aggregation queries. *Adjacency* queries are used to determine the neighborhood of a vertex while *reachability* queries identify if and how two vertices are connected. Reachability queries are also used to find all vertices reachable from a start vertex within a certain number of traversal steps or via vertices and edges meeting given traversal constraints. *Pattern matching* retrieves subgraphs (embeddings) isomorphic to a given pattern graph. Pattern matching is an important operator for data analytics as it requires no specific start point but can be applied to the whole graph. Figure 3a shows an example pattern graph representing an analytical question about social network data. Finally, *aggregation* is used to derive aggregated, scalar values from graph structures. In contrast to Angles, we use the term aggregation instead of *summarization*, as the latter is also used to denote structural summaries of graphs [106]. Such summarization queries are not supported by any of the considered systems.

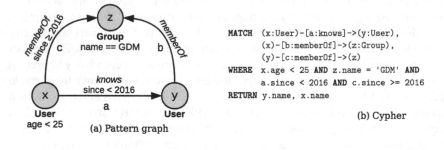

```
MATCH  (x:User)-[a:knows]->(y:User),
       (x)-[b:memberOf]->(z:Group),
       (y)-[c:memberOf]->(z)
WHERE  x.age < 25 AND z.name = 'GDM' AND
       a.since < 2016 AND c.since >= 2016
RETURN y.name, x.name
```

(b) Cypher

(a) Pattern graph

```
PREFIX rdf: <http://www.w3.org/1999/02/22-rdf-syntax-ns#>
PREFIX : <http://example.schema#>
SELECT ?y_name ?x_name
{
   ?x rdf:type :User;  :name ?x_name; :age ?x_age. FILTER(?x_age < 25).      (c) SPARQL
   ?y rdf:type :User;  :name ?y_name.
   ?z rdf:type :Group; :name 'GDM'.
   ?a rdf:subject ?x; rdf:predicate :knows;    rdf:object ?y; :since ?a_since. FILTER(?a_since < 2016).
   ?b rdf:subject ?y; rdf:predicate :memberOf; rdf:object ?z .
   ?c rdf:subject ?x; rdf:predicate :memberOf; rdf:object ?z; :since ?c_since. FILTER(?c_since >=2016).
}
```

```
g.V().match(
       __.as('x').has(label, 'User').and().has('age', lt(25)),
       __.as('y').has(label, 'User'),
       __.as('z').has(label, 'Group').and().has('name', 'GDM'),
       __.as('x').outE('knows').has('since', lt(2016)).inV().as('y'),    (d) Gremlin
       __.as('y').out('memberOf').as('z'),
       __.as('x').outE('memberOf').has('since', gte(2016)).inV().as('z')
).select('y', 'x').by('name')
```

Fig. 3 Comparison of pattern matching queries

Most of the recent graph database systems either support SPARQL for RDF or TinkerPop Gremlin for the property graph model. Both query languages support adjacency, reachability, pattern matching and aggregation queries. Figure 3c, d show example pattern matching queries equivalent to the pattern graph of Fig. 3a expressed in SPARQL and Gremlin. The result are pairs of Users who are member of the same Group with name GDM. Further on, one User should be younger than 25, member since 2016 and already knew the other user before 2016. The query was chosen to highlight syntactical differences and involves predicates related to labels and properties of vertices and edges. To support edge predicates, the SPARQL query relates to edge properties expressed by standard reification. While such complex graph patterns in SPARQL are expressed by a composition triple patterns and literal predicates (FILTER), the Gremlin equivalent is a composition of traversal chains, similar to the syntax of object-oriented programming languages.

Beside this, there are also some vendor-specific query languages or vendor-specific SQL extensions. However, these languages miss pattern matching. A notable exception is Neo4j Cypher [26]. In Cypher, pattern graphs are described by ASCII characters where predicates related to vertices and edges are separated within a WHERE clause. Cypher is currently exclusively available for Neo4j but it is planned to make it an open industry standard similar to Gremlin. Participants of the respective openCypher[6] project are i.a. Oracle and databricks (Apache Spark), which could make Cypher available to more graph database and graph processing systems in future. A common limitation of SPARQL, Gremlin and Cypher is the representation of pattern matching query results in the form of tables or single graphs (SPARQL CONSTRUCT). In consequence, it is not possible to evaluate the embeddings in more detail, e.g., by visual comparison, and to execute any further graph operations on query results. A recently proposed solution to this problem is representing the result of pattern matching queries by a collection of graphs (see Sect. 5).

3 Graph Processing

Many algorithms for graph analytics such as *pagerank*, *triangle counting* or *connected components* need to iteratively process the whole graph while other algorithms such as *single source shortest path* might require access to a large portion of it. Graph databases excel at querying graphs but usually cannot efficiently process large graphs in an iterative way. Such tasks are the domain of distributed graph processing frameworks.

In this section, we focus on dedicated distributed graph processing systems such as Pregel [71] and its derivates. More general dataflow systems like Apache Flink or Apache Spark, which also provide graph processing capabilities, will be discussed in the next section. Our presentation focuses on the popular vertex-centric processing model and its variations like partition- or graph-centric processing. To illustrate

---

[6]http://www.opencypher.org/.

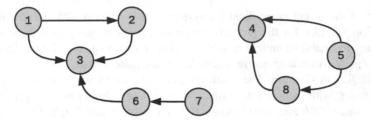

Fig. 4 Directed graph with two weakly connected components

different programming models, we show their use to compute weakly connected components (WCC) of a graph. A connected component is a subgraph where each pair of vertices is connected via a path. For weakly connected components the edge direction is ignored, i.e., the graph is considered to be undirected. Figure 4 shows an example graph with two weakly connected components $V_{C_1} = \{1, 2, 3, 6, 7\}$ and $V_{C_2} = \{4, 5, 8\}$.

3.1   General Architecture

The different programming models are based on a general architecture of a distributed graph processing framework. The architecture uses a *master node* for coordination and a set of *worker nodes* for the actual distributed processing. The input graph is partitioned among all worker nodes, typically using hash or range-based partitioning on vertex labels. In the vertex-centric model, a worker node stores for each of its vertices the vertex value, all outgoing edges including their values and vertex identifiers (ids) of all incoming edges. Figure 5a shows our example graph partitioned across four worker nodes A, B, C and D. Different frameworks extend upon this structure such as Giraph++ [107] where each worker node also stores a copy of each vertex that resides on a different worker but has a connection to a vertex on the worker node (Fig. 5b).

All graph processing systems discussed in this section use a directed generic multigraph model as introduced in Sect. 2. Vertices have a unique identifier K, e.g., of type 64bit-integer. Vertices and edges may store a generic value further referred to as VV (vertex value) and EV (edge value). All frameworks allow the exchange of messages passed along edges and denoted by M.

3.2   Think Like a Vertex

The "Think Like a Vertex" or vertex-centric approach has been pioneered by Google Pregel in 2010 [71]. Ever since many frameworks have adopted or extended it [10, 41, 61, 67, 97, 102]. To write a program in a Pregel-like model, a so called *vertex*

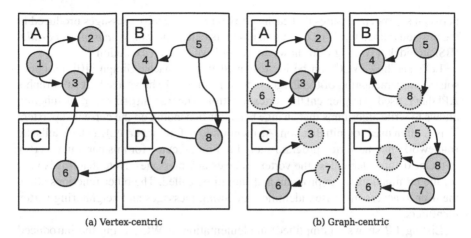

(a) Vertex-centric   (b) Graph-centric

Fig. 5 Partitioned input graph for different computation models

```
 1   long getSuperstep(); // returns the current iteration
 2   void sendMsg(K id, M msg);
 3   void sendMsgToAllEdges(M msg);
 4   void voteToHalt();
 5   K getVertexId();
 6   VV getVertexValue();
 7   void setVertexValue(VV vertexValue);
 8   Iterator<K> getNeighbors();
 9   Iterable<M> getMessages();
10   void aggregateValue(aggregatorName, aggregatedValue);
11   AV getAggregatedValue(aggregatorName);
```

Listing 1.1: Subset of the Apache Giraph API used to write a vertex function.

compute function[7] has to be implemented. This function consists of three steps: Read all incoming messages, update the internal vertex state (i.e., its value) and send information (i.e., messages) to its neighbors. Note that each vertex only has a local view of itself and its immediate neighbors. Any other information about the graph necessary for computation has to be sent along the edges. This paradigm is similar to the actor-based programming model [2] as implemented by Akka [3] or Quasar [87].

Vertex functions are executed in synchronized *supersteps*. In each superstep each worker node executes the compute function for all of its active vertices, marks them inactive if the voteToHalt() function is called and gathers their output messages. When all workers have finished, the gathered messages are delivered synchronously. Vertices that receive messages are then marked active. This is repeated until there is no active vertex at the end of a superstep. Note that the synchronization barrier

---

[7]We use vertex compute function and vertex function interchangeably throughout this section.

between supersteps ensures that each vertex will only receive messages produced in the previous superstep. This execution model is called the *bulk synchronous parallel* (BSP) model [110]. Figure 6 shows an example of such an execution.

Let's see how WCC can be implemented using Apache Giraph [10], an open-source implementation of the Pregel model. Listing 1.1 shows a subset of Giraph's API that is used to implement the vertex function. The getSuperstep() function allows to write algorithms that change behavior depending on the current superstep. This is often used for initialization. As mentioned before, voteToHalt() tells the framework that the vertex program should not be executed for this particular vertex in the next superstep unless the vertex receives any messages. Note that this is vital for the termination of the program and should be called. The other functions allow the user to access the vertex identifier, incoming messages and neighboring vertex identifiers.

Listing 1.2 shows a (simplified) implementation of WCC using the introduced API. The basic idea is that vertices propagate their label along the edges until convergence. After termination, each vertex stores a component id which will be equal to the smallest vertex id that can be reached from this vertex. This value will be the same for each vertex in a component and thus identifies a component. In superstep 0, we initialize the component id with the vertex id and send the value to all neighbors. In each subsequent superstep, each vertex computes the smallest component id among all received messages; if it is smaller than the currently stored value, it is replaced and the new value is sent to all neighbours. Each vertex always votes to halt at the end of each superstep. As a result, no message will be sent, if no vertex has changed its component id within a superstep and the algorithm terminates. Figure 6

```
1   void compute(Vertex v) {
2     if (getSuperstep() == 0)
3       v.setValue(v.getVertexID())
4       sendMessageToAllEdges(v.getVertexValue())
5     else
6       minValue = min(v.getMessages())
7       if (minValue < v.getVertexValue())
8         v.setVertexValue(minValue)
9         sendMessageToAllEdges(v.getVertexValue())
10    v.voteToHalt();
11  }
12
13  void combine(M message1, M message2) {
14    return min(message1, message2)
15  }
```

Listing 1.2: WCC in Apache Giraph. The vertex function will be executed for each vertex in the graph. Messages sent by the vertices are stored at the worker and delivered at the end of the current superstep. The execution loops until no vertex has received any message.

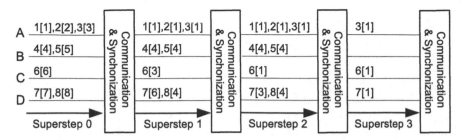

Fig. 6 Vertex-centric WCC computation for the graph of Fig. 4. We show the vertex value at the end of each superstep. Initially, the vertices use their ids as initial vertex values. In any superstep, each vertex changes its value to the minimum among all messages and its own value. However, a vertex function will only be executed if the vertex has received any messages in the previous superstep. Looking at vertex 1 and 7, one can see how vertex id 1 is propagated through the component. Note that we omitted the 4th superstep that solely consists of vertex 6 processing a message received from vertex 7

shows the WCC execution for the graph shown in Fig. 5a resulting in two connected components represented by the identifiers 1 and 4.

Variants Various vertex-centric graph processing systems provide specific features and optimization techniques, for example, to mutate the graph or to reduce network traffic and computation time. In the following, we will discuss the most differentiating features as shown in Table 2.

Aggregation: Certain graph algorithms need global knowledge in terms of aggregated values such as the number of vertices in the graph or the total sum of all vertex values. In the basic model, this can be achieved by creating a vertex that is connected to all other vertices. However, this approach creates vertices with a huge amount of incident edges that will take longer to process than a regular graph vertex. This will decrease performance since workers have to wait for each other at the end of each superstep. Additionally, these special purpose edges and vertices require specific programming logic in the vertex program which increases complexity. Many frameworks (see column aggregation in Table 2) require the user to provide a function that is run on the master node between *supersteps* for this purpose. For example, to calculate the sum of all vertex values, each vertex would send its value to the master node (API method `aggregateValue()`, Listing 1.1) which aggregates them and makes the results accessible in the next superstep (`getAggregatedValue()` in Listing 1.1). Note, that for associative and commutative operations, such as counting or summation, this can be done in an aggregation tree where the worker node will aggregate the values of all its vertices before sending the aggregated value to the master, therefore reducing communication costs.

Reducing network communication: A technique, similar to the one used for aggregation, can also be used to reduce the number of messages between different worker nodes. If a worker node has multiple messages addressing the same vertex, they can potentially be combined into a single message. In some of the frameworks (Table 2) the user can define a combiner, a dedicated function that takes two messages as input

Table 2 Key features of the discussed graph processing systems (n.a., not applicable)

	Language	Programming model	BSP	Asynchronous	Generic scheduler	Aggregation	Add vertex/edge	Remove vertex/edge	Combiner
Pregel [71]	C++	Pregel	✓						
Giraph [10]	Java	Pregel	✓			✓	✓	✓	✓
GPS [97]	Java	Pregel	✓			✓			✓
Mizan [61]	C++	Pregel	✓			✓			✓
GraphLab [69]	C++	GAS	✓	✓	✓	✓	✓		n.a.
GraphChi [67]	C++, Java	Pregel		✓	✓	✓	✓	✓	n.a.
Signal/Collect [102]	Java	Scatter–Gather	✓	✓					n.a.
Chaos [94]	Java	Scatter–Gather	✓	✓					n.a.
Giraph++ [107]	Java	Partition-Centric	✓			✓	✓	✓	✓
GraphX [42]	Scala, Java	GAS	✓			✓			n.a.
Gelly [39]	Scala, Java	GSA, Scatter–Gather	✓			✓	✓	✓	n.a.

and combines them into one. Listing 1.2 includes the combine function for WCC as implemented in Giraph. In our WCC implementation, we are only interested in the smallest value, so the combiner can discard the larger message. With this combiner, no vertex will receive more messages than the number of worker nodes.

Powergraph [41] further extended the idea of the combiner by introducing the *Gather-Apply-Scatter* (GAS) model. Instead of a single vertex compute function, the user has to provide a gather, apply and scatter function. The gather function has the same functionality as the combiner: it aggregates messages addressing the same vertex on the sending worker nodes. The apply function has the incoming messages as input and updates the vertex state. The scatter function has the vertex state as input and produces the outgoing messages. Similar to the gather function, the scatter function can be executed on the worker nodes. Instead of sending multiple messages from one vertex to vertices on the same worker node, only the vertex value is send and the messages are then created locally. This execution is transparent to the user which only has to provide the three functions. The GAS model is especially effective on graphs with highly skewed degree distributions. It not only reduces the amount of network traffic, but also helps balancing the workload between worker nodes by spreading out the computation. One downside of the GAS model is that all information about messages that should eventually be sent needs to be part of the vertex value. In case of WCC, we need to extend the vertex value by a boolean field that reflects if the vertex value has changed or not to decide if messages should be sent.

The systems Signal/Collect [102] and Chaos [94] introduced the *Scatter–Gather* *model*. This model requires the user to provide an edge and a vertex function. The vertex function has all incoming messages as input and can modify the vertex value. The edge function takes the vertex value as input and can then generate a message. Compared to the GAS model, in the scatter–gather model, the computation is parallelized across the vertices, which may lead to unbalanced load, if the edge degree distribution is skewed. Depending on the computation, the execution time for high-degree vertices increases as they need to process more messages and thus the synchronization barrier is eventually delayed.

Asynchronous execution: Looking at Fig. 6, one can see that worker node A takes longer to compute the vertex function on all of its vertices. In consequence, the faster working nodes B, C, and D have to wait. Not all algorithms require the strict synchronization offered by the BSP execution model. Our WCC implementation in Listing 1.2 tries to find the minimum vertex id in each component. Finding the minimum of a set does not require a specific execution order and can be executed without synchronization. If a worker node is a superstep behind and does not deliver its messages in time, the minimum of each component will eventually be found once the delayed messages are delivered. The overall execution time can be potentially reduced since worker nodes do not spend time waiting for other workers to finish. Furthermore, some algorithms [16, 66] converge much faster on an asynchronous execution model up to the point where running them in a BSP model will not converge in reasonable time. Other graph algorithms such as Ja-be-Ja [89], a peer-to-peer inspired graph

```
boolean containsVertex(K id);
boolean isInternalVertex(K id);
boolean isBoundaryVertex(K id);
Vertex<K, VV, EV, M> getVertex(K id);
Collection<Vertex<K, VV, EV, M>> internalVertices();
Collection<Vertex<K, VV, EV, M>>
activeInternalVertices();
Collection<Vertex<K, VV, EV, M>> boundaryVertices();
Collection<Vertex<K, VV, EV, M>> allVertices();
```

Listing 1.3: Additional functions in the Giraph++ API.

partitioning algorithm, can only be implemented using an asynchronous execution model. To address these challenges GraphLab [69], Signal/Collect and GraphChi [67] allow for asynchronous execution. Instead of waiting for a synchronization barrier, in these models, messages produced by a vertex will be delivered to the target vertex directly. Each worker node processes its vertices in order, thus, within a partition, each vertex will be executed with the same frequency. However, in Fig. 5a, vertex 6 on worker node C might already have executed ten times while vertex 1 on worker node A has only executed once. GraphLab and GraphChi also allow the user to provide a scheduler function that changes the execution order, for example, to prioritize vertices with a high value. This allows to focus an algorithm on a certain part of the graph, which can lead to faster convergence in some cases.

Note, that optimizations such as combiners or the GAS model cannot use their full potential when executed asynchronously since messages are not necessarily batched together. As a result, asynchronous execution generally uses more network resources. Performance gains are hard to quantify since the speedup highly depends on the graph structure and to which degree work is equally distributed between worker nodes. For our WCC example, each superstep might be faster due to the removal of the synchronization barrier but the algorithm might require more steps to terminate. In the BSP execution, the number of required steps is equal to the longest shortest path in the graph since each vertex processes the data from all its neighbors in each superstep. In an asynchronous execution it is possible that the message with the true minimum is delayed so that there are additional steps finding the minimum between larger values before the true minimum is found.

Graph mutation: Transformational algorithms such as graph coarsening or computing the minimum spanning tree need to modify the graph structure during execution. This is a non-trivial task since it may lead to load imbalances, performance loss and memory overflow. Currently, only few frameworks support these operations. For example, while Giraph supports adding and removing vertices and edges, GraphLab only allows addition. Vertices are added or removed from inside the vertex function and the changes to the graph, similar to messages, become visible in the next superstep. Newly created vertices are always marked as active in the superstep they appear in and are therefore guaranteed to be executed.

```
1   void compute() {
2     if (getSuperstep() == 0)
3       sequentialCC();
4       for (bV in boundaryVertices())
5         sendMsg(bV.getVertexId(), bV.getVertexValue())
6     else
7       equiCC = new MultiMap;
8       for (iV in activeInternalVertices())
9         minValue = min(iV.getMessages())
10        if (minValue < iV.getVertexValue())
11          equiCC.add(iV.getVertexValue(), minValue)
12          for (v in allVertices())
13            minValue = equiCC.getMinFor(v.getVertexValue())
14            if (minValue < v.getVertexValue())
15              v.setVertexValue(minValue)
16              if (isBoundaryVertex(v.getId())
17                sendMsg(v.getVertexId(), vertex.getVertexValue())
18      allVoteToHalt()
19  }
20
21  void combine(M message1, M message2) {
22    return min(message1, message2)
23  }
```

Listing 1.4: WCC in Giraph++. First each worker node finds all internal connected components. Then it iteratively shares the information with other worker nodes that have connected vertices.

3.3   Think Like a Graph

Instead of writing a compute function executed on each vertex, in a graph-/partition-centric model, the user provides a compute function that takes all vertices managed by a worker node as input. These functions are then executed using the BSP model. This approach requires additional support structures when distributing the graph. The input graph is distributed across worker nodes in the same way as for vertex-centric computations. The vertices of worker node n are called *internal vertices* to n. On each worker node n we then create a copy of each vertex that is not internal to n, but is directly connected to an internal vertex of n. These vertices are called *boundary vertices* and represent the cached vertex values of copied vertices. Every internal vertex may have up to one of these boundary vertices on each worker node. Figure 5b shows the distributed graph with internal and boundary vertices on the four worker nodes.

Listing 1.3 shows the additional methods of the Giraph++ [107] API. Having a partition compute function instead of a vertex compute function allows direct access to all internal and local boundary vertices and thus computing the entire subgraph. Each worker node executes its user-defined function and afterwards sends messages from all boundary vertices to their internal representation. The partition-centric model can mimic a vertex centric execution by iterating through all active internal nodes once in each superstep. Listing 1.4 shows a partition-centric implementation of WCC. In the

Table 3 Vertex states/values in vertex-centric iteration

Vertex	Step 0	Step 1	Step 2	Step 3
1	1	1	1	1
2	2	1	1	1
3	3	1	1	1
4	4	4	4	4
5	5	4	4	4
6	6	3	1	1
7	7	6	3	1
8	8	4	4	4

initialization step, a sequential connected component algorithm is executed to find all local connected components. The locally computed component label for each boundary vertex is then sent to its corresponding internal vertex. In each of the subsequent supersteps, the algorithm processes all the incoming messages and merges labels representing the same component. Although the implementation of this approach is more complex, it can reduce the amount of iterations and thus improve performance. The number of steps required to converge is smaller or equal to the longest shortest path in the graph. The precise number of saved iteration steps depends on the graph structure, in particular on how vertices are distributed among the worker nodes.

Tables 3 and 4 show the convergence in vertex- and graph-centric iterations respectively. One can see, that it takes four supersteps for a vertex-centric iterations whereas using a graph-centric approach, the components can be computed in only two supersteps. Notice that the reduction in supersteps depends on the partitioning of the input graph. A partitioning where each component resides on a single worker node requires zero supersteps, while the worst case partition would require the same amount of supersteps as a vertex centric program. The performance gain can be hard to predict and cannot justify the additional complexity of the program in all cases.

Table 4 Vertex state/values in graph-centric iteration

Vertex	Step 0	Step 1
1	1	1
2	1	1
3	1	1
4	4	4
5	4	4
6	6	1
7	7	1
8	8	4

In this section we gave an overview about the different dedicated graph processing frameworks available. We summarized the most common programming models and have shown their variants. In real-world scenarios, graph processing is often only a single step of a longer pipeline consisting of data transformations. Therefore modern processing frameworks such as Apache Spark and Apache Flink provide graph processing libraries that can be directly integrated into a larger program. These libraries support vertex-centric graph processing with additional graph operations that can be combined with general-purpose data operations on structured and unstructured data.

4 Graph Dataflow Systems

In the previous section, we introduced specialized systems providing tailored programming abstractions for the fast execution of a single iterative graph algorithm on large graphs with billions of vertices and edges. However, complex analytical problems often require the combination of multiple techniques, for example, to create combined graph structures based on unstructured or structured data originated from different sources (e.g., distributed file systems, database systems) or to combine graph algorithms and non-graph algorithms (e.g., for machine learning). In such cases, using dedicated systems for each part of an analytical program increases the overall complexity and leads to unnecessary data movement between systems and respective data duplication [42, 116].

By contrast, distributed in-memory dataflow systems such as Apache Spark [42, 115, 116, 118], Apache Flink [4] or Naiad [76, 79] provide general-purpose operators (e.g., map, reduce, filter, join) to load and transform unstructured and structured data as well as specialized operators and libraries for iterative algorithms (e.g., for machine learning and graph analysis). Using such a system for the implementation of complex analytical programs reduces the overall complexity for the user and may lead to performance improvements since the holistic view on the whole program enables optimizations, such as operator reordering or caching of intermediate results.

In this section, we will discuss graph analytics on distributed dataflow systems using Apache Flink as a representative system. We briefly introduce Apache Flink and its concept for iterations and will then focus on Gelly, a graph processing library integrated into Apache Flink. Gelly implements the Scatter–Gather and Gather-Sum-Apply programming abstractions for graph processing and provides additional operators for graph transformation and computation. We will finish the section with a brief comparison to GraphX, a graph library on Apache Spark.

```
1   ExecutionEnvironment env = getExecutionEnvironment();
2   DataSet<String> text = env.readTextFile("hdfs:///text");
3
4   DataSet<Tuple2<String, Integer>> wordCounts = text
5       // splits the line and outputs (word, 1) tuples
6       .flatMap(new LineSplitter())
7       // group tuples by word
8       .groupBy(0)
9       // add together the "1"s in all tuples per group
10      .sum(1);
11
12  wordCounts.print();
```

Listing 1.5: Word Count in Flink

4.1 Apache Flink

Apache Flink is the successor of the former research project Stratosphere [4] and supports the declarative definition and distributed execution of analytical programs on batch and streaming dataflows.[8] The basic abstractions of such programs are *datasets* and *transformations*. A dataset is a collection of arbitrary data objects and transformations describe the transition of one dataset to another one. For example, let X,Y be datasets, then a transformation could be seen as a function $t : X \rightarrow Y$. Example transformations are *map*, where for each input object $x_i \in X$ there is exactly one output object $y_i \in Y$, and *reduce*, where all input objects are aggregated to a single one. Further transformations are well known from relational databases, e.g., *join*, *group-by*, *project*, *union* and *distinct*. To express application logic, transformations are parameterized with user-defined functions. A Flink program may include multiple chained transformations. When executed, Flink handles program optimization as well as data distribution and parallel execution across a cluster of machines.

We give an exemplary introduction to the dataset API using a simple word count program to compute the frequency of each word in an input text (Listing 1.5). We first create a Flink execution environment (Line 1), which abstracts either a local machine (e.g., for developing and testing) or a cluster. In Line 2, we define an input data source, here a file from HDFS, the Hadoop Distributed File System.[9] The resulting dataset contains strings whereas each string represents a line in our input file. In Line 6, we use *flatMap* to declare the first transformation on our input dataset. This transformation allows us to output an arbitrary number of objects for each input object. Here, the user-defined function `LineSplitter` is applied on each line in the input dataset and splits it into words. For each word, the function outputs a tuple containing the word and the frequency 1, for example, the line *"graphs are*

---

[8]In its core, Flink is a distributed streaming system and provides streaming as well as batch APIs. We focus on the batch API, as Gelly is currently implemented on top of that.

[9]Flink supports further systems as data source and sink, e.g., relational and NoSQL databases or queuing systems.

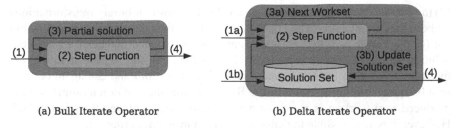

(a) Bulk Iterate Operator   (b) Delta Iterate Operator

Fig. 7 Iteration operators in Apache Flink [9]

everywhere" results in the tuples (*"graphs"*, 1), (*"are"*, 1) and (*"everywhere"*, 1). In Line 8, we perform a *group-by* transformation on the output dataset of the previous *flatMap* transformation to gather all tuples that represent the same word. In Line 10, we add together the single frequencies to get the total frequency for each word using *sum*, a predefined aggregation transformation. Flink programs are executed lazily, i.e., program execution needs to be started explicitly. Here, we trigger the execution by printing the dataset to system console (Line 12). When triggered, Flink analyzes the program, optimizes it and executes it in the specific environment. Data lines are read in parallel from the data source and "flow" through the transformations which are scaled-out to all workers in the cluster.

Iterations in Apache Flink Our word count example represents a dataflow whose execution graph is a directed acyclic graph of transformations. However, iterative or recursive graph and machine learning algorithms require cyclic execution graphs. To support cyclic dataflows, Flink offers two specialized operators: Bulk and Delta Iteration [34, 35].

With *Bulk Iteration* (Fig. 7a), each iteration computes a new solution based on the previous iteration result which is then used as input for the next iteration. Conceptually, Flink's Bulk Iteration can be separated into four phases: (1) the *iteration input* is the initial dataset for the first iteration; (2) the *step function* takes the output of the previous iteration as input and executes an acyclic dataflow containing arbitrary transformations on that dataset to create a new dataset; (3) the result of the step function is the *next partial solution*, which is used as input for the next iteration; (4) the *iteration result* is the dataset created by the last iteration and can be used in subsequent dataflows. The convergence criterion for the Bulk Iteration is either a maximum number of iterations or a custom convergence criterion.

With *Delta Iteration*, each iteration computes only incremental updates for an evolving global solution set instead of a completely new solution set. The motivation for this approach are algorithms where an update on one element has a direct impact only on a small number of other elements, such that different parts of the solution may converge at different speeds [34]. When applicable, this leads to faster convergence as large parts of the solution are computed in the first iterations so that later iterations compute on much smaller subsets. Figure 7b shows the phases of Flink's Delta Iteration: (1) In contrast to Bulk Iteration, we now have two input datasets: (a) the *initial workset* and (b) an *initial solution set* which evolves with each iteration;

(2) the step function again performs an acyclic dataflow of arbitrary transformations on both the current workset and the solution set; (3) the outputs of the step function are (a) the *update solution set*, which contains incremental updates for the initial solution set and (b) the *next workset*, which is the input for the next iteration; (4) the solution set after the last iteration is the *iteration result* and can again be used in subsequent dataflows. In contrast to Bulk Iteration, the iteration terminates if the produced next workset is empty or a maximum number of iterations is reached. However, it is also possible to define a custom convergence criterion.

With reference to the introduced programming models for graph processing in Sect. 3, each iteration in the Bulk and Delta Iteration can be seen as a superstep in a synchronous BSP process. Multiple instances of the step function are executed in parallel and synchronized at the end of each iteration. In Sect. 3, we also showed that for specific graph algorithms, for example, connected components or single-source-shortest-path, not all vertices are necessarily active in each superstep.[10] The Delta Iteration is a good foundation for this class of algorithms which is why Gelly uses it to implement vertex-centric programming abstractions, which we will discuss next.

4.2 Apache Flink Gelly

Flink Gelly [39] is a graph library integrated into Apache Flink and implemented on top of its dataset API. Besides dedicated graph processing abstractions, Gelly provides a wide set of additional operators to simplify the definition of graph analytical programs. The provided data model is a directed, labeled multigraph where vertex and edge labels are generic, i.e., vertices and edges can carry arbitrary user-defined payload ranging from basic data types such as numbers and strings to complex domain objects. In the following, we will discuss the graph representation on Flink's dataset API, transformation methods and how graph processing abstractions are mapped to the Delta Iteration.

Graph Representation Gelly uses two classes to represent the elements of a graph: `Vertex` and `Edge`. A `Vertex` comprises a comparable, unique identifier (id) and a value, an `Edge` consists of a source vertex id, a target vertex id and an edge value. Identifiers and values are generic and need to be declared upon graph creation. Internally, a graph is represented by a dataset of vertices and a dataset of edges as shown below:

```
class Graph<K, VV, EV> {
    DataSet<Vertex<K, VV>> vertices
    DataSet<Edge<K, EV>> edges
}
```

The generic type K represents the vertex id type, VV the vertex value type and EV the edge value type. Since Gelly offers methods to return the vertex and edge

---

[10]When implemented using a synchronous graph-processing system.

datasets, an analytical program can combine those datasets with any other library in Flink (e.g., for machine learning) as well as third-party libraries that are implemented on the dataset API (e.g., GRADOOP in Sect. 5). A Gelly graph provides basic methods for creating graphs and returning simple metrics such as vertex count, edge count or in- and out-degrees of vertices, which result in new datasets for further processing.

Graph Transformations Graph transformation methods are applied on an input graph and return a new, possibly modified graph, hence enabling the composition of multiple graph transformations in an analytical program. Internally, Gelly translates each graph transformation to a series of transformations on the vertex and edge datasets. Similar to other graph dataflow frameworks, e.g., GraphX [42, 115, 116], Gelly offers the following transformation methods:

- **Mutation** methods enable adding and removing of vertices and edges. The result is a new graph with an updated vertex and edge dataset respectively.
- **Map** allows the modification of vertex and edge values by applying user-defined transformation functions on all elements in the corresponding datasets.
- **Subgraph** enables the extraction of a new graph based on user-defined vertex and edge predicates. If an element in the input graph satisfies the predicate, it is contained in the output graph.
- **Join** allows the combination of vertex and edge datasets with additional input datasets. The transformation applies a user-defined function on each matching pair and returns a graph with a updated datasets. This can be useful to attach external data, e.g., from a relational database, to the graph.
- **Undirected** can be used to transform a directed graph into an undirected graph by cloning and reversing all edges.
- **Union/Difference/Intersect** enable merging of two graphs into a new graph based on the respective set-theoretical method applied on vertex and edge datasets.

Neighborhood Methods Neighborhood methods are applied on all incident edges and adjacent vertices of each vertex and can be used to aggregate edge and vertex values (e.g., average/min/max values, vertex degree, etc.). Gelly provides two variants of neighborhood methods:

- **reduceOnEdges/Neighbors** allow the aggregation of edge and vertex values by providing a user-defined, associative and commutative function on pairs of values. The methods result in a new dataset containing exactly one aggregate per vertex.
- **groupReduceOnEdges/Neighbors** allow the aggregation of edge and vertex values by providing a user-defined, non-associative, non-commutative function on all respective values. This is useful, if one needs to have all values available in the function or if more than one aggregate needs to be computed per neighborhood. The methods result in new datasets containing an arbitrary amount of aggregates for each vertex.

Graph Processing In Sect. 3, we introduced various programming abstractions for graph processing. Gelly currently adopts two variants of vertex-centric iterations:

```
class WCCMessenger extends MessagingFunction {
  void sendMessages(Vertex<K, VV> v) {
    sendMessageToAllNeighbors(vertex.getValue())
  }
}
class WCCUpdater extends VertexUpdateFunction {
  void updateVertex(Vertex<K, VV> v, Iterator<VV>
      messages) {
    VV current = v.getValue()
    VV min = current
    for (Message message in messages)
      if (message < min) min = m
    if (current != min) v.setValue(min)
  }
}
```

Listing 1.6: Scatter/Messaging and Gather/Update functions for WCC. `MessagingFunction` and `VertexUpdateFunction` are provided by Gelly and need to be extended by the user.

Scatter–Gather and Gather-Sum-Apply. Both are implemented using the Delta Iteration operator and are thus executed in synchronous supersteps. In the following, we will discuss both abstractions in further detail.

The Scatter–Gather abstraction is adopted from the Signal/Collect model [102] and divides a superstep in two phases. In the Scatter (or messaging) phase, the messages sent to other vertices are being produced, while in the Gather (or update) phase each vertex updates its value using the received messages. The user needs to implement both, a messaging and an update function, which are applied during the computation. Picking up the running example of Sect. 3, Listing 1.6 shows a WCC implementation using the Scatter–Gather abstraction. While the Scatter function sends the updated vertex value to all neighbors, the Gather function searches for the smallest value among all messages and updates the vertex value if necessary.

Figure 8a illustrates the implementation of the Scatter–Gather abstraction using Delta Iteration. Here, the initial workset and solution set is the vertex dataset. In the step function, Scatter and Gather functions are applied using Flinks *coGroup* transformation.[11] First, an adjacency list is built by grouping each vertex with all of its incident edges. For each row in that adjacency list, Gelly applies the Scatter function to create new messages. That messages are again grouped with the vertex values (solution set) and fed into the Gather function. The output of that transformation is a dataset containing all vertices that changed their value. This dataset is then used to update the solution set and also as workset for the next iteration.

In contrast to Scatter–Gather, where information is pushed to a vertex, in the Gather-Sum-Apply (GSA) abstraction, each vertex instead pulls information from

---

[11] The `coGroup` transformation groups each input dataset on one or more fields and then joins the groups.

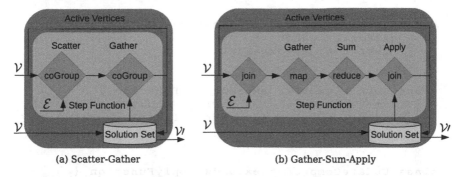

(a) Scatter-Gather   (b) Gather-Sum-Apply

Fig. 8 Scatter–Gather and Gather-Sum-Apply abstraction using Delta Iteration [39]. Input for both iterations are the vertex dataset $\mathcal{V}$ (initial working and solution set), the edge dataset $\mathcal{E}$ and the respective user-defined functions. In both cases, the output dataset $\mathcal{V}'$ contains the updated vertex values

its neighbors.[12] One iteration is divided into three phases: In the Gather phase, a user-defined function is applied on the neighborhood of each vertex. Here, each pair of incident edge value and corresponding adjacent vertex value produces a partial value. In the Sum phase, a second user-defined function aggregates the partial values for each neighborhood to a single value. In the final Apply phase, the aggregated value and the current vertex value are used to produce a new vertex value. For a WCC computation, the user-defined functions are presented in Listing 1.7. In the Gather function, we select the value stored at each adjacent vertex.[13] After that, we compute the minimum among those values by reducing them pair-wise in the Sum function. In the Apply function, we finally update the vertex value if the reduced value is smaller than the current vertex value.

Figure 8b illustrates the GSA abstraction implemented using Delta Iteration. In contrast to the Scatter–Gather implementation, vertices are first joined with their incident edges to construct neighbors as input for the Gather function. The latter is applied using a *map* transformation and returns a value for each neighbor. Those values are reduced for each neighborhood by applying the Sum function and finally joined with the vertices to update their values using the Apply function. As with Scatter–Gather, the result is a dataset of updated vertices which is used to evolve the solution set and as workset for the next iteration.

As denoted in Sect. 3, the main difference between Scatter–Gather and GSA computations is that in the Gather phase of GSA, the computation is parallelized over the edges, while in the Scatter phase, it is parallelized over the vertices. Through this, GSA is advantageous if the Gather phase contains expensive computation or if the graph shows a skewed degree distribution. Also, since the Sum phase of a GSA computation exploits a *reduce* transformation, the results computed on a single worker are internally combined before they are sent to other workers which decreases network

---

[12]GSA is a variant of the GAS abstraction introduced by PowerGraph [41] and discussed in Sect. 3.
[13]The `Neighbor` class allows access to the incident edge value and the adjacent vertex value.

```
class GatherNeighborValues extends GatherFunction {
  VV gather(Neighbor n) {
      return n.getVertexValue()
  }
}
class GetMiniumValue extends SumFunction {
  VV sum(VV newValue, VV currentValue) {
    return (newValue < currentValue) ? newValue :
        currentValue
  }
}
class UpdateComponent extends ApplyFunction {
  void apply(VV sumValue, VV originalValue) {
    if (sumValue < originalValue) setResult(sumValue)
  }
}
```

Listing 1.7: Gather, Sum and Apply functions for WCC in Gelly.

traffic and computation times [39]. However, in contrast to Scatter–Gather, the GSA composition prohibits the communication between vertices that are not adjacent in the graph.

4.3 Comparison to Other Graph Dataflow Frameworks

Another prominent implementation of a graph dataflow framework is GraphX [42, 115, 116] which is integrated into Apache Spark [118]. GraphX provides a similar API for graph transformation and neighborhood methods that can be composed with other Spark libraries. For iterative graph processing, GraphX implements the Gather-Apply-Scatter abstraction introduced by Powergraph [41] and discussed in Sect. 3. Like Gelly, GraphX is built on top of the underlying batch API and uses two distributed collections, so-called Resilient Distributed Datasets (RDD), to manage vertices and edges. RDDs are similar to the concept of a dataset in Flink and support transformations (e.g., map, reduce, join) which result in new RDDs. However, in contrast to Gelly, GraphX offers various optimizations tailored for graph analytics. One important optimizations is the partitioning of edges based on vertex-cut algorithms like 2D hash partitioning. Here the edge collection is equally distributed across all workers by minimizing the number of times each vertex is cut. A second optimization is the reduction of network traffic between workers by introducing so called mirror vertices in combination with multicast joins [42]. Here, a join operation between vertex and edge RDD transfers only those vertices to edge partitions that are incident to the contained edges.

5 Gradoop

The distributed graph processing and graph dataflow approaches presented in the preceding sections are well suited for scalable graph analytics, especially to execute iterative graph algorithms on large graphs. The graph dataflow approaches also support a flexible combination of graph processing with general data transformation operators provided by the underlying frameworks. However, the implemented graph data models are largely generic and do not meet the requirements posed in the introduction, in particular schema-flexible support for semantic graph data with vertices and edges of different types and varying attributes. Without this support, graph operators such as evaluations on vertex or edge attributes need to be user-defined making the analysis of heterogeneous real-world data a laborious programming task. Moreover, none of the graph systems discussed so far has built-in support to manage collections of graphs, e.g., application-specific subgraphs such as communities in social networks. Finally, the graph data model should provide a set of declarative operators on graphs and graph collections that can be used for the simplified development of advanced graph analysis programs.

The GRADOOP framework (*Gra*ph data management and analytics with Ha*doop*) [56, 57] aims at meeting these requirements and improving current graph dataflow systems. It is built on the so-called **Extended Property Graph Model** [57] supporting semantically rich, schema-free graph data within many distinct graphs. A set of high-level operators is provided for analyzing both single graphs and collections of graphs. These operators fulfill the closure property[14] as they take single graphs or graph collections as input and result in single graphs or graph collections thus enabling their composition to complex analytical programs. GRADOOP is GPLv3-licensed and publicly available.[15] In the following subsections, we will first introduce the architecture of GRADOOP and then focus on the data model including its operators. Finally, we illustrate the capabilities of GRADOOP with an exemplary analytical dataflow program.

5.1 Architecture

GRADOOP aims at providing a framework for scalable graph data management and analytics on large, semantically expressive graphs. To achieve horizontal scalability of storage and processing capacity, GRADOOP runs on shared nothing clusters and utilizes existing Hadoop-based software for distributed data storage and processing.

Figure 9 shows the high-level architecture of GRADOOP. Analysts declare graph analytical programs using a domain specific language, called Graph Analytical Language (GrALa). The language contains analytical operators for single graphs and

---

[14]An operator fulfills the closure property if the execution of that operator on members of an input domain results in members of the same domain.

[15]http://www.gradoop.com.

Fig. 9 Gradoop high-level
architecture

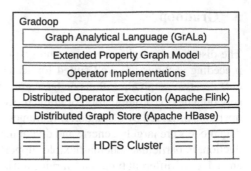

graph collections as well as general operators to read and write graph data from and
to data stores. GrALa has been developed on top of the Extended Property Graph
Model (EPGM) that will be discussed in the next section.

To execute analytical programs in a distributed environment, the EPGM and
GrALa are implemented on top of Apache Flink. This way, GRADOOP provides
new features for graph analytics while benefiting from existing Flink capabilities for
large-scale data and graph processing. Flink handles program optimization as well
as data distribution and parallel execution across a cluster of machines. Furthermore,
GRADOOP can be easily integrated with other Flink libraries, like Gelly or Machine
Learning.

The distributed graph store offers the possibility to manage a persistent graph
database structured according to the EPGM and is implemented in Apache HBase,[16]
a distributed, non-relational database running on the Apache HDFS (Hadoop Distributed File System). The graph store offers basic methods to read and write a database
and therefore serves as data source and sink for graph analytical programs. Additionally, GRADOOP allows reading from and writing to any data store which is supported
by Apache Flink (e.g., HDFS files, relational databases, NoSQL databases).

5.2   Extended Property Graph Model

The EPGM extends the popular property graph model [93] (Sect. 2.2) by supporting graph collections and composable analytical operators. Graph collections are a
natural way to represent logical partitions of a graph, e.g., communities in a social
network [36] or business process executions [83]. Further on, graph collections are
the result of certain graph algorithms, e.g., embeddings found by graph pattern matching [37] or frequent subgraph mining [55]. Using GrALa, the EPGM operators for
graphs and graph collections can be used together within analytical programs. In the
following, we present the EPGM graph representation and operators in more detail.

---

[16]http://hbase.apache.org.

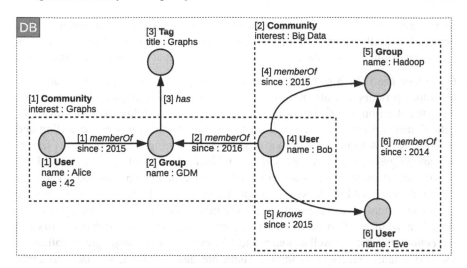

Fig. 10 Example EPGM database representing a simple social network containing two logical graphs. Each logical graph describes a community inside the social network, for example, people that are member of a group related to graphs form the *Graphs* community

Graph Representation A property graph is a directed, labeled and attributed multi-graph. To express heterogeneity, *type labels* can be defined for vertices and edges (e.g., `Person` or `likes`). Attributes have the form of key-value pairs (e.g., `name:Alice` or `age:42`) and are referred to as *properties*. Such properties are set at the instance level without an upfront schema definition. In an *extended* property graph, a database consists of multiple property graphs which are called *logical graphs*. These graphs are application-specific subsets from shared sets of vertices and edges, i.e., may have common vertices and edges. Additionally, not only vertices and edges but also logical graphs have a type label and can have different properties.

Figure 10 shows an example EPGM database DB of a simple social network. Formally, DB consists of the vertex set $V = \{v_1, .., v_6\}$ and the edge set $\mathcal{E} = \{e_0, .., e_6\}$ where each element has a unique identifier (e.g., [1]). Vertices represent users, groups and interest tags, denoted by corresponding type labels (e.g., `User`) and are further described by their properties (e.g., `name:Alice`). Edges describe the relationships between vertices and also have type labels (e.g., `memberOf`) and properties. Type labels do not determine a schema, as elements with the same type label may have different property keys, e.g., v_1 and v_4.

The sample database contains the set of logical graphs $\mathcal{L} = \{G_1, G_2\}$, where each graph represents a community inside the social network, in particular specific interest groups (e.g., `Graphs`). Each logical graph has a dedicated subset of vertices and edges, for example, $V(G_1) = \{v_1, v_2, v_4\}$ and $E(G_0) = \{e_1, e_2\}$. One can see that vertex (and also edge sets) of logical graphs may overlap since $V(G_1) \cap V(G_2) = \{v_4\}$. Note that also logical graphs have type labels (e.g., `Community`) and may have properties to annotate the graph with specific metrics or descriptive information (e.g., `interest:Big Data`). Logical graphs, such as those of our example, are either declared explicitly or output of a graph algorithm, e.g., community detection or graph pattern matching. In both cases, they can be used as input for subsequent operators.

Operators The EPGM provides operators for single logical graphs and graph collections; operators may also return single logical graphs or graph collections. Here, a graph collection $\mathcal{G} \in \mathcal{L}^n$ is a n-tuple of logical graphs and thus may contain duplicate elements. Collections are ordered to support application-specific sorting and position-based selection of logical graphs. In the following, we use the terms *collection* and *graph collection* as well as *graph* and *logical graph* interchangeably. Table 5 lists the analytical operators together with their corresponding pseudocode syntax for calling them in GrALa. The syntax adopts the concept of higher-order functions for several operators (e.g., to use aggregate or predicate functions as operator arguments). Based on the input of operators, GrALa distinguishes between *graph operators* and *collection operators* as well as *unary* and *binary operators* (single graph/collection vs. two graphs/collections as input). There are also *auxiliary operators* to apply graph operators on collections or to call specific graph algorithms. In addition to the listed ones GrALa provides operators to create graphs, vertices and edges including respective labels and properties. In the following, we will present a subset of available operators, a detailed discussion of all operators and their implementation can be found in [56].

Aggregation An operator often used in analytical applications is aggregation, where a set of values is mapped to a single value of significant meaning. The EPGM supports aggregation at the graph level. Formally, the operator maps an input graph G to an output graph G' and applies the user-defined aggregate function $\alpha : \mathcal{L} \rightarrow A$. Thus, the resulting graph is a modified version of the input graph with an additional property k. In the following, we show a simple vertex count example:

```
alpha = (g => g.V.count())
outGraph = inGraph.aggregate('vertexCount', alpha)
```

Here, a user-defined aggregate function `alpha` computes the cardinality of the vertex set `g.V` of an input graph `g`. The aggregation operator is called on the logical graph referred to by the variable `inGraph`. The operator takes property key `vertexCount` and aggregate function `alpha` as arguments. The resulting logical graph is assigned to the variable `outGraph` and provides a property `vertexCount` storing the result of `alpha`. Basic aggregate functions such as *count*, *sum*, *min* and *max* are predefined in GrALa and can be applied to vertex and edge collections.

Pattern Matching A fundamental operation of graph analytics is the retrieval of subgraphs isomorphic to a user-defined pattern graph [37]. The operator results in a graph collection containing all embeddings of that pattern graph in the input graph. For example, in Fig. 11a, a simple pattern graph describes the membership relation between an arbitrary user and an arbitrary group. Applied on our example graph in Fig. 10, the operator returns the collection shown in Fig. 11b. Each logical graph in that collection represents a subgraph that is isomorphic to the pattern graph. To support such queries, GrALa provides the pattern matching operator, where a pattern

Table 5 Overview of operators provided by the domain specific language GrALa

	Graph Analytical Language		
	Operator	Operator Signature	Output
Unary	Aggregation	`Graph.aggregate(propertyKey, aggregateFunction)`	Graph
	Transformation	`Graph.transform(graphFunction, vertexFunction, edgeFunction)`	Graph
	Pattern Matching	`Graph.match(patternGraph)`	Collection
	Subgraph	`Graph.subgraph(vertexPredicateFunction, edgePredicateFunction)`	Graph
	Grouping	`Graph.groupBy(vertexGroupingKeys, vertexAggregateFunction, edgeGroupingKeys, edgeAggregateFunction)`	Graph
	Selection	`Collection.select(predicateFunction)`	Collection
	Distinct	`Collection.distinct()`	Collection
	Limit	`Collection.limit(n)`	Collection
	Sorting	`Collection.sortBy(propertyKey, [:asc|:desc])`	Collection
Binary	Equality	`Graph.equals(otherGraph, [:identity|:data])`	Boolean
	Combination	`Graph.combine(otherGraph)`	Graph
	Exclusion	`Graph.exclude(otherGraph)`	Graph
	Overlap	`Graph.overlap(otherGraph)`	Graph
	Equality	`Collection.equals(otherCollection, [:identity|:data])`	Boolean
	Difference	`Collection.difference(otherCollection)`	Collection
	Intersect	`Collection.intersect(otherCollection)`	Collection
	Union	`Collection.union(otherCollection)`	Collection
Aux.	Apply	`Collection.apply(unaryGraphOperator)`	Graph
	Reduce	`Collection.reduce(binaryGraphOperator)`	Graph
	Call	`[Graph|Collection].callForGraph(algorithm, parameters)`	Graph
		`[Graph|Collection].callForCollection(algorithm, parameters)`	Collection

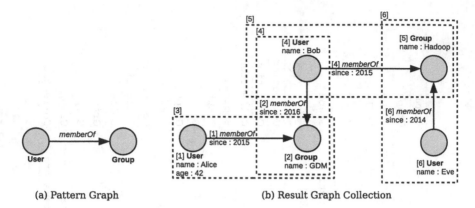

(a) Pattern Graph   (b) Result Graph Collection

Fig. 11 Example of a pattern matching execution where **a** represents the pattern which is applied on the graph of Fig. 10 and **b** shows the resulting graph collection containing all subgraphs that match the pattern

graph G^* and a predicate $\varphi : \mathcal{L} \to \{true, false\}$ are the operator arguments. Pattern matching is applied to a graph G and returns a graph collection $\mathcal{G}' = \{G' \subseteq G \mid G' \simeq G^* \land \varphi(G') = true\}$ containing all matches, for example:

```
matches = db.G.match("(a:User)-[e:memberOf]-(b:Group))")
```

The shown pattern graph reflects our membership query. GrALa adopts the basic concept of describing graph patterns using ASCII characters from Neo4j Cypher [26], where (a)-[e]->(b) denotes an edge e from vertex a to vertex b. The predicate function φ is embedded into the pattern by defining type labels and properties. In the example, we describe a pattern of two vertices and one edge, which are assigned to variables (a,b for vertices; e for the edge). Variables are optionally followed by a label (e.g., a:User) and properties (e.g., {name = 'Alice'}). The operator is called for the logical graph representing the whole database DB (db.G) of Fig. 10 and returns a collection assigned to variable matches and containing four new logical graphs.

Grouping The *groupBy* operator determines a structural grouping of vertices and edges to condense a graph and thus helps to uncover insights about patterns hidden in the graph. Let G' be the grouped graph of G, then each vertex in $V(G')$ represents a group of vertices in $V(G)$; edges in $E(G')$ represent a group of edges between the vertex group members in $V(G)$. Vertices are grouped based on selected property values (including their type label) while edges are grouped along their incident vertices and optionally by selected property values. Vertices and edges in the grouped graph are called *super vertices* and *super edges*, respectively. Additionally, the vertex and edge aggregate functions can be used to compute aggregated property values for super vertices and edges, e.g., the average age of users in a group or the number of group members. The aggregate value is stored at the super vertex and super edge, respectively. The following example shows the application of the grouping operator:

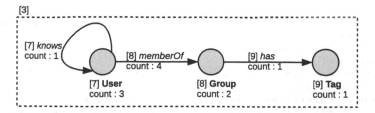

Fig. 12 Example of the grouping operator applied on the graph of Fig. 10. The graph is grouped by vertex and edge label and a count aggregate is used to compute the number of elements represented by each resulting super vertex/edge

```
1  outGraph = db.G.groupBy(
2    [:label],
3    (suVertex, vertices => suVertex['count'] = vertices.
        count()),
4    [:label],
5    (suEdge, edges => suEdge['count'] = edges.count()))
```

The goal of this example is to group vertices and edges in the graph of Fig. 10 by their corresponding type label. Furthermore, we want to count the number of vertices and edges represented by each label. In line 2, we define the vertex grouping keys. Here, we want to group vertices by their type label. However, it is also possible to define property keys which are used to select property values for grouping (e.g., to group users by their age). Edges are also grouped by type label (line 4). In lines 3 and 5, we define the vertex and edge aggregate functions. Both receive the super entity (i.e., `suVertex`, `suEdge`) and the set of group members (i.e., `vertices`, `edges`) as input. Both functions apply the aggregate function `count()` on the set of grouped entities to compute the group size. The resulting value is stored as a new property `count` at the super vertex and super edge respectively. Figure 12 shows the resulting logical graph of the grouping example.

Analytical Example Finally, we illustrate the capabilities of GRADOOP using an exemplary analytical program based on social network data. We assume a heterogeneous network including various vertex and edge types including users and their mutual friendship relations similar to Fig. 10. Vertices and edges have properties, for example, user vertices store the corresponding name, gender and the city the user lives in.

The graph analytical program used for our example is shown in Listing 1.8 and includes several operators from Table 5 not discussed before. The input is an entire social network represented as a single logical graph. Here, the graph is stored in HBase and distributed across a cluster of machines. In line 1, we load the graph and refer to it using the variable `socialNetwork`. Starting from Line 2, we define our analytical program as a composition of GrALa operators. First, we extract the *subgraph* containing only users and their mutual friendship relationships by applying user-defined vertex and edge predicate functions. The vertices of the resulting graph are then *transformed* to a representation which is limited to information necessary

```
1   Graph socialNetwork = EPGMDatabase.fromHBase().getDBGraph()
        Graph
2   result = socialNetwork
3     .subgraph(
4       (vertex => vertex[:label] == 'User'),
5       (edge => edge[:label] == 'knows'))
6     .transformVertices(currentVertex, transformedVertex => {
7         transformedVertex['city']   = currentVertex['city']
8         transformedVertex['gender'] = currentVertex['gender'])
9     .callForCollection(:LabelPropagation, [:id, 5]))
10    .apply(g => g.aggregate('vertexCount', (h => h.V.count())))
11    .select(g => g['vertexCount'] > 50_000)
12    .reduce(g, h => g.combine(h))
13    .groupBy(['city','gender'],
14      (suVertex, vertices => suVertex['count'] = vertices.count
          ()),
15      [], (suEdge, edges => suEdge['count'] = edges.count()))
16    .aggregate('vertexCount', (g => g.V.count()))
17    .aggregate('edgeCount', (g => g.E.count()))
18  result.writeAsJSON('hdfs:///output/')
```

Listing 1.8: Analytical program which shows the combination of EPGM operators.

for subsequent operators. The user-defined transformation function takes the current vertex and a copy of that vertex with omitted label and properties as input and determines, which data gets transferred from the current to the copied vertex. Here, we adopt only the gender and city properties. The transformed subgraph is then used as input for the *call* operator in line 9. That operator allows us to call specific graph algorithms on logical graphs (e.g., pagerank) or graph collections (e.g., common subgraph detection). Here, we use Label Propagation [88], a community detection algorithm that is already implemented in Flink Gelly. The algorithm propagates the value associated with a given property key (we use the vertex id) through the graph in five iterations. The result is a graph collection containing all found communities. In line 10, we *apply* the *aggregate* operator on each of these communities to compute their respective vertex counts. Then, we use the *selection* operator to filter communities that have more than 50 K users. The filtered graphs are then combined to a single logical graph by applying the *reduce* operator on the filtered collection. The result is a single logical graph containing all vertices and edges from all graphs in the collection. We further *group* this graph by the vertex properties city and gender to see the relations between those groups. Edges are grouped along their incident vertices. By applying group-wise counting, we can find out how many vertices and edges are represented by their respective super entities. In lines 16 and 17, we use *aggregation* to compute how many super entities are contained in the resulting logical graph. As GRADOOP is build on top of Apache Flink, program execution needs to be triggered explicitly. In the last line, we start the program by writing the resulting logical graph to HDFS using a dedicated JSON output format.

The example illustrates that GRADOOP allows the combined application of graph queries and transformations as well as the execution of graph algorithms such as for community detection within a compact dataflow program. The entire program can be automatically executed in parallel on distributed clusters since all operators are implemented using Flink operators.

6 Comparison

In our introduction we stated various requirements for flexible and efficient management and analysis of big graph data. In the previous sections, we discussed three system categories in detail: graph database, graph processing and graph dataflow systems. We now want to compare these categories based on the stated requirements. Table 6 highlights the features of the respective categories.

- *Powerful graph data model*: The need to process graphs with heterogeneous vertices and edges of varying types and with different attributes is currently addressed best by graph database systems that offer schema-free, flexible data models like the PGM or RDF. From the considered distributed frameworks for graph analytics, only GRADOOP supports such a graph data model. Its EPGM is the only data model with versatile support for graph collections.

Table 6 Feature comparison of different approaches to graph data management and analytics

	Graph database systems	Graph processing systems	Graph dataflow systems	
Examples	Neo4j, Marklogic	Pregel, Giraph	Gelly, GraphX	Gradoop
Data model	PGM/RDF	Generic graph	Generic graph, Datasets	EPGM, Datasets
Graph collections	No	No	No	Yes
Query approach	Query languages	Vertex-/Graph-centric computation models	Vertex-centric computation, dataflow programs	
Scope	OLTP/Queries	Analytics	Analytics	Analytics
Scalability	Up/(Out)	Out	Out	Out
Persistency	Yes	No	No	Yes
Transactions	Yes	No	No	No
Graph visualization	Interactive traversal	No	No	No

- *Powerful query and analysis capabilities*: We saw that each system category has its own approach for querying and analyzing the graph. While declarative languages, like Cypher or SPARQL, are unique for graph database systems, graph processing systems provide vertex- and graph-centric programming abstractions that simplify the implementation of distributed graph algorithms. In contrast, graph dataflow systems combine vertex-centric computation with additional libraries and general-purpose data operators for pre- and post-processing. While Gelly and GraphX provide transformation and aggregation methods for single graphs, GRADOOP in addition offers operators that exploit the expressiveness of the underlying graph data model.
- *High performance and scalability*: The main focus of graph database systems are OLTP applications with demand for very low query execution times. To achieve that, those systems focus on query optimization, indexing, efficient physical storage and data replication. Graph processing and dataflow systems on the other hand, focus on analytical programs involving graphs that span an entire cluster of machines. Here, the focus is on balanced load distribution, reducing network traffic and fault-tolerance in case of system-failures during long running programs. While the architecture of graph processing as well as dataflow systems is built with data distribution in mind, only a subset of available graph databases provides that feature. The actual system performance depends on many implementation decisions as well as on the data and workload characteristics - an extensive benchmarking could help clarify differences between the different approaches (see Sect. 7.2).
- *Persistent graph storage and transaction support*: As stated before, graph databases focus on OLTP applications, hereby offering support for ACID compliant transactions on persistent data. Graph processing and dataflow systems solely focus on reading the graph from data sources, process it in a distributed manner and write the results back to an arbitrary data sink. GRADOOP offers rudimentary support for managing graphs in a persistent database. Those graphs can be either used in further analytical programs or be queried directly in the graph store. However, some graph databases, for example Titan [108], already provide APIs to execute ACID compliant graph processing algorithms.
- *Ease of use/graph visualization*: The growing interest in graph-based data systems indicates that the use of graphs is intuitive for many use cases. However, if it comes to meaningful visualization of graphs, there is only limited support in some graph database systems for navigating through the graph. Hence, support for versatile visualization and interactive exploration for large graphs or the results of graph analytics is still missing and a topic for future research and development (see Sect. 7.5).

7 Current Research and Open Challenges

The development of systems for graph analytics has made great progress in the past decade but there are still several areas requiring significant further improvement and research. In the following, we discuss some of these areas together with a brief outline of initial results that have already been achieved.

7.1 Graph Data Allocation and Partitioning

The efficiency of distributed graph processing substantially depends on a suitable data allocation (partitioning) of the graph data among all nodes of the processing system. This data allocation should enable graph processing with a minimum of inter-node communication and data transfer while at the same time ensure a good load balancing such that all nodes can be effectively utilized. The associated optimization objective is to find a balanced distributed of vertices and their edges such that the each partition includes about the same number of vertices while the sum of edges crossing partitions is minimized ("vertex cut"). The graph partitioning problem is known to be NP-hard and has attracted a large amount of research, in particular in graph theory [20]. The most promising approximate solutions are multilevel approaches such as METIS [59] that include steps for coarsening graphs to find partitions for condensed graphs and uncoarsening to the detailed graph while keeping the partitioning from the coarse graph [20]. Although these approaches can be run in parallel they are still expensive and thus likely of limited scalability to very large graphs [75]. A further problem is that even a near-perfect static data allocation is not sufficient since during the execution of long-running analysis, e.g., in a Pregel-like system, the processing of some partitions may have already terminated while others still have to be processed. Furthermore, some graphs may quickly change, e.g., in social networks, so that the data allocations needs to be quickly adapted without causing a completely new static data allocation [51].

Current distributed graph data systems mostly follow a simple hash-based partitioning of the vertices across all nodes. This approach achieves an even distribution of vertices and thus good load balancing and does not require a data structure to locate vertices. On the other hand, it frequently assigns neighboring vertices to different partitions leading to poor locality of processing and high communication overhead for many algorithms. A number of proposals has been made to address these limitations and also support adaptive data allocation to deal with changing graphs or load imbalances during analytical processing. Stanton and Kliot [100] propose a locality-aware data allocation that incrementally assigns vertices to the partition where most of the already assigned neighbors have been placed without causing significant load imbalances. This approach is also suitable for changing graphs to allocate new vertices. PowerGraph [41] as well as Spark GraphX [42] support an *edge-based partitioning* rather than a vertex-driven approach so that the number of edges is balanced across partitions and vertices are replicated along with their edges ("vertex cut"). Such an approach is especially valuable for graphs with a highly uneven distribution of vertex degrees such that a large fraction of the edges is associated with few vertices that could easily lead to load imbalances with a vertex-based data partitioning. The replication of vertices has also been proposed for an adaptive data allocation for dynamically evolving graphs [51, 78]. Furthermore, graph repartitioning approaches based on the dynamic migration and replication of vertices have been proposed to deal with load imbalances during analytical processing in graph processing systems such as GPS and XPregel, but the associated overhead has not always resulted in significantly improved execution time [75].

The discussion shows that graph data allocation is a challenging problem that is not sufficiently solved with a single solution such as hash partitioning. This is because a good data allocation depends on the characteristics of the graph data as well on the intended kinds of graph analytics. It would be desirable to have support of a spectrum of allocation approaches from which the system can automatically choose depending on the graph and workload characteristics (similar as for data allocation in parallel database systems [80, 91]). Furthermore, it would be desirable to find a data allocation that can deal with mixed workloads including different kinds of complex analytical tasks as well as diverse kinds of interactive queries.

7.2   Benchmarking and Evaluation of Graph Data Systems

The large number of existing systems for analyzing graph data poses the question for potential users of such systems which of the systems performs best for which kind of analysis tasks, datasets and platforms. This asks for a comprehensive and comparative performance evaluation or benchmarking of the different implementations under comparable conditions. Such evaluations are also expected to help identify existing bottlenecks that may be addressed in the further development of systems. A large number of studies has addressed these issues by comparing the performance of selected systems, in particular for graph database and graph processing systems, e.g., [31, 44, 47, 70, 74, 98, 121]. While these studies have been insightful, their results are mostly not comparable as each study has chosen a different set of systems, different sets of real or synthetically generated graph datasets, different sets of queries and analytical tasks (ranking from pagerank to collaborative filtering and graph coloring) as well as different environments in terms of number of worker nodes and their characteristics such as memory size and number of cores. The different studies thus have only few observations in common such that graph processing systems significantly outperform MapReduce-based implementations [31, 44] and that Giraph is mostly slower than other graph processing systems such as GraphLab or GPS [44, 47, 70, 98, 121].

A few studies also considered Apache Spark [121] and Stratosphere (the predecessor of Flink) [31, 44], but comparative performance evaluations for GraphX and Gelly are still missing. Here, it is of great interest, how the underlying dataflow systems behave for graph analytical workflows involving multiple graph and non-graph transformations as well as iterative algorithms. For example, Flink optimizes dataflow programs with cost-based query-optimization techniques that are similar to the ones used in database systems, e.g., to reorder transformations. As the optimizer assumes independent datasets, such generic program optimizations may, however, cause problems for the processing of graph data with strongly interrelated vertex and edge datasets. It would thus be interesting to evaluate the performance of different kinds of graph dataflow programs in more detail and develop optimization techniques customizable for graph processing and analysis, e.g., to automatically repartition the graph during program execution.

In addition to the individual performance studies there are also several recent attempts for the definition of graph analysis benchmarks, namely Linkbench [12], LDBC [8, 33] and gMark [13]. These benchmarks specify the synthetic generation of datasets of different sizes, the workloads and performance metrics together with rules on how to perform the evaluation in order to achieve comparable results between different systems. Furthermore, there are proposals for the synthetic generation of graph datasets, e.g., for business intelligence [84]. From the mentioned benchmarks, the LDBC (Linked Data Benchmark Council) effort is the most ambitious as it consists of two benchmarks for semantic publishing and social network benchmarking (SNB) and different sets of query and analysis workloads exhibiting several "choke points" to stress-test the systems. Unfortunately, there are only few evaluations so far for all benchmarks so that they could not yet demonstrate their usefulness.

7.3   Analysis of Dynamic Graphs

Previous approaches for graph analytics focus on static graphs that remain stable. Most graphs, e.g., social networks, however are constantly changing so that the results of analytical processes, e.g., for community detection or on metrics such as pagerank or centrality, need to be updated or refreshed. Furthermore, there is a need for fast, one-pass graph analysis in data streams, e.g., to quickly identify new topics and correlations in Twitter data, to determine online recommendation for users based on their current website usage (clickstream) or to identify potentially criminal acts such as credit card misuse or planned terror attacks.

According to [1], dynamic graphs fall into two categories: slowly evolving graphs (e.g., co-authorship networks) and streaming networks. In the first case, it is possible to maintain different snapshots of the graph as the basis for an offline analysis while in the second case a near real-time analysis is necessary. The analysis can further focus on understanding the evolution, e.g., by comparing different snapshots, or on refreshing previous analysis results for the new graph data. A large amount of research has already dealt with these topics as surveyed in [1]. Typical observations show that the number of edges grows stronger than the number of vertices leading to increasingly denser networks (reduced distances between vertices). Many studies focused on analyzing the evolution of communities, e.g., by applying a clustering-based community detection on different snapshots and analyzing the cluster changes. Graph analysis for streamed data has also found interest already, e.g., to detect outliers such as a new co-author link between authors of different communities (linkage anomaly). There is also some work to incrementally update complex graph metrics such as betweenness centrality[17] for streamed data, e.g., using approximation techniques and specific index structures [50]. The Kineograph system [25] supports the dynamic graph-based analysis of Twitter data (correlations between users and hashtags) by continuously creating new in-memory graph snapshots that can then be evaluated by conventional mining approaches for static graphs, e.g., for ranking or community detection.

Despite the relatively large body of previous theoretical and experimental work on dynamic networks, little work has been done for big graph data utilizing current distributed graph data platforms as discussed in this chapter. Analyzing massive amounts of changing graph data in a distributed way poses many new algorithmic and data management challenges including the need for adaptive data allocation (as discussed in Sect. 7.1 above). Data management and graph analytics is challenging on a sequence of large graph snapshots as well as for streaming data and needs much further research. Most studies for graph evolution and dynamic graph analytics focused on structural changes such as the addition of new vertices and edges; more work is needed for considering both changes in structure and content, e.g., new publication topics or changing interests of users in social networks. Furthermore, the graph changes may have to be associated with information in different data sources, e.g., to better understand certain changes or identify potential criminal acts. The latter aspects might imply the need to develop application-specific approaches to take the specific kinds of changes and additional information to correlate with into account.

7.4   Graph-Based Data Integration and Knowledge Graphs

Before graph data can be analyzed it is necessary to construct and store the graphs for further processing. As for big data analysis in general, the graph data typically needs to be extracted from the original data sources (e.g., from social networks, web pages, tweets, relational databases, etc.), transformed and cleaned. Furthermore, it is often necessary to combine and interrelate data from multiple sources into the combined graph. These steps are typically carried out within so-called ETL (extract-transform-load) workflows that may be performed in parallel on Hadoop platforms, e.g., using MapReduce or other frameworks such as Hive, Spark or Flink [15, 45, 62]. A particularly important and expensive step is the matching of equivalent entities (users, products, etc.) from different sources so that they can be fused together, e.g., within one graph vertex. Map-reduce-based tools such as Dedoop [64] have been

---

[17]The betweenness centrality of a vertex is defined as the number of shortest paths in a network pathing through the vertex. A high value thus indicates that a vertex is centrally located so that it plays an important role in a network.

developed for scalable entity matching. So far, relatively little work has focused on ETL for graph data, although there are new challenges in all steps of a typical ETL pipeline, e.g., to extract graphs from certain data sources such as relational databases, for data cleaning and for data integration. GraphBuilder is one of the few tools for graph ETL [54]. It utilizes MapReduce jobs to extract data from sources based on user-defined parsers and to generate vertices and edges. It also provides different options for distributed storage of the resulting graph data. The BIIIG system supports the extraction of graph data from several relational databases to support a graph-based business intelligence [83, 85].

A particularly challenging kind of graph-based data integration becomes neces-sary for the generation and continuous maintenance of so-called *knowledge graphs* [30, 81, 90] providing a large amount of interrelated information about many real-world entities (persons, locations, ...) and their describing metadata concepts, typi-cally extracted and combined from several other sources. Non-commercial knowl-edge graph projects include YAGO,[18] DBpedia,[19] Freebase and its successor Wiki-data.[20] Companies such as Yahoo! [15], Google, Microsoft or Facebook utilize even larger knowledge graphs [81] combining information from more resources including web pages and search queries. Most of the systems make use of the RDF data model to express the contained knowledge.

A massive problem is the typically low data quality, high diversity and large volume of the automatically extracted information to be integrated into knowledge graphs. Dealing with these issues requires scalable and largely automatic (learning-based) approaches for information extraction, cleaning, classification and matching [15, 30, 90].

Low data quality including incomplete and contradicting information from the information to be integrated into a knowledge graphs is a huge challenge to deal with requiring scalable and largely automatic (learning-based) approaches for information extraction, cleaning, classification and matching [15, 30, 90].

7.5 *Interactive Graph Analytics*

Interactive graph analytics supported by suitable visualizations is highly desirable to put the human in the loop for exploring and analyzing graph data. However, inter-active graph analysis is currently only supported for query processing with graph databases (Sect. 2) while graph analytics with the discussed distributed frameworks is largely batch-oriented. For example, Neo4j allows such an interactive and visual exploration of the immediate neighborhood of selected vertices.[21] Screen size and human recognition capabilities limit this approach to inspecting only tens to a few

---

[18]www.mpi-inf.mpg.de/yago-naga/yago/.

[19]http://dbpedia.org/.

[20]www.wikidata.org.

[21]http://neo4j.com/graph-visualization-neo4j/.

hundreds of vertices at a time. More promising is the exploration and visualization of summarizing graph data, similar to multidimensional OLAP queries for data warehouses. Several approaches for such graph summaries [106, 119], graph OLAP [24, 40, 112, 120] and grouping (Sect. 5.2) have already been proposed and can potentially be applied for large graphs. For example, k-SNAP [106] automatically creates summarized graphs with k vertices, where the change of parameter k enables an OLAP-like *roll-up* and *drill-down* within a dimension hierarchy [24]. However, the approach is not yet fully interactive as it depends on a pre-determined parameter.

To improve ease-of-use there is a strong need for extending interactive and visual analysis to more kinds of graph analysis, from OLAP-style aggregations for single large graphs and graph collections to exploring evolution in dynamic graphs. Furthermore, it should be possible to interactively evaluate the results of expensive graph analytics, e.g., to inspect parts of the graphs with a high centrality, certain communities of interest, etc. The currently existing separation between interactive query processing with graph databases and batch-oriented graph analytics should thus be overcome by providing all kinds of analysis in a unified, distributed platform with support for interactive and visual analysis. Some of the graph databases of Sect. 2, e.g., Blazegraph, System G and Titan, try to go into this direction, but there are still many open issues in finding suitable visualizations and interaction forms for the different kinds of analysis. Furthermore, the combined processing of mixed workloads with queries and heavy-weight graph algorithms should also be possible with the graph processing frameworks for Hadoop-based clusters.

8 Conclusions and Outlook

The analysis of graph data has become of great interest in many applications and a major focus of big data platforms. We have posed major requirements for big data graph analytics and surveyed current systems in three categories: graph database systems, distributed graph processing systems and distributed graph dataflow systems. The summarizing comparison of these system categories with respect to the posed requirements in Sect. 6 showed that there are still big differences between the query-focused graph database systems and the distributed platforms focusing on large-scale iterative graph analysis. While distributed graph analysis platforms generally lack an expressive graph data model, the distributed dataflow approach GRADOOP provides an extended property graph model with powerful support for analyzing collections of graphs.

Despite the significant advances made in the last few years, the development and use of distributed graph data systems are still in an early stage. Hence, the posed requirements are not yet fully achieved and there are many opportunities for improvement and future research. As discussed in Sect. 7, this is especially the case for evaluating and improving the performance and scalability of graph data systems, for graph data partitioning and load balancing, for the analysis of dynamic graph data, for graph-based data integration, and for interactive and visual graph analytics.

Acknowledgements This work is partially funded by the German Federal Ministry of Education and Research under project ScaDS Dresden/Leipzig (BMBF 01IS14014B).

References

1. C. Aggarwal, K. Subbian, Evolutionary network analysis: a survey. ACM Comput. Surv. (CSUR) **47**(1), 10 (2014)
2. G.A. Agha, *Actors: a model of concurrent computation in distributed systems* Technical report, DTIC Document (1985)
3. Akka. http://www.akka.io. Accessed 10 Mar 2016
4. A. Alexandrov et al., The stratosphere platform for big data analytics. VLDB J. **23**(6) (2014)
5. AllegroGraph. http://franz.com/agraph/allegrograph/. Accessed 10 Mar 2016
6. R. Angles, A comparison of current graph database models, in *Proceedings of ICDEW* (2012)
7. R. Angles, C. Gutierrez, Survey of graph database models. ACM Comput. Surv. (CSUR) **40**(1) (2008)
8. R. Angles et al., The linked data benchmark council: a graph and RDF industry benchmarking effort. Proc. SIGMOD **43**(1) (2014)
9. Apache Flink Iteration Operators. https://ci.apache.org/projects/flink/flink-docs-master/apis/batch/index.html#iteration-operators. Accessed 09 Mar 2016
10. Apache Giraph. http://www.giraph.apache.org. Accessed 10 Mar 2016
11. Apache Jena - TBD. https://jena.apache.org/documentation/tdb/. Accessed 09 Mar 2016
12. T.G. Armstrong et al., Linkbench: a database benchmark based on the facebook social graph (2013)
13. G. Bagan et al. gMark: Controlling Diversity in Benchmarking Graph Databases. CoRR abs/1511.08386 (2015)
14. O. Batarfi et al., Large scale graph processing systems: survey and an experimental evaluation. Clust. Comput. **18**(3) (2015)
15. K. Bellare et al., Woo: a scalable and multi-tenant platform for continuous knowledge base synthesis. PVLDB **6**(11) (2013)
16. D.P. Bertsekas, J.N. Tsitsiklis, Parallel and distributed computation: numerical methods, vol. 23 (1989)
17. Big Data Spatial and Graph User's Guide and Reference. http://docs.oracle.com/cd/E69290_01/doc.44/e67958/toc.htm. Accessed 16 Mar 2016
18. H. Bolouri, Modeling genomic regulatory networks with big data. Trends Genet. **30**(5) (2014)
19. D. Brickley, L. Miller, Foaf vocabulary specification 0.98. Namespace document 9 (2012)
20. A. Buluç et al., Recent advances in graph partitioning. CoRR (2013)
21. M. Canim, Y.C. Chang, System G data store: big, rich graph data analytics in the cloud, in *IEEE Cloud Engineering (IC2E)* (March 2013)
22. G. Carothers, RDF 1.1 N-Quads: a line-based syntax for RDF datasets. W3C Recommendation (2014)
23. R. Cattell, Scalable SQL and NoSQL data stores. Proc. SIGMOD **39**(4) (2011)
24. C. Chen et al., Graph OLAP: towards online analytical processing on graphs, in *IEEE Data Mining (ICDM)* (2008)
25. R. Cheng et al., Kineograph: taking the pulse of a fast-changing and connected world, in *Proceedings of EuroSys* (2012)
26. Cypher Query Language. http://neo4j.com/docs/stable/cypher-query-lang.html. Accessed 16 Mar 2016
27. S. Das et al., A Tale of two graphs: property graphs as RDF in Oracle, in *EDBT* (2014)
28. R. Diestel, Graph theory, *Graduate Texts in Mathematics*, vol. 173, 4th edn. (2012)
29. Y. Ding, Scientific collaboration and endorsement: network analysis of coauthorship and citation networks. J. Inform. **5**(1) (2011)

30. X. Dong et al., Knowledge Vault: a web-scale approach to probabilistic knowledge fusion, in *Proceedings of SIGKDD* (2014)
31. B. Elser, A. Montresor, An evaluation study of bigdata frameworks for graph processing, in *IEEE Big Data* (2013)
32. O. Erling, I. Mikhailov, RDF support in the Virtuoso DBMS, in *Networked Knowledge-Networked Media* (2009)
33. O. Erling et al., The ldbc social network benchmark: interactive workload, in *Proceedings of SIGMOD*(2015)
34. S. Ewen et al., Spinning fast iterative data flows. PVLDB **5**(11) (2012)
35. S. Ewen et al., Iterative parallel data processing with stratosphere: an inside look, in *Proceedings of SIGMOD* (2013)
36. S. Fortunato, Community detection in graphs. Phys. Rep. **486**(3–5) (2010)
37. B. Gallagher, Matching structure and semantics: a survey on graph-based pattern matching. AAAI FS 6 (2006)
38. J. Gao et al., Glog: a high level graph analysis system using mapreduce, in *Proceedings of ICDE* (2014)
39. Gelly: Flink Graph API. https://ci.apache.org/projects/flink/flink-docs-master/apis/batch/libs/gelly.html. Accessed 15 Mar 2016
40. A. Ghrab et al., A framework for building OLAP cubes on graphs, in *Advances in Databases and Information Systems* (2015)
41. J.E. Gonzalez et al., Powergraph: distributed graph-parallel computation on natural graphs, in *Proceedings of OSDI* (2012)
42. J.E. Gonzalez et al., GraphX: graph processing in a distributed dataflow framework, in *Proceedings of OSDI* (2014)
43. GraphDB: At Last, the Meaningful Database. http://ontotext.com/documents/reports/PW_Ontotext.pdf. Whitepaper July 2014
44. Y. Guo et al., How well do graph-processing platforms perform? An empirical performance evaluation and analysis, in *Proceedings of Parallel and Distributed Processing Symposium* (2014)
45. D. Haas et al., Wisteria: nurturing scalable data cleaning infrastructure. PVLDB **8**(12) (2015)
46. T. Haerder, A. Reuter, Principles of transaction-oriented database recovery. ACM Comput. Surv. **15**(4) (1983)
47. M. Han et al., An experimental comparison of pregel-like graph processing systems. PVLDB **7**(12) (2014)
48. S. Harris, A. Seaborne, E. Prudhommeaux, SPARQL 1.1 query language. W3C Recommendation 21 (2013)
49. O. Hartig, B. Thompson, Foundations of an alternative approach to reification in RDF. Technical Report. arXiv:1406.3399 (2014)
50. T. Hayashi, T. Akiba, Y. Yoshida, Fully dynamic betweenness centrality maintenance on massive networks. PVLDB **9**(2) (2015)
51. J. Huang, D.J. Abadi, LEOPARD: lightweight edge-oriented partitioning and replication for dynamic graphs. PVLDB **9**(7) (2016)
52. InfiniteGraph: The Distributed Graph Database. http://www.objectivity.com/wp-content/uploads/Objectivity_WP_IG_Distr_Benchmark.pdf. Whitepaper 2012
53. B. Iordanov, HyperGraphDB: a generalized graph database, in *Web-Age Information Management* (2010)
54. N. Jain, G. Liao, T.L. Willke, Graphbuilder: scalable graph ETL framework, in *International Workshop on Graph Data Management Experiences and Systems* (2013)
55. C. Jiang et al., A survey of Frequent Subgraph Mining algorithms. Knowl. Eng. Rev. **28**(1) (2013)
56. M. Junghanns et al., GRADOOP: Scalable Graph Data Management and Analytics with Hadoop. Technical Report. arXiv:1506.00548 (2015)
57. M. Junghanns et al., Analyzing extended property graphs with apache flink, in *Proceedings of SIGMOD Workshop on Network Data Analytics* (2016)

58. Z. Kaoudi, I. Manolescu, RDF in the clouds: a survey. VLDB J. **24**(1) (2015)
59. G. Karypis, V. Kumar, Multilevel k-way partitioning scheme for irregular graphs. J. Parallel Distrib. Comput. **48**(1) (1998)
60. Key Features - ArangoDB. https://www.arangodb.com/key-features/. Accessed 10 Mar 2016
61. Z. Khayyat et al., Mizan: a system for dynamic load balancing in large-scale graph processing, in *Proceedings EuroSys* (2013)
62. Z. Khayyat et al., Bigdansing: a system for big data cleansing, in *Proceedings SIGMOD* (2015)
63. G. Klyne, J.J. Carroll, Resource description framework (RDF): concepts and abstract syntax (2006)
64. L. Kolb, A. Thor, E. Rahm, Dedoop: efficient deduplication with Hadoop. PVLDB **5**(12) (2012)
65. L. Kolb, Z. Sehili, E. Rahm, Iterative computation of connected graph components with MapReduce. Datenbank-Spektrum **14**(2) (2014)
66. D. Koller, N. Friedman, Probabilistic graphical models: principles and techniques (2009)
67. A. Kyrola, G. Blelloch, C. Guestrin, GraphChi: large-scale graph computation on just a PC, in *Proceedings OSDI* (2012)
68. J. Lin, M. Schatz, Design patterns for efficient graph algorithms in MapReduce, in *Proceedings of 8th Workshop on Mining and Learning with Graphs* (2010)
69. Y. Low et al., Distributed GraphLab: a framework for machine learning and data mining in the cloud. PVLDB **5**(8) (2012)
70. Y. Lu, J. Cheng, D. Yan, H. Wu, Large-scale distributed graph computing systems: an experimental evaluation. PVLDB **8**(3) (2014)
71. G. Malewicz et al., Pregel: a system for large-scale graph processing, in *Proceedings of SIGMOD* (2010)
72. MarkLogic Semantics. http://www.marklogic.com/resources/marklogic-semantics-datasheet/. Datasheet March 2016
73. N. Martinez-Bazan, S. Gomez-Villamor, F. Escale-Claveras, DEX: a high-performance graph database management system, in *Proceedings of ICDEW* (2011)
74. R. McColl et al., A performance evaluation of open source graph databases, in *Proceedings of PPAAW* (2014)
75. R.R. McCune, T. Weninger, G. Madey, Thinking like a vertex: a survey of vertex-centric frameworks for large-scale distributed graph processing. ACM Comput. Surv. (CSUR) **48**(2) (2015)
76. F. McSherry et al., Composable incremental and iterative data-parallel computation with naiad. Technical Report MSR-TR-2012-105 (October 2012)
77. J.J. Miller, Graph database applications and concepts with Neo4j, in *Proceedings of Southern Association for Information Systems Conference*, vol. 2324 (2013)
78. J. Mondal, A. Deshpande, Managing large dynamic graphs efficiently, in *Proceedings of SIGMOD* (2012)
79. D.G. Murray et al., Naiad: a timely dataflow system, in *Proceedings of 24th ACM Symposium on Operating Systems Principles. SOSP '13* (2013)
80. R. Nehme, N. Bruno, Automated partitioning design in parallel database systems, in *Proceedings of SIGMOD* (2011)
81. M. Nickel, K. Murphy, V. Tresp, E. Gabrilovich, A review of relational machine learning for knowledge graphs. Proc. IEEE **104**(1) (2016)
82. Oracle Spatial and Graph: Advanced Data Management. http://www.oracle.com/technetwork/database/options/spatialandgraph/spatial-and-graph-wp-12c-1896143.pdf. Whitepaper September 2014
83. A. Petermann et al., BIIIG: enabling business intelligence with integrated instance graphs, in *Proceedings of ICDEW* (2014)
84. A. Petermann et al., FoodBroker-generating synthetic datasets for graph-based business analytics, in *Big Data Benchmarking* (2014)

85. A. Petermann et al., Graph-based data integration and business intelligence with BIIIG. PVLDB **7**(13) (2014)
86. A. Poulovassilis, M. Levene, A nested-graph model for the representation and manipulation of complex objects. ACM Trans. Inform. Syst. (TOIS) **12**(1) (1994)
87. quasar. http://www.paralleluniverse.co/quasar. Accessed 10 Mar 2016
88. U.N. Raghavan et al., Near linear time algorithm to detect community structures in large-scale networks. Phys. Rev. E **76**, 036106 (2007)
89. F. Rahimian et al., Distributed vertex-cut partitioning, in *Distributed Applications and Interoperable Systems* (2014)
90. E. Rahm, The case for holistic data integration, in *Advances in Databases and Information Systems* (2016)
91. J. Rao et al., Automating physical database design in a parallel database, in *Proceedings of SIGMOD* (2002)
92. M.A. Rodriguez, The gremlin graph traversal machine and language (invited talk), in *Proceedings of 15th Symposium on Database Programming Languages* (2015)
93. M.A. Rodriguez, P. Neubauer, Constructions from dots and lines. Bull. Am. Soc. Inform. Sci. Technol. **36**(6) (2010)
94. A. Roy et al., Chaos: scale-out graph processing from secondary storage, in *Proceedings of 25th Symposium on Operating Systems Principles* (2015)
95. M. Rudolf et al., The graph story of the SAP HANA database, in *Proceedings of BTW* (2013)
96. S. Sakr, A. Liu, A.G. Fayoumi, The family of mapreduce and large-scale data processing systems. ACM Comput. Surv. (CSUR) **46**(1) (2013)
97. S. Salihoglu, J. Widom, GPS: a graph processing system, in *Proceedings of 25th International Conference on Scientific and Statistical Database Management. SSDBM* (2013)
98. N. Satish et al., Navigating the maze of graph analytics frameworks using massive graph datasets, in *Proceedings of SIGMOD* (2014)
99. K. Shim, MapReduce algorithms for big data analysis. PVLDB **5**(12) (2012)
100. I. Stanton, G. Kliot, Streaming graph partitioning for large distributed graphs, in *Proceedings of SIGKDD*
101. Stardog 4 - The Manual. http://docs.stardog.com/. Accessed 10 Mar 2016
102. P. Stutz, A. Bernstein, W. Cohen, Signal/collect: graph algorithms for the (semantic) web, in *ISWC* (2010)
103. W. Sun et al., SQLGraph: an efficient relational-based property graph store, in *Proceedings of SIGMOD* (2015)
104. C. Teixeira et al., Arabesque: a system for distributed graph mining, in *Proceedings of 25th Symposium on Operating Systems Principles* (2015)
105. The bigdata RDF Database. https://www.blazegraph.com/whitepapers/bigdata_architecture_whitepaper.pdf. Whitepaper May 2013
106. Y. Tian, R.A. Hankins, J.M. Patel, Efficient aggregation for graph summarization, in *Proceedings of SIGMOD* (2008)
107. Y. Tian et al., From "Think Like a Vertex" to "Think Like a Graph". PVLDB **7**(3) (2013)
108. TITAN: Distributed Graph Database. http://thinkaurelius.github.io/titan/. Accessed 10 Mar 2016
109. N.B. Turk-Browne, Functional interactions as big data in the human brain. Science **342**(6158) (2013)
110. L.G. Valiant, A bridging model for parallel computation. CACM **33**(8) (1990)
111. X.H. Wang et al., Ontology based context modeling and reasoning using owl, in *Pervasive Computing and Communications Workshops* (2004)
112. Z. Wang et al., Pagrol: parallel graph olap over large-scale attributed graphs, in *Proceedings of ICDE* (2014)
113. Why OrientDB? http://orientdb.com/why-orientdb/. Accessed 10 Mar 2016
114. Y. Xia et al., Graph analytics and storage, in *IEEE Big Data* (2014)
115. R.S. Xin et al., GraphX: a resilient distributed graph system on spark, in *First International Workshop on Graph Data Management Experiences and Systems. GRADES '13* (2013)

116. R.S. Xin et al., GraphX: Unifying Data-Parallel and Graph-Parallel Analytics. Technical Report. arxiv:1402.2394 (2014)
117. P. Yuan et al., Triplebit: a fast and compact system for large scale rdf data. PVLDB **6**(7) (2013)
118. M. Zaharia et al., Spark: cluster computing with working sets, in *Proceedings of 2Nd USENIX Conference on Hot Topics in Cloud Computing. HotCloud'10* (2010)
119. N. Zhang, Y. Tian, J.M. Patel, Discovery-driven graph summarization, in *Proceedings of ICDE* (2010)
120. P. Zhao et al., Graph cube: on warehousing and OLAP multidimensional networks, in *Proceedings of SIGMOD* (2011)
121. Y. Zhao et al., Evaluation and analysis of distributed graph-parallel processing frameworks. J. Cyber Secur. Mobil. **3**(3) (2014)

Similarity Search in Large-Scale Graph Databases

Peixiang Zhao

Abstract Graphs are ubiquitous and play an essential role in modeling and representing complex structures in real-world networked applications. Given a graph database that comprises a large collection of graphs, it is fundamental and critical to enable fast and flexible search for structurally similar graphs. In this paper, we survey recent graph similarity search techniques and specifically focus on the work based on the graph edit distance (GED) metric. State-of-the-art approaches for the GED based similarity search typically adopt a *pruning and verification* framework. They first take advantage of some easy-to-compute lower-bounds of graph edit distance, and use novel graph indexing structures to efficiently evaluate such lower-bounds between graphs in the graph database and the query graph. This way, graphs that violate the GED lower-bound constraints can be identified and filtered from the graph database from further investigation. Then, the costly GED verification is performed only for the graphs that pass the GED lower-bound evaluation. We examine existing GED lower-bounds, graph index structures, and similarity search algorithms in detail, and compare different similarity search methods from multiple aspects including index construction cost, similarity search performance, and applicability in real-world graph databases. In the end, we envision and discuss the future research directions related to similarity search and high-performance query processing in large-scale graph databases.

1 Introduction

Today's highly networked world is facing numerous challenges raised in particular by the abundance of massive, complex, and structurally correlated data, which, without loss of generality, are often modeled and interpreted as graphs [1, 16]. The ubiquity of graphs and information networks has ignited intensive interest in enabling efficient access mechanisms and versatile querying functionalities in large collections

P. Zhao (✉)
Department of Computer Science, Florida State University, 1017 Academic Way,
James Love Building, Tallahassee, FL 32306, USA
e-mail: zhao@cs.fsu.edu

© Springer International Publishing AG 2017
A.Y. Zomaya and S. Sakr (eds.), *Handbook of Big Data Technologies*,
DOI 10.1007/978-3-319-49340-4_15

507

of graphs, which, in the database context, is to search relevant graphs from within large-scale graph databases [4, 28]. For example, in chemistry, chemical molecules are often represented as graphs that are recorded and compared for new material discovery and synthesis [5, 57]. In computer vision, graphs representing hand-written symbols, fingerprints, or medical images are retrieved and matched approximately for identity discovery, object detection, and scene identification [6, 15]. In bioinformatics, similarity search tools are implemented on graph-structured data for biological pathway enumeration and protein interaction detection [40]. The principles and methodologies of graph search have accordingly been studied in a wide range of real-world application domains, including pattern recognition [49], drug design [5, 40, 42, 44], social network analysis [14, 32, 33], program analysis [12], and business intelligence [48, 53], to name a few.

In order to search relevant graphs from a graph database given a user-specified, graph-structured query, we need to evaluate the "structure relevancy" between data graphs in the graph database and the query graph, the process of which is often referred to as *graph matching* [11, 15, 20, 50]. Graph search methods based on the *exact* graph matching principle are to retrieve data graphs that are isomorphic (or subgraph/supergraph isomorphic) to the query graph [13, 25–27, 35, 56, 63]. Although exact graph search methods offer a rigorous way for relevant graph retrieval in graph databases, they are too restrictive and thus only applicable in very few real-world scenarios due primary to the following reasons:

1. Real-world graphs are typically noisy with distorted or inconsistent information arising in different phases of graph modeling, acquisition, storage, management, and processing [33, 57]. Exact search upon noisy, error-prone graphs oftentimes leads to meaningless search results in graph databases, and incurs a lot of wasteful computation during query processing;
2. For graph database users who are not equipped with a mastery of domain knowledge, they may have vague information need when formulating graph-structured queries. In many occasions, they have to continuously modify and reformulate queries against graph databases, which typically results in a tedious, time-consuming query refinement process if the rigid, exact graph search mechanism is adopted [5, 40].

On the other hand, the flexible graph similarity search methods based on the *inexact, fault-tolerant* graph matching principle are able to cope with strong distortions present in real-world graphs. Meanwhile, the query results turn out to be a series of data graphs with *graded similarities* to the query graph. Such a relaxed, approximate search mechanism can greatly ease the pain of graph database users when they formulate complex, structure-enriched graph queries with little or no prior knowledge of the graph databases. As a result, flexible and efficient graph similarity search methods become essential and highly desirable especially for exploratory studies and rank-based applications in real-world, large-scale graph databases [48, 53, 54, 64, 66, 67].

There has been a rich literature on the modeling and quantification of pairwise similarity/distance between graphs, including, but not limited to, graph edit distances [23,

43, 59], maximum common subgraphs [10, 48], edge/feature misses [57, 58, 60, 68], graph alignment [50], graph kernels [7, 39], and graph simulations [18, 38]. In this chapter, we consider the *graph similarity search problem* that is defined on the *graph edit distance* (GED) constraint: given a graph database $\mathscr{G} = \{g_1, g_2, \ldots, g_n\}$, and a query graph q, to search as output $g_i \in \mathscr{G}$ $(1 \le i \le n)$ whose graph edit distance from q is within a user-specified GED threshold, τ. The graph edit distance, $\mathrm{GED}(g, q)$, is the minimum number of *graph edit operations* required to transform g to q (or vise versa), and an graph edit operation can be either vertex/edge insertion, deletion, or relabeling. GED has proven to be one of the most significant and intuitive graph proximity functions due to its generality, high degree of flexibility, and interpretability, and thus has been employed in widely varying graph-based applications [23, 39]. Specifically, GED has a series of advantages for pairwise graph similarity modeling and formulation, as follows,

1. GED is a metric applicable to virtually any type of graphs. It allows a complete set of interpretable graph edit operations that can precisely capture both structure and content differences between graphs [21];
2. GED defines a generic theoretical framework for graph similarity quantification, within which many graph proximity measures or distance functions, such as maximum common subgraphs [8] and edge misses [57], are just its special cases.

Unfortunately, the GED computation is NP-hard [22], which makes the graph similarity search problem challenging especially in large-scale graph databases. As a result, a systematic exploration of existing graph similarity search methods becomes essential to real-world large graph databases and big networked data. Furthermore, the solutions to the GED-based graph similarity search problem can be extended to help address a family of graph similarity search problems that are defined on other types of graph proximity measures.

In this chapter, we will provide an introduction to, and an overview of, the state-of-the-art GED-based graph similarity search solutions in large-scale graph databases. In Sect. 2, we will brief the preliminary concepts and core definitions for GED-based graph similarity search. We will discuss a basic *pruning-verification* framework underpinning existing graph similarity search solutions, together with a cost-based graph similarity search model in Sect. 3. In Sect. 4, we will survey the state-of-the-art solutions to the graph similarity search problem in large-scale graph databases. We will discuss the further work and potential research directions in Sect. 5, followed by concluding remarks in Sect. 6.

2 Preliminaries

In the chapter, we focus our discussion on *simple*, *undirected*, *labeled*, and *connected* graphs. It turns out that the definitions below are sufficiently flexible for a large variety of graphs in real-world applications, and the graph similarity search methods discussed in this chapter can be extended to other types of graphs with minor revision.

Definition 1 (*Graph*) A graph g is a 4-tuple (V, E, Σ, l), where V is a finite set of vertices; $E \subseteq V \times V$ is a set of edges; Σ is a set of labels of vertices and edges; $l : V \cup E \to \Sigma$ is a labeling function for vertices and edges of g.   ∎

Given a graph g, we denote the number of vertices of g to be $n = |V|$, and the number of edges of g to be $m = |E|$. For a vertex $u \in V$, we denote all its neighboring vertices as $N(u) = \{v | (u, v) \in E, v \in V\}$. The label of the vertex u is denoted as $l(u)$, and the label of an edge $e = (u, v)$ is denoted as $l(e)$ or $l(u, v)$. Labels in Σ represent the vertex/edge attributes or contents, such as tags in XML documents, atoms and bonds in chemical compounds, gene ontology (GO) terms in biological networks, or object descriptors of images. In practical applications, labels of vertices and edges are often defined as multidimensional vectors depicting a series of attributes affiliated with vertices and edges of graphs.

To perform graph search in graph databases, a common operation is to determine whether two graphs are equal or not with respect to both graph structure and vertex/edge labels. The problem of graph equality is often formulated as *graph isomorphism*, defined as follows,

Definition 2 (*Graph Isomorphism*) Given two graphs $g_1 = (V_1, E_1, \Sigma, l_1)$ and $g_2 = (V_2, E_2, \Sigma, l_2)$, A graph isomorphism between g_1 and g_2 is a bijective function $f : V_1 \to V_2$ satisfying (1) $\forall u \in V_1, l_1(u) = l_2(f(u))$, (2) $\forall (u, v) \in E_1$, there exists an edge $(f(u), f(v)) \in E_2$ such that $l_1(u, v) = l_2(f(u), f(v))$, and (3) $\forall (u, v) \in E_2$, there exists an edge $(f^{-1}(u), f^{-1}(v)) \in E_1$ such that $l_1(f^{-1}(u), f^{-1}(v)) = l_2(u, v)$.   ∎

Two graphs g_1 and g_2 are isomorphic if there exists a graph isomorphism between them. Although Laszlo Babai has proven that graph isomorphism can be evaluated in quasipolynomial time [2], it is still time-demanding in practical applications. Typically, one has to adopt a straightforward approach to examining all the possible vertex-to-vertex correspondences, the time complexity of which is exponential in the number of vertices of both graphs [52].

Similar to graph isomorphism, we can define *subgraph isomorphism* from one graph g_1 to another graph g_2, denoted as $g_1 \subseteq g_2$, if there exists a subgraph g of g_2 such that g_1 is graph isomorphic to g. If $g_1 \subseteq g_2$, we say g_1 is *contained* in g_2, or g_2 *contains* g_1. Subgraph isomorphism has proven to be NP-complete [22].

To account for the pairwise proximity/distance of real-world graphs, we typically take into consideration the differences of graphs from both perspectives of graph structures and vertex/edge labels. Graph Edit distance (GED) offers an intuitive way to integrate structure/label differences into the graph similarity modeling process and is thus applicable to virtually any types of real-world graphs. The key idea of GED is to model graph differences by a series of *graph edit operations* reflecting the structure/label variations between two graphs. Specifically, a graph edit operation is one of the following six basic operations that transform one graph to another [9, 19, 45],

1. insert an *isolated labeled* vertex u into a graph;
2. delete an *isolated labeled* vertex u from a graph;
3. change the label $l(u)$ of a vertex u;
4. insert a *labeled* edge (u, v) into a graph;
5. delete a *labeled* edge (u, v) from a graph;
6. change the label $l(u, v)$ of an edge (u, v).

Given two graphs g_1 and g_2, the sequence of graph edit operations performed on one graph in order to get the other is called an *edit path*. Formally, an edit path $\mathscr{P} = \{p_1, \ldots, p_k\}$ is a series of k graph edit operations that transform g_1 to g_2. The *edit cost* of such a transformation is $\sum_{i=1}^{k} c(p_i)$, where $c(p_i)$ is the cost of the particular graph edit operation, p_i. Taking the unit cost $c(p_i) = 1$ for each graph edit operation, an edit path of minimal length is called an *optimal edit path*.

Definition 3 (*Graph Edit Distance (GED)*) Given two graphs g_1 and g_2, the graph edit distance between g_1 and g_2, denoted as $\mathrm{GED}(g_1, g_2)$, is the length of an optimal edit path between g_1 and g_2. Namely, it is the minimum number of graph edit operations required to transform g_1 to g_2, or vice versa. ∎

Example 1 Figure 1 presents a toy graph database containing two sample graphs g_1 and g_2, and a query graph q. The graph edit distance between q and g_1, $\mathrm{GED}(q, g_1) = 5$, indicates that 5 graph edit operations are required to transform q to g_1: inserting an isolated vertex P, inserting an edge (P, C_1), inserting an edge (P, C_2), relabeling the vertex label N to S, and relabeling the edge label of (C_1, C_3) from single bond to double bond. Similarly, $\mathrm{GED}(q, g_2) = 2$. ∎

With the basic operations such as insertion, deletion, and label substitution, any graph can be transformed to another graph by iteratively applying graph edit operations. Consequently, GED can be used as an elegant and flexible similarity/distance measure between graphs, and it has been proven that GED is a metric [21]. However, the GED computation is NP-hard [22]. The state-of-the-art GED computational methods are based on the best-first search paradigm [19], and are only feasible for graphs of very small sizes [23].

We finally formulate the graph similarity search problem in a graph database as follows,

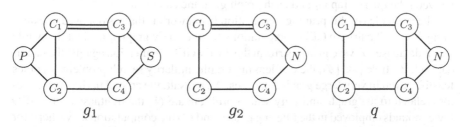

$$g_1 \qquad\qquad g_2 \qquad\qquad q$$

Fig. 1 Two sample graphs g_1, g_2 and a query graph q. Vertex labels represent atom symbols, and edge labels are either single bond or double bond. Subscripts of labels differentiate the vertices that share identical labels

Definition 4 (*Graph Similarity Search*) Given a graph database $\mathscr{G} = \{g_1, g_2, \ldots, g_N\}$, a query graph q, and a graph edit distance threshold τ, the graph similarity search problem is to search as output all the graph $g_i \in \mathscr{G}$ such that $\text{GED}(g_i, q) \leq \tau$. ∎

Example 2 Consider a sample graph database $\mathscr{G}$ consisting of two graphs g_1 and g_2, a query graph q, as shown in Fig. 1, and the GED threshold $\tau = 2$. The graph g_2 is returned as output for the similarity search because $\text{GED}(q, g_2) = 2 \leq \tau$. ∎

Because the computation of $\text{GED}(g_i, q)$ is NP-hard, it is straightforward to know that the graph similarity search problem is also NP-hard.

3 The Pruning-Verification Framework

A naive solution to the graph similarity search problem is to iteratively examine for each graph g_i in the graph database $\mathscr{G}$, whether the GED constraint, $\text{GED}(g_i, q) \leq \tau$, satisfies or not. However, the NP-hardness of GED computation poses serious algorithmic challenges to the graph similarity search problem. Furthermore, real-world graph databases may contain a large number N of graphs, thus leading to N costly GED computations. To ameliorate this computational bottleneck, the state-of-the-art methods typically adopt a *pruning-verification* algorithmic approach. First of all, we consider some light-weight, easy-to-compute GED lower-bounds, $\underline{\text{GED}}(g_i, q)$, to identify the graphs from the graph database $\mathscr{G}$ such that $\underline{\text{GED}}(g_i, q) > \tau$. It is easy to verify that

$$\text{GED}(g_i, q) \geq \underline{\text{GED}}(g_i, q) > \tau.$$

As a result, the graphs that satisfy $\underline{\text{GED}}(g_i, q) > \tau$ can be safely pruned from $\mathscr{G}$ without the costly GED verification. On the other hand, the graphs that pass the pruning phase, *i.e.*, $\underline{\text{GED}}(g_i, q) \leq \tau$, are further considered in a *candidate set*, $\mathscr{C}$. That is,

$$\mathscr{C} = \{g_i | g_i \in \mathscr{G}, \underline{\text{GED}}(g_i, q) \leq \tau\}. \tag{1}$$

At the second verification phase, we verify the pairwise GED constraint, $\text{GED}(g_i, q)$, between the query graph q and each graph g_i in the candidate set, $\mathscr{C}$.

The main idea of this pruning-verification framework is that, as opposed to examining $N = |\mathscr{G}|$ pairwise GED computations for the query graph q against the whole graph database, $\mathscr{G}$, we can just perform the costly GED verification against the candidate set, $\mathscr{C}$. If $|\mathscr{C}| \ll |\mathscr{G}|$, the challenging graph similarity search problem becomes feasible, even in very large graph databases. As a result, the crucial performance factors related to the graph similarity search problem are (1) the tightness of the GED lower bounds employed in the filtering phase, and (2) the computational overhead for the identification and filtering of graphs that fail the GED lower-bound constraints.

Formally, we consider the runtime cost, T, for graph similarity search, which is dependent on the following critical factors: (1) T_C: the runtime cost for graph pruning and candidate generation, (2) $|\mathscr{C}|$: the size of the candidate set, and (3) T_{GED}: the runtime cost for GED computation. Upon the pruning-verification framework, it is straightforward to know that

$$T = T_C + |\mathscr{C}| * T_{GED} \tag{2}$$

As a result, in order to improve the query performance for graph similarity search in large-scale graph databases, the most significant issue is to reduce the size of the candidate set, $|\mathscr{C}|$, because the runtime cost T_{GED} for pairwise GED computation is high, and it is unlikely to make breakthroughs in this aspect unless P = NP. To this end, powerful GED lower-bounds that can effectively identify and filter unqualified graphs from the graph database will result in a small value of $|\mathscr{C}|$, and thus bring significant speedup for graph similarity search. Meanwhile, another importance performance factor is T_C. To reduce T_C, state-of-the-art solutions typically adopt efficient and cost-effective graph indexing techniques to facilitate the process of GED lower-bound evaluation and graph pruning before the costly GED verification is performed upon the candidate set, $\mathscr{C}$.

4 State-of-the-Art Approaches

In this section, we elaborate on the state-of-the-art solutions to the similarity search problem in large-scale graph databases. All the methods to be discussed employ some graph-structured patterns as cost-effective graph indexing features, and take advantage of count-based GED lower-bounds based on the index features for effective graph identification and filtering.

To give readers a clear overview of these graph indexing based approaches, we will first consider q-gram indexes as an analogy, which have been extensively studied for the string edit distance computation [24, 36, 37, 41, 51, 61]. The main idea of q-gram is as follows: For each letter t in a string S of length $|S|$, we keep explicitly its small continuous q-grams, each of which is a substring of length q with t as its central character. This way, the string S can be decomposed into a collection of q-grams. As a result, we have the following important properties: (1) two identical strings have the same q-gram collections, and (2) any string edit operation will only affect a limited number of q-grams. These properties make it possible to perform string edit distance based similarity search upon a large set of strings, because *the number of matched q-grams* (based on string identity) is correlated with the exact string edit distance between strings, and therefore can be effectively leveraged as a lower-bound of string edit distance for effective pruning.

By analogy, we can adapt the same reasoning behind the q-gram methods to the problem of similarity search in graph databases, and indeed, existing graph similarity search solutions are primarily inspired by this idea. Essentially, the costly graph edit distance constraint is relaxed to a weaker, count-based constraint on the number of common graph-structured q-grams shared by both graphs, g, in the graph database and the query graph, q. These methods differ, however, primarily in (1) the q-grams adopted and indexed from the graph database, (2) the count-based GED lower-bound derived based on the q-grams, and (3) the graph indexing based filtering process for candidate graph generation and false-positive graph pruning. Broadly, the existing graph similarity search methods can be categorized as *tree-based approaches, star-based approaches, path-based approaches*, and *partition-based approaches*, each of which will be discussed in detail in the following sections.

4.1 A Tree-Based Approach: K-Adjacent Tree

Guoren Wang *etc.* proposed a graph indexing approach, *k-Adjacent Tree* (k-AT), for similarity search in large sparse graph databases [53]. They first decompose graphs into a series of tree-structured features, k-Adjacent Tree (k-AT for short), and then use the number of common k-AT features derived from $g \in \mathcal{G}$ and q, respectively, as a GED lower-bound for graph pruning. Formally, we define k-AT as follows,

Definition 5 (*k-AT*) Given a graph $g = (V, E, \Sigma, l)$ and a vertex $u \in V$, the adjacent tree of u in g is a breath-first search tree originated from u with the children of each tree node being sorted by the vertex labels. The k-adjacent tree of u, denoted as k-AT(u), is the top k-level subtree of the adjacent tree of u in g. The k-adjacent tree of g, denoted as k-AT(g), is a multi-set of k-ATs for each vertex $u \in V$. That is

$$k\text{-AT}(g) = \{k\text{-AT}(u) \mid u \in V\} \tag{3}$$

■

It is straightforward to decompose a graph g into k-AT(g), and each k-AT(u) can be properly maintained into an index structure for efficient lookup and comparison. Here the tree-structured k-AT serves as the basic "q-gram" for GED estimation. Intuitively, if the number of graph edit operations exerted on g is small, there must be a considerable number of its k-ATs remain unchanged. Formally, we can use the number of common k-ATs of two graphs to estimate the GED lower-bound between them, which is detailed in the following theorem.

Theorem 1 *Given two graphs g_1 and g_2, let $\delta(g_1)$ and $\delta(g_2)$ be the maximum degree of g_1 and g_2, respectively. If $\delta(g_1) > 1$ and $\delta(g_2) > 1$, then the number of common k-ATs of g_1 and g_2, $|k\text{-AT}(g_1) \cap |k\text{-AT}(g_2)|$, and their graph edit distance, $GED(g_1, g_2)$, satisfy the following inequality:*

$$|k\text{-AT}(g_1) \cap k\text{-AT}(g_2)| \geq |V_{g_1}| - \text{GED}(g_1, g_2) \cdot 2(\delta(g_1) - 1)^{k-1} \tag{4}$$

□

Namely, if q and $g \in \mathcal{G}$ are similar ($\mathrm{GED}(q, g) \leq \tau$), the number of their common k-ATs has to be no smaller than $|V(g)| - 2\tau(\delta(g) - 1)^{k-1}$. In other words, if the following inequality satisfies,

$$|k\text{-AT}(q) \cap k\text{-AT}(g)| < |V_q| - 2\tau(\delta(q) - 1)^{k-1} \qquad (5)$$

g is definitely an unqualified graph, and thus can be safely pruned from the graph database from further inspection.

During index construction, we generate all the k-ATs of each graph g from the graph database, $\mathcal{G}$, and maintain the k-ATs into an hierarchical index structure, k-**AT lattice**. To further compact the k-AT lattice, each k-AT is associated with a unique identifier, and the identifier of a small-size k-AT is reused to represent a large-size k-AT in the lattice. Such tree-structured k-ATs are further sequentialized for compact storage and efficient retrieval. When end-users issue a graph query q, we decompose it into k-ATs in an analogous way. Afterwards, we examine the number of common k-ATs between q and every graph $g \in \mathcal{G}$ in order to evaluate the inequality in Eq. 5. The graphs fail in the evaluation are put into the candidate set, $\mathcal{C}$, for further pairwise GED verification.

There are significant issues in this k-AT approach. First, the GED lower-bound is loose if graphs of the graph database are dense, and it is difficult to further tighten this bound [53]. Interestingly, the GED lower-bound may become equal to or even less than zero if there is a vertex with a high degree in the graph. Such phenomenon is often called *underflowing*. Second, this approach require enumerating all k-ATs exhaustively from the graph database, which will introduce significant time and space overheads. Third, it is hard to choose a proper value of k to guarantee both the filtering capability and the compactness of the index for a large graph database. Empirically, small values of k usually lead to limited pruning ability because small-size k-ATs fail to capture the global structure of graphs, which are crucial in GED computation, while large values of k may result in large-size k-ATs, within which many graph edit operations may occur and overlap, thus yielding poor pruning power and extremely high query processing cost.

4.2 A Star-Based Approach: SEGOS

Xiaoli Wang *etc.* proposed another graph indexing method, **SEGOS** (SEarching similar Graphs based On Sub-units), for similarity search in graph databases [54, 59]. In this approach, each graph $g \in \mathcal{G}$ is decomposed into specialized features called *stars*, each of which contains a vertex and discriminative information about its neighboring vertices and edges.

Definition 6 (*Star*) A star is a labeled, single-level and rooted tree represented as a 3-tuple $s = (r, L, l)$, where r is the root, L is the set of leaves and l is a labeling function. For each u in the graph g, a star can be built as $s_u = (u, L_u, l)$, where L_u is the label set of u's neighboring vertices. A graph g can be decomposed into its star representation, $S(g)$, which is a multiset of stars originating from each vertex u of g as a root,

$$S(g) = \{s_u | u \in V_g\} \tag{6}$$

■

The edit distance between two stars can be define as follows,

Definition 7 (*Star Edit Distance*) Given two stars s_1 and s_2, the edit distance between s_1 and s_2 is

$$\lambda(s_1, s_2) = T(r_1, r_2) + d(L_1, L_2) \tag{7}$$

where

$$T(r_1, r_2) = \begin{cases} 0 & \text{if } l(r_1) = l(r_2) \\ 1 & \text{otherwise} \end{cases}$$

and

$$d(L_1, L_2) = ||L_1| - |L_2|| + M(L_1, L_2)$$

$$M(L_1, L_2) = \max\{|\Psi_{L_1}|, |\Psi_{L_2}|\} - |\Psi_{L_1} \cap \Psi_{L_2}|$$

where Ψ_L is the multiset of vertex labels in L. ■

We then consider quantifying an estimated distance between two graphs g_1 and g_2 in terms of their star representations, $S(g_1)$ and $S(g_2)$, respectively. Such a distance is defined as a *mapping distance*, as follows,

Definition 8 (*Mapping Distance*) Given two star representation $S(g_1)$ and $S(g_2)$ with the same cardinality, and a bijective function $\mathscr{P} : S(g_1) \to S(g_2)$, the mapping distance between them is

$$\mu(g_1, g_2) = \min_{\mathscr{P}} \sum_{s_i \in S(g_1)} \lambda(s_i, \mathscr{P}(s_i)) \tag{8}$$

where the function $\lambda(s_i, s_j)$ is the start edit distance, as defined in Eq. 7. ■

The computation of the mapping distance is equivalent to finding an optimal mapping between two star representations of graphs, which can be implemented based on the Hungarian algorithm in cubic time [34]. The weight between two stars is the star edit distance. If two graphs are of different size, a special node ϵ is inserted for normalization.

Example 3 Figure 2 illustrates two graph g_1 and g_2 together with their star representations $S(g_1)$ and $S(g_2)$. The bottom left matrix $M(S(g_1), S(g_2))$ is the weight matrix between $S(g_1)$ and $S(g_2)$. Cells in gray denote the optimal matching between $S(g_1)$ and $S(g_2)$. Namely, $\mu(g_1, g_2) = 2 + 0 + 2 + 0 + 0 + 5 = 9$. To have a clear view, two sets of stars in $S(g_1)$ and $S(g_2)$ are shown on the right, and the optimal matching is marked with solid arrows. ■

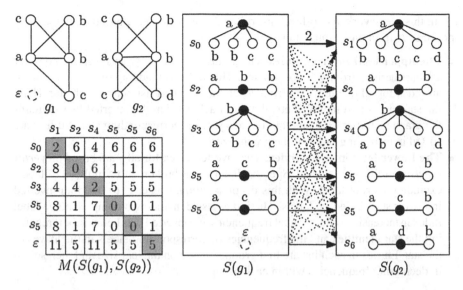

Fig. 2 Two graphs g_1 and g_2 are decomposed into their star representations $S(g_1)$ and $S(g_2)$, respectively. The mapping distance computation between g_1 and g_2 is also illustrated based on the Hungarian algorithm [54]

The mapping distance of graphs can be used effectively to derive a lower-bound, $\underline{GED}(g, q)$, and a upper-bound, $\overline{GED}(g, q)$ of the graph edit distance [59].

Theorem 2

$$\underline{GED}(g, q) = \frac{\mu(g, q)}{\max\{4, [\max\{\delta(g_1), \delta(g_2)\} + 1]\}} \leq GED(g, q) \qquad (9)$$

where $\delta(\cdot)$ is the maximum degree of a graph, and

$$\overline{GED}(g, q) = C(g_1, g_2, \mathcal{P}) \geq GED(g, q) \qquad (10)$$

where $C(g_1, g_2, \mathcal{P})$ is the overall cost of transforming g_1 to g_2 using the optimal mapping $\mathcal{P}$. □

The authors further proposed a two-level inverted index framework, **SEGOS**, to facilitate the computation of mapping distances and the evaluation of GED lower-bounds for similarity search. To enhance the filtering power, the decomposed stars from the graph database $\mathcal{G}$ are compared against the stars generated from the query graph q using the Hungarian algorithm. Formulated as a weighted bipartite matching problem, each star from $\mathcal{G}$ can have a partial matching with each star in q. This need arises in order to find the stars that not only match exactly but also are highly similar with each other.

In this framework, a two-level inverted index is constructed based on the decomposed stars from the graph database, $\mathscr{G}$:

- **The upper-level inverted index**. Given any graph $g \in \mathscr{G}$, an inverted index for all stars derived from g is constructed. This index is made up of two main parts: an index for all distinct stars from $\mathscr{G}$, and an inverted list below each star. Here, the stars are sorted in alphabetical order. Each entry in the inverted lists contains the graph identifier in $\mathscr{G}$ and the frequency of the corresponding star. All lists are sorted in increasing order of the graph sizes;
- **The Lower-level inverted index**. The inverted index for all stars based on *vertex labels* is further constructed. Specifically, a star is broken into a multiset of labels excluding its root label. With this decomposition, it is easy to build an inverted index for stars based on labels. The index also contains two components: a label index in increasing order of label frequencies and inverted lists below labels recording the star identifies and the frequencies of corresponding labels in the leaves of the star. Entries in each list are first grouped based on the leaf size and then sorted in decreasing frequencies within each group.

This two-level inverted index is preprocessed to maintain a global order for both stars and graphs in $\mathscr{G}$. This order ensures that stars or graphs can be accessed in increasing dissimilarity to stars decomposed from the query graph q. Therefore, it is convenient to search similar stars for a star derived from the query graph based on the star edit distance.

Given a query graph q, the similar search follows a novel, *cascaded* framework: in the lower level of the inverted index, top-k similar stars to each star of the query can be retrieved and returned quickly; in the upper level of the inverted index, graph pruning is carried out based on the top-k results returned from the lower level. Two search algorithms based on the paradigm of the TA (threshold algorithm) and the CA (combined algorithm) [17] are proposed to retrieve stars and graphs. Using the summation of star edit distances as the aggregation function, sorted lists can be easily constructed to guarantee the global orders on the increasing dissimilarity for graphs. The CA based method can enhance similarity search by avoiding access to graphs with high dissimilarity. It is clear that the top-k stars returned from the lower-level star search can be automatically used as the input to the upper-level graph search. Therefore, these two search stages can be easily pipelined to support continuous graph pruning.

The star structure considered in SEGOS is exactly the same feature as the 1-AT defined in the k-AT method, where $k = 1$. Unlike k-AT, SEGOS computes the new lower and upper GED bounds, which have proven to be tighter than that of k-AT. However, SEGOS has to invoke the Hungarian algorithm for the weighted bipartite matching between star representations of two graphs, which is time-consuming. Furthermore, star structures still lead to poor similarity search performance because they are typically frequent features in a graph database and exhibit limited pruning capabilities, thus yielding a very large candidate set, $\mathscr{C}$, for similarity search.

4.3   A Path-Based Approach: GSimJoin

Xiang Zhao *etc.* proposed a different, path-based indexing approach, **GSimJoin**, for *similarity join* in graph databases, while their methods can be directly applied to solve the similarity search problem as well [65]. Distinct from the tree-based approaches, they explore a novel perspective of utilizing *paths* as the basic q-grams and index features:

Definition 9   *(q-Path)* A q-path is a simple path of length q, in terms of the number of edges, in a graph g.   ∎

Here "simple" means that there is no repeated vertex in the path. Since a path has two ends, namely, start vertex and end vertex, two sequences can be formed by concatenating the vertex and edge labels from both ends. Here we only consider the lexicographically smaller one as a q-path. Since the length of a path can be zero for the case of a single vertex, a 0-path is just a single vertex.

The path-based GED lower-bound can be derived accordingly, as follows,

Theorem 3   *Consider two graphs g_1 and g_2, if $GED(g_1, g_2) \leq \tau$, then g_1 and g_2 must share at least*

$$\max(|P_{g_1}| - \tau \cdot D_{path}(g_1), |P_{g_2}| - \tau \cdot D_{path}(g_2)) \tag{11}$$

common q-paths. Here P_g denotes the multiset of q-paths of g, and $D_{path}(g)$ denotes the maximum number of q-paths that can be affected by one particular graph edit operation.   □

Compared with the tree-based approaches, such as k-AT and SEGOS, this path-based method, GSimJoin, has the advantage of presenting tighter GED lower-bounds. As a result, it will deliver the chance of using longer q-paths in seek of greater pruning capabilities and better similarity search performance.

GSimJoin further makes use of an inverted index to evaluate the path-based GED lower-bounds during the candidate generation. First of all, all q-paths from the graph database $\mathcal{G}$ are generated. The inverted index maps each q-path, p, to a list of identifiers of graphs in $\mathcal{G}$ that contain p. Once given a query graph, we can decompose it into a series of q-paths and probe the inverted index to produce the candidate set $\mathcal{C}$ that contains all graphs sharing common q-paths with the query and satisfying the path-based GED lower-bound, as formulated in Eq. 11.

To address the main performance bottleneck in accessing long inverted lists in the index, GSimJoin further adopts a *prefix filtering* idea that if two multisets of q-paths meet the GED lower-bound constraint, they must share at least one common q-path in their prefixes. In other words, if P_{g_1} and P_{g_2} have no common q-gram in their prefixes, the number of their common q-grams is no more than the GED lower-bound in Eq. 11. As a result, the prefix filtering can be employed to further reduce the candidate set, $\mathcal{C}$. In practice, in order to achieve a small candidate size and fast execution speed, rare q-paths are favored in prefixes. Therefore the multiset of q-paths of each graph can be sorted in ascending order of the number of graphs of $\mathscr{G}$ that contain the q-paths.

Another novelty of GSimJoin is to exploit the valuable information provided by mismatching q-paths that cannot be matched between graphs and the query. Two filtering conditions, minimum prefix length and local labels, are accordingly proposed so that the size of candidates can be substantially reduced.

GSimJoin still suffers from significant performance issues for similarity search because the basic index features, k-paths, may overlap with a lot of other q-paths within a graph, and thus a single graph edit operation will affect many k-paths. This is especially true in dense graphs with high-degree vertices. As a result, the path-based GED lower-bound becomes loose and thus fails to filter unqualified graphs from the graph database during the candidate generation.

4.4 A Partition-Based Approach: Pars

The tree-based, star-based, and path-based approaches make use of some fixed-size, overlapping substructures of graphs as basic index features for GED lower-bound evaluation and graph filtering. As a consequence, these approaches suffer from the following drawbacks for similarity search: (1) They fail to take full advantage of the global topological structure of graphs for candidate generation. The fixed-size substructures significantly limit the selectivity of corresponding graph indexes, and therefore become nonadaptive to large-scale graph databases and queries; (2) There exist a lot of structure redundancy among fixed-size substructures. As a result, their corresponding GED lower-bounds are primarily established in a pessimistic way for false-positive graph identification and pruning, and thus are very sensitive to large vertex degrees or large GED threshold, τ.

Xiang Zhao *etc.* proposed another novel graph indexing method, **Pars**, by decomposing graphs into *variable-size*, *non-overlapping* half-edge graph partitions, which constitute the basic index features for similarity search [64]. Such partition-based indexing scheme is not prone to large vertex degrees, and can accommodate large distance thresholds in practice.

Definition 10 *(Half-edge Graph)* A half-edge graph $g = (V, E, \Sigma, l)$, where $E \subseteq V \times (V \cup \{*\})$, is a graph with the possible half edge $(u, *) \in E$, which has a definite one-end vertex $u \in V$, while the other end (and its label) is not explicitly specified, represented as $*$. ∎

A half-edge graph is a general case of graphs, as defined in Definition 1. Given a graph g, it can be partitioned into a series of half-edge graphs. This graph partitioning can be formalized as follows,

Definition 11 *(Graph Partitioning)* A graph g can be decomposed into a set $\mathscr{P}$ of collective exhaustive, mutually exclusive, and non-empty half-edge graphs, denoted

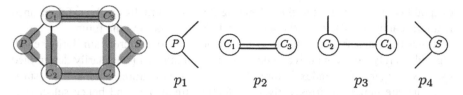

Fig. 3 The graph g on the *left* is partitioned into four half-edge graphs: $\mathscr{P}(g) = \{p_1, p_2, p_3, p_4\}$ as shown on the *right*

as $\mathscr{P}(g) = \{p_i \mid \bigcup_i V_{p_i} = V_g, \ \bigcup_i E_{p_i} \subseteq E_g \cup V_g \times \{*\}, \ p_i \cap p_j = \emptyset, \ \forall i, j, \ i \neq j\}$. $\mathscr{P}$ is called a *partitioning* of g. ∎

Example 4 As shown in Fig. 3, the graph g is partitioned into four half-edge graphs, $p_1, p_2, p_3,$ and p_4. So $\mathscr{P} = \{p_1, p_2, p_3, p_4\}$ is one partitioning, among others, of g. ∎

Partitioning a graph g into a set of half-edge graph partitions has a clear advantage in GED computation: for any graph edit operation that can be applied upon g, it can only appear in, and thus affect, at most one graph partition. As a result, the following partition-based GED lower bound can be derived,

Theorem 4 *Consider a query graph q and a graph $g \in \mathscr{G}$ with a partitioning $\mathscr{P}(g)$ of $\tau + 1$ partitions. If $GED(g, q) \leq \tau$, at least one of the $\tau + 1$ partitions is subgraph isomorphic to q.* □

A partition $p_i \subseteq g$ is called a *matching partition* if it is subgraph isomorphic to the query q, or otherwise a *mismatching partition*. Based on Theorem 4, if the number of mismatching partitions of g is larger than τ, then $GED(g, q) > \tau$. Namely, g is a false-positive graph and thus can be safely pruned. In order to leverage this partitioned based GED lower-bound for effective graph filtering and candidate generation, a partition-based similarity search framework, **Pars**, is developed which contains two main phases: an index construction phase that can be performed offline and a query processing phase that support online similarity search in graph databases.

In the offline index construction phase, each graph $g \in \mathscr{G}$ is partitioned into $(\tau + 1)$ half-edge subgraphs. An inverted index is further built to maintain for each partition p all the graphs of $\mathscr{G}$ that contain p as a subgraph. The graph-structured partitions are sequentialized into their canonical DFS-code representations [55] to facilitate the lookup and maintenance of the indexed half-edge subgraph features during similarity search.

Once the partitioned based index has been properly built, similarity search can be supported in a straightforward way: Given a query graph q, we start probing the inverted index for candidate generation. At first, we maintain a data structure, *states*, which are set to *uninitialized* for all graphs g in the graph database $\mathscr{G}$. Then, for each partition p in the inverted index, we examine whether p is contained as a subgraph in the query q or not. If so, each graph g with an uninitialized state in the postings list of p are set to *true* and become candidate graphs in $\mathscr{C}$, while the states of the disqualified graphs are set to *false* denoting that these graphs can be safely pruned from the candidate set, $\mathscr{C}$, without costly GED verification in the future.

It has been observed that the filtering performance of different similarity search algorithms relying on inclusive logic over inverted indexes is primarily determined by the *selectivity* of the index features. Fixed-size index features, such as trees, stars, or paths, are generated irrespectively of feature frequency, and hence selectivity; while variable-size, half-edge graph partitions offer more flexibility in constructing the feature-based inverted index. Meanwhile, partition-based features can capture the global structural information of graphs in the graph database, and thus can obtain statistically more selective index features than the previous, fixed-sized feature based approaches. Furthermore, partition-based half-edge subgraphs generated from the graph database are non-overlapping. This property restricts that an graph edit operation can affect at most one index feature, and thus, the number of features affected by τ graph edit operations can be tremendously reduced. As a result, unlike previous approaches, the partition-based approach does not suffer from the drawback of loose GED bounds when handling large GED thresholds and graphs/queries with large vertex degrees.

A critical component in Pars is how to partition a graph g into $(\tau + 1)$ half-edge graphs. As expected, even assigned with a trivial cost function, this graph partitioning problem is NP-hard [64]. In order to address this problem, a practical *randomized-refinement* graph partitioning algorithm is proposed for a good, if not the optimal, partitioning. First of all, it randomly picks $(\tau + 1)$ distinct vertices of g as the initial partition p_i, where $1 \leq i \leq \tau + 1$, respectively. This ensures that every graph partition p_i is non-empty and contains at least one vertex. Then, for each p_i, we further extend it with a neighboring vertex that has not been assigned to any existing partition. This offers each p_i a chance to expand in a way that the sizes and the selectivities of all the partitions are balanced. Finally, for all the remaining edges $e = (u, v)$ whose end vertices are assigned to different partitions, it randomly assigns e to either the partition containing u or v as a half-edge. If the information of historical query workload is available, the randomized graph partitioning can be further refined by selecting the best option of moving vertices from one partition to another in order to minimize the size of the candidate set, $\mathscr{C}$.

Another novelty of Pars is to dynamically rearrange graph partitions to adapt the online query by *recycling* and making use of the information in mismatching partitions. The basic idea of *dynamic partition filtering* is to leverage the mismatching partitions and to dynamically add, if possible, additional vertices and edges to a partition having been tested to be contained by the query q. The recycled vertices and edges are used once the subgraph isomorphism test between the graph partition p and the query graph q returns true. In particular, for each graph g in p's postings list, we append g's recycled vertices and edges to p and perform another subgraph isomorphism test. Only if the new partition is contained by q, g becomes a candidate and is allocated to $\mathscr{C}$. Note that if the new subgraph isomorphism test fails, the vertices and edges can be recycled again for further evaluation.

Although Pars has claimed to be one of the best approaches to similarity search in large-scale graph databases, there are still several weaknesses. First, the partition-based GED lower bound is still not tight to filter a majority of false-positive graphs from the graph database. Second, Pars adopts a randomized partitioning method to generate half-edge graph partitions as basic index features, the selectivity of which, however, are actually quite limited, and in some occasions, their pruning power will be significantly poor because some randomly generated index features might be matching partitions of a lot of queries. Although a sophisticated partitioning refinement process is devised to remedy this issue, it still suffers from a laborious process that involves a large amount of subgraph isomorphism computation and index reorganization, which poses another significant performance bottleneck to similarity search.

5 Future Research Directions

Although there have been numerous GED lower-bounds and similarity search techniques proposed thus far, the topic of enabling efficient, versatile, and high-quality search functionalities and graph access methods still stays in its infant stage, and there exist extensive research interest and real-world need for Google-like search functions and tools that are well tailored and highly optimized for large-scale graph databases. In this section, we envision the potential research directions and future frontiers that are closely related to similarity search in graph databases.

5.1 New GED Bounds and Search Algorithms

As discussed above, there have been quite a few similarity search algorithms proposed for similarity search in large-scale graph databases [23, 53, 54, 59, 64, 65, 67]. However, state-of-the-art solutions still suffer from severe performance issues. As analyzed in Sect. 3, the key to boost the similarity search performance is to leverage some powerful GED lower-bound that can effectively pre-prune false-positive graphs from the graph database $\mathscr{G}$ in order to avoid the costly GED verification. However, existing GED lower-bounds have demonstrated limited filtering capabilities, thus resulting in the corresponding similarity search algorithms hard to deploy in real-world graph databases. Furthermore, there is no theoretical results about the tightness guarantee of the proposed GED lower-bounds. Therefore, the resultant similarity search performance becomes unstable in the presence of different graph datasets and queries of varying structures and types. More theoretical breakthroughs about graph edit distances and new GED lower-bounds and upper-bounds are expected in order to reduce the size of the candidate set, $\mathscr{C}$. An interesting exploration will be examining other graph-structured patterns, like trees, paths, or half-edge partitions, which will bring forth more powerful pruning capability and better selectivity for similarity search.

Meanwhile, graph indexing has somehow become the de facto standard to evaluate and crosscheck different GED lower-bounds for false-positive graph identification and pruning. In this aspect, all index-related issues, such as novel graph indexing structures, cost-effective index organizations, efficient index lookup and maintenance mechanisms, and dynamic index updates become utterly important and play an essential role in similarity search in graph databases.

5.2 Rich Semantics of Similarity Search

As discussed above, most similarity search problems in graph databases are formulated as retrieving graphs g from $\mathscr{G}$ in a way that $\text{GED}(g, q) \leq \tau$, where τ is a user-specified GED threshold. In the traditional database terminology, this is often termed as a *range query*, where τ determines the range of graph edit distances tolerated by end-users. In some occasions, however, it is hard for users to choose the appropriate values of τ, and if τ is not carefully determined, a tedious and time-consuming process of query re-formulation has to be undertaken. Furthermore, the graph indexes have to be rebuilt as most existing index structures are strongly dependent on the exact values of τ. A possible and promising research direction is to support GED-based top-k similarity search in graph databases. Namely, the objective is to retrieve k graphs from $\mathscr{G}$ that have the smallest graph edit distances from the query q. This way, the query results can be naturally ranked, which are way more interpretable to users who have little or no prior knowledge of the underlying graph databases.

Another possibility of extension to the existing similarity search semantics lies in the definition of GED. Here we assume all different graph edit operations play an equal role in formulating the differences between two graphs. In practice, however, graph edit operations may introduce varied significance in quantifying distances between graphs. For example, the cost of relabelling a vertex might be different from that of relabelling an edge in a graph. It means that, we can assigned weights to graph edit operations, and such weights are primarily determined in real-world application domains. In this new setting, the similarity search problem turns out to be a new, weighted version, which, unfortunately, is hard to address using existing approaches to the primitive, unweighted version. New GED lower-bounds and graph indexing techniques are thus required in order to address the generalized, weighted GED similarity search problem in graph databases.

Besides graph edit distances, there have been a great many proximity/distance measures proposed thus far to quantify the similarity/distances between graphs, including, but not limited to, maximum common subgraphs, graph alignment, kernels, edge misses, and graph simulations. It will be of great interest to examine the theoretical correlation between GED and other graph similarity measures, and it is also beneficial to adapt existing similarity search solutions to addressing the similarity search problem that are defined based on different graph proximity measures in graph databases.

Similarity search is not only confined within graph databases. With the advent and popularity of social networks, communication networks, road networks, biological networks, and the Web, there is growing interest in searching graph-structured patterns from within such large-scale networked data. Here we want to support similarity search in a single graph of massive size. Namely, we need directly address the challenging GED problem in very large graphs, which has proven to be much harder than similarity search in a graph database. Although there have been related work that support *exact* search in large-scale graphs or information networks [28, 62], the similarity search problem in a single large graph is yet to be explored thoroughly and a lot of research can be carried out in this direction.

5.3 Graph Query Formulation and Understanding

Due to its great power in representing and formulating complex structures in real-world applications, graph has gradually become the first-class citizen in the data management and mining communities. However, there still lack some high-level query languages or tool-chains that help formulate user's structured information need into graph-shaped queries. Although SPARQL has been proposed to query semantic webs [3, 46, 47], it is primarily employed to support search functionalities in graphs formulated in the RDF format. Meanwhile, some visual query formulation techniques have been studied to ease the pain of constructing graph-structured queries [29–31]. However, they are primarily focused on exact queries in graph databases. As a result, there is an urgent need to support some user-friendly graph query languages or graph query formulation mechanisms that can enable similarity search semantics in graph databases or large-scale networks. Consequently, a series of research topics, such as graph query completion, suggestion, enrichment, and diversification can be examined in order to support a full-fledged graph querying interface for real-world graph databases.

It is interesting to note that, because graph queries can be relaxed approximately based on the GED constraint, an immediate result is that there might be a lot of graphs from the graph database $\mathscr{G}$ returned as feasible answers to a given query q, which can easily overwhelm the end-users. On the other hand, it would be highly desirable to pick k representative graphs that are more comprehensible and manageable than the complete set of query results. This problem has proven to be NP-hard [43]. New algorithms and novel definitions of "graph representativeness" will be of special interest in enhancing the understanding of similarity search results from within large-scale graph databases.

6   Summary

The similarity search problem plays a fundamental and critical role in managing, accessing, and analyzing graph-structured data, and has found widely varying applications in real-world graph databases. In this chapter, we examined the similarity search problem under the graph edit distance (GED) constraint, and surveyed state-of-the-art graph indexing approaches to addressing the similarity search problem in real-world, large-scale graph databases. We envisioned and explored the potential research directions that are closely related to the similarity search problem, which is expected to fuse a series of fundamental and practical research in large graph databases and real-world social and information networks.

References

1. C.C. Aggarwal, H. Wang, *Managing and Mining Graph Data* (Springer, US, 2010)
2. L. Babai, Graph isomorphism in quasipolynomial time. in *Proceedings of the 48th Annual ACM SIGACT Symposium on Theory of Computing (STOC'16)* (2016), pp. 684–697
3. D.F. Barbieri, D. Braga, S. Ceri, E.D. Valle, M. Grossniklaus, Querying rdf streams with c-sparql. SIGMOD Rec. **39**(1), 20–26 (2010)
4. P. Barceló Baeza, Querying graph databases. in *Proceedings of the 32nd Symposium on Principles of Database Systems (PODS'13)* (2013), pp. 175–188
5. H.M. Berman, J. Westbrook, Z. Feng, G. Gilliland, T.N. Bhat, H. Weissig, I.N. Shindyalov, P.E. Bourne, The protein data bank. Nucleic Acids Res. **28**, 235–242 (2000)
6. S. Berretti, A. Del Bimbo, E. Vicario, Efficient matching and indexing of graph models in content-based retrieval. IEEE Trans. Pattern Anal. Mach. Intell. **23**(10), 1089–1105 (2001)
7. K.M. Borgwardt, H.-P. Kriegel, Shortest-path kernels on graphs. in *Proceedings of the Fifth IEEE International Conference on Data Mining (ICDM'05)* (2005), pp. 74–81
8. H. Bunke, On a relation between graph edit distance and maximum common subgraph. Pattern Recogn. Lett. **18**(9), 689–694 (1997)
9. H. Bunke, Error correcting graph matching: on the influence of the underlying cost function. IEEE Trans. Pattern Anal. Mach. Intell. **21**(9), 917–922 (1999)
10. H. Bunke, K. Shearer, A graph distance metric based on the maximal common subgraph. Pattern Recogn. Lett. **19**(3–4), 255–259 (1998)
11. X. Chen, K.S. Candan, M.L. Sapino, P.Shakarian, KSGM: Keynode-driven scalable graph matching. in *Proceedings of the 24th ACM International on Conference on Information and Knowledge Management (CIKM'15)* (2015), pp. 1101–1110
12. H. Cheng, D. Lo, Y. Zhou, X. Wang, X. Yan, Identifying bug signatures using discriminative graph mining. in *Proceedings of the Eighteenth International Symposium on Software Testing and Analysis (ISSTA'09)* (2009), pp. 141–152
13. J. Cheng, Y. Ke, W. Ng, Efficient query processing on graph databases. ACM Trans. Database Syst. **34**(1), 2:1–2:48 (2009)
14. S. Choudhury, L. Holder, G. Chin, A. Ray, S. Beus, J. Feo, Streamworks: a system for dynamic graph search. in *Proceedings of the 2013 ACM SIGMOD International Conference on Management of Data (SIGMOD'13)* (2013), pp. 1101–1104
15. D. Conte, P. Foggia, C. Sansone, M. Vento, Thirty years of graph matching in pattern recognition. Int. J. Pattern Recognit. Artif. Intell. **18**(3), 265–298 (2004)
16. D.J. Cook, L.B. Holder, *Mining Graph Data* (Wiley, New Jersey, 2006)

17. R. Fagin, A. Lotem, M. Naor, Optimal aggregation algorithms for middleware. in *Proceedings of the Twentieth ACM SIGACT-SIGMOD-SIGART Symposium on Principles of Database Systems (PODS'01)* (2001), pp. 102–113

18. W. Fan, J. Li, S. Ma, N. Tang, Y. Wu, Y. Wu, Graph pattern matching: from intractable to polynomial time. Proc. VLDB Endow. **3**(1–2), 264–275 (2010)

19. S. Fankhauser, K. Riesen, H. Bunke, Speeding up graph edit distance computation through fast bipartite matching. in *Proceedings of the 8th International Conference on Graph-based Representations in Pattern Recognition (GBRPR'11)* (2011), pp. 102–111

20. B. Gallagher, Matching structure and semantics: a survey on graph-based pattern matching. in *American Association for Artificial Intelligence (AAAI'06)*, vol. 6 (2006), pp. 45–53

21. X. Gao, B. Xiao, D. Tao, X. Li, A survey of graph edit distance. Pattern Anal. Appl. **13**(1), 113–129 (2010)

22. M.R. Garey, D.S. Johnson, *Computers and Intractability; A Guide to the Theory of NP-Completeness* (W. H. Freeman & Co., New York, 1990)

23. K. Gouda, M. Arafa, An improved global lower bound for graph edit similarity search. Pattern Recogn. Lett. **58**, 8–14 (2015)

24. L. Gravano, P.G. Ipeirotis, H.V. Jagadish, N. Koudas, S. Muthukrishnan, D. Srivastava, Approximate string joins in a database (almost) for free. in *Proceedings of the 27th International Conference on Very Large Data Bases (VLDB'01)* (2001), pp. 491–500

25. W.-S. Han, J. Lee, J.-H. Lee, Turboiso: towards ultrafast and robust subgraph isomorphism search in large graph databases. in *Proceedings of the 2013 ACM SIGMOD International Conference on Management of Data (SIGMOD'13)* (2013), pp. 337–348

26. W.-S. Han, M.-D. Pham, J. Lee, R. Kasperovics, J.X. Yu, Igraph in action: performance analysis of disk-based graph indexing techniques. in *Proceedings of the 2011 ACM SIGMOD International Conference on Management of Data (SIGMOD'11)* (2011), pp. 1241–1242

27. H. He, A.K. Singh, Closure-tree: an index structure for graph queries. in *Proceedings of the 22nd International Conference on Data Engineering (ICDE'06)* (2006), pp. 38–49

28. H. He, A.K. Singh, Graphs-at-a-time: query language and access methods for graph databases. in *Proceedings of the 2008 ACM SIGMOD International Conference on Management of Data (SIGMOD'08)* (2008), pp. 405–418

29. H.H. Hung, S.S. Bhowmick, B.Q. Truong, B. Choi, S. Zhou, Quble: blending visual subgraph query formulation with query processing on large networks. in *Proceedings of the 2013 ACM SIGMOD International Conference on Management of Data (SIGMOD'13)* (2013), pp. 1097–1100

30. N. Jayaram, S. Goyal, C. Li, VIIQ: Auto-suggestion enabled visual interface for interactive graph query formulation. Proc. VLDB Endow. **8**(12), 1940–1943 (2015)

31. C. Jin, S.S. Bhowmick, X. Xiao, J. Cheng, B. Choi, GBLENDER: towards blending visual query formulation and query processing in graph databases. in *Proceedings of the 2010 ACM SIGMOD International Conference on Management of Data (SIGMOD'10)* (2010), pp. 111–122

32. A. Khan, N. Li, X. Yan, Z. Guan, S. Chakraborty, S. Tao, Neighborhood based fast graph search in large networks. in *Proceedings of the 2011 ACM SIGMOD International Conference on Management of Data (SIGMOD'11)* (2011), pp. 901–912

33. A. Khan, Y. Wu, C.C. Aggarwal, X. Yan, NeMa: Fast graph search with label similarity. Proc. VLDB Endow. **6**(3), 181–192 (2013)

34. H.W. Kuhn, B. Yaw, The hungarian method for the assignment problem. Naval Res. Logist. Quart. 83–97 (1955)

35. J. Lee, W.-S. Han, R. Kasperovics, J.-H. Lee, An in-depth comparison of subgraph isomorphism algorithms in graph databases. in *Proceedings of the 39th International Conference on Very Large Data Bases (PVLDB'13)* (2013), pp. 133–144

36. C. Li, J. Lu, Y. Lu, Efficient merging and filtering algorithms for approximate string searches. in *Proceedings of the 2008 IEEE 24th International Conference on Data Engineering (ICDE'08)* (2008), pp. 257–266

37. C. Li, B. Wang, X. Yang, VGRAM: improving performance of approximate queries on string collections using variable-length grams. in *Proceedings of the 33rd International Conference on Very Large Data Bases (VLDB'07)* (2007), pp. 303–314

38. S. Ma, Y. Cao, W. Fan, J. Huai, T. Wo, Strong simulation: Capturing topology in graph pattern matching. ACM Trans. Database Syst. **39**(1), 4:1–4:46 (2014)

39. M. Neuhaus, H. Bunke, *Bridging the Gap Between Graph Edit Distance and Kernel Machines* (World Scientific Publishing, Singapore, 2007)

40. H. Ogata, S. Goto, K. Sato, W. Fujibuchi, H. Bono, M. Kanehisa, KEGG: kyoto encyclopedia of genes and genomes. Nucleic Acids Res. **27**(1), 29–34 (1999)

41. J. Qin, W. Wang, Y. Lu, C. Xiao, X. Lin, Efficient exact edit similarity query processing with the asymmetric signature scheme. in *Proceedings of the 2011 ACM SIGMOD International Conference on Management of Data (SIGMOD'11)* (2011), pp. 1033–1044

42. S.A. Rahman, M. Bashton, G.L. Holliday, R. Schrader, J.M. Thornton, Small molecule subgraph detector (SMSD) toolkit. J. Cheminform. **1**, 1–12 (2009)

43. S. Ranu, M. Hoang, A. Singh, Answering top-k representative queries on graph databases. in *Proceedings of the 2014 ACM SIGMOD International Conference on Management of Data (SIGMOD'14)* (2014), pp. 1163–1174

44. S. Ranu, A.K. Singh, Indexing and mining topological patterns for drug discovery. in *Proceedings of the 15th International Conference on Extending Database Technology (EDBT'12)* (2012), pp. 562–565

45. K. Riesen, S. Emmenegger, H. Bunke, A novel software toolkit for graph edit distance computation. in *9th International Workshop on Graph-Based Representations in Pattern Recognition* (2013), pp. 142–151

46. S. Sakr, S. Elnikety, Y. He, G-SPARQL: A hybrid engine for querying large attributed graphs. in *Proceedings of the 21st ACM International Conference on Information and Knowledge Management (CIKM'12)* (2012), pp. 335–344

47. M. Schmidt, M. Meier, G. Lausen, Foundations of SPARQL query optimization. in *Proceedings of the 13th International Conference on Database Theory (ICDT'10)* (2010), pp. 4–33

48. H. Shang, X. Lin, Y. Zhang, J.X. Yu, W. Wang, Connected substructure similarity search. in *Proceedings of the 2010 ACM SIGMOD International Conference on Management of Data (SIGMOD'10)* (2010), pp. 903–914

49. A. Tefas, C. Kotropoulos, I. Pitas, Using support vector machines to enhance the performance of elastic graph matching for frontal face authentication. IEEE Trans. Pattern Anal. Mach. Intell. **23**(7), 735–746 (2001)

50. Y. Tian, R.C. Mceachin, C. Santos, D.J. States, J.M. Patel, SAGA: a subgraph matching tool for biological graphs. Bioinformatics **23**(2), 232–239 (2007)

51. E. Ukkonen, Approximate string-matching with q-grams and maximal matches. Theor. Comput. Sci. **92**(1), 191–211 (1992)

52. J.R. Ullmann, An algorithm for subgraph isomorphism. J. ACM **23**(1), 31–42 (1976)

53. G. Wang, B. Wang, X. Yang, G. Yu, Efficiently indexing large sparse graphs for similarity search. IEEE Trans. Knowl. Data Eng. **24**(3), 440–451 (2012)

54. X. Wang, X. Ding, A.K.H. Tung, S. Ying, H. Jin, An efficient graph indexing method. in *Proceedings of the 2012 IEEE 28th International Conference on Data Engineering (ICDE'12)* (2012), pp. 210–221

55. X. Yan, J. Han, gSpan: graph-based substructure pattern mining. in *Proceedings of the 2002 IEEE International Conference on Data Mining (ICDM'02)* (2002), pp. 721–724

56. X. Yan, P.S. Yu, J. Han, Graph indexing: a frequent structure-based approach. in *Proceedings of the 2004 ACM SIGMOD International Conference on Management of Data (SIGMOD'04)* (2004), pp. 335–346

57. X. Yan, P.S. Yu, J. Han, Substructure similarity search in graph databases. in *Proceedings of the 2005 ACM SIGMOD International Conference on Management of Data (SIGMOD'05)* (2005), pp. 766–777

58. Y. Yuan, G. Wang, J.Y. Xu, L. Chen, Efficient distributed subgraph similarity matching. VLDB J. **24**(3), 369–394 (2015)

59. Z. Zeng, A.K.H. Tung, J. Wang, J. Feng, L. Zhou, Comparing stars: On approximating graph edit distance. Proc. VLDB Endow. **2**(1), 25–36 (2009)
60. S. Zhang, J. Yang, W. Jin, SAPPER: Subgraph indexing and approximate matching in large graphs. Proc. VLDB Endow. **3**(1–2), 1185–1194 (2010)
61. Z. Zhang, M. Hadjieleftheriou, B.C. Ooi, D. Srivastava, Bed-tree: an all-purpose index structure for string similarity search based on edit distance. in *Proceedings of the 2010 ACM SIGMOD International Conference on Management of Data (SIGMOD'10)* (2010), pp. 915–926
62. P. Zhao, J. Han, On graph query optimization in large networks. Proc. VLDB Endow. **3**(1–2), 340–351 (2010)
63. P. Zhao, J.X. Yu, P.S. Yu, Graph indexing: tree + delta ≥ graph. in *Proceedings of the 33rd International Conference on Very Large Data Bases (VLDB'07)* (2007), pp. 938–949
64. X. Zhao, C. Xiao, X. Lin, Q. Liu, W. Zhang, A partition-based approach to structure similarity search. PVLDB **7**(3), 169–180 (2013)
65. X. Zhao, C. Xiao, X. Lin, W. Wang, Efficient graph similarity joins with edit distance constraints. in *Proceedings of the 2012 IEEE 28th International Conference on Data Engineering (ICDE'12)* (2012), pp. 834–845
66. X. Zhao, C. Xiao, X. Lin, W. Wang, Y. Ishikawa, Efficient processing of graph similarity queries with edit distance constraints. VLDB J. **22**(6), 727–752 (2013)
67. W. Zheng, L. Zou, X. Lian, D. Wang, D. Zhao, Graph similarity search with edit distance constraint in large graph databases. in *Proceedings of the 22nd ACM International Conference on Conference on Information & Knowledge Management (CIKM'13)* (2013), pp. 1595–1600
68. G. Zhu, X. Lin, K. Zhu, W. Zhang, J.X. Yu, TreeSpan: efficiently computing similarity all-matching. in *Proceedings of the 2012 ACM SIGMOD International Conference on Management of Data (SIGMOD'12)* (2012), pp. 529–540

Big-Graphs: Querying, Mining, and Beyond

Arijit Khan and Sayan Ranu

Abstract Graphs are a ubiquitous model to represent objects and their relations. However, the complex combinations of structure and content, coupled with massive volume, high streaming rate, and uncertainty inherent in the data, raise several challenges that require new efforts for smarter and faster graph analysis. With the advent of complex networks such as the World Wide Web, social networks, knowledge graphs, genome and scientific databases, Internet of things, medical and government records, novel graph computations are also emerging, including graph pattern matching and mining, similarity search, keyword search, and graph query-by-example. These workloads require both topology and content information of the network; and hence, they are different from classical graph computations such as shortest path, reachability, and minimum cut, which depend only on the structure of the network. In this chapter, we shall describe the emerging graph queries and mining problems, their applications and resolution techniques. We emphasize the current challenges and highlight some future research directions.

1 Introduction

Recent advances in social and information science have shown that linked data is pervasive in the natural world around us [88]. Examples include communication and computer systems, the World Wide Web, online social networks, biological networks, transportation systems, epidemic networks, chemical networks, and hidden terrorist networks. All these systems can be modeled as graphs, in which individual components interact with a specific set of components, resulting in massive, interconnected, and heterogeneous networks.

A. Khan (✉)
School of Computer Science and Engineering, Nanyang Technological University,
Singapore, Singapore
e-mail: arijit.khan@ntu.edu.sg

S. Ranu
Department of Computer Science and Engineering, IIT Madras, Chennai, India
e-mail: sayan@cse.iitm.ac.in

© Springer International Publishing AG 2017
A.Y. Zomaya and S. Sakr (eds.), *Handbook of Big Data Technologies*,
DOI 10.1007/978-3-319-49340-4_16

Given a network modeled as a *graph*, there have been several commonly-used graph computations, such as breadth-first (BFS) and depth-first searches (DFS) [27], shortest path [124, 129], reachability [139], subgraph isomorphism [10], PageRank [14], graph partitioning [62, 78, 116], and clustering [4] — their algorithms are also implemented using a number of graph processing libraries, e.g., LEDA [106], BLAS [17], Parallel Boost Graph Library (PBGL) [46], and MultiThreaded Graph Library (MTGL) [11]. Graph workloads can be broadly classified into two categories [83]. (1) *Offline graph analytics* usually perform computation over the entire graph, and they require high throughput. PageRank computation, diameter estimation, partitioning, and clustering belong to this category. (2) *Online graph queries*, on the other hand, require very fast response time, and they often explore only a small fraction of the entire graph dataset. Examples include reachability, shortest path, and sub-graph isomorphism queries.

We note that the aforementioned commonly-used workloads depend only on the graph structure, and thus may not be able to capture the rich semantics associated with nodes, edges, and structures in the network. With the advent of complex networks, novel graph computations are emerging, which combine both contents and topology information of the graph. Typical examples of these queries include graph pattern matching [38, 40, 41] and mining [47, 85, 149], similarity search [84, 135, 136], anomaly detection [137], influence maximization [81], graph skyline [162] and OLAP [20], ranking and expert finding [130, 131], and graph aggregation [154], among others. In this chapter, we shall discuss graph pattern matching as the representative of emerging online queries, and graph pattern mining as the representative of emerging offline analytics.

As these novel queries integrate both the structure and attribute information of the network, existing algorithms and techniques may not directly apply. Besides, when graphs are complex and large, efficiency and scalability become an issue. Therefore, novel indexing, pruning, sampling, and summarization techniques, as well as distributed algorithms and systems are necessary. Finally, due to lack of fixed schema, missing type information, incomplete knowledge and uncertainty about the structure and contents of the real-life information networks, it might be infeasible to use conventional database techniques to query these graphs. Therefore, we emphasize on user-friendly, and often approximate, query answering and pattern mining techniques.

We carefully limit the scope of this chapter. We consider emerging queries over static, deterministic graphs. Querying of uncertain graphs were discussed in [82], whereas graph stream algorithms were surveyed in [104]. In earlier tutorials [37, 95], Faloutsos et al. discussed large graph mining using matrix based methods. In [52], Han et al. systematically introduced data mining and knowledge discovery algorithms for information networks including graph clustering, ranking, and graph OLAP. Algorithmic developments for managing and mining of graph data are discussed in [4, 86]. There are many tutorials and surveys about recent advances on distributed graph processing systems [6, 83, 103] and graph databases [7]. Different from those literature, in this chapter we present an overview of the emerging graph computations in the context of novel graph pattern matching and and mining techniques, including approximate algorithms, their user-friendliness, and scalability. We intend to give a first impression on their challenges, current solutions, and related future topics.

2 Graph Data Models

The network data models have been studied since 1970s in the context of object oriented databases, semi-structured data, knowledge base, hypertext, and semantic web [7]. Their influence gradually died out with the popularity of other database models, including relational, geographical, spatial, and XML. However, with the recent emergence of complex social and information networks, the need to manage graph data has been re-established. In the following we discuss two popularly used graph data models: Resource Description Framework (RDF) [102] and property graphs [112].

2.1 *RDF*

RDF allows descriptions of Web resources in a machine understandable format. A Web resource can be any object with a Uniform Resource Identifier (URI), e.g., http://www.ietf.org/rfc/rfc2396.txt. RDF databases are collections of *triples*: ⟨subject, predicate, object⟩. Each triple models the binary relation between the subject and object which are RDF terms, e.g., URIs, literals, or blank nodes. This linking structure forms a directed, labeled graph, where an edge depicts the relation between two resources, represented by graph nodes. Figure 1 shows the graph structure for the following RDF triples.

```
Person1 isNamed "John Waters"
Person2 isNamed "Stephen Spielberg"
Person3 isNamed "Darren E. Burrows"
Movie1 hasTitle "Cry-Baby"
Movie1 hasActor Person3
Movie1 hasDirector Person1
Movie2 hasTitle "Amistad"
Movie2 hasActor Person3
Movie2 hasDirector Person2
```

One major advantage of the RDF data model is its semi-structured nature in comparison to the relational model where the entities, their attributes, and relationships to other entities are *strictly* defined. In RDF, the schema may evolve over time, which facilitates data merging even if the underlying schemas differ.

Fig. 1 An example RDF
graph

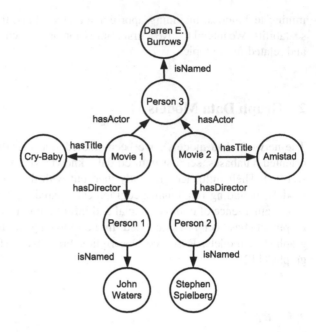

SPARQL [117] is the official W3C standard query language for RDF. A simple
SPARQL query has the form:

```
select ?variable1 ?variable2 ...
where { pattern1. pattern2. ... }
```

Here, every pattern is a triple: ⟨subject, predicate, object⟩, and each of subject,
predicate, and object can be either a variable or an RDF term. The variables specify
the unknowns in the query, which may occur in multiple patterns, thereby requiring
join operations. The query processor finds all possible variable bindings that satisfy
the given patterns. Below we show a SPARQL query over the example RDF graph in
Fig. 1, and it retrieves all actors who appeared in both a "John Waters" movie and a
"Steven Spielberg" movie. As one may observe, in order to write a SPARQL query,
the user needs to know how the entities are connected to each other in the dataset,
e.g., the user must know that the actors and directors are connected via the movie
entities.

```
select ?p1
where {
  ?m1 hasActor ?p1 .
  ?m1 hasDirector ?p2 .
  ?m2 hasActor ?p1 .
  ?m2 hasDirector ?p3 .
```

```
    ?p2 isNamed"John Waters" .
    ?p3 isNamed "Stephen Spielberg" .
  }
```

In addition to the simple SPARQL query illustrated above, operators akin to relational join, union, left outer join, filter, selection, and projection can be combined to construct more expressive queries. SPARQL 1.1 included path queries that extends matching of triple pattern to arbitrary length paths.

One may note that SPARQL is a query language for RDF data. In RDF stores with Relational database back-end (e.g., SW-Store [1]), SPARQL queries are translated into SQL queries, which are optimized and processed by the Relational database management system (RDBMS). On the other hand, graph-based RDF querying techniques (e.g., gStore [115, 163]) convert the SPARQL query into a query graph, and perform exact or approximate subgraph matching to evaluate the query. We shall discuss more about graph querying techniques in Sect. 3.

2.2 Property Graph

A property graph contains connected entities (i.e., nodes), which can hold any number of attributes (also called labels). Nodes can be also tagged with types representing their different roles. Relationships (e.g., edges) provide directed, named connections between two nodes. Therefore, an edge can also have any number of types and properties. We present an example property graph in Fig. 2. Property graphs are

Fig. 2 An example property graph

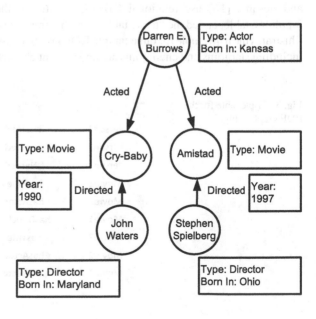

often referred to as entity-relationship graphs, labeled graphs, attributed graphs, information networks, or knowledge graphs.

There are various studies on the theoretical foundation of graph query languages [147], such as conjunctive queries (CQ), e.g., [60], regular path queries (RPQ), e.g., [108], conjunctive regular path queries (CRPQ), e.g., GraphLog [24], Lorel [2], StruQL [42] and UnQL [18], and extended conjunctive regular path queries (ECPRQ), e.g., SPARQLeR [89]. Various declarative (e.g., Cypher [112]) and domain-specific (e.g., Green-Marl [64]) APIs and languages also exist. One may note that these are query languages, and the underlying query answering mechanisms can vary, e.g., SQL and NoSQL-based techniques, or exact and approximate graph pattern matching, which we shall discuss in Sect. 3.

3 Pattern Matching Techniques Over Big-Graphs

3.1 SQL and NoSQL Approaches

Given the success of Relational database management systems (RDBMS) in dealing with wide varieties of data such as structured, spatiotemporal, and XML data, there have been several attempts to store and query graphs using relational approaches [122]. In this section, we shall provide an overview of two relational stores, namely RDF-3X [113] and SW-Store [1], and one NoSQL graph database — Neo4J [112], and how they answer graph pattern matching queries.

RDF-3X. Many RDF data stores, including 3store [57], Oracle [22], Jena [146], and Sesame [16] use relational DBMS, by storing the RDF triples in a giant 3-column table, called the "triple table", i.e., having one row for each triple. Figure 3 illustrates the triple table for our example RDF graph in Fig. 1. Since large join queries including star joins and chain joins are an inherent characteristics of searching RDF,

Fig. 3 Triple table for the RDF graph in Fig. 1

Subject	Predicate	Object
Person1	isNamed	"John Waters"
Person2	isNamed	"Stephen Spielberg"
Person3	isNamed	"Darren E. Burrows"
Movie1	hasTitle	"Cry-Baby"
Movie1	hasActor	Person3
Movie1	hasDirector	Person1
Movie2	hasTitle	"Amistad"
Movie2	hasActor	Person3
Movie2	hasDirector	Person2

triple tables incur too many expensive self-joins. Neumann and Weikum designed the RDF-3X [113] system to overcome this difficulty of triple tables.

The RDF-3X system is workload-independent and eliminates the need for a physical-design tuning by performing "aggressive" indexing over the RDF triples. In particular, to ensure that it can answer every possible pattern triple by performing a single index scan, the system maintains all six permutations (SPO, SOP, OSP, OPS, PSO, POS) of subject (S), predicate (P), and object (O) in six separate indexes. Additionally, indexes over count-aggregated variants for all three two-dimensional and all three one-dimensional projections are created. As RDF-3X compresses the original RDF triples, it can store all the indexes in a space smaller than the size of the original data. The triples in an index are sorted lexicographically, and the query processor uses the full set of indexes on the triple tables to rely mostly on efficient merge joins over these sorted index lists.

The query optimizer focuses on join order and applies dynamic programming to generate the lowest-cost execution plan. In general, the semi-structured nature of RDF data makes the query optimization problem more challenging. RDF-3X considers the two following statistics for query optimization. (1) Histograms for all six permutations of triples: Histograms are generic, and they can handle any kind of triple patterns and joins. However, its disadvantage is that it assumes independence between predicates. (2) Frequent join paths: Frequent star and chain join paths are identified and their exact join statistics are kept. They are more accurate compared to histogram-based statistics, but are also expensive to compute.

One major disadvantage of "aggressive" indexing in RDF-3X is that it is difficult to update the existing RDF triples.

SW-Store. Due to the disadvantage of triple store in large join queries processing, alternative RDF stores were designed. Jena2 [146] and Oracle [22] adopted "property table", each containing a cluster of properties that can be defined together. Figure 4 models the properties of a movie entity for the RDF graph in Fig. 1. One can observe that property tables are good at speeding up queries that can be answered from a single property table, for example, subject-subject self-joins over triple tables reduces to a simple selection operation if they can be answered from the same property table.

Though property tables look similar to relational tables, they suffer from two main disadvantages. First, due to the semi-structured nature of RDF data, not every subject listed in a property table will have all the properties defined. In other words, the property tables can be very sparse with many NULL values, which incur storage

ID	Name	Actor(s)	Director(s)
Movie1	Cry-Baby	Person3	Person1
Movie2	Amistad	Person3	Person2

Fig. 4 Property table for the movie entity in RDF graph of Fig. 1

Acted	
Movie	Actor
Movie1	Person3
Movie2	Person3

Directed	
Movie	Director
Movie1	Person1
Movie2	Person2

Fig. 5 Vertical partition of the RDF graph in Fig. 1

overhead and additional complexities in the query processing. Second, due to the abundance of multi-valued attributes in RDF data, it is difficult to store them in a tabular format, unless one uses lists, sets, or bags of attributes, which also complicate the storage and query processing techniques.

Due of the aforementioned disadvantages of a property table, Abadi et al. introduced SW-Store [1] by vertically partitioning the dataset, that is, partitioning the triple table into m two-column tables, where m is the number of unique properties in the data. The first column is the subject which defines the property, and the second column contains the object values for those subjects. Each RDF triple is inserted into one of these m two-column tables. Figure 5 demonstrates the vertical partitioning of the RDF graph in Fig. 1.

The authors in [1] implemented SW-Store by using a column oriented DBMS (column-store). Since a column-store vertically partitions the attributes of a relational table, it is well-suited to store vertical partitioning of the RDF data, in addition to inheriting the advantages of a column oriented DBMS. In summary, SW-store can easily support multi-valued attributes, and it does not suffer due to NULL values because NULL data does not need to be explicitly stored. It also supports efficient merge joins. However, SW-store usually requires more number of joins than the property table approach; and insertion can be slower, since multiple tables need to be accessed for statements about the same subject.

Neo4J. Neo4J [112] is a NoSQL graph database that models network data in the form of a property graph. Neo4j is disk-based, where edges are stored as first-class objects similar to nodes. Besides, each node uses direct memory links to adjacent nodes rather than requiring joins or key lookups. Hence, the system is efficient for graph traversals. Neo4j also has partial ACID support for graph transactions. The API is in Java, and it supports Java object storage. Cypher is the query language for Neo4J, which is a declarative language for describing patterns in graphs. It allows a user to describe what she wants to select, insert, update, or delete from a graph database without requiring to describe exactly how to do it. Neo4J currently does not support distributed graph processing, that is, it requires the full graph dataset to be stored on a single machine.

The performance of Neo4J is usually better compared to RDBMS for graph traversal queries [142]. However, one must note that the performance of Neo4J depends heavily on the amount of cache available. Neo4J uses two different types of caches. The file buffer cache caches the storage file data in the same format as it is stored on the disk. The object cache caches the nodes, edges, and properties in a format suitable for efficient graph traversals. It was indeed shown that an in-memory graph processing system can outperform Neo4J for traversal queries [145]. Besides, queries that access node and edge properties can be slower in Neo4J compared to that in RDBMS [142]. This is because Neo4J uses Lucene indexing, and Lucene, by default, treats all data as text. Thus, equivalence and inequality comparison over numerical attributes are not very fast since a conversion to text must be done.

Two other popular NoSQL graph databases are InfiniteGraph [71] and Sparksee [31]. Interested readers may find various graph databases in recent surveys [7, 19]. From the usability perspective, however, it is clear that a strict, structured language (e.g., SPARQL, Cypher, and those alike) would be difficult to use for querying graphs. Therefore, the focus now-a-days is to devise more user-friendly query answering techniques for graph data. In this context, we shall discuss keyword search, graph pattern matching, and graph query-by-example.

3.2 Keyword Search

Keyword search is a commonly used information retrieval technique for text data and the World Wide Web. The query consists of a collection of keywords, and it retrieves the documents containing those keywords. Due to its simpler query interface and user-friendly nature, keyword search is also popular over structured and semi-structured data. As opposed to more sophisticated query languages (e.g., XQuery for XML, SQL for Relational data, and SPARQL for RDFs), the user does not require to have a full knowledge of the underlying schema for constructing a keyword query.

However, answering keyword queries over structured and semi-structured data is different from that over text, because for the first two categories of data, one needs to assemble information from various locations that are inter-connected and collectively relevant to the query. In the following, we shall introduce keyword search over XML, Relational data, and graphs.

Keyword Search over XML. Keyword search over XML is simpler than that over graphs and relational data, which is due to the tree-like structure of XML. This property of XML makes the semantics of a keyword query simpler and the query answering technique more efficient. In the literature, there are many semantics on what qualifies as an answer to a keyword search query. Possibly the simplest of them is the "lowest common ancestor" (LCA) which is defined as follows. Given a set of keywords $\{k_1, k_2, \ldots, k_m\}$, assume that L_i be the list of nodes in the XML tree which have the keyword k_i, $1 \leq i \leq m$. Then, an answer to the query is denoted as $LCA(n_1, n_2, \ldots, n_m)$, which is the lowest common ancestor of

Fig. 6 Keyword search
in XML

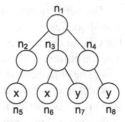

nodes $n_1, n_2, \ldots, n_m$, where each $n_i \in L_i$. As an example, both nodes n_1 and n_3 are
answers to the keyword query $\{x, y\}$ over the XML tree in Fig. 6. This is because
$L_x = \{n_5, n_6\}$, $L_y = \{n_7, n_8\}$, and $n_1 = LCA(n_5, n_7)$, $n_3 = LCA(n_6, n_7)$.

Several other works [67, 148] instead opted for the "Smallest LCA" (SLCA)
semantics, which requires that an answer (i.e., a least common ancestor of nodes
that contain all the keywords) does not have any descendent that is also an answer.
In Fig. 6, node n_1 is not an SLCA for the keyword query $\{x, y\}$, because n_3 is a
descendent of n_1 and n_3 is an SLCA for the same keyword query.

XRank [48] adopted a different semantics called the "Exclusive Lowest Common
Ancestor" (ELCA). A node is said to be an ancestor of a keyword if that node or one
of its descendants has the keyword. According to ELCA, an answer is a node that
is ancestor of at least one occurrence of all the query keywords, after excluding its
descendent nodes that are also ancestors of all the query keywords. For example, in
Fig. 7, node n_3 is ancestor of both the query keywords $\{x, y\}$. While n_3 is a descendent
of n_1, even after removing n_3, the node n_1 is still ancestor of the query keywords x (at
node n_5) and y (at node n_8). Therefore, according to ELCA semantics, both n_1 and
n_5 qualify to be answers. More generally, one can derive the following relationship.

$$SLCA(k_1, k_2, \ldots, k_m) \subseteq ELCA(k_1, k_2, \ldots, k_m) \subseteq LCA(k_1, k_2, \ldots, k_m) \quad (1)$$

Based on the above query semantics, many answers could be found. However,
all answers might not be equally relevant to the user. Therefore, several ranking
methods are also proposed in the literature, e.g., XRank [48] and XSEarch [23].
Below we shall demonstrate the ranking method designed in XRank. Assume that
node v_{t+1} has the keyword k, and there is a simple path from node v_0 to v_{t+1} as
follows: $v_0, v_1, \ldots, v_t, v_{t+1}$. Then, the rank of node v_i, $0 \le i \le t$, with respect to
the occurrence of keyword k at node v_{t+1} is given by:

$$r(v_i, k) = ElemRank(v_t) \times decay^{t-i} \quad (2)$$

Here, $ElemRank(v_t)$ denotes the importance of node v_t in the XML tree, and
is calculated in a way analogous to the PageRank computation. $decay$ is a value
between 0 and 1, and therefore, it scales $ElemRank(v_t)$ based on the distance of v_i
from v_t. In order to compute $r(v_i, k)$, usually the closest occurrence of k from v_i is

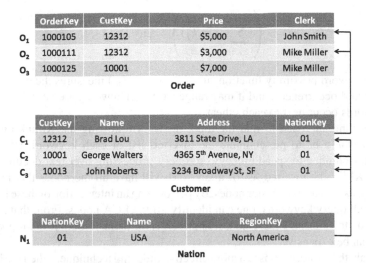

	OrderKey	CustKey	Price	Clerk
O_1	1000105	12312	$5,000	John Smith
O_2	1000111	12312	$3,000	Mike Miller
O_3	1000125	10001	$7,000	Mike Miller

Order

	CustKey	Name	Address	NationKey
C_1	12312	Brad Lou	3811 State Drive, LA	01
C_2	10001	George Walters	4365 5th Avenue, NY	01
C_3	10013	John Roberts	3234 Broadway St, SF	01

Customer

	NationKey	Name	RegionKey
N_1	01	USA	North America

Nation

(a) Lyrics Database Example

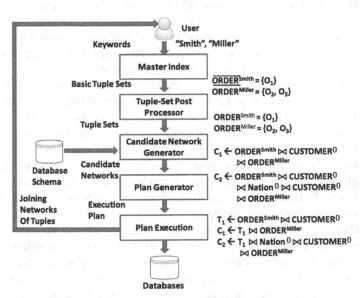

(b) Architecture of DISCOVER

Fig. 7 DISCOVER keyword search over relational databases

considered. Finally, if there are multiple keywords in the query $Q = \{k_1, k_2, \ldots, k_m\}$, the overall ranking of a node v is defined as follows.

$$R(v, Q) = \left(\sum_{1 \leq j \leq m} r(v, k_j) \right) \times p(v, k_1, k_2, \ldots, k_m) \qquad (3)$$

The keyword proximity function $p(v, k_1, k_2, \ldots, k_m)$ measures the proximity of the keyword occurrences, and it may range from 0 (keywords are very far apart) to 1 (keywords occur next to each other).

A baseline approach to find the top-k answers following the XRank objective function would be to pre-compute the inverted list of each keyword, which is a standard technique to speed up keyword query processing over text documents. In case of XML, the inverted list of a keyword will have all the nodes that have the keyword, as well as its ancestor nodes. By performing an intersection of these inverted lists for all query keywords, one can identify all the LCA nodes. From that set, one still needs to find the ELCA nodes, and then rank them to output the top-k ones, which can be expensive.

XRank, therefore, proposed a more effective indexing technique, called the Dewey Inverted List (DIL). The nodes in the XML tree are first numbered in a BFS order, starting at the root as 0. The Dewey ID of a node n is then constructed as the concatenation of the numbers assigned to the nodes on the path from root to n. In this way, one can easily determine the ancestor-descendent relationships between two nodes: Node n_1 is an ancestor of node n_2 if the Dewey ID of n_1 is a prefix of that of node n_2. In XRank, the DIL inverted list of a keyword k consists of only those nodes that directly have that keyword. Therefore, the DIL inverted index structure in XRank occupies less space than that of the baseline inverted list approach discussed earlier. The nodes in a DIL inverted list are ordered based on their *Elem Rank*. Given a keyword query $\{k_1, k_2, \ldots, k_m\}$, an algorithm similar to the Threshold algorithm (TA) [36] is designed to find the top-k ECLA answers. The nodes from the m sorted DIL inverted lists corresponding to m query keywords are accessed in parallel (i.e., round-robin manner). For each such node, its closest ancestor (i.e., ancestor with the longest Dewey ID) is determined which contains all the query keywords. To find this ancestor node efficiently, each inverted list is further indexed with a B+ tree over the Dewey ID of its constituent nodes. Therefore, finding the closest ancestor of a node containing all the query keywords becomes equivalent to a range search query over B+ trees of the other query keywords' inverted lists. The stopping criterion to ensure that the top-k ECLA nodes have been found directly follows from the TA algorithm.

The authors in [48] have empirically shown that the XRank algorithm is quite efficient — for a query containing 5 keywords, it takes less than 2 s to identify the top 10 answers from the DBLP XML dataset (143 MB) using a 2.8 GHz Pentium IV processor with 1 GB of main memory and 80 GB of disk space.

Keyword Search over Relational Data. Keyword search in relational databases follows the join semantics across tables defined by foreign key relationships. Given a keyword query $\{k_1, k_2, \ldots, k_m\}$, it looks for all potential joins and verifies if there is a tuple in the join results that contains all the query keywords. Usually, the number of join operations is upper-bounded by a pre-defined threshold, because the relations that are far apart might not be meaningful, and also to avoid a large number of self-joins.

In this section, we shall highlight two systems, DBXplorer [5] and DISCOVER [65]. Given a set of keywords, DBXplorer identifies the columns in database tables that contain the keywords. This can be computed efficiently by leveraging database indexes on the columns, or by keeping an inverted index over the rows (called the symbol table in [5]). The algorithm then enumerates all possible join trees in order to join the corresponding tables and retrieves the relevant tuples from those join results. One may note that several join trees as constructed above could have some common join structures. DISCOVER, therefore, considers the problem of finding an optimal join execution order that maximizes the reuse of common subexpressions. Since the problem is NP-hard, the DISCOVER system applies greedy techniques.

In particular, the entire pipeline of DISCOVER [65] can be expressed as in Fig. 7b for the example database in Fig. 7a. A database consists of a set of relations $R_1, R_2, \ldots, R_n$. The schema graph G is a directed graph that captures the primary key to foreign key relationships in the database schema. It has a node for each relation of the database, and an edge $R_i \rightarrow R_j$ for each primary key to foreign key relationship from a set of attributes of R_i to a set of attributes of R_j. The undirected version of G is denoted as G_u. A *joining network J of tuples* is a tree of tuples, where for each pair of adjacent tuples $t_i, t_j \in J$, with $t_i \in R_i$, $t_j \in R_j$, there is an edge (R_i, R_j) in G_u and $(t_i \bowtie t_j) \in (R_i \bowtie R_j)$. As an example, $O1 \bowtie C1 \bowtie O2$ is a joining network of tuples from Fig. 7a. A keyword query is a set of keywords $k_1, k_2, \ldots, k_m$. The result of the keyword query is a set of all possible joining networks of tuples that are both *Total* and *Minimal*.

- *Total*: Every keyword is contained in at least one tuple of the joining network.
- *Minimal*: One cannot remove any tuple from the joining network, and still have a total joining network of tuples.

We refer to such joining networks as *Minimal Total Joining Networks of Tuples* (MTJNT). Clearly, the result of a keyword query, which consists of all MTJNTs, is unique. For example, $O1 \bowtie C1 \bowtie O2$ and $O1 \bowtie C1 \bowtie N1 \bowtie C1 \bowtie O2$ are two MTJNTs of the keyword query: "Smith", "Miller".

Given a keyword query $k_1, k_2, \ldots, k_m$, the master index component of DISCOVER (Fig. 7b) outputs a set of *basic tuple sets* $\overline{R_i}^{k_j}$, for $i = 1, 2, \ldots, n$ and $j = 1, 2, \ldots, m$. The basic tuple set $\overline{R_i}^{k_j}$ consists of all tuples of relation R_i that contain the keyword k_j. For example, $\overline{ORDER}^{Miller} = \{O_2, O_3\}$. Next, the tuple set post-processor takes the basic tuple sets as above and produces *tuple sets* R_i^K for all subsets K of $k_1, k_2, \ldots, k_m$, where

$$R_i^K = \bigcup_{k \in K} \overline{R_i}^k - \bigcup_{k \in \{k_1, k_2, ..., k_m\} - K} \overline{R_i}^k \qquad (4)$$

These tuple sets, along with the database schema graph, are passed to the candidate network generator, which outputs the *candidate networks* as follows. A candidate network is a joining network of tuple sets that can produce one or more MTJNTs for the keyword query. For example, $ORDER^{Smith} \bowtie CUSTOMER^{\{\}} \bowtie ORDER^{Miller}$ is a candidate network that produces the MTJNT: $O1 \bowtie C1 \bowtie O2$ for the keyword query: "Smith", "Miller". Finally, the plan generator inputs a set of candidate networks and creates an execution plan to evaluate them. Since these candidate networks often share common subexpressions, an efficient execution plan stores the common join expressions as intermediate results and reuse them in evaluating the candidate networks. However, as stated earlier, the problem of finding an optimal join execution order that maximizes the reuse of common subexpressions is NP-hard. Therefore, the DISCOVER system applies greedy techniques. Based on empirical evidences, the authors have shown that their greedy algorithm creates a near-optimal execution plan to evaluate the set of candidate networks.

One may also note that both DISCOVER and DBXplorer do not provide any ranking of the answer tuples. Later, Hristidis et al. [66] and Liu et al. [98] proposed IR-style ranking methods for keyword search query over relational databases.

Keyword Search over Graphs. Keyword search over graphs can be classified into two broad categories based on the answer semantics: (1) The result is a connected *sub-tree* containing all query keywords, and (2) the result is a connected *sub-graph* having all query keywords. Various ranking criteria also exist to find the top-k answers, e.g., sum of all edge weights in the resulting tree/ graph, sum of all path weights from root to each keyword in the tree, maximum pairwise distance among nodes, etc.

In the first category, notable algorithms are BANKS [3], bidirectional search [76], and BLINKS [61]. BANKS follows a backward graph exploration method starting from the query keywords, which is similar to the XRank algorithm, until the root of the answer sub-tree is found. At any point during the backward exploration, let us denote by E_i the set of nodes that we know they can reach the query keyword k_i. We call E_i the cluster for k_i. The BANKS technique uses the two following strategies for choosing which node to visit next.

- *Equi-distance expansion in each cluster*: This strategy decides which node to visit for expanding a cluster of a keyword. Intuitively, the algorithm expands a cluster by visiting nodes in order of increasing distance from the cluster origins, i.e., the nodes that directly contain the keyword.
- *Distance-balanced expansion across clusters*: This strategy decides the frontier of which keyword will be expanded. Specifically, the algorithm attempts to balance the distance between each cluster's origins to its frontier, across all clusters. Therefore, it selects the keyword cluster that has the lowest distance between its origins and the frontier.

It was later shown in [61] that while the equi-distance expansion in each cluster is a necessary criterion for an optimal backward search algorithm, the second strategy,

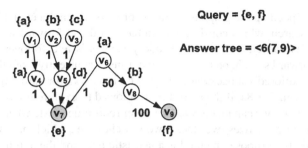

(a) Graph Keyword Search Example

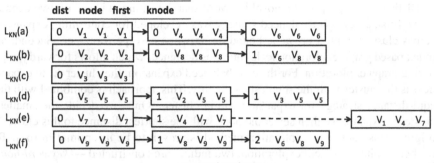

(b) BLINKS index

Fig. 8 BLINKS keyword search over graphs

i.e., distance-balanced expansion across clusters could result in sub-optimal performance. Figure 8a shows such an example. Assume that nodes v_7 and v_9 are the two cluster origins for the keyword query: $\{e, f\}$. There are many nodes that can reach v_7 with shorter paths, but only one edge into v_9 with a large weight (100). With distance-balanced expansion across clusters, BANKS would not expand the second cluster (for keyword f) along this edge until we have visited all nodes within distance 100 to node v_7. It would have been unnecessary to visit many of these nodes had the algorithm chosen to expand the second cluster earlier.

In order to speed up the graph exploration process, Kacholia et al. [76] proposed bidirectional search, which consists of both backward and forward searches. Backward search, as earlier, starts from all query keywords and follows reverse edge directions. Forward search, on the other hand, runs concurrently from the nodes that were already explored via backward search, and it follows forward edge directions. As an example, in Fig. 8a, if the algorithm is allowed to explore forward from node v_6 towards node v_9, we can identify v_6 as answer roots much faster. The bidirectional search algorithm prioritizes over these two searches, i.e., which one will be called next. However, the proposed method is a heuristic one, and the authors do not provide any worst-case performance guarantee. In fact, without additional connectivity information, the forward search could be as ineffective as backward search.

BLINKS [61] proposed a novel keyword search technique based on two central ideas. First, a new, cost-balanced strategy was developed for controlling expansion across clusters, with a provable bound on its worst-case performance. Second, an index based graph exploration technique was designed, which allows forward jumps during graph exploration. For the cost-balanced expansion, the cluster E_i to expand next is the cluster with the smallest cardinality. This approach is combined with the equi-distance strategy for expansion within clusters: Once we decide the smallest cluster to expand, we then select the node with the shortest distance to this cluster's origins. This backward exploration method is proven to be m-optimal in [61]. To further facilitate efficient exploration, two indices are constructed — *keyword-node lists* and *node-keyword map* (Fig. 8b), as defined below.

- *Keyword-node lists*: For each keyword, the shortest distances from every node to that keyword (more precisely, to any node containing the keyword) is precomputed. This results in a collection of keyword-node lists. For a keyword w, $L_{KN}(w)$ denotes the list of nodes that can reach keyword w, and these nodes are ordered by their distances to w. Each entry in the list has four fields $(dist, node, first, knode)$, where $dist$ is the shortest distance between that node and a node containing w; $knode$ is a node containing w for which this shortest distance is realized; $first$ is the first node on the shortest path from node to $knode$.
- *Node-keyword map*: Blinks pre-computes, for each node u, the shortest graph distance from u to every keyword, and organizes this information in a hash table called node-keyword map, denoted as M_{NK}. Given a node u and a keyword w, $M_{NK}(u, w)$ returns the shortest distance from u to w, or ∞ if u cannot reach any node that contains w. The hash entry for (u, w) also contains $first$ and $knode$, which are defined identically as in L_{KN} and used for the same purposes.

The keyword-node list supports the equi-distance expansion order in each cluster. Across clusters, we pick a cluster to expand next in a round-robin manner, which implements cost-balanced expansion among clusters. These two techniques together ensure an m-optimal backward search. In addition, as soon as we visit a node, we look up its distance to the other keywords using the node-keyword map. Using this information we can immediately determine if we have found the root of an answer. The stopping criterion that we found the top-k answer root nodes follows from the Threshold algorithm (TA) [36]. While these two indexes can speed up the online search mechanism, they can be expensive in terms of space requirement and offline computation time. To address this problem, BLINKS first partitions the data graph into multiple subgraphs, or blocks. A bi-level index is designed, which consists of a top-level block index that stores the mapping between keywords and nodes to blocks, and an intra-block index for each block that stores more detailed information within a block. The authors empirically demonstrated that the total size of the bi-level index is a fraction of that of a single-level index discussed earlier.

Under the second category of keyword search over graphs, [77, 93, 96] find answers that are connected subgraphs containing all the query keywords. In particular, Kargar et al. [77] introduced the notion of r-clique, where the nodes in an r-clique are separated by a distance at most r. Any r-clique having all the query keywords can be an answer, and the authors further rank them based on the aggregation of all node-pair distances in an r-clique. However, the problem of finding the minimum-weight r-clique is NP-hard. Therefore, [77] designed approximation algorithms with theoretical performance guarantees.

Keyword Query Reformulation. While the aforementioned approaches directly evaluate a keyword query, the query reformulation technique converts the keyword query into a more structured format, e.g., SPARQL [53] or graph query [155]. Given the set of keywords, the top-k structured queries are identified by considering term similarity, their co-occurrences, and semantic relationships in the input graph.

3.3 Graph Matching Query

In graph matching query, the query is a graph — thereby allowing a user to input the connectivity information among query entities, in addition to specifying keywords on query nodes and edges. Compared to keyword queries, a query graph may provide more information (i.e., both structure and keywords) about the query. Therefore, the later often retrieves better quality top-k answers [79, 87]. On the other hand, query graphs are more complex to formulate than keyword queries, they require more sophisticated query interfaces, and the query answering technique gets more difficult. In fact, the exact subgraph isomorphism problem and many of its variants are NP-complete. Nevertheless, as one may only look for some approximate matches with respect to the query graph, the user does not need to know how exactly various entities are connected in the original data graph. In other words, the query graph constructed by a user can be slightly different than the exact answers that are present in the data graph. This is why graph search is more flexible compared to writing a strict SQL or SPARQL query, and it provides an intermediary querying technique between keyword search and RDBMS.

A vast majority of literature in graph matching considers labeled (or, attributed) graphs, i.e., graphs with labels on nodes, edges, or both. This reduces the size of the search space, because a query node (resp. a query edge) can only be matched with data nodes (resp. data edges) having similar labels. In general, more selective labels and structures in the query graph reduces the number of possible matches. Therefore, with effective indexing, pruning, and query optimization techniques, one can significantly speed up the query evaluation process for labeled graphs [84].

We next provide some formal definitions which are widely used in the literature of graph matching. An undirected labeled graph G is defined as a triple $G = (V, E, L)$ where V is the set of nodes, $E \subseteq V \times V$ is the set of undirected edges, and L is a labeling function that maps a node or an edge to a label (or a set of labels). The following definitions, however, can easily be generalized over directed graphs.

Graph Isomorphism. Given a data graph $G = (V, E, L)$ and a query graph $Q = (V', E', L')$, a graph isomorphism is a *bijective function* $M : V' \to V$ such that (1) $\forall v \in V', L'(v) = L(M(v))$, and (2) $\forall (v_1, v_2) \in E', (M(v_1), M(v_2)) \in E$, and $L'(v_1, v_2) = L(M(v_1), M(v_2))$.

Subgraph Isomorphism. Given a data graph $G = (V, E, L)$ and a query graph $Q = (V', E', L')$, a subgraph isomorphism is an *injective function* $M : V' \to V$ such that (1) $\forall v \in V', L'(v) \subseteq L(M(v))$, and (2) $\forall (v_1, v_2) \in E', (M(v_1), M(v_2)) \in E$, and $L'(v_1, v_2) \subseteq L(M(v_1), M(v_2))$.

Graph isomorphism and subgraph isomorphism are depicted in Fig. 9.

Graph Homomorphism. Given a data graph $G = (V, E, L)$ and a query graph $Q = (V', E', L')$, a graph homomorphism is a *function (not necessarily bijective)* $M : V' \to V$ such that (1) $\forall v \in V', L'(v) = L(M(v))$, and (2) $\forall (v_1, v_2) \in E', (M(v_1), M(v_2)) \in E$, and $L'(v_1, v_2) = L(M(v_1), M(v_2))$.

Graph homomorphism is illustrated in Fig. 10a. In case of graph homomorphism, usually the query graph is larger than the data graph.

Graph Automorphism. An automorphism of a graph $G = (V, E, L)$ is a *permutation* M of the node set V such that (1) $\forall v \in V, L(v) = L(M(v))$, and (2) $\forall (v_1, v_2) \in E, (M(v_1), M(v_2)) \in E$, and $L(v_1, v_2) = L(M(v_1), M(v_2))$. In other words, an automorphism of a graph is a graph isomorphism with itself, that is, a mapping from the nodes of the given graph G back to nodes of G such that the resulting graph is isomorphic with G.

Figure 10b demonstrates a graph automorphism. It can be used to measure self-similarity in the graph.

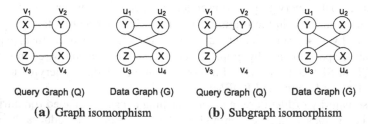

(a) Graph isomorphism   (b) Subgraph isomorphism

Fig. 9 **a** Graph isomorphism: $v_1 \to u_2, v_2 \to u_1, v_3 \to u_3, v_4 \to u_4$. **b** Subgraph isomorphism: $v_1 \to u_2, v_2 \to u_1, v_3 \to u_3$. Labels on graph nodes

Graph Simulation. Graph simulation considers relations instead of functions from one graph to another. A graph $Q = (V', E', L')$ is said to be simulated by graph $G = (V, E, L)$ if there exists a *binary relation* R between the nodes of Q and the nodes of G such that:
(1) for each node v in Q, there exists a node u in G, such that $(v, u) \in R$, and
(2) for each node pair $(v, u) \in R$, (a) $L(v) = L(u)$, and (b) for each edge (v, v_1) in Q, there is an edge (u, u_1) in G such that (v_1, u_1) is also in R, and $L(v, v_1) = L(u, u_1)$.

 If we require R to be a function, then graph simulation becomes graph homomorphism.

Graph Bisimulation. Given two graphs Q and G, a bisimulation is a *binary relation* R between the nodes of Q and the nodes of G such that both R and R^{-1} are graph simulations.

 We depict graph simulation and bisimulation in Fig. 11. Note that both Fig. 11a, b are graph simulations, as Q can be simulated by G in both. However, Fig. 11a is not a bisimulation because G cannot be simulated by Q. For example, if (u_2, v_2) is in the binary relation, then there is an edge (u_2, C, u_4) in G, but there is no corresponding edge from v_2 in Q. Thus, G cannot be simulated by Q in Fig. 11a. In Fig. 11b, since Q can be simulated by G and G can be simulated by Q, it is a bisimulation.

 The decision versions of subgraph isomorphism and graph homomorphism problems are NP-complete [25], while graph isomorphism is not known to be in polynomial or NP-complete. Graph simulation and bisimulation, on the other hand, can

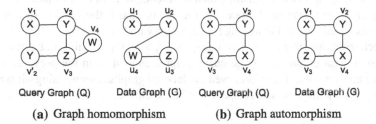

Query Graph (Q)   Data Graph (G)   Query Graph (Q)   Data Graph (G)

(a) Graph homomorphism   (b) Graph automorphism

Fig. 10 **a** Graph homomorphism: $v_1 \to u_1$, $v_2 \to u_2$, $v_2' \to u_2$, $v_3 \to u_3$, $v_4 \to u_4$. **b** Graph automorphism: $v_1 \to v_4$, $v_2 \to v_2$, $v_3 \to v_3$, $v_4 \to v_1$. Labels on graph nodes

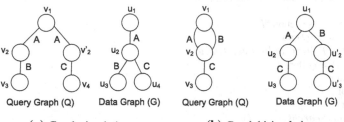

Query Graph (Q)   Data Graph (G)   Query Graph (Q)   Data Graph (G)

(a) Graph simulation   (b) Graph bisimulation

Fig. 11 **a** Graph simulation: (v_1, u_1), (v_2, u_2), (v_2', u_2), (v_3, u_3), (v_4, u_4). **b** Graph bisimulation: (v_1, u_1), (v_2, u_2), (v_2, u_2'), (v_3, u_3), (v, u_3'). Labels on graph edges

be computed within polynomial time [32, 63]. It is important to note that the above definitions can be extended for unlabeled graphs by assuming that all nodes and edges have the same label.

Graph matching queries, as demonstrated in Fig. 12, can be broadly classified into three categories: (1) Subgraph/supergraph containment query, (2) Graph similarity search, and (3) Graph pattern matching. In the first type of query, the input data is a graph database consisting of many graphs — however, each graph in the database is usually small in size — and the results are all those graphs from the database which are subgraph/ supergraph of the query graph [151]. In the second category, the input data also consists of a graph database with several graphs, and the results are those graphs that are graph-isomorphic to the query graph [161]. In the third category, the input is only one large graph, and the query identifies all occurrences of the query graph in that data graph [126]. It is important to note that both exact and approximate versions of these three queries were studied in the literature, e.g., exact subgraph containment (gIndex [151]) versus approximate subgraph containment (Grafil [152]). While the semantics of "exact" queries are rigid and usually follows the definitions in the earlier paragraph, the semantics of "approximate" queries are often application specific, and they differ significantly in the past studies. Since our focus is primarily on large-scale graphs processing, we shall discuss graph pattern matching queries in the following. Interested readers may look at [55, 143] for surveys and experimental comparisons on graph containment and similarity search queries.

Exact Graph Pattern Matching. The exact graph pattern matching requires checking for the subgraph isomorphism, which is NP-complete. In case of labeled graphs, if the query graph has total k nodes $v_1, v_2, \ldots, v_k$, and if the number of candidate data nodes based on label matching for each query node v_i is $|C(v_i)|$, then the overall search space has size $|C(v_1)| \times |C(v_2)| \times \ldots \times |C(v_k)|$. This can be quite large for massive data graphs and large query graphs, as well as in the presence of less selective query nodes. Therefore, even for labeled graphs, enumerating all possible answer graphs within the search space can be expensive.

Algorithm 1 GenericQueryProg: Exact Graph Pattern Matching

Require: query graph Q, data graph G
Ensure: all subgraph isomorphisms of Q in G
 1: $M = \phi$
 2: /* Candidate Selection */
 3: **for all** $u \in V(Q)$ **do**
 4:     $C(u) = FilterCandidates(Q, G, u)$
 5:     **if** $C(u) = \phi$ **then**
 6:         return
 7:     **end if**
 8: **end for**
 9: $SubgraphSearch(Q, G, M)$

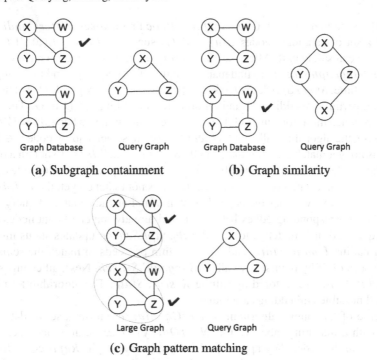

(a) Subgraph containment   **(b) Graph similarity**

(c) Graph pattern matching

Fig. 12 Graph matching queries: Labels on graph nodes

Algorithm 2 Subroutine *SubGraphSearch(Q,G,M)*

1: **if** $|M| = |V(Q)|$ **then**
2:    output M
3: **else**
4:    $u = NextQueryVertex()$
5:    $C_R = RefineCandidates(M, u, C(u))$
6:    **for all** $v \in C_R$ such that v is not yet matched **do**
7:      **if** $IsJoinable(Q, G, M, u, v)$ **then**
8:        $UpdateState(M, u, v)$
9:        $SubgraphSearch(Q, G, M)$
10:       $RestoreState(M, u, v)$
11:      **end if**
12:    **end for**
13: **end if**

Ullmann proposed the first practical algorithm for subgraph isomorphism search in graphs [140]. It is a backtracking algorithm which finds exact matches by incrementing partial solutions or abandoning them when it determines that they cannot be completed. Algorithm 1 shows a generic subgraph isomorphism procedure, *GenericQueryProc* [94]. It inputs a query graph Q and a data graph G, and it output all subgraph isomorphic mappings (or embeddings) of Q in G. To represent an embedding, we use a list M of pairs of a query node and a corresponding data node. For each node u in Q, $GenericQueryProc$ first invokes $FilterCandidates$ to find a set of candidate nodes $C(u) \subseteq V(G)$ such that $L(u) \subseteq L(v)$. If $C(u)$ is empty, we can safely exit. After that, $GenericQueryProc$ invokes a recursive subroutine, $SubgraphSearch$, to find mapping pairs of a query node and matching data nodes at a time. $SubgraphSearch$ takes as parameters a query graph Q, a data graph G, and a partial embedding M and reports all embeddings of Q in G. The recursion stops when the algorithm finds the complete solution (i.e., when $|M| = |V(Q)|$). Otherwise, the algorithm calls $NextQueryVertex$ to select a query node $u \in V(Q)$ which is not yet matched. After that, it calls $Refine\ Candidates$ to obtain a refined candidate node set C_R from $C(u)$ by using algorithm-specific pruning rules. Next, for each candidate data node $v \in C_R$ such that v is not matched yet, the $IsJoinable$ subroutine checks whether the edges between u and already matched query nodes of Q have corresponding edges between v and already matched data nodes of G. If v is qualified, it is matched to u, and $SubgraphSearch$ updates status information by calling $UpdateState$, and the algorithm proceeds to match the remaining query nodes of Q by recursively calling $SubgraphSearch$. Next, all changes done by $UpdateState$ are restored by calling $RestoreState$. The algorithm terminates when all possible embeddings are found.

In case of Ullmann's algorithm, $FilterCandidate$ returns a set of data graph nodes with a matching label with u. $NextQueryVertex$ returns one query node at a time, in the order they appear in the input query graph. $RefineCandidates$ prunes all candidate nodes $v \in C(u)$ that have a smaller degree than u. $IsJoinable$ iterates through all adjacent query nodes of u. If an adjacent query node u' is already matched, i.e., $(u', v') \in M$, then it checks whether there is a corresponding edge (v, v') in the data graph. Finally, $UpdateState$ appends a pair (u, v) to M, while $RestoreState$ restores M by removing the pair (u, v) from M.

Recently, many algorithms such as VF2 [26], QuickSI [126], GraphQL [60], GADDI [156], and SPath [159] are proposed for subgraph isomorphism search in large graphs. These algorithms follow the same underlying backtracking principle of Algorithm 1; however, they improve the efficiency by exploiting different join orders and with smart pruning techniques (e.g., with more advanced methods for $NextQueryVertex$ and $RefineCandidates$). Lee et al. empirically compared these techniques [94] and found that QuickSI is often the most efficient method for both small and large graphs.

Approximate Graph Pattern Matching. In bioinformatics, approximate graph matching has been extensively studied, e.g., PathBlast [80], SAGA [135], NetAlign [97], and IsoRank [128]. However, these algorithms target smaller biological networks. It is difficult to apply them in large social and information networks.

There have been significant studies on inexact subgraph matching in large graphs. Tong et al. [138] proposed G-Ray, which aims to maintain the shape of the query. In contrast, NESS [84] and NeMa [87] identify the optimal matches in terms of proximity among entities rather than the shape of the query graph. Tian et al. [136] proposed an approximate subgraph matching tool, called TALE, with efficient indexing. Mongiovi et al. introduced a set-cover-based inexact subgraph matching technique, called SIGMA [109]. Both these techniques use edge misses to measure the quality of a match. There are other works on inexact subgraph matching. An incomplete list (see [44] for surveys) includes homomorphism based subgraph matching [39], belief propagation based net alignment [10], edge-edit-distance based subgraph indexing technique [158], subgraph matching in billion node graphs [132], regular expression based graph pattern matching [9], schema [107] and unbalanced ontology matching [160]. There are also works on simulation and bisimulation-based graph pattern matching, e.g., [38, 100].

3.4   Graph Query by Example

Graph data is not easier than relational data in either query language or data model. This largely has to do with the sheer volume and complexity of such data. As of March 2012, the Linking Open Data community had interlinked over 52 billion RDF triples spanning over several hundred datasets. Freebase alone has over 22 million entities and 350 million relationships in about 100 domains. Before users and developers can do anything meaningful with the data, they are often overwhelmed by the daunting task of attempting to even digest and understand it. While a number of graph database systems and RDF stores have emerged in recent years, usability has not been the focus of innovation. In retrieving data from these databases, the norm is often to use structured query languages such as SQL, SPARQL, and those alike. In the literature on graph querying, the starting point is virtually always a query graph, which is a graphical representation of structured query. However, writing structured queries is hard. It requires extensive experiences in query language and data model and good understanding of particular datasets.

For this very reason, query by example systems over graph data has received a considerable attention lately. Query by example (QBE) has a positive history in relational databases [90], HTML tables [49], and entity sets [144]. Exemplar query [110] and GQBE [73] were the first to adapt similar ideas over knowledge graphs. To illustrate this point, assume that a journalist is interested in preparing an article on university professors who have designed a programming language and also won an award in Computer Science. She may find it difficult how to write this query, but she might know a few relevant professor-university-award triples, such as ⟨Donald Knuth, Stanford University, Turing Award⟩. A query by example system will allow her to input the above triple as a query, and will return other similar tuples that are present in the database, such as ⟨John McCarthy, Stanford University, Turing Award⟩ and ⟨Alan Perlis, Yale, Turing Award⟩.

In particular, GQBE follows a two-step approach. Given the input example tuple(s), GQBE first identifies the query graph that captures the user's query intent. Then, it evaluates the query graph to find other relevant answer tuples. The GQBE framework is shown in Fig. 13. Let us consider the knowledge graph in the figure and let ⟨$JerryYang, Yahoo!$⟩ be the input query tuple. Neighbors of the query entities up to length d are captured to form a neighborhood graph

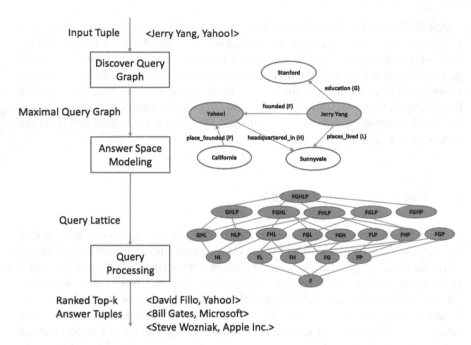

Fig. 13 The GQBE framework

H_t, which can be large. Hence, its unimportant edges are pruned. We rank the edges by weighting them using several distance-based and frequency-based heuristics. The weight $w(e)$ of an edge $e = (u, v)$ is (1) directly proportional to its inverse edge frequency that captures how rare a relationship is globally in the data graph, (2) inversely proportional to its participation that determines the number of edges in the data graph that share the same label and one of e's end nodes (u or v), and (3) inversely proportional to the distance that captures the distance of edge e from the query entities. A greedy heuristic is used to choose the maximal query graph MQG_t, which is an m-edged weakly connected subgraph of H_t containing all query entities, while maximizing the aggregated edge weight.

We model the space of possible query graphs by a query lattice. Each query graph in the lattice is a connected subgraph of MQG_t and contains all query entities. The bottom-most nodes in the lattice are called the minimal query trees (nodes F and HL in Fig. 13), which together capture all relationships between the input entities. The top-most node ($FGHLP$ in Fig. 13) is the MQG_t, and other lattice nodes have exactly one edge more than its children. Answer graphs to these query graphs are also subgraphs of the data graph and are structurally isomorphic to the query graph. The score of a query graph Q is equal to the sum of all its edge weights. Given an answer graph, nodes corresponding to the query tuple entities are projected as its answer tuple. Thus the answer tuples are approximate answers to MQG_t.

The data graph is stored in a relational database via vertical partitioning [1], that is, we maintain a table for each property with two columns ($subj$, obj), for the edges' source and destination nodes, respectively. For efficient query processing, two in-memory hash tables are created on each table, using $subj$ and obj as the hash keys. A query graph can be evaluated using a multi-way join query. We use the right-deep hash-joins to process such a query. Let us consider the topmost join operator in a join tree for query graph Q. Its left operand is the build relation which is one of the two in-memory hash tables for an edge e. Its right operand is the probe relation which is a hash table for another edge or a join subtree for $Q' = Q - e$ (i.e., the resulting graph of removing e from Q). GQBE uses a best-first exploration strategy of the lattice to obtain the top-k answers. It explores the query lattice in a bottom up way, starting with the minimal query trees. After a query graph is processed, its answers are materialized in files. To process a query Q, at least one of its children $Q' = Q - e$ must have been processed. The materialized results for Q' form the probe relation and a hash table on e is the build relation. The best-first strategy always chooses to evaluate the most promising lattice node Q_{best} from a set of candidate nodes. Q_{best} is the candidate with the highest upper-bound score. If processing Q_{best} does not yield any answer graph, Q_{best} and all its ancestors are pruned and the upper-bound scores of other candidate nodes are recalculated. The algorithm terminates when it has obtained at least k answer tuples with scores higher than the highest possible upper-bound score among all unevaluated nodes. The quality of the answer tuples obtained by GQBE can be further improved by using multiple input tuples as query to the system.

4   Mining Techniques Over Big-Graphs

The field of graph mining has emerged as one of the hottest topics in recent years due to its wide applications in a variety of fields including social networks, computational biology, software bug localization and transportation networks. Graph mining covers a variety of topics such as clustering and classification of graphs, anomaly detection, link analysis, mining subgraph patterns, graph anonymization and privacy preservation, etc. In this chapter, we discuss the problem of mining subgraph patterns.

Mining subgraph patterns has received considerable interest in the graph mining community due to its wide-ranging applications. For example, frequent subgraph mining has been widely used for drug discovery [91, 114, 149]. Specifically, given a dataset of molecules that are active against a particular disease, chemists are often interested in identifying the molecular substructures that are frequent in this set. This same line of reasoning evolved into discriminative [133] and statistically significant subgraph mining [59, 74, 120, 153], where the goal is to mine subgraphs that are "over-represented" in the active dataset. Both discriminative subgraph mining and significant subgraph mining have shown good performance in molecular activity prediction [74, 133]. Beyond drug-discovery, subgraph mining has also been used for bug localization [21] and predicting disease susceptibility from gene expression data in protein-protein interaction networks [33, 119].

A subgraph pattern mining problem generally has three pieces of input information. A graph dataset $\mathbb{D}$, a significance function $\phi(g)$ that quantifies the significance of a subgraph g, and a threshold θ. Given these inputs, the goal is to identify all subgraphs that have a significance value greater than the threshold. Mathematically, we need to compute the set of subgraphs $\mathbb{A}$

$$\mathbb{A} = \{g \text{ is a subgraph from dataset } \mathbb{D} | \phi(g) \geq \theta\} \tag{5}$$

The significance function could model a variety of application specific properties such as frequency of a subgraph, statistical significance, representative power, etc.

The main challenge in subgraph mining is to explore the exponential subgraph search space in an efficient manner. A graph with n nodes could potentially contain 2^n subgraphs and evaluating each possible subgraph is not scalable. Hence, the primary goal of all existing techniques is to develop strategies that are effective in pruning the search space. These strategies however depend on the significance function being used and the type of graph dataset being mined.

Generally, there are two kinds of graph datasets. In the first kind, we have a database of objects, where each object is a graph. This setup is common while mining molecular repositories [120, 149] and is popularly known as the *transactional graph database*. The second type of scenario is where we have a *single large graph*. Such graphs are routinely used to model protein-protein interaction networks, social networks, and road networks [33, 34, 119]. Generally, the size of the graphs are small in the transactional setting. However, the database may contain millions of graphs. On the other hand, in the single large graph scenario, although there is only one graph, the graph may span millions of nodes. Typically, a graph mining algorithm targets only one of these dataset types since each dataset type brings in their own unique challenges.

4.1 Frequent Subgraph Mining

Frequent subgraph mining has several applications. One of the most prominent applications of frequent subgraph mining is to characterize graphs as feature vectors. In this conversion, each frequent subgraph corresponds to a dimension and the dimension value represents whether the subgraph occurs in the graph that is being converted to a feature vector. In other words, a feature vector representation of a graph G encodes the presence or absence of the frequent subgraph in G. Owing to this conversion to a feature space, higher order operations such as classification, clustering, and querying of graphs can be performed using traditional tools built for feature vectors. For example, to perform graph search efficiently, Yan et al. [151] used frequent subgraph as indexing features. Deshpande et al. [30] classified molecules by considering frequent patterns as features.

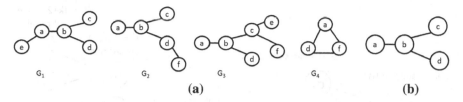

Fig. 14 **a** A sample graph database. **b** A subgraph with a frequency of 75%

Beyond conversion to feature vectors, frequent subgraphs can also explain why a graph demonstrates certain properties. For example, Borgelt et al. [13] studied HIV-screening molecular dataset and found active chemical structures in it by contrasting the support of frequent subgraphs between various classes. Huan et al. [70] studied protein structure families by applying frequent graph mining techniques.

Frequent subgraph mining has been studied both in transactional graph databases as well as in the single large graph setting. We first discuss the problem in transactional databases.

To formulate the problem, we first introduce some definitions. Let $\mathbb{D} = \{G_1, \cdots , G_n\}$ be a database of n graphs. The *support set*, D_g, of a subgraph g is the number of graphs in the database $\mathbb{D}$ to which g is subgraph isomorphic. Mathematically,

$$D_g = \{g \subseteq G | \, G \in \mathbb{D}\} \tag{6}$$

where $g \subseteq G$ denotes that g is subgraph isomorphic to G. The frequency, $freq(g)$, of a subgraph g is the proportion of database graphs that are included in $g's$ support set. Specifically,

$$freq(g) = \frac{|D_g|}{|\mathbb{D}|} \tag{7}$$

Given a threshold θ, a subgraph g is *frequent* if $freq(g) \geq \theta$. Therefore, the problem of mining frequent subgraphs is to identify all subgraphs that are frequent.

Example. Fig. 14a shows a graph database. The frequency of the subgraph shown in Fig. 14b is 75% since it is present in three database graphs.

The baseline algorithm to mine frequent subgraphs is to enumerate all unique subgraphs of the database, compute their support sets, and output those that are frequent. This approach however does not scale due to the sheer number of subgraphs that exist in the search space. A crucial property of frequent subgraphs that is employed to prune the search space is the apriori property of frequency.

Apriori property. The apriori property of frequent subgraphs means that a subgraph g is frequent only if all subgraphs of g are also frequent. Conversely, if g is not frequent, none of g's supergraphs are frequent either.

Many frequent subgraph mining techniques have been proposed [13, 68, 72, 91, 114, 141, 149]. These techniques can broadly be grouped into two categories based on the searching strategy they follow: *join-based approach* [72, 91, 141] and *pattern-growth approach* [13, 68, 114, 149].

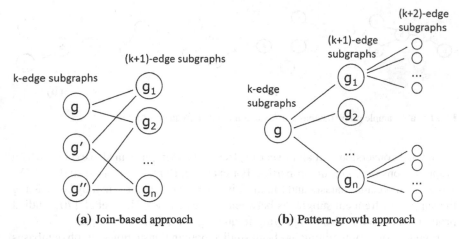

(a) Join-based approach   (b) Pattern-growth approach

Fig. 15 Overview of the two search exploration strategies for frequent subgraph mining

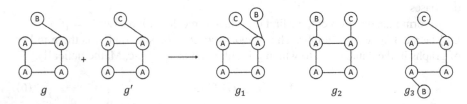

Fig. 16 Joining two k-edge subgraphs to form $k+1$-edge subgraph candidates in FSG

Join-based approach. We illustrate this approach using FSG [91] as the representative technique. As outlined in Fig. 15a, FSG explores the search space in a bottom-up, iterative manner. FSG starts by enumerating all k-edge frequent subgraphs for some small k. These k-edge frequent subgraphs are next *joined* to construct $k+1$-edge subgraph candidates. These candidates are then verified to check if they are indeed frequent. After this verification, FSG proceeds to the next iteration where $k+1$-edge frequent subgraphs are joined to form $k+2$-edge candidates. This iteration stops when no further candidates remain to be evaluated.

An important step in this approach is to join two k-edge subgraphs. Two k-edge subgraphs can be joined if and only if they contain a common $k-1$ edge subgraph. An example of this operation is shown in Fig. 16. Note that the join of two k-edge subgraphs may create more than one $k+1$-edge subgraph candidates. The large number of candidates generated in the join-based approach is in fact a major computational overhead. To overcome this weakness, the pattern-growth approach was developed.

Pattern-growth approach. Figure 15b outlines the search exploration strategy of the pattern-growth approach. We explain this approach by using gSpan [149] as the representative technique. As in FSG, gSpan initiates search by first enumerating all k-edge frequent subgraphs for some small value of k. Next, it picks a k-edge

Fig. 17 **a** The growth-rate of the running times of FSG [91], gSpan [149], and Gaston [114] against the frequency threshold

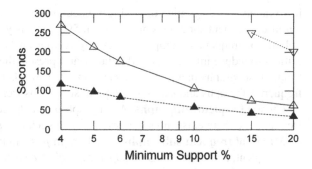

frequent subgraph and extends it by one more edge in every possible position to create $k + 1$-edge frequent subgraphs. Each of these $k + 1$-edge frequent subgraphs are then recursively extended in the same manner till it cannot be extended any further to construct a larger frequent subgraph. In other words, the search space is explored in a depth-first manner. The pattern-growth approach does not perform expensive subgraph joins.

One potential problem in the pattern-growth approach is that a subgraph may be discovered multiple times. To avoid this redundant operation, gSpan develops the idea of *right-most extension*. In right-most extension, edge extensions take place only on the *right-most path*. A right-most path for a given graph is the shortest path from the starting vertex v_0 to the last vertex v_n, according to a depth-first search on the graph.

Generally, the pattern-growth approach has been shown to perform better than the join-based approach. Among the existing techniques to mine frequent subgraphs, Gaston [114] produces the best performance. Gaston also follows the pattern-growth approach like gSpan. However, it also employs some additional heuristics to prune the search space. Figure 17 shows a comparison of running times in the AIDS antiviral molecular dataset, which contains 42,689 graphs. On average, the graphs contain 26 nodes and 30 edges. As can be seen in, Gaston is almost twice as fast as gSpan. Gaston has been shown to perform well on graph datasets containing more than 250,000 graphs. It is safe to assume that Gaston and gSpan will also scale to million sized datasets since the running time grows linearly with dataset size. The primary weakness of these techniques however lie in their scalability with graph size. Typically, the frequent subgraph mining techniques developed for the transactional setting fail to scale on graphs containing more than 100 nodes. This weakness is addressed by the techniques that are built to mine frequent subgraphs from a single large graph [34].

Closed and Maximal frequent subgraph mining. Due to the apriori property, all subgraphs of a frequent subgraph g are also frequent. In the worst case, a frequent graph of size n contains 2^n frequent subgraphs. Consequently, the number of frequent subgraphs in a database grows exponentially as the frequency threshold decreases. This ultimately results in mining a huge number of frequent subgraphs. For example,

in a database of 422 graphs containing molecules active against the HIV virus, close to
a million frequent subgraphs are mined at 5% frequency threshold [150]. This large
volume of frequent subgraphs makes any further analysis difficult. It is therefore
critical to reduce information redundancy and present the information embedded in
frequent subgraphs in a more concise manner. Towards that goal the idea of *closed*
frequent subgraphs and *maximal* frequent subgraphs were proposed.

Closed frequent subgraphs. A subgraph g is a closed frequent subgraph if g is
frequent and there exists no supergraph $g' \supseteq g$ such that g' is as frequent as g.

Maximal frequent subgraphs. A subgraph g is a maximal frequent subgraph if
g is frequent, and there exists no supergraph $g' \supseteq g$ such that g' is frequent.

The set of closed frequent subgraphs is a subset of frequent subgraphs and the
set of maximal subgraphs is a subset of closed subgraphs. Note that given the set
of closed frequent subgraphs, one can generate the set of frequent subgraphs. Such
an operation is not possible from maximal subgraphs. Thus, no information is lost
when one retains only the closed patterns. CloseGraph [150] mines closed frequent
subgraphs, and Spin [69] and Margin [134] mine maximal frequent subgraphs.

Mining frequent subgraphs in a single-large graph. The problem of mining frequent subgraphs from a single large graph has also been studied [15, 34, 43, 92].
The major challenge in mining frequent subgraphs from a single-large graph is the
violation of apriori property of subgraph frequency. Figure 18a illustrates the issue.
As visible, g_1 is a subgraph of g_2. Despite this relationship between them, g_1 has one
embedding within the graph compared to three embeddings of g_2. In other words, the
apriori property is violated. The apriori property is violated due to overlap among
instances of a subgraph pattern. For example, although g_2 occurs thrice in the graph,
the same ABC component overlaps in all of its instances.

Without the apriori property, pruning the subgraph search space is hard. Hence,
multiple definitions of subgraph frequency were proposed. Kuramochi et al. proposed
a definition of subgraph frequency based on the idea of *maximum independent sets*
(MIS). Specifically, for any subgraph g, let $I_1, \cdots, I_n$ be all instances of g. A network
N is created using these instances where each instance is a node and two instances I_i
and I_j are connected by an edge if the instances overlap. The frequency of g is defined
as maximum independent set of N. It has been shown that the MIS-based frequency
follows the apriori property [43]. Kuramochi et al. proposed two algorithms to mine

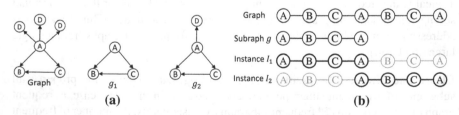

Fig. 18 **a** Illustration of how the apriori property is violated while mining frequent subgraphs in a
single-large graph. **b** Counting the frequency of a subgraph

frequent subgraphs based on this definition [92], namely HSIGRAM and VSIGRAM. While HSIGRAM explores the search space in breadth-first manner, VSIGRAM follows a depth-first exploration.

Subsequently, Fiedler et al. [43] proposed a definition of frequency called *Harmful Overlap (HO)*. In HO, a network is constructed among the instances of a subgraph just like in [92]. However, the definition of overlap is different. Specifically, two instances I_1, I_2 of a graph g overlap if there exists at least one vertex in g that is mapped to the same node in both I_1 and I_2. Figure 18b illustrates the idea. Instances I_1 and I_2 overlap based on the definition proposed in [92]. However, they do not based on HO principle and hence g has a frequency of 2 according to HO and 1 based on the definition in [92]. HO follows the apriori property as well and is closer to the true frequency than [92].

Both definitions of frequency discussed above are NP-complete to compute. To address this weakness, Bringmann et al. defined a simpler version of frequency called *Minimum Image based frequency (MNI)* [15]. In MNI, first a frequency is defined for each vertex of a subgraph g within a graph G. The frequency of a vertex $v \in g$ is the least number of unique vertices in G to which v is mapped in some instance of g. The frequency of g is the minimum of all vertex frequencies in g. Revisiting Fig. 18b, g has a frequency of 2 based on MNI. MNI is not only more computationally efficient, but also retains the apriori property.

GRAMI. The state-of-the-art technique for mining frequent subgraphs in a single-large graph is GRAMI [34]. GRAMI uses MNI as subgraph frequency. GRAMI models the problem of frequency evaluation of a subgraph as a *constraint satisfaction problem (CSP)*. A CSP is represented as a tuple $(\mathcal{X}, \mathcal{D}, \mathcal{C})$, where $\mathcal{X}$ is an ordered set of variables, $\mathcal{D}$ is a set of domains corresponding to variables $\mathcal{X}$, and $\mathcal{C}$ is a set of constraints among the variables in $\mathcal{X}$. A solution for a CSP is an assignment of values to the variables in $\mathcal{X}$ from their corresponding domains such that none of the constraints in $\mathcal{C}$ is violated. GRAMI maps the subgraph isomorphism problem to a CSP as follows. Each node in a subgraph S corresponds to a variable, the domain of each variable is the set of nodes in the target graph G on which S is being searched, and the constraints are the following:

1. $x_v \neq x'_v$, for all distinct variables $x_v, x'_v \in \mathcal{X}$
2. the labels of v and x_v are same
3. the labels of all edges (x_v, x'_v) in G are same as edges (v, v') in S

GRAMI explores the search space in the depth-first manner that is used by gSpan. Specifically, it first computes all frequent edges and then tries to extend them by one more edge. The only difference from traditional depth-first approaches is that the subgraph enumeration is performed using CSP, which allows higher efficiency. GRAMI first checks the number of valid assignments in the domain of each node of the extended subgraph. Only if the number of valid assignments for all nodes (or variables in the CSP) is larger than the support threshold, the extended subgraph can be frequent.

GRAMI employs two heuristics to further speed-up the frequency computation of a subgraph: *push-down pruning* and *lazy search*. In push-down pruning, GRAMI exploits the property that any assignment that was invalid for a subgraph S will remain invalid for all of its children (supergraphs of S) as well in the depth-first search tree. The intuition behind lazy search is that if the search for an assignment of variable values to a CSP takes a long time, then most likely a solution does not exist for the partial assignments that have already been made in this search. It might be better to abandon this partial assignment and look for an alternative solution. To incorporate this intuition, GRAMI imposes a time limit on the search for any assignment to a CSP. If the search times out, a different assignment of variable values is initiated to solve the CSP. The searches that are timed out are kept in memory and re-visited only if assignments that did not time out have not yet provided θ solutions to the CSP, where θ is the frequency threshold.

A notable feature of GRAMI is that corresponding to a frequent subgraph S, it only stores a template of S based on the solutions to the CSP of S, and not all instances of S itself. This allows GRAMI to scale to low frequency thresholds and large graphs. GRAMI has been shown to perform well on a snapshot of the Twitter network, which contains ≈ 11 million nodes and ≈ 85 million edges. In this network, each node represents a user, and an edge represents an interaction between users. Compared to the basic depth-first approach, GRAMI is shown to be up to 2 orders of magnitude faster on the Twitter network. Quite naturally, the running time grows exponentially as the frequency threshold is lowered. However, even at a low frequency threshold of 3000, GRAMI takes less than two and half hours to complete. Recently, the problem was also studied in dynamic networks [50], where the idea of *temporal subgraph isomorphism* was proposed.

4.2   Mining Discriminative Subgraphs

Frequent subgraphs are often used as features for graph classification. The problem has been studied in both transactional setting as well as in a single graph. We first discuss the problem in transactional databases.

In graph classification, each graph G in the dataset is also tagged with a class label. To classify graphs, frequent subgraphs have been used to convert each graph into a feature vector. Specifically, each frequent subgraph g_i corresponds to dimension i, and G is represented as an n-dimensional binary feature vector where n is the total number of frequent subgraphs used as features. The value in dimension i of G is set as follows.

$$G^i = \begin{cases} 1, & \text{if } g_i \subseteq G \\ 0, & \text{otherwise} \end{cases} \tag{8}$$

Owing to this transformation of graphs into feature vectors, any standard classifier such as decision trees, support vector machines, etc., can be used. However, using frequent subgraphs as features has several issues. First, since the number of frequent subgraphs is often huge, converting graphs to feature vectors is computationally expensive. Second, when the number of dimensions is larger than the number of training samples, there is a chance of over-fitting. It is therefore important to reduce the number of dimensions without losing too much information. While closed or maximal frequent subgraphs can be used, they do not reduce the number of features enough to address the issues outlined above. Furthermore, a high frequency does not necessarily indicate a high discriminative power. To overcome these weaknesses, the problem of mining *discriminative subgraphs* was proposed [133].

In [133], the authors develop a technique called CORK to mine subgraphs that can discriminate between two classes of graphs. Specifically, the graph dataset $\mathbb{D} = D_+ \cup D_-$ contains a *positive* set of graphs and a *negative* set of graphs. The goal is to select subgraphs such that they are present in graphs predominantly from either the positive set or the negative set. Towards that goal, CORK defines the idea of *correspondence*. Two graphs $G_1 \in D_+$, $G_2 \in D_-$ form a correspondence if for all dimensions i, $G_1^i = G_2^i$. Clearly, higher the correspondence, lower is the quality of the features. Hence, CORK first mines all frequent subgraphs using a low frequency threshold. It then chooses a subset of these frequent subgraphs through a greedy, iterative approach, where in every iteration, the chosen subgraph provides maximum reduction in correspondence.

Essentially, a subgraph that reduces correspondence by a high margin is discriminative in nature. Subsequent works LEAP [153] and GraphSig [120] highlighted that for a subgraph to be discriminative, it does not need to be frequent. This observation resulted in the problem of mining statistically significant subgraphs [59, 74, 120, 153]. However, before discussing this problem, we overview mining discriminative subgraphs from single-large graphs.

In a single-large graph, mining discriminative subgraphs has been studied for a class of networks called *global-state networks* [28, 29, 119]. A global-state network has multiple *snapshots*. While the structure of each snapshot is identical, the node labels are snapshot specific. Furthermore, each snapshot is associated with a global class label. The goal is to predict the class label of a snapshot using the node labels and the structure of the graph. In addition, it is desirable to predict using minimal amount of information. Thus, instead of using the entire structure and all node labels to predict, if a subgraph can accomplish the same task, then it is a better solution. Towards that goal, given an accuracy threshold θ, a subgraph g is defined as *discriminative* if a classification model can be learned from the just the structure and node labels of g to predict the class label of any given snapshot with an accuracy of at least θ. A subgraph g is *minimally discriminative*, if g is discriminative and there does not exist any subgraph of g that is as discriminative.

The problem was formulated by Ranu et al. in [119] and an algorithm called *MINDS* was developed. MINDS employs a sampling based framework to identify minimally discriminative subgraphs. Subsequently, the solution was further improved by Dang et al. [28, 29].

4.3 Mining Statistically Significant Subgraphs

The applications of mining significant patterns is same as mining discriminative subgraphs; the only difference being it decouples frequency from the discriminative power (or statistical significance) of a subgraph. Instead of refining frequent subgraphs to identify discriminative subgraphs, this group of techniques mine statistically significant subgraphs directly [59, 74, 120, 153]. Thus, the invalid assumption that for a subgraph to be discriminative, it must be frequent as well is removed. At the same time, since statistical significance of a subgraph does not follow the apriori property, pruning the search space becomes a more difficult problem. In the following discussion, we will introduce some of the seminal works in the space of significant subgraph mining. Unlike frequent subgraph mining, these techniques cannot be grouped into two broad classes. However, they have the common goal of avoiding the enumeration of the complete set of frequent subgraphs while presenting only a compact set of significant subgraph patterns. Consequently, both scalability and higher efficacy in discriminative power are achieved.

In transactional graph databases, statistically significant subgraph mining has been studied under two scenarios. Figure 19a depicts the two cases. In the first scenario, the graphs are tagged with a class label. Here, the goal is to mine the subgraphs that are *significantly more* frequent in the positive graphs than in the negative graphs and vice versa. Since a high difference in frequencies indicates bias towards a particular class, these subgraphs are discriminative in nature and can be used for graph classification. In the second scenario, class labels are absent. Here, the goal is to identify subgraphs that are significantly more frequent than their *expected frequency*. The expected frequency is computed based on some *background model*.

The first step in mining statistically significant subgraphs is to identify the function that quantifies the significance of a subgraph. Any of the known statistical measures such as p-value [120], G-test score [153], and chi-square value [8] could be used. Since statistical significance does not follow the apriori property, we need to enumerate all possible subgraphs of the database, compute their significance, and return those that exceed the threshold. Clearly, this two-step baseline approach is not

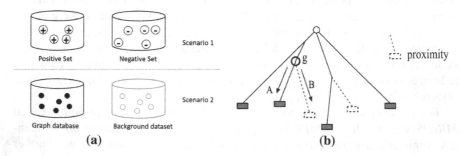

Fig. 19 a The two problem settings for statistically significant subgraph mining. **b** Illustration of structural leap search

scalable. The bottleneck lies in the subgraph enumeration step where we generate an exponential search space. Thus, it is critical to mine significant subgraphs without enumerating all possible subgraphs. We discuss two of the techniques here, namely LEAP [153] and GraphSig [120], that allows us to perform this task in a scalable manner.

LEAP. LEAP is one of the first techniques to directly mine statistically significant subgraphs [153] without enumerating all subgraphs. Let $F(g) = f(p(g), q(g))$ be the statistical significance of subgraph g, where $p(g)$ denotes the frequency of g in the positive set and $q(g)$ denotes the frequency in the negative set. LEAP allows the usage of any $f(\cdot, \cdot)$ as long as $f(\cdot, \cdot)$ increases with increase in the difference between the positive and the negative frequencies of g. One example of such a function is the G-test score [153]. Consequently, for any supergraph g' of g, the following upper bound $\widehat{F}(g')$ on $F(g')$ can be provided.

$$\widehat{F}(g') = \max\{f(p(g), 0), f(0, q(g))\} \tag{9}$$

In other words, all super graphs of g can be skipped if the above upper bound is less than the significance threshold. $\widehat{F}(g')$ however is often not tight and therefore not much effective in pruning the search space. To overcome this bottleneck, LEAP proposes two different pruning strategies, namely *structural leap search* and *frequency-descending mining*, to mine the subgraph with the highest G-test score. The technique can easily be generalizes to top-k as well.

Figure 19b shows the search space of significant subgraphs. Consider the two branches A and B originating from the subgraph g. Branch A contains all supergraphs of $g \diamond e$, i.e., g extended by some edge e. On the other hand, B contains all supergraphs of g except those of $g \diamond e$. LEAP observes that in a graph dataset if g and $g \diamond e$ often occur together, then it is likely that some subgraph $g' \supset g$ in branch B occurs with $g'' = g' \diamond e$, where g'' is in branch A. In other words, $p(g'') \sim p(g')$ and $q(g'') \sim q(g')$. This behavior in turn would result in $F(g'') \sim F(g')$. Extending this same line of reasoning, when g and $g \diamond e$ co-occur often, the significance scores of subgraphs in branch A and branch B are likely to be similar. Thus, if branch A has already been searched, then B can be skipped for faster searching. Note that this strategy does not guarantee optimality.

To formalize the above intuition, let $I(G, g, g \diamond e)$ be an indicator function of a graph G where $I(G, g, g \diamond e) = 1$ if for any $g' \supseteq g$, if $g' \subseteq G \exists g'' = g' \diamond e$, such that $g'' \subseteq G$; otherwise, $I(G, g, g \diamond e) = 0$. Essentially, $I(G, g, g \diamond e) = 1$ means if a supergraph g' of g has an embedding in G, then $g' \diamond e$ is also present in G. Let $D_+(g, g \diamond e) = \{G \in D_+ | I(G, g, g \diamond e) = 1\}$ be the set of positive graphs where this event occurs. In $D_+(g, g \diamond e)$, $g' \supset g$ and $g'' = g' \diamond e$ have the same frequency. Therefore, the maximum frequency difference between g' and g'' in the positive set can be defined as follows.

$$\Delta_+(g, g \diamond e) = p(g) - \frac{|D_+(g, g \diamond e)|}{|D_+|} \tag{10}$$

Input: Graph dataset D, difference threshold σ
Output: Optimal graph pattern candidate g^*

```
1: S = {1 - edge graph};
2: g* = ∅; F(g*) = -∞;
3: while S ≠ ∅ do
4:     S = S \ {g};
5:     if g was examined then
6:         continue;
7:     if ∃g ◊ e, g ◊ e ≺ g, (2Δ₊(g,g◊e))/(p(g◊e)+p(g)) ≤ σ, (2Δ₋(g,g◊e))/(q(g◊e)+q(g)) ≤ σ then
8:         continue;
9:     if F(g) > F(g*) then
10:        g* = g;
11:    if F̂(g) ≤ F(g*) then
12:        continue;
13:    S = S ∪ {g'|g' = g ◊ e};
14: return g*;
```

Fig. 20 The pseudocode for structural leap search

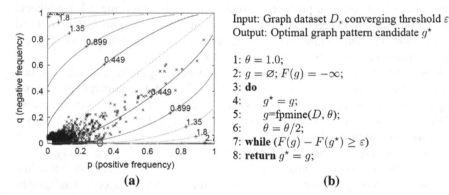

Input: Graph dataset D, converging threshold ε
Output: Optimal graph pattern candidate g^*

```
1: θ = 1.0;
2: g = ∅; F(g) = -∞;
3: do
4:     g* = g;
5:     g=fpmine(D, θ);
6:     θ = θ/2;
7: while (F(g) - F(g*) ≥ ε)
8: return g* = g;
```

(a)   (b)

Fig. 21 a Relationship between subgraph frequency and G-test score. **b** The pseudocode of frequency-descending search

where D_+ is set of positive graphs. If the difference in frequency is smaller than a threshold σ, then the entire branch B can be "leaped" over. The pseudocode of the structural leap search is provided in Fig. 20. Line 7 encodes the leaping condition. The denominators in the fractions act as the normalizing factors.

Structural leap search utilizes the correlation between structural similarity and statistical significance. LEAP notices that subgraph frequency is also correlated to statistical significance. Consider Fig. 21a, which is a contour plot to show the correlation between frequency and G-test score for subgraphs in the AIDS-antiviral dataset [153]. The x-axis represents the frequency in the negative dataset and the y-axis depicts the frequency in the positive dataset. The curves show the G-test scores. As can be seen, subgraphs with high G-test scores have a high frequency in either the positive set or the negative set. For example, the subgraph enclosed in the red circle has the highest G-test score. As can be seen, this subgraph has a frequency in the positive set that is higher than most.

LEAP utilizes the above observation to design an iterative frequency-descending mining of significant subgraphs. Figure 21b presents the pseudocode of the frequency-descending search. It starts by mining frequent subgraphs at a threshold of 1 (line 5), and evaluates the G-test scores of these graphs. Next, it repeats the same process with a lower frequency threshold and this process continues till the G-test score converges.

Finally, LEAP integrates structural-leap search and frequency-descending search into a single framework.

1. Perform structural-leap search only on subgraphs with frequency threshold of 1.
2. Repeat the above procedure with a lower threshold iteratively till the significance of the best subgraph found so far, g^*, converges.
3. Perform structural leap search without frequency-descending mining using g^*. Specifically, in addition to structural leap, search in a branch only if it can lead to a subgraph with a score higher than $F(g^*)$. This can be determined using $\widehat{F}(g)$.

Leap has been benchmarked on a series of graph datasets from the biology domain. Specifically, each graph corresponds to a molecule, and the graph is also labeled with a tag denoting whether it is active or inactive against a disease such as Aids or cancer. The sizes of these datasets range up to $\approx 80,000$ graphs. Compared to frequent subgraph mining techniques, LEAP is considerably slower. This is expected since the apriori property cannot be used to prune the search space. However, the proposed heuristics of structural-leap and frequency descending mining results in a running time that is up to two orders of magnitude faster than the naive depth-first approach. lEAP shares several properties with the frequent subgraph mining techniques built for transactional databases. First, the running time of LEAP grows linearly with dataset size. Second, LEAP does not scale to graph sizes beyond 100 since the search space explodes exponentially.

GraphSig. GraphSig uses p-value to quantify the statistical significance of a subgraph. Given a dataset of graphs $\mathbb{D}$, and a p-value threshold θ, the goal is to mine the following answer set.

$$\mathbb{A} = \{g | p - value(g) \leq \theta, g \subseteq G, G \in \mathbb{D}\} \qquad (11)$$

Note that a lower p-value indicates higher statistical significance.

Figure 22 outlines the pipeline of GraphSig. GraphSig first converts each graph G into a set of feature vectors. Each feature vector represents a certain subgraph within G. Owing to the conversion of G into a feature space, the problem of mining significant subgraphs translates to the problem of mining significant *sub-feature vectors*. These significant sub-feature vectors represent potential significant subgraphs in the

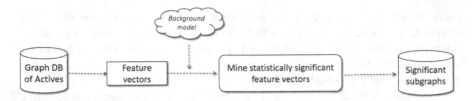

Fig. 22 The pipeline of GraphSig

Sample Graph	Random Walk Results			

ID	Starting Node	O-2-C	C-1-C	C-1-N
h_1	O	4	2	2
h2	C	2	3	3
h3	C	2	4	2
h4	N	2	2	4

Fig. 23 Conversion of a graph into feature vectors

feature space. Thus, GraphSig develops a technique to mine these significant sub-feature vectors and convert them back to the graph space to verify if they are indeed statistically significant. Those that pass this check are returned as the final answer set of significant subgraphs. GraphSig derives its pruning power from the feature space analysis following which only a small portion of the exponential subgraph search space is accessed for further analysis. We next explain the major steps of GraphSig.

To convert a graph into a set of feature vectors, GraphSig performs random walk with restarts on each node. This simulates sliding a window across the graph. RWR simulates the trajectory of a random walker that starts from a target node and jumps from one node to a neighbor. Each neighbor has an equal probability of becoming the new station of the walker. At each jump, GraphSig updates the edge-type traversed. Further, with a restart probability α, the walker teleports to the starting node and restarts the walk. For example, at $\alpha = 0.25$, on average, after every four jumps, the walker comes back to the starting node. The random walk continues till the feature distribution converges. The feature values are finally discretized into 10 bins. An example is shown in Fig. 23. RWR can therefore be visualized as placing a window at each node of a graph and capturing a feature vector representation of the subgraph within it. As a result, a graph of n nodes is represented by n feature vectors.

Next, GraphSig develops a model to compute the p-value of a feature vector. GraphSig models the occurrence of a feature vector $\underline{x}$ in a vector generated from a random graph. The frequency distribution of a vector is generated using the prior probabilities of features in the background dataset. The p-value is then calculated by comparing the observed support of $\underline{x}$ with its expected support.

For two feature vectors $\underline{x} = [x_1, \cdots , x_n]$ and $\underline{y} = [y_1, \cdots , y_n]$, $\underline{x}$ is a *sub-feature vector* of $\underline{y}$ if and only if $x_i \leq y_i$ for $i = 1, \cdots , n$. The relation is denoted as $\underline{x} \subseteq \underline{y}$.

The probability of $\underline{x}$ occurring in a random feature vector $\underline{y} = [y_1, \cdots, y_n]$ is expressed as follows.

$$P(\underline{x}) = P(y_1 \geq x_1, \cdots, y_n \geq x_n) \tag{12}$$

where each event is the probability of a feature in the random vector $\underline{y}$ having a higher or equal value than the same feature in $\underline{x}$.

GraphSig assumes independence of the features. As a result, Eq. 12 is expressed as a product of the individual probabilities.

$$P(\underline{x}) = \prod_{i=1}^{n} P(y_i \geq x_i) \tag{13}$$

In other words, Eq. 13 gives us the probability of finding $\underline{x}$ in a random feature vector.

Once we know $P(\underline{x})$, the support of $\underline{x}$ in a database of random feature vectors is modeled as a binomial distribution. Specifically, a random vector can be viewed as a trial and $\underline{x}$ occurring in it as "success". A database consisting m feature vectors will involve m trials. The support of $\underline{x}$ in the database is the number of successes. Therefore, the probability of $\underline{x}$ having a support μ is

$$P(\underline{x}; \mu) = \binom{m}{\mu} P(\underline{x})^{\mu} (1 - P(\underline{x}))^{m-\mu} \tag{14}$$

Therefore, given an observed support μ_0 of $\underline{x}$, its p-value is the area under the pdf in the range $[\mu_0, m]$.

$$p\text{-}value(x, \mu_0) = \sum_{i=\mu_0}^{m} P(\underline{x}; i) \tag{15}$$

GraphSig makes the following observations on monotonicity of p-values of feature vectors, which is later used to prune the search space.

1. Given two feature vectors $\underline{x}$ and $\underline{y}$, if $\underline{x} \subseteq \underline{y}$, then $p\text{-}value(x, \mu) \geq p\text{-}value(y, \mu)$ for any μ.
2. Given a feature vector $\underline{x}$, if $\mu_1 \geq \mu_2$, then $p\text{-}value(x, \mu_1) \leq p\text{-}value(x, \mu_2)$ for any $\underline{x}$.

With the ability to compute p-value of a feature vector, GraphSig next focuses on mining the set of statistically significant sub-feature vectors. Towards that goal, GraphSig develops the significant sub-feature vector mining algorithm. Figure 24 presents the pseudocode. GraphSig explores closed sub-vectors in a bottom-up, depth-first manner. At each step, the floor of the supporting set is evaluated for significance. Next, it moves to a state with a smaller supporting set along a branch and repeats the evaluation process. The algorithm stops branching from a state when

Require: x is current sub-feature vector
Require: $\mathbb{S}$ supporting set of x
Require: b current starting position
Ensure: $\mathbb{A}$ is the set of all significant sub-feature vectors
1: **if** $p\text{-}value(x) \leq maxPvalue$ **then**
2:    $\mathbb{A} \leftarrow \mathbb{A} + x$
3: **for** $i = b$ **to** m **do**
4:    $\mathbb{S}' \leftarrow \{y | y \in \mathbb{S}, y_i > x_i\}$
5:    **if** $|\mathbb{S}'| < minSup$ **then**
6:       **continue**
7:    $x' = floor(\mathbb{S}')$
8:    **if** $\exists j < i$ such that $x'_j > x_j$ **then**
9:       **continue**
10:   **if** $p\text{-}value(ceiling(\mathbb{S}'), |\mathbb{S}'|) \geq maxPvalue$ **then**
11:      **continue**
12:   $FVMine(x', \mathbb{S}', i)$

Fig. 24 The pseudocode of the feature vector mining algorithm

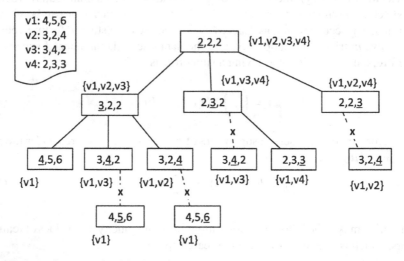

Fig. 25 A running example of the feature vector mining algorithm

all its descendants are guaranteed to have a p-value higher than the p-value threshold (lines 10–11), or produces a duplicate state (lines 8–9). Figure 25 demonstrates a running example in a database of four vectors with a support and p-value threshold of 1.

Given the knowledge that each significant sub-feature vector could describe a significant subgraph, GraphSig scans the database to identify the regions where the current sub-feature vector occurs. This involves finding all nodes described by a feature vector f, such that f is a super-vector of the current sub-feature vector. Each of these nodes is located in a region of interest. GraphSig isolates the subgraph centered at each node by using a user-specified radius. This produces a candidate set of subgraphs for each significant sub-feature vector. The cutoff radius could be selected based on some prior knowledge about the typical size of subgraphs that one wants to study. In the worst case, one can select the entire graph where the node occurs. Finally, GraphSig performs *maximal* frequent subgraph mining on each candidate set with a high frequency threshold since each set is expected to contain a common subgraph corresponding to the significant sub-feature vector. To mine maximal subgraphs, any of the existing techniques could be used [69, 91, 134]. This step also prunes out false positives where dissimilar subgraphs are grouped into a set due to similar vector representations. Due to the absence of a common subgraph, when frequent subgraph mining is performed on the set, no frequent subgraph will be produced and as a result the set will be filtered out.

In a subsequent work, GraphSig has been extended to mine graphs with proba-bilistic class labels [118].

In terms of efficiency, GraphSig is faster than Leap and this is largely due to processing graphs in feature space. However, the answer sets of the two techniques cannot directly be compared since their models to quantify statistical significance are different. GraphSig has been applied on datasets containing $\approx 80,000$ graphs. Due to the linear growth rate with running time, GraphSig can scale to million-sized graph databases. However, the limitation to scale on larger graphs remains. Like LEAP, the significant subgraphs mined by GraphSig have been employed as features to classify graphs and have shown higher accuracy than LEAP [121].

Mining statistically significant subgraphs on a single large graph. So far, we have discussed the problem of mining significant subgraphs only on transactional databases. Recently, the problem has also been studied on a single large network [8].

Given an undirected, vertex-labeled graph G, Arora et al. propose a technique to mine the top-k most statistically significant subgraphs. The statistical significance of a subgraph is quantified using its Chi-square value. Arora et al. assume that the vertex labels are drawn randomly and independently from a multinomial distribution. Based on this null hypotheses, the expected support of a subgraph is computed. By comparing the deviation from the observed support, the chi-square value of a subgraph is calculated.

The naive approach is to enumerate all possible subgraphs and compute their significance one by one. However, due to the exponential search space, this approach is not scalable. Arora et al. overcome this bottleneck through a data compression based strategy where vertices are merged into a *super-vertex* and a graph is built over these super-vertices alone. Since this *super-graph* is significantly smaller in size, the brute force approach is scalable.

Figure 26 outlines the pipeline of the algorithm. The mining process initiates by first constructing a super-graph from the original graph by collapsing the vertices along *contracting* edges. If the vertex labels are discrete, then an edge is contracting if the connecting vertices have the same label. On the other hand, if the vertex labels are continuous, an edge is contracting if the chi-square value of the merged vertex is larger than the individual vertices. The merging process ensures that any vertex

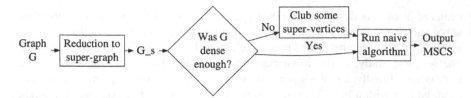

Fig. 26 Pipeline of the technique to mine statistically significant subgraphs from a single large graph [8]

that is part of a significant subgraph in the original graph, continues to be part of a significant subgraph even in the super-graph. The reduction in the size of the super-graph depends on the density of the original graph. If the original graph is dense, the resulting super-graph is small enough such that the brute-force algorithm of finding all connected subgraphs from the supergraph is feasible. Otherwise, if the original graph is not dense enough, then the super-graph is not small enough either and further reduction in the size of the super-graph is required. This reduction is performed by merging vertices that have low chi-square values. Once the super-graph is reduced to a small-enough size, the brute-force algorithm is again applied to find the statistically significant subgraphs.

4.4 Mining Representative Subgraphs

As discussed earlier, due to the exponential subgraph search space, the number of subgraph patterns mined is also large. Thus, it is often required to represent all patterns through a more concise set of representative subgraphs. This problem has been studied in the context of frequent subgraph mining by ORIGAMI [58] and RING [157].

ORIGAMI. ORIGAMI proposes the idea of α-orthogonal, β-representative subgraph patterns. Given a graph dataset $\mathbb{D}$ and a similarity threshold $\alpha \in [0, 1]$, a subset $\mathbb{R} \subset \mathbb{D}$ is α-orthogonal if and only if for any $G_1, G_2 \in \mathbb{R}$, $sim(G_1, G_2) \leq \alpha$ and for any $G_i \in \mathbb{D} - \mathbb{R}$, $\exists G \in \mathbb{R}, sim(G, G_i) > \alpha$. Given a similarity threshold $\beta \in [0, 1]$, $G \in \mathbb{D}$ is represented by $\mathbb{R}$ if $\exists G' \in \mathbb{R}$ such that $sim(G, G') \geq \beta$. Let $\mathbb{S} = \{G \in \mathbb{D} | \exists G' \in \mathbb{R}, sim(G, G') \geq \beta\}$ be the set of all represented graphs. The *residue set*, or the set of unrepresented graphs, is therefore $\mathbb{D} - \mathbb{S}$. The goal of ORIGAMI is to mine the α-orthogonal, β-representative subgraphs from the set of maximal frequent subgraphs such that the size of the residue set is minimized. ORIGAMI shows that this problem is NP-Hard. Hence, it follows a two step approach. First, it mines a sample of maximal frequent subgraphs. Second, it constructs a network among the sampled maximal frequent subgraphs where each subgraph is a node, and two subgraphs are connected by an edge if their similarity is bounded by α. In this network, ORIGAMI identifies a random maximal clique and improves it iteratively till the size of the residue set converges.

RING. Given a graph database $\mathbb{D}$, a support threshold θ and an integer N, RING aims to find up to N representative frequent subgraph patterns and a minimized distance upper bound R. Specifically, for any frequent subgraph pattern g', there must exist a representative frequent subgraph pattern g in the representative set such that the distance between g and g' is no larger than R.

RING initiates the mining process by first computing the subgraph pattern distribution. For that task, RING mines a random set of frequent subgraph patterns, clusters the patterns, and selects the centers of the clusters as the initial representative patterns. The cluster centers provide an approximate distribution of how the frequent patterns are distributed in the subgraph search space. Clustering requires RING to perform a large number of distance computations between subgraphs. Common subgraph distance functions such as graph edit distance and maximum common connected subgraphs are expensive to compute. To overcome this computational bottleneck, RING converts subgraphs into feature vectors. This operation is done by first identifying a set of *graph invariants*. A graph invariant is some function $f(g)$ of a graph g such that if g_1 and g_2 are isomorphic, then $f(g_1) = f(g_2)$. Examples include number of vertices, diameter of the graph, minimum spanning tree, etc. An invariant is good if the probability of $f(g_1) = f(g_2)$ is low when g_1 and g_2 are not isomorphic to each other and the invariant is fast to compute. RING indexes the feature space representations of the cluster centers using R-Tree [51]. This complete the first stage of the mining process.

In the second stage, RING performs a depth-first exploration of the search space. For each enumerated frequent subgraph g, RING checks if it is already represented.

1. If g is not represented, RING either assigns g as a new representative or extends the radius of one of the existing representatives so that g is now represented.
2. If g already belongs to a representative and a pattern belonging to the same representative has been reached, then RING stops growing from g; if g is the first pattern belonging to this representative, RING continues growing in a depth-first manner.

5 Open Problems

With the advent of big-graphs and their applications in real world, the last ten years has seen an unprecedented interest and research in designing algorithms and systems for big-graphs. There are yet many questions that need to be investigated. We conclude this chapter by highlighting some future research directions.

5.1 Large-Scale Graph Processing Systems

While the topology of today's large-scale graphs may often fit on a single server [105], adding all data associated with the graph nodes and edges can overwhelm the main memory and computation capacity of one server. In order to achieve low latency and high throughput over massive graph datasets, various scale-out systems, such as Pregel [101], GraphLab [99], GraphX [45], and Horton [125] (for a survey, see [54, 83]) are designed in which the graph and its data are partitioned horizontally across cheap commodity servers in the cluster. Keeping with the modern database trends to support low-latency operations, the system designers often target fully in-memory systems, and disks are used only for durability. Nevertheless, existing systems for big-graphs processing have some unique challenges as follows.

- Many of the existing large-scale graph processing systems follow the node centric computation model, which severally restricts the usability and expressibility of these systems. In particular, node centric programs are difficult to write and they are procedural in nature (i.e., the user needs to take care of query optimization, in addition to scheduling and locking for asynchronous node centric computation). Besides, it is difficult to design a node centric algorithm for many graph problems (e.g., between-ness centrality).
- Due to the interconnected nature of graph data, graph computations are irregular: It is difficult to extract parallelism by partitioning. Unbalanced computational workload resulting from poor partitioning and synchronization overheads reduces scalability. For example, several graph workloads (e.g., reachability and shortest path queries) are performed in an iterative manner and they show a varying degree of parallelism over the course of execution. This brings to the following question. What could be the ideal graph-partitioning scheme? Does one size fit all? Do we need to vary the partitioning and re-partitioning strategy based on the graph data, systems, algorithms, and even at various stages of an algorithm?
- What will be the roles of modern hardware in accelerating big-graphs processing? In the past, there have been a few attempts to build graph-processing systems using GPU, FPGA, and FlashSSD [56]. However, they are not widely accepted.
- Which one is a better design choice for queries on big-graphs — "scaling out" on cheap, commodity clusters (distributed memory) versus "scaling up" with more cores and more memory (shared memory)? Perhaps, for online graph queries, scaling up is a better option due to their lower communication cost [127].
- Do we need stand-alone systems only for graph processing, such as Horton [125] and GraphLab [99]? Can they be integrated with the existing big-data and dataflow systems? There is a recent trend for such integration and several hybrid systems have been proposed, e.g., GraphX [45] and Naiad [111].

5.2 Graph Databases, Languages, and Query Interfaces

With the emergence of complex social and information networks, many big-graph startups came out in the last few years, such as Gephi, FlockDB, GraphBuilder, HypergraphDB, AllegroGraph, Gremlin, and Neo4J. They usually provide their own open-source tools, APIs, and graph databases. However, unlike the classical RDBMS vendors, these startups are not popular yet. There are indeed many fundamental problems in the area of graph databases that we need to resolve.

- Is the relational algebra expressive and efficient enough for querying graph databases? Alternatively, do we require a separate graph algebra? Some recent works, e.g., Horton [125] and G-SPARQL [123] proposed an operator "graph traverse", in addition to the classical relational algebra operators. In the presence of such additional operators, how can one perform efficient indexing and query optimization?
- It is important to analyze what types of transactions and consistency features a graph database requires. Social and information networks are temporal in nature; they evolve over time. Communication networks and social media data, on the other hand, arrive as a continuous graph stream.
- While privacy and security is an important feature in today's commercial RDBMS, the aforementioned big-graph startups hardly ensure any privacy guarantee for their graph databases. Due to outsourcing of graph data, we also need to design efficient provenance techniques to verify the correctness and completeness of the answers.
- An important aspect of graph databases is the query and the data manipulation languages. The complexity of graph query languages makes them unsuitable for ordinary users in a variety of domains. A usable graph querying technique could borrow ideas from the HCI domain, such as visual [12, 75], sensory, and natural-language interfaces, auto-query completion, personalized feedback, and re-ranking of query results, etc.

5.3 Datasets and Benchmarks

It is important to identify standard datasets and benchmarks [35] for various graph queries and systems. While there are several algorithmic advances in this area, lack of standard datasets, open-source softwares, and benchmarks is often a challenging problem in terms of comparing with previous works.

6 Conclusions

In this chapter, we discussed graph pattern matching and mining as the representative of emerging online queries and offline analytics, respectively. Due to lack of fixed schema, complexity of graph query languages, and also because of efficiency reasons, the focus now-a-days is on user-friendly, and often approximate, query answering and pattern mining techniques, that we also emphasized in this work. We concluded by highlighting future research directions in the context of usable graph systems, databases, and benchmarks.

7 About Authors

Arijit Khan is an assistant professor in the School of Computer Engineering at Nanyang Technological University, Singapore. His research interests span in the area of big-data, big-graphs, and graph systems. He received his PhD from the Department of Computer Science, University of California, Santa Barbara, and did a post-doc in the Systems group at ETH Zurich. Arijit is the recipient of the prestigious IBM PhD Fellowship in 2012–2013. He published several papers in premier database and data-mining conferences and journals including SIGMOD, VLDB, TKDE, ICDE, SDM, EDBT, and CIKM. Arijit co-presented tutorials on emerging graph queries, big-graph systems, and uncertain graphs at ICDE 2012, VLDB 2014, VLDB 2015, and served in the program committee of KDD, SIGMOD, VLDB, ICDM, EDBT, WWW, and CIKM. Arijit served as the co-chair of Big-O(Q) workshop co-located with VLDB 2015.

Sayan Ranu is an assistant professor in the department of Computer Science and Engineering at IIT Madras. His research interests include spatio-temporal data analytics, graph indexing and mining, and bioinformatics. Prior to joining IIT Madras, he was a researcher at IBM Research. He obtained his PhD from the Department of Computer Science, University of California, Santa Barbara (UCSB) in March 2012. He was a recipient of the "Distinguished Graduate Research Fellowship" at UCSB. He obtained his Bachelor of Science from Iowa State University, where he received the "Presidents top 2% of the class" award. He has published several papers in premier database and data-mining conferences including SIGMOD, VLDB, ICDE, KDD, ICDM, and WWW. Sayan regularly serves in the program committees and review panels of prestigious conferences and journals including KDD, SDM, TKDE, VLDB Journal.

References

1. D.J. Abadi, A. Marcus, S.R. Madden, K. Hollenbach, SW-Store: a vertically partitioned DBMS for semantic web data management. VLDB J. **18**(2), 385–406 (2009)
2. S. Abiteboul, D. Quass, J. McHugh, J. Widom, J.L. Wiener, The lorel query language for semistructured data. Int. J. Digit. Libr. **1**(1), 68–88 (1997)
3. B. Aditya, G. Bhalotia, S. Chakrabarti, A. Hulgeri, C. Nakhe, P. Parag, S. Sudarshan, BANKS: browsing and keyword searching in relational databases, in *VLDB* (2002)
4. C. Aggarwal, H. Wang, *Managing and Mining Graph Data* (Springer, Berlin, 2010)
5. S. Agrawal, S. Chaudhuri, G. Das, DBXplorer: a system for keyword-based search over relational databases, in *ICDE* (2002)
6. D. Ajwani, M. Karnstedt, A. Sala, Processing large graphs: representations, storage, systems, and algorithms, in *WWW* (2015)
7. R. Angles, C. Gutierrez, Survey of graph database models. ACM Comput. Surv. **40**(1), 1:1–1:39 (2008)
8. A. Arora, M. Sachan, A. Bhattacharya, Mining statistically significant connected subgraphs in vertex labeled graphs, in *SIGMOD* (2014)
9. P. Barceló, L. Libkin, J.L. Reutter, Querying graph patterns, in *PODS* (2011)

10. M. Bayati, M. Gerritsen, D.F. Gleich, A. Saberi, Y. Wang, Algorithms for large sparse network alignment problems, in *ICDM* (2009)
11. J. Berry, B. Hendrickson, S. Kahan, P. Konecny, Software and algorithms for graph queries on multithreaded architectures, in *IPDPS* (2007)
12. S.S. Bhowmick, B. Choi, S. Zhou, VOGUE: towards a visual interaction-aware graph query processing framework, in *CIDR* (2013)
13. C. Borgelt, M.R. Berthold, Mining molecular fragments: finding relevant substructures of molecules, in *ICDM* (2002)
14. S. Brin, L. Page, The anatomy of a large-scale hypertextual web search engine. Comput. Netw. **30**(1–7), 107–117 (1998)
15. B. Bringmann, S. Nijssen, What is frequent in a single graph? in *PAKDD* (2008)
16. J. Broekstra, A. Kampman, F.v. Harmelen, Sesame: a generic architecture for storing and querying RDF and RDF schema, in *ISWC* (2002)
17. A. Buluç, J.R. Gilbert, The combinatorial BLAS: design, implementation, and applications. Int. J. High Perform. Comput. Appl. **25**(4), 496–509 (2011)
18. P. Buneman, M.F. Fernandez, D. Suciu, UnQL: a query language and algebra for semistructured data based on structural recursion. VLDB J. **9**(1), 76–110 (2000)
19. M. Bureli, The Current State of Graph Databases (2012). http://bigbe.su/lectures/2014/16.3.pdf
20. C. Chen, X. Yan, F. Zhu, J. Han, P.S. Yu, Graph OLAP: towards online analytical processing on graphs, in *ICDM* (2008)
21. H. Cheng, D. Lo, Y. Zhou, X. Wang, X. Yan, Identifying bug signatures using discriminative graph mining, in *ISSTA* (2009)
22. E.I. Chong, S. Das, G. Eadon, J. Srinivasan, An efficient SQL-based RDF querying scheme, in *VLDB* (2005)
23. S. Cohen, J. Mamou, Y. Kanza, Y. Sagiv, XSEarch: a semantic search engine for XML, in *VLDB* (2003)
24. M.P. Consens, A.O. Mendelzon, Expressing structural hypertext queries in graphlogm, in *HYPERTEXT* (1989)
25. S. Cook, The complexity of theorem-proving procedures, in *STOC* (1971), pp. 151–158
26. L.P. Cordella, P. Foggia, C. Sansone, M. Vento, A (sub)graph isomorphism algorithm for matching large graphs. IEEE Trans. Pattern Anal. Mach. Intell. **26**(10), 1367–1372 (2004)
27. T.H. Cormen, C. Stein, R.L. Rivest, C.E. Leiserson, *Introduction to Algorithms* (McGraw-Hill Higher Education, New York, 2001)
28. X.H. Dang, A. Singh, P. Bogdanov, H. You, B. Hsu, Discriminative subnetworks with regularized spectral learning for global-state network data, in *ECML PKDD* (2014)
29. X.H. Dang, H. You, P. Bogdanov, A. Singh, Learning predictive substructures with regularization for network data, in *ICDM* (2015)
30. M. Deshpande, M. Kuramochi, N. Wale, G. Karypis, Frequent substructure-based approaches for classifying chemical compounds. IEEE Trans. Knowl. Data Eng. **17**, 1036–1050 (2005)
31. DEX/Sparksee, http://sparsity-technologies.com/
32. A. Dovier, C. Piazza, The subgraph bisimulation problem. TKDE **15**(4), 1055–1056 (2003)
33. J. Dutkowski, T. Ideker, Protein networks as logic functions in development and cancer. PLoS Comput. Biol. **7**, 09 (2011)
34. M. Elseidy, E. Abdelhamid, S. Skiadopoulos, P. Kalnis, GraMi: frequent subgraph and pattern mining in a single large graph, in *VLDB* (2014)
35. O. Erling, A. Averbuch, J. Larriba-Pey, H. Chafi, A. Gubichev, A. Prat, M.-D. Pham, P. Boncz, The LDBC social network benchmark: interactive workload, in *SIGMOD* (2015)
36. R. Fagin, A. Lotem, M. Naor, Optimal aggregation algorithms for middleware, in *PODS* (2001)
37. C. Faloutsos, G. Miller, C. Tsourakakis, Large graph mining: power tools and a practioner's guide, in *KDD* (2009)
38. W. Fan, J. Li, S. Ma, N. Tang, Y. Wu, Y. Wu, Graph pattern matching: from intractable to polynomial time, in *VLDB* (2010)

39. W. Fan, J. Li, S. Ma, H. Wang, Y. Wu, Graph homomorphism revisited for graph matching, in *VLDB* (2010)
40. W. Fan, J. Li, J. Luo, Z. Tan, X. Wang, Y. Wu, Incremental graph pattern matching, in *SIGMOD* (2011)
41. W. Fan, J. Li, S. Ma, N. Tang, Y. Wu, Adding regular expressions to graph reachability and pattern queries, in *ICDE* (2011)
42. M.F. Fernandez, D. Florescu, A.Y. Levy, D. Suciu, Declarative specification of web sites with STRUDEL. VLDB J. **9**(1), 38–55 (2000)
43. M. Fiedler, C. Borgelt, Subgraph support in a single large graph, in *ICDM Workshops, 2007* (2007)
44. B. Gallagher, Matching structure and semantics: a survey on graph-based pattern matching, in *AAAI FS* (2006)
45. J.E. Gonzalez, R.S. Xin, A. Dave, D. Crankshaw, M.J. Franklin, I. Stoica, GraphX: graph processing in a distributed dataflow framework, in *OSDI* (2014)
46. D. Gregor, A. Lumsdaine, The parallel BGL: a generic library for distributed graph computations, in *POOSC* (2005)
47. Z. Guan, J. Wu, Q. Zhang, A. Singh, X. Yan, Assessing and ranking structural correlations in graphs, in *SIGMOD* (2011)
48. L. Guo, F. Shao, C. Botev, J. Shanmugasundaram, XRANK: ranked keyword search over XML documents, in *SIGMOD* (2003)
49. R. Gupta, S. Sarawagi, Answering table augmentation queries from unstructured lists on the web, in *VLDB* (2009)
50. S. Gurukar, S. Ranu, B. Ravindran, COMMIT: a scalable approach to mining communication motifs from dynamic networks, in *SIGMOD* (2015)
51. A. Guttman, R-trees: a dynamic index structure for spatial searching, in *SIGMOD* (1984)
52. J. Han, Y. Sun, X. Yan, P.S. Yu, Mining knowledge from databases: an information network analysis approach, in *SIGMOD* (2010)
53. L. Han, T. Finin, A. Joshi, GoRelations: an intuitive query system for dbpedia, in *JIST* (2011)
54. M. Han, K. Daudjee, K. Ammar, M.T. Özsu, X. Wang, T. Jin, An experimental comparison of pregel-like graph processing systems, in *VLDB* (2014)
55. W.-S. Han, J. Lee, M.-D. Pham, J. Yu, iGraph: a framework for comparisons of disk-based graph indexing techniques, in *VLDB* (2010)
56. W.-S. Han, S. Lee, K. Park, J.-H. Lee, M.-S. Kim, J. Kim, H. Yu, TurboGraph: a fast parallel graph engine handling billion-scale graphs in a single PC, in *KDD* (2013)
57. S. Harris, N. Gibbins, 3store: efficient bulk RDF, in *PSSS* (2003)
58. M.A. Hasan, V. Chaoji, S. Salem, J. Besson, M.J. Zaki, ORIGAMI: mining representative orthogonal graph patterns, in *ICDM* (2007)
59. M.A. Hasan, M.J. Zaki, Output space sampling for graph patterns, in *VLDB* (2009)
60. H. He, A. Singh, Graphs-at-a-time: query language and access methods for graph databases, in *SIGMOD* (2008)
61. H. He, H. Wang, J. Yang, P.S. Yu, BLINKS: ranked keyword searches on graphs, in *SIGMOD* (2007)
62. B. Hendrickson, R. Leland, A multilevel algorithm for partitioning graphs, in *Supercomputing* (1995)
63. M.R. Henzinger, T.A. Henzinger, P.W. Kopke, Computing simulations on finite and infinite graphs, in *FOCS* (1995)
64. S. Hong, H. Chafi, E. Sedlar, K. Olukotun, Green-Marl: a dsl for easy and efficient graph analysis, in *ASPLOS* (2012)
65. V. Hristidis, Y. Papakonstantinou, Discover: keyword search in relational databases, in *VLDB* (2002)
66. V. Hristidis, L. Gravano, Y. Papakonstantinou, Efficient IR-style keyword search over relational databases, in *VLDB* (2003)
67. V. Hristidis, N. Koudas, Y. Papakonstantinou, D. Srivastava, Keyword proximity search in XML trees. TKDE **18**(4), 525–539 (2006)

68. J. Huan, W. Wang, J. Prins, Efficient mining of frequent subgraphs in the presence of isomorphism, in *ICDM* (2003)
69. J. Huan, W. Wang, J. Prins, J. Yang, Spin: mining maximal frequent subgraphs from graph databases, in *KDD* (2004)
70. J. Huan, W. Wang, D.Bandyopadhyay, J. Snoeyink, J. Prins, A. Tropsha, Mining spatial motifs from protein structure graphs, in *Proceedings of the 8th Annual International Conference on Research in Computational Molecular Biology (RECOMB04)* (2004), pp. 308–315
71. InfiniteGraph, http://www.objectivity.com/products/infinitegraph/
72. A. Inokuchi, T. Washio, H. Motoda, An apriori-based algorithm for mining frequent substructures from graph data. Princ. Data Min. Knowl. Discov. **1910**, 13–23 (2000)
73. N. Jayaram, A. Khan, C. Li, X. Yan, R. Elmasri, Querying knowledge graphs by example entity tuples. TKDE **27**(10), 2797–2811 (2015)
74. N. Jin, C. Young, W.Wang, 0010. GAIA: graph classification using evolutionary computation, in *SIGMOD* (2010)
75. C. Jin, S.S. Bhowmick, X. Xiao, B. Choi, S. Zhou, GBLENDER: visual subgraph query formulation meets query processing, in *SIGMOD* (2011)
76. V. Kacholia, S. Pandit, S. Chakrabarti, S. Sudarshan, R. Desai, H. Karambelkar, Bidirectional expansion for keyword search on graph databases, in *VLDB* (2005)
77. M. Kargar, A. An, Keyword search in graphs: finding R-cliques, in *VLDB* (2011)
78. G. Karypis, *METIS and ParMETIS, in Encyclopedia of parallel computing* (Springer, Berlin, 2011)
79. Z. Kefato, M. Lissandrini, D. Mottin, T. Palpanas, Keyword Query to Graph Query. Technical report DISI-14-003, University of Trento (2013)
80. B.P. Kelley, B. Yuan, F. Lewitter, R. Sharan, B.R. Stockwell, T. Ideker, PathBLAST: a tool for alignment of protein interaction networks. Nucleic Acids Res. **32**, 83–88 (2004)
81. D. Kempe, J.M. Kleinberg, E. Tardos, Maximizing the spread of influence through a social network, in *KDD* (2003)
82. A. Khan, L. Chen, On uncertain graphs modeling and queries, in *VLDB* (2015)
83. A. Khan, S. Elnikety, Systems for big-graphs, in *VLDB* (2014)
84. A. Khan, N. Li, Z. Guan, S. Chakraborty, S. Tao, Neighborhood based fast graph search in large networks, in *SIGMOD* (2011)
85. A. Khan, X. Yan, K.-L. Wu, Towards proximity pattern mining in large graphs, in *SIGMOD* (2010)
86. A. Khan, Y. Wu, X. Yan, Emerging graph queries in linked data, in *ICDE* (2012)
87. A. Khan, Y. Wu, C. Aggarwal, X. Yan, NeMa: fast graph search with label similarity, in *VLDB* (2013)
88. J. Kleinberg, Navigation in a small world. Nature **406**, 845 (2000)
89. K. Kochut, M. Janik, SPARQLeR: extended sparql for semantic association discovery, in *ESWC* (2007)
90. R. Krishnamurthy, S.P. Morgan, M. Zloof, Query-by-example: operations on piecewise continuous data, in *VLDB* (1983)
91. M. Kuramochi, G. Karypis, Frequent subgraph discovery, in *ICDM* (2001)
92. M. Kuramochi, G. Karypis, GREW-a scalable frequent subgraph discovery algorithm, in *ICDM* (2004)
93. T. Lappas, K. Liu, E. Terzi, Finding a team of experts in social networks, in *KDD* (2009)
94. J. Lee, W.-S. Han, R. Kasperovics, J.-H. Lee, An in-depth comparison of subgraph isomorphism algorithms in graph databases, in *VLDB* (2013)
95. J. Leskovec, C. Faloutsos, Tools for large graph mining: structure and difference, in *WWW* (2008)
96. G. Li, B.C. Ooi, J. Feng, J. Wang, L. Zhou, EASE: an effective 3-in-1 keyword search method for unstructured semi-structured and structured data, in *SIGMOD* (2008)
97. Z. Liang, M. Xu, M. Teng, L. Niu, NetAlign: a web-based tool for comparison of protein interaction networks. Bioinformatics **22**(17), 2175–2177 (2006)

98. F. Liu, C. Yu, W. Meng, A. Chowdhury, Effective keyword search in relational databases, in *SIGMOD* (2006)
99. Y. Low, D. Bickson, J. Gonzalez, C. Guestrin, A. Kyrola, J.M. Hellerstein, Distributed graphlab: a framework for machine learning and data mining in the cloud, in *VLDB* (2012)
100. S. Ma, Y. Cao, W. Fan, J. Huai, T. Wo, Capturing topology in graph pattern matching, in *VLDB* (2012)
101. G. Malewicz, M.H. Austern, A.J.C. Bik, J.C. Dehnert, I. Horn, N. Leiser, G. Czajkowski, Pregel: a system for large-scale graph processing, in *SIGMOD* (2010)
102. F. Manola, E. Miller, RDF Primer, W3C Recommendation (2004). http://www.w3.org/TR/REC-rdf-syntax/
103. R.R. McCune, T. Weninger, G. Madey, Thinking like a vertex: a survey of vertex-centric frameworks for large-scale distributed graph processing. ACM Comput. Surv. **48**(2), 25:1–25:39 (2015)
104. A. McGregor, Graph stream algorithms: a survey. SIGMOD Rec. **43**(1), 9–20 (2014)
105. F. McSherry, M. Isard, D.G. Murray, Scalability! but at what COST? in *HotOS* (2015)
106. K. Mehlhorn, S. Naher, LEDA, a platform for combinatorial and geometric computing. Commun. ACM **38**(1), 96–102 (1995)
107. S. Melnik, H.G.-Molina, E. Rahm, Similarity flooding: a versatile graph matching algorithm and its application to schema matching, in *ICDE* (2002)
108. A.O. Mendelzon, P.T. Wood, Finding regular simple paths in graph databases. SIAM J. Comput. **24**(6), 1235–1258 (1995)
109. M. Mongiovì, R.D. Natale, R. Giugno, A. Pulvirenti, A. Ferro, R. Sharan, Sigma: a set-cover-based inexact graph matching algorithm. J. Bioinform. Comput. Biol. **8**(2), 199–218 (2010)
110. D. Mottin, M. Lissandrini, Y. Velegrakis, T. Palpanas, Exemplar queries: give me an example of what you need, in *VLDB* (2014)
111. D.G. Murray, F. McSherry, R. Isaacs, M. Isard, P. Barham, M. Abadi, Naiad: a timely dataflow system, in *SOSP* (2013)
112. Neo4j, https://neo4j.com/
113. T. Neumann, G. Weikum, The RDF-3X engine for scalable management of RDF data. VLDB J. **19**(1), 91–113 (2010)
114. S. Nijssen, J.N. Kok, The gaston tool for frequent subgraph mining, in *Proceedings of the International Workshop on Graph-Based Tools* (2004)
115. M.T. Özsu, A survey of rdf data management systems (2015). http://arxiv.org/abs/1601.00707
116. F. Pellegrini, J. Roman, SCOTCH: a software package for static mapping by dual recursive bipartitioning of process and architecture graphs, in *HPCN* (1996)
117. E. Prud'hommeaux, A. Seaborne, SPARQL query language for RDF. W3C Recommendation (2008)
118. S. Ranu, B.T. Calhoun, A.K. Singh, S.J. Swamidass, Probabilistic substructure mining from small-molecule screens. Mol. Inform. **30**(9), 809–815 (2011)
119. S. Ranu, M. Hoang, A. Singh, Mining discriminative subgraphs from global-state networks, in *KDD* (2013)
120. S. Ranu, A.K. Singh, GraphSig: a scalable approach to mining significant subgraphs in large graph databases, in *ICDE* (2009)
121. S. Ranu, A.K. Singh, Mining statistically significant molecular substructures for efficient molecular classification. J. Chem. Inf. Model. **49**, 2537–2550 (2009)
122. S. Sakr, G. Al-Naymat, Relational processing of RDF queries: a survey. SIGMOD Rec. **38**(4), 23–28 (2010)
123. S. Sakr, S. Elnikety, Y. He, G-SPARQL: a hybrid engine for querying large attributed graphs, in *CIKM* (2012)
124. H. Samet, J. Sankaranarayanan, H. Alborzi, Scalable network distance browsing in spatial databases, in *SIGMOD* (2008)
125. M. Sarwat, S. Elnikety, Y. He, M.F. Mokbel, Horton+: a distributed system for processing declarative reachability queries over partitioned graphs, in *VLDB* (2013)

126. H. Shang, Y. Zhang, X. Lin, J. Yu, Taming verification hardness: an efficient algorithm for testing subgraph isomorphism, in *VLDB* (2008)
127. J. Shun, G.E. Blelloch, Ligra: a lightweight graph processing framework for shared memory, in *PPoPP* (2013)
128. R. Singh, J. Xu, B. Berger, Global alignment of multiple protein interaction networks with application to functional orthology detection. PNAS **105**(35), 12763–12768 (2008)
129. C. Sommer, Shortest-path queries in static networks. ACM Comput. Surv. **46**(4), 45:1–45:31 (2014)
130. H. Sun, M. Srivatsa, S. Tan, Y. Li, L.M. Kaplan, S. Tao, X. Yan, Analyzing expert behaviors in collaborative networks, in *KDD* (2014)
131. Y. Sun, J. Han, X. Yan, P.S. Yu, T. Wu, PathSim: meta path-based top-K similarity search in heterogeneous information networks, in *VLDB* (2011)
132. Z. Sun, H. Wang, H. Wang, B. Shao, J. Li, Efficient subgraph matching on billion node graphs, in *VLDB* (2012)
133. M. Thoma, H. Cheng, A. Gretton, J. Han, H.-P. Kriegel, A. Smola, L. Song, P.S. Yu, X. Yan, K. Borgwardt, Near-optimal supervised feature selection among frequent subgraphs, in *SDM* (2009)
134. L.T. Thomas, S.R. Valluri, K. Karlapalem, MARGIN: maximal frequent subgraph mining. ACM Trans. Knowl. Discov. Data **4**(3), 10:1–10:42 (2010)
135. Y. Tian, R. McEachin, C. Santos, D. States, J. Patel, SAGA: a subgraph matching tool for biological graphs. Bioinformatics **23**(2), 232–239 (2006)
136. Y. Tian, J.M. Patel, TALE: a tool for approximate large graph matching, in *ICDE* (2008)
137. H. Tong, C.-Y. Lin, Non-negative residual matrix factorization with application to graph anomaly detection, in *SDM* (2011)
138. H. Tong, C. Faloutsos, B. Gallagher, T. Eliassi-Rad, Fast best-effort pattern matching in large attributed graphs, in *KDD* (2007)
139. S. Trißl, U. Leser, Fast and practical indexing and querying of very large graphs, in *SIGMOD* (2007)
140. J.R. Ullmann, An algorithm for subgraph isomorphism. J. ACM **23**, 31–42 (1976)
141. N. Vanetik, E. Gudes, Mining frequent labeled and partially labeled graph patterns, in *ICDE* (2004)
142. C. Vicknair, M. Macias, Z. Zhao, X. Nan, Y. Chen, D. Wilkins, A comparison of a graph database and a relational database: a data provenance perspective, in *ACMSE* (2010)
143. S.V.N. Vishwanathan, N.N. Schraudolph, R. Kondor, K.M. Borgwardt, Graph Kernels. J. Mach. Learn. Res. **11**, 1201–1242 (2010)
144. R.C. Wang, W. Cohen, Language-independent set expansion of named entities using the web, in *ICDM* (2007)
145. A. Wlc, R. Raman, Z. Wu, S. Hong, H. Chafi, J. Banerjee, Graph analysis: do we have to reinvent the wheel? in *GRADES* (2013)
146. K. Wilkinson, C. Sayers, H. Kuno, D. Reynolds, Efficient RDF storage and retrieval in Jena2, in *SWDB* (2003)
147. P.T. Wood, Query languages for graph databases. SIGMOD Rec. **41**(1), 50–60 (2012)
148. Y. Xu, Y. Papakonstantinou, Efficient keyword search for smallest LCAs in XML databases, in *SIGMOD* (2005)
149. X. Yan, J. Han, gSpan: graph-based substructure pattern mining, in *ICDM* (2002)
150. X. Yan, J. Han, Closegraph: mining closed frequent graph patterns, in *KDD* (2003)
151. X. Yan, P.S. Yu, J. Han, Graph indexing: a frequent structure-based approach, in *SIGMOD* (2004)
152. X. Yan, F. Zhu, P.S. Yu, J. Han, Feature-based similarity search in graph structures. ACM Trans. Database Syst. **31**(4), 1418–1453 (2006)
153. X. Yan, H. Cheng, J. Han, P.S. Yu, Mining significant graph patterns by scalable leap search, in *SIGMOD* (2008)
154. X. Yan, B. He, F. Zhu, J. Han, Top-K aggregation queries over large networks, in *ICDE* (2010)

155. J. Yao, B. Cui, L. Hua, Y. Huang, Keyword query reformulation on structured data, in *ICDE* (2012)
156. S. Zhang, S. Li, J. Yang, GADDI: distance index based subgraph matching in biological networks, in *EDBT* (2009)
157. S. Zhang, J. Yang, S. Li, RING: an integrated method for frequent representative subgraph mining, in *ICDM* (2009)
158. S. Zhang, J. Yang, W. Jin, SAPPER: subgraph indexing and approximate matching in large graphs, in *VLDB* (2010)
159. P. Zhao, J. Han, On graph query optimization in large networks, in *VLDB* (2010)
160. Q. Zhong, H. Li, J. Li, G. Xie, J. Tang, L. Zhou, Y. Pan, A Gauss function based approach for unbalanced ontology matching, in *SIGMOD* (2009)
161. Y. Zhu, L. Qin, J. Yu, H. Cheng, Finding top-k similar graphs in graph databases, in *EDBT* (2012)
162. L. Zou, L. Chen, M.T. Özsu, D. Zhao, Dynamic skyline queries in large graphs, in *DASFAA* (2010)
163. L. Zou, J. Mo, L. Chen, M.T. Özsu, D. Zhao, gStore: answering SPARQL queries via subgraph matching, in *VLDB* (2011)

Link and Graph Mining in the Big Data Era

Ana Paula Appel and Luis G. Moyano

Abstract Graphs are a convenient representation for large sets of data, being complex networks, social networks, publication networks, and so on. The growing volume of data modeled as complex networks, e.g. the World Wide Web, and social networks like Twitter, Facebook, has raised a new area of research focused in complex networks mining. In this new multidisciplinary area, it is possible to highlight some important tasks: extraction of statistical properties, community detection, link prediction, among several others. This new approach has been driven largely by the growing availability of computers and communication networks, which allow us to gather and analyze data on a scale far larger than previously possible. In this chapter we will give an overview of several graph mining approach to mine and handle large complex networks.

1 Introduction

Over the past years the amount of data collected has increased substantially, especially with the growing availability of the World Wide Web, expansion not only for text but also with images and video. For instance, Facebook estimates that video exhibition move from 1 billion in 2015 to 8 billion in 2016.

Social networks and social media are becoming a regular part of people lives, as a person spends, in average, almost 2 hours per day in a social network. These kind of data was studied in the past by social scientists, however in a much smaller scale. They usually worked with hundreds of nodes to answer questions such as which person is the most connected in the network, or which one, if removed, could break the connection among all the individuals. Today, social networks are composed by hundreds of million users and the analysis is different not only in the class of

A.P. Appel (✉)
IBM Research, São Paulo, Brazil
e-mail: apappel@br.ibm.com

L.G. Moyano
CONICET and Facultad de Ciencias Exactas y Naturales,
Universidad Nacional de Cuyo, Mendoza, Argentina

© Springer International Publishing AG 2017
A.Y. Zomaya and S. Sakr (eds.), *Handbook of Big Data Technologies*,
DOI 10.1007/978-3-319-49340-4_17

techniques but also in which type of questions we want to answer. Asking which person will break the connection of the network does not make sense anymore, since there could be no one to cause this kind of damage in a network of such size. This crucial change in scale made these type of problems much more interesting and lead to emergence of a new research field: **graph mining**.

These data are naturally mapped in complex networks that are represented using graphs. Under this model, nodes are entities (e.g. people or groups of people), and edges represent some kind of interaction between entities (e.g. friendship). Another example is that of information networks, in which the nodes are information resources such as Web pages or documents, and edges represent logical connections such as hyperlinks, citations, or cross-references and so on.

These large volume of data makes it easy to study global phenomena that are not discernible in smaller networks, for example, how a community is born, how a network evolves over time, understand the importance of single node or an edge, among many examples that were almost impossible to tackle in small networks.

Graphs are very a powerful tool to express and model mathematically complex network structures, appearing in many domains, whenever it is useful to represent how things are either physically or logically linked to one another in a network structure.

A graph $\mathcal{G}$ is mathematically represented as $\mathcal{G} =< \mathcal{V}, \mathcal{E} >$, on which $|\mathcal{V}| = n$ represents the number of nodes (or vertex set), and $|\mathcal{E}| = m$ represents the number of edges (or links), and a relation that associates with each edge two vertexes [109]. We say that two nodes are neighbors if they are connected by an edge.

The graph-based representation used for data has substantial non-trivial topological features, with patterns of connections between their elements that are neither purely regular nor purely random. For this reason this representation is usually referred to as a Complex Network.

The study of complex networks brought to light important properties such as power-law degree distributions [47], the Small World phenomenon [104], among several others [109]. These patterns help us understand the interaction of people in social networks [91, 99] as well as the dissemination of information and diseases [36], and has other practical applications such as anomaly detection [7] and so on.

In this chapter we will navigate through several graph mining task, such as pattern discovery using statistical properties (Sect. 2), community detection (Sect. 5), link prediction (Sect. 4). We also will talk about how to represent networks with weight (Sect. 7), multiple edges (Sect. 8) and temporally (Sect. 3). We also we show that networks can be use to represent knowledge and map knowledge bases (Sect. 6.2). We also present several platforms to store and process large graphs (Sect. 6). We finish this chapter present what are the big challenges and open issues of this chapter (Sect. 9) and than we conclude (Sect. 10).

2 Definitions

A graph is a useful way to specify relationships among a collection of items. A graph consists of a set of objects called *nodes*, with certain pairs of these objects connected by links called *edges*. We say that two nodes are neighbors if they are connected by an edge. A complex network is modeled as a graph $\mathcal{G} = \langle \mathcal{V}, \mathcal{E} \rangle$, on which $\mathcal{V}$ represents the number of nodes or verticals $N = |\mathcal{V}|$, and $\mathcal{E}$ represents the number of edges or links $M = |\mathcal{V}|$. The traditional way of represent a graph $\mathcal{G}$ computationally is the adjacency matrix, which is a square matrix $\mathbf{A} = N \times N$ where $\mathbf{A}_{i,j} = 1$ is $(v_i, v_j) \in \mathcal{E}$ and 0 otherwise [109].

A graph is undirect if $(v_i, v_j) \in \mathcal{E} \Leftrightarrow (v_j, v_i) \in \mathcal{E}$, that is, the edges are unordered pairs.

However, in many cases, it is desirable to express asymmetric relationships and other attributes such as weights, time or multiple relations in the links. Thus, the graph can become directed where the presence of $v_j, v_i \in \mathcal{E}$ does not imply that $v_i, v_j \in \mathcal{E}$. For example, A points to B but not vice versa, which means that edges are ordered pairs. It is possible to also have weighted networks, $\mathcal{G} = \langle \mathcal{V}, \mathcal{E}, \mathcal{W} \rangle$ where $w_i \in \mathcal{W}$, which means that each edge will have a weight w_i associated to it (weights are discussed in more detail in Sect. 7).

The *node degree* $\tau(v_i)$, also called *neighborhood* of a node, can be defined by the amount of incident edges on node v_i (Table 1).

Another important measure in real complex network is called the *clustering coefficient*, which is proportional to the total number of triangles that a node or a network has.

A *triangle* Δ of a graph $\mathcal{G}$ is a set of three completely connected nodes where $(u, v, w) \in \mathcal{V}$ and edges $(u, v), (v, w), (w, u) \in \mathcal{E}$.

In many networks it is found that if vertex A is connected to vertex B and vertex B to vertex C, then there is a high probability that vertex A will also be connected to vertex C. In the language of network theory, a friend's friend is likely also to be

Table 1 Symbols used in this chapter

Symbols	Description
$\mathbf{A}$	Adjacency matrix
$\mathcal{G}$	Graph
$\mathcal{E}$	Set of edges
$\mathcal{V}$	Set of nodes
v_i	Node
e_k	Edge
Δ	Triangle
$\tau(v_i)$	Degree of node v_i
$C(v_i)$	Cluster coefficient of node v_i
$C(\mathcal{G})$	Cluster coefficient of $\mathcal{G}$

a friend. In terms of network topology, transitivity means the presence of a high number of triangles in the network [146].

Formally, the cluster coefficient can be expressed by the following equation:

$$C(v_i) = \frac{2 * \Delta(v_i)}{d(v_i) * (d(v_i) - 1)} \tag{1}$$

A node v_i with degree $|\tau(v_i)|$ has at most $\tau(v_i) * (\tau(v_i) - 1)/2$ edges that could exist among them, being $\Delta(v_i)$ the fraction of edges that really exist (the number of triangles). Thus, the cluster coefficient in Eq. 1 $C(v_i)$ of a node v_i is the proportion of edges among nodes that are at distance 1 if the neighborhood of v_i, divided by the total number of edges that could exist among them. Also, $C(v_i)$ is the fraction of triangles centered at node v_i which $(d(v_i) * (d(v_i) - 1))/2$ triangles could exist.

The *global cluster coefficient* $C(\mathcal{G})$ is the average of the clustering of all nodes $C(v_i)$ from graph $\mathcal{G}$, divided by the total number of nodes n.

$$C(\mathcal{G}) = \frac{1}{N} * \sum_{i=1}^{N} C(v_i) \tag{2}$$

Nodes are also important in real graphs to analyze *degree probability distribution* [47]. We define *pk* to be the fraction of nodes in the network that have degree *k*. Equivalently, *pk* is the probability that a node chosen uniformly at random has degree *k*. A plot of *pk* for any given network can be formed by making a histogram of the degrees of nodes. This histogram is the degree distribution for the network. An example of degree distribution from DBLP[1] is presented in Fig. 1.

In a random graph of the type studied by Erdõs and Rényi [109], each edge is present or absent with some constant probability *p*. As a result, the degree distribution is, as mentioned earlier, binomial, or Poisson in the limit of large graph size.

Real-world networks are mostly found to be very unlike the random graph in their degree distributions [2]. Many of them asymptotically follow power-laws in their tails: $pk \approx k^{\alpha}$ for some constant exponent α. The degrees of the nodes in most networks are strongly right-skewed, meaning that their distribution has a long right tail of values that are far above the mean. In the context of social networks, for example, power-law distribution means that most of people has few friends and few people has a lot of friends. There are several metrics that also follows a power-law in real networks, as nodes and edges during evolution [90], triangle distribution [142] and weights [100].

Another important characteristic observed in real networks is the small diameter. In graph theory, the *diameter* is defined as been the longest path among all the shortest paths. However, in this definition the diameter is susceptible to outliers if the graph has a long chain. Also, to calculate the diameter in real large graphs is computationally very expensive.

---

[1]http://dblp.uni-trier.de.

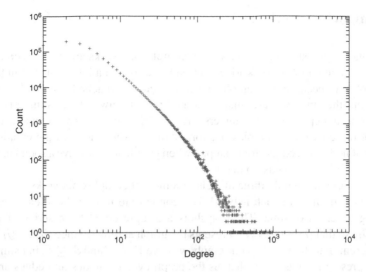

Fig. 1 Degree distribution from co-authorship network extract from DBLP

One of the traditional metrics for diameter is the effective diameter [79, 117], defined as $f(\mathcal{G})$ of a graph $\mathcal{G}$. It is the minimum number of hops (steps or links) in which 90% of all connected pairs of nodes can reach each other. One way to calculate the effective diameter is build what we call a *Hop Plot*. The hop plot shows each distance, starting in one, until a distance where the number of nodes reach does not change over a very small threshold. An example of this plot is shown in Fig. 2. As we can see, after distance 7 the number of nodes reached (y axis) do not longer increase.

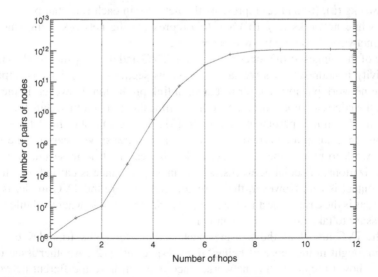

Fig. 2 Hop Plot from co-authorship network extract from DBLP

3 Temporal Evolution

One of many interesting characteristics of complex networks are the ones connected
with temporal evolution. Network evolution has attracted a lot of interest in the last
years, not only because the complex network field has reached a stable knowledge
point about the simplest case, that is the undirected, unweighted, simple network.
Thus, the next step is to start to understand special cases of networks and evolution
is one of these cases. A complex network evolves over time with the creation and
deletion of nodes and edges, for example, when people join or leave a social network,
they create or break friendship ties.

Today, there are mainly three models of temporal complex networks.

The most traditional such network is a condensate in snapshots, each one representing a period of time. The snapshots are represented by a series of graphs
$\mathcal{G}_1, ..., \mathcal{G}_T$, so that $\mathcal{G}_t = (\mathcal{V}_t, \mathcal{E}_t)$ represents the graph at time t. Since $\mathcal{G}_1, ..., \mathcal{G}_T$ represent different snapshots of the same graph, we have $\mathcal{V}_t \subseteq \mathcal{V}$ and $\mathcal{E}_t \subseteq \mathcal{E}$. For simplicity
most of presentation assumes that, as the graph evolves, nodes and edges are only
added and never deleted, that is, $\mathcal{V}_1 \subseteq \mathcal{V}_2 \subseteq ...\mathcal{V}_T$ and $\mathcal{E}_1 \subseteq \mathcal{E}_2 \subseteq ...\mathcal{E}_T$.

The second one follows the data stream model, where a large number of edges
representing interactions are continuously received over time and are superposed
over a much larger network. An example of such a scenario would be a Twitter post
stream, in which several posts are continuously received over time [3].

Another way to represent temporal networks is through Time-Varing Graphs models (TGVs) also known in the literature as temporal or time-dependent networks [29].

TGVs graphs can be easily converted in snapshots by creating a graph with an
edge between nodes, if and only if there is connection between nodes during a time
interval. A example is shown in Fig. 3 where we illustrate a network in time t_0, t_1 and
t_2. Networks (a), (b) and (c) represents the network in each timestamp using TGV
model while networks (a'), (b') and (c') represents the network using snapshots
model, network (c') being the whole network.

One of the biggest differences between the TVG and the snapshot models is how
transitivity is addressed. As we saw in previous sections, transitivity is important
in some network phenomena, for instance, in link prediction. However, in the TVG
model, the edges are not carried on from one timestamp to another, thus transitivity
can be used within a particular timestamp [81]. In the snapshot model, the edges
from one timestamp are carried to the next one, so transitivity can use the whole
past network to predict the future network. This can be true in several scenarios,
as organizations, or social networks, where an acquaintance is carried for life even
if the contact is lost. However, there are situations where the TVG model is more
suitable, as is the case of the air-transport network for example where to build a route
is necessary to take time into account [71].

In the TVG model, all the measures such as diameter, paths, triangles, etc., need
to be rethought in the sense of being "time-respectful" [72]. Another issue in this
model is how to represent the network, since one will have a different network in
each timestamp and sometimes one wants to represent an edge that repeats from one

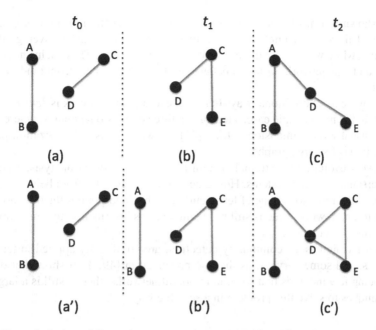

Fig. 3 Networks in three different timestamps, using model TVG and Snapshots to represent them in each timestamp. Network (c') represents the whole network using snapshots model

timestamp to another. In [147], the authors propose a unifying model for representing TVG graphs. In this model a square adjacency matrix of $n * t$ (where n is the number of nodes and t is the number of timestamps) is created and follows the same principles of traditional adjacency matrix.

The third model considers the network as a graph data stream. These networks are based on transient interactions, such as email or telecommunication networks. To process a graph data stream is usually necessary real-time analytical methods [3] as the one presented in [4].

Depending on what it is the purpose of the task, the choice of the model could have a big impact. Also, graph data streams are far more challenging than graph snapshots, since the first one is in general not possible to store in memory (even in disk) for analysis.

Many interesting properties about network evolution have been studied over the years using the snapshot model. In one case, the network presents the characteristic of a shrinking diameter, i.e., in most cases the effective diameter of a network decreases over time while the network grows. Another one is the network densification which means the network becomes denser and the average degree increases as it evolves [90].

$$|\mathcal{E}_t| \propto |\mathcal{V}_t|^{(\alpha)} \tag{3}$$

As shown in Eq. 3 the number of nodes grows as function of number of edges. For $\alpha = 1$ there is a complex network that grows with a constant average degree over time, while with $\alpha = 2$ there is a much denser network. These helps not only to understand real network evolution but also to discriminate them from random graph models.

One of the most traditional ways to model evolution in network is *Tensor Factorization*. In the most simple case, a tensor is a three dimensional matrix, where in our contect usually one dimension is time [46]. However, tensor factorization present scaling issues for large graphs.

Network evolution is extremely useful for complex network analysis, especially to understand dynamic systems. How a network evolves is far from being answered simply. Link prediction is one of few techniques that incorporates the time notion in its definition, however time is still not being used as information that may improve the method's accuracy.

Other techniques, as community detection, are not directly applied in temporal networks, and some surveys, as the one presented in [49, 129] show that despite there being few methods that work in temporal networks, there's still is a large gap in techniques to solve this problem in a scalable way.

4   Link Prediction

Social networks are one of the most clear and well-known examples of complex networks. These applications typically need to recommend connections among users, such as the *"People you may know"* feature. Thus, the problem of how people get connected is relevant not only for social network but also for a large number of use cases such as organizations [42].

There are many reasons, sometime exogenous to the social network, why two individuals will become friends: they may happen to be geographically close if one moves to a city near the other's neighborhood, or they may attend to same party, or go to the same school, and so on. Such type of interaction can be hard to predict.

Commonly, two nodes are more likely to be connected if they are more similar. Similarity may be based only on network structure. Thus, a large number of new interactions are hinted at by the topology of the network: two individuals who are close in the network will have friends in common, and this suggests that they are more likely to become a friend in the near future.

Understanding the mechanisms by which people get connected and how networks evolve have been addressed by what is known as *the link prediction task*. Link prediction methods are based on graph snapshots as a model to support evolution. Thus, in a formal way, link prediction can be defined as: Given a snapshot of a graph $\mathcal{G}$ at time t, predict accurately which edges will appear in the network in time $t + 1$ [93]. Translated to the context of social networks, friends of your friends are likely also to be your friends.

In terms of network topology, transitivity means the presence of a high number of triangles in the network [146].

Most of traditional linking prediction methods are based on graph structural properties as assigning a connection value, called $score(u, w)$, to pairs of unconnected nodes $< u, w >$ based on a desired graph $\mathcal{G}$. The scores are ranked in a list in decreasing order of $score(u, w)$ and afterwards predictions are made according to this list.

Let $\tau(u) = (u \in \mathcal{V} : \exists (v, u) \in \mathcal{E})$ of node u be defined as the set of nodes in $\mathcal{V}$ that are adjacent to u. For a node u, let $\tau(u)$ denote the set of neighbors of u in $\mathcal{G}$. A number of link prediction approaches are based on the idea that two nodes u and w are more likely to form a link, in the future, if their sets of neighbors $\tau(u)$ and $\tau(w)$ have large overlap. The most direct implementation of this idea for the link-prediction problem is the common-neighbors predictor, which may be defined as follows:

$$score(u, w) = |\tau(u) \cap \tau(w)|$$

The common-neighbors predictor captures the notion that a friend may introduce two strangers who have a common friend. This introduction has the effect of "closing a triangle" in the graph and feels like a common mechanism in real life. In this sense some other measures, also neighborhood based, were propose to rank nodes that are likely to have a link in a near future.

The *Jaccard coefficient* is a similarity metric that is commonly used in information retrieval. It is used to measure the probability that both u and w have a feature f, for a randomly selected feature f that either u or w has. In the case of networks, the feature f can be a list of friends that a node has. Formally, the Jaccard predictor uses the following measure:

$$score(u, w) = |\tau(u) \cap \tau(w)| / |\tau(u) \cup \tau(w)|$$

The *Adamic/Adar predictor* [1] evaluates the degrees of the common neighbors and emphasizes the nodes that share neighbors with small degree. This is because a high degree node has a higher chance to be in the common neighborhood of other nodes. This method computes features of the nodes, and defines the similarity between two nodes to be the following:

$$score(u, w) = \sum_{(V \in \tau(u) \cap \tau(w))} \frac{1}{\log \tau(v)}$$

While the neighborhood-based measures provide a robust estimation of the likelihood of a link forming between a pair of nodes, they are not quite as effective when the number of shared neighbors between a pair of nodes is small. A particular walk-based measure that is used commonly to measure the link-prediction strength is the Katz [80] measure, which is arguably one of the best link predictors available because it has been shown to outperform many other methods as showed in [93].

$$score(u, w) = \sum_{l=1}^{\infty} \beta^l |path^l_{(u,w)}|$$

Another path measured in link prediction task is Random Walking which is a path that consisted of succession of steps chosen randomly. As showed in [94] one difficulty with all Random-walk to link prediction is their sensitive dependence to parts of the network far away from target nodes. For example, in a random walk from x to y, the walker has a certain probability to go too far away from both x and y although they may be close to each other. This may lead to a low prediction accuracy since in most real networks nodes tend to connect with the ones nearby rather than far away. Another algorithm for link prediction in social network based on Random Walks is presented in [17]. As we will present in Sect. 6.2, PRA is a random walk-based algorithm to predict relations in a Knowledge base mapped as a network.

Link prediction can be applied in other scenarios than social networks, for example predict interactions and collaborations among people in organizations can help manage companies in a productive way. The task of recommending unknowns but similar people is quite different from possible friend recommendation tasks, which focus on recommending individuals who have friends in common [67].

In [145], the author finds that the similarity between individuals' movements, their social connectedness and the strength of interactions between them are strongly correlated with each other. Thus, the authors also reports that human mobility could serve as a good predictor for the formation of new links, yielding comparable predictive power to traditional network-based measures.

Another challenge which remains largely open in link prediction methods is how to effectively combine the information from the network structure with rich node and edge attribute data. Social ties could improve link prediction metrics as the authors show in [17] where a supervised random walk that naturally combines the information from the network structure with node and edge level attributes.

There are many more metrics used to produce the score between two unconnected nodes as presented in [69, 97].

The link-prediction problem can also be related to the task of inferring missing links in complex networks: in many domains, one constructs a network and then tries to infer additional links that, while not directly visible, are likely to exist, as *Prophet* [13] presented in Sect. 6.2 [44].

This line of work differs from link prediction problem formulation in that it works with a static snapshot of a network, rather than considering network evolution. It also tends to take into account specific attributes of the nodes in the network, rather than evaluating the power of prediction methods based purely on the graph structure.

The metrics presented above for link prediction can actually be called in a more specific way as Link Existence Prediction. The link existence problem is defined as the problem of predict whether a link will or not exist in undirected networks, which is the same of link prediction, since most of studies emphasize only unweighted undirected networks.

However the link prediction problem could be extended for other problems related to discover a link, such as direction, multiplicity and weight. Follow we define each one of these problems are.

The link direction problem can be viewed as a link-prediction extension on directed networks, where the link and direction will be predicted. An example of the importance would be in a phone call network someone may want predict who-calls-whom. However, most of the work developed predicts the direction of an existent link [10].

Another possible extension is to apply the link prediction task in multiplex networks, where not only links between unconnected nodes could be predicted but also new link between node already connected (more in Sect. 8). This is a big challenge because links could have different meanings. Thus, most of work in this line needs extra information other than topology. Multiplex network commonly are seen in academic, companies and social networks where relationships among individuals can have different roles, as friend, family, co-work and so on [24].

In weighted networks, the link prediction problem can be viewed as the prediction of both the link and the weight associated with it. The most common use of weights in link prediction is to help predict the existence of links by combining them with the observed links [95]. In this case, most works adjust the metrics presented (common neighbor, Jaccard, Adamic/Adar) from unweighted to weighted networks. Yet, how weights improve the accuracy of a link prediction task and how to predict the weights together (or not) with the links has not been well studied. Example of problems that could benefit from weighted link prediction is urban or air transportation [138]. One of the few studies, and a very interesting one, on the link prediction problem in weighted networks is [96], where the authors find that weak links may play a more important role than strong links.

The works presented until now are based on homogeneous networks, meaning links are all from the same type and there are only static snapshots. When we say the network is dynamic, we are implying that new links are constantly being added to the network. Such new links may also arrive in the context of new nodes being added to the network, or they may correspond to edges between already existing nodes.

Recent works focus on heterogeneous networks where link prediction are extended for it, as co-author network [45], Location-based social networks [150], information network [83, 133, 134].

Link prediction problem becomes extremely challenging when it is addressed to dynamic massive heterogeneous network because of the challenges associated with the dynamic nature of the network, and the different types of nodes and attributes in it.

In [4] the authors present a method called "DYNALINK", an algorithm for dynamic link inference in temporal and heterogeneous networks. The algorithm is able to construct link inference models for online and heterogeneous networks which are continuously evolving over time.

Time can have a big influence in link prediction, since old links are less important than the recent ones. For example, in a co-authorship network, new co-authors are more important than the oldest ones in terms of indicate new co-authors. The authors in [46] show that Katz metric can be improved adding a weight in the link reflecting how recent or how old the link is.

Another information that could be used to improve link prediction techniques is community structure information, as in traditional data mining, where cluster detection can be used as a pre-processing technique. In [127] the authors use the community structure to help in link prediction. The same technique is used in [4].

An interesting type of research area that may extend research for weighted link prediction is in signed networks. A signed network consists of a network composed by positive and negative links, which could mean friends and foes [84], trusted and distrusted peers [99]. A method to predict the signs of links (positive or negative) is proposed in [89, 126], however the prediction of both the existence of a link and its sign simultaneously has not been addressed yet.

5 Community Detection

A very important and rich research area in network theory is that of community detection. The basic idea behind community detection is the possibility to group nodes into larger groups with some criterion of similarity. The goal is to have a way to capture mesoscopic structures and in some way decrease the complexity of the original graph. This fertile research area has produced many community finding methods and algorithms [50]. Many of these methods rely in the optimization of a special function of the edges of the graph, usually called *modularity*, as we will se next.

5.1 Modularity Maximization

Many real networks have some type of inner structure beyond local edges, but which at the same time is different from and contained within the complete graph. For instance, a social graph may be though locally by studying ego-networks (i.e. just the immediate connections in a node), or may be analyzed globally, may be expressing scale-free structure which is evidenced by the whole set of nodes and edges. But in a social network it is also common to find friendship groups, or work groups which are larger than ego networks and smaller than the whole graph. The aim of community algorithms is finding these kind of mesoscopic structures, usually refered to as communities or modules [27, 50, 110].

Girvan and Newman [54, 112] propose an elegant way of finding these structures. They reasoned that by analyzing edges and assuming that a set of nodes with larger number of edges between them (compared to what would be expected if the edges were randomly placed) could be thought to form a community. They defined an objective function, the *modularity Q*, which represents the fraction of edges inside communities minus the fraction of edges in groups *if they were randomly assigned*.

In a network with n nodes, one can propose a given partition of nodes between just two communities, so as to assign $s_i = 1$ if node i belongs to one community and $s_i = -1$ if belongs to the other community. One can express the modularity of such setting as follows:

$$Q = \frac{1}{4m} \sum_{ij} \left(A_{ij} - \frac{k_i k_j}{2m} \right) s_i s_j, \tag{4}$$

where A_{ij} is the adjacency matrix of the network, k_i is the degree of node i, and $m = \frac{1}{2} \sum_i k_i$ is the number of edges in the network. The second term in the parenthesis represents the expected fraction of nodes if the edges where randomly assigned. Note that this assumes a certain *null-model*, i.e. an idea of which structure would the network have is it was randomly generated. This is an important point that may influence the outcome of the final set of communities, and that needs discussion in any kind of generalization of the concept, as will be further discussed in Sect. 8.2. The above framework may be repeated iteratively to the found subgraphs to subsequently find smaller communities, taking care of modifying Eq. 4 to correctly account for all edges [112]. Whenever a proposed split gives a contribution non-positive to the total modularity, then the algorithm is carried no further. Thus, the definition of community in the Girvan-Newman method is a subgraph that is not further divisible for maximizing its modularity Q. The modularity method has been widely successful and even though it has been extended in many ways it continues to be the basis of the most robust methods for community finding [110].

5.2 The Louvain Method for Community Detection

Another successful method for finding communities in very large networks is known as the Louvain community detection method [26], a very efficient method that has proved extremely useful in a number of big data graphs, taking a few minutes in regular hardware to compute communities for graphs of hundredths of millions of nodes and a few billion edges [16]. It was originally proposed by Lefebvre,[2] and later further developed by a group of researchers led by Blondel, all which at some point had worked at the Universit Catholique de Louvain, hence its name. The Louvain method is a modularity optimization algorithm based in the same principle that of the Girvan-Newman algorithm developed in [110, 112] (see previous Sect. 5.1) It was originally defined for weighted networks by allowing the adjacency matrix A_{ij} to contain weights w_{ij} (as is also the case of Eq. 4).

The algorithm is a heuristic greedy optimization method performed in two steps. First the method acts locally finding small communities. Initially every node is assigned to its own community. Next, for each node i, the node is assigned to the community C of each of its neighbors and the corresponding change in modularity ΔQ is computed. The expression for ΔQ may be expressed as follows:

---

[2]https://perso.uclouvain.be/vincent.blondel/research/louvain.html.

$$\Delta Q = \left[\frac{\sum_{in} + k_{i,in}}{2m} - \left(\frac{\sum_{tot} + k_i}{2m} \right)^2 \right] - \left[\frac{\sum_{in}}{2m} - \left(\frac{\sum_{tot}}{2m} \right)^2 - \left(\frac{k_i}{2m} \right)^2 \right], \quad (5)$$

where $\sum_{in}$ is the sum of weights of edges in C, $\sum_{tot}$ is the sum of weights of edges incident to nodes in C, k_i is the sum of weights of edges incident to node i, $k_{i,in}$ is the sum of weights of edges from i to nodes in C and finally m is the total sum of weights of edges in the network. Node i is moved to the community that contributes the most to the total modularity (with a breaking rule in case of ties) or stays in its original community in case no positive gain in Q is possible. This procedure is repeated until no improvement in Q is possible.

In the second part of the algorithm a new network is built by now grouping all nodes belonging to the communities found in the first part as new nodes in a new network. The edges between two nodes in this new network have weights equal to the sum of weights in edges between the original communities from the first stage. Self-loops appear whose weights are equal to the sum of weights of all edges within the community in the first stage of the algorithm corresponding to the new node. One this new network is created, another pass (stages one and two) is applied to the new network.

The Louvain community finding method is by no means the only community finding method for large networks [39, 92], but it has proved a nice example of an efficient and successful algorithm for networks by-product of rapidly growing— and increasingly common—big datasets.

6   Graphs in Big Data

Graphs and networks are a fundamental concept in the context of big data, as big data business is being driven by the possibility of quantifying relationship data. Indeed, much of the value provided by the availability of vast quantities of data resides in the ability to spot and quantify these relationships. Users that express an interest, friends that stay in touch, clients that spend in a given item, all these have in common that represent a relationship between two entities. And as we have seen in the previous sections, relationships are susceptible to be efficiently described by networks or graphs.

Google and Weibo funded their businesses linking users to topics and interest by search and advertisement. Facebook linked people and of course interests, LinkedIn connects people and professional opportunities. They all gather an enormous amount of information from their data, so the ability to extract useful insights, in the form of distilled bits of data, is crucial.

Many data wealthy businesses, different from social networks, are also starting to see the fact that they could benefit greatly by the possibilities offered by graph analytics and graph methodologies in general. For instance, there is increasing interest in graph methods applied to healthcare [118, 135].

In this section we will lay out some important examples and use cases, and we will mention the most important methods, applications and tools for the kind of networks typically found in Big Data business.

6.1 Graphs in the Big Data Era

Google's PageRank One of the most interesting examples of graph methods applied to big data was the PageRank algorithm [116], a cornerstone of Google's early success. PageRank is the ranking algorithm originally used by Google Search to present an ordered set of web pages to the user, and even though today has been considerably extended, today continues to play a relevant part in Google search results.[3] PageRank represented a success factor for Googles search engine, as it performed quite accurately for search results ranking. The ranking algorithm computes the "importance" of webpages with simple notion: based on the structure of the web page graph, use links from other pages as a proxy for the importance of the page. In essence, PageRank computes the probability that a random walk will end in a given node of the network. The algorithm is iterative, and can compute the rank of all nodes in a graph of arbitrary size. At every iteration s, the algorithm computes the (unnormalized) probability PR for every node i in the network:

$$PR^{t+1}(i) = r + (1 - r) \sum_j \frac{PR^t(j)}{k_j^o}, \qquad (6)$$

where r is the probability of a step for the random walk and k_j^o is the our-degree of node j. Even though there has been many variations to the algorithm since its introduction, it is undeniable that PageRank still remains important to Google's business.

Facebook's Graph Search Beginning 2013, Facebook introduced its Graph Search product,[4] as "a new way to navigate (the graph's) connections and make them more useful", i.e. with the aim of improving the efficient exploration of the wealth of data produced by their social network. Facebook Graph Search is designed for users to be able to make *semantic* searches for entities and their relationships. The tool uses a battery of techniques to deliver search results, such as named entity queries as well as structured queries, but it also relies heavily in graph-related quantities such as graph distance (which is fundamental to the result) in tight combination with attributes of nodes and edges such as friendship relationships, age, gender, number of friends, celebrity status, among others [130].

---

[3] https://www.google.com/insidesearch/howsearchworks/algorithms.html.
[4] http://newsroom.fb.com/news/2013/01/introducing-graph-search-beta/.

6.2 Knowledge Graphs

Tom Gruber defines an ontology as follows: *"An ontology is a description (like a formal specification of a program) of the concepts and relationships that can formally exist for an agent or a community of agents. This definition is consistent with the usage of ontology as set of concept definitions, but more general. And it is a different sense of the word than its use in philosophy"* [64].

The RDF (https://www.w3.org/RDF/) data model, also known as RDF triples composed by subject-predicate-object, is a standard model for data interchange on the Web. The subject denotes the resource, and the predicate denotes traits or aspects of the resource and expresses a relationship between the subject and the object. For example, one way to represent the notion "Messi plays soccer" in RDF is as the triple: a subject denoting "Messi", a predicate denoting "plays", and an object denoting "soccer". RDF data model is naturally suited to knowledge representation and collection of RDF statements can be represents a labeled, directed multi-graph. Mapping a Knowledge Base as graph, categories become nodes and relations are edges.

Over the last few years, many research projects focused on building large scale ontological knowledge bases (OKB) have been developed, such as Google Knowledge Graph based on Freebase [28], YAGO [131], DBpedia [15], Elementary/Deep-Dive [114], Walmart [43], Microsoft Satori and a continuously learning program called NELL (Never Ending Language Learner) [34]. These projects store their knowledge using what we call Knowledge Bases (KBs) with millions of facts about the world, such as information about people, places and things referred as entities.

Traditionally, an knowledge base (KB) organizes and stores knowledge in two different parts, namely: (i) an ontological model, where categories (*city, company, person*, etc.) and relations (**worksFor**(*person, company*), **headQuarteredIn**(*company, city*))) are defined, and (ii) a set of facts which are instances of categories (**city** (*New York*), **company**(*Disney*), **person**(*Walt Disney*)) and relations.

Despite their size, KB are far from complete. As showed in [44] 71% of people in Freebase have no known place of birth, and 75% have no known nationality. Furthermore, coverage for less common relations can be even lower.

Therefore, a new approach is necessary to further scale up knowledge base construction. Such an approach should automatically extract facts from the whole Web, to augment the knowledge we collect from human input and structured data sources. Unfortunately, standard methods for this task often produce very noisy, unreliable facts. To alleviate the amount of noise in the automatically extracted data, the new approach should automatically leverage already-cataloged knowledge to build prior models of fact correctness.

One of the biggest problems in knowledge bases is extending it by inferring new relations. The ability to infer new knowledge may be straightforward for humans, but is tipically very hard be done automatically by a machine, as learning programs populates the KB from corpora or the Web, which may be a difficult task.

Mapping a KB as a network allows us to apply graph mining techniques to infer new relations. Thus, one of task that are mainly used is link prediction, which is applied to find implicit information so as to populate the KB. There are several projects that use graph mining, such as *Prophet* [13] in NELL, or Knowledge Vault [44] from Google. NELL also uses Random Walks to infer relations [86] and PageRank for search [144].

Prophet [13] was created to be one of NELL's components, to apply link-prediction techniques into NELL's KB maped to a graph. It executes a link-prediction task using a metric called extra-neighbors to extend NELL's ontology by finding new possible relations and also it finds new instances of these relations and some possible misplaced facts present on the KB (Fig. 4).

Knowledge Vault used PRA [86] to extend their KB that is based on Freebase. Similar to distant supervision, PRA begins with an instance of a relation such as (*Basketball, MichaelRedd*), i.e. a pair of entities, then it performs a random walk on the graph, starting at all the subject (source) nodes. Paths that reach the object (target) nodes are considered successful. For example [44], PRA learns that pairs (X, Y) which are connected by a *marriedTo* edge often also have a path of the form $X \xrightarrow{\text{childOf}} Z \xleftarrow{\text{childOf}} Y$, since if two people share a common child, they are likely to be married. The paths that PRA learns can be interpreted as rules.

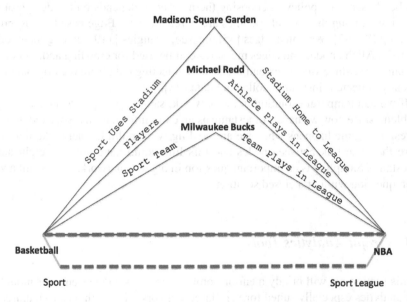

Fig. 4 An example of rule (*Sport, SportLeague*) and instance (*Basketball, NBA*) infer by *Prophet* based on three independent paths (Madison Square Garden, Michael Redd, Milwaukee Bucks)

6.3 Graph Sampling

One way to work with massive amounts of data is sampling. Data sampling has been applied over years in large data sets to extract representative small portions from data allowing us to apply data mining techniques as clustering and classification algorithms, usually computational expensive in the complete dataset. However, in the context of complex networks, the question of how to extract a representative small network from the original dataset is nontrivial. Also, one issue with graph sampling is if one should focus on nodes, edges or both. Additionally, in non-linked data, random sampling usually performs reasonably, however in complex network random sampling, be it in nodes or edges, the process usually produces a disconnected graph [88].

It can be quite difficult to find a method to reduce the size of network and keep all the important measures, useful for graph mining techniques as cluster and link prediction [143].

Another area where graph sampling techniques are widely applied is in crawling. Collecting data for analysis is a very important task and how these data are collected is extremely important to the post analyze that will be done. The crawling process of a graph starts from selecting one (or multiple) node(s) called seed(s). After a node has been visited, the edges incident in this node are known and the next node can be chosen. The policy of choosing the next node depends on the design of the crawling. Among the possible policies are: Breadth-First, Edge based [5], Random Walking [87, 87], Weighted edges [85], degree, triangles [119], among many other [9, 153]. All the nodes and edges measures can be used for crawling and, of course, for sampling, since we can extend the idea of selecting data from the original graph (already collected) instead of collecting data [73].

If we add temporal evolution in the network, sampling becomes an even harder problem, since now we have a timestamp as an extra information associated with the edges. The same happens when we are dealing with multiplex networks that have more than one link between nodes and this link could have both a weight and a timestamp. Sampling is an important question in big data networks, but is still a very open question without a closed solution.

6.4 Graph Analytics Tools

In this section we will briefly mention some graph tools and systems for modeling and analytics especially suited for very large networks, i.e. with tens to hundredths of millions of nodes and up to hundredths of billions of links. These systems are fundamental to efficiently support Big Data applications, such as Natural Language Processing tasks or targeted advertising.

Apache Giraph [12] is an open source distributed system for large scale graph processing, based on Google's proprietary Pregel [98]. It is an iterative graph system designed for high scalability (based on Apache Hadoop's MapReduce implementation), and it extends Pregel with a number of features such as edge-oriented output, master computation, sharded aggregators among others. Giraph is used by Facebook to analyze its vast social graph, and is able to process trillions of edges [37], and new, faster extensions are being developed based on it [137, 139] as well as dedicated machine learning libraries [59].

Another graph analytics example is PowerGraph [58], a distributed graph placement and representation that exploits a know feature of social networks: their power-law degree probability distribution. PowerGraph was shown to process PageRank and other tasks such as LDA in data from the Twitter social network, containing 41 million nodes and 1.5 billion edges. GraphLab, a CMU initiative, and afterwards GraphLab Create [62], an open source framework for distributed, high-performance computation over graphs, stemmed originally from PowerGraph.

Project Pegasus [120] is another CMU-based open-source, big graph-mining system designed for high scalability. In [78], the authors make an interesting comparison between Pegasus, Pregel, GraphLab and Microsoft's contribution, names Trinity at the time (now GraphEngine) [123, 124]. They compare system performance in a number of graph-oriented tasks over two big datasets, a snapshot of the World Wide Web (2002), crawled by Yahoo! with 1.4 billion nodes (web pages) and 6.6 billion links, and a Twitter who-follows-whom graph (2009), containing 63 million nodes (users) and 1.8 billion links.

Another Big Graph system is Twitter's Cassovary [35], a processing library for the Java Virtual Machine, which is now open source. Cassovary, written in Scala, is designed to handle large graphs such as Twitter's and also to be space-efficient. In [66], the authors describe some variants of recommender systems implemented in Cassovary, and a very interesting take on the architecture design, as the entire graph is put in a single server for optimization purposes, contrary to the mainstream tendency of distributed architectures. There are several other initiatives for big graph analytics and processing systems, ranging from industrial tools such as IBM System G [75], DataStax/Aurelius Faunus [48] or Teradata's SQL-GR Graph Analytics engine [14], to less production-oriented such as Microsoft's GraphEngine [61] and even more academic research-oriented systems such as the Stanford Network Analysis Project (SNAP) [125], Galois [51], from University of Texas, GUESS [65] as well as iGraph [76] (in its three flavors, R, C/C++ and Python), Gephi [21, 53] and Python-based NetworkX [70], among several others.

It is also worth mentioning a different but important class of systems, graph databases. Graph databases are generally not relational databases and exploit graph structure to optimize searches and semantic queries, most commonly by keeping track of relationships among nodes, among other things. Even though their purpose may be diverse (some are operational, while others are for analytics or development, and so on), we will focus on these differences and instead we will just present some of the most known solutions as of 2016. Some examples include Titan [141] (which Amazon integrates through their NoSQL database Amazon DynamoDB), Neo4j [108], OrientDB [115], Sparksee [128], IBM Graph [74], and GraphX [63] (Apache Spark's API for graphs and graph-parallel computation), among several others. Some of these have in common that they use the Apache TinkerPop graph computing framework [140], in particular, they are able to process instructions from the Gremlin traversal language [122], a cross-platform virtual machine and language that supports imperative and declarative querying for graph databases and graph analytical engines.

7 Weighted Networks

Single networks represent their connections as binary entities, i.e. an edge is present or not. Usually, edges do not provide more information than if they are present or not. However, links between nodes may have some describing attribute, reflecting their intensity, capacity, duration, intimacy or exchange of services [19, 60], which may be encoded in some variable usually known referred to as *weight* of the link.

The study of weighted networks has not been thorough in the last decade. Some initial work was done in [18, 149]. Many methods developed for single networks are not trivial to extend for weighted networks. A weighted network can be treated as a multiplex network where the weight becomes the number of edges between two verticals, as shown in Fig. 5.

The weights in the edges of a network help model and define week and strong ties. One way to measure this is the so-called *strength* of a node V_i (Eq. 7) defined as the sum of all weights of neighbors of node v_i. The strength of a node integrates the information both with its connectivity and the importance of the weights of its links, and can be considered as the natural generalization of the connectivity [102].

$$S_{v_i} = \sum_{v_j \in V(V_i)} w_{V_i, V_j} \tag{7}$$

In [6] the authors present *OddBall*, a fast, unsupervised method to detect abnormal nodes in weighted graphs. They also show that the total weight W_{v_i} and the number of edges $\mathcal{E}_{v_i}$ of $\mathcal{G}$ follow a power-law probability distribution. In another interesting work [8], the authors study the weights associate to the reciprocity in mobile phone calls and describe several patterns found. In the particular case of link prediction, models using

Fig. 5 Weighted network W and multiplex network W

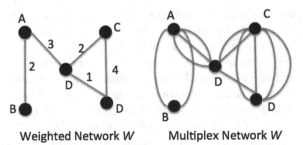

Weighted Network *W*   Multiplex Network *W*

weighted networks are usually simpler than models that use the multiplex network framework, since it has been shown that weights can also help in the algorithm accuracy to predict new edges [95].

8 Extending Graph Models: Multilayer Networks

There have been constant research efforts to extend network models to describe more adequately complex real life networks. As we will see, many of this efforts involve generalizing the definition of the basic blocks that constitute a graph. In this section we will briefly describe one of the most interesting attempts, which is to consider graphs with several types of edges. As we will see this allows for very rich networks models.

8.1 The Layered Point of View: Multilayer Networks

As we have mentioned Sect. 1, graphs are an extraordinarily useful representation of real networks, i.e. any collection of entities with a given relationship between each other. In some practical settings, it is interesting to model a given type of relationship, and the definition of edge is clear. This is the case, for instance, of a co-authorship network, where we are interested in which authors published together, so the edge definition is simply if there exists any publication with both authors' names.

On the other hand, other settings are much more complex, given that the two entities may have more than one type of relevant relationship. For instance, if we are interested in modeling a social network, it could be useful to distinguish among family and work relationships. One way of dealing with this situation is to count everything as a link and try to keep the nature of the edge somehow, for instance as weights in the link. In this way, we can use these weights to restrict or filter some calculations.

Another equivalent approach is to think about these multiple types of relationships as defining different networks or subnetwork, and then connect these accordingly, as they are defined by the same set of nodes. By this point of view, each network is characterized as a layer, so a network may be represented by a number of interconnected layers [82].

This way of framing the structure of interactions among nodes has gained a lot of track in particular in the community of complex networks, which has termed the concept generally as *multiplex networks* [56, 57, 107] or *networks of networks* [40, 52].

One example of this concept in very practical scenario is the case of overlay networks [11]. In the field of engineering and computer science, and specifically in the context of network virtualization, an overlay network is a virtual network created on top of an existing "substrate" network, where only some of its nodes and links are used for the virtual network. An overlay network can be used to share infrastructure and simplify topology, defining a network with different properties than the underlying network, in terms of routing, security, caching, or other network functionalities. Thus, an overlay network can be though of a multiplex network where certain nodes have special types of *virtual* links [38]. The Internet itself started out as an overlay network over the telephone network, and currently many services such as VoIP applications are also defined over the Internet, and so are also work on top of overlays.

Indeed, over the last 40 years many fields of research have in some way or another turned their attention to the same problem under different names: multiplex networks, networks of networks, multidimensional networks, multislice networks, etc. In [82], the authors make a thorough review of different approaches and a provide a complete historic perspective, as well as the different technical aspects of the state of the art. In the following sections we will address some points directly related to the Big Data scenario.

8.2 Models, Methodologies and Other Tools

To get some intuition, we will consider that a multilayer network is a set of nodes belonging to a set of layers, and where any node in any layer may be connected by an edge [82]. (There are even more general and elaborate models of multilayer networks [41], but for our purposes we will set with this description.)

In Fig. 6 we show a basic representation of such a construction, with only two layers and a few nodes. Not every layer must have the same nodes, and edges may connect any pair of nodes between layers. Layers may have any meaning, for instance they could define a type od interaction, so edges have a particular meaning according to the layer, or they may signify a particular time frame, in fact describing the dynamics of the network. These and many other choices fit in this very general definition. Additional constraints may also be put in place, for instance, considering that the set

Fig. 6 Multilayer network representation, showing two layers and two single-layer networks, with some edges connecting nodes between layers

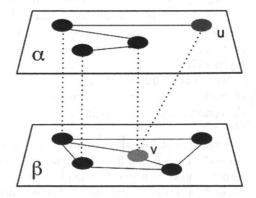

of nodes is fixed for all layers (node alignment), or that edges across layers can only connect the same node in each layer.

Generalization of single-layer algorithms Naturally, research efforts in multilayer networks have focused on extending tools and methods successfully used to study and diagnose single-layer networks. New definitions have been developed for generalizing basic concepts such as node degree, neighborhoods, clustering coefficients, etc. as well as methods and models such as community detection or diffusion dynamics. We explore a few of these ideas next.

In the same way that single-layer networks may be represented by and adjacency matrix, a natural generalization for multi-layer networks is by describing them through tensor representations [41]. Thus, one way to represent a multi-layer network is by providing the tensor $\mathcal{A}_{uv\alpha\beta}$, whose elements take value 1 if node u of layer α is connected with node v of layer β, and 0 otherwise. This generalization corresponds to multi-layer models with node alignment, but it is possible to go beyond to the mode general case where this constraint is relaxed.

One of the most basic concepts in networks is that of degree. As we have seen, this is simply the number of edges incident in a given node. Is it possible to generalize the degree by having into account the weight of the node (weight degree or strength) as well as discriminating by edge direction (*in* or *out*), for directed networks.

Generally speaking, a general procedure to transform a multi-layer network into a single-layer network, so as to apply known (single-layer) techniques is by aggregating layers in some way. This *network aggregation* may be done in several ways according to the study being made.

In the case of the degree, one way to extend this concept is precisely by aggregating through all the layers, i.e. summing all (or some) type of edges to get a value for the degree, e.g. $k_u = \sum_{v\alpha\beta} \mathcal{A}_{uv\alpha\beta}$. There are of course other approaches, for instance, considering thresholds to the quantity of edges that contribute to the degree, and also different normalizing methods [41]. The authors in [30–32] describe various definitions of degree as well as diverse methods to compute degree centrality, which we will comment in detail later, and other related concepts such as neighborhood.

As we have seen in Sect. 2 other central concept in network theory is that of clustering coefficients, and, generally speaking, the notion of transitivity [111]. The extension of these ideas to multi-layer networks has been quite challenging as there are inherent ambiguities in the different possible definitions of these quantity. One well known interpretation of the clustering coefficient for single-layer networks is that is equal to the ratio of closed triples to connected triples. But the definition of triple in multi-layer networks requires some care, as a triple may be defined in multiple ways in this case, depending on the set of constraints that may be imposed regarding the layers (i.e., the type of edges) allowed for consideration in the definition. Great effort has been made in defining a suitable generalization of the clustering coefficient (see [30, 32, 82] among many others). One important conclusion that one may reach from these efforts is that an adequate definition of the clustering coefficient is dependant on the domain on is trying to describe, as the notion of neighborhood or path may differ for different type of networks.

A related and also central concept in single-layer networks is that of communities or modules [111], i.e. the fact that a set of nodes within the network may be more related between them than with other nodes not in the set (or rather, compared to what is expected in a randomly connected network). Despite the numerous research efforts involving some variant of multi-layer networks, few community detection algorithms have been put forward compared to the huge number of such algorithms for single-layer networks. One again, the additional degrees of freedom introduced by these models makes previous definitions ambiguous and non-trivial to generalize.

One important concept in the are of community detection is the choice of a null-model to which compare the network in order to precisely quantify the meaning of "expected in a randomly connected network". The authors in [20, 107] study this problem in the case of multislice modularity, which has been used particularly for the case of temporal networks [106] (i.e. layers representing time).

Centrality measures seek to inform about the relative importance of a given element in the network [111]. There are several versions of centrality measures for different contexts. Central of network science, many of these measures have already been generalized for the case of multi-layer centrality. For instance, PageRank centrality (introduced in Sect. 6.1) has been extended for the more general case through random walker able to also traverse across layers, with suitable definitions of rank also for the layers [113], or differentiating probabilities inside layers from those across layers [152], or with the introduction of biased walkers [68].

8.3   Theoretical Models, Empirical Applications and Other Examples

The possibilities open by the multi-layer formalism are indeed enormous [82]. Many networks from different fields of study fit very well with this formalism.

For instance, Morris et al. [105] develop a two layer theoretical model for transportation in spacial networks and show that different transport regimes may be found. In turn, Cardillo et al. [33] apply the multiplex theoretical framework to the European Air Transportation Network, and they claim that the topology of each layer affects the emergence of structural properties in the aggregate network.

In [106, 107], the authors show how a multislice framework may help understand the communities developed as a function of time in a dataset of the U.S. Senate roll call voting, from 1798 to 2008.

On a more theoretical note, the authors in [56] explore the diffusion properties in the context of (node-aligned) multiplex networks, and explain the dynamics of the diffusion process through the mathematical properties of the system, more specifically, the spectrum of eigenvalues of a matrix built with the Laplacians of each layer.

In [25], Bianconi proposed a statistical mechanics framework to study mutliplex networks, on the premise that a given link between nodes in a layer may be highly correlated to another link between in another layer. The author develops entropy expressions for the multiplex system that may be useful tools for inference problems.

Another interesting application of multiplex networks is in the field of evolutionary game-theory, in particular the study of cooperation in the context of interacting agents. The authors in [57] explore the Prisoner's Dilemma (PD) game in a multiplex setting with random (Erdös-Rényi) networks in each layer, where layers are coupled via the payoff parameter of the PD game, which is the sum of the payoffs in all layers. They show that the resilience of the fraction of agents which stay in the cooperation state is boosted by the introduction of interaction between layers in the system, an important result indicating that the multiplex character of, for instance, social networks, could influence favorably for the emergence of stable of cooperation.

9 Open Challenges

In this chapter, we have shown the relevance and ubiquity of graphs and networks in big data systems and applications. Graph analytics and mining provide value to these big data systems, and the field is really starting to emerge, driven by ever new technologies and applications [55, 148]. Naturally, this remarkable growth pushes the state-of-the-art of current systems until limitations are reached. Next we mention some of the future challenges of the area and some efforts to push the capabilities of today's systems to meet tomorrow's needs.

Streaming Analytics The data stream model may be useful in many situations where data is constantly produced. Many non-trivial challenges arise when dealing with graph data streams, such as the trade-off between data size and accuracy in the computation of graph measured destined to summarize the data. In this way, streaming methologies have become central in many applications where the real-time nature of information flow is relevant or needed [55, 151], as well as in other issues of a more technological origin such as how to compute graph quantities in a distributed or parallel setting [101]. Usually, graph streaming is done typically by providing a stream of edge information to add or subtract, and make some computation in the resulting graph [77]. The challenge is to maintain a precise or at least approximate picture of the network or associated summary variables. A research field strongly connected with this type of issues is graph visualization. For example, in streaming visualization of power-grid networks, the accurate and quick description of the network may be crucial at times of failure where responsibles have minutes or even seconds to respond [148]. In general, there is currently a research effort in this field is focusing in algorithms for directed edges, which may be more general as most practical examples are directed [101].

Representation learning for networks Graphs are intrinsically a non-linear combination of data, not always readily summarizable in every aspect by a few parameters. One interesting research area in network theory is how to decrease the dimensionality of a given graph, while at the same time preserving useful characteristics or informative traits we are in the first place interested in. This dimension reduction techniques play an important part in prediction tasks, such as link prediction (see Sect. 4) or prediction of node attributes such as user interests or functional labels in biological networks.

On the other hand, a manifest characteristic in a graph dataset is its sparseness. Depending on the task or application one is trying to accomplish, this sparseness may be of help or, on the contrary, become a computational burden, leading to inefficient algorithms. For instance, statistical learning in a graph may be hindered by the inherent sparsity of some graphs, especially as they turn into very large graphs, typical of big data domains. Traditional dimensionality reduction methods, such as Principal Component Analysis or Multi-Dimensional Scaling, have been studied widely in the literature [22]. But these techniques usually involve finding eigenvalues of the adjacency matrix (or another equivalent matrix), which normally does not scale well for large graphs.

A way to go around this problem is to find *latent representations* of the network, which encode what is interesting or useful from the graph but are defined in a space with much lower dimensionality [23]. Versions of such latent representations based in deep neural networks have been quite successful in the context of Natural Language Processing [103] and have lead the way for applications in many research areas, including graph analytics and social networks.

In [121], the authors apply this ideas to social networks with the aim of encode social relationships in a continuous vector space, which are then easily exploited by statistical models. The authors propose a random walk algorithm which is used to capture neighborhood similarity, in the sense that nodes with similar neighborhoods will finally present similar representation in vector space. In [136], the authors propose finding latent representations as an optimization problem, by carefully devicing an objective function to capture both local and global characteristics of network structure.

With these type of methods, the learned representations may be used to perform classification tasks or link prediction tasks, which proves to be much more efficient due to the decreased dimensionality. However there is still much research needed as, to date, representation learning have been mainly proposed as heuristic methods aiming to automatically capture useful features from networks, with not much study as to the general validity of the results, both in types of networks and in types of features learned.

All-pairs **computation** On a related note, graphs have the inevitable characteristic of scaling as $O(n^2)$ when taking into account all pairs of nodes. Due to this fact, some graph analytics methods in big data recurrently find limitations whenever computations involve computing over all pairs of nodes, e.g. all-pair shortest paths, or any other type of similar quantity. This is the case, for instance, discussed in Sect. 4 for the Katz measure, which is based in a sequence of matrix-matrix computations [80, 132]. The usual way of bypassing this scaling limitation is to provide estimates or approximations, which depending on the problem and the size of the data may no longer be a satisfactory solution.

10 Conclusions

Many data sources in big data scenarios represent *relationship data*. Being social data, Internet-of-things data, or even semi-structured data such as Twitter posts or document corpora, relationships among entities are usually present, thus making it viable to represent the data as graphs. The graph representation will not always be necessary, but most of the times will be convenient and useful.

In this chapter we have covered many aspects of graph algorithms and network analysis which are important, especially in the case of very large graphs. We have shown that there is an increasing confluence of efforts towards the practical use of graph analytics in the context of big data. One the one hand, the research community continually provides powerful graph algorithms and methods, some of outstanding business success such as PageRank, or the link prediction algorithms in Facebook to grow their social base. On the other hand, a relentless developer community yields ever more powerful software, libraries, as well as commercial and open-source tools, focusing in the implementation of improved graph algorithms and in providing more efficient ways to capture, handle and process large quantities of data as graph information.

Graphs and networks are at the core of the big data era. The advent of big data tools and systems has changed radically the access to real data, and both businesses and the research community have benefited from this by leveraging graph analytics and methods to produce new sources of wealth and information.

References

1. L.A. Adamic, E. Adar, Friends and neighbors on the web. Soc. Network. **25**(3), 211–230 (2003)
2. L.A. Adamic, B.A. Huberman A. Barabási, R. Albert, H. Jeong, G. Bianconi, Power-law distribution of the world wide web. Science **287**(5461):2115a+ (2000)
3. C. Aggarwal, K. Subbian, Evolutionary network analysis: a survey. ACM Comput. Surv. **47**(1), 10:1–10:36 (2014)
4. C. Aggarwal, Y. Xie, P.S. Yu, *On Dynamic Link Inference in Heterogeneous Networks, chap. 35*, pp. 415–426
5. N. Ahmed, J. Neville, R.R. Kompella, Network sampling via edge-based node selection with graph induction (2011)
6. L. Akoglu, M. McGlohon, C. Faloutsos, Oddball: spotting anomalies in weighted graphs, in *Advances in Knowledge Discovery and Data Mining*, ed. by M.J. Zaki, J.X. Yu, B. Ravindran, V. Pudi (Springer, Heidelberg, 2010), pp. 410–421

7. L. Akoglu, H. Tong, D. Koutra, Graph based anomaly detection and description: a survey. Data Min. Knowl. Discov. **29**(3), 626–688 (2015). May
8. L. Akoglu, P.O.S. Vaz de Melo, C. Faloutsos, Quantifying reciprocity in large weighted communication networks, in *Proceedings of the 16th Pacific-Asia Conference on Advances in Knowledge Discovery and Data Mining - Volume Part II, PAKDD'12* (Springer, Heidelberg, 2012), pp. 85–96
9. M. Al Hasan, M.J. Zaki, Output space sampling for graph patterns. Proc. VLDB Endow. **2**(1), 730–741 (2009)
10. U. Alon, Network motifs: theory and experimental approaches. Nat. Rev. Genet. **8**(6), 450–461 (2007)
11. D. Andersen, H. Balakrishnan, F. Kaashoek, R. Morris, Resilient overlay networks (ACM, 2001)
12. Apache Giraph, an iterative graph processing system. http://giraph.apache.org/. Accessed 10 March 2016
13. A.P. Appel, E.R.H. Junior, Prophet – a link-predictor to learn new rules on nell, in *2011 IEEE 11th International Conference on Data Mining Workshops (ICDMW)*, Dec 2011, pp. 917–924
14. Aster SQL-GR Big Data Parallel Graph Analytics. http://www.teradata.com/SQL-GR-Engine/. Accessed 10 March 2016
15. S. Auer, C. Bizer, G. Kobilarov, J. Lehmann, R. Cyganiak, Z. Ives, DBpedia: a nucleus for a web of open data, in *The Semantic Web: 6th International Semantic Web Conference, 2nd Asian Semantic Web Conference, ISWC 2007 + ASWC 2007, Busan, Korea, 11–15 November 2007. Proceedings* (Springer, Heidelberg, 2007), pp. 722–735
16. T. Aynaud, V.D. Blondel, J.-L. Guillaume, R.Lambiotte, Multilevel local optimization of modularity, in *Graph Partitioning* (2013), pp. 315–345
17. L. Backstrom, J. Leskovec, Supervised random walks: predicting and recommending links in social networks, in *Proceedings of the Fourth ACM International Conference on Web Search and Data Mining, WSDM'11* (ACM, New York, 2011), pp. 635–644
18. A. Barrat, M. Barthélemy, R. Pastor-Satorras, A. Vespignani, The architecture of complex weighted networks. Proc. National Acad. Sci. **101**, 3747–3752 (2004)
19. M. Barthélemy, A. Barrat, R. Pastor-Satorras, A. Vespignani, Characterization and modeling of weighted networks. Physica A **346**, 34–43 (2005)
20. D.S. Bassett, M.A. Porter, N.F. Wymbs, S.T. Grafton, J.M. Carlson, P.J. Mucha, Robust detection of dynamic community structure in networks. J. Nonlinear Sci. **23**(1), 013142 (2013)
21. M. Bastian, S. Heymann, M. Jacomy et al., Gephi: an open source software for exploring and manipulating networks. ICWSM **8**, 361–362 (2009)
22. M. Belkin, P. Niyogi, Laplacian eigenmaps and spectral techniques for embedding and clustering. NIPS **14**, 585–591 (2001)
23. Y. Bengio, A. Courville, P. Vincent, Representation learning: a review and new perspectives. IEEE Trans. Pattern Anal. Mach. Intell. **35**(8), 1798–1828 (2013)
24. M. Berlingerio, M. Coscia, F. Giannotti, A. Monreale, D. Pedreschi, Multidimensional networks: foundations of structural analysis. World Wide Web **16**(5), 567–593 (2012)
25. G. Bianconi, Statistical mechanics of multiplex networks: entropy and overlap. Phys. Rev. E **87**(6), 062806 (2013)
26. V.D. Blondel, J.-L. Guillaume, R. Lambiotte, E. Lefebvre, Fast unfolding of communities in large networks. J. Stat. Mech. Theory Experiment **2008**(10), P10008 (2008)
27. S. Boccaletti, V. Latora, Y. Moreno, M. Chavez, D.-U. Hwang, Complex networks: structure and dynamics. Phys. Rep. **424**(4), 175–308 (2006)
28. K. Bollacker, C. Evans, P. Paritosh, T. Sturge, J. Taylor, Freebase: a collaboratively created graph database for structuring human knowledge, in *Proceedings of SIGMOD* (2008)
29. D. Braha, Y. Bar-Yam, Time-dependent complex networks: dynamic centrality, dynamic motifs, and cycles of social interactions, in *Adaptive Networks: Theory, Models and Applications* (Springer, Heidelberg, 2009), pp. 39–50
30. P. Bródka, K. Musial, P. Kazienko, A method for group extraction in complex social networks, in *Knowledge Management, Information Systems, E-Learning, and Sustainability Research,*

ed. by M.D. Lytras, P. Ordonez De Pablos, A. Ziderman, A. Roulstone, H. Maurer, J.B. Imber (Springer, Heidelberg, 2010), pp. 238–247

31. P. Bródka, K. Skibicki, P. Kazienko, K. Musiał, A degree centrality in multi-layered social network, in *2011 International Conference on Computational Aspects of Social Networks (CASoN)* (IEEE, 2011), pp. 237–242

32. P. Bródka, P. Kazienko, K. Musiał, K. Skibicki, Analysis of neighbourhoods in multi-layered dynamic social networks. Int. J. Comput. Intell. Syst. **5**(3), 582–596 (2012)

33. A. Cardillo, J.Gómez-Gardeñes, M. Zanin, M. Romance, D. Papo, F. del Pozo, S. Boccaletti, Emergence of network features from multiplexity. Sci. Rep. **3** (2013)

34. A. Carlson, J. Betteridge, B. Kisiel, B. Settles, E.R. Hruschka Jr., T.M. Mitchell, Toward an architecture for never-ending language learning, in *Proceedings of AAAI* (2010)

35. Cassovary. https://github.com/twitter/cassovary. Accessed 10 March 2016

36. D. Chakrabarti, Y. Wang, C. Wang, J. Leskovec, C. Faloutsos, Epidemic thresholds in real networks. ACM Trans. Inf. Syst. Secur. **10**(4), 1–26 (2008)

37. A. Ching, S. Edunov, M. Kabiljo, D. Logothetis, S. Muthukrishnan, One trillion edges: graph processing at facebook-scale. Proc. VLDB Endow. **8**(12), 1804–1815 (2015)

38. N.M.K. Chowdhury, R. Boutaba, A survey of network virtualization. Comput. Network. **54**(5), 862–876 (2010)

39. A. Clauset, M.E. Newman, C. Moore, Finding community structure in very large networks. Phys. Rev. E **70**(6), 066111 (2004)

40. G. D'Agostino, A. Scala, *Networks of Networks: The Last Frontier of Complexity*, vol. 340 (Springer, Heidelberg, 2014)

41. M. De Domenico, A. Solé-Ribalta, E. Cozzo, M. Kivelä, Y. Moreno, M.A. Porter, S. Gómez, A. Arenas, Mathematical formulation of multilayer networks. Phys. Rev. X **3**(4), 041022 (2013)

42. R.A. de Paula, A.P. Appel, C.S. Pinhanez, V.F. Cavalcante, C.S. Andrade, Using social analytics for studying work-networks: a novel, initial approach, in *2012 Brazilian Symposium on Collaborative Systems (SBSC)*, Oct 2012, pp. 146–153

43. O. Deshpande, D.S. Lamba, M. Tourn, S. Das, S. Subramaniam, A. Rajaraman, V. Harinarayan, A. Doan, Building, maintaining, and using knowledge bases: a report from the trenches, in *Proceedings of the 2013 ACM SIGMOD International Conference on Management of Data, SIGMOD'13* (ACM, New York, 2013), pp. 1209–1220

44. X. Dong, E. Gabrilovich, G. Heitz, W. Horn, N. Lao, K. Murphy, T. Strohmann, S. Sun, W. Zhang, Knowledge vault: a web-scale approach to probabilistic knowledge fusion, in *Proceedings of the 20th ACM SIGKDD International Conference on Knowledge Discovery and Data Mining, KDD'14* (ACM, New York, 2014), pp. 601–610

45. Y. Dong, J. Tang, S. Wu, J. Tian, N.V. Chawla, J. Rao, H. Cao, Link prediction and recommendation across heterogeneous social networks, in *Proceedings of the 2012 IEEE 12th International Conference on Data Mining, ICDM'12* (IEEE Computer Society, Washington, DC, 2012), pp. 181–190

46. D.M. Dunlavy, T.G. Kolda, E. Acar, Temporal link prediction using matrix and tensor factorizations. ACM Trans. Knowl. Discov. Data **5**(2), 10:1–10:27 (2011)

47. M. Faloutsos, P. Faloutsos, C. Faloutsos, On power-law relationships of the internet topology, in *ACM SIGCOMM Computer Communication Review*, vol. 29 (ACM, 1999), pp. 251–262

48. Faunus: Graph Analytics Engine. http://thinkaurelius.github.io/faunus/. Accessed 10 March 2016

49. S. Fortunato, Community detection in graphs. Phys. Rep. **486**(3–5), 75–174 (2010)

50. S. Fortunato, C. Castellano, Community structure in graphs, in *Computational Complexity*, ed. by R.A. Meyers (Springer, Heidelberg, 2012), pp. 490–512

51. Galois: The University of Texas at Austin. http://iss.ices.utexas.edu/?p=projects/galois. Accessed 10 March 2016

52. J. Gao, S.V. Buldyrev, S. Havlin, H.E. Stanley, Robustness of a network of networks. Phys. Rev. Lett. **107**(19), 195701 (2011)

53. Gephi: The Open Graph Viz Platform. https://gephi.org/. Accessed 10 March 2016

54. M. Girvan, M.E. Newman, Community structure in social and biological networks. Proc. National Acad. Sci. **99**(12), 7821–7826 (2002)
55. D.F. Gleich, M.W. Mahoney, Mining large graphs, in *Handbook of Big Data* (2016), p. 191
56. S. Gomez, A. Diaz-Guilera, J. Gomez-Gardeñes, C.J. Perez-Vicente, Y. Moreno, A. Arenas, Diffusion dynamics on multiplex networks. Phys. Rev. Lett. **110**(2), 028701 (2013)
57. J. Gómez-Gardeñes, I. Reinares, A. Arenas, L.M. Floría, Evolution of cooperation in multiplex networks. Sci. Rep. **2** (2012)
58. J.E. Gonzalez, Y. Low, H. Gu, D. Bickson, C. Guestrin, Powergraph: distributed graph-parallel computation on natural graphs, in *Presented as part of the 10th USENIX Symposium on Operating Systems Design and Implementation (OSDI 12)* (2012), pp. 17–30
59. Grafos.ML - Empowering Giraph. http://grafos.ml/index.html. Accessed 10 March 2016
60. M. Granovetter, The strength of weak ties. Am. J. Sociol. **78**(6), 1360–1380 (1973)
61. GraphEngine: serving big graphs in real-time. http://www.graphengine.io/. Accessed 10 March 2016
62. GraphLab Create - an extensible machine learning framework. https://dato.com/products/create/. Accessed 10 March 2016
63. GraphX: Apache Spark's API for graphs and graph-parallel computation. http://spark.apache.org/graphx/. Accessed 10 March 2016
64. T. Gruber, What is an ontology (1993). WWW Site http://www-ksl.stanford.edu/kst/whatis-an-ontology.html. Accessed 07 Sep 2004
65. GUESS: The graph exploration system. http://graphexploration.cond.org. Accessed 10 March 2016
66. P. Gupta, A. Goel, J. Lin, A. Sharma, D. Wang, R. Zadeh, Wtf: the who to follow service at twitter, in *Proceedings of the 22nd International Conference on World Wide Web Conferences Steering Committee* (2013), pp. 505–514
67. I. Guy, S. Ur, I. Ronen, A. Perer, M. Jacovi, Do you want to know?: recommending strangers in the enterprise, in *Proceedings of the ACM 2011 Conference on Computer Supported Cooperative Work, CSCW'11* (ACM, New York, 2011), pp. 285–294
68. A. Halu, R.J. Mondragón, P. Panzarasa, G. Bianconi, Multiplex pagerank. PloS One **8**(10), e78293 (2013)
69. M.A. Hasan, M.J. Zaki, A survey of link prediction in social networks, in *Social Network Data Analytics*, ed. by C.C. Aggarwal (Springer, Boston, 2011), pp. 243–275
70. High-productivity software for complex networks. https://networkx.github.io/. Accessed 10 March 2016
71. P. Holme, C. Edling, F. Liljeros, Structure and time-evolution of an internet dating community. Soc. NetworK. **26**, 155 (2004)
72. P. Holme, J. Saramäki, Temporal networks. Phys. Rep. **519**(3), 97–125 (2012)
73. P. Hu, W.C. Lau, A survey and taxonomy of graph sampling. arXiv preprint arXiv:1308.5865 (2013)
74. IBM Graph: easy-to-use, fully-managed graph database service. https://new-console.ng.bluemix.net/catalog/services/ibm-graph/. Accessed 10 March 2016
75. IBM System G. http://systemg.research.ibm.com/. Accessed 10 March 2016
76. igraph: The network analysis package. http://igraph.org/. Accessed 10 March 2016
77. M. Jha, C. Seshadhri, A. Pinar, A space efficient streaming algorithm for triangle counting using the birthday paradox, in *Proceedings of the 19th ACM SIGKDD International Conference on Knowledge Discovery and Data Mining* (ACM, 2013), pp. 589–597
78. U. Kang, C. Faloutsos, Big graph mining: algorithms and discoveries. ACM SIGKDD Explor. Newslett. **14**(2), 29–36 (2013)
79. U. Kang, C.E. Tsourakakis, A.P. Appel, C. Faloutsos, J. Leskovec, Hadi: mining radii of large graphs. ACM Trans. Knowl. Discov. Data (TKDD) **5**(2), 8 (2011)
80. L. Katz, A new status index derived from sociometric analysis. Psychometrika **18**(1), 39–43 (1953). March
81. D. Kempe, J. Kleinberg, A. Kumar, Connectivity and inference problems for temporal networks, in *Proceedings of the Thirty-second Annual ACM Symposium on Theory of Computing, STOC'00* (ACM, New York, 2000), pp. 504–513

82. M. Kivelä, A. Arenas, M. Barthelemy, J.P. Gleeson, Y. Moreno, M.Λ. Porter, Multilayer networks. J. Complex Network. **2**(3), 203–271 (2014)
83. X. Kong, J. Zhang, P.S. Yu, Inferring anchor links across multiple heterogeneous social networks, in *Proceedings of the 22Nd ACM International Conference on Information & Knowledge Management, CIKM'13* (ACM, New York, 2013), pp. 179–188
84. J. Kunegis, A. Lommatzsch, C. Bauckhage, The slashdot zoo: mining a social network with negative edges, in *Proceedings of the 18th International Conference on World Wide Web, WWW'09* (ACM, New York, 2009), pp. 741–750
85. M. Kurant, M. Gjoka, C.T. Butts, A. Markopoulou, Walking on a graph with a magnifying glass: stratified sampling via weighted random walks, in *Proceedings of the ACM SIGMET-RICS Joint International Conference on Measurement and Modeling of Computer Systems* (ACM, 2011), pp. 281–292
86. N. Lao, T. Mitchell, W.W. Cohen, Random walk inference and learning in a large scale knowledge base, in *Proceedings of the 2011 Conference on Empirical Methods in Natural Language Processing* (Association for Computational Linguistics, Edinburgh, 2011), pp. 529–539
87. C.-H. Lee, X. Xu, D.Y. Eun, Beyond random walk and metropolis-hastings samplers: why you should not backtrack for unbiased graph sampling, in *ACM SIGMETRICS Performance Evaluation Review*, vol. 40 (ACM, 2012), pp. 319–330
88. J. Leskovec, C. Faloutsos, Sampling from large graphs, in *Proceedings of the 12th ACM SIGKDD International Conference on Knowledge Discovery and Data mining* (ACM, 2006), pp. 631–636
89. J. Leskovec, D. Huttenlocher, J. Kleinberg, Predicting positive and negative links in online social networks, in *Proceedings of the 19th International Conference on World Wide Web, WWW'10* (ACM, New York, 2010), pp. 641–650
90. J. Leskovec, J. Kleinberg, C. Faloutsos, Graph evolution: densification and shrinking diameters. ACM Trans. Knowl. Discov. Data **1**(1) (2007)
91. J. Leskovec, L. Backstrom, R. Kumar, A. Tomkins, Microscopic evolution of social networks, in *Proceedings of the 14th ACM SIGKDD International Conference on Knowledge Discovery and Data Mining, KDD'08* (ACM, New York, 2008), pp. 462–470
92. J. Leskovec, K.J. Lang, A. Dasgupta, M.W. Mahoney, Community structure in large networks: natural cluster sizes and the absence of large well-defined clusters. Internet Math. **6**(1), 29–123 (2009)
93. D. Liben-Nowell, J. Kleinberg, The link prediction problem for social networks, in *Proceedings of the Twelfth International Conference on Information and Knowledge Management, CIKM'03* (ACM, New York, 2003), pp. 556–559
94. W. Liu, L. Lü, Link prediction based on local random walk. EPL (Europhysics Letters) **89**(5), 58007 (2010)
95. L. Lü, T. Zhou, Role of weak ties in link prediction of complex networks, in *Proceedings of the 1st ACM International Workshop on Complex Networks Meet Information & Knowledge Management, CNIKM'09* (ACM, New York, 2009), pp. 55–58
96. L. Lü, T. Zhou, Link prediction in weighted networks: the role of weak ties. EPL (Europhysics Letters) **89**(1), 18001 (2010)
97. L. Lü, T. Zhou, Link prediction in complex networks: a survey. Physica A **390**(6), 1150–1170 (2011)
98. G. Malewicz, M.H. Austern, A.J. Bik, J.C. Dehnert, I. Horn, N. Leiser, G. Czajkowski, Pregel: a system for large-scale graph processing, in *Proceedings of the 2010 ACM SIGMOD International Conference on Management of Data* (ACM, 2010), pp. 135–146
99. P. Massa, P. Avesani, Controversial users demand local trust metrics: an experimental study on epinions.com community, in *Proceedings of the 20th National Conference on Artificial Intelligence - Volume 1, AAAI'05* (AAAI Press, 2005), pp. 121–126
100. M. McGlohon, L. Akoglu, C. Faloutsos, Weighted graphs and disconnected components: patterns and a generator, in *Proceedings of the 14th ACM SIGKDD International Conference on Knowledge Discovery and Data Mining, KDD'08* (ACM, New York, 2008), pp. 524–532

101. A. McGregor, Graph stream algorithms: a survey. ACM SIGMOD Rec. **43**(1), 9–20 (2014)
102. G. Menichetti, D. Remondini, P. Panzarasa, R.J. Mondragón, G. Bianconi, Weighted multiplex networks. CoRR, abs/1312.6720 (2013)
103. T. Mikolov, I. Sutskever, K. Chen, G.S. Corrado, J. Dean, Distributed representations of words and phrases and their compositionality, in *Advances in Neural Information Processing Systems* (2013), pp. 3111–3119
104. S. Milgram, The small world problem. Psychol. Today **2**(1), 60–67 (1967)
105. R.G. Morris, M. Barthelemy, Transport on coupled spatial networks. Phys. Rev. Lett. **109**(12), 128703 (2012)
106. P.J. Mucha, M.A. Porter, Communities in multislice voting networks. Chaos **20**(4), 041108 (2010)
107. P.J. Mucha, T. Richardson, K. Macon, M.A. Porter, J.-P. Onnela, Community structure in time-dependent, multiscale, and multiplex networks. Science **328**(5980), 876–878 (2010)
108. Neo4j: The World's Leading Graph Database. http://neo4j.com/. Accessed 10 March 2016
109. M.E.J. Newman, The structure and function of complex networks. SIAM Rev. **45**(2), 167–256 (2003)
110. M.E. Newman, Modularity and community structure in networks. Proc. National Acad. Sci. **103**(23), 8577–8582 (2006)
111. M. Newman, *Networks: An Introduction* (Oxford University Press, Oxford, 2010)
112. M.E. Newman, M. Girvan, Finding and evaluating community structure in networks. Phys. Rev. E **69**(2), 026113 (2004)
113. M.K.-P. Ng, X. Li, Y. Ye, Multirank: co-ranking for objects and relations in multi-relational data, in *Proceedings of the 17th ACM SIGKDD International Conference on Knowledge Discovery and Data Mining* (ACM, 2011), pp. 1217–1225
114. F. Niu, C. Zhang, C. Ré, J. Shavlik, Elementary: large-scale knowledge-base construction via machine learning and statistical inference. Int. J. Semant. Web Inf. Syst. **8**(3), 42–73 (2012). July
115. OrientDB: Distributed Graph Database. http://orientdb.com/. Accessed 10 March 2016
116. L. Page, S. Brin, R. Motwani, T. Winograd, The pagerank citation ranking: bringing order to the web (1999)
117. C.R. Palmer, P.B. Gibbons, C. Faloutsos, Anf: a fast and scalable tool for data mining in massive graphs, in *Proceedings of the Eighth ACM SIGKDD International Conference on Knowledge Discovery and Data Mining* (ACM, 2002), pp. 81–90
118. Y. Park, M. Shankar, B.-H. Park, J. Ghosh, Graph databases for large-scale healthcare systems: a framework for efficient data management and data services, in *2014 IEEE 30th International Conference on Data Engineering Workshops (ICDEW)* (IEEE, 2014), pp. 12–19
119. A. Pavan, K. Tangwongsan, S. Tirthapura, K.-L. Wu, Counting and sampling triangles from a graph stream. Proc. VLDB Endow. **6**(14), 1870–1881 (2013)
120. PEGASUS - Peta-scale graph mining system. http://www.cs.cmu.edu/~pegasus/. Accessed 10 March 2016
121. B. Perozzi, R. Al-Rfou, S. Skiena, Deepwalk: online learning of social representations, in *Proceedings of the 20th ACM SIGKDD International Conference on Knowledge Discovery and Data Mining* (ACM, 2014), pp. 701–710
122. M.A. Rodriguez, The gremlin graph traversal machine and language (invited talk), in *Proceedings of the 15th Symposium on Database Programming Languages* (ACM, 2015), pp. 1–10
123. B. Shao, H. Wang, Y. Li, The trinity graph engine. Microsoft Res., 54 (2012)
124. B. Shao, H. Wang, Y. Li, Trinity: a distributed graph engine on a memory cloud, in *Proceedings of the 2013 ACM SIGMOD International Conference on Management of Data* (ACM, 2013), pp. 505–516
125. SNAP: Stanford Network Analysis Platform. http://snap.stanford.edu/. Accessed 10 March 2016
126. D. Song, D.A. Meyer, D. Tao, Efficient latent link recommendation in signed networks, in *Proceedings of the 21th ACM SIGKDD International Conference on Knowledge Discovery and Data Mining, KDD'15* (ACM, New York, 2015), pp. 1105–1114

127. S. Soundarajan, J. Hopcroft, Using community information to improve the precision of link prediction methods, in *Proceedings of the 21st International Conference on World Wide Web, WWW'12 Companion* (ACM, New York, 2012), pp. 607–608
128. Sparkse: Scalable high-performance graph database. http://www.sparsity-technologies.com/. Accessed 10 March 2016
129. M. Spiliopoulou, Evolution in social networks: a survey, in *Social Network Data Analytics*, ed. by C.C. Aggarwal (Springer, Heidelberg, 2011), pp. 149–175
130. N.V. Spirin, J. He, M. Develin, K.G. Karahalios, M. Boucher, People search within an online social network: large scale analysis of facebook graph search query logs, in *Proceedings of the 23rd ACM International Conference on Information and Knowledge Management* (ACM, 2014), pp. 1009–1018
131. F.M. Suchanek, G. Kasneci, G. Weikum, Yago: a core of semantic knowledge, in *Proceedings of WWW* (2007)
132. X. Sui, T.-H. Lee, J.J. Whang, B. Savas, S. Jain, K. Pingali, I. Dhillon, Parallel clustered low-rank approximation of graphs and its application to link prediction, in *Languages and Compilers for Parallel Computing* (Springer, 2012), pp. 76–95
133. Y. Sun, R. Barber, M. Gupta, C.C. Aggarwal, J. Han, Co-author relationship prediction in heterogeneous bibliographic networks, in *Proceedings of the 2011 International Conference on Advances in Social Networks Analysis and Mining, ASONAM'11* (IEEE Computer Society, Washington, DC, 2011), pp. 121–128
134. Y. Sun, J. Han, C.C. Aggarwal, N.V. Chawla, When will it happen?: relationship prediction in heterogeneous information networks, in *Proceedings of the Fifth ACM International Conference on Web Search and Data Mining, WSDM'12* (ACM, New York, 2012), pp. 663–672
135. J. Sun, C.K. Reddy, Big data analytics for healthcare, in *Proceedings of the 19th ACM SIGKDD International Conference on Knowledge Discovery and Data Mining* (ACM, 2013), pp. 1525–1525
136. J. Tang, M. Qu, M. Wang, M. Zhang, J. Yan, Q. Mei, Line: large-scale information network embedding, in *Proceedings of the 24th International Conference on World Wide Web Conferences Steering Committee* (2015), pp. 1067–1077
137. S. Tasci, M. Demirbas, Giraphx: parallel yet serializable large-scale graph processing, in *Euro-Par 2013 Parallel Processing*, ed. by F. Wolf, B. Mohr, D. an Mey (Springer, Heidelberg, 2013), pp. 458–469
138. T.T. Tchrakian, B. Basu, M. O'Mahony, Real-time traffic flow forecasting using spectral analysis. IEEE Trans. Intell. Transp. Syst. **13**(2), 519–526 (2012)
139. Y. Tian, A. Balmin, S.A. Corsten, S. Tatikonda, J. McPherson, From think like a vertex to think like a graph. Proc. VLDB Endow. **7**(3), 193–204 (2013)
140. TinkerPop: an Apache2 licensed graph computing framework for both graph databases (OLTP) and graph analytic systems (OLAP). http://tinkerpop.apache.org/. Accessed 10 March 2016
141. Titan: Distributed Graph Database. http://thinkaurelius.github.io/titan/. Accessed 10 March 2016
142. C.E. Tsourakakis, Fast counting of triangles in large real networks without counting: algorithms and laws, in *ICDM'08* (IEEE Computer Society, Washington, DC, 2008), pp. 608–617
143. T. Wang, Y. Chen, Z. Zhang, T. Xu, L. Jin, P. Hui, B. Deng, X. Li, Understanding graph sampling algorithms for social network analysis, in *Proceedings of the 2011 31st International Conference on Distributed Computing Systems Workshops, ICDCSW'11)* (IEEE Computer Society, Washington, DC, 2011), pp. 123–128
144. W.Y. Wang, K. Mazaitis, W.W. Cohen, Programming with personalized pagerank: a locally groundable first-order probabilistic logic, in *Proceedings of the 22nd ACM International Conference on Information and Knowledge Management (CIKM 2013)* (2013, to appear)
145. D. Wang, D. Pedreschi, C. Song, F. Giannotti, A.-L. Barabasi, Human mobility, social ties, and link prediction, in *Proceedings of the 17th ACM SIGKDD International Conference on Knowledge Discovery and Data Mining, KDD'11* (ACM, New York 2011), pp. 1100–1108
146. D.J. Watts, S.H. Strogatz, Collective dynamics of 'small-world' networks. Nature **393**(6684), 409–10 (1998)

147. K. Wehmuth, A. Ziviani, E. Fleury, A unifying model for representing time-varying graphs. In *2015 IEEE International Conference on Data Science and Advanced Analytics, DSAA 2015, Campus des Cordeliers, Paris, France, 19–21 October 2015* (2015), pp. 1–10, 2015
148. P.C. Wong, C. Chen, C. Gorg, B. Shneiderman, J. Stasko, J. Thomas, Graph analyticslessons learned and challenges ahead. IEEE Comput. Graph. Appl. **5**, 18–29 (2011)
149. S.H. Yook, H. Jeong, A.L. Barabasi, Weighted evolving networks. Phys. Rev. Lett. **86**(25), 5835–5838 (2001)
150. J. Zhang, X. Kong, P.S. Yu, Transferring heterogeneous links across location-based social networks, in *Proceedings of the 7th ACM International Conference on Web Search and Data Mining, WSDM'14* (ACM, New York, 2014), pp. 303–312
151. Y. Zhao, Mining Large Graphs. Ph.D. thesis, University of Illinois at Chicago (2013)
152. D. Zhou, S.A. Orshanskiy, H. Zha, C.L. Giles, Co-ranking authors and documents in a heterogeneous network, in *Seventh IEEE International Conference on Data Mining, 2007. ICDM 2007* (IEEE, 2007), pp. 739–744
153. R. Zou, L.B. Holder, Frequent subgraph mining on a single large graph using sampling techniques, in *Proceedings of the Eighth Workshop on Mining and Learning with Graphs* (ACM, 2010), pp. 171–178

Granular Social Network: Model and Applications

Sankar K. Pal and Suman Kundu

Abstract Social networks are becoming an integral part of the modern society. Popular social network applications like Facebook, Twitter produces data in huge scale. These data shows all the characteristic of Big data. Accordingly, it leads to a deep change in the way social networks were being analyzed. The chapter describes a model of social network and its applications within the purview of information diffusion and community structure in network analysis. Here fuzzy granulation theory is used to model uncertainties in social networks. This provides a new knowledge representation scheme of relational data by taking care of the indiscernibility among the actors as well as the fuzziness in their relations. Various measures of network are defined on this new model. Within the context of this knowledge framework of social network, algorithms for target set selection and community detection are developed. Here the target sets are determined using the new measure granular degree, whereas it is granular embeddedness, together with granular degree, which is used for detecting various overlapping communities. The resulting community structures have a fuzzy-rough set theoretic description which allows a node to be a member of multiple communities with different memberships of association only if it falls in the (rough upper - rough lower) approximate region. A new index, called normalized fuzzy mutual information is introduced which can be used to quantify the similarity between two fuzzy partition matrices, and hence the quality of the communities detected. Comparative studies demonstrating the superiority of the model over graph theoretic model is shown through extensive experimental results.

1 Introduction

Social network is a collection of social ties among friends and acquaintances. After a child is born, (s)he gets immediately connected with the members of the family. Over the course of time (s)he develops connections with larger networks like village,

S.K. Pal · S. Kundu (✉)
Center for Soft Computing Research, Indian Statistical Institute,
Kolkata 700108, India
e-mail: suman@sumankundu.info

© Springer International Publishing AG 2017
A.Y. Zomaya and S. Sakr (eds.), *Handbook of Big Data Technologies*,
DOI 10.1007/978-3-319-49340-4_18

school, and office. Due to the technological advancement, distance travel, global communication and digital interaction have been growing in numbers and in effect social networks are also growing steadily in complexity. This complex "connectedness" of modern society got the attention of different fields of studies.

The term "social network" was coined by the social scientists. The network was considered as a theoretical construct to study the relationships between individuals, groups, organizations or even the entire society. However, the recent boom in online services related to social networks, viz Facebook, Twitter, WhatsApp, LinkedIn, provides new research opportunities to the scholars of computer science, because the data available from these networks are dynamic, large, diverse and complex. That is, it shows all the characteristics of Big Data [69] such as Velocity, Volume, and Variety. Accordingly, recent algorithms [43, 53, 61, 85] are addressing the Big Data issues related to social networks.

Since its inception in early 20th century, social networks are represented using graphs [58], and graph analysis has become crucial to understand the features of these networks [24]. Due to the recent revolution in computing (processing) power, one can now handle relatively larger real networks [67] potentially reaching millions of vertices. Accordingly, it leads to a deep change in the way social networks were being analyzed.

In contrast to random network, social networks shows fascinating patterns and properties [57]. The degree distribution follows power law [5, 21] or truncated geometric distribution [8]. Diameter of the network is very small compared to the size of the network, and the network possesses high concentration of edges in its certain parts forming groups. Such groups with high internal edge density within themselves and low between them characterizes the community structure (or clusters) of the network.

Two of the major research areas in Social Network Analysis (SNA) are (a) analysis of network values [16, 39, 96], and (b) community detection [9, 65]. The objective of the former is to analyze the relative importance of a node in the network. One of the major research application of this area is **target set selection**. In this problem, one seeks to find a set of influential nodes for which the information diffusion over the network is maximum. It is effectively used in viral marketing [81] through online social networks. In addition, this can be used for finding the top stories from a news network, spreading of social awareness or combat with deceptions spreading over social media. Other applications of network values include study on epidemic spreading, diffusion of innovations, homophily analysis and optimal price-setting in market.

Several attempts [16, 35, 37, 38, 40, 76, 81, 95] were made to solve the target set selection problem. However, these are very restrictive either in terms of performance or in execution time, specifically for large scale social networks. For example, greedy hill climbing algorithm of [37] provides approximation within a factor of $(1 - \frac{1}{e} - \epsilon)$ to the optimal solution. Here e is the base of natural logarithm, and ϵ depends on the accuracy of Monte-Carlo estimate of influence spread. But it takes days to compute the set of seeds, even for a moderate sized social network. In contrast to this, heuristic methods (e.g., [10, 11]) are fast but provides sub-optimal output as compared to the greedy method of [37].

Community detection, on the other hand, deals with the problem of identifying virtual groups in a network. A community is formed when a group of nodes are more densely connected with each other compared to rest of the network. In addition to the social implication study of such groups, the solution to this problem has broad application in different fields. For example, in world wide web it will help to optimize the Internet infrastructure [42], in a purchase network it can boost the sell by recommending the appropriate products [78], and in computer network it will help to optimize the routing table creation [84].

Scientists from several disciplines studied the community detection problem for a long time [28, 54, 62, 63, 77, 80, 89]. These involve mainly two strategies for finding different communities in a network. The first approach considers a partition of the whole network into disjoint communities (i.e., a node belongs to only one community). The second strategy, on the other hand, allows a node to be a member of multiple communities with equal membership. However, for large-scale networks, it is possible that a node may belong to more than one community with different degrees of association.

Beside these, highly overlapping neighborhoods in real life big social networks enforce uncertainties in decision making. Although the graph modeling has been in use for social networks since its inception in 1934 [58], a better modeling to deal with these uncertainties is in need. The new modeling may lead to a deep change in the way social networks were being analyzed.

2 Preliminaries

2.1 Social Network Analysis

At a more precise level, a network is any collection of objects in which some pairs are connected by links [17]. Based on configuration, different forms of relationships or connections may be used to define links. Due to this flexible options, it is easy to find network in different domains. Graph based modeling is a typical way to represent social networks. Let us first explain some of the basic elements of graphs before providing a review on modeling social network.

Graph, Nodes and Edges: Conceptually, a graph is formed by nodes (vertices) and edges (links) connecting the nodes. Formally, a graph is an ordered pair (V, E) where V is the set of nodes and E is the set of edges, formed by pairs of nodes.

Undirected and Directed Graphs: Edges can be symmetric such as in Fig. 1a, or asymmetric like in Fig. 1b. The former is referred as undirected graph or simply graph and the latter is called directed graph.

Graphs as Models of Networks: Graphs are useful in social network study as they serve as mathematical models of network structure. Let us now replace aforesaid toy example Fig. 1 with a real social network of Fig. 2. It is popularly known as Zachary karate club [92]. This network shows the friendship relations between 34 members of a US karate club in 1970s. People are represented by nodes and edges are constructed where two people shows friendship outside the context of club. Note that the actual placement of nodes is immaterial. All that matters is which node is connected with which others. Statistics about the network is shown in Table 1.

Paths and Cycles: A path is a sequence of nodes where each consecutive pair in the sequence is connected by an edge. For example, in the Zachary karate club we have a path from node 1 to 34 as 1, 14, 3, 34. A path can repeat nodes such as, 1, 4, 13, 1, 12. Cycle is a specific kinds of path which forms a ring like structure. For example, in Zachary karate club 11, 5, 7, 6, 11 is a cycle.

Connectivity: Whether we are dealing with small or large scale social networks, it is natural to check if every node can reach every other node via a path. We say a graph is connected if for every pair of nodes there exists a path between them. For any social network it may happen that two persons are not reachable via a valid path. This then leads to a disconnected network. For example, Fig. 3 shows a disconnected network of metabolic cellular network.

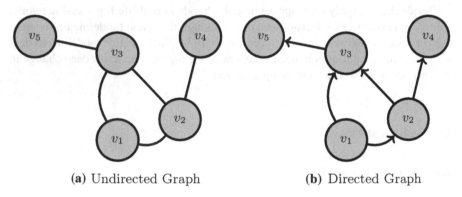

(a) Undirected Graph   (b) Directed Graph

Fig. 1 Example: graphs, nodes, edges

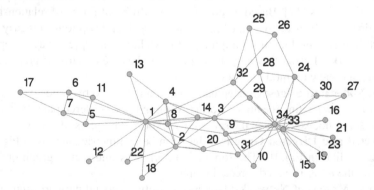

Fig. 2 Zachary karate club

Table 1 Statistics of Zachary karate club network

Nodes	34
Edges	78
Nodes in largest Weakly Connected Component(WCC)	34
Edges in largest WCC	78
Nodes in largest Strongly Connected Component(SCC)	34
Edges in largest SCC	78
Diameter	5
Avg. clustering coefficient	0.570638

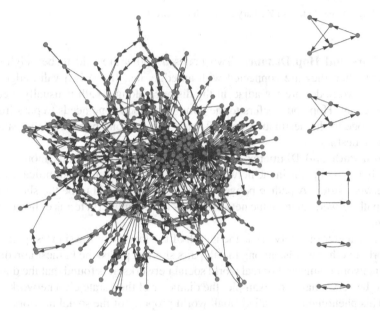

Fig. 3 Metabolic cellular network data for Oryza Sativa [33]

Components (Weakly Connected versus Strongly Connected) If a graph is not connected it breaks apart naturally. These separate subsets are called components. Each of the components when considered separately represents a connected graph. For example the disconnected network in Fig. 3 has 6 connected component.

For directed social network the notion of connectivity can be expressed in two different forms, namely, weakly connected component and strongly connected component. A weakly connected component is a subgraph of a directed graph such that for every pair of nodes u, v in the subgraph, there is an undirected path from u to v and a directed path from v to u. On the other hand, a strongly connected component is a subgraph of a directed graph such that for every pair of nodes u, v in the subgraph, there is a directed path from u to v and a directed path from v to u.

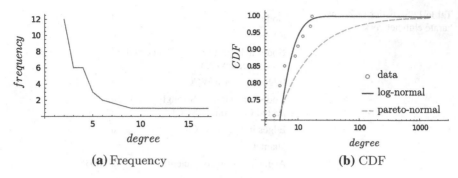

(a) Frequency   (b) CDF

Fig. 4 Degree distribution of Zachary karate club network

Neighbors and Hop Distance: Two nodes u and v are said to be neighbors or adjacent when they are connected with an edge, i.e., (u, v) is a valid edge in the graph. If two nodes are not adjacent the distance along a path is usually measured by hop count. Hop count refers to the number of nodes one needs to pass from the source node to the destination node. That is a hop is one portion of the path from source to destination.

Shortest Path and Diameter: One may reach to a node u from another node v through different paths in the network. Shortest among them has significant value in the network study. A path p between nodes u and v is said to be the shortest if no other path between them in the network holds lesser length (in terms of hop distance) than p.

The diameter of a network is the length of the longest of the shortest paths in the network. In other words, among the all pairs shortest paths, the highest hop distance is the network diameter. For real world social networks, it is found that the diameters tend to be very small. For example, the diameter of the karate club network (Fig. 2) is 5. This phenomenon is called small world property of the social network.

Degree and Degree Distribution: Degree of a node is measured by the number of incident edges on it. It is denoted by $d(v)$. For directed graphs, a node has two different degrees, the in-degree, which is the number of incoming edges, and the out-degree, which is the number of outgoing edges.

Degree distribution refers to the frequency distribution of the degrees of a network. Degrees are usually plotted in x-axis and the frequencies are plotted in y-axis. Figure 4a shows the degree distribution of the karate club data. Similarly we can plot cumulative distribution function (CDF) as shown in Fig. 4b.

An observation can be made from the degree distribution of the karate club data that the number of nodes with higher degree is low as compared to the number of nodes with lower degree values. Similar long tail can be found in most of the real world networks. This is different from random graphs and due to this, social networks are referred as scale free network.

2.2 Fuzzy Sets

Traditional set theory deals with whether an element "belongs to" or "does not belong to" a set. Fuzzy set theory [93], on the other hand, concerns with the continuum degree of belonging, and offers a new way to observe and investigate the relation between sets and its members. It is defined as follows:

Let X be a classical set of objects, called the universe. A fuzzy set A in X is a set of ordered pairs $A = \{(x, \mu_A(x)) | x \in X\}$, where $\mu_A : X \to M$ is called the membership function of x in A which maps X to membership space M. Membership $\mu_A(x)$ indicates the degree of similarity (compatibility) of an object x to an imprecise concept, as characterized by the fuzzy set A. The domain of M is $[0, 1]$. If $M = \{0, 1\}$, i.e., the members are only assigned either 0 or 1 membership value, then A possesses the characteristics of a crisp or classical set.

The set of all elements having positive memberships in fuzzy set A constitutes its support set, i.e.,

$$Support(A) = \{x | \mu_A(x) > 0\}. \tag{1}$$

The cardinality of the fuzzy set A is defined as

$$|A| = \sum_{x \in X} \mu_A(x). \tag{2}$$

Union and intersection of two fuzzy sets A and B are also fuzzy sets and we denote them as $A \cup B$ and $A \cap B$ respectively. The membership functions characterizing the union and intersection of A and B are as follows:

$$\mu_{A \cup B}(x) = \max(\mu_A(x), \mu_B(x)), x \in X \tag{3}$$

$$\mu_{A \cap B}(x) = \min(\mu_A(x), \mu_B(x)), x \in X. \tag{4}$$

2.3 Rough Sets

Let X be a classical set of objects, in a universe of discourse U. Under situations when relations exist among elements of U, X might not be exactly definable in U as some elements of U that belong to the set X might be related to some elements of U that do not belong to set X.

When a relation, say R, exists among elements of U, limited discernibility draws elements of U together governed by the relation R resulting in the formulation of granules in U. Here, a set of elements in U that are indiscernible from or related to each other is referred to as a granule. Let us represent granules using Y and the family of all granules formed due to the relation R using U/R.

As mentioned earlier, the relation R among elements of U might result in an inexact definition of X. To tackle such cases, in rough set theory, X is approximately represented by two exactly definable set $\underline{R}X$ and $\overline{R}X$ in U given as

$$\underline{R}X = \bigcup \{Y \in U/R | Y \subseteq X\} \tag{5}$$

$$\overline{R}X = \bigcup \{Y \in U/R | Y \cap X \neq \} \tag{6}$$

In the above, the set $\underline{R}X$ is defined by the union of all granules that are subsets of the set X and the set $\overline{R}X$ is defined by the union of all granules that have non-empty intersection with the set X. The sets $\underline{R}X$ and $\overline{R}X$ are respectively called the lower approximation and upper approximation of X with the imprecise concept R.

Fuzzy set and rough set are reputed to handle uncertainities arising from overlapping concepts (or characters) and granularity in the domain of discourse respectively. While the former uses the notion of class membership of an element, the latter hinges on the concept of approximating from lower and upper side of a set defined over a granular domain.

2.4 Granular Computing

Granular computing (GrC) is a problem solving paradigm with the basic element, called granules. The construction of granules is a crucial process, as their sizes and shapes are responsible for the success of granular computing based models. Further, the inter and intra relationships among granules play an important role. A granule may be defined as the clump of elements that are drawn together, for example, by indiscernibility, similarity and functionality. Each of the granules according to its shape and size, and with a certain level of granularity may reflect a specific aspect of the problem. Granules with different granular levels may represent a system differently.

Granulation is the process of construction, representation and interpretation of granules. It involves the process of forming larger objects into smaller and smaller into larger based on the problem in hand. According to Zadeh [94], "granulation involves a decomposition of whole into parts. Conversely, organization involves an integration of parts into whole."

One of the realizations behind GrC is that - precision is sometimes expensive and not very meaningful in modeling and controlling complex systems. When a problem involves incomplete, uncertain and vague information, it may sometimes become difficult to differentiate the individual elements, and one may find it convenient to consider granules to represent a structure of patterns evolved by performing operations on the individual patterns [26]. Accordingly, GrC became an effective framework in designing efficient and intelligent information processing systems for various real life decision-making applications. The said framework can be modeled, for example, with the principles of fuzzy sets, rough sets, neural networks, power algebra, interval analysis [73]. For further details on the significance and various applications of GrC, one may refer to [7, 66, 68, 70, 72, 91].

3 Literature Review

3.1 Modeling Social Networks

As mentioned in Sect. 2.1, network structures with actors and their relationships are usually modeled as graphs. In sociology, this representation is sometime referred as sociogram. In a sociogram, actors are represented by vertices of a graph, and relations by edges. Graphs appear naturally here as it is useful to represent how things are either physically or logically linked together. Sociogram was developed by Moreno [58] to analyze the choices of preferences within a group. It was used to diagram the structure and patterns of group interactions.

Social network data, sometime represented in two-way matrices, is termed as sociomatrices [88]. The two dimensions of a sociomatrix are indexed by the senders (rows) and the receivers (column) of relationships. Usually the matrix has n rows and n columns, where n represents the number of actors in the network. Thus a basic sociomatrix is square. Sociomatrices were first used together with sociogram by Moreno [58] who showed how social relationship can be pictured through these.

The same network can also be represented using the relational form. Relational algebras (also called role algebras) are used to analyze the structure of social roles by emphasizing multiple relations rather than actors. Harrison White and his students [6, 90] pioneered this approach as an extension to block modeling. A block model is a representation of objects in groups based upon patterns that occur in the relations between these objects [3]. The structure of a block model is a matrix in which the $(i, j)^{th}$ entry denotes the number of directed edges from nodes in cluster i to nodes in cluster j. A block models can represent any pattern that arises in the relations between objects, such as bipartite relations, hierarchies, rings, bridges, and other unique aggregate connectivity patterns between groups of vertices.

Another approach to model social networks is based on statistical modeling. The idea of statistical modeling of network is to represent the main features of the social network by a few parameters and express the uncertainty of those estimates by standard error, p-value, posterior distribution etc. There are two ways for statistical modeling of network, viz. model-based inference and design-based inference. When a sample is drawn from a larger graph, design-based method can be used. Link-tracing technique [83] is one kind of design based method. Examples of this technique are snowball design and random walk design. On the other hand, in model-based inference, it is required to construct a probability model with the assumption that the observed data can be regarded as the outcome of a random draw from this model [25, 27]. Multiple linear regression models are an example.

Thus several models for describing social network exist starting from 1930s. Recently, the development on modeling social network problems using multi-agent theory and/or game theory has been observed. In their paper [41], Kleinberg et. al. modeled a network with n distinct agents who build link to one another based on a strategic game. The payoff to an agent arises as a difference of costs and benefits. Narayanam et. al. [60], on the other hand, mapped the information diffusion process of social network to the formation of coalitions in an appropriately defined cooperative game. In [34], authors modeled the user interactions of a network to explore the dynamic evolutionary process of knowledge sharing among users using the agent-based computational approach. But the focus of these researches is mostly problem centric.

Fuzzy set theory has also received attention on social network analysis in recent years. In their work, Nair and Sarasamma [59] analyzed multi-modal social networks using fuzzy graphs and referred it as fuzzy social network. Later in 2008, Davis and Carley [14] used a stochastic model to identify fuzzy overlapping groups in social networks. Here they modeled the fuzzy overlapping group detection using an optimization problem. Another area where fuzzy sets have been used by different scientists is positional analysis (finding similarities between actors in the network) of social networks [22]. Instead of a general framework, these recent developments of fuzzy set theoretic approach in social network are more focused on a particular type of the network or particular application of the network.

Beside these, an attempt was made to use the concept of granular computing to model relational database for association discovery [32]. The technique is a specialized version of the general relational data mining framework which efficiently provides the search space for association discovery. Also, there were several research investigations focused on a problem oriented modeling of social network using different soft computing tools. For example, Chen and Li [9] proposed evolutionary computing based algorithm to detect community structures in complex networks. Genetic algorithm based diffusion model for information cascade in a social network is used in [46, 52]. For target set selection problem, Wang et al. [86] proposed a set-based coding genetic algorithm. However, none of these techniques provides any general framework which can serve as a generic platform, similar to sociogram or sociomatrices, to analyze social network data in view of different problems in the field.

Algorithm 1: Greedy Hill Climbing Algorithm

---

    input   : A Social Network $G(V, E)$ and k
    output   : Set $S \in 2^V$ having cardinality k

    initialization: $S := \emptyset$
    while $|S| \neq k$ **do**
        $v^* \leftarrow \arg\max_{v \in V \setminus S} \hat{\sigma}(S \cup \{v\})$; /* $\hat{\sigma}(.)$ returns the estimated influence   */
        $S \leftarrow S \cup \{v^*\}$;

3.2 Target Set Selection

In the area of information diffusion, finding a target set is to find the influential nodes mainly in terms of the total influence in the network. The natural solution to the problem will be to select those persons having higher numbers of neighbors. That is, select the nodes based on their degree centrality scores. Domingos and Richardson were the first to study the problem [16, 81] in the algorithmic aspect and proposed probabilistic methods to solve it. Later, Kempe et al. formulated it as a discrete optimization problem [37] and showed that the problem is NP hard. They proposed a greedy hill climbing algorithm shown in Algorithm 1. In each iteration of the algorithm, marginal contribution of every non seed node (i.e., nodes in $V \setminus S$) to the information diffusion is separately estimated and the highest contributor is selected as the next seed. Thus the algorithm maximizes the influence contribution during seed selection. Hence it is able to find higher quality seeds. However, for the same reason it leads to high computational time, specially for large scale networks. The main drawback of the algorithm comes from the marginal contribution estimation. There is no deterministic methods available till date to get the marginal contribution of a node. In their paper, Kempe et al. [37] uses Monte Carlo simulation for the estimation os such contribution. As the process of information diffusion is highly stochastic, the simulation needs to be performed for a sufficiently large number of times to obtain more accurate results. It may take days to identify top 50 seeds even on a graph of moderate size of 30 K nodes [12]. To overcome this drawback, several algorithms were proposed in last few years [11, 18, 30, 49]. Notably, in [49] Leskovec et al. presented a *cost-effective lazy forward* (CELF) method which exploits the sub-modularity property of the influence function. For any given set function $\sigma(.)$, sub-modularity property confirms that the effect of v to a subset is always higher than that of the super set. That is, $\sigma(S \cup \{v\}) > \sigma(T \cup \{v\})$ if $S \subset T$. Authors argued in [49] that most of the realistic outbreak detection objectives are *sub-modular*. Their experiments with blog network and water network show that CELF runs 700 times faster than the greedy algorithm of [37]. However, CELF method still takes hours to generate 50 seeds [11]. Improvement in execution time was also sought by considering the properties of the underlying diffusion model. One of such popular diffusion models is Independent cascade model. In this model of information diffusion, information propagates in discreet time steps. In each time t, one node with the information tries to influence one of its neighbors who does not have the information already. Success depends on a probability called propagation probability. Irrespective of the success, the same node will never get a chance to influence the same neighbor again. In [11], authors provide two new greedy algorithms designed on independent cascade model of information diffusion. One of them, NewGreedyIC, uses a random removal of edges instead of Monte Carlo simulation to estimate a node's marginal contribution. The random removal uses the propagation probability to identify the edges to be removed. This process leads to an improvement in execution time. Further they integrated the idea of the CELF inside the NewGreedyIC and proposed improved MixGreedyIC. Goyal et al. [30] suggested an improved version of CELF as CELF++ and showed empirically that the algorithm is faster than CELF with insignificant amount of additional memory usage. In CELF++, authors maintained a heap with intermediary results of the Monte Carlo simulation, which reduces the execution time of the subsequent iterations. A greedy sketch-based influence maximization (SKIM) was described in [13] very recently and it is reported that it may be scaled to large social network data.

In contrary, several heuristic algorithms [10–13] were proposed which improve the performance compared to the centrality measures while the execution time remains lower than that of the greedy. One such algorithm is degree discount heuristic of [11], which runs with the following principle. If a node u is already considered as a seed then in later iterations a node v's degree is calculated after discounting the edge $e(u, v)$. This algorithm works very well for undirected social networks. In 2012, Wang et al. [87], reported a heuristic method named prefix excluding maximum influence path (PMIA) where the propagation probability of a path is calculated and used to identify a node's contribution in the diffusion. These heuristic approaches used underlying diffusion principles to improve the performance. Some of the heuristic algorithms, on the other hand, are designed to perform well on specific social networks. For example, Chen et al. [12] proposed a liner time algorithm for directed acyclic graphs, and Gomez- Rodriguez and Schölkopf [29] proposed probability based methods to identify influential nodes for continuous time diffusion networks. Similarly, Aslay et al. [4] described a target set selection algorithm for topic-aware influence maximization queries and Li et al. [51] reported a location-aware target set selection method using spacial-based indexes.

3.3 Community Detection

Community detection is to identify virtual groups of a network. The main challenge is to identify the groups and possibly their hierarchical organization by only using the network topology. One of the first studies on community identification was carried out by Rice [80]. In the work, clusters were identified in a small political body based on their voting patterns. Later in 1955, Jacobson [89] studied community structure within a government agency [89]. They have separated work-groups by removing those people who work with different groups. This idea of removing edges is the basis of several algorithms in recent times. One such algorithm, presented by Girvan and Newman [28], aims at the identification of the edges lying between two communities for possible removal. These edges were identified based on their centrality values. The concept is considered as the start of modern era in community detection. In [63], Newman [63] proposed the modularity measure to quantify the quality of the identified community structure. The modularity is, up to a multiplicative constant, the number of edges falling within groups minus the expected number in an equivalent network with edges placed at random [64]. Modularity value can be either positive or negative. Positive value of modularity indicates the presence of community structures. So, one may partition the network with the aim to maximizing the modularity value of the community structure. This idea of optimizing modularity using some optimization technique is used to identify the community structure by Newman [62]. On the contrary, Raghavan et al. [77] described a near liner localized community detection algorithm based on label propagation which does not optimize any similar measures of community strength. In this method, initially each node is assigned with a unique label. At every iteration of the algorithm, each node adopts the label which is used by maximum number of its neighbors. Ties are broken randomly. At the end of the algorithm, nodes with the same label are grouped together to form a community. Density based graph partitioning algorithm is also available in the literature. Example of one such algorithm is by Falkowski et al. [20]. More traditional methods such as hierarchical [31] and partition based clustering, where vertices are jointed into groups as per their mutual similarities, are also used for identifying communities in a social network.

All these algorithms discussed above create a crisp partition in the network. That is, a node belongs to a single community only. However in a real life a person may belong to multiple groups, i.e., the existence of overlapping community structures. In [36, 79], authors showed that overlapping is indeed a significant feature of many real world social networks. One of the most popular overlapping community detection algorithms, namely, clique percolation method (CPM) of [71], detects overlapping communities by searching of adjacent cliques. The algorithm first searches for all the cliques of size k and constructs another graph by considering a k-clique as a node. A link is added when two cliques share $k - 1$ edges. Each connected component on this new graph is considered to be a community in the network, and k-cliques belonging to a component are considered to be in the same community. Overlap is possible because a node can be a member of multiple cliques. A version of the same algorithm for weighted network was proposed by Farkas et al. [23]. Here k-cliques with weight greater than a threshold are considered for the community. Another approach to get overlapping community structure is to partition links instead of nodes. Ahn et al. [2] used hierarchical clustering to partition edges of the network. In this algorithm, each edge belongs to a unique cluster but nodes may naturally belong to different clusters. Evans and Lambiotte [19], on the other hand, constructed a new weighted line graph by considering links of the original graph as a node and then partition this new graph using disjoint community detection algorithm. Although the link partitioning for overlapping detection seems conceptually to be natural, there is no guarantee that it provides higher quality than the node based detection does [24]. Readers may refer to [24, 50] for review on different Community detection algorithms.

4 Fuzzy Granular Social Networks (FGSN)

Social network is nothing but a collection of relations between social actors and their interactions. Social actors often form closely operative groups among themselves, which are often indistinguishable. A granule is a clump of objects (points) in the universe of discourse, drawn together, for example, by indistinguishability, similarity, proximity or functionality [94]. So, the characteristic of indistinguishability among closely operative groups of the social actors may be modeled using the concepts of granules for further processing.

Further, the basic concepts of conceptual similarities between nodes, cluster of nodes, relations between nodes and their interactions etc. do not lend themselves to precise definition, i.e., they have ill-defined boundaries. So, it is appropriate and natural if a social network is viewed in terms of a collection of *fuzzy granules*. Based on these notions, a new unified framework to model social networks effectively and efficiently in the framework of granular computing is developed [44, 45]. In this model a granule is constructed around a node with fuzzy boundary. The membership function for computing the degree of belonging of a node to the said granule is determined depending upon the problem in hand. Within this framework, some of the popularly known network measures are redefined [44].

4.1 The Model

Global phenomenon of a social network always ensembles the local behaviors of individuals as well as their closely related neighborhoods. While the concept of neighborhoods in the network can be modeled in terms of granules, the vagueness in term "closeness" can be quantified using fuzzy set theory. In this section, we provide the description of the model *fuzzy granular social network* (FGSN).

Knowledge Representation: Let us consider the graph $G(V, E)$ represents a social network, where V is the set of all nodes (or vertices) and E represents the relationships (or edges). If I is the unit interval $[0, 1]$, a fuzzy granular neighborhood defined over V is a function $\phi : V \rightarrow A(V)$, which assigns every node $v \in V$ to a fuzzy set $A \in I^V$. When $\phi(v)$ is non empty, we call it the fuzzy neighborhood of the node v, i.e., $\phi(v)$ is the granule defined around the node v. Due to the complex nature of social networks a node can be a member of different such neighborhood sets reflecting its different degrees of association. Let family of fuzzy sets associated with the node $v \in V$ be $\Phi(v)$. $\Phi(v)$ represents the neighborhood sets of node v. A fuzzy granular social network is represented by a triple:

$$S = (C, V, G) \text{ where} \tag{7}$$

- V is a finite set of nodes of the network
- $C \subseteq V$ is a finite set of granule representatives
- G is the finite set of all granules,

  i.e., $G = \{\bigcup \Phi(c) | c \in C\}$

A granule $g \in G$ around a representative node $(c \in C)$ is constructed by assigning fuzzy membership values to its neighborhood nodes. Due to the overlapping nature of the neighborhoods, a node may belong to more than one granule. Their association with different granules may have different degrees as well. However, in case of directed social network, two different granules may be constructed around one single node. One for inbound relations and other for the outbound relationships [44].

4.2 Network Measures of FGSN

A social network is analyzed based on social measures defined over its graph representation. Similarly, several equivalent granular measures available for FGSN are provided in this section.

Let us first see the construction of FGSN of our example network shown in Fig. 2. Our objective here is to model the graph representation $G(V, E)$ by a *fuzzy granular social network* representation $S(C, V, G)$. So, we need to define three sets C, V and G from the network $G(V, E)$.

We consider preserving the maximum information of the network inside the FGSN. So, we constructed granules around every nodes in the network. Following is the definition of $S(C, V, G)$ for Zachary karate club data.

- $V = \{v | \forall v \in V\}$
- $C = \{c | \forall c \in V\}$
- $G = \{A_c | \forall c \in C, A_c \equiv \sum_{v \in V} \tilde{\mu}_c(v)/v\}$.

Normalized membership value $\tilde{\mu}_c(v)$ is the degree of belonging of node v in the granule (A_c) around node c. $\tilde{\mu}_c(v)$ is calculated based on the Eq. 8 with minimum *hop* distance as the distance metric and $r = D$, the network diameter.

$$\tilde{\mu}_c(v) = \frac{\mu_c(v)}{\sum_{i \in C} \mu_i(v)} \text{ such that } \sum_{i \in C} \tilde{\mu}_i(v) = 1 \qquad (8)$$

where,

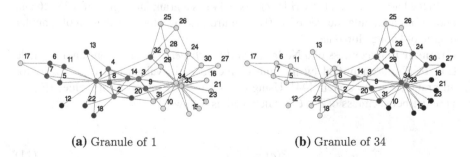

(a) Granule of 1   (b) Granule of 34

Fig. 5 Color coded granules of Zachary karate club

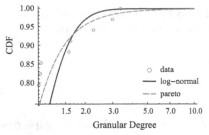
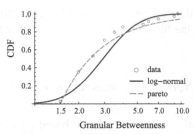

(a) Granular degree distribution   **(b)** Granular betweenness distribution

Fig. 6 FGSN features of Zachary karate club

$$
\mu_c(v) = \begin{cases} 0 & \text{for } d(c, v) > r \\ \dfrac{1}{1 + d(c, v)} & \text{otherwise} \end{cases} \tag{9}
$$

where $d(c, v)$ is the distance between node v and the center c.

Two such granules around nodes 1 and 34 are shown in Fig. 5. Here darker shades of brown represent higher values of membership. As we used normalized membership values, the nodes in less overlapping region may turn to have higher membership than the center nodes of the granules. This indicates that those nodes belong only to a fewer number of granules as compared to the centers. This is intuitively appealing as the former ones have higher possibilities of 'definitely belonging' to a granule than the latter ones

Granular Degree of a Node: *Granular degree* of a node in FGSN is equivalent to the degree measure of a node in graph representation. Granular degree of a node c is the cardinality of the granule constructed around the node c [44]. Here each granule is represented by a fuzzy set, so we use Eq. 2 to compute the granular degree of a node c as

$$
\mathcal{D}(c) = |A_c| = \sum_{v \in \mathcal{V}} \tilde{\mu}_c(v) \tag{10}
$$

In the karate club example (Fig. 2), node 34 has a granular degree of 3.38026 and node 1 has a granular degree of 3.0044. Figure 6a shows the distribution of granular degree of karate club data.

Granular Betweenness of a Node: *Granular betweenness* of a representative node c in FGSN is quantified by the sum of membership values that c possesses for all granules in the system [44]. Using the normalized membership values (Eq. 8), granular betweenness of $c \in C$ is calculated as follows.

$$
\mathcal{B}(c) = \frac{1}{\max\limits_{i \in C}(\tilde{\mu}_i(c))} \tag{11}
$$

$\mathcal{B}(c)$ takes values in $[0, |\mathcal{C}|]$. In our example karate club network, granular betweenness of node 1 and node 34 is 9 and 9.5, respectively. The distribution of granular betweenness of karate club data is shown in Fig. 6b.

Granular Embeddedness of a Pair of Nodes: *Granular embeddedness* for any pair of nodes defines how much a granule centered at one node is embedded inside that of the other [44]. It may be measured by the cardinality of the intersection of granules centered by the pair of points. Using Eqs. 4 and 2, granular embeddedness of a pair of nodes a and b is defined as

$$\mathcal{E}(a, b) = |A_a \cap A_b| = \sum_{v \in \mathcal{V}} \min(\tilde{\mu}_a(v), \tilde{\mu}_b(v)) \tag{12}$$

where A_a and A_b are the fuzzy sets representing the granules having the center nodes a and b, respectively.

In the example of karate club, the embeddedness of 1 and 34 is found to be 0.610714 when $r = 2$, and 0.959073 when $r = D(= 5)$, the diameter of the network.

4.3 Uncertainties in FGSN

Uncertainties in a social network arises due to the presence of vaguely defined *closeness* between nodes. Each relationship has a degree of togetherness. The presence of a relational link in a network does not imply that both the nodes are 100% committed towards each other. Similarly, the absence of a link does not necessarily mean they are not following each other. Let us now define two measures of uncertainties in FGSN in terms of fuzziness, as follows:

Energy Measure of a Granule in FGSN: Let us consider a monotonically increasing mapping $e : [0, 1] \rightarrow [0, 1]$ with the boundary conditions $e(0) = 0$ and $e(1) = 1$. An energy measure of a granule $A_c \in \mathcal{G}$, denoted by $\mathbb{E}(A_c)$, is a function of its characterizing membership values, represented as

$$\mathbb{E}(A_c) = \sum_{x \in \mathcal{V}} e[\tilde{\mu}_c(x)] \tag{13}$$

This measure quantifies the energy associated with the granule A_c. The energy increases as the membership values of its supporting nodes increase. The energy measure of A_c reduces to its cardinality if we use the identity mapping $e(x) = x \; \forall x \in \mathcal{V}$, i.e.,

$$\mathbb{E}(A_c) = \sum_{x \in \mathcal{V}} \tilde{\mu}_c(x) = |A_c| \tag{14}$$

One can also think of a different functional for e other than the identity mapping, for example, $e(x) = x^a, a > 0$ or $e(x) = sin(\frac{\pi}{2}x)$.

Entropy Measure of FGSN: Given a FGSN $\mathcal{S}(\mathcal{C}, \mathcal{V}, \mathcal{G})$, each granule $A_c \in \mathcal{G}$ represents a fuzzy equivalence class under the attribute set $\mathcal{C}$. If we have n objects in the universe $\mathcal{V}$ then the fuzzy relative frequency [56] of a granule will be

$$\rho(A_c) = \frac{|A_c|}{n} \tag{15}$$

where $|A_c|$ is the cardinality of the granule A_c. Based on this relative frequency of granules, one can find the information gain of the FGSN through its entropy, using Shannon's logarithmic function, as

$$H(\mathcal{S}) = -\sum_{A_c \in \mathcal{G}} \rho(A_c) log_\beta(\rho(A_c)) \tag{16}$$

where β represents the base of logarithm. Applying Eq. 15 into Eq. 16 we get

$$H(\mathcal{S}) = -\frac{1}{n} \sum_{A_c \in \mathcal{G}} |A_c| log_\beta(\frac{|A_c|}{n}). \tag{17}$$

The value of $H(\mathcal{S})$ can vary in $[0, log_\beta(|\mathcal{C}|)]$. $H(\mathcal{S}) = 0$ means the FGSN is least uncertain, while its value equal to $log_\beta(|\mathcal{C}|)$ signifies the highest uncertainty. Readers may refer [82] for generalized entropy measures in the granular space.

4.4 Granular Degree Heuristic for Target Set Selection in FGSN

The section report the experimental results demonstrating the applicability of fuzzy granular social network for target set selection problem.

Problem Statement: Let us consider an influence function $\sigma : 2^\mathcal{V} \to \mathbb{N}$, defined for a social network $\mathcal{S}(\mathcal{C}, \mathcal{V}, \mathcal{G})$, such that given a set of initial active nodes $K \in 2^\mathcal{V}$, $\sigma(K)$ returns the expected number of active nodes at the end of information cascade. The problem of target set selection is to find the k number of influential nodes for which influence in $\mathcal{S}$ is maximum. So, this is a maximization problem defined as

$$\max_K \quad \sigma(K)$$
$$\text{subject to} \quad |K| = k, k > 0.$$

Data Sets: In the experiments, we used three data sets, namely Zachary karate club [92], Dolphin social network [55], and Political blog network [1]. We already described the details of Zachary karate club in Sect. 2.1. Properties of Dolphin social graph and Political blog network are shown in Figs. 7 and 8 respectively.

Results: We first selected the top-k nodes (that is, the centers of the granules) from a given FGSN, in descending order of granular degree value. We refer this algorithm

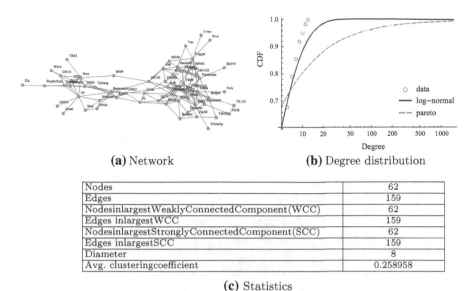

(a) Network   **(b)** Degree distribution

Nodes	62
Edges	159
Nodes in largest Weakly Connected Component (WCC)	62
Edges in largest WCC	159
Nodes in largest Strongly Connected Component (SCC)	62
Edges in largest SCC	159
Diameter	8
Avg. clustering coefficient	0.258958

(c) Statistics

Fig. 7 Dolphin social graph

as Granular degree heuristic. Then we pass these top k nodes, as the set of seeds, in the Monte Carlo simulation of information diffusion (independent cascade model [37]). The output of the simulation process represents the number of total nodes influenced due to the said set of input seeds. We have varied the value of k from 1 to 15. These results are reported graphically in Fig. 9. Here X-axis shows the value of k and the Y-axis presents the total number of nodes influenced. As the Monte Carlo process is a stochastic process, we executed each experiment for 10000 trials and reported here the average values. It is clear from the figure that, for Zachary karate club and Dolphin social networks, results obtained with the proposed granular degree heuristic on FGSN outperform those obtained by other graph theoretic algorithms (High degree heuristic, Random and Diffusion degree heuristic [67]) for most values of k. This signifies that the set of seeds selected using the FGSN based method is able to determine the superior top k influential nodes. For Political blogs, the performance is at par with the High Degree Heuristic, superior to random and inferior to Diffusion Degree Heuristics.

Execution time (in seconds) of different algorithms for 1000 runs is shown in Table 2. As expected, the random selection method needs least time for all the data sets. Diffusion degree heuristics, on the other hand, takes longest time for all the cases. The proposed Granular degree heuristic requires much lower execution time as compared to diffusion degree for all the data sets. For Zachary karate club and Dolphin social graph, it is almost as fast as the high degree heuristics. For Political blog network, however, the proposed algorithm takes longer time compared to high degree heuristics. Further the algorithm is seen to perform best for $r = 2$. For other values e.g., $r = 1, 3, 4, 5$, the performance deteriorates [44].

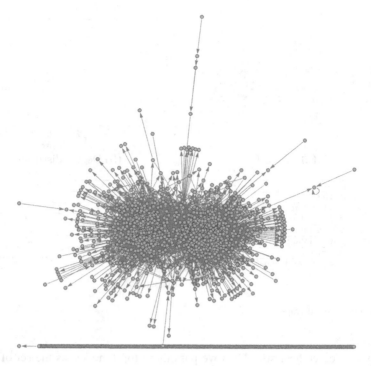

(a) Network

Nodes	1490
Edges	16718
Nodes in largest Weakly Connected Component (WCC)	1222
Edges in largest WCC	16717
Nodes in largest Strongly Connected Component (SCC)	1
Edges in largest SCC	0
Diameter	10
Avg. clustering coefficient	0

(b) Statistics

Fig. 8 Political blogs network

The computation complexity of the granular degree heuristic is $O(|V| + |E| + |\mathcal{C}| + kn)$ as reported in [44]. Here k is the number of desire seeds and n is the number of granules having granular degree greater than 1.

4.5 Fuzzy-Rough Community (FRC) Detection

A new community detection algorithm within the new knowledge representation scheme of FGSN is described in this section. Communities detected here show fuzzy-

Fig. 9 Variation of total influence with k for different algorithms ($r = 2$)

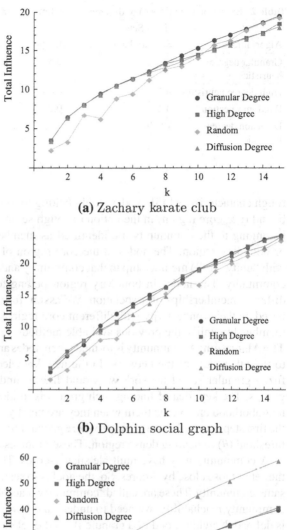

(a) Zachary karate club

(b) Dolphin social graph

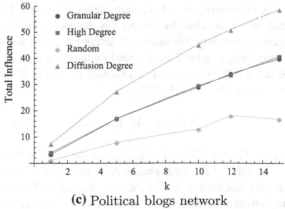

(c) Political blogs network

Table 2 Execution time (in sec) of different algorithms for 1000 runs

Algorithms	Data Sets		
	Zachary karate club	Dolphin social graph	Political blogs network
Granular degree heuristics	0.311	0.52	27.26
High degree heuristics	0.3	0.48	3.7
Random selection	0.2	0.2	0.5
Diffusion degree heuristics	12.19	16.532	9.29×10^4

rough characteristics [45]. Nodes surely belong to a community constitute its lower bound (i.e., core region) in the notion of rough set theory while the others possibly belonging to the community are identified as members of "upper - lower" bound or boundary region. The nodes in the core region of the community are assigned with "unity" (full) membership to that community and "zero" (no) for the remaining community. The nodes in boundary region belong to multiple communities with different memberships of association. We assign fuzzy membership to these nodes based on their connectivity with different core regions, thereby resulting in unequal membership unlike the previous available methods.

The Algorithm: A community is formed when nodes are densely connected, compare to the other parts of the network. In the new knowledge representation scheme of fuzzy granular social network we would like to find out such densely connected groups. The key idea of finding such groups is to identify the granules with dense neighborhood and merge them when they are nearby (merging dense regions). Thus the first step is to find those granules where *granular degree* (Eq. 10) exceeds a certain threshold (θ) indicating dense region. These granules are referred as θ-Core.

A community may have multiple such θ-cores. The algorithm needs to identify the set of those close by θ-cores. So, the goal is to search for θ-cores which belong to same community. These are called 'community reachable cores' [45]. To understand community reachability, we need to understand how the neighborhood of a granule is defined. Neighborhood of a granule A_c is the set of all granules whose centers lie in the support set of A_c [45], i.e.,

$$\Gamma(A_c) = \{A_i | A_i \in \mathcal{G} \text{ and } i \in Support(A_c) \forall i \neq c\}$$

where $Support(A_c) = \{v | \tilde{\mu}_c(v, r) > 0\}$ and r is the radius of the granule.

Based on the neighborhood, thus defined, we can find the θ-cores which are community reachable to each other, i.e., belong to the same community. There are three notions of community reachability. Two θ-cores are said to be (1) directly community reachable when one of them is in the neighborhood of the other, (2) indirectly community reachable, when one is reachable via a chain of directly community reachable θ-cores to the other and (3) r-connected community reachable when both of them are indirectly community reachable to a third θ-core r.

In a network, there might be nodes, which reside at the boundary regions and have neighborhood spread over multiple groups. To represent the notion of this overlapping, a normalized granular embeddedness measure [45] is introduced as

$$\mathcal{E}(A_p, A_q) = \frac{|A_p \cap A_q|}{|A_p \cup A_q|}.$$

$\mathcal{E} = 0$ implies no overlapping between granules A_p and A_q. $\mathcal{E} = 1$ signifies complete overlapping.

On the basis of community reachable θ-cores one may define community as follows.

Definition 1 (*Community*) Given a social network $\mathcal{S} = (\mathcal{C}, \mathcal{V}, \mathcal{G})$, and θ and ϵ, a community $\mathbb{C}$ is a non empty subset of granules $\mathcal{G}$ satisfying the following conditions:

- $\forall A_p, A_q \in \mathbb{C}$, A_p and A_q are community reachable cores
- $\forall A_p \in \mathbb{C}, \mathcal{E}(A_p, \bigcup_{A_q \in \mathbb{C} \setminus A_p} A_q) > \frac{1}{\epsilon}$

θ and ϵ are referred as density and coupling co-efficient of the community respectively [45]. One may note that the communities, thus identified, have fuzzy (ill defined) boundaries. These communities can further be viewed in terms of lower and upper approximations in the framework of rough set theory. That is, each community has a lower approximate region reflecting nodes definitely belonging to, and a boundary (i.e., upper - lower) region reflecting the nodes possibly belonging to. Therefore it may be appropriate to assign fuzzy membership values in $(0, 1)$ to only those nodes which lie within the said (upper - lower) region, and assign unity (1) value to those of lower approximation. The fuzzy-rough communities are accordingly defined (Definition 2).

Definition 2 (*Fuzzy-Rough Community*) Let the n communities found for a social network be $\mathbb{C}_1, \mathbb{C}_2, ..., \mathbb{C}_n$, and the upper and lower approximation of the ith community be $\overline{\mathbb{C}_i \theta}$ and $\underline{\mathbb{C}_i \theta}$ respectively. Then

$$\underline{\mathbb{C}_i \theta} = \{x | x \in Support(A_p) \wedge x \notin Support(A_q);$$
$$\forall A_p \in \mathbb{C}_i \text{ and } A_q \in \mathbb{C}_j; i \neq j\} \tag{18}$$
$$\overline{\mathbb{C}_i \theta} = \{x | x \in Support(A_p); A_p \in \mathbb{C}_i\}$$

Fuzzy-Rough membership function characterizing the community $\mathbb{C}_i$ is defined as,

$$\delta_{\mathbb{C}_i}^{\theta}(x, r) = \begin{cases} 1 & \text{if } x \in \underline{\mathbb{C}_i \theta} \\ \sum_{c \in \underline{\mathbb{C}_i \theta}} \tilde{\mu}_c(x, r) & \text{if } x \in \overline{\mathbb{C}_i \theta} \setminus \underline{\mathbb{C}_i \theta} \\ 0 & \text{Otherwise} \end{cases} \tag{19}$$

where $\tilde{\mu}_c(x, r)$ is defined in Eq. 8.

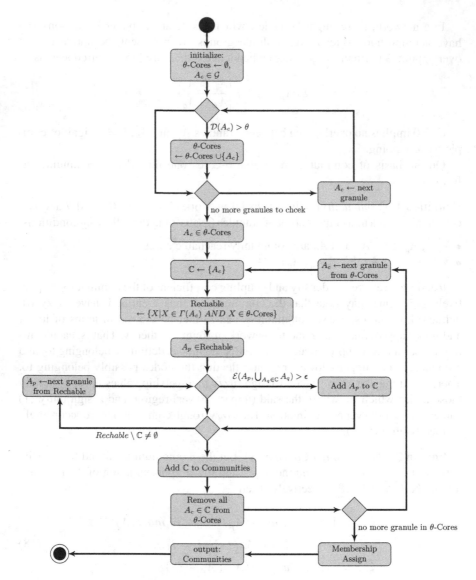

Fig. 10 Block diagram of FRC-FGSN algorithm

Orphans: A node is said to be orphan if it is not a member of any identified community.

Given a social network, the algorithm (FRC-FGSN) finds its various communities (Definition 1) with fuzzy-rough description (Eq. 19) defined over the granular model (Eq. 9) of knowledge representation. Nodes not included as a part of any community are designated as orphans. A block diagram of the algorithm is shown in Fig. 10 [45].

LFR Benchmark Data: LFR benchmark data is one of the popular benchmark data for comparing community detection algorithms [48]. Later, it was modified to accommodate more properties of network and communities viz. directed, weighted network and overlapping communities [47]. The idea is to generate network graphs based on various parameters. These parameters are

- Size of the network N
- Size of the communities (within C_{min} to C_{max})
- Mixing parameter, i.e., the average ratio of edges within community and edges with other communities (η)
- Fraction of overlapping nodes (O_n) and
- Number of overlapping communities (O_m)

With LFR data, we compare the identified community structures with the output of three popular graph theoretic algorithms. These are, centrality based community detection method [28], Modularity optimization method [62] and k-clique percolation method (CPM) [71]. A point to note here is that, CPM can identify overlapping communities whereas the other two comparing methods identify non-overlapping partitions of the network.

Normalized fuzzy mutual information [45] is used to compare different community detection algorithms. For two community structures $\mathbb{C}^X$ and $\mathbb{C}^Y$ the NFMI value can be calculated as

$$NFMI(\mathbb{C}^X : \mathbb{C}^Y) = \frac{1}{2} \left[\frac{H(\mathbb{C}^X) - H(\mathbb{C}^X|\mathbb{C}^Y)}{H(\mathbb{C}^X)} + \frac{H(\mathbb{C}^Y) - H(\mathbb{C}^Y|\mathbb{C}^X)}{H(\mathbb{C}^Y)} \right] \tag{20}$$

where $H(\mathbb{C}^X|\mathbb{C}^Y)$ (or $H(\mathbb{C}^Y|\mathbb{C}^X)$) is the conditional information measure in terms of lack of information of $\mathbb{C}^X$ (or $\mathbb{C}^Y$) given $\mathbb{C}^Y$ (or $\mathbb{C}^X$). Here, $H(\mathbb{C}^X)$ (or $H(\mathbb{C}^Y)$) is the information contained in $\mathbb{C}^X$ (or $\mathbb{C}^Y$) and is defined as

$$H(\mathbb{C}^X) = - \sum_{P \in \mathbb{C}^X} \lambda_P^X \log_2(\lambda_P^X) \tag{21}$$

where $\lambda_P^X = \sum_i^n m_P^X(i)$ is the fuzzy relative frequency of community $P \in \mathbb{C}^X$.

In the experiments, the size of the network is fixed to 1001 and we vary the other variables, and analyze the algorithms and their performance. The benchmark data generated by LFR algorithm for overlapping communities is far from the reality. It considers a fixed number of overlaps for the nodes which is unusual for real world networks. Furthermore, for nodes in overlapping region, we are assigning different memberships for belonging to different communities, but the network generated by LFR assigns unity value to these nodes. So, it is not the perfect data set to test our algorithms, yet results are convincing, as described below.

Fig. 11 Comparative results
with different values of
mixing parameter. Network
size: 1001; Min community
size: 150; Max community
Size: 250

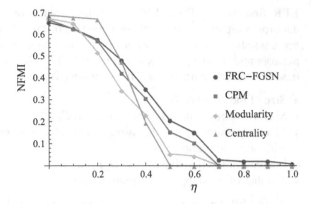

Fig. 12 Comparison
showing variation of NFMI
for different fraction of
overlapping community.
Network size: 1001; Mixing
parameter: 0.4; Min
community size: 150; Max
community size: 250

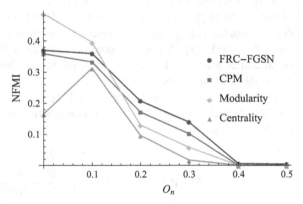

First, we vary the mixing parameter η from 0.0 to 1.0 by fixing the fraction of overlap to 0.15 and run all the four algorithms. We measure NFMI of each output with the ground truth. Figure 11, shows the variation of NFMI with respect to η for these algorithms. As expected, NFMI decreases when η increases in all the cases. For lower values of η, modularity and centrality based algorithms show better results, but for $\eta \geq 0.3$ the proposed FRC-FGSN shows prominent improvement over all other methods.

In another experiment, we vary the fraction of overlapping nodes (O_n) from 0.0 to 0.5 by fixing the mixing parameter at 0.4. Results are reported in Fig. 12. It shows that the proposed FRC-FGSN produces superior performance for O_n ranging from 0.2 to 0.4 and second best for $O_n < 0.2$.

One may restrict the number of granules to reduce the execution time to a tolerable range. We perform an experiment to observe this phenomenon. The result in this regard for one of the benchmark networks is shown in Fig. 13. Here, x-axis shows the percentage of granules corresponding to the number of nodes in the network. The blue curve with square points shows the time taken by the proposed FRC-FGSN and the red curve with circular points shows its accuracy in terms of NFMI. As expected, the time and accuracy both decrease as we reduce the number of granules from 100

Fig. 13 Plot showing the performance on number of granules for LFR data

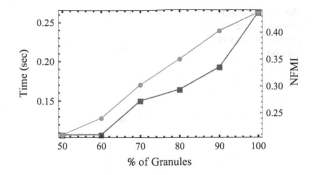

Table 3 Characteristics of generated data sets

Properties	Datasets				
	Dataset 1	Dataset 2	Dataset 3	Dataset 4	Dataset 5
Nodes	8404	16998	25761	34328	42965
Edges	163397	396178	660909	934708	1233418
Closed triangles	349156	879612	1526011	2178342	2952935
Open triangles	22228753	66714578	125632613	183297754	256587753
Approx. full diameter	4	5	5	5	5
90% effective diameter	2.875267	2.900036	2.931684	2.932677	2.935486

to 50%. Interestingly, the rate of drop in execution time is higher than that of the accuracy. This shows that by reducing the number of granules in FGSN one may obtain execution benefits in the algorithm.

4.6 Scalability of FGSN

We conducted experiments to understand the performance of FGSN with the growing number of links in the network. We used LDBC DATAGEN [75] to generate social network data of different scale. LDBC DATAGEN is a synthetic graph data generator which internally uses S3G2 [74] algorithm to generate social network data. DATAGEN generates realistic social networks based on the link distributions found in a real social network such as Facebook [75]. DATAGEN follows the MapReduce [15] paradigm, allowing for the generation of large data sets on commodity clusters.

With the help of DATAGEN, we generated five data sets. Characteristics of these networks are listed in Table 3. We observed the time required to convert these networks into FGSN model. Python modules NetworkX and Pandas are used for graph-

Fig. 14 Variation of conversion time with the number of links in the network

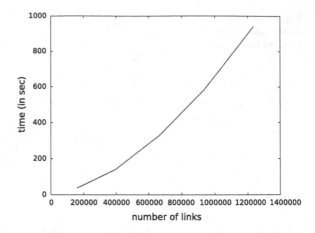

Fig. 15 Variation of execution time with number of links in the network

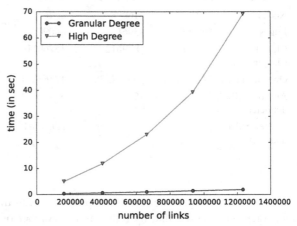

based operations and granule based operations respectively. Figure 14 shows the results in graphical format. The x-axis shows the number of links in the network and y-axis shows the time required for the conversion. As expected, the time increases with the number of links in the network. Clearly the curve shows a quadratic pattern. It is seen that for a network with over 1.23 millions of edges require slightly over 15 min of time for the conversion. A point to note here is that this time is required only once and the granular social network thus produced can be saved in flat files or database for future use. After the conversion, one may efficiently execute algorithms designed for the granular social networks. For example, the granular degree heuristic algorithm for the problem of target set selection is magnitude faster than the high degree heuristic. Figure 15 shows the execution time for extracting 50 seeds with aforesaid two algorithms on different data sets. The time here corresponds to 100 runs. The granular degree heuristics take only 1.93 s for the network with 1.23 millions edges, where as for the same network high degree heuristic took 69.14 s.

5 Discussions and Conclusions

In this chapter, we described a model of social network based on fuzzy granulation theory. Granules in the model characterize the closely operative groups formed within the highly overlapping neighborhood of social networks. The presence of vaguely defined closeness in relationships is modeled through the fuzzy set theory. The model is named *fuzzy granular social network* (FGSN).

In the graph representation of network, an individual node is used as an actor, whereas in FGSN, a granule is considered as an actor. A granule is constructed around each node in the network. This enables to capture the maximum information of the network inside the FGSN model. Under this granular framework, characteristics of a network are described using various measures defined over one or more granules. These measures include granular degree, granular betweenness and granular embeddedness.

The FGSN framework assumed the same role for all the actors in a network. This means, the model is valid for any social network as long as the roles of all the actors in the network remain the same. However, if a network has different roles for its different actors, then a modification may be required to accommodate such characteristics.

The data used in the experiment are collected with the view of graph representation. Hence, we had to convert such graph networks into the new knowledge representation of FGSN. Time taken for these conversions in seconds is seen to be 3.61, 12.54 and 7.09×10^3 for the Zachary karate club network, Dolphin social graph and Political blog network respectively. Once the modeling is complete, algorithms for different tasks of network analysis can be formulated.

A point to note here is that FGSN only encodes the structural information of the network. However for online social networks, many other contents (like posts, images, tags and profile) are also available to attribute with the actors. How to encode these information inside the granular social network model is not addressed in the current study.

Two major tasks concerning social network analysis are provided in this article. These are target set selection and community detection. Granular degree heuristic algorithm described for target set selection on FGSN uses granular degrees to rank the influencing nodes. Top k nodes selected from this ranked list are then used as the seed for the problem of target set selection. This selected target set is seen to perform better for most of the test cases in the undirected social networks of karate club network and Dolphin social graph. For directed network it is at par with the high degree heuristic but lower than that of the diffusion degree heuristics.

The output communities found by FRC-FGSN are characterized with crisp lower and fuzzy upper memberships, and are designated as "fuzzy-rough communities". A fuzzy membership is assigned only to those nodes which fall into the boundary (upper - lower) region of a community signifying that a node in that region can belong to multiple communities with different degrees of association. Nodes in the lower approximate region are assigned unity membership reflecting the certainty in belonging. In the process orphan (nodes with zero membership to all communities) are detected automatically.

Normalized fuzzy mutual information (NFMI) quantifies well the goodness of the identified communities. Larger is the value of NFMI between two community structures, higher is their similarity. It is shown that the FRC-FGSN algorithm produces superior outcome as compared to other popularly known community detection algorithms when the network contains overlapped communities.

Social networks available from popular mobile and Internet applications produce data in huge scale. These data show all the characteristics of Big data. Scalability is one of the important issues for Big data analysis. In case of big social networks, FGSN has the following two advantages over the graph modeling. First, in FGSN, the network properties of a node are embedded inside the granule constructed around it. If an algorithm demands to work on fewer nodes rather than the full network then one may avoid feeding the full network into the algorithm and yet can get the network properties from the granular characteristics. Even for the global property analysis, for reducing the execution time of data processing one may restrict the number of granules either based on a threshold, decided over the cardinality of the granule, or with human intervention. Experimentally, we found that with the reduction in number of granules, the rate of improvement in execution time is exponential while the rate of drop in accuracy is linear. Second, FGSN supports asynchronous nature of distributed computing better than the graph modeling. Two of the major challenges involving the distributed computing are, (1) coping with the intrinsic asynchrony between the different entities, and (2) coping with the spatial distribution of these computing entities. Granules may be more effectively fed into such asynchronous distributed systems where one computing unit will only deal with a subset of granules. Whereas feeding a graph model in such system is difficult. With graphs an additional care needs to be taken to work with synchronous algorithms in a distributed environment.

It is seen experimentally that the algorithm scales well with the growing number of links in the network. DATAGEN, a Hadoop/MapReduce based data generator is used to generate synthetic data for scalability analysis. The growth in execution time with the number of links for granular degree heuristics is found to be linear and the slope is also very low.

The model of FGSN is seen here to perform effectively and efficiently for two of the major applications in the domain of social network analysis. There are other applications in social network analysis e.g., link prediction and evolution of social network which are also very important to study. For example existing algorithms on link prediction through graph implementation find the similarity between two nodes which in turn provides the plausibility that a link may form between them in future. Similarities can easily be identified using the normalized embeddedness measure in FGSN. If two granules are highly embedded to each other, then there is a high possibility that there would be a link between the centers of the granules in the network.

Although, some of the algorithms available in the domain might provide better solutions as compared to the proposed methodologies, the way of modeling a network with FGSN opens a new avenue and provides directions on using the established granular computing theory and other efficient data mining techniques into the demanding dynamics of social networks and related problems with a scope of newly defined measures and efficient algorithms.

Acknowledgements The work was completed while S.K. Pal held the J.C. Bose National Fellowship, Indian National Academy of Engineering Chair professorship, and DAE Raja Ramanna Fellowship.

References

1. L.A. Adamic, N. Glance, The political blogosphere and the 2004 US election: divided they blog, in *Proceedings of the 3rd International Workshop on Link Discovery, LinkKDD '05* (ACM, Chicago, 2005), pp. 36–43
2. Y.Y. Ahn, J.P. Bagrow, S. Lehmann, Link communities reveal multiscale complexity in networks. Nature **466**(7307), 761–764 (2010)
3. A. Anthony, S. Biesan, Block Modeling in Large Social Networks with Many Clusters. Technical report (2012)
4. C. Aslay, N. Barbieri, F. Bonchi, R. Baeza-Yates, Online topic-aware influence maximization. Proc. VLDB Endow. **8**(6), 666–677 (2015)
5. A.L. Barabási, R. Albert, Emergence of scaling in random networks. Science **286**(5439), 509–512 (1999)
6. S.A. Boorman, H.C. White, Social structure from multiple networks. ii. role structures social structure from multiple networks. Am. J. Sociol. **81**(6), 1384–1446 (1976)
7. D. Chakraborty, B.U. Shankar, S.K. Pal, Granulation, rough entropy and spatiotemporal moving object detection. Appl. Soft Comput. J. **13**(9), 4001–4009 (2013)
8. S. Chattopadhyay, C.A. Murthy, S.K. Pal, Fitting truncated geometric distributions in large scale real world networks. Theor. Comput. Sci. **551**, 22–38 (2014)
9. S. Chen, Y. Li, Dynamic grade on the major hazards using community detection based on genetic algorithm, in *Proceedings of 2009 International Conference on Signal Processing Systems* (IEEE, Singapore, 2009), pp. 713–717
10. W. Chen, C. Wang, Y. Wang, Scalable influence maximization for prevalent viral marketing in large-scale social networks, in *Proceedings of the 16th ACM SIGKDD International Conference on Knowledge Discovery and Data Mining* (ACM, New York, 2010a), pp. 1029–1038
11. W. Chen, Y. Wang, S. Yang, Efficient influence maximization in social networks, in *Proceedings of the 15th ACM SIGKDD International Conference on Knowledge Discovery and Data Mining* (ACM Press, Paris, 2009), pp. 199–208
12. W. Chen, Y. Yuan, L. Zhang, Scalable influence maximization in social networks under the linear threshold model, in *2010 IEEE International Conference on Data Mining* (IEEE, New Jersey, 2010b), pp. 88–97
13. E. Cohen, D. Delling, T. Pajor, R.E. Werneck, Sketch-based influence maximization and computation: scaling up with guarantees, in *Proceedings of the 23rd ACM International Conference on Information and Knowledge Management (CIKM '14)* (ACM Press, New York, 2014), pp. 629–638
14. G.B. Davis, K.M. Carley, Clearing the FOG: fuzzy, overlapping groups for social networks. Soc. Netw. **30**(3), 201–212 (2008)
15. J. Dean, S. Ghemawat, mapreduce: simplified data processing on large clusters. Commun. ACM **51**(1), 107 (2008)
16. P. Domingos, M. Richardson, Mining the network value of customers, in *Proceedings of the 7th ACM SIGKDD International Conference on Knowledge Discovery and Data Mining* (ACM, San Francisco, CA, 2001), pp. 57–66

17. D. Easley, J. Kleinberg, Networks, Crowds, and Markets: Reasoning about a Highly Connected World (Cambridge, Cambridge University Press, 2010)
18. P.a. Estevez, P. Vera, K. Saito, Selecting the most influential nodes in social networks, in *Proceedings of 2007 International Joint Conference on Neural Networks* (IEEE, New Jersey, 2007), pp. 2397–2402
19. T.S. Evans, R. Lambiotte, Line graphs of weighted networks for overlapping communities. Eur. Phys. J. B **77**(2), 265–272 (2010)
20. T. Falkowski, A. Barth, M. Spiliopoulou, DENGRAPH: a density-based community detection algorithm, in *Proceedings of IEEE/WIC/ACM International Conference on Web Intelligence (WI'07)* (IEEE, Washington, 2007), pp. 112–115
21. M. Faloutsos, P. Faloutsos, C. Faloutsos, On power-Law relationships of the internet topology, in *Proceedings of the Conference on Applications, Technologies, Architectures, and Protocols for Computer Communication, SIGCOMM '99* (ACM, New York, 1999), pp. 251–262
22. T.F. Fan, C.J. Liau, T.Y. Lin, Positional analysis in fuzzy social networks, in *Proceedings of 2007 IEEE International Conference on Granular Computing (GRC 2007)* (IEEE, Silicon Valley, 2007), pp. 423–428
23. I.J. Farkas, D. Ábel, G. Palla, T. Vicsek, Weighted network modules. New J. Phys. **9**(6), 180 (2007)
24. S. Fortunato, Community detection in graphs. Phys. Rep. **486**(3–5), 75–174 (2010)
25. O. Frank, Estimation and sampling in social network analysis, in Encyclopedia of Complexity and Systems Science, by R.A. Meyers (ed.) (Springer, New York, 2009), pp. 8213–8231
26. A. Ganivada, S. Dutta, S.K. Pal, Fuzzy rough granular neural networks, fuzzy granules, and classification. Theor. Comput. Sci. **412**(42), 5834–5853 (2011)
27. K.J. Gile, M.S. Handcock, Respondent-driven sampling: an assessment of current methodology. Sociol. Methodol. **40**(1), 285–327 (2010)
28. M. Girvan, M.E.J. Newman, Community structure in social and biological networks. Proc. Natl. Acad. Sci. U.S.A **99**(12), 7821–7826 (2002)
29. M. Gomez-Rodriguez, B. Schölkopf, Influence maximization in continuous time diffusion networks, in *Proceedings of the 29th International Conference on Machine Learning (ICML-12)* (Edinburgh, 2012), pp. 313–320
30. A. Goyal, W. Lu, L. Lakshmanan, CELF++: optimizing the greedy algorithm for influence maximization in social networks, in *Proceedings of the 20th International Conference Companion on World Wide Web* (ACM, New York, 2011), pp. 47–48
31. T. Hastie, R. Tibshirani, J. Friedman, *The Elements of Statistical Learning: Data Mining, Inference, and Prediction*, 2nd edn. (Springer Series in Statistics, Springer, New York, 2009)
32. P. Hońko, Association discovery from relational data via granular computing. Inf. Sci. **234**(2), 136–149 (2013)
33. H. Jeong, B. Tombor, R. Albert, Z.N. Oltvai, aL Barabási, The large-scale organization of metabolic networks. Nature **407**(6804), 651–654 (2000)
34. G. Jiang, F. Ma, J. Shang, P.Y. Chau, Evolution of knowledge sharing behavior in social commerce: an agent-based computational approach. Inf. Sci. **278**, 250–266 (2014)
35. L.j. Kao, Y.P. Huang, Mining influential users in social network, in *Proceedings of IEEE International Conference on Systems, Man, and Cybernetics (SMC), 2015* (Hong Kong, 2015), pp. 1209–1214
36. S. Kelley, M. Goldberg, M. Magdon-Ismail, K. Mertsalov, A. Wallace, Defining and discovering communities in social networks, in *Handbook of Optimization in Complex Networks* (2012), pp. 139–168
37. D. Kempe, J. Kleinberg, É. Tardos, Maximizing the spread of influence through a social network, in *Proceedings of the 9th ACM SIGKDD International Conference on Knowledge Discovery and Data Mining* (ACM Press, New York, NY, 2003), p. 137
38. D. Kempe, J. Kleinberg, É. Tardos, Influential nodes in a diffusion model for social networks. Autom. Lang. Program. **3580**, 1127–1138 (2005)
39. Y.A. Kim, R. Phalak, A trust prediction framework in rating-based experience sharing social networks without a Web of Trust. Inf. Sci. **191**, 128–145 (2012)

40. J. Kleinberg, Cascading behavior in networks: algorithmic and economic issues, in *Algorithmic Game Theory*, by eds. N. Nisan, T. Roughgarden, E. Tardos, V.V. Vazirani (Cambridge, Cambridge University Press, 2007), pp. 613–632

41. J. Kleinberg, S. Suri, É. Tardos, T. Wexler, Strategic network formation with structural holes, in *Proceedings of the 9th ACM Conference on Electronic Commerce - EC'08* (ACM Press, New York, USA, 2008), pp. 284–293

42. B. Krishnamurthy, J. Wang, On network-aware clustering of web clients, in *Proceedings of of the Conference on Applications, Technologies, Architectures, and Protocols for Computer Communication, SIGCOMM '00* (ACM, New York, Stockholm, 2000), pp. 97–110

43. L. Kuang, X. Tang, M. Yu, Y. Huang, K. Guo, A comprehensive ranking model for tweets big data in online social network. EURASIP J. Wireless Commun. Netw. **2016**(1), 46 (2016)

44. S. Kundu, S.K. Pal, FGSN: fuzzy granular social networks - model and applications. Inf. Sci. **314**, 100–117 (2015a)

45. S. Kundu, S.K. Pal, Fuzzy-rough community in social networks. Pattern Recognit. Lett. **67**(2), 145–152 (2015b)

46. M. Lahiri, M. Cebrian, The genetic algorithm as a general diffusion model for social networks, in *Proceedings of the 24th AAAI Conference on Artificial Intelligence* (Atlanta, Georgia, 2010), pp. 494–499

47. A. Lancichinetti, S. Fortunato, Benchmarks for testing community detection algorithms on directed and weighted graphs with overlapping communities. Phys. Rev. E. **80**(1), 016118 (2009)

48. A. Lancichinetti, S. Fortunato, F. Radicchi, Benchmark graphs for testing community detection algorithms. Phys. Rev. E. **80**(1), 016118 (2008)

49. J. Leskovec, A. Krause, C. Guestrin, C. Faloutsos, J. VanBriesen, N. Glance, N, Cost-effective outbreak detection in networks, in *Proceedings of the 13th ACM SIGKDD International Conference on Knowledge Discovery and Data Mining* (ACM Press, San Jose, 2007), pp. 420–429

50. J. Leskovec, K.J. Lang, M. Mahoney, Empirical comparison of algorithms for network community detection, in *Proceedings of the 19th International Conference on World Wide Web - WWW '10* (Raleigh, 2010), p. 631

51. G. Li, S. Chen, J. Feng, K.J. Tan, W.s. Li, Efficient location-aware influence maximization, in *Proceedings of the 2014 ACM SIGMOD international conference on Management of data (SIGMOD'14)* (Snowbird, 2014), pp. 87–98

52. L. Li, S. Li, X. Chen, A new genetics-based diffusion model for social networks, in *Proceedings of 2011 International Conference on Computational Aspects of Social Networks (CASoN)* (IEEE, Salamanca, Spain, 2011), pp. 76 81

53. O. Liu, K.L. Man, W. Chong, C.O. Chan, Social network analysis and big data, in *Proceedings of the International Multi Conference of Engineers and Computer Scientists*, vol. II (Hong Kong, 2016), pp. 6–7

54. W, Liu., X, Jiang, M, Pellegrini, X, Wang, Discovering communities in complex networks by edge label propagation. Sci. Rep. **6** (2016)

55. D. Lusseau, K. Schneider, O.J. Boisseau, P. Haase, E. Slooten, S.M. Dawson, The bottlenose dolphin community of doubtful sound features a large proportion of long-lasting associations. Behav. Ecol. Sociobiol. **54**(4), 396–405 (2003)

56. P. Maji, S.K. Pal, Fuzzy-rough sets for information measures and selection of relevant genes from microarray data. IEEE Trans. Syst. Man Cybern. Part B **40**(3), 741–52 (2010)

57. F.D. Malliaros, M. Vazirgiannis, Clustering and community detection in directed networks: a survey. Phys. Rep. **533**(4), 95–142 (2013)

58. J.L. Moreno, *Who Shall Survive? A New Approach to the Problem of Human Interrelations*, Nervous and Mental Disease Monograph Series (Nervous and Mental Disease Publishing co., New York, 1934)

59. P.S. Nair, S.T. Sarasamma, Data mining through fuzzy social network analysis, in *Proceedings of the 26th International Conference of North American Fuzzy Information Processing Society* (IEEE, San Diego, California, 2007), pp. 251–255

60. R. Narayanam, Y. Narahari, A Shapley value-based approach to discover influential nodes in social networks. IEEE Trans. Autom. Sci. Eng. **8**(1), 130–147 (2011)
61. M. Narayanan, A. Cherukuri, A study and analysis of recommendation systems for location-based social network (LBSN) with big data. IIMB Manag. Rev. **28**(1), 25–30 (2016)
62. M. Newman, Fast algorithm for detecting community structure in networks. Phys. Rev. E **69**(6), 066133 (2004)
63. M. Newman, M. Girvan, Finding and evaluating community structure in networks. Phys. Rev. E **69**(2), 1–15 (2004)
64. M. Newman, Modularity and community structure in networks. Proc. Natl. Acad. Sci. U.S.A **103**(23), 8577–8582 (2006)
65. G.K. Orman, V. Labatut, The effect of network realism on community detection algorithms, in *Proceedings of the 2010 International Conference on Advances in Social Networks Analysis and Mining* (IEEE, Odense, Denmark, 2010), pp. 301–305
66. S.K. Pal, Granular mining and rough-fuzzy pattern recognition: a way to natural computation. IEEE Intell. Inf. Bull. **13**(1), 3–13 (2012)
67. S.K. Pal, S. Kundu, C.A. Murthy, Centrality measures, upper bound, and influence maximization in large scale directed social networks. Fundam. Inf. **130**(3), 317–342 (2014)
68. S.K. Pal, S.K. Meher, Natural computing: a problem solving paradigm with granular information processing. Appl. Soft Comput. J. **13**(9), 3944–3955 (2013)
69. S.k. Pal, S.K. Meher, A. Skowron, Data science, big data and granular mining. Pattern Recognit. Lett. **67**(2), 109–112 (2015)
70. S.K. Pal, P. Mitra, *Pattern Recognition Algorithms for Data Mining* (CRC Press, Boca Raton, 2004)
71. G. Palla, I. Derényi, I. Farkas, T. Vicsek, Uncovering the overlapping community structure of complex networks in nature and society. Nature **435**(7043), 814–818 (2005)
72. W. Pedrycz, *Granular Computing: Analysis and Design of Intelligent Systems* (CRC Press, Boca Raton, 2013)
73. W. Pedrycz, A. Skowron, V. Kreinovich (eds.), *Handbook of granular computing* (Wiley, Sussex, 2008)
74. M.D. Pham, P. Boncz, O. Erling, S3G2: a scalable structure-correlated social graph generator, in *Selected Topics in Performance Evaluation and Benchmarking: 4th TPC Technology Conference, TPCTC 2012, Istanbul, Turkey, August 27, 2012, Revised Selected Papers*, ed. by R. Nambiar, M. Poess (Springer, Berlin, 2013), pp. 156–172
75. A. Prat, DATAGEN: Data Generation for the Social Network Benchmark (2014). http://ldbcouncil.org/blog/datagen-data-generation-social-network-benchmark
76. Y. Qin, J. Ma, S. Gao, Efficient influence maximization under TSCM: a suitable diffusion model in online social networks. Soft Comput. 1–12 (2016)
77. U. Raghavan, R. Albert, S. Kumara, Near linear time algorithm to detect community structures in large-scale networks. Phys. Rev. E **76**(3), 36106 (2007)
78. P.K. Reddy, M. Kitsuregawa, P. Sreekanth, S.S. Rao, A graph based approach to extract a neighborhood customer community for collaborative filtering, in *Proceedings of the Second International Workshop on Databases in Networked Information Systems*, DNIS '02 (Springer, London, 2002), pp. 188–200
79. F. Reid, A. McDaid, N. Hurley, Partitioning breaks communities, in *Proceedings of 2011 International Conference on Advances in Social Networks Analysis and Mining, ASONAM 2011* (Kaohsiung City, Taiwan, 2011), pp. 102–109
80. S.A. Rice, The identification of blocs in small political bodies. Am. Polit. Sci. Rev. **21**(3), 619–627 (1927)
81. M. Richardson, P. Domingos, Mining knowledge-sharing sites for viral marketing, in *Proceedings of the 8th ACM SIGKDD International Conference on Knowledge Discovery and Data Mining* (ACM Press, Edmonton, Alberta, 2002), pp. 61–70
82. D. Sen, S. Pal, Generalized rough sets, entropy, and image ambiguity measures. IEEE Trans. Syst. Man Cybern. Part B **39**(1), 117–128 (2009)

83. M. Spreen, Rare Populations, Hidden Populations, and Link-Tracing Designs: What and Why? Bulletin de Méthodologie Sociologique **36**, 34–58 (1992)
84. M. Steenstrup, Cluster-based networks, in *Ad Hoc Networking, Chap*, vol. 4 (Addison-Wesley Longman Publishing Co., Inc, Boston, MA, USA, 2001), pp. 75–138
85. Z. Su, Q. Xu, Q. Qi, Big data in mobile social networks: a QoE-oriented framework. IEEE Netw. **30**(1), 52–57 (2016)
86. C. Wang, L. Deng, G. Zhou, M. Jiang, A global optimization algorithm for target set selection problems. Inf. Sci. **267**, 101–118 (2014)
87. C. Wang, W. Chen, Y. Wang, Scalable influence maximization for independent cascade model in large-scale social networks. Data Mining Knowl. Discov. **25**(3), 545–576 (2012)
88. S. Wasserman, K. Faust, *Social Network Analysis: Methods and Applications* (Cambridge University Press, Cambridge, 1994)
89. R.S. Weiss, E. Jacobson, A method for the analysis of the structure of complex organizations. Am. Sociol. Assoc. **20**(6), 661–668 (1955)
90. H.C. White, S.A. Boorman, R.L. Breiger, Social structure from multiple networks. I. Block-models of roles and positions. Am. J. Sociol. **81**(4), 730–780 (1976)
91. J. Yao, A.V. Vasilakos, W. Pedrycz, Granular computing: perspectives and challenges. IEEE Trans. Cybern. **43**, 1977–1989 (2013)
92. W. Zachary, An information flow model for conflict and fission in small groups. J. Anthropol. Res. **33**(4), 452–473 (1977)
93. L. Zadeh, Fuzzy sets. Inf. Control **8**, 338–353 (1965)
94. L.A. Zadeh, Toward a theory of fuzzy information granulation and its centrality in human reasoning and fuzzy logic. Fuzzy Sets Syst. **90**, 111–127 (1997)
95. Y. Zeng, X. Chen, G. Cong, S. Qin, J. Tang, Y. Xiang, Maximizing influence under influence loss constraint in social networks. Expert Syst. Appl. **55**, 255–267 (2016)
96. T. Zhu, B. Wang, B. Wu, C. Zhu, Maximizing the spread of influence ranking in social networks. Inf. Sci. **278**, 535–544 (2014)

Part IV
Big Data Applications

Big Data, IoT and Semantics

Beniamino di Martino, Giuseppina Cretella and Antonio Esposito

Abstract Big data and the Internet of things are two parallel universes, but they are so close that in most cases they blend together. The amount of devices that connect to the internet grows day by day and they bring millions of data. The IoT generates unprecedented amounts of data and this impacts on the entire big data universe. The IoT and big data are clearly growing apace, and are set to transform many areas of business and everyday life. Semantic technologies play a fundamental role in reducing incompatibilities among data formats and providing an additional layer on which applications can be built, to reason over data and extract new meaningful information. In this chapter we report the most common approaches adopted in dealing with Big Data and IoT problems and explore some of the semantic based solutions which address such problematics.

Keywords Big data · IoT · Semantics

1 Introduction

With the diffusion of Smart Sensors and mobile Sensor devices communicating with each other and with remote servers, the number of connections established over the Internet has exponentially grown, and with that the volume of data continuously exchanged over such connections. The advent of the Internet of Things (IoT) era has deeply shaken the already thriving IT ecosystem, and has given birth to new potentialities, challenges and issues. In particular Big Data methodologies and technologies, already at a test due to the huge amount of information generated by not

B. di Martino · G. Cretella · A. Esposito (✉)
Second University of Naples, Via Roma 29, Aversa, Caserta, Italy
e-mail: antonio.esposito@unina2.it

B. di Martino
e-mail: beniamino.dimartino@unina.it

G. Cretella
e-mail: giuseppina.cretella@unina2.it

© Springer International Publishing AG 2017
A.Y. Zomaya and S. Sakr (eds.), *Handbook of Big Data Technologies*,
DOI 10.1007/978-3-319-49340-4_19

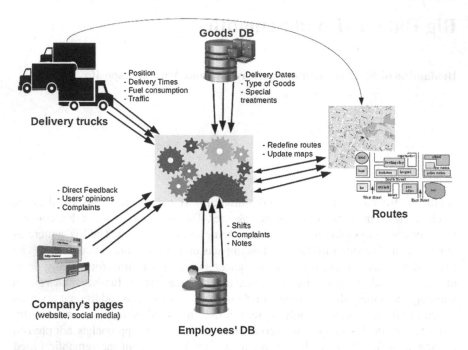

Fig. 1 The system used by a delivery company to update its delivery routes

sensor-related resources (Social Media, Business Processes, e-mails and so on) are further stressed and need to be adapted to the new sensor-centred scenarios.

Nevertheless, existing techniques are a necessary background for data analysts approaching the IoT field, as in many cases sensors' outputs come in the form of data streams which can be analysed, filtered and classified via well-known algorithms. Also, Smart Sensors are often integrated in Big Data applications. As an instance, consider the simple application scenario proposed in Fig. 1. In such example, a delivery company collects data from different sources in order to optimize the routes which the company's delivery trucks have to follow. In particular, the data come from:

- A database containing information on all the goods to deliver: expiration day, address of sender and receiver of the item, treatments for special goods and so on.
- A database of the company's employees, which collects information on their shifts, ordinary routes, residence.
- Sensors placed on the delivery trucks, which continuously feed the system with data regarding the trucks' position, the traffic conditions, the fuel consumption, etc.
- Feedback from customers, directly provided via explicit forms, or extracted from comments on social networks or on the company's web-site.

Now, let's suppose the system detects a high number of complaints or negative feedbacks for goods delivered in a specific town. It thus follows a series of steps in order to determine the cause of such complaints and eventually avoid future ones.

1. First of all, the system tries to understand what the complaints are about: let's suppose that customers complain about damaged goods.
2. Then, the system checks the goods' database to understand if the affected items belong to some specific category of products: it finds out that only items categorized as *fragile* are affected. Also, only items delivered in a specific area of the town, or whose route passes through that area, are involved in the complaints.
3. By analysing the data coming from delivery trucks following the identified routes, the system determines that, when passing through a specific road, all trucks sensibly slow down. However there are no problems with other roads, so the items are always delivered on-time.
4. At this point, the system re-arrange the routes in order to avoid that specific road.
5. In re-arranging the routes, the system also tries to assign them to employees living in or close to the interested areas of the town.

At the end of the day, it turned out that a very poorly maintained road was the cause of the damages to the goods: even if trucks' drivers slowed down significantly while driving along such a road, the shaking and vibrations caused by trucks passing over potholes ruined fragile items.

Considering the dimension of the data to be analysed to follow the simple steps described above (just consider how many goods a delivery company has to take care about every day), Big data analysis techniques are needed.

1. In order to analyse the feedbacks from users, and to understand if they are positive or negative and the topic of the reports, **Text Mining** and **Sentiment Analysis** techniques are applied. Section 4 reports an overview of such techniques, starting with the more general topic of **Social Mining**.
2. In order to rapidly query the goods' database and understand which items are affected by the reports, for which reasons and what are their main characteristics, **Graph** based representations can be employed. Graphs are a very efficient and convenient instrument for data representation, especially when several relevant relationships exist among data, or indexing capabilities are highly desired. Section 5 presents and overview of **Graph Mining** algorithms, with some application examples.
3. The data coming delivery trucks are actually sent by Smart Sensors applied to them. In order to handle the data traffic generated by such kind of sensors, specific protocols have been standardized (see Sect. 3), and Data Stream analysis techniques need to be applied (see Sect. 6 for such techniques). Also, the data collected from such sensors are almost all geo-referenced, as they specify position, speed and acceleration of the trucks. Section 7 briefly introduces such a topic.

4. The data coming from all of these different sources need to be efficiently integrated in order to be of some use to the system elaborating the routes. Semantic-web based models and technologies are the most suited to provide a homogeneous representation for data, which can be used as a base to infer new information and solve interoperability and portability issues which may arise. Section 2 provides and insight on semantic technologies and their application to Big Data.

2 Semantics for Big Data

Big Data is a relatively emerging field, in which innovative technologies are exploited to deal with issues often arising when dealing with huge amounts of data. In particular, new technologies and methodologies are investigated to reuse and extract value from information often coming from heterogeneous, unstructured or semi-structured and dynamic sources. Whilst the main issue referred to by researchers is the problem of processing very large sources of data in affordable time, analysts are equally busy in defining ways to rapidly and effortlessly normalize, integrate and transform the data coming from different sources into the format required by their own systems and tools. This depends on the extreme heterogeneity displayed by data coming from different resources, which is caused by the different perspectives on the reality the original developers of each of the considered source systems had. This problem is even more evident when considering data silos produced within the same company or organization, as a result of the use of different systems (CMRs, ERPs and so on) which too often are not natively integrated.

In such a situation semantic based technologies come in handy, as they can help solving several incompatibility errors related to data definitions and standard differences, and they can provide further context information which, in some cases, are fundamental to correctly analyse them. In short, the contribution of semantics to Big Data can be roughly summarized into these fields:

- Data contextualization. Information on the context the analysed set of data is referred to is sometimes fundamental for a correct analysis. Often the same term can refer to very different objects, places or even people: for example if we use the word "President", are we referring to the President of United States, to a film or to a restaurant with that name? The outcome of a term-based research will strongly depend on the context we are moving in (the time also can affect such search).
- Data reconnection. Since the same concepts can be represented in very different ways, according to the descriptors' perspectives, different terms could be used to address the same data. Using a common and shared semantic representation, it is possible to resolve eventual inconsistencies and "reconnect" data from different database silos.
- Management of overlapping representations. There are situations in which data models simply overlap, that is they could share some definition and differ on some other. In such a case, it can be difficult for an analyst to integrate them, or

to discard one in favour of another. Semantic technologies provide the means to operate such an integration effortlessly, retaining the essential information from all the interested models.

- Knowledge discovery. Discovery of new information represents a foundational element of Big Data. Semantic technologies provide many instruments, from logical rules to query languages, which can be used to assess and retrieve new information from an existing knowledge base.

2.1 Semantic Representation of Things, People and Web

Semantic web technologies include a plethora of standards and languages for knowledge representation, querying and inferencing. Some of these technologies find immediate application in the Big Data scenery, as they provide a better support to the storage of huge and complex amount of data, than traditional relational databases. Resource Description Framework [32] (RDF) is a framework for the description of knowledge, defined and maintained by the W3C Group, which provides a structure for describing identified "things" or resources and their relationships. RDF uses URIs (Uniform Resource Identifiers) to unambiguously point to resources in statements. RDF statements are represented by triples in the form **Subject-Predicate-Object**. These elements are all defined as unique resources, but only objects can be primitive types. An RDF model can be always serialized into an XML document, or it can be represented through an oriented graph in which nodes describe resources or primitive types and arcs are predicates. Using RDF triples, it is possible to easily represent knowledge, as even very complex concepts can be reduced to Subject-Predicate-Object instances. Also, RDF enables the inferencing of new knowledge from existing corpora. Figure 2 reports a very simple example of an RDF triple in which a person *Mary* (the Subject) is connected to a city *Rome* (the Object) via the property *Works_In* (the Predicate). Since *Rome* is also the Subject of another triple, with Predicate *Is_in* and Object *Italy*, it is possible to infer a new predicate between *Mary* and *Italy*. An extension to RDF, known as RDF-Schema (RDFS) is used to provide native predicates to define basic relationships among the described resources: **rdf:type** for example is used to state that a resource is an instance of a class, while **rdfs:subClassOf** is used to define a class hierarchy. Triples stored in an RDF database can be easily accessed and retrieved via SPARQL Protocol And RDF Query Language [45] (SPARQL). SPARQL can be used to express queries across different

Fig. 2 Simple RDF triple example

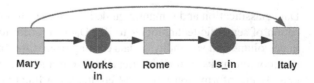

Mary   Works   Rome   Is_in   Italy
            in

data sources, provided that such data are expressed in or can be seen as RDF triples through a middleware. SPARQL queries use a syntax which is very similar to SQL, and these can be written in the form of triple patterns, conjunctions and disjunctions. SPARQL enables analysis and querying of large RDF graphs and different studies have been carried out to define multiple scalable and optimized queries [18, 24, 36]. SPARQL queries can be used to retrieve RDF described resources, filtered according to selected constraints. While RDF enables a very simple, effective and straightforward approach to resource description, it lacks the expressive capabilities needed to add useful semantic information to such descriptions. So, we opted for a much more expressive formalism for semantic description of resources, known as Web Ontology Language [37] (OWL). OWL is completely derived from RDF, of which it retains all the native predicates and properties and the triples based description. OWL was born to develop ontologies (definition and classification of concepts and entities, together with their relationships) that are compatible with the World Wide Web. Since OWL is based on RDF, SPARQL queries can be still executed on produced ontologies. OWL adds new properties to the basic RDF, allowing to:

- put restrictions on defined classes (for example, allValuesFrom selects resources with properties that only have values that meet a specific criteria);
- define complex relationships among classes (disjointWith states that resources belonging to one class cannot belong to another;
- state equality between classes through the sameAs attribute;
- enrich defined relationships (a Symmetric property between A and B affirms that a relationship between A and B is also true between B and A).

Adding semantic information to resource description allows the execution of automatic inferences on the existing knowledge base, provided inference rules are defined and an inferential engine is used to enforce them. In order to define such rules, languages like the Semantic Web Rule Language [22] (SWRL) can be used. SWRL can be used to express rules as well as logic, combining OWL DL or OWL Lite dialects with a subset of the Rule Markup Language. Rules in SWRL are described as an implication between an antecedent (body) and consequent (head). Whenever the conditions specified in the antecedent hold, then the conditions specified in the consequent also hold. Using SWRL rules it is possible, for example, to automatically assess equivalence among semantically described data and documents, or to interpret the associated meta-data and produce new knowledge.

2.2 Semantic Based Classification and Learning

Data classification and semantic-guided learning is an immediate example of application of semantic technologies to Big Data. In order to effectively respond to the huge volume and variability of data coming from different sources (just consider the data continuously coming in streaming from a sensor network), it is necessary to classify the information and act differently according to its interpretation. One of the

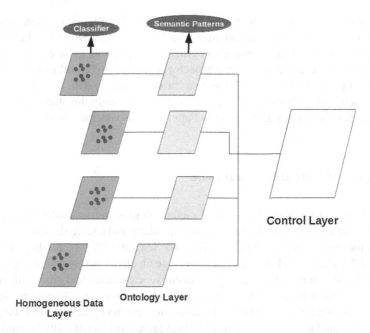

Fig. 3 Multilayer semantic classification

most followed approach to classify data from heterogeneous sources is based on the development of multi-layer classification models, which refine and abstract data, by identifying more specific "buckets" in which to put them at each layer. An example of such approach is described in [1], and the basic model followed for data classification is reported in Fig. 3. Here the data, coming from different sources (in the specific case, a sensor network), are firstly roughly classified according to the generating source: since the system knows the origin of the collected data, some knowledge on them is already owned and can be used to create coarse grained groups. One of the main characteristics of the proposed approach consists in associating each data to one or more possible initial categories: in this way, even if some redundancy is created as some data will be replicated in several classes, there will be more precision when applying semantics. Semantics are applied at the **Ontology layer**, in which data are associated to different domain ontologies. Such ontologies are chosen according to the previous classification operated at the lower layer, and are used to add context-specific information. However, the ontologies used at this level are not isolated, as they are connected via properties which enable reasoning. At the highest level an **Access/Control layer** can be found, which acts as an interface to query the created knowledge base and provide access to classified data. In order to correctly associate semantic annotations to incoming data, classification systems need a set of rules which identify data features and then create annotations accordingly. Several languages can be used to express such rules: SWRL is a classic example of these. However, it is also important to collect the semantic features in order to not hard-code them into the rules: that is why **Semantic Patterns** [21, 49] can be used. Such Patterns provide a structured representation of semantic features which can be identified in a dataset: once such features are recognized, it is possible to assess that a certain kind of data has been identified. One aspect which needs to be addressed regards the **training** of the system to "teach" it how to correctly classify data and annotate them. Training sets can be defined and used in order to both verify the effectiveness of the defined recognition rules and Semantic Patterns or to contribute to their design.

2.3 *Linked Data and Open Data*

Linked Data can be defined as a set of methodologies and technologies to publish structured data on the web, in order to interlink them and enable the use of semantic queries for their retrieval. Standard Web and Semantic Web technologies such as HTTP, RDF, URIs, OWL are needed to provide an environment where applications can query data, draw inferences by accessing vocabularies and share information seamlessly. The semantic web technologies do not aim to serve human readers, but to extends classic web pages to support automatic reasoning processes and computers. The use of a standard format, which can be interpreted and managed by Semantic Web tools, is a necessary condition to make the Linked Data paradigm a reality. Also, not only access to data must be continuously ensured, but also the relationships among them need to be always available: otherwise, we would obtain only a mere collection of fairly annotated data. As also stated by Berners-Lee in [8], a set of four simple rules need to be applied in order to correctly define Linked Data and enable their availability on the web:

- In order to identify and refer to entities (things), URIs need to be used to name them.
- To look-up entities, HTTP URIs need to be used. This ensures both uniqueness of the references item and its availability on the web.
- Beside the name of the referred entities, other useful information need to be provided regarding it, by using standard Semantic Web formats and technologies. Just putting names on the Web does not help in building a Linked Data framework: descriptions and relationships are needed too.
- Always refer to other entities by using their HTTP URI-based names when publishing data on the Web.

Graph notations are widely used to represent linked data, as also shown by Fig. 4 where the Linked Open Data dataset is graphically represented. Linked Open Data is linked data that is also open content. Practically, it follows the same rules as above, but the connected information must be open to everyone. The goal of the W3C Semantic Web Education and Outreach group's Linking Open Data community project is to extend the Web by publishing several open datasets as RDF on the Web and by setting RDF links between data items from different data sources. Large linked open

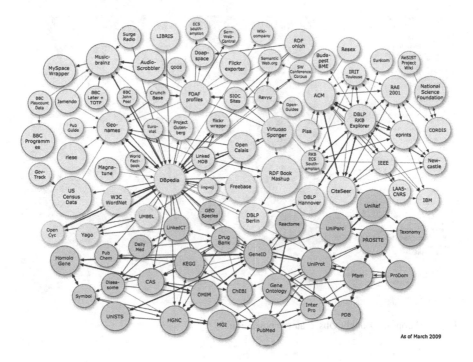

Fig. 4 Linked Open Data collection, march 2009

data sets include DBpedia [4] and Freebase [9]. The development of Linked Open Data databases and datasets is strongly favoured by the European Union, which has funded several research projects and currently supports the development of the Eu Open Data Portal [10]. Such portal includes a very wide variety of high-value open data across EU policy domains, as also more recently identified by the G8 Open Data Charter. Data are currently provided by a growing number of agencies, such as Eurostat, the European Environment Agency and the Joint Research Centre continues to grow. The portal also provides access to a wide range of visualization applications which both enable the retrieval of information and the displaying of examples of possible practical applications. Figure 5 reports a screen-shot of an application for the monitoring of temperatures and precipitations in Europe, as part of a drought monitoring project. Among the European funded research projects and communities, we find:

- **PlanetData** [16] which aims to establish a European community of researchers that supports organizations in exposing their data in new and useful ways. I n particular, it focuses on the mining and analysis of the huge amount of data continuously published online, including data streams, (micro)blog posts, digital archives, eScience resources, public sector data sets, and the Linked Open Data Cloud.
- **DaPaaS** [12] research project has developed a set of tools which enable the use, re-use, definition and sharing of Linked Open Data. The provide **DataGraft**

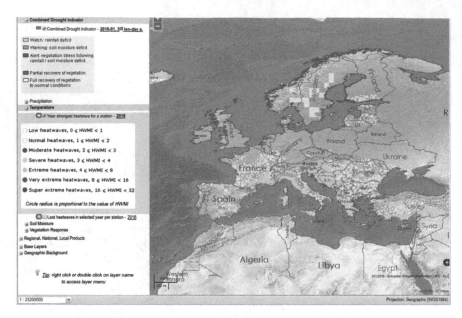

Fig. 5 European Drought Observatory application

platform consists of an integrated suite of four elements: **Grafter**, an open source collection of tools for tabular data transformation and processing; **Grafterizer**, the graphical fronted for the Grafter collection; a **Database-as-a-Service structure** completely built upon RDF; a **Data Portal** providing a catalogue of various datasets and reusable data transformation services.

• **LOD2** [5] is a large-scale integrating project co-funded by the European Commission within the FP7 framework. The project aims at advancing the technologies and standards for Linked Open Data management and representation, by developing tools, methodologies, testbeds and standards to support interaction among enterprises and the sharing of new datasets. The project focuses on the development of trustful standards which can reliably track provenance, ensure privacy and security of data.

2.4 Reasoning over Big Data

For a long time Business Intelligence tools and Data Warehouses have helped companies to understand more about their operations and to take decisions over future development strategies. However, such tools work best on structured and limited (in dimension) data, whilst they are very limited when it come to unstructured data, such as those considered in Big Data scenarios. That depends on the fact that BI tools work by having a clear model already defined and a set of well described **Patterns** to identify in the data source. Due to the high heterogeneity and volume of data to analyse in a Big Data scenario, it would be impossible to identify a set of Patterns to recognize, since they would change dynamically as new information is digested by the system. In order to identify relationships and patterns among different types of data, companies are turning to **Predictive analytics**. This term covers a large number of analysis techniques, such as:

- **Social media analytics** (see Sect. 4), which collects and analyzes information gathered from a variety of social media sources;
- **Text mining**, focused on the deep analysis of text-based documents;
- **Sentiment analysis** (see Sect. 4.2), which aims to identify a user's desires, thinking and priorities, by analyzing electronic text from files, reports, surveys, forms, e-mail and more;
- **Geospatial analysis** (see Sect. 7), focused on data coming from satellites, global navigation systems, aerial surveys, sensor networks and radar.

Predictive analytic makes large use of **Inductive** reasoning, which does not make presumptions of patterns or relationships and is more about data discovery. Several approaches can be used to explore data and discover interrelationships and patterns: machine learning, neural networks, computational mathematics and so on.

Semantics plays a key role in providing the means to correctly understand the meaning of the analysed data: frameworks like Hadoop can store huge amounts of raw documents and files, through the Hadoop Distributed File System (HDFS), but it is difficult to assess what they exactly mean, and to enable several users to understand them without ambiguities. Semantic technologies can help solving the problem, by adding a semantic layer on top of the unstructured data. However, classical semantic-based reasoning approaches are not suitable to support scalable analysis of Big Data: the graph structures, often adopted when representing knowledge in the Semantic Web, are not the best choice for high-bandwidth applications which require speed and scalability.

3   Big Data and Semantics in the Internet of Things

Big data and the Internet of things are two parallel universes, but they are so close that in most cases they blend together. The amount of devices that connect to the internet grows day by day and they bring millions of data. The IoT generates unprecedented amounts of data and this impacts on the entire big data universe. The IoT and big data are clearly growing apace, and are set to transform many areas of business and everyday life.

3.1   Impact of IoT on Big Data

The Internet of Things has deeply influenced the Big Data ecosystem, driving companies towards the research of new, highly efficient data collection and elaboration mechanisms. Since the idea at the base of the IoT is the possibility to connect devices to each other via an IP address, and also considering the fact that millions of devices are and will be connected together, thus generating a huge volume of data, the need for new methodologies and technologies to handle them has become urgent. Along with the necessity to actually handle the enormous volumes of data continuously exchanged and elaborated by smart devices, there are two important issues to consider: **Security** issues, due to the difficulties to apply traditional security mechanisms to such a dynamic and ever growing collection of data; **Data Usefulness**, companies will need to distinguish between useful and redundant data, which will translate in the research for new and efficient analysis algorithms and techniques.

Collecting huge volumes of data from sensors The task of collecting all and only the data that are really relevant to their business can be extremely challenging for companies, as they need not only to actually collect them, but also to filter them by eliminating redundant data, and protect them against security threats. Several software and protocols have been developed for such tasks, and some of them are currently employed with success: among these, the **Message Queue Telemetry Transport** (MQTT) [26] and **Data Distribution Service** [44] (DDS) are two of the most comprehensive protocols.

The MQTT, which is currently an OASIS standard, collects data from several sensor fitted devices, and then feeds them to the IT infrastructure which will elaborate them. It represents, in simple words, a lightweight message queueing and transport protocol, suited for Mobile to Mobile (M2M), Wireless Sensor Networks (WSN) and IoT applications. Its implementation follows a clear **Observer** Design Pattern. A **Sensor Node** sends data to an MQTT **Broker** which sorts them in different message queues, also referred to as **Topics**. An **Actor Node**, or **Subscriber**, which has previously registered to one or more Topics, will receive only those messages containing data of specific interest. Figure 6 shows the general idea behind the protocol model, while Fig. 7 describes the general sequence of steps sensor nodes, subscribers and broker have to follow. In general, when a Publisher sends data to the Broker,

Fig. 6 Model of the MQTT protocol

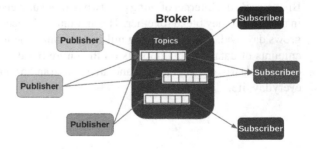

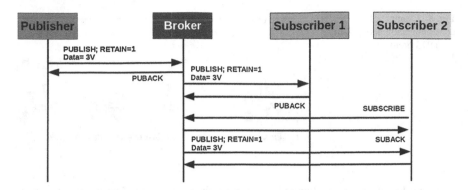

Fig. 7 Example of usage of the MQTT protocol

or the Broker pushes such data to a Subscriber, an ack is sent by the receiver. This feature is enabled when the protocol works with a Quality of Services of level one (QS1), for which a message has to be delivered at least once. Instead, a QS0 can be chosen (with no ack messages), or the level can be raised to two (QS2), in which messages are delivered exactly once. In the last case, additional ack messages are exchanged between senders and receivers. When the Broker receives data, and the message containing them is marked with the "Retain = 1" option, then the data has to be stored on the Broker and will eventually be updated. When such data is sent to the Subscriber, the same option points out that the data is old and the stored value could not match the current one. In any case, when a new Subscriber requests update from a topic, the last stored data is immediately sent to it.

In order to keep the protocol simple and reduce overhead in communication, the mandatory header for MQTT messages is just two bytes long. Additional and optional fields can be used, especially to reduce retransmissions and data losses.

The DDS protocol, while based on exactly the same publisher/subscriber paradigm, is more suited for the distribution of data among several devices. This derives from the distributed, rather than centralized, architecture of a DDS implementation, which does not imply the use of a central broker for message delivery. The concepts which are defined in the DDS standard are very similar to those found in the MQTT. The concept of **Topic** still exists in DDS, with more or less the same meaning. IN particular, DDS's topics are a class of streams, which are associated to a user defined extensible type and a set of QoS policies. They can be defined locally, or discovered by devices. As in the MQTT case, we have Subscribers called **DataReaders** and Publishers called **DataWriters**. The main difference between MQTT and DDS relies in the complete lack of a central broker in the latter approach. Instead, the DDS protocol implementations exploit a distributed Peer to Peer network, in which devices can act as readers or writers indifferently and at the same time. As shown in Fig. 8, no broker is used in the DDS approach and, while clusters of devices/sensors are possible and supported, the nodes act independently. DDS also offers a better approach to QoS than MQTT, as several properties regarding delivery time, dimen-

Fig. 8 Example of
communication network in
the DDS protocol

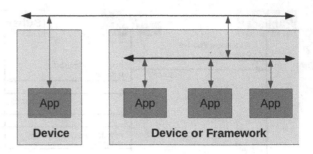

sion and length of the exchanged messages can be enforced. The approaches used
by the two protocols are not mutually exclusive: indeed, they can be used together in
order to solve issues in data communication over very large and distributed sensor
networks. The most common approach is to exploit a **Gateway**, located on a sensor
device or on the MQTT broker, between a DDS and MQTT network.

3.2 Ongoing Research Efforts

The Internet-of-Things vision provides a large set of opportunities to users, manufacturers and companies [39]. In fact, IoT technologies will find wide applicability
in many productive sectors including, e.g., environmental monitoring, health-care,
inventory and product management, workplace and home support, security and surveillance. From a user point of view, the IoT will enable a large amount of new
always responsive services, which shall answer to users needs and support them in
everyday activities. Before the IoT and big data coalition can deliver on its promise,
there are a number of barriers to overcome.

The first challenge is the worldwide adoption of shared standards. The use of
standards ensures interoperable and cost-effective solutions, opens up opportunities
in new areas and allows the market to reach its full potential. For the IoT to work,
there must be a framework within which devices and applications can exchange data
securely over wired or wireless networks. In this area there are a lot of player: M2M
[41], AllJoyn [3], OIC [42].

M2M (Machine to Machine) refers to technologies that allow both wireless and
wired systems to communicate with other devices of the same type. M2M is a broad
term as it does not pinpoint specific wireless or wired networking, information and
communications technology. This broad term is particularly used by business exec-
utives. M2M is considered an integral part of the Internet of Things (IoT) and brings
several benefits to industry and business in general as it has a wide range of applications such as industrial automation, logistics, Smart Grid, Smart Cities, health,
defense etc. mostly for monitoring but also for control purposes. Open Machine to
Machine (OM2M) [2] provides an open source service platform for M2M interoperability based on the ETSI-M2M standard. OM2M follows a RESTful approach with

open interfaces to enable developing services and applications independently of the underlying network. It proposes a modular architecture making it highly extensible via plugins. It supports multiple protocol bindings such as HTTP and CoAP. Various interworking proxies are provided to enable seamless communication with vendor-specific technologies such as Zigbee and Phidgets devices. OM2M implements the SmartM2M [13] standard. It provides a horizontal Service Capability Layer (SCL) that can be deployed in an M2M network, a gateway, or a device. Each SCL provides Application Enablement, Generic Communication, Reachability, Addressing and Repository, Interworking proxy, Entity Management, etc. It includes several primitive procedures to enable machines authentication, resources discovery, applications registration, containers management, synchronous and asynchronous communications, access rights authorization, groups organisation, re-targeting, etc.

The AllSeen Alliance [3] is a cross-industry consortium dedicated to enabling the interoperability of billions of devices, services and apps that comprise the Internet of Things. AllJoyn, a framework created by the AllSeen Alliance, is an open, universal, secure and programmable software connectivity and services framework that enables companies and enterprises to create interoperable products that can discover, connect and interact directly with other AllJoyn-enabled products. AllJoyn is agnostic respect to the transport layer, the OS, the platform and brand, enabling the emergence of a broad ecosystem of hardware manufacturers, application developers and enterprises that can create products and services that easily communicate and interact. It consists of an open source SDK and code base of service frameworks that enable such fundamental requirements as discovery, connection management, message routing and security, ensuring interoperability among even the most basic devices and systems. The initial planned set of service frameworks include: device discovery to exchange information and configurations (learning about other nearby devices); onboarding to join the users network of connected devices; user notifications; a common control panel for creating rich user experiences; and audio streaming for simultaneous playback on multiple speakers. In addition, the Alliance is producing developer tools and verifying correct implementation through a compliance program.

The Open Interconnection Consortium (OIC) is defining a common communication framework based on industry standards to wirelessly connect and manage the flow of information among IoT devices. It sponsors the IoTivity Project [29], an open source software framework for device-to-device connectivity. The IoTivity architectural goal is to create a new standard by which billions of wired and wireless devices will connect to each other and to the internet. The goal is an extensible and robust architecture that works for smart and thin devices. The IoTivity framework APIs expose the framework to developers, and are available in several languages and for multiple operating systems. The APIs are based on a resource-based, RESTful architecture model. The framework operates as middleware across all operating systems and connectivity platforms and has four essential building blocks: discovery, data transmission, data management and device management. IoTivity Services, which are built on the IoTivity base code, provide a common set of functionalities to application development. IoTivity Services are designed to provide easy and scalable access to applications and resources and are fully managed by themselves. There are six

IoTivity Services in v1.0, each with its own unique functionality: Resource Encapsulation, Resource Container, Things Manager, Resource Hosting, Resource Directory and MultiPHY EasySetup. Resource Encapsulation abstracts common resource function modules. It provides functionalities for both the client and server side functions to IoTivity Service developers. For client side, it provides resource Cache and Presence Monitoring functions. On the other hands, for the server side, it provides the simple, direct way to create the resource and to set the properties, attributes of resources. Resource Container provides a way to integrate non-OIC resources into OIC ecosystem by creating, registering, loading and unloading resource bundles. It also provides common resource templates and configuration mechanism for resource bundles. It deals with OIC specific communication features, and provides common functionalities in a generic way. Things Manager creates Groups, finds appropriate member things in the network, manages member presence, and makes group action easy. The goal of Resource Hosting is to off-load the request handling works from the resource server where original resource is located to reduce the power consumption of resource constrained devices. A resource directory is a server that acts on behalf of the thin-client. The thin-client after it publishes their resources, resource-directory will respond on behalf of these devices. The device acting as a resource directory could itself hold resources. MultiPHY Easysetup is an IoTivity primitive service to enable different sensor devices (with different connectivity support) to be easily connected to the end user's IoTivity network seamlessly. Thus enabling Sensor devices to be part of the IoTivity network in a user friendly manner.

The big data universe can provide infrastructure and tools for handling, processing and analyzing deluge of the IoT data. However, there is a lack of efficient methods and solutions that can structure, annotate, share and make sense of the IoT data and facilitate transforming it to actionable knowledge and intelligence in different application domains [6]. The issues related to interoperability, automation, and data analytics naturally lead to a semantic-oriented perspective towards IoT. Semantic technologies based on machine-interpretable representation formalism have shown promise for describing objects, sharing and integrating information, and inferring new knowledge together with other intelligent processing techniques.

Semantic technologies are necessary to integrate data from heterogeneous data sources. In fact ontologies and related semantic technologies, such as ontology merging and mapping, could offer a simple and powerful mean to provide not only a formats unifier but also a semantic translator by providing an unified interpretation of different data sources. Moreover, by means of semantic annotation the data will be self explanatory information carriers and thus enabling the dynamic discovery of relevant data sources and data. Semantic technologies enable the development of extensible context models which can be adapted to different application domain and that can be easily enriched to accommodate the continuous evolution of IoT systems. There are many efforts in creating common models for describing and representing the IoT data and resource descriptions, among many IOT-A [7], SSN [11], OpenIoT [31].

4 Social Mining

Modern Social media is designed as a group of Internet-based applications that build on the ideological and technological foundations of Web 2.0. Such technologies allow the creation, sharing and exchanges of huge volumes of user-generated content. Vast amounts of data are created on social media sites every day. **Social data mining** can be defined as the systematic analysis and extraction of valuable information from the Social media. Since the Social media data are largely user-generated. This means that the analysed content is:

- **Vast**, as the volume of data produced and exchanged exceeds the computational capabilities of standard algorithms and machines.
- **Noisy**, since people exchange information about everything, often biased by their own opinions and experiences. The really useful information could be hidden behind a large amount of useless data.
- **Distributed** because the source of data cannot be found in just one place, such as in a single database. Data are being generated everywhere, by people connecting to their preferred social networks from work, school, home or in the middle of a street, thanks to their mobile phones.
- **Unstructured**, as data produced by people interacting on Social Networks come in the form of chats, comments, messages, e-mails, which do not have a fixed structure or standard to follow.
- **Dynamic**, since data change at a dramatic pace, following people's opinions, moods, movements and so on.

Despite the complexity behind the analysis of such social generated data, there are many reasons which drive companies to invest money on this topic. The main reason is represented, however, by **Marketing**. By mining people's opinions and mood, market researchers try to determine the best sale trends and to propose, according to it, the right product at the right moment. By analysing customer experiences, which can be extracted from dedicated blogs or web-sites (not considering Facebook pages), companies can try to optimize their products and satisfy their clients. There are, of course, nobler applications of social mining. First of all, by analysing the position of people, their movements and their opinion on public transports, institutions can try to identify problems and issues in the public transportation system (very crowded/almost empty trains could represent the symptoms of a problem in the circulation system). The analysis of messages exchanged by people during or immediately after major incidents (fire out-brakes, earthquakes, floods) can be used to direct rescuers towards the most hit areas and rapidly organize first-aid operations.

As in all social relationships and communities, Social Media evolve and gradually acquire a well defined structure, which reflect the interaction ad relations among the community members, and impose rules of well behaving within that community. As an instance, online forums often have criteria to determine if a member is an expert on a specific topic, and provide means to give the user credit for that. Social applications are generally designed to support the definition and creation of such structures and criteria. Very common social structure to be found in Social Media are:

- Hierarchical structures, which follow the hierarchical nature of human relationships, in which a limited set of members has higher privileges and can enforce rules.
- Conversational structures, which are built upon the messages and replies exchanged by members. Such structures are present in all networks which imply or require a direct interaction among users.

Knowing the structure of the target Social Media is essential to analyse it, since the value of the information generated by a member of the network strongly depends on its source within the community and the influence such a source has on other participants. In order to study a social media and apply mining techniques, Social Media Graph Mining Graphs (or networks) have been largely applied, since they constitute a dominant data structure and appear essentially in all forms of information. Typically, the communities correspond to groups of nodes, where nodes within the same community (or **clusters**) tend to be highly similar sharing common features, while on the other hand, nodes of different communities show low similarity. Useful knowledge (patterns, outliers, etc.) can be extracted from structured data represented as a graph. Using a graph representation of social interactions and behaviours, it is possible to run several algorithms which help in mining new information as required. A classical example is represented by Google Page Rank algorithm, used as one of many predictors for the relevance of a web page. The link structure in the world-wide-web network provides valuable contextual information about which pages are deemed most relevant by the web page creators and visitors. The link-based graph structure is used to predict relevance for a user's query, but Social media hosts also use it to understand if a specific group within their network will grow or be disbanded in the near future. Section 5 focuses more on graph mining techniques.

4.1 Text Mining

Text Mining is an emerging technology that attempts to extract meaningful information from unstructured textual data. Since Social networks are based on messages exchanging, comments publication, links to posts, blogs or other news articles, most of the data created and passed from one user to another come in for of text. Text mining techniques to discover information from unstructured text are part of many of today's applications, some of which we use everyday:

- Automatic processing of e-mails, for the automatic classification of text and recognition of "junk mail" is used in order to fight spam. Using text based filters it is possible to recognize specific terms or words that are not likely to appear in proper and legitimate messages.
- Mining of medical records is used to improve care of patients' health.
- Cyber Security applications, also to fight terrorism and crime, often analyse online plain text sources such as news, blogs, feeds, to monitor suspicious activities and detect malicious intentions.

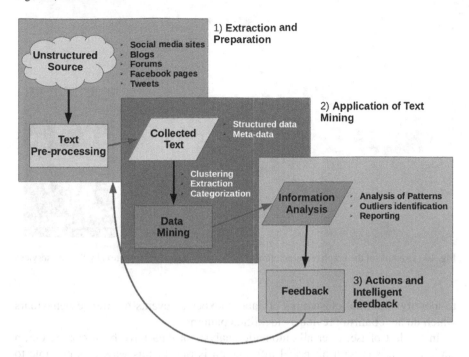

Fig. 9 Phases of a text mining process

- Web pages "crawling" is often used by companies to analyse competitors' portfolios and identify competitive strategies. Crawling techniques are also used by search engines to determine connections among web pages and to index new content, also making researches faster.

In order to carry out an effective Text Mining process, three basic steps have to be followed, as also shown in Fig. 9.

First of all, the data need to be correctly collected from one or more social platforms. Raw data can be stored into Big Data oriented databases, especially if their volume need particular care. However, some pre-processing need to be done in order to reduce the volume od data to be analysed and to remove redundant and noisy text. This is done by applying Data Modelling techniques which enforce specific requirements on the data (which have to be determined in the application design phase) and strongly depended on the scope of the application. According to this pre-processing, meta-data are generated and associated to the raw data, which are then sorted and arranged in order to be better classified during the second stage.

In the Mining stage, different Data Mining techniques are applied to the semi-structured data which have been previously collected. Cluster analysis, as an instance, is applied to identify unknown and interesting patterns which glue together groups of data records. Anomaly and outliers detection is also applied in this phase in order

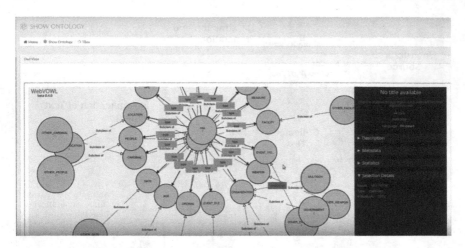

Fig. 10 Example of the graph representation of information extracted from text by Watson services

to identify abnormal clusterings of data, unexpected events or strange behaviours which do not confirm previously identified patterns.

In the last phase, after all automatic analysis of data have been carried out, a static comparison with historical information is run. In this way, it is possible to compare past and present situations, in order to understand if some major change has happened, if the previous model can be confirmed or need to be update on the base of the new information. Business companies often use such comparison to perform market analysis, to determine future trends or correct/avoid manufacturing problems. Among the most recent advancements in the State of the Art on text mining, Cloud based solutions surely have an important role. As an instance, the IBM Bluemix Cloud Platform has provided a set of very interesting services, based on the Watson framework, which enable the online analysis of text and web-sites. In particular, via a **Relation extraction service** [27], which accepts as an input both textual documents and URLs of web-pages, it is possible to retrieve semantic relationships among entities described in the text and polarity information on the analysed terms. The output produced by the Watson framework includes parse trees, grammatical and logical information on the text, and semantic relationships which can be used to derive entire ontologies. Figure 10 shows the graph-based representation of entities and their relationships which have been detected in a simple text analysed via Watson services. By selecting one of the bubbles, whose dimension depends on the number of occurrences and relevance of the referenced term, it is possible to obtain additional information on that term (relationships with other entities, meta-data and so on).

4.2 Sentiment Analysis

One recurring argument in Social and Text Mining scenarios, is represented by **Sentiment Analysis** (also Opinion Mining) which is referred to as the use of Natural Language Processing (NLP) techniques, statistics, or machine learning methods to extract, identify, or otherwise characterize the sentiment content of a text unit. Mostly adopted to identify people's reaction to particular events, such as the launch on the market of a new product, upcoming elections, a change in the marketing advertisement and so on, it is applied to several possible sources: customer forums, tweets, product reviews. However, sentiment analysis is also used for other purposes, such as during information extraction for discarding subjective opinions from a text and identifying biased news sources, or when trying to summarize the content of a specific page, taking in consideration multiple opinions. Moreover, it finds application in flame detections, or in identifying the content of public media from the comments left by users.

While Sentiment Analysis deal with far less categories than Text Mining, as the opinions can be positioned into a limited and strictly dependent set of classes (positive or negative, good or bad and similar), it is still extremely complex to actually understand under which category a specific opinion falls. This depends on the fact that humans express their opinions in very complex ways, making use of expressions where lexical meaning can be misleading, or using rhetorical modes such as irony, sarcasm implication and so on.

The first question to answer when trying to analyse the sentiment of a text unit regards the exact subject of the classification: according to what we are interested in (users, small sentences, tweets and updates, smileys) different techniques can be applied. Since words represent the building block of language, most techniques start from them to classify the sentences they compose. Short sentences (two to three words) can also be taken in consideration, since they can compose some particular meaningful expression. When working with small texts, tweets or comments, a common approach is to analyse smileys, since they portray a lot of information in a very few coded characters.

Whatever part of the text we want to analyse, there are different techniques and approaches which can be applied. The most known and followed approach tries to identify the **Semantic orientation** [20] of the text and the **Polarity** of words contained in it. Examining the Semantic Orientation of a text towards a features tries to determine if the opinion expressed by that text is positive, negative or neutral. It represents a real-number measure of positive or negative sentiment in a phrase, and it is calculated on the Polarity of the relevant words composing it, which is in turn a binary value either positive o negative. In order to assess if a specific word has a positive or negative meaning, we can rely on **Heuristic** methodologies, which start from some already owned knowledge and apply it to solve the new problem: as an instance, we already own dictionaries which classify words and know their meaning. Some tools support this kind of approach: **General Inquirer** [50] and **WordNet** [38].

General Inquirer is a "Computer-assisted approach for content analyses of textual data" which has been developed over several years and continuously updated to support modern programming languages and run on today's machines. The approach it uses consists in dividing words into more than 180 categories (like a dictionary with more classes), which are used to tag and map terms and phrases. In order to classify words by polarity, it is possible to apply positive and negative categories, which is very fast and easy using the instruments provided by the Inquirer. However, there are major drawbacks represented by the binary-based approach (no gradations or weighting of the polarity) and the blindness to the context. Since the categorization is only lexical, the context is ignored and sometimes interpretation can be misleading.

WordNet creates an extended thesaurus in which words are part of a network, connecting them on the base of synonymy in order to create a rich semantic organization. Synonyms are grouped into **synsets**, which are in turn connected via different kind of relationships:

- Hyponym relationship connects a term with a specific meaning to a more general one: *butterfly* is hyponym of *insect*;
- Hypernym is the opposite of Hyponym;
- Has-member connects a group with one or more of its components: *book* has-member *page*;
- Has-stuff connects a term with one or more elements describing its composing materials: *book* has-stuff *paper*;
- Entail express an implication or natural involvement: *dream –> sleep*;
- Cause-to express a cause-effect relationship: *time passes –> become older*
- Attribute connects and adjective to a base word: *hypocritical –> insincerity*

Other relationships connect words directly. The work presented in [23] demonstrates the use of WordNet to determine adjectives' polarity: the authors apply a machine learning methodology, based on a set of rules which, starting from a set of "seed" words (whose polarity is already well known), label new terms by measuring their proximity via synonymy/antonymy relations to seed adjectives. Despite the good results obtained via such a technique (the authors documented an average sentiment orientation accuracy of 84%), the WordNet approach is limited when it comes to word with multiple senses (context issues again) and sentences with very few or no adjectives.

Another approach which is used in combination with WordNet or other systems is based on the **Theory of Semantic Differentiation** [43], which assigns three different values to words, according to the emotive meaning of adjectives: **Potency** (strong or weak), **Activity** (active or passive) and **Evaluative** (good or bad). The work presented in [30] used WordNet to calculate such values, by comparing the minimal path length (MPL) between two words representing the factor's range (as an instance, the MPL between a word and the good and bad terms, for Evaluative). This approach has the evident limitation that if two adjectives are not related to the ranges' extremes by a synonymy relationship, they will never get a score.

4.3 Social and Political Trends

One of the application of Sentiment Analysis techniques is represented by the inspection of informal texts regarding political discussions and opinions. The subject of such an analysis can be quite heterogeneous, as the arguments of interest vary significantly. In general, government and politicians are strongly interested in knowing the Public opinion, and their general attitude towards policies, parties, government agencies and politicians. The knowledge of the political trends and opinions expressed by people, active voters in particular, can be used to drive elections by influencing them. Targeting advertising and communications such as donation requests, petitions, and notices can represent a formidable weapon to gain consent. Also, it is possible to understand if a particular information source, such as a news paper, is biased towards a specific opinion more than another, and choose/discard it to release a positive interview.

The main source of informal political discussions is represented by Newsgroups, Blogs, Online publications allowing readers to write their feedback, or even Social Networks. What characterizes such sources is informal register which is often used in the text. This depends on the fact that people express their opinion by quickly writing short sentences, often fragmented, which contain expressions the author (but not necessarily the readers) is very familiar with. To this, one should add the high number of misspellings and grammatical errors which often affect such textual interventions, the extensive use of jargon, and the occurrence of non-existing words which are satirical re-spellings of known ones.

Differently from Sentiment Analysis categorizations, political opinion does not easily fall in binary classifications, which are way too simple representations. Political attitudes cover a wide range of possible judgements, whose relationships are not always clear. As instance, the "Pro-life" anti-abortion groups in the USA political scenario are often affiliated with "Pro-death" penalty organizations.

In [40] and subsequent works, the authors have applied a classification methodology to identify public opinion trends, using "quote" patterns to determine the affiliation of writers to a political party. In particular, political attitudes were divided in three coarse grain categories *left, right and other*, where the "other" contained noisy and uncertain data. Starting from the discussions posted on the www.politics.com website, the textual data has been divided in "chunks", and three (or more) words long chunks, which appeared in more than one post, were considered as quotes. By analysing chunks and quotes, and by comparing them with lexical terms extracted from other sources (such as on-line newspapers), text is labelled with manual identified classes, which then are associated to the original categories. In order to correctly identify the afferent categories, Naive Bayesian techniques were used, associated to a **User Citation Graph**, in which nodes represented users and edges were quotes. Among the results of the study, it was evident that users from one wing mostly quoted users from the opposing party. In order to reduce the set of data to analyse and fasten the computation, only posts and quotes from frequent users were taken in consideration.

The selection of the starting set of terms, sentences and textual fragments to analyse is an important task, as the dimension of the data to be evaluated strongly influences the time to elaborate them and the needed computational power. In the approach followed in [35], which analyses sentiments expressed in tweets on USA President approval, the selection of posts to analyse was reduced to one per day per poster, and all comments left on other topics were simply discarded. Subjectivity Lexicon [52] and SentiStrength [51] were used to determine the polarity of words and the strength of the expressed sentiments.

5 Graph Mining

When trying to model complex structures, Graphs based representations can provide the optimal solution to store all the required information regarding elements of the structure and their relationships. Circuits, biological and social networks, the Internet can all be represented through graphs. The use of graph search algorithms is documented in several fields, such as chemical informatics, computer vision, video indexing, and text retrieval. Because of the extreme flexibility of graphs, their capability to describe complex situations and the intense research on algorithms to manage them, they have been naturally adopted to support many data mining applications. Recent studies have developed several graph mining methods and applied them to the discovery of interesting patterns in various applications. In particular, such methods are based on the discovery of **Frequent sub-structures**, that is basic patterns which repeat themselves in several graphs and show peculiar and distinguishable characteristics. The presence of such Frequent sub-structures in graphs can help in discriminating, classifying and clustering sets of graphs, building indexes or speeding researches in graph databases.

Graph sub-structure mining In order to understand how to mine for sub-structures, we need to define the main components of a graph. In particular, a graph is composed of: a set of **Vertexes**, which represent entities to be modelled; a set of **Edges** which denote relationships among entities and connect vertexes; a **Labelling function** which associates vertexes and edges to a label used to annotate it with information (meta-data). Given a set D of N graphs, we will say that a sub-graph **g** is frequent, if it is sub-graph of at least m graphs from the set, where m represents a **threshold**. The definition of sub-graph relies on that of **isomorphism** on which a very rich literature exists [14].

The sub-graph mining problem is generally solved by following a two-step approach: in the first step, a set of sub-graph candidates is generated, while in the second their frequency is calculated. Most algorithms provide optimized methodologies to create the candidate set.

The **Apriori-based Approach** starts from a small structure (the definition of size depends on the specific graphs analysed), and define new candidates by expanding it. The expansion can require the addition of a vertex, and edge or a path to the original

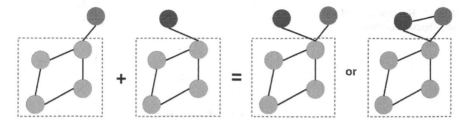

Fig. 11 Functioning of an Apriori procedure (AGM in particular)

structure. In general, the algorithms based on this approach follow a bottom-up procedure: starting from a set of already known frequent sub-graphs, either already known or built and evaluated during the first phases (thus being very small and simple), they merge such sub-graphs and check if they are frequent enough to become part of the solution set. By means of this iterative procedure, larger sub-graphs are generated at each iteration: the algorithm stops when the new larger structures do not meet the frequency threshold. Figure 11 shows how the basic procedure works, by adding one vertex at a time.

   Implementations of such procedure are, as an instance:

- the **Apriori-based Graph Mining** (AGM) algorithm [28], which increases the size of graphs by adding one vertex at a time. In particular, in order to obtain a graph of dimension k (size here is the number of vertexes), it joins only two graphs of dimension k − 1 sharing the same sub-graph. The new graph will share the common elements of the joined graphs, plus two more vertexes which can be or not be connected: thus, two new candidates are defined. Figure 11 makes reference to this kind of approach.
- the **Frequent Sub-Graph** (FSG) algorithm [34] follows an edge-based approach, as the expansion is made not by adding vertexes, but one edge at a time. Two k − 1 graphs sharing the same substructure are joined to form a new k sized graph.
- the **Edge-disjoint path** method [19] defines the size of a graph as the number of **Disjoint paths** (paths not sharing edges) they include. Two graphs of dimension k, sharing a set of disjoint paths, are combined together to obtain a k + 1 sized graph.

Apriori-based approaches all have to evaluate all k-size sub-graphs before increasing their dimension: thus, they all follow a Breadth-first search kind of approach which, in some cases, can slow down the process since the number of k-size graph to be evaluated before moving to k + 1 can be considerable. **Pattern Growth** approaches use a completely different way of visiting the solution space and, being more flexible, they can be implemented in a BFS or DFS (Depth-first search) fashion. The idea is to add to a sub-graph g, one edge at a time. This edge can add a vertex or not, that is not important. Instead of adding edges to the graph g by keeping the size constant, this approach builds all the g' graphs including g which can be generated by adding edge after edge. The algorithm stops when no more edges can be added, so when the

Fig. 12 Expansions used for
the gSpan algorithm

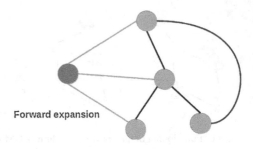

Forward expansion

maximum size possible, starting from g, has been reached. This approach rapidly determines sub-graphs of big size, but has a potential bottleneck: it can produce a high number of duplicate solutions, which need to be continuously checked and can represent an unwanted overhead.

The **gSpan** algorithm provides a solution to such a problem. It adopts a DFS search, by selecting a random start vertex which is then expanded to add new edges. Each edge will provide a connection to a new vertex, which will be expanded itself. Already visited vertex are not expanded. The algorithm stops when no more expansion are possible. A criterion for expansion order is needed, as it is necessary to choose which vertex among new added ones the algorithm has to visit. When the DFS visit has been completed, we will have a single path from the root note and the last visited vertex, which is also called **Right-most Vertex**, while the path connecting the root the such vertex is called the **Right-most path**. In order to avoid duplicate graphs, edges can be added only in two cases:

- **backward extension**: the edge connects the right-most vertex with another vertex on the right-most path;
- **forward extension**: the new edge introduces a new vertex on the right-most path.

Figure 12 reports how the two expansion methods work.

Since different DFS trees can be defined, depending on the root node selected and the ordering criterion, one and only one base DFS has to be chosen to expand the sub-graph by using the two extension mechanisms.

Since each node in the examined graphs can be labelled, it is possible to generate a **code** which identifies all the vertexes and edges from the root to the last added one: in this way, if a visit has generated two identical signatures, it is possible to simply **prune** one of them, and proceed to explore the remaining. In this way, the algorithm avoid expanding the same nodes twice and being capped in a loop.

Other graph mining algorithms The Apriori and Pattern Growth mining algorithms are among the most used ones to determine frequent sub-graphs in a graph set. However, there are other algorithms and methods which try to reduce the search space by imposing some limitation criteria on the possible sub-graph candidates. As an instance, the **Closed Frequent Substructure** method only researches for **Closed Graphs**, where we define a graph G as closed if there is no super-Graph G' with the same frequency in the graph set: that is, by expanding the graph G we have a reduction in its frequency. With one of the previous approaches the new graph G' would have been considered as a valid candidate, if it had a minimum frequency of course. With the Closed graph approach, it is ruled out of the solution set, even if it has a frequency above the threshold. The result set would obviously be smaller than in the base Apriori or Pattern Growth approaches, but the missing solutions can still be calculated from the discovered Closed graphs.

If we consider the gSpan algorithm, we can see that it is limited in regards to the graphs it can handle: the original algorithm considers only labelled, connected, undirected graphs. So, if one or more of these properties is not present, the algorithm cannot be applied. However, these situations can be addressed and eventual issues solved.

- If a graph is not labelled or partially labelled, it is possible to define a new label set, comprehending the old one and a new "empty" symbol, which is applied to all unlabelled nodes and edges. This symbol can match with all other symbols or only with other empty ones, depending on the application.
- In the case of **nonsimple graphs** which may contain self loops (a vertex pointing to itself), or multiple edges from vertexes, a change to the expansion rules must be introduces. In particular, the order to be followed during the expansion needs to be backward edges, self-loops, forward edges. Multi-edge graphs can be handled by simply defining a correct order in the nodes' codes: the classic lexicographic order can handle multiple edges without problems.
- Directed graphs only differ from undirected because edges have a starting and target node. It is possible to represent an edge with a tuple o 6 elements (i, j, d, l_i, $l_{(i,j)}$, l_j), where i and j are nodes, l_i, $l_{(i,j)}$, l_j are the labels associated to such nodes and the edge existing between them, while d is the direction (1 for forward, -1 for backward). So the expansion algorithm just has to consider the direction when adding new edges to the solution set.
- Disconnected graphs are handled in two ways, depending on the source of disconnection. If the graphs in the data set are disconnected, then it is possible to a add virtual vertex to eliminate the disconnection. If the disconnection is in the graph patterns, then we can see each of them as a set of connected graphs, and impose and order on their visit.

Applications of Graph mining Among the several applications in which graph mining techniques can be adopted, **Indexing** represents is particularly important. All modern databases exploit indexing algorithms to speed-up queries and the retrieval of information, but graph-based databases cannot be handled by standard algorithms, because of the exponential number of sub-graphs to manage. A **Path-based** approach is often applied, where the path (a list of visited vertexes) represents the indexing unit. The idea is to enumerate all possible paths in a graph-based database, limiting the path length to a maximum number L. When a query is issued, graphs are identified by matching them with a graph index extracted from the query. Since, despite the length limitation, the index can be quite long as several paths are considered, other approaches choose different indexes. The **Discriminative frequent substructure** approach uses frequent substructures to index graphs in the database, but it only considers discriminative substructures. A substructure is discriminative if its frequency is not approximated by the intersection of the graph sets containing one of its sub-graphs. The **gIndex** [53] algorithm is an implementation of this approach.

Classification and cluster analysis is another application of graph and sub-graph mining algorithms. By discovering frequent graph patterns and their variants, it is possible to associate discriminative features to graphs and, as a consequence, to classify them according to such features. Thresholds like frequency, graph connectivity, discriminativeness can be tweaked to determine the classification accuracy. Once a set of features has been identified, classic classification techniques can be applied: vector machines, naive Bayesian, associative classification and so on. Clustering can also be based on mined graph patterns. Graphs sharing a large sub-set of similar graph patterns can be labelled as similar and thus be grouped in the same cluster, while graphs not belonging to any of the determined clusters or not similar enough to other graphs, are considered as outliers.

5.1 Link Mining

In Social Networks relationships among entities are represented by links in a graph: however, traditional machine learning and data mining algorithms and techniques may not be efficient and appropriate in the Social Network analysis. The data produced and exchanged within a social network tend to be very heterogeneous, multi-relational and semi-structured: in order to define new mining techniques, **Link Mining** has emerged as a new research field. In particular, while traditional data classification methods, based on graphs, identify objects as possible graph's nodes based on their attributes only, Link Mining takes in consideration the attributes of existing links and the properties eventually associated to the whole object-relation-object triples. This is used, as an instance, for **Link-based object classification**. A typical application of link based classification is represented by Web pages analysis, which tries to predict the category of a web page on the base of two attributes: the words occurrences in the text of the page and the **anchor** text that is, the words used for hyperlinks. Classification also benefits from the knowledge of links existing between pages and the attributes of the linked pages. Strictly connected to link-based object classification are **Object type prediction** and **Link type prediction**, which try to predict an Object and Link type, based on the links and objects they are connected to. Link prediction is particularly interesting, since in most analysis it is necessary to understand the nature of the relationship existing among entities. If we want to assess if a certain link exists, that is we do not know if two or more objects are connected or not, we talk about **Predicting link existence**. As an instance, we may want to know if two web pages are connected, if an article cites another one and so on. In some cases, knowing the **Cardinality** of the links is necessary to correctly weight them, and optimize the pages' categorization. In general, an object authoritativeness is reflected by the number of other objects linking to it. A highly cited paper, for an instance, is more likely to have a high impact on the research field it addresses. Conversely, if an object acts as a **hub**, that is it links to a high number of different authoritative objects (out-links), than it can be seen just as a collector of links and obtain scarce consideration. Another form of link cardinality estimation calculates the number of object touched while visiting a link-based graph, moving along the path between two nodes. As an instance, while crawling a web page, it is useful to predict how many in and out links that page own and, as a consequence, to choose the path which will require less computational power to complete the crawling. A common problem which can be solved by analysing the attributes of the links connecting object is the **Recognition of Duplicates**, that is the identification of two or more objects as exactly the same. This is an important task in web pages classification, in order to identify page mirrors, or in the bibliography field, to understand if two citations actually refer to the same paper. **Clustering** of objects is also another interesting task which can be addressed via link mining: by analysing objects' properties and links existing among them, it is possible to place them in particular groups or clusters, thus inferring similarities among different entities. The identification of Web or research Communities is an example of Cluster detection. If we are more interested in identifying particular structures based on the connections existing among objects, we talk about **Sub-graph** detection, which identify recurring sub-graphs within networks. Information on objects and links can be stored as Meta-data, which provide semi-structured data on otherwise unstructured ones.

6 Big Stream Data Mining

Most of the algorithms described in literature to mine data assume that the information is permanently stored in some kind of database and that it is always available. However, when considering Big Data Streams, data arrive in a stream or streams, and needs to be immediately processed or stored somewhere, or they are lost. In most cases, the rapidity with which data arrive is so high that it is not even possible, or feasible, to actually store them. Stream mining algorithms tend to apply one of two different approaches: **Summarization** and **Windowing**. In Summarization only a part of the stream, the one considered relevant and desirable, is actually elaborated or stored temporarily, in order to reduce storing and computational costs. In Windowing, the stream is divided in chunks of fixed length (generally large), and data are elaborated in batches. Since the windows tend to be very large themselves, summarization may be applied on a smaller scale. The reduced windows can be then queried like a standard database.

Models used to manage Data Streaming all follow a standard architecture, which is summarised in Fig. 13 in its main components. The data streams, which can be of different nature and come from very different sources, feed a streaming processor. Among the possible data sources, we have:

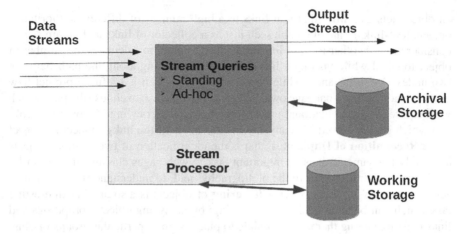

Fig. 13 Data Streaming Model

- Sensors and Sensor networks, which produce vast amounts of data regarding several aspects of the world (temperature, devices' positions, measures of voltage or current).
- Image data, sent for example from satellites to home and mobile devices for reproduction.
- Internet and Web-traffic, which represent a very relevant part of the data exchanged among human and artificial net users.

Such sources are very different from one another, but they a Big Data Stream Analysis framework has often to deal with all of them at the same time and to integrate the knowledge extracted from it.

The Streaming Processor is in charge of executing queries on the collected data, regardless of the fact that it has been summarized or windowed (or both). **Standing Queries** are always executed on the stream, that is they are defined statically once and are always used to retrieve information from the stream. As an instance, a standing query could determine, for a specific reading, the medium value of the preceding x hours, for every time that measurement arrives. Instead, ad-hoc queries are executed once and only if the system is asked to. Since it is not feasible to store all the data obtained by the streams in order to answer arbitrary queries, the designer of the stream analysis system need to foresee which ad-hoc queries could be formulated in the future and organize the **Working Storage** accordingly: in such a storage only a limited sub-set of the entering data can be archived, as they will fade after a set amount of time (sliding-window approach). On the other hand, **Archival Storage** is used to preserve some particular data over a longer period of time: generally it stores aggregate data (max, minimum and similar) extracted from the streams.

Several operations can be done on the data stream: here we will consider two basic operations which are necessary to analyse data efficiently.

6.1 Data Sampling

Data Sampling is one of such operations. As we have already stated, it is not feasible to keep track of all the data flowing into the analysis system, and to permanently store them somewhere. Sop, in order to answer both static and ad-hoc query, the most suitable samples must be selected from the data stream. The general idea which is applied to solve the sampling problem, is to choose a **Hash** function to apply to the incoming data and, based on the value returned by such a function, to keep or discard the data. In general, the data flowing in via the stream come in form of tuples, with one or more of the tuple's components representing the **key** which will be used by a query to retrieve the required data. The hash function idea is to choose a **threshold value**, generally high enough to not discard too much data, and to store all and only the tuples whose key correspond to a hash value below the threshold. Furthermore, if the hash value is used to label the tuples and put them in discreet buckets, the data is also partially sorted. Since the storage space is not free and sooner or later the system will need to erase some data, it is possible to lower the threshold and discard tuples stored in buckets with hashing value above it.

6.2 Data Filtering

Another operation which can be done on the data stream is represented by **Filtering**, that is remove/discard those tuples which do not meet certain criteria. Note that this is different from Sampling, since the hash function adopted in that case does not consider the characteristics of the data. Among the techniques used to enforce a filtering rule on a set of tuples, the **Bloom Filter** represents one outstanding tool. The idea behind this filter is represented by combining several hash functions, which are all evaluated on the incoming tools. In particular, the Bloom Filters can be built by using a simple bit array of N elements, initially all set to 0, and a collection of K hash function which, operating on the tuples' key(s), map them to one of N buckets. Finally a set S of M keys, which represent tuples meeting the filter criteria, is selected and evaluated: the N bit array is initialized by calculating the hash functions of the S set, and setting to 1 all and only the bits corresponding to the calculated hash values. After the training, when a new tuple arrives, the set of hash functions is evaluated on its key(s) and, if the resulting values correspond to the 1 bits of the N array, then it is accepted. Otherwise, it does not respect the acceptance criterion and it is rejected. Obviously, while there is no possibility of having a false negative, it is still possible to have a false positive, and to accept a tuple which does not meet all the criteria.

7 Geo-Referenced Data Mining

Geographic information science moves within an increasingly data-rich and computation consuming environment. The digital geographic datasets are very extensive,
and they provide an impressive coverage: remote sensing systems and environmental
monitoring devices have gathered in the past years vast amounts of geo-referenced
images, videos and sounds. Location aware technologies, such as the GPS and other
modern tools such as cell phones, in-vehicle navigation systems and so on have
enable the possibility to associate spatial information to practically all of the data we
produce, exchange and consume everyday. While information infrastructure initiatives, such as the U. S. National Spatial Data Infrastructure, have provided the means
to enhance data sharing and interoperability, and the growth in computer power has
allowed for faster computations on large datasets, there are still issues related to the
available algorithms and methodologies for mining huge volumes of data. Traditional
statistical methods, particularly spatial statistics, have high computational burdens.

   Geographic knowledge discovery (GKD) and **Geographic data mining** start
from the presupposition that novel and useful geographic knowledge is hidden in the
huge volumes and scopes of digital geo-referenced data being collected, archived
and shared by researchers, public agencies, private organizations and web users.
Traditional methods fail to reveal such knowledge, as they cannot handle massive
data. Also, the very nature of geo-references data is different from the classic aspatial
data. Non spatial data analysed in typical Knowledge Mining applications can be
often reduced to points in a n-dimensional space, without the risk of losing important
information. On the other hand, for some (most of) geographical objects such a
reduction causes a significant information loss, which cannot be neglected by mining
applications. A problem which is often related to geographic data is the sensitivity
shown to spatial measurement units, especially when treating aggregate variables
(density of population over an area, census districts and so on). This is known as
the **Modifiable Area Unit Problem** (MAUP) [17]. As a consequence, discovered
patterns have to be checked for robustness, as they may emerge because of unit
conventions.

   In spatial and geo-reference mining, the pattern types which can be discovered,
such as classes, associations, rules, clusters, outliers and trends all have spatial expressions, since their are heavily influences by the morphology as well as spatial relationships among these objects. Several techniques are available to examine these kind of
patterns.

- **Spatial classification** techniques map spatial objects into meaningful categories
  that take in consideration several characteristics, such as distance, direction, morphological aspects and so on. Spacial buffers are used in the work presented in
  [33]. The approach classify objects on the base of morphological similarity and
  proximity. A similar approach, but which uses path relationships in a defined
  neighbourhood of a target object is used in [15].
- Spatial association rules are association rules that also contain spatial predicates
  in their precedent or antecedent. Such rules are used to associate spatial data by

exploiting tree search techniques, which explore hierarchical structured data. Co-location patterns are a particular type of association rules that determine subsets of spacial objects which are often found together in the same location. As a top-down tree search technique that exploits background knowledge in the form of a geographic concept hierarchy. A specific type of association rule is a co-location pattern: these are subsets of spatial objects that are frequently located together. Huang, Shekhar and Xiong [25] develop a multi-resolution filtering algorithm for discovering co-location patterns in spatial data.

- **Spatial prediction** algorithms try to infer new information on the characteristics of a set of spatial objects via inductive learning techniques. The target is generally represented by general purpose collections of data, such as topographic maps produced by cartographic organizations. As an instance the work presented in [46] used a tree-based induction algorithm to extract knowledge about complex soil-landscape processes, by combining background knowledge from soil surveys with data coming from environmental sensors.

- **Spatial clustering** algorithms comprehend a vast number of heuristic approaches for the grouping of spacial objects, following some similarity criteria. The heuristic of such approaches is necessary since it is not possible to automatically decide the optimal set of k clusters in which to divide the data, without a previous knowledge of the problem. Traditional partitioning methods such as k-means and the "expectation maximization" method can be used to cluster data on the base of simple distance relationships, and are therefore applicable to large spatial datasets.
  Density-based methods are used to define regions of space where a large number of spatial objects reside. The definition of density will depend of the particular case examined and discriminates the dimension of the regions to be considered. Using density as a parameter, it is possible to determine clusters of arbitrary shapes. Conversely, grid-based methods divides the space into a very well defined tessellated structure, and cluster objects accordingly. Model-based methods define functions and rules which are reflected in region shapes which include or exclude subsets of the spatial database. The curves drawn by these functions represent the border between adjacent regions. Constraint based methods are used to capture spatial restrictions: they are particularly useful to take in account for geographical obstacles such as rivers, borders and mountains [47].

- **Spatial outlier analysis** can be focused on both spatial and non-spatial characteristics of the examined objects. In the work presented in [48] spatial outliers are geo-referenced object which show non-spatial characteristics which are not consistent with other objects in the neighbourhood. This means that the object would be inconsistent with the local region it would belong spatially, but not with the entire database. Obviously, more generic approaches can take in consideration spatial characteristics (i.e. distance, speed, geometry, position) as well to determine outliers.

8 Conclusion

In the steadily evolving IT environment, strong emphasis has been given to Big Data
and the Internet of Things. These two topics are strongly related to each other (see
Sect. 3): indeed, IoT strongly influences Big Data research, since the huge volumes
of data continuously produced by sensors need to be managed in a high efficient
way. That is why it is important for data analysts to own knowledge of Big Data
methodologies, techniques and technologies, which are also applied in the IoT field.
In this chapter we have presented the most prominent Big Data algorithms and techniques, in order to provide an insight on Big Data thematics and fields of application,
stress its connection to IoT, and show how Semantics can support reasoning and
classification over collected data (Sect. 2). In particular, we have focused on mining
techniques, applied to Social contexts (Sect. 4), Graphs (Sect. 5) and Data Streams
(Sect. 6), which are particularly relevant to IoT developers, since the information
coming from sensors networks are often in the form of streams.

References

1. M.G. Al Zamil, S. Samarah, The application of semantic-based classification on big data,
   in *2014 5th International Conference on Information and Communication Systems (ICICS)*
   (IEEE, 2014), pp. 1–5
2. M.B. Alaya, Y. Banouar, T. Monteil, C. Chassot, K. Drira, Om2m: extensible etsi-compliant
   m2m service platform with self-configuration capability. Procedia Comput. Sci. **32**, 1079–1086
   (2014)
3. AllJoyn, https://allseenalliance.org/
4. S. Auer, C. Bizer, G. Kobilarov, J. Lehmann, R. Cyganiak, Z. Ives, Dbpedia: A Nucleus for a
   Web of Open Data (Springer, Berlin, 2007)
5. S. Auer, V. Bryl, S. Tramp, Linked Open Data–Creating Knowledge Out of Interlinked Data:
   Results of the LOD2 Project, vol. 8661 (Springer, 2014)
6. P. Barnaghi, W. Wang, C. Henson, K. Taylor, Semantics for the internet of things: early progress
   and back to the future. Int. J. Semant. Web Inf. Syst. (IJSWIS) **8**(1), 1–21 (2012)
7. R. Beneficiary, I. FhG, S.H. SAP, E.H. HSG, C. Jardak, A.O. CEA, A. Serbanati, M.T. SAP,
   J.W. Walewski, Internet of things-architecture iot-a deliverable d1. 3–updated reference model
   for iot v1. 5
8. T. Berners-Lee, Linked data-design issues. https://www.w3.org/DesignIssues/LinkedData.
   html (2006). Accessed 01 Apr 2016
9. K. Bollacker, C. Evans, P. Paritosh, T. Sturge, J. Taylor, Freebase: a collaboratively created
   graph database for structuring human knowledge, in *Proceedings of the 2008 ACM SIGMOD
   International Conference on Management of Data* (ACM, 2008), pp. 1247–1250
10. E.U. Commission, European union open data portal (2012). https://open-data.europa.eu.
    Accessed 01 Apr 2016
11. M. Compton, P. Barnaghi, L. Bermudez, R. GarcíA-Castro, O. Corcho, S. Cox, J. Graybeal, M.
    Hauswirth, C. Henson, A. Herzog et al., The ssn ontology of the w3c semantic sensor network
    incubator group. Web Semant. Sci. Serv. Agents World Wide Web **17**, 25–32 (2012)
12. D. Consortium, A data- and platform-as-a-service approach to efficient. open data publication
    and consumption (2006). http://project.dapaas.eu/. Accessed 01 Apr 2016
13. S.K. Datta, C. Bonnet, Smart m2m gateway based architecture for m2m device and endpoint
    management, in *Internet of Things (iThings), 2014 IEEE International Conference on, and*

Green Computing and Communications (GreenCom), IEEE and Cyber, Physical and Social Computing (CPSCom), IEEE (IEEE, 2014), pp. 61–68

14. D. Eppstein, Subgraph isomorphism in planar graphs and related problems. SODA **95**, 632–640 (1995)

15. M. Ester, H.P. Kriegel, J. Sander, Spatial data mining: a database approach, in *Advances in spatial databases* (Springer, 1997), pp. 47–66

16. A. Fensel, D. Fensel, E. Simperl, R. Studer, Planetdata: a european network of excellence on large-scale data management (2012)

17. A.S. Fotheringham, D.W. Wong, The modifiable areal unit problem in multivariate statistical analysis. Environ. Plan. A **23**(7), 1025–1044 (1991)

18. R. Gomathi, C. Sathya, D. Sharmila, Efficient optimization of multiple sparql queries. IOSR J. Comput. Eng. (IOSR-JCE) **8**(6), 97–101 (2013)

19. V. Guruswami, S. Khanna, R. Rajaraman, B. Shepherd, M. Yannakakis, Near-optimal hardness results and approximation algorithms for edge-disjoint paths and related problems. J. Comput. Syst. Sci. **67**(3), 473–496 (2003)

20. V. Hatzivassiloglou, K.R. McKeown, Predicting the semantic orientation of adjectives, in *Proceedings of the 35th Annual Meeting of the Association for Computational Linguistics and Eighth Conference of the European Chapter of the Association for Computational Linguistics* (ACL '98, Association for Computational Linguistics, Stroudsburg, PA, USA, 1997), pp. 174–181. http://dx.doi.org/10.3115/976909.979640

21. M.A. Hearst, Automatic acquisition of hyponyms from large text corpora, in *Proceedings of the 14th conference on Computational linguistics-Volume 2* (Association for Computational Linguistics, 1992), pp. 539–545

22. I. Horrocks, P.F. Patel-Schneider, H. Boley, S. Tabet, B. Grosof, M. Dean et al., Swrl: a semantic web rule language combining owl and ruleml 21, 79 (2004)

23. M. Hu, B. Liu, Mining and summarizing customer reviews, in *Proceedings of the Tenth ACM SIGKDD International Conference on Knowledge Discovery and Data Mining* (ACM, 2004), pp. 168–177

24. J. Huang, D.J. Abadi, K. Ren, Scalable sparql querying of large rdf graphs. Proc. VLDB Endow. **4**(11), 1123–1134 (2011)

25. Y. Huang, S. Shekhar, H. Xiong, Discovering colocation patterns from spatial data sets: a general approach, in *IEEE Transactions on Knowledge and Data Engineering*, vol. 16, no. 12 (2004), pp. 1472–1485. doi:10.1109/TKDE.2004.90

26. U. Hunkeler, H.L. Truong, A. Stanford-Clark, Mqtt-sa publish/subscribe protocol for wireless sensor networks, in *3rd International Conference on Communication Systems Software and Middleware and Workshops, 2008. Comsware 2008* (IEEE, 2008), pp. 791–798

27. Ibm watson relationship extraction service, http://www.ibm.com/smarterplanet/us/en/ibmwatson/developercloud/doc/sireapi/

28. A. Inokuchi, T. Washio, H. Motoda, An apriori-based algorithm for mining frequent substructures from graph data, in *Principles of Data Mining and Knowledge Discovery* (Springer, 2000), pp. 13–23

29. IoTivity, https://www.iotivity.org/

30. J. Kamps, M. Marx, R.J. Mokken, M. de Rijke, Words with attitude. Citeseer (2001)

31. J. Kim, J.W. Lee, Openiot: an open service framework for the internet of things, in *2014 IEEE World Forum on Internet of Things (WF-IoT)* (IEEE, 2014), pp. 89–93

32. G. Klyne, J.J. Carroll, Resource description framework (rdf): Concepts and abstract syntax (2006)

33. K. Koperski, J. Han, J. Adhikary, Mining knowledge in geographical data. Commun. of ACM (accepted) (1998)

34. M. Kuramochi, G. Karypis, An efficient algorithm for discovering frequent subgraphs. IEEE Trans. Knowl. Data Eng. **16**(9), 1038–1051 (2004)

35. P. Lai, Extracting strong sentiment trends from twitter (2010)

36. W. Le, A. Kementsietsidis, S. Duan, F. Li, Scalable multi-query optimization for sparql, in *2012 IEEE 28th International Conference on Data Engineering (ICDE)* (IEEE, 2012), pp. 666–677

37. D.L. McGuinness, F. Van Harmelen et al., Owl web ontology language overview **10**(10), 2004 (2004)
38. G.A. Miller, Wordnet: a lexical database for english. Commun. ACM **38**(11), 39–41 (1995)
39. D. Miorandi, S. Sicari, F. De Pellegrini, I. Chlamtac, Internet of things: vision, applications and research challenges. Ad Hoc Netw. **10**(7), 1497–1516 (2012)
40. T. Mullen, R. Malouf, A preliminary investigation into sentiment analysis of informal political discourse, in *AAAI Spring Symposium: Computational Approaches to Analyzing Weblogs* (2006), pp. 159–162
41. D. Niyato, L. Xiao, P. Wang, Machine-to-machine communications for home energy management system in smart grid. Commun. Mag. IEEE **49**(4), 53–59 (2011)
42. Open Connectivity Foundation, http://openconnectivity.org/
43. C. Osgood, G. Suci, P. Tannenbaum, *The Measurement of Meaning* (University of Illinois Press, Urbana, 1957)
44. G. Pardo-Castellote, Omg data-distribution service: architectural overview, in *Proceedings 23rd International Conference on Distributed Computing Systems Workshops, 2003* (IEEE, 2003), pp. 200–206
45. E. PrudHommeaux, A. Seaborne et al., Sparql query language for rdf 15 (2008)
46. F. Qi, A.X. Zhu, Knowledge discovery from soil maps using inductive learning. Int. J. Geogr. Inform. Sci. **17**(8), 771–795 (2003)
47. S. Sekhar, C.T. Lu, P. Zhang, R. Liu, Data mining for selective visualization of large spatial datasets, in *Proceedings 14th IEEE International Conference on Tools with Artificial Intelligence, 2002.(ICTAI 2002)* (IEEE, 2002), pp. 41–48
48. S. Shekhar, C.T. Lu, P. Zhang, A unified approach to detecting spatial outliers. GeoInformatica **7**(2), 139–166 (2003)
49. S. Staab, M. Erdmann, A. Maedche, Semantic patterns. Technical report, AIFB, University of Karlsruhe (2001)
50. P.J. Stone, D.C. Dunphy, M.S. Smith, The general inquirer: a computer approach to content analysis (1966)
51. M. Thelwall, K. Buckley, G. Paltoglou, D. Cai, A. Kappas, Sentiment strength detection in short informal text. J. Am. Soc. Inform. Sci. Technol. **61**(12), 2544–2558 (2010)
52. T. Wilson, J. Wiebe, P. Hoffmann, Recognizing contextual polarity in phrase-level sentiment analysis, in *Proceedings of the Conference on Human Language Technology and Empirical Methods in Natural Language Processing* (Association for Computational Linguistics, 2005), pp. 347–354
53. X. Yan, P.S. Yu, J. Han, Graph indexing: a frequent structure-based approach, in *Proceedings of the 2004 ACM SIGMOD International Conference on Management of Data* (ACM, 2004), pp. 335–346

SCADA Systems in the Cloud

Philip Church, Harald Mueller, Caspar Ryan, Spyridon V. Gogouvitis,
Andrzej Goscinski, Houssam Haitof and Zahir Tari

Abstract SCADA (Supervisory Control And Data Acquisition) systems allow users
to monitor (using sensors) and control (using actuators) an industrial system remotely.
Larger SCADA systems can support several 100,000 sensors, sending and storing
hundreds of thousands of messages per second, generating large amounts of data. As
these systems are critical to industrial processes, they are often run on highly reliable
and dedicated hardware. This is in contrast to the current state of computing, which
is moving from running applications on internally hosted servers to cheaper, internal
or external cloud environments. Clouds can benefit SCADA users by providing the
storage and processing power to analyse the collected data. The goal of this chapter
is twofold; provide an introduction to techniques for migrating SCADA to clouds,
and devise a conceptual system which supports the process of migrating a SCADA
application to a cloud resource while fulfilling key SCADA requirements (such as;
support for big data storage).

P. Church · C. Ryan · A. Goscinski · Z. Tari (✉)
School of CS and IT, RMIT University, Melbourne, Australia
e-mail: zahir.tari@rmit.edu.au

P. Church
e-mail: philip.church@rmit.edu.au; philip.church@research.deakin.edu.au

C. Ryan
e-mail: caspar.ryan@rmit.edu.au

A. Goscinski
e-mail: andrzej.goscinski@rmit.edu.au; ang@deakin.edu.au

H. Mueller · S.V. Gogouvitis · H. Haitof
Corporate Technology, Siemens AG, Munich, Germany
e-mail: h.mueller@siemens.com

S.V. Gogouvitis
e-mail: gogouvitis@siemens.com

H. Haitof
e-mail: houssam.haitof@siemens.com

P. Church · A. Goscinski
School of IT, Deakin University, Geelong, Waurn Ponds, Australia

© Springer International Publishing AG 2017   691
A.Y. Zomaya and S. Sakr (eds.), *Handbook of Big Data Technologies*,
DOI 10.1007/978-3-319-49340-4_20

Keywords SCADA · Migration · Clouds

1 Introduction

Industry enterprise these days are located over huge areas; individual components of a flood prediction system, power generation and distribution complex, and mineralogical plant are distributed and distances between components are huge. The problem is; how to control, co-ordinate and manage all the components, identifying the current state of a production process and changing it to satisfy production requirements. This implies the collection of a huge amount of data, to be stored and processed. To make the best decisions there is a need for a lot of data which describe the current state and predict the future state of all components; Enterprise IT is of the most crucial importance.

While cloud computing has gained significant inroads to enterprise IT, Operational Technology (OT) still relies on dedicated on-premise hardware in most cases. OT systems utilize hardware and software that can monitor and/or control physical devices, processes, and events in an industrial enterprise or in the public infrastructure. They show the following characteristics: connectivity to physical devices, soft real-time monitoring and/or control of physical processes, control of assets essential to the function of a society and/or economy (mission-critical systems), high security, and high availability. In many cases OT systems collect large amounts of data about the monitored and controlled infrastructure and processes in order to make decisions based on this data. Much of this data is still unused today, which gives the opportunity to open up new fields of big data applications. A very prominent example of Operational Technology are Supervisory Control and Data Acquisition (or SCADA) systems.

SCADA systems are instrumental to a wide range of mission-critical industrial systems, from infrastructure installations like gas pipelines or water control facilities to industrial plants. SCADA systems allow a user to monitor (using sensors) and control (using switches/actuators) an industrial system remotely. As SCADA systems are critical to industrial processes, they are often run on highly reliable and dedicated hardware. This is in contrast to the current pattern in Enterprise IT, which is moving from running applications on internally hosted servers to private or public cloud resources.

For users, the main benefits of moving applications such as SCADA to the cloud lie in potential cost savings, shifting from upfront investment to operational cost, reduced setup and maintenance efforts, and faster access to new functionality. Cloud resources are purchased and accessed on-demand, at a price cheaper than buying hardware; furthermore, as there is no need to install and/or maintain hardware and manage software, the need for technical staff is reduced. Clouds provide storage for the big data generated by SCADA systems, and the processing power and tools to carry out analytics. For the vendor, offering cloud-based SCADA can lead to new, pay-per-use business models, instead of a traditional one-off hardware cost and software licensing fee. Recurring charges can be based on time of use, the amount of computing resources (compute, storage, bandwidth) needed, or any number derived from the size, complexity or quality of the managed industrial process or system.

One way to provide a cloud-based SCADA system would be the cloud native approach, which means to develop it from scratch, ideally using an appropriate architectural design to make use of cloud inherent features. Another option is to take an existing SCADA implementation and migrate it to a cloud. Often this latter approach is an initial step towards the cloud, given that solid and well proven implementations are available.

The goal of this chapter is twofold; provide a tutorial covering cloud migration techniques applied to SCADA systems, and propose a conceptual framework that supports the process of migrating a SCADA application from on-premise hardware to a cloud-based infrastructure.

The chapter is structured as follows: Sect. 2 presents exsisitng SCADA system architectures that have been deployed in the cloud environment. Section 3 presents an overview of SCADA systems and their characteristics, deriving a generalized architecture. Section 4 describes advantages and challenges to build cloud-based SCADA systems and presents options, methodologies and technologies for cloud migration. The remaining sections describe our work simplifying the migration process through a SCADA cloud orchestration framework. Section 5 presents a design of SCADA cloud services and a SCADA cloud orchestration framework. Section 6 presents the time taken to transfer sensor data, in order to validate the use of cloud to support SCADA real-time requirements. The chapter concludes with Sect. 7 which provides a summary of the proposed framework, and the recommendations derived from it.

2 Related Work

There are a number of papers, which discuss how to build cloud-based SCADA architectures. They focus on implementing a solution from the ground up, as opposed to utilizing pre-existing SCADA solutions.

- Liu, et al. presents a generalized overview of clouds and SCADA, and propose the possibility of running SCADA in the cloud [1].
- Gligor and Turc recommend exposing each SCADA component as a service and deploying them through a Local Directory Service (LDS) [2]. The LDS stores a description of available SCADA resources, access methods, and description. The use of a broker allows some components can be replaced by cloud services; for example the database service can utilize Data Center as a Service (DaaS). This approach is very flexible; allowing users to extending the SCADA system by adding new functionalities to existing services or defines new ones in accordance with needs and formulated requirements. Based on these concepts, a web-based SCADA system is implemented on Rackspace cloud resources. Data was transferred using a simple protocol, consisting of a few numerical and process

monitoring variables. Data transfer rates were measured from a local database to a cloud database, results measured were between 125 to 156 ms.

- Goose et al. present a secure SCADA cloud framework called SKYDA [3]. This SCADA system is designed to take advantage of the scalability and reliability offered by a cloud-based infrastructure. This paper focuses on providing a high level understanding of SCADA replication using clouds, moving all SCADA components (except the field devices) as a single service. Field devices are connected to the cloud based SCADA system directly or (for legacy devices) via a proxy. The framework utilizes multiple cloud providers, running multiple copies of the SCADA Master application in multiple clouds to provide fault tolerance.

There also are solutions provided by commercial cloud-based SCADA providers; the two major commercial solutions are Ignition and XiO's Cloud SCADA.

- Ignition SCADA is a SCADA solution that has been built from the ground up using Java to take advantage of cloud features [3]. Ignition interfaces with most Programmable Logic Controllers (PLC), allowing users to take advantage of existing sensors/actuators. Ignition users do not maintain hardware themselves; instead they access systems remotely via web interfaces. Users are charged based on the number of servers used instead of via software licensing fees. Ignition allows users to customize their architecture by choosing to deploy components individually.
- XiO Cloud SCADA [4] consists of two components, a local (customizable) hardware module called a Soft-I/O, which contains the sensors and actuators, and the cloud component, the SCADA application which runs on secure commercial servers. Users subscribe to a Cloud Service, for a monthly fee, giving them access to the SCADA system through web and mobile apps. Users can customize the priorities of their XiO SCADA system, for example to priorities energy efficiency.

In general, there is a trend to develop SCADA systems specifically for clouds. The fact that existing SCADA solutions have not been used alludes to potential issues with migrating SCADA systems. However, it is not clear if issues are performance or deployment related. The solution presented by Gligor and Turc addresses performance through the use of a simple data transfer protocol, while the local directory service addresses deployment issues. The SKYDA cloud framework only addresses deployment issues through the use of automated replication. Common open-source SCADA systems include EPICS, TANGO, EclipseSCADA and IndigoSCADA. EPICS (Experimental Physics and Industry Control System) [5] is a SCADA system designed to operate devices such as particle accelerators, large experiments, and major telescopes. TANGO [6] is an object orientated distributed control system supported by a consortium of European Synchrotrons in Germany, Spain, Italy, Poland, and France. EclipseSCADA [7] is a key eclipse foundation project used commercially, the details of which have not been made public. IndigoSCADA [8] is a light weight SCADA system for Linux and Windows. This paper focuses on the process of migrating existing open SCADA solutions to run on a cloud.

3 An Overview of SCADA

3.1 Generalized SCADA Architecture

SCADA describes applications, which aim to control and monitor remote equipment via a communication channel [9]. There have been a number of attempts to generate a generalized SCADA framework and architecture. Boye [10] defines a simple SCADA architecture which consists of sensors, switches and/or actuators (field devices), connected and read by a device server (see Fig. 1). Data is transfered across a network to a control server, which handles events (informing a user if sensor data exceeds set boundaries). A user of a SCADA system accesses data via the master server using a workstation. This setup could be extended with a Historian which stores sensor data and events for analysis.

In contrast, IEEE [11] defines an in-depth standard describing the components that make up a SCADA framework. Based on this IEEE standard, study of SCADA systems, their characteristics, and requirements (see Sect. 3.2) we present a generalized SCADA architecture (shown in Fig. 2).

As defined by the IEEE standard, the system is divided into a remote site and master station. The remote site consists of field devices connected to a device server. Communication between the field device and device server makes use of a SCADA communication standard (used by the driver). Collected information is stored in real-time and historical databases on the master station. Communication between components of the master station uses an internal communication protocol. The Master Terminal Unit (MTU) contains a number of tools that interact with the data stored in the databases including:

- An event handler, which reacts to changes to the real time database;
- A device manager, which can modify the behaviour of field devices;
- An alarm manager, which allows users to setup monitoring rules and notify a user if rules have been broken;

Fig. 1 Simple SCADA system

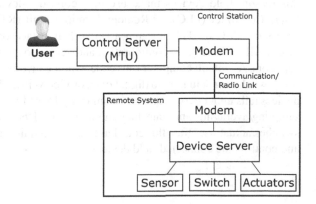

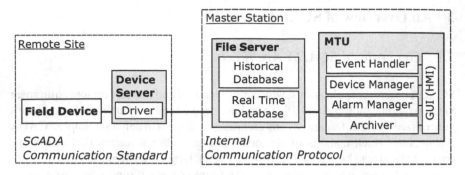

Fig. 2 General SCADA software architecture

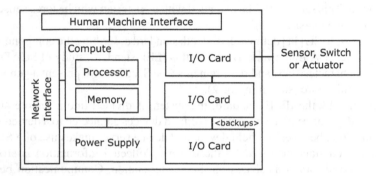

Fig. 3 Field device architecture

- An archiver, which provides analytics of stored data; and
- A GUI or Human Machine Interface, which provides the user with a graphical representation of the remote site.

Every SCADA system uses a large number of field devices. A field device must be able to understand and collect data from sensors, switches and actuators. For this reason, field devices make use of a compute device called a Programmable Logic Controller (PLC), or Remote Terminal Unit (RTU) (see Fig. 3). Users can access the PLC or RTU through a network interface or a Human Machine Interface (HMI), allowing for configuration and access to connected sensors, switches or actuators. An inbuilt computer runs code which converts signals from connected sensors, switches or actuators to digital data, or vice versa. The code which runs on field devices falls under two categories: monitoring loops for sensors that may incorporate sampling/averaging, and state diagrams that control the state of output devices such as switches/actuators. Often this code has real time requirements, as it directly monitors and controls the connected field devices.

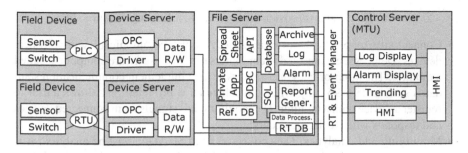

Fig. 4 General SCADA software architecture (Detailed)

Field devices are designed to be reliable, often incorporating backup power and redundancy in the form of backup I/O cards. If a device is connected to an I/O card that fails, the device automatically gets connected to the redundant I/O. In the absence of this feature, if an input card fails, signals would be lost until the card is replaced. Through the use of SCADA and Field Devices, it becomes possible to monitor and control large scale systems (such as gas pipelines which cover very large distances) cheaply and efficiently.

A realization of the generalized SCADA architecture is shown in Fig. 4 [12]. This SCADA system consists of field devices, device servers, a file server, and a control server (MTU). Field Devices consist of sensors, switches or actuators connected to Programmable Logic Controllers (PLCs) or Remote Terminal Units (RTU). Two Device Servers are used, each interfacing with a field device through the Driver or Open Platform Component (OPC). Information collected from these devices are collected and stored in a real time (RT) database hosted on the File Server. Data from the RT database is used by a number of software components running on the File Server. The File Server runs a Report Generation component that queries the RT database using SQL in order to produce report summarizing information generated by connected devices. The Alarm component is programmed to recognize and react to behaviour patterns found in the RT database. Both the Log and Archive component are responsible for moving data from the RT database to historical database for medium or long term storage. The archive database is made accessible to external programs through methods such as: Open Data Base Connectivity (ODBC) or Application Interfaces (API). This data is provided to users through a number of tools hosted on the Master Terminal Unit.

3.2 SCADA Characteristics

There are numerous SCADA systems available (both open-source and commercial), each with different features, architectures, hardware and standard support [13, 14]. Both open-source and commercial SCADA systems provide similar functionality and requirements.

Current SCADA systems provide users with the following functionalities:

- Connectivity to physical devices: SCADA must communicate with physical devices. Devices are connected to computers running a software component called a monitor. The monitor integrates drivers, which implement communication standards such as ISA-95 [15]. Often a mechanism is provided that allows a user to register a device with a monitor. The monitor reads signals from connected devices and responds when a change occurs.
- Monitoring and Analytics: Data collected from SCADA must be analysed. It is common to provide short term monitoring and long term analytics to identify system problems. Short term monitoring is often provided in the form of alarms, which notify a user or trigger actions if a threshold has been exceeded. Historical data is presented to users in the form of graphs.
- Data Storage: Collected data is stored by the SCADA system for monitoring and analysis. Often two databases are used, a historical database for long term storage and a real-time database for monitoring. Due to the different requirements of these databases, different data storage technologies are used. Real-time databases must store data at real-time, so these databases often run in memory. Historical databases needs to store large amounts of data and is typically written to the disk.

SCADA systems must be designed and implemented to support the following non-functional requirements:

- Real-time: Communication between the device and SCADA system must be at real time. The acceptable delay between communications depends on the SCADA application. Internal communication standards are used to reduce the traffic on the network. These standards often use: polling techniques which requests sensor data on a timer, or event driven techniques which sends data when sensor signals change.
- High availability: SCADA systems collect data in real-time, and therefore need to be available at all times. Furthermore, users must be able to control attached hardware via these SCADA systems at any time. Availability is considered one of the most important requirements for SCADA systems [16]. SCADA solutions often provide availability through replication of software and hardware (hot standby).
- Reliability: SCADA systems are responsible for controlling critical infrastructure, which is often distributed geographically or in places difficult to maintain (underground). For these reasons, SCADA systems are designed to be fault-tolerant, running in adverse environmental conditions and operating for long periods of time without human intervention. Some SCADA solutions utilize hardware that is ruggedized to withstand temperature, vibration, and voltage extremes.
- High Security: Using SCADA systems, users can monitor and control critical infrastructure. If an intruder got access to a SCADA system, core infrastructure could be attacked and real time information collected by SCADA systems could be used maliciously. Solutions such as TDMS-Plus utilize the OpenVMS operating system, which is known for extensive security protection.

3.3 SCADA and Big Data

During operation, SCADA systems generates big data which need to be stored and analysed. A typical SCADA system consists of 100,000 sensors polling at a rate of 1 s. Assuming a data size of 8 bytes, over the course of an hour, this SCADA system will generate 2.8 GB of data. Some SCADA systems can generate even larger amounts of data, with large numbers of sensors polling at rates of 0.01 s, and even shorter periods.

Big data is a paradigm applied to datasets of these large sizes. Therefore, there is a need for new software tools and infrastructure to collect, store, manage, and process these datasets in an acceptable time period. These datasets are of the 4 V type: Variety: from various sources, structured and unstructured; Volume: of large size; Velocity: with fast data in/out; and Value: high value.

In the case of SCADA, big data are characterised by Volume, as they are collected by thousands of sensors; Velocity, as they are collected with very high frequency, data in/out occuring at soft- real time (with 1 ms polling rates and control signals being sent to actuators with a matching frequency); and Value, although individual signals are not of a critical value, a set of them could form a pattern that might be critically important, for example, to predict failure of a industrial system.

Various systems are proposed and used to support the handling of big data: scalable storage systems; parallel processing databases; cloud computing systems, in general distributed systems. In this chapter we deal with porting a SCADA system into clouds, for lowering costs, improving reliability and availability, and increasing performance.

Although we do not address the issues of industry complex coordination and management carried out based on data provided and stored within a SCADA system, it is worth mentioning, from the big data point of view, data integration, which covers: data preservation, information integration, spatiotemporal data management, and computational modelling. The following tools are usually used to deal with these operations: analytics tools for big data, visualisation tools for big data, big data acquisition, integration and cleaning tools, and big data pre-processing tools. Reliability is of the most crucial importance to industry complexes. When using SCADA based big data there is a need for analytics to deal with: duplicate, erroneous, and inconsistent data, data trustworthiness (e.g., integrity, reputable source), and provide data validation. Clouds can be used to support big data storage and analytics.

4 Moving SCADA to the Cloud

4.1 Benefits of Cloud-Based SCADA Systems

Today's SCADA systems usually consist of software packages installed and running on on-premise servers. When moving SCADA systems to a cloud, we propose a combined cloud consisting of infrastructure and data storage. Infrastructure as a

Service (IaaS) is used to provide the computational power to monitor sensors, while Data as a Service (DaaS) provides flexible and cheap storage for the generated data.

Providing such a cloud-based SCADA system has advantages for the customer as well as for the vendor. For the customer, the main benefits are savings in the total cost of ownership, focus on core competences, and faster access to new functionality. For the vendor, the cloud can inherently provide support for redundancy, scalability, and increased uptime [17, 18]. Cloud features can also help support core SCADA functionality;

- Connectivity to Physical Devices: SCADA systems are distributed systems consisting of many connected devices (sensors/actuators). Each device is connected to a monitor, a piece of software which integrates drivers and event handlers, in order to provide control and system status to the user. As sensors are added to the SCADA system, more monitors must also be added. A growing SCADA system can take advantage of the resource on demand provided by Infrastructure as a Service (IaaS) cloud solutions.
- Data Storage: As discussed in Sect. 3.3, SCADA systems can generate large amounts of data. Data as a Service (DaaS) cloud solutions provide flexible storage.
- Monitoring and Analytics: To monitor and analyse the large amounts of data generated by a SCADA system, computational resources are required. Commonly analysis is carried out in two phases, short-term and long-term. Short-term analysis is carried out on recent data (often a day to week) in order to predict failures of a system. Long-term analysis is carried out on months and years worth of data, in order to monitor trends which can be used to improve the system. Infrastructure as a Service (IaaS) cloud solutions can provide the resources required to carry out this analysis. In the case of short-term analysis, resources are required to constantly analyse data. In the case of long-term analysis, a large amount of resources are required for a short period, making this analysis suitable to take advantage of cloud resources on demand.

While cloud characteristics can benefit SCADA systems, moving SCADA to the cloud also introduces problems which must be resolved.

4.2   SCADA Requirements Versus Cloud Solutions

When moving an existing SCADA system to the cloud, there is a shift from private hardware to shared public cloud infrastructure. Due to the change in computational environment, SCADA software may not be designed to address specific cloud issues such as: network security and shared network connections. We propose the following core requirements must be supported (see Table 1).

SCADA systems have high security requirements; for this reason they currently often operate on isolated networks and servers. On the other hand, one of the most publicized issues in public clouds is security. Public clouds are open to multiple users, and hardware is often shared. Security must be addressed when moving SCADA to a

Table 1 Main SCADA requirements and Cloud features

SCADA requirements	Exemplary possible cloud solutions
High security requirements	Private clouds, data encryption
High availability	Service level agreements (Cloud and Internet Provider), high availability architectures, redundancy concepts
Real time communication	SLAs, region selection, private clouds, traffic policing
Ease of use	Data, hardware, software transparency and abstraction
Reliability	Replication of application components and databases, geographical isolation

public cloud. In a simple case, this can be done by using concepts like virtual private clouds, that provide virtual separation of the used resources. A more rigorous solution is to use less shared environments like private clouds, or to only run those parts of the SCADA system in the public cloud that are non-security critical. Databases and data communication can also be encrypted, to protect data during transfer and storage. This does not solve all security issues; data must be decrypted when carrying out analysis, reacting to alarms, or providing event handling leading to potential vulnerabilities. Cloud-based SCADA system are also vulnerable to denial of service attacks.

SCADA requires high availability; public clouds provide Service Level Agreements (SLA) made between the customer and the provider. As SCADA systems are typically distributed, there is also a need to have a service level agreement with the providers of the communication networks. High availability could also be achieved by having stand-by or redundancy concepts implemented in the cloud or between the cloud and the remote sites. The XiO cloud solution [4], for example, gets around the need to provide an SLA by having backup SCADA implemented at the remote site.

SCADA systems require (soft) real-time response, typically requiring worst case latencies in the order of sub second to several seconds. In order to enable (soft) real-time applications to run in cloud environments certain guarantees need to be provided [19]. As an example, the speed of network communication needs to be guaranteed. Common solutions to reduce latency include hosting data and processing on the same cloud, and ensuring the cloud resources are in the same geographical location as the remote site [20]. Other solutions are targeted at optimising clouds for communication [21]. iLAND provides a virtualized middleware based on the Data Distribution Service (DDS) standard, implementing a publish/subscribe model for sending and receiving data [22]. ISONI improves the flow of traffic on the network by isolating the traffic of individual virtual machines through a virtual address space and policing network traffic [23].

As key parts of critical infrastructure, SCADA systems need to be reliable, able to continue operating correctly with an expected level of performance at all times. Often, SCADA systems replicate key components (see Sect. 3.2). This approach can be supported by clouds; IaaS clouds allow users to replicate their virtual machines on demand. Users can replicate data stored on DaaS clouds. Clouds can further improve reliability by replicating across different geographical regions. Once a SCADA system is replicated, synchronization must be performed to ensure that the states of all SCADA components are consistent. A common method to ensure data consistency is by keeping track of and carrying out events on each replicated database, in the order in which they occur. Many of these problems can be solved when implementing a cloud-friendly design, keeping service components stateless, and storing state in distributed databases. Several cloud-based databases of this kind are available.

4.3   Overview of Cloud Migration

Migration is the process of moving an application between clouds and/or on-premise environments. A re-occurring operation, migration is often automated through the use of resource selection and scheduling algorithms.

When migrating SCADA to the cloud, three different approaches can be considered: re-hosting, re-factoring, and revising. The quickest and simplest approach is achieved by simply re-hosting an application in the cloud ("lift and shift" approach). Re-hosting is the process of installing an existing application in a cloud environment and mainly relying on IaaS offerings. This can be a first step in a gradual approach, enabling to perform initial analysis and to improve the application in multiple iterations.

To better benefit from the characteristics of cloud computing, e.g., making an application redundant or scalable, it is usually required to re-engineer the application. This can be a simple refactoring, i.e. a modification of one or a few features. An example is adding monitoring capabilities, which enable elastic behaviour by reporting unused resources and react to it. It might also require revising an application, i.e. making major modifications at its core; examples would be using a cloud-based database offering or changing an application into a multi-tenancy SaaS offering. To fully benefit from the cloud, it might also be necessary to rebuild an application and integrate, for example automation for horizontal scaling.

Before carrying out the migration process, there is a need to understand the ramifications of running SCADA on a cloud. The outcomes of a study of advantages of moving SCADA to the cloud, how clouds can fulfil SCADA requirements, and how existing cloud-based SCADA solutions utilize cloud resources are of the most critical importance. An examination of cloud-based SCADA solutions demonstrate a range of different deployment options. In general, moving an application to a cloud is broken into three stages: planning, execution, and evaluation.

- The planning stage incorporates preliminary migration studies and decision making. Preliminary studies are carried out to determine if the application can be moved to the cloud, and to define migration requirements. Decisions are made about the cloud provider to decide, what sub-systems to move, what cloud services to use, and how the moving procedure is to be carried out.

- The execution stage covers the tasks to port and migrate the selected application to the cloud. The system architecture may be adapted to the cloud. If necessary, code is modified to take advantage of cloud services (in the case of legacy applications, code wrapping is used). Data from the pre-existing system is extracted and moved to the cloud. In order to take full advantage of life cycle improvements and elasticity of the cloud, a special focus needs to be taken on automating the steps of migrating the ported application to different cloud environments and setups and operating it.
- The evaluation stage involves validating the application now running on the cloud. Testing is carried out to ensure that the system requirements (defined in the planning stage) are fulfilled. Migration validation is carried out to ensure the behaviour of the system matches the on-premise solution.

Of these three stages, the approaches used during the execution stage are the most varied. Decisions are made based on an understanding of the specific SCADA system architecture, and selected cloud provider and services. The first part of the execution stage is adapting the system architecture to the cloud, through the partitioning of the target application. This is followed by the implementation of the adapted architecture, which is based on the selected cloud model, and by the automated migration and operation steps.

4.4 Cloud Migration Technologies

The task of migrating an application to the cloud can be supported by different migration technologies, that can be structured in open standards, cloud abstraction, and cloud orchestration.

Open Standards There are several attempts to standardize the different interfaces that are relevant in cloud computing. Open standards address all layers of clouds, from defining a common architecture to file formats. The most prominent standard is the Open Cloud Computing Interface (OCCI) from the Open Grid Forum, which is aimed at supporting remote management of mainly IaaS cloud infrastructure [24]. However, the OCCI standard is not supported by many cloud providers and cloud technologies, or it is only available as an add-on (e.g. for OpenStack).

Another standard is Open Virtual Format (OVF), ratified by ANSI and adopted by ISO and IEC. OVF defines a virtual machine format, which is an "open, secure, portable, efficient and extensible format for the packaging and distribution of software to be run in virtual machines" [25]. OVF is supported by a number of cloud and virtualization technologies, including VMware, VirtualBox, OpenNode and Microsoft. However, this does not cover all cloud technologies.

Docker is a light weight Linux container, which can package an application and its dependencies in a virtual container that can run on almost any Linux server [26]. Docker can be used as a lightweight virtualization technology and can run on bare metal servers. However, it lacks some functionalities of IaaS like virtualizing and managing storage and networking. There are numerous ways how Docker can be used in conjunction with clouds. Although Docker is not an official standard, it is widely supported by different cloud providers and technologies.

Cloud Abstraction As described in the previous section, clouds providers and technologies use different interfaces for managing and accessing their services. Several cloud abstraction technologies have emerged, that provide access to a multitude of cloud stacks through a generic API. Current abstraction technologies cover the functionality of IaaS clouds. Using them allows for writing cloud-agnostic code to manage infrastructure resources.

Apache jclouds is a multi-cloud toolkit written in Java, which is compatible with over 30 cloud providers, including Amazon, OpenStack, the Google Cloud Platform, and Docker [27]. Apache delta-cloud is a REST-based API for simple any-platform access; it supports a range of IaaS cloud providers and cloud storage providers [28].

Cloud orchestration Orchestration can be referred to as the automated deployment, configuration, coordination, and management of complex computing configurations, spanning infrastructure, services and applications. Orchestration techniques become especially important in the context of cloud computing, with its core characteristics of self-service and elasticity. Furthermore, cloud-based solutions get more and more complex due to their modular and distributed nature.

Usually, orchestration technologies provide some sort of formal description of the required configuration and its dependencies. In the simplest case, this can be a form of scripting. In more powerful technologies, more advanced concepts are used, that include hierarchical and functional descriptions, or concepts of autonomic computing.

Orchestration technologies can be classified according to two main criteria: scope of managed resources - from infrastructure to applications, and proprietary versus open. Most cloud technologies come with proprietary orchestration technologies tailored to their service offerings: e.g. Amazon Cloud Formation [29], Microsoft Azure Automation [30], OpenStack Heat [31]. Usually they cover the full set of services offered by the respective cloud. However, using them will provide specific configuration descriptions that can only be used for the specific cloud.

Another type of orchestration is related to deploy and manage containers. With Docker having established as one of the most wide-spread container technologies, many tools have emerged to handle Docker containers.

Kubernetes is an open source container cluster manager by Google. It provides capabilities to deploy and manage containers among a cluster of Kubernetes nodes, which are physical or virtual machines managed by Kubernetes. Kubernetes also provides services like service discovery or load balancing [32]. Many cloud service providers offer container management services based for running and managing Docker containers on their infrastructure and integrate with other services. One example is the Amazon EC2 Container Service (ECS) [33].

Open orchestration technologies provide users with the ability to deploy and manage applications across many clouds, often built on top of cloud abstraction libraries. Uncinus provides automated provisioning and deployment of applications, built on top of the euca2ools library [34]. Aoleus is a cloud management tool written in Ruby; it is built atop the delta-cloud API [35].

Cloudify is an open source cloud orchestration technology, using so-called blueprints defined in YAML. It uses an abstraction layer to interface with different cloud stacks. The blueprints describe the execution plans for the lifecycle of the application for installing, starting, monitoring, and terminating the complete application stack [36].

The open source orchestration technology Brooklyn also uses YAML blueprints to describe configurations. Brooklyn blueprints can be composed from simpler blueprints in a hierarchical manner. Another description element is the so-called entity, that can use Java to implement functional behaviour. Brooklyn implements the concept of autonomic computing, that uses sensors and actuators in the entities together with policies to result in self-managed components and systems. Brooklyn uses jclouds as cloud abstraction layer and can thus provide cloud-agnostic orchestration supporting multiple cloud providers and technologies, as well as container-based and even physical server infrastructure [37].

Existing standards such as OCCI and OVF have not been widely adopted and are only available in a few cloud management tools. OCCI is available with OpenStack as an add-on, while OVF is supported by the IBM SmartCloud and OpenNode cloud platforms. In practice, cloud abstraction libraries have to be used to stay independent from specific cloud providers and technologies. In order to automate the deployment and management of the typically complex configurations of cloud-based applications, several orchestration technologies are available. Again, using open orchestration technologies helps to keep the automation descriptions independent from the underlying cloud infrastructure and to flexibly migrate software in multi-cloud environments.

5 Conceptual SCADA Cloud Orchestration Framework

5.1 Migration Recommendations

In principle, there are two ways to implement a cloud-based SCADA system: migration of an existing non-cloud implementation, or cloud-native development of a SCADA system. While cloud-native development has the opportunity to design the system for the cloud and to make maximum use of cloud inherent features, it usually implies major implementation effort to match the functionality of existing SCADA systems. For this reason, the approach of migrating existing solutions is an important one, allowing to make use of available functionality very fast, and optimize the migrated system incrementally afterwards.

In order to migrate a SCADA system to the cloud, different technologies can be used. Virtualization and containerization can be used to simplify the movement of the SCADA system by implementing the same resource stack. Cloud abstraction techniques can be used to deploy SCADA across different cloud providers. Orchestration technologies can be used to support automated deployment of SCADA on cloud resources. Based on the study of SCADA (see Sect. 3) and cloud migration techniques and technology (see Sect. 4), the following recommendations can be made:

- SCADA components (the device server, master server and historian) should be replicated locally due to: real-time constraints and reliability in case connection to the cloud is lost.
- SCADA components should be exposed as services. Services would consist of SCADA components deployed on virtual machines or Docker containers, and exposed through APIs or graphical interfaces.
- Virtualization or containerization techniques (such as Docker) should be used to ensure a consistent environment.
- Cloud abstraction libraries such as Apache jclouds should be used to ensure compatibility with many cloud systems.
- Orchestration should be used to automate deployment and management of SCADA components exposed as services.

Following these recommendations, a SCADA system can be moved to the cloud in an efficient and open manner. However, it still needs to take into account user and application requirements (see Sect. 3.2). From the view point of a user: SCADA must be reliable, available at any time, provide real-time data and control over connected devices, and be accessible through easy to understand interfaces. From the view point of the company: SCADA must be reliable (and easy to repair) and secure so that data is protected and the connected devices are not taken over by hackers. SCADA running on a cloud should also guarantee user satisfaction. Clouds provide some features targeted at these requirements. For example by using Service Level Agreements that can be utilized to ensure availability of the SCADA system.

An ideal cloud-based SCADA system should be flexible, allowing a user control over what parts of their SCADA system is running on the cloud. Based on these suggestions, the SCADA cloud orchestration framework will enable two deployment strategies: the first is aimed at users with static SCADA requirements and no knowledge of SCADA design; the second is aimed at technical users with dynamic SCADA requirements and knowledge of SCADA design (see Table 2).

- Pre-configured SCADA system: This variant deploys a specified SCADA architecture to the cloud. An entire SCADA system is exposed as a service, which can be deployed to a cloud (via an orchestrator). While the SCADA system is treated as a single service, the deployed SCADA service is to be partitioned across multiple computer resources. Implementation of this approach focuses on building mechanisms to transform a local SCADA system to the defined cloud based architecture. Each migrated SCADA system must be based on a scenario (with a well defined number of sensors, switches and actuators, network bandwidth and regions) and finding an optimal solution.
- Customizable SCADA system: This variant gives users control over where to move SCADA components (no assumptions are made about the user's SCADA requirements). SCADA components are exposed as services, which can be deployed, duplicated, and linked together. Implementation of this approach must support

Table 2 Advantages and Disadvantages of SCADA migration approaches

Approaches	Advantages	Disadvantages
Preconfigured SCADA	Optimised for a specific scenario, standardized solution so less effort	Low flexibility to adapt to different SCADA scenarios
Customizable SCADA	Customizable and could be extended to other clouds and SCADA packages	Requires an understanding of system's requirements, more complex solution

the dynamic nature of this solution; the migrated SCADA system must be able to recognize and integrate new services. Time must be spent to ensure that services are constructed in such a way that communication can be redirected, and services can interact with each other over local and cloud networks (supporting one-to-one, one-to-many, and many-to-many relationships). As control is given to the user, a sub-optimal SCADA solution could be deployed. Migration tools could be extended to enable automatic selection and deployment of a SCADA system based on user requirements.

In the following paragraphs we describe how a SCADA system can be exposed as a service and provide a framework that can deploy a SCADA system on different clouds. Deployment on different clouds could be carried out to improve reliability, availability and performance. The last improvement being the outcome of migration of the SCADA system close to data sources and sinks.

5.2 SCADA as a Service

Typically, access to cloud resources is offered as service (usually through web interfaces), encapsulating software, and providing resource management. Offering a SCADA system as a service allows the non-technical user access to SCADA running on cloud resources without having to purchase and operated hardware or install specific software. There are many ways to deploy a cloud service; commonly this requires packaging software in a virtual machine or container, and offering it as a web-based service (through an API and graphical interface) or through an application marketplace [38].

From the generalized SCADA architecture, three core SCADA components have been defined: a device service that converts data between formats, a file service that stores data, and a control service that carries out event management. According to the general migration approach presented in Sect. 4, in order to devise a portable SCADA system there is a need to: expose the three SCADA components as cloud services, and construct an environment in which these services can be deployed.

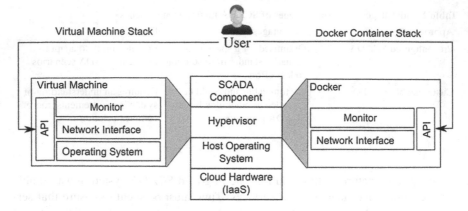

Fig. 5 Cloud service deployment enviroment

The service deployment environment contains the middleware required to host and run services. Deployed on top of the environment are the three defined SCADA components which are encapsulated in either a VM or Docker container. Users can access and configure these services using an Application Programing Interface (API).

5.3 Cloud Service Deployment Environment

The Cloud Service Deployment Environment (see Fig. 5) is designed to ensure that services can be run on a range of IaaS cloud resources.

There are two possible implementations of the deployment environment; both run similar services using different virtual environments. The first uses a virtual machine image, while the second uses a Docker container. In the case of the virtual machine, the operating system is encapsulated into the service. This increases the size of the service, but gives users the flexibility to specify the operating system type and version. This is useful when running legacy or windows-based SCADA systems. In the case of Docker, the container runs on top of the host operating system and the SCADA system must be Linux compatible.

Running in the virtual machine or Docker container are applications that expose key OS functionalities. A network interface application enables changes to be made to the network interface, and enables collection and modification of network traffic. A monitor application collects performance data of each service (such as CPU, and RAM usage). The network interface and monitor are accessible through an API, which can be accessed by a user remotely. Also running in the virtual machine or Docker container shown in Fig. 5, are SCADA components and their dependencies.

Fig. 6 Device SCADA service

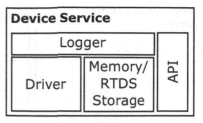

Fig. 7 File SCADA service

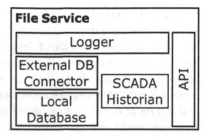

5.4 SCADA Service Design

The implementation of a SCADA service depends on the type of SCADA component being deployed. We have proposed and developed three different SCADA services. They provide the core functionality of the device server, file server, and control server (MTU). The device service provides the core functionality of the device server (see Fig. 6). The device service consists of drivers which enable communication with connected field devices, and memory storage which provides fast storage for device data. Each component is exposed through an API which enables configuration of drivers and memory storage. Users can modify the driver component to support different devices and specify how often data is collected from connected devices. Users can also configure the memory/Real Time Database storage to configure what type of data is collected.

The file service implements the core functionality of the file server (see Fig. 7). The file service consists of: a SCADA Historian, which collects and stores device data; a local database for internal data storage; an external DB connector to enable optional usage of cloud storage (DaaS); and a logger, which stores events and errors. Each component is exposed through an API, which enables access to stored data, and configuration of the data storage environments. Users can link the file service to a device service and a control service, configure the SCADA historian to use either the local or a external database as storage, and query the historic data stored in the local or external databases.

The control service implements the core functionality of the MTU (see Fig. 8). The control service consists of: a device manager, which connects to and consolidates data from device services; an event manager, which defines when events are triggered by connected devices; an alarm handler, which implements logic, which

Fig. 8 Control SCADA
service

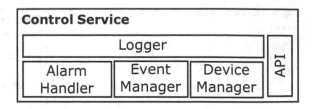

is executed when events are triggered; and a logger, which stores events and errors. Each component is exposed through an API, which enables access to alarm and event handling, and consolidation of devices. Through the API, users connect to the alarm handler, device manager and event manager. Users can access the device manager to setup connections to device services, and to classify connected devices in the form of groups. Users can access the event manager to define the rules which indicate success and failure of devices or groups of devices. Users can access the alarm handler to link operational and recovery logic to defined events.

Once implemented, the Cloud Service Deployment Environment (see Sect. 5.3) and the presented SCADA services can be manually deployed on IaaS cloud resources. To achieve migration, there is a need to automate deployment and manage these SCADA services. A framework which supports the orchestration of SCADA services, is proposed; which serves as a proxy between SCADA services and IaaS cloud providers.

5.5 SCADA Cloud Orchestration Framework

The SCADA cloud orchestration framework is designed to deploy the SCADA cloud services (described in Sect. 5.2) to IaaS cloud resources and manage the SCADA services. This SCADA orchestration framework supports: the allocation and termination of IaaS cloud resources, the deployment of SCADA services to IaaS resources, and the management and evaluation of SCADA services.

Framework Design The proposed SCADA cloud orchestration framework consists of SCADA services (see Sect. 5.2), a SCADA orchestration service (see Sect. 4) and field devices (see Sect. 3). In default operation, the SCADA services, SCADA orchestration service, and field devices run locally (see Fig. 9). SCADA services are stored and managed by the SCADA orchestration service, where running SCADA services connect to field devices.

At any time, a user can connect to the SCADA orchestration service and request that deployment of SCADA services be carried out. During this process the user can either: create a SCADA system configuration from the available SCADA services or choose a pre-configured SCADA system. Upon selection of the SCADA services, the Orchestration Service selects and requests cloud resources from a cloud provider, and carries out service deployment.

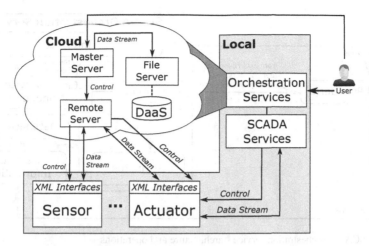

Fig. 9 SCADA cloud orchestration framework

Using the SCADA cloud orchestration framework, two possible scenarios for SCADA deployment are supported (see Sect. 5.1). In the case of the pre-configured SCADA approach, the SCADA orchestration service stores instructions for deploying a full SCADA system, deployment and configuration of multiple SCADA services occurring as a single operation. In the case of the customizable SCADA system, users deploy and configure each SCADA service individually either manually or through a scripting language.

Regardless of the deployment scenario, once SCADA services are deployed to the cloud, users can connect to the cloud-based SCADA system through the control service. In this framework, a Human Machine Interface (HMI) can be used to simplify access to the control service API, enabling users to access and monitor data from connected field devices.

Architecture and Operations of the SCADA Orchestration Service As discussed in Sect. 5.5, the SCADA orchestration service serves as an intermediary between a cloud provider and user. The SCADA orchestration service helps users accomplish a range of tasks including service publication, selection, deployment, and management. The proposed design of the SCADA orchestration service consists of a cloud service manager, which interacts with an open library and database, and is made accessible through a graphical interface (see Fig. 10).

The cloud service manager is divided into two parts, the "Application Broker", which implements algorithms to store and discover cloud services; and the "Cloud Interface Services", which implement the algorithms to select, deploy, and manage cloud resources and services (see Fig. 11).

- The application broker facilitates storage and retrieval of service information from a database. The database provides persistent storage of deployable services, which can be discovered by users, and information about running services.

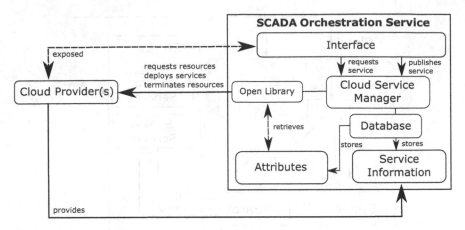

Fig. 10 SCADA orchestration service - architecture and operations

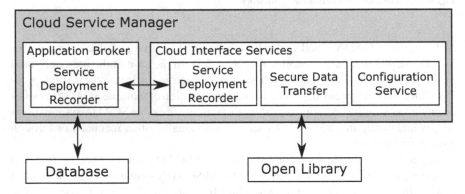

Fig. 11 Cloud service manager

- The cloud interface services facilitate communication with cloud providers using an abstraction mechanism like jclouds. Services are divided into three categories: "Cloud Resource Allocator", which requests and terminates IaaS resources; "Secure Data Transfer", which manages transfer of files to IaaS resources; and "Configuration Services", which interacts with deployed services and configures their operation. Each of these services use a library, which abstracts the API of many cloud providers, so that common operations such as starting and terminating VMs is carried out through a single interface, despite the lower level differences in cloud providers.

When a user publishes services, the Application Broker is used to store those services into the database. Each service consists of a Docker or VM image and related metadata, or attributes [34]. The following attributes are stored for each service: service name, service description, service type, minimal CPU and RAM requirements (see Table 3).

Table 3 SCADA service attributes

Attribute	Description
Name	Name of service
Description	Description of service, used for service discovery
Type	Type of service (VM or Docker)
CPU	Minimal CPU requirements used for resource selection
RAM	Minimal RAM requirements used for resource selection

When an end-user wants to deploy a SCADA system on cloud resources, the user first accesses the "Application Broker" and retrieves the name and description of all stored services. The user selects from the available SCADA services to port to the cloud. "Secure Data Transfer" is used to move the selected SCADA service files to the cloud. Attributes of each selected service is retrieved from the database and given to the "Cloud Interface Services". Using the CPU and RAM attributes, the "Cloud Resource Allocator" selects and contacts the cloud provider, requesting resources that meet the service requirements. If resources have been allocated successfully, the cloud provider returns service allocation information such as the deployment location, type of service, and authentication details. This information is sent to the "Service Deployment Recorder" and stored in the database. Lastly, the "Configuration Service" accesses each service's API, and uses the service allocation information customizing each service so they interact with each other. This procedure commonly involves: adding devices to the control service through modifications of the device manager, configuring the historian to link to the control server and data storage, and configuring the device server to support sensors as indicated by the user.

Supporting SCADA requirements The above conceptual SCADA cloud orchestration framework is designed in such a way that the four main requirements of SCADA systems: security, availability, real time communication, and reliability are supported (see Sect. 4.2).

The SCADA cloud orchestration framework supports SCADA security by allowing the user to choose parts of the SCADA system to run on the cloud. Depending on security constraints, a private cloud could be used. The inclusion of a cloud abstraction layer into the cloud middleware allows support for private clouds such as OpenStack. Security can also be implemented at the base service level, where the network interface service could be modified to support data encryption.

The SCADA cloud orchestration framework provides SCADA systems with availability through replication. By default, the entire SCADA system is replicated with SCADA services running both locally and on the cloud in order to implement hot-standby. Furthermore, individual SCADA services can be replicated; for example deploying multiple historians can allow for both local and external data storage (noSQL databases) to be provided.

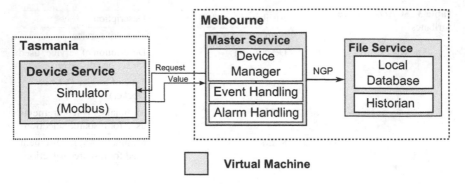

Fig. 12 EclipseSCADA service enviroment

The SCADA cloud orchestration framework helps SCADA systems support real time communication through geographical cloud region selection. Cloud abstraction libraries allow for SCADA systems to take advantage of cloud providers that are close to field devices.

The SCADA cloud orchestration framework provides SCADA systems with reliability through replication and monitoring. As stated above, the SCADA cloud orchestration framework supports replication of services. Monitoring is used to ensure that the SCADA system is operational. The middleware provides a configuration component, which communicates with the monitor in each SCADA service on a regular basis. In this way, the broker is aware of what services are running and their status. End-users are notified if their service has failed, and automated recovery can be carried out. Local SCADA services take over from the cloud, until a connection to the cloud is once again made available.

It should be noted, that while SCADA requirements can be supported by the framework, they are not ensured by the proposed framework alone. Ideally, security, availability, real time communication, and reliability should be also provided at lower levels through system architecture and underlying resource configuration.

6 Results

Using the framework, EclipseSCADA was converted into services and deployed on virtual machines hosted on the NeCTAR research cloud [39]. Each virtual machine (m1.medium) had 2 virtual CPUs, 8 GB RAM and 60 GB secondary disk.

The EclipseSCADA system was deployed across Melbourne and Tasmania cloud regions, where the device service was deployed in Tasmania, and File and Control services were deployed in Melbourne. The Device Service was configured to use Modbus/TCP simulators, where the Control service will request data from the Device service on a timed loop of 1 ms. Data which triggers alarms will be converted to NGP format and sent to the File service for long term storage (see Fig. 12) [40].

Table 4 Modbus Transfer Time between Service (ms)

Origin	Destination	Time (ms)
Control service	Device service	4
Device service	Control service	7
Control service	File service	0.7

The time taken to transfer data between each service was measured (see Table 4).

Results show that the time taken to transfer a single Modbus/TCP sensor of data between regions differs deepening on the service. During operation of the migrated SCADA system, 4 ms are spent requesting data, while 7 ms are spent recieving sensor data. The Device service will store data via the File service, each operation taking 0.7 ms. For each connected sensor, 10 bytes of sensor data was generated every 11 ms, for a total of 909 kbyte per hour. These measurements are faster than those reported by Gligor and Turc [2]. Results confirm that the soft real-time requirements of a SCADA system can be met using cloud resource.

7 Conclusion

SCADA systems are of fundamental importance to a wide range of mission-critical industrial systems in the scope of gas pipelines, dams, to industrial plants. SCADA systems allow users to monitor and control systems remotely and in real time, based on huge amount of data. The ability to control, co-ordinate and manage all the components, in addition to identifying the current state of a production process and changing it to satisfy production requirements, is of crucial importance.

Migrating existing SCADA systems to clouds can allow users to benefit from new business and technological models where users are charged for resources they use, and on-demand resources allow for reduced setup and deployment time. Users can also take advantage of flexible storage and clouds set up for big data processing. For this reason a study and experience report was presented that covered SCADA, clouds and services, and SCADA migration approaches techniques and technologies.

We propose the development of a SCADA cloud orchestration framework that combines different techniques and solutions to access IaaS clouds. Through the proposed system, SCADA will be made available to developers and users by automating two main scenarios: (i) exposing a SCADA application running on a local computer as a service; (ii) deploying a SCADA service to a cloud.

Based on the study presented in this chapter, the following recommendations were made (see Sect. 5.1):

- SCADA components (the device server, master server and historian) should be replicated locally due to: real-time constraints and reliability in case connection to the cloud is lost.

- SCADA components should be exposed as services. Services would consist of SCADA components deployed on virtual machines or Docker containers, and exposed through APIs or graphical interfaces.
- Virtualization or containerization techniques (such as Docker) should be used to ensure a consistent environment.
- Cloud abstraction libraries such as Apache jclouds should be used to ensure compatibility with many cloud systems.
- Orchestration should be used to automate deployment and management of SCADA components exposed as services.

From these recommendations a conceptual SCADA cloud orchestration framework was devised, which aimed to support the deployment and management of SCADA systems to Infrastructure as a Service clouds, thereby fulfilling the core SCADA requirements.

As a framework, the methodology had to be abstract enough to be applied to a range of SCADA systems. Therefore a generalized SCADA architecture was developed to serve as the basis of the SCADA cloud orchestration framework. This generalized SCADA architecture consists of three servers: a remote server that collects data from field devices, a file server that stores data, and a master server that performs alarm and event handling. Each SCADA server is converted into a service, resulting in the creation of a device service, a remote service, and a control service. The device service is a cloud enabled remote server with customizable drivers. The file service is a cloud enabled file server with customizable storage. The control service is a cloud enabled master server with customizable event and alarm handling. These services are integrated with a SCADA orchestration service to form the SCADA cloud orchestration framework.

The SCADA Orchestration Service is designed to help users accomplish a range of tasks including service publication, selection, deployment, and management. A service manager implements algorithms, which fall under two categories: service storage and service deployment. By providing service storage, it is possible to re-use deployment information and reduce the time taken to deploy SCADA services. By automating service deployment, non-computing specialists can deploy and access SCADA running on cloud infrastructures. Results show that SADA deployed through this framework can meet the soft-real time requirments of SCADA systems.

In conclusion, a SCADA cloud orchestration framework like the one presented in this chapter can simplify the migration procedure. By treating each SCADA component as an individual service, which can be deployed separately, this framework fulfills the four requirements of SCADA systems. Users can deploy each SCADA service based on their requirements. Components with high security requirements could be deployed on private networks and hardware. Components with tight real time communication requirements could be deployed on appropriate infrastructure, private clouds or even on-premise hardware. Availability and reliability requirements can be met using replication; by default the SCADA system runs locally in case of failure of the cloud system. Additionally it is possible to deploy multiple SCADA file services to support both local and remote big data storage.

References

1. M. Liu, C. Guo, M. Yuan, in *The Framework of SCADA System Based on Cloud Computing, in Cloud Computing*, vol. 133, ed. by V.C.M. Leung, M. Chen (Springer International Publishing, 2014), pp. 155–163
2. A. Gligor, T. Turc, Development of a service oriented SCADA system. Procedia Econ. Finance **3**, 256–261 (2012)
3. S. Goose, J. Kirsch, D. Wei, SKYDA: cloud-based, secure SCADA-as-a-service. Int. Trans. Electr. Energy Syst. **25**(11), 3004–3016 (2014)
4. XiO, XiO Cloud SCADA Control System (2015). http://www.xioio.com/wp/
5. A. Johnson, EPICS - Experimental Physics and Industrial Control System (2014). http://www.aps.anl.gov/epics/index.php
6. TANGO Consortium, "The TANGO Controls website," (2014)
7. IBH Systems GmbH, openSCADA | We are the good guys (2014). http://openscada.org
8. apaatsf, IndigoSCADA (2015). http://sourceforge.net/projects/indigoscada/
9. K. Barnes, B. Johnson, R. Nickelson, Review Of Supervisory Control And Data Acquisition (SCADA) Systems (U.S. Department of Energy) (pp. 57). Idaho Falls, Idaho: Idaho National Engineering and Environmental Laboratory (2004)
10. S.A. Boye (ed.), *SCADA: Supervisory Control and Data Acquisition* (International Society of Automation, Research Triangle Park, 2010)
11. IEEE Power Engineering Society. IEEE Standard for SCADA and Automation Systems (2008)
12. P. Kumar, SCADA SYSTEMS Introduction, architecture, functionality, and other aspects (28 July, 2011). http://purnendukumar.wordpress.com/2011/12/12/scada-systems-introduction-architecture-functionality-and-other-aspects/
13. O. Barana, P. Barbato, M. Breda, R. Capobianco, A. Luchetta, F. Molon, M. Moressa, P. Simionato, C. Taliercio, E. Zampiva, Comparison between commercial and open-source SCADA packages: a case study. Fusion Eng. Des. **85**(3–4), 491–495 (2010). doi:10.1016/j.fusengdes.2010.02.004
14. C. Queiroz, A. Mahmood, Z. Tari, SCADASim - a framework for building SCADA simulations. IEEE Trans. Smart Grid **2**(4), 589–597 (2011). doi:10.1109/tsg.2011.2162432
15. American National Standards Institute/International Society of Automation, Enterprise-Control System Integration, Part 3: Activity Models of Manufacturing Operations Management (ISA The Instrumentation, Systems, and Automation Society, Research Triangle Park, NC, 2005)
16. ETM professional control GmbH, 1200 survey participants judge SCADA systems and PVSS (2015). http://www.etm.at/index_e.asp?seite_id=91
17. K. Wilhoi, "SCADA in the Cloud - A Security Conundrum?" (Trend Micro 2013)
18. Inductive Automation, Cloud-Based SCADA systems: The benefits and Risks (2011). http://www.inductiveautomation.com
19. S. Gogouvitis, K. Konstanteli, S. Waldschmidt, G. Kousiouris, A. Katsaros, A. Menychtas, D. Kyriazis, T. Varvarigou, Workflow management for soft real-time interactive applications in virtualized environments. Future Gener. Comput. Syst. **28**(1), 193–209 (2012)
20. A. Goscinski, M. Brock, P. Church, *High Performance Computing Clouds* (CRC, Taylor & Francis group, Boca Raton, 2011)
21. M. Garcia-Valls, T. Cucinotta, C. Lu, Challenges in real-time virtualization and predictable cloud computing. J. Syst. Archit. **60**(9), 726–740 (2014). doi:10.1016/j.sysarc.2014.07.004
22. M. Garcia-Valls, I.R. Lopez, L.F. Villar, iLAND: an enhanced middleware for real-time reconfiguration of service oriented distributed real-time systems. IEEE Trans. Ind. Inf. **9**(1), 228–236 (2013). doi:10.1109/TII.2012.2198662
23. T. Voith, M. Kessler, K. Oberle, D. Lamp, A. Cuevas, P. Mandic, A. Reifert, ISONI Whitepaper v2.0 (2009)
24. Open Cloud Computing Interface, Open Cloud Computing Interface (2015). http://occi-wg.org/

25. American National Standards Institute, Information technology – Open Virtualization Format (OVF) specification, **17203**, 37 (2011)
26. Docker Inc, Docker - Build, Ship and Run Any App, Anywhere (2014). https://www.docker.com/
27. Apache, jCloud (2015c). https://jclouds.apache.org/
28. Apache, Apache delta-cloud (2015b). https://deltacloud.apache.org/
29. Amazon, AWS CloudFormation (2015). http://aws.amazon.com/cloudformation/
30. Microsoft, Automation - Cloud process (2015a). http://azure.microsoft.com/en-us/services/automation/
31. OpenStack Project, Heat - OpenStack (2015). https://wiki.openstack.org/wiki/Heat
32. Google, Kubernetes (2015b). http://kubernetes.io/
33. J. Jackson, Amazon embraces Docker with new customer tool (2014). http://www.itnews.com/virtualization/86050/amazon-embraces-docker-new-customer-tool?page=0,0
34. P. Church, A. Goscinski, C. Lefere, Exposing HPC and sequential applications as services through the development and deployment of a SaaS cloud. Future Gener. Comput. Syst. **43–44**, 24–37 (2015)
35. Aeolus, Aeolus (2015). http://www.aeolus-project.org/
36. GigaSpaces Technologies, Cloudify (2015). http://getcloudify.org/
37. Apache Software Foundation, Apache Brooklyn (2016). https://brooklyn.apache.org/
38. F. Curbera, M. Duftler, R. Khalaf, W. Nagy, N. Mukhi, S. Weerawarana, Unraveling the Web services web: an introduction to SOAP, WSDL, and UDDI. IEEE Internet Comput. **6**(2), 86–93 (2002)
39. J. Kirby, NeCTAR - Australian Research Cloud (2012). http://www.nectar.org.au/
40. P. Church, H. Mueller, C. Ryan, S. Gogouvitis, A. Goscinski, H. Haitof, Z. Tari, Moving SCADA systems to IAAS clouds, in *Proceedings of the 5th International Symposium on Cloud and Service Computing (SC2 2015)*, Chengdu, China, 19–21 Dec 2015 (2015), pp. 908–914

Quantitative Data Analysis in Finance

Xiang Shi, Peng Zhang and Samee U. Khan

Abstract Quantitative tools have been widely adopted in order to extract the massive information from a variety of financial data. Mathematics, statistics and computers algorithms have never been so important to financial practitioners in history. Investment banks develop equilibrium models to evaluate financial instruments; mutual funds applied time series to identify the risks in their portfolio; and hedge funds hope to extract market signals and statistical arbitrage from noisy market data. The rise of quantitative finance in the last decade relies on the development of computer techniques that make processing large datasets possible. As more data is available at a higher frequency, more researches in quantitative finance have switched to the microstructures of financial market. High frequency data is a typical example of big data that is characterized by the 3V's: velocity, variety and volume. In addition, the signal-to-noise ratio in financial time series is usually very small. High frequency datasets are more likely to be exposed to extreme values, jumps and errors than the low frequency ones. Specific data processing techniques and quantitative models are elaborately designed to extract information from financial data efficiently. In this chapter, we present the quantitative data analysis approaches in finance. First, we review the development of quantitative finance in the past decade. Then we discuss the characteristics of high frequency data and the challenges it brings. The quantitative data analysis consists of two basic steps: (i) data cleaning and aggregating; (ii) data modeling. We review the mathematics tools and computing technologies behind the two steps. The valuable information extracted from raw data is represented by a group of statistics. The most widely used statistics in finance are expected return and volatility, which are the fundamentals of modern portfolio theory. We further introduce some simple portfolio optimization strategies as an example of the application of financial data analysis. Big data has already changed financial industry

X. Shi · P. Zhang (✉)
Stony Brook University, Stony Brook, NY 11794, USA
e-mail: peng.zhang@stonybrook.edu

X. Shi
e-mail: xiang.shi@stonybrook.edu

S.U. Khan
North Dakota State University, Fargo, ND 58108, USA
e-mail: samee.khan@ndsu.edu

© Springer International Publishing AG 2017
A.Y. Zomaya and S. Sakr (eds.), *Handbook of Big Data Technologies*,
DOI 10.1007/978-3-319-49340-4_21

fundamentally; while quantitative tools for addressing massive financial data still have a long way to go. Adoptions of advanced statistics, information theory, machine learning and faster computing algorithms are inevitable in order to predict complicated financial markets. These topics are briefly discussed in the later part of this chapter.

1 Introduction

1.1 History of Quantitative Finance

The modern quantitative finance or mathematical finance is an important field of applied mathematics and statistics. The major task of it is to model the finance data, evaluate and predict the value of an asset, identify and manage the potential risk in a highly scientific way. One can divide the area of quantitative finance into two distinct branches based on its tasks [43]. The first one is called the "$\mathbb{Q}$" area, which serves to price the derivatives and other assets. The character "$\mathbb{Q}$" denotes the risk-neutral probability. The other one is the "$\mathbb{P}$" area, which are developed to predict the future movements of the market. The character "$\mathbb{P}$" denotes the "real" probability of the market.

The first influential theory in quantitative finance is the Black–Scholes option pricing theory. Unlike public equities that are frequently traded in the market, derivatives like options often lack liquidity and are hard to be evaluated. The theory was initiated by Merton [42] who applied continuous-time stochastic models to get the equilibrium price of equity. Black and Scholes [7] derive an explicit formula for option pricing based on the idea of arbitrage free market. This formula, as Duffie [19] called, is "the most important single breakthrough" of the "golden age" of the modern asset pricing theory. Following works by Cox and Ross [15], Cox et al. [16] and Harrison and Kreps [29] form the footstone of the "$\mathbb{Q}$" area. The theory is most widely applied in sell-side firms and market makers like large investment banks. Today the Black–Scholes formula is the core curriculum of any quantitative finance programs in university. The fundamental mathematical tools in this area are Ito's stochastic calculus, partial differential equation and modern probability measure theory developed by Kolmogorov. The security and the derivatives are often priced individually, thus high dimensional problems are often not considered in classical "$\mathbb{Q}$" theories.

Unlike the "$\mathbb{Q}$" theory which focuses on measuring the present; the goal of the "$\mathbb{P}$" area is to predict the future. Financial firms who are keen on this area are often mutual funds, hedge funds or pension funds. Thus the ultimate goal of the "$\mathbb{P}$" area is portfolio allocation and risk management. The foundation of the "$\mathbb{P}$" world is the modern portfolio theory developed by Markowitz [37]. The idea of Markowitz's theory is that any risk-averse investor tends to maximize the expected returns (alpha) of his portfolio while the risk is under control. Other important contributions to this area are the capital asset pricing models (CAPM) introduced by Treynor [58], Sharpe [54], Lintner [35] and Mossin [46].

Financial data is fundamentally discrete in nature. In the "$\mathbb{Q}$" area, asset prices are usually approximated by a continuous-time stochastic process so that one can obtain a unique equivalent risk-neutral measure. The continuous-time process, however, has difficulties in capturing some stylized facts in financial data such as mean-reverting, volatility clustering, skewness and heavy-tailness unless highly sophisticated theories are applied to these models. Thus the "$\mathbb{P}$" area often prefers discrete-time financial econometric models that can address these problems more easily than their continuous-time counterparties. Rachev et al. [50] suggest that there are three fundamental factors that make the development of financial econometrics possible, which are: *"(1) the availability of data at any desired frequency, including at the transaction level; (2) the availability of powerful desktop computers and the requisite IT infrastructure at an affordable cost; and (3) the availability of off-the-shelf econometric software."*

Furthermore, most problems in the "$\mathbb{P}$" area are high dimensional. Portfolio managers construct their portfolios from thousands of equities, ETFs or futures. Dependence structure among these risky assets is one of the most important topics in the "$\mathbb{P}$" world. Traditional statistics are challenged by these high dimensional financial data and complicated econometric models.

Thus the big data together with related techniques is the foundation of the "$\mathbb{P}$" world, just like coal and petroleum that make the industrialization possible. And the technologies behind big data become more important as the development of high frequency trading. Just a decade ago, the major research in the "$\mathbb{P}$" area was based on the four prices: Open, High, Low, Close (OHLC) that are reported at the end of each day. Data at higher frequency was not provided or even kept by most of the exchanges. For example, commodity trading floors did not keep intraday records for more than 21 days until 6 years ago [2]. Comparing to the low frequency OHLC data, the high frequency data is often irregularly spaced, and exhibits stronger mean-reverting and periodic patterns. A number of researches in econometrics have switched to the high frequency area. As an example, we use the keywords "financial econometrics" and "high frequency" to search related publications on Google Scholar®. To compare we also search the results of "financial econometrics" only. Figure 1 plots the number of the publications during each period.

One can observe that there is a tremendous growth of financial econometrics publications over the past decade. The percentage of the papers related to high frequency data is about 13% in 1990–1994 periods. This number increases to about 34 and 32% in 2005–2009 and 2010–2014 periods. Figure 1 is also an evidence of the growing importance of the big data in finance; since the high frequency data is a typical example of big data that is characterized by the 3V's: velocity, variety and volume. We discuss these concepts in depth in the following sections.

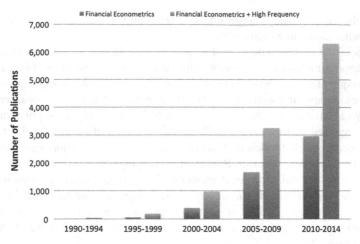

Fig. 1 Number of publications related to high frequency econometrics on Google Scholar® (Data source: Google Scholar®)

1.2 Compendium of Terminology and Abbreviations

Briefly, we summarize the terminology and abbreviations in this chapter (Table 1):

Algorithmic trading strategy refers to a defined set of trading rules executed by computer programs.

Quantitative data analysis is a process of inspecting, cleaning, transforming, and modeling data based on mathematical models and statistics.

Moore's law is the observation that the number of transistors in a dense integrated circuit doubles approximately every two years.

Equity is a stock or any other security representing an ownership interest. In this chapter, the term "equity" only refers to the public traded ones.

High frequency data refers to intraday financial data in this chapter.

ETF refers to exchange traded fund, is a marketable security that tracks an index, a commodity, bonds, or a basket of assets like an index fund.

Derivative refers to a security with a price that is dependent upon or derived from one or more underlying assets.

Option refers to a financial derivative that represents a contract sold by one party (option writer) to another party (option holder). The contract offers the buyer the right, but not the obligation, to buy (call) or sell (put) a security or other financial asset at an agreed-upon price (the strike price) during a certain period of time or on a specific date (exercise date).

Buy side is the side of the financial industry comprising the investing institutions such as mutual funds, pension funds and insurance firms that tend to buy large portions of securities for money-management purposes.

Table 1 List of abbreviations

TAQ data	Trade and quote data
OHLC	Traditional open, high, low, close price data
HFT	High frequency trading
MLE	Maximum likelihood estimator
QMLE	Quasi-maximum likelihood estimator
PCA	Principle component analysis
EM	Expectation maximization
FA	Factor analysis
ETF	Exchange traded fund
NYSE	New York stock exchange
AR	Autoregressive model
ARMA	Autoregressive moving average model
GARCH	Generalized autoregressive conditional heteroscedasticity model
ACD	Autoregressive conditional duration

Sell side is the part of the financial industry involved with the creation, promotion, analysis and sale of securities. Sell-side individuals and firms work to create and service stock products that will be made available to the buy side of the financial industry.

Bid price refers to the maximum price that a buyer or buyers are willing to pay for a security.

Ask price refers to the minimum price that a seller or sellers are willing to receive for the security. A trade or transaction occurs when the buyer and seller agree on a price for the security.

2 The Three V's of Big Data in High Frequency Data

Big data is often described by the three V's: velocity, variety and volume, all of which are the basic characteristics of high frequency data. The three V's bring both opportunities and difficulties to practitioners in finance [26]. In this section we introduce the concept, historical development and challenges of high frequency data.

2.1 Velocity

Telling about the velocity of the high frequency data seems to be tautology. Over the past two decades, the financial markets adopt computer technologies and electronic systems. This leads to a dramatic change of the market structure. Before 1970s, the traditional market participates usually negotiate their trading ideas via phone calls.

Today most of jobs of the traditional traders and brokers are facilitated by computers, which are able to handle tremendous amount of information in an astonishing speed. For example, the NYSE TAQ (Trade and Quote) data was presented in seconds' timestamp when it was first introduced in 1997. This was already a huge advance comparing to the pre 1970s daily data. Now the highest frequency of the TAQ data is in millisecond, which is a thousand of a second. Furthermore, a stock can have about 500 quote changes and 150 trades in a millisecond. No one would be surprised if the trading speed would grow even faster in the near future because of Moore's law. As a result, even traditional low frequency traders may need various infrastructures, hardware and software techniques to reduce their transaction costs in their transactions. The high frequency institutions, on the other side, are willing to invest millions of dollars not only on computer hardware but also on real estate; since 300 miles closer to the exchange will provide about one millisecond advantage in sending and receiving orders.

2.2 Variety

With the help of electronic systems the market information can be collected not only in higher frequency but also in a greater variety. Traditional price data of a financial instrument usually consists of only 4 components: open, high, low, close (OHLC). The microstructure of the price data is fundamentally different with the daily OHLC, which are just 4 numbers out of about ten thousands trade prices of equity in a single day. For example, the well-known bid-ask spread which is the difference between the highest bid price and the lowest ask price is the footstone of many high frequency trading strategies. The level-2 quote data also contains useful information can be used to identify buy/sell pressure. Another example is the duration, which measures how long it takes for price change, can be used to detect the unobservable good news in the market. Diamond and Verrecchia [17] and Easley and O'hara [21] suggest that the lower the durations, the higher probability of the presence of the good news when the short selling is not allowed or limited. Together with the trade volume, the duration can also be a measurement of market volatility. Engle and Russell [25] first found the intraday duration curve that indicated the negative correlation with the U-shaped volatility pattern.

2.3 Volume

Both velocity and variety contributes to the tremendous volume of the high frequency data. And that amount is still growing. The total number of transactions in the US market has been increased by 50 times in the last decade. If we assume that there are about 252 trading days in each year, then the number of quotes observed on November 9, 2009 for SPY alone would be greater than 160 years of daily OHLC and volume data points, Aldridge [1]. Not only the number of records, but also the accuracy is increasing. The recent TAQ prices are truncated to five implied decimal places comparing to the two decimal digits of the traditional daily price data. The size of one-day trade data is about 200 MB on average; while the quote data is about 30 times larger than the trade data. Most of these records are contributed by the High Frequency Trading (HFT) companies in US. For example, in 2009 the HFT accounted for about 60–73% of all US equity trading volume while the number of these firms is only about 2% overall operating firms [26].

2.4 Challenges for High Frequency Data

Like most Big Data, high frequency data is a two-sided sword. While it carries a great amount of valuable information; it also brings huge challenges to quantitative analyst, financial engineers and data scientists. First of all, most high frequency data are inconsistent. These data are strongly depended on the regulations and procedures of the institution that collects them, which varies for different periods and different exchanges. For example, the bid-ask spreads in NYSE are usually smaller than the ones in other exchanges. Moreover, a higher velocity in trading means a larger likelihood that the data contains wrong records. As a result, some problematic data points should be filtered out from the raw data; and a fraction of the whole data can be used in practice.

Another challenge is the discreteness in time and price. Although all financial data are discrete, many of them can be approximately modeled by a continuous stochastic process or a continuous probability distribution. The classical example of Black Scholes formula is based on the assumption of geometric Brownian motion price process. However this is not the case for high frequency data. The tick data usually falls on a countable set of values. Figure 2 plots the histogram of the trade price changes of IBM on Jan 10, 2013. There are about 66% of the prices are the same as the previous one. And about 82% of the price changes fall in −1 to 1 cent. Similar observation can be found in Russell et al. [53]. Another property of high frequency data is the bid-ask bounce. Sometimes it can be observed that the prices frequently back and forth between the best bid and ask price. This phenomenon introduces a jump process that differs with many traditional models. Furthermore, the irregularly spaced data makes it difficult to be fitted by most continuous stochastic processes that are widely used in modeling daily returns. The problem becomes even harder in high dimension, since the duration pattern varies in different assets.

3 Data Cleaning, Aggregating and Management

Cleaning data is the first step of any data analysis, modeling and prediction. The raw data provided by data collectors is referred as dirty data, since it contains inaccurate

Fig. 2 Histogram of the trade price changes of IBM on Jan 10, 2013

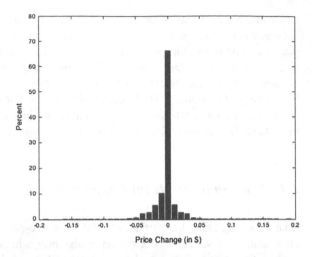

FILE	FORMAT	RECORD SIZE (BYTES)	FTP SIZE (COMPRESSED)	NUMBER OF ROWS	FILE TIME AVAILABILITY (EST)
TAQ Master	ASCII	251	360 KB	8000	8pm (20:00)
TAQ Master Beta	ASCII	Variable – Pipe Delimited	Approximately: *.txt - 750kb *.xls - 2.6mb	8000	Midnight (00:00)
TAQ Quotes	ASCII	133	6 GB	550 million	11pm (23:00)
TAQ Trades	ASCII	108	200 MB	24 million	9pm (21:00)
TAQ NBBO	ASCII	182	1.2 GB	110 million	11pm (23:00)
TAQ Quote Admin Messages	ASCII	Variable – Pipe Delimited	TBC	TBC	2am (02:00)
TAQ Trade Admin Messages	ASCII	Variable – Pipe Delimited	TBC	TBC	2am (02:00)

Fig. 3 Daily TAQ file details (*Source* https://www.nyxdata.com/doc/243156)

or even incorrect data point almost surely. In addition data cleaning is sometimes followed by data aggregation that generates data with a desired frequency. The size of data is often significantly reduced after the two steps. Thus one can extract useful information from the cleaned data in a great efficiency.

In this section we take NYSE TAQ data as an example. Figure 3 lists the details of daily TAQ files. The information is available on http://www.nyxdata.com/Data-Products/Daily-TAQ.

Fig. 4 The trade prices of
IBM on Jan 10, 2013

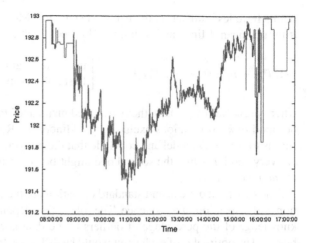

3.1 Data Cleaning

As we have discussed in the previous section, most of high frequency data contains
certain errors. Some of them can be detected simply by plotting all the data points.
Figure 4 plots all the trade prices of IBM on Jan 10, 2013. The trades not happened
in regular market hours (9:30 AM to 4:00 PM) are also included in the dataset. This
kind of data lacks liquidity and contains more outliers than the others; and therefore
they are not considered in most data analysis. But one can also observe that there are
several abnormal outliers within the regular hours.

We introduce several numerical approaches for cleaning high frequency data. The
first step is to filter out the data that potentially have lower quality and accuracy. For
example, Brownlees and Gallo [9] suggest removing non-NYSE quotes in TAQ data;
since NYSE records usually have less outlier than the non-NYSE ones as shown by
Dufour and Engle [20]. In addition, the data record that were corrected or delayed
should also be removed. These kinds of information about data condition and location
are listed in COND, CORR and EX columns in the TAQ data, see Yan [59] for details.

Consider a price sequence p_i where $i = 1, 2, \ldots, N$ with length N. Brownlees
and Gallo [9] propose the following algorithm for removing outliers:

$$\text{If } (|p_i - \bar{p}_i(k)| < 3s_i(k) + \phi) = \begin{cases} \text{true, observation } i \text{ is kept} \\ \text{false, observation } i \text{ is removed/truncated} \end{cases}$$

where $\bar{p}_i(k)$ and $s_i(k)$ are the α-trimmed mean and standard deviation of a neighborhood of k observations and ϕ is a positive number called granularity parameter. ϕ is to
prevent p_i to be removed when $s_i(k) = 0$. As we have seen in Fig. 2 high frequency
data often contains many equal prices. α is a percentage number. For example, a
10%-trimmed mean and standard deviation are the average of the sample excluding
the smallest 10% and the largest 10% numbers. Thus outliers and unreasonable data

points have less impact on the trimmed statistics. Median can be viewed as a fully trimmed mean. Mineo and Romito [44] propose a slightly different algorithm:

$$\text{If } (|p_i - \bar{p}_{i-}(k)| < 3s_{-i}(k) + \phi) = \begin{cases} \text{true, observation } i \text{ is kept} \\ \text{false, observation } i \text{ is removed/truncated} \end{cases}$$

where $\bar{p}_{i-}(k)$ and $s_{-i}(k)$ are the α-trimmed mean and standard deviation of a neighborhood of k observations excluding p_i. Mineo and Romito [45] apply both algorithms to the ACD model and conclude that the performances of the two algorithms are very similar, while the second one might be better in modeling the correlations of model residuals.

The α-trimmed mean and standard deviation are the robust estimates of the location and dispersion of a sequence. The robustness depends on the choice of α. Prior knowledge of the percentage of outliers in the data is required in order to find the best α. The optimal α of each asset would be different. In some cases the α-trimmed mean and the standard deviation can be replaced by the following statistics:

$$\bar{p}_i(k) = median\{p_j\}_{j=i-k,\dots,i+k}$$
$$s_i(k) = c \cdot median\{|p_j - \bar{p}_i(k)|\}_{j=i-k,\dots,i+k}$$

where c is a positive coefficient. Outlier detecting algorithms with above statistics are sometimes called Hampel filter that is widely used in engineering. The second equation can be generalized by replacing the median by quartile with certain level. The median based $\bar{p}_i(k)$ and $s_i(k)$ are also more robust than the trimmed ones.

A very important issue in the data cleaning approaches is that the volatility of the cleaned data depends on the choice of methods and corresponding parameters. The volatility of many high frequency data, including equity and currency, exhibits strong periodic patterns. The outlier detection algorithms with moving window can potentially diminish or remove these patterns that are important in prediction and risk control. Thus it is crucial to consider the periodic behavior before using above algorithms directly. One way is to apply robust estimates of volatility to raw data and then remove this effect via certain adjustment. We discuss this problem in Sect. 4.1.

3.2 Data Aggregating

Most econometric models are developed for equally spaced time series, while most high frequency data are irregular spaced and contain certain jumps. In order to apply these models to the high frequency data, some aggregating techniques are necessary for generating equally spaced sequence from the raw data. Consider a sequence $\{(t_i, p_i)\}$ where $i = 1, \dots, N$, t_i is time step and p_i is trade or quote price. Given an equally-spaced time stamps $\{\tau_j\}$ where $j = 1, \dots, M$ and $\tau_j - \tau_{j-1} = \tau_{j+1} - \tau_j$ for all j, a simple but useful way to construct a corresponding price series $\{q_j\}$ where

$j = 1, \ldots, M$ is to take the previous data point:

$$q_j = p_{i_{\text{last}}},$$

where $i_{\text{last}} = \max\{i | t_i \leq \tau_j, i = 1, \ldots, N\}$. This approach is called last point interpolation. It assumes that the price would not change before the new data come in. Gençay [27] propose a linear interpolation approach:

$$q_j = p_{i_{\text{last}}} + (p_{i_{\text{next}}} - p_{i_{\text{last}}}) \frac{\tau_j - t_{i_{\text{last}}}}{t_{i_{\text{next}}} - t_{i_{\text{last}}}}$$

where $i_{\text{next}} = \min\{i | t_i \geq \tau_j, i = 1, \ldots, N\}$. The second method is potentially more accurate than the first one, but one should be very careful when use it in practice, especially in back-testing models or strategies; since it contains the future information $p_{i_{\text{next}}}$ which is not available at τ_j.

There are several ways to deal with the undesirable jumps caused by bid-ask bounce. The most widely used approach is to replace the trade prices by the mid-quote prices. Let $\{(t_{i^b}, p_{i^b})\}$ where $i = 1, \ldots, N^b$ and $\{(t_{i^a}, p_{i^a})\}$ where $i = 1, \ldots, N^a$ be the best bid and ask prices together with their time stamps. The mid-quote price is given by

$$p_i = \frac{1}{2}(p_{i^b}^b + p_{i^a}^a),$$

where

$$t_i = \max\{t_{i^a}, t_{i^b}\}$$
$$i_b = \min\{t_{i^b} > t_{i-1}, i = 1, \ldots, N^b\}$$
$$i_a = \min\{t_{i^a} > t_{i-1}, i = 1, \ldots, N^a\}.$$

Another approach is to weight the bid and ask by their sizes s_{i^b} and s_{i^a}

$$p_i = \frac{s_{i^b}^b p_{i^b}^b + s_{i^a}^a p_{i^a}^a}{s_{i^b}^b + s_{i^a}^a}.$$

Once we get an equal time spaced price series $\{q_j\}$ where $j = 1, \ldots, M$, we are able to calculate the log returns of the asset:

$$r_j = \log \frac{q_j}{q_{j-1}}.$$

In high frequency data, the price difference is usually very small. Thus the log returns would be very close to the linear returns

$$r_j = \frac{q_j - q_{j-1}}{q_{j-1}}.$$

There are several good reasons to consider the log returns instead of the linear returns in financial modeling. First it is symmetric with respect to the up and down of the prices. If the price increases 10% and decreases 10% in terms of the log return, then it will remain the same. The linear return can exceed 100% but cannot be lower than -100% while the log return does not have this limit. Furthermore the cumulative log returns can be simply represented as the sum of the log returns; this fact would be very helpful in applying many linear models to the log returns.

The last thing we want to mention here is that the size of overnight returns in equity market is often tremendous comparing to the size of intraday returns. The currency market does not have that problem. Overnight returns in equity market are often considered as outliers and removed from the data in most applications. One can also rescale these returns since they may contain useful information. But different methods in rescaling overnight returns might affect the performance of model and strategy.

3.3   Scalable Database and Distributed Processing

Cleaning and aggregating high-volume data always needs a big data infrastructure that combines a data warehouse and a distributed processing platform. To address the challenges of such a big data infrastructure with emerging computing multi-source platforms such as heterogeneous architectures and Hadoop with emphasis on addressing data-parallel paradigms, people have extensively been working on various aspects, such as scalable data storage and computation management of big data, multisource streaming data processing and parallel computing, etc.

Database is an essential datastore for high-volume finance data such long-term historical market data sets. In data management, the column-based database like NoSQL and in-memory database are replacing the traditional relational database management system (RDBMS) in financial data-intensive applications. RDBMS is database based on the relational model and it has been used for decades in industry. Although it is ideal for processing general transactions, RDBMS is less efficient in processing enormous structured and unstructured data, for examples, for market sentiment analysis, real-time portfolio and credit scoring in modern financial sector. Usually, these financial data are seldom modified but their volume is overwhelmed and they need to be queried frequently and repeatedly. In this, a column based database often stores time series based metadata with support of data compression and quick read. In this regard, the columnar databases are preferably suitable for time series of financial metadata. For example, when a financial engineer pulls out a time series of only a few specified metrics with a specific point, a columnar database is faster for reading than a row-based database since only specified metrics such as OHLC are needed. In this case, a columnar database is more efficient because of the cache efficiency and it has no need for scanning all rows like in a row based database. Beyond the columnar database, the in-memory data-base is another emerging datastore solution when performing analytics. That is, if the data set is frequently used and its size fits into memory, the data should persist in the memory for sake of data retrieving, eliminating the need for accessing disk-mediated databases. In practice, what solution is favorable should depend on the practitioners application and available computing facilities.

In addition to data warehouse, distributed processing is equally important. Hadoop often works on Big Data for financial services [26]. Hadoop refers to a software platform for distributed datastore and distributed processing on a distributed computing platform such as a computer cluster. Hadoop is adopted for handling the big data sets for some financial services such as fraud detection, customer segmentation analysis, risk analytics and assessment. In these services, the Hadoop framework helps to enable a timely response. As a distributed data infrastructure, Hadoop does not only include a distributed data storage known as HDFS, Hadoop Distributed File System, but it also offers a data-parallel processing scheme called as MapReduce. However, Hadoop, as a tool, is not a complete big data solution and it has its limitations like everything. For example, it is inefficient to connect structured and unstructured data, unsuitable for real-time analytics, unable to prioritize tasks when multiple tasks are running simultaneously in distributed computing platforms, and its performance closely depends on the scalability of a distributed file system which in turn limits this architecture. Apache Spark, on the other hand, is a data-processing tool and it operates on distributed data storage. Spark does not provide a distributed data storage like HDFS so it needs to be integrated with one distributed data platform. It can run on top of HDFS or it can process structured data in Hive. Spark is an alternative to the traditional map/reduce model that is used by Hadoop and it supports real-time stream data processing and fast queries. Generally, Sparks needs more RAM instead of network and disk-backed I/O and thus it is relatively faster than Hadoop. Spark often completes the full real-time data analytics in memory. However, as it uses large RAM, Spark needs a high-end machine with a large memory capacity. In the code development, Spark is a library for parallel processing through function calls and a Hadoop MapReduce program can be written by inheriting Java classes.

4   Modeling High Frequency Data in Finance

In this section we discuss the mathematical models for high frequency data. There are a number of quantitative models with different features in financial econometrics. The purpose of majority of these models is to estimate expected returns and volatility of a risky asset or portfolio. As we have discussed in the first section, expected return and volatility are the two footstones of the modern portfolio theory. Expected return, sometimes called alpha, is the prediction of profit and loss in the future. It is the most crucial statistics for a portfolio manager. Volatility measures variation of value change for a financial instrument or portfolio. The behavior of a portfolio whose volatility is controlled properly is more consistent than the ones with large volatility. Thus Markowitz's theory states that a portfolio may generate relatively stable revenues by maximizing its expected return and minimizing the volatility. Other useful statistics and performance measures such as skewness, kurtosis, VaR or drawdown can also be estimated by some of the following models. There a number of literatures consider portfolio selection and risk management based on these statistics. We will not discuss them in this chapter.

4.1 Volatility Curve

The intraday market exhibits a more clearly periodic pattern especially in volatility comparing to the low frequency financial data. There a number of papers propose different approaches to modeling the volatility of the high frequency data. The most common idea is to separate the volatility into deterministic seasonal part and stochastic part. The deterministic part is usually fitted by a smooth function, as Andersen and Bollerslev [4] and Andersen et al. [5] suggest. The stochastic part can be modeled by ARCH type models, since Engle and Manganelli [24] discover volatility clustering effect in high frequency market.

The volatility is often considered as a hidden factor of the market. The most common way to extract seasonal volatility from the data is to compute the norms of the absolute returns. To make it clear, let an integer $K > 0$ be the period length and $r_1, r_2, \ldots, r_{KN}$ be a sequence of equally time-spaced log returns in N periods. Then the seasonal realized volatility can be defined as:

$$v_i = \left(\frac{1}{N} \sum_{j=1}^{N} |r_{K(j-1)+i}|^p \right)^{\frac{1}{p}}, i = 1, 2, \ldots, K,$$

where the exponent p is usually set to be 1 or 2. However the above representation is sensitive to the outliers. The seasonal structure could be destroyed by a single abnormal extreme value. A more robust way is to consider the quartiles of the absolute returns:

$$v_i = quartile_\alpha\{|r_{K(j-1)+i}|\}_{j=1,2,\ldots,N}, i = 1, 2, \ldots, K,$$

where $0 \leq \alpha \leq 1$. Seasonality with different periods can be observed from the high frequency data. As an example, Dong [18] considers 1-min log returns of all the stocks in Russell 3000 on 2009. The period K is set to be 390 that is the number of minutes in each trading day. Figure 5 plots the volatility curve together with the aggregated volume curves of NYSE and NASDAQ against 390 min.

In addition Dong [18] discovers that there exist 5-min spikes on the curve. This phenomena are more clear when we plot the volatility curve when $K = 60$ min (see

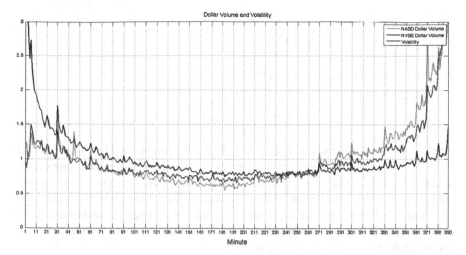

Fig. 5 The volatility curve together with the aggregated volume curves of NYSE and NASDAQ against 390 min (*Credit* [18])

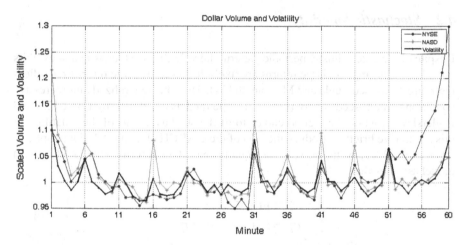

Fig. 6 The volatility curve together with the aggregated volume curves of NYSE and NASDAQ against 60 min (*Credit* [18])

Fig. 6). Both volatility and volume exhibit the U-shape pattern but they are different at tails. The volatility is relatively higher at market opening and lower at the end.

To fit the volatility curve above one can use a smooth rational function, for example:

$$f(x) = \frac{ax^2 + bx + c}{dx + 1}.$$

The coefficients a, b, c, d can be fitted by least square approach:

$$\min_{a,b,c,d} \sum_{i=1}^{K} (v_i + div_i - ai^2 - bi - c)^2,$$

and the de-seasonal log returns can be:

$$\hat{r}_{K(j-1)+i} = \frac{r_{K(j-1)+i}}{f(i)},$$

where $i = 1, 2, \ldots, K$ and $j = 1, 2, \ldots, N$. As we have mentioned before, the volatility patterns may not be preserved if we apply the outlier cleaning techniques introduced in Sect. 3.1 before computing the realized volatility. The quartile-based realized volatility, which is a robust estimator, can be applied directly to un-cleaned data. Thus instead of removing outliers in the price, one can first aggregate data and get an equal spaced return series with abnormal outliers. Then the data cleaning approach can be applied to the de-seasonal returns.

4.2 Stochastic Volatility

Despite of the deterministic periodic pattern, the volatility is stochastic and exhibits volatility clustering, i.e. large returns are likely followed by large returns regardless their directions, see Engle and Manganelli [24]. Thus the generalized autoregressive conditional heteroscedasticity (GARCH) type models developed by Engle [22] and Bollerslev [8] would be a good choice to fit the stochastic part of the volatility. In this section we briefly introduce the idea of the GARCH (1,1) model. For simplicity let r_i, $i = 1, 2, \ldots$ be the de-seasonal equally spaced log returns. The GARCH (1,1) model assumes that:

$$r_i = \mu_i + \sigma_i \epsilon_i$$
$$\sigma_i^2 = \omega + \alpha r_{i-1}^2 + \beta \sigma_{i-1}^2$$

where ω, α, β are positive real numbers, and ϵ_i where $i = 1, 2, \ldots$ are i.i.d normally distributed with zero mean and unit variance. The drift term μ_i is the conditional expectation of r_i given all the information up to time i. There are a lot of approaches in modeling μ_i that is often called "alpha" in finance. We discuss several examples in Sect. 4.4. The parameters ω, α, β should satisfy the constraint $\alpha + \beta \leq 1$ in order to make the process to be stationary. The estimation of the model is usually performed by the maximum likelihood estimator (MLE). We refer to McNeil et al. [40] for details. Scientific programming languages including Matlab and R have matured packages for fitting the GARCH model. In practice, the ω is often a small number close to zero; β ranges from 0.7 to 0.9 and $\alpha + \beta \approx 1$. α is usually much smaller than β, but it plays a key role in measuring the volatility sensitivity to the market impact.

4.3 Multivariate Volatility

The simplest approach to model the dependence structure of multi-assets is to compute the covariance of their returns. However, the traditional sample covariance is usually ill- conditioned when the dimension is relatively high comparing to the sample size. An ill-conditioned covariance matrix may lead huge errors in risk forecasting and portfolio optimization. The simplest way to improve the conditions of the sample covariance is to adjust its eigenvalues. Another method is to shrink the covariance to some well-conditioned matrix. The most famous shrink-age estimator is proposed by Ledoit and Wolf [33].

The third approach, which might be the most widely used, is to impose certain structure on the covariance. For example, one can assume that a d by d covariance matrix has the expression:

$$\Sigma = FF' + D,$$

where F is a d-by-n matrix, D is a d-by-d diagonal matrix and $n < d$. The rationale of the above formula is that the asset return follows the linear factor model:

$$r = Fx + \epsilon,$$

where r is the d-dimensional vector of log returns, x is the vector of uncorrelated risky factors with unite variance in a lower dimension n, and ϵ is the uncorrelated errors with covariance D. Unlike the traditional factor models, the factor x does not come from real data, which are usually correlated. In this model x is some un-correlated statistical factors that are hidden from the market. The well-known principle component analysis (PCA) is one way to extract x from the original data. Let $\hat{\Sigma}$ be the sample covariance matrix; by the singular value decomposition it can be written as:

$$\hat{\Sigma} = U\Lambda U',$$

where U is a d-by-d unitary matrix, i.e. $UU' = U'U = I$, and Λ is a diagonal matrix with eigenvalues $\lambda_1 \geq \lambda_2 \geq \cdots \geq \lambda_d$. Then we can set

$$x = \Lambda_n^{-\frac{1}{2}} U_n' r$$

$$F = U_n \Lambda_n^{\frac{1}{2}},$$

where the d-by-n matrix U_n consists the first n columns of U, and the diagonal matrix Λ_n is the first n-by-n block of Λ. In fact one can show that F is the solution of:

$$\min_{F} \| \hat{\Sigma} - FF' \|_2,$$

where $\| \cdot \|_2$ is the induced 2-norm of a matrix. The residual matrix D can be simply written as:

$$D = diag(\hat{\Sigma} - FF').$$

The PCA is a simple standard statistical tool for dimension reduction. A potentially more precise approach to fit F and D is to apply the expectation maximization (EM) algorithm to the log returns. This approach is also known as the factor analysis (FA). The standard EM algorithm for FA proposed by Rubin and Thayer [52] is an iterative algorithm. Let $\{r_i\}$ where $i = 1, \ldots, N$ be a sequence of vectors of log returns, and $F^{(0)}$, $D^{(0)}$ be the initial inputs. Then the k-th iteration of the EM is given by:

E Step: Re-compute the conditional expectations:

$$E[x|r_i] = F^{(k-1)'}(D^{(k-1)} + F^{(k-1)}F^{(k-1)'})^{-1}r_i$$
$$E[xx'|r_i] = I - F^{(k-1)'}(D^{(k-1)} + F^{(k-1)}F^{(k-1)'})^{-1}F^{(k-1)} + E[x|r_i]E[x|r_i]'.$$

M Step: Update F and D:

$$F^{(k)} = \left(\sum_{i=1}^{N} r_i E[x|r_i] \right) \left(\sum_{i=1}^{N} E[xx'|r_i] \right)^{-1}$$

$$D^{(k)} = \frac{1}{N} diag\left(\sum_{i=1}^{N} r_i r_i' - F^{(k)} E[x|r_i]r_i' \right).$$

The above algorithm will converge to the maximum likelihood estimator of F and D given that x and ϵ are independently Gaussian distributed. There are some variations of the classical EM algorithm that may improve the convergence speed, for example, the ECM algorithm proposed by Meng and Rubin [41], Donald B the ECME algorithm proposed by Liu and Rubin [36], the GEM algorithm pro-posed by Neal and Hinton [48] and the α-EM algorithm proposed by Matsuyama [39]. Jia [31] applies the α-EM algorithm together with conjugate gradient method to the FA and shows a significant improvement in the speed.

4.4 Expected Return

The high frequency data usually have a stronger cross-sectional dependency than the low frequency one. This fact can be observed not only in the volatility but also in the expected returns or alphas. Thus the classical autoregressive (AR) models may have a better performance in the high frequency market. Let $\{r_i\}$, $i = 1, 2, \ldots$ be a sequence of de-seasonal log returns equally spaced in time. The AR(p) model can be written as:

$$r_i = h_0 + h_1 r_{i-1} + h_2 r_{i-2} + \cdots + h_p r_{i-p} + x_i,$$

where $i = p + 1, p + 2, \ldots$; $h_0, h_1, \ldots, h_p$ are called AR coefficients or impulse response in electronic engineering and x_i are often assumed to be i.i.d zero mean normally distributed noises. Given the information up to time $i - 1$, the expectation of r_i, which is given by $h_0 + \sum_{j=1}^{p} h_j r_{i-j}$, is the alpha prediction of the AR(p) model.

The estimation of AR(p) model can be performed by the least square method. Suppose that we have data samples with length $N > p$, the least square method solves the following optimization problem:

$$\min_{h_0, h_1, \ldots, h_p} \sum_{i=p+1}^{N} \left(r_i - h_0 - \sum_{j=1}^{p} h_j r_{i-j} \right)^2,$$

which can be solved explicitly:

$$\hat{h} = (R'R)^{-1} R' r,$$

where

$$r = \begin{pmatrix} r_{p+1} \\ r_{p+2} \\ \vdots \\ r_N \end{pmatrix},$$

and

$$R = \begin{pmatrix} 1 & r_p & r_{p-1} & \cdots & r_1 \\ 1 & r_{p+1} & r_p & \cdots & r_2 \\ 1 & r_{p+2} & r_{p+1} & \cdots & r_3 \\ \vdots & \vdots & \vdots & \ddots & \vdots \\ 1 & r_{N-1} & r_{N-2} & \cdots & r_{N-p} \end{pmatrix},$$

The naïve least square method is simple, but it is not the most numerically efficient approach in the estimation of AR(p). A better alternative called Burg's method is usually considered a standard approach for estimating AR(p) systems. We refer readers to Marple [38]. Some software such as Matlab also provides build-in function for Burg's algorithm.

A generalization of the AR(p) model is so-called autoregressive moving average (ARMA) model. Similar as AR(p), the ARMP(p,q) model can be represented as

$$r_i = h_0 + \sum_{j=1}^{p} h_j r_{i-j} + \sum_{j=1}^{q} q_j x_{i-j} + x_i.$$

In fact one can show that ARMA(p,q) is also a special case of AR(∞) process. However methods like least squares or Burg's algorithm cannot be applied to the estimation of ARMA(p,q) model. Instead the general maximum likelihood estimator is the standard approach for fitting ARMA(p,q) with normally distributed residuals x_i. The ARMA process often works together with GARCH model. In that case the estimations of ARMA and GARCH can be done separately. This approach is called quasi maximum likelihood (QMLE). A comprehensive introduction of the ARMA-GARCH type models can be found in McNeil et al. [40]. Beck et al. [6] apply the ARMA-GARCH model to intraday data with frequency ranged from 75 to 300 s; and discover the heavy-tailness in the residuals of the model.

The financial data often has mean-reverting pattern. For example, the estimated h_1 of AR(p) model is usually negative. Roughly speaking, the scales of the rest parameters h_j, $(j > 1)$ are small comparing to h_1, and become smaller as j increases, since the impact of historical values to the present will diminish as time goes. However, this does not mean that h_j with large j should be ignored. The aggregation of small impulse responds may have a strong impact to the prediction; since it contains information of long-term trends. Sun et al. [56] find that the intraday equity data may have long-range dependence, i.e. the decay of h_j with respect to j is very slow. Kim [32] applies an ARMA-GARCH model with fractional heavy-tailed distributions to model high frequency data. Although neither ARMA(p,q) nor AR(p) processes can capture the long-range dependency of the data, one may approximate a long-range dependent time series by an AR(p) with large p in a finite amount of time. However as the number of parameters increases, the error of the least squares estimator or Burg's method grows tremendously, due to the Cramér-Rao bound. Thus similar as the covariance matrix estimation, one may need some biased estimators like shrinkage. Mullhaupt and Riedel [47] impose a specific structure called triangular input balanced form on the AR process. They show that the estimation error can be significantly reduced by adding small bias to the estimator.

4.5   Duration

Up to now we introduce how to transfer data into equal spaced series. However the frequency of the data would be reduced and certain information would be lost in the aggregation. The original data with irregular time stamps are called ultra-high frequency data in Engle [23]. Consider a sequence of ultra-high frequency data $\{(t_i, p_i)\}$ where $i = 1, \ldots, N$, the number of trades that occur before time t is given by $N(t) = \sup\{i | t_i \leq t, i = 1, \ldots, N\}$. The simplest way is to fit $N(t)$ by a homogeneous Poisson process, i.e. the probability that there $N(t)$ events happen between t and $t + \Delta t$ is

$$P(N(t + \Delta t) - N(t) = k) = \frac{(\lambda \Delta t)^k}{k!} e^{-\lambda \Delta t}, k = 0, 1, 2, \ldots,$$

where λ is the instantaneous arrival rate of an event

$$\lambda = \lim_{\Delta t \to 0} \frac{P(N(t + \Delta t) - N(t) \geq 0)}{\Delta t}.$$

The Poisson process implies that the durations $\Delta t_i = t_i - t_{i-1}$ are i.i.d exponentially distributed with constant rate λ

$$P(\Delta t \leq s) = 1 - e^{-\lambda s}.$$

However the Poisson process might over simplify the problem. Similar as the volatility, duration exhibits periodicity and heteroskedasticity. Engle [23] shows that the duration of mid-quote prices has an n-shape curve in contrast to the volatility. The periodicity can be removed using the same approach in Sect. 4.1. The heteroscedasticity, however, contradicts to the assumption that λ is constant. Engle and Russell [25] propose an autoregressive conditional duration (ACD) model as follows:

$$\Delta t_i = \phi_i \epsilon_i$$
$$\phi_i = \omega + \alpha \Delta t_{i-1} + \beta \phi_{i-1},$$

where ϵ_i are i.i.d positive random variables. The ACD model looks very similar to the GARCH model. The distribution of residuals ϵ_i is often set to be the exponential or Weibull distribution. It is clear that the instantaneous arrival rate λ of the ACD model is not a constant. Simple calculation shows that given $\phi_{N(t)}$ and exponentially distributed $\epsilon_{N(t)}$:

$$\lambda(t) = \frac{1}{\phi_{N(t)}}.$$

Similar as the GARCH model, the parameters of the ACD model can be fitted via QMLE, see Engle and Russell [25], Engle [23].

4.6 Scalable Parallel Algorithms on Supercomputers

As we have seen, all of the computations in previous sections are based on high dimensional matrix operations. For example, multivariate least squares method is applied to fit volatility curves and AR models. Eigenvalues are important in estimating the covariance matrix.

Of these methods, matrix multiplication is the core problem as a basis for most of other methods such as least square, eigenvalue and matrix factorization. Matrix multiplication (MM) is the simplest yet most difficult problem in mathematics [60]. The standard algorithm for MM is $O(n^3)$ but in mathematics, researchers never stop in pursuit of faster approached for multiplying matrices. For example, Strassen

Matrix Multiplication Algorithms:

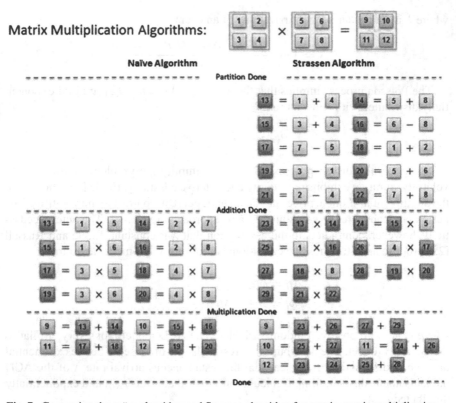

Fig. 7 Comparing the naïve algorithm and Strassen algorithm for matrix-matrix multiplication

reduced the computing complexity to $O(n^{2.8})$ in 1969 and another breakthrough is the Coppersmith-Winograd algorithm that per-forms MM in $O(n^{2.4})$ operations. In addition to theoretical studies, the complex architectures of computing facilities have further escalated the difficulty for the MM implementation. For example, the task mapping in parallel computers and the task scheduling in hybrid CPU-GPU computers made the MM implementations even harder. In this regard, some data-oriented schedule paradigm is proposed and it has been applied to the MM problem on todays high-performance computing facilities [62]. Of experiments, the best-practice matrix-multiplication approach is found [60]. Figure 7 compares the naïve and Strassen algorithms for tile-based matrix-matrix multiplication.

Cholesky inversion method is to compute the inverse for a positive-definite matrix. In finance, the covariance matrix is positive semidefinite. Cholesky inversion is more challenging than matrix multiplication and it consists of three successive steps: Cholesky factorization, inversion for lower triangular matrix and product of lower triangular matrices. A naïve approach is to perform three steps sequentially but its performance is very poor. To deliver better parallelism, one has to interweave these steps by adhering to the complex data dependencies. This goal could be achieved through a thorough critical path approach [57] or a dynamic data-oriented schedule approach Zhang et al. [61, 62].

5 Portfolio Selection and Evaluation

Data cleaning, aggregation and modeling can all be viewed as searching valuable information from the massive data. The amount of data would be significantly reduced after each step. Expected return, volatility and other statistics are the gold extracted from raw ore. The final steps are developing trading ideas, constructing portfolios and testing strategies. Although data volume in this procedure is relatively small, there is a great need of computing speed from high frequency investors who want to execute their strategies faster than their opponents. In this section we review two different classes of strategies: Markowitz's mean variance portfolio selection and on-line portfolio selection. The first one is relatively slow but mature and well developed. The second one is simple but fast which can be potentially applied to ultra-high frequency trading.

5.1 Markowitz Portfolio Optimization with Transaction Costs

Suppose that there are d risky assets with expected return μ and covariance Σ. A self-financing portfolio is represented by a d-dimensional weight vector w that satisfies $\sum_{i=1}^{d} w_i = 1$. The well-known Markowitz portfolio states that a rational risk averse investor wants to maximize the utility function

$$\max_{w} w'\mu - \frac{\lambda}{2}w'\Sigma w$$
$$\text{subject to: } \mathbf{1}'w_i = 1,$$

where $\mathbf{1}_d$ is a d-dimensional vector with all ones, $w'\mu$ is the expected return of the portfolio, $w'\Sigma w$ is the variance of the portfolio and $\lambda > 0$ reflects the degree of risk aversion. In high frequency market, the log return and the real return are very close, so $w'r$ with log return r can be an approximation of the real portfolio return in a short period. Thus μ and Σ in the optimization problem can be log return based mean and covariance. But this would not be true for long-term prediction. The above optimization problem can be solved explicitly. The optimal portfolio weight together with its expectation and variance changes as the risk aversion parameter λ varies. By plotting the expected return against the variance with all possible λ then we obtain the famous efficient frontier.

There are many variations of the Markowitz mean-variance portfolio strategy. One can replace the variance term $w'\Sigma w$ by other risk measures like the value-at-risk

(VaR), conditional value-at-risk (CVaR) or maximum drawdown. These risk measures are often considered to be superior than the variance since they are able to capture the tail-risk. Rockafellar and Uryasev [51] show that the mean-CVaR problem can be transferred to a linear programming with a higher dimension. Chekhlov et al. [11] propose a similar approach for drawdown measures. However, the trade-off of these approaches is that the dimension of the problem increases tremendously by introducing auxiliary variables. CVaR for example, is often calculated via Monte Carlo; and the dimension of the auxiliary variables in the equivalent linear programming is the same as the number of Monte Carlo scenarios. Regular computers may fail to deal with this kind of problem efficiently due to the memory limitation. Under some special cases the mean-risk problem can be solved easily. For example, Shi and Kim [55] show that the dimension of any mean-risk problem with coherent risk measures and a subclass of normal mixture distributions can be reduced to two. In general, however, the mean-risk problem is usually very hard to solve.

The most important problem within the above strategies is that they assume no transaction cost. Transaction cost is usually ignored in the low frequency finance, but it grows dramatically as the trading frequency increases. Broker commissions, exchange fees and taxes are all major sources of the transaction cost. But the most significant one is the portfolio turnovers. For example, if the current best ask price of equity is $10, it does not mean you are able to buy 500 shares at $5000. The size of the best asks might just be 200 shares. The next best ask price might be $10.1 with 300 shares. Overall the average price you paid grows almost proportionally as objective shares increases. Thus a high frequency trader may not choose to change his current position even when he observes a signal. Even you have a perfect prediction of the expected returns and variance, the optimal mean-variance portfolio may be completely different with the current ones; and the profit would be dwarfed by the huge transaction cost in rebalancing the portfolio. Thus a constraint on the portfolio turnover is necessary in portfolio optimization problems. Suppose that your current portfolio weight is given by a d-dimensional vector $\tilde{w}$; then the turnover is usually modeled by the 1-norm of the weight change:

$$\|w - \tilde{w}\|_1 := \sum_{i=1}^{d} |w_i - \tilde{w}_i|.$$

Thus the mean-variance problem with transaction cost can be rewritten as:

$$\max_{w} w'\mu - \frac{\lambda}{2}w'\Sigma w - c\|w - \tilde{w}\|_1$$
$$\text{subject to: } \mathbf{1}'w = 1,$$

where $c > 0$ is the degree of the turnover. The object function is neither quadratic nor smooth at the point $\tilde{w}$. But we are able to convert it to a quadratic programming problem:

$$\max_{v} v'\tilde{\mu} - \frac{\lambda}{2}v'\tilde{\Sigma}v$$

$$\text{subject to: } \tilde{\mathbf{1}}'v = 0, v \geq 0,$$

where

$$\tilde{\mu} = \begin{pmatrix} \mu - \lambda\tilde{w}'\Sigma + c\mathbf{1} \\ -\mu + \lambda\tilde{w}'\Sigma + c\mathbf{1} \end{pmatrix}, \tilde{\Sigma} = \begin{pmatrix} \Sigma & -\Sigma \\ -\Sigma & \Sigma \end{pmatrix},$$

and $\tilde{\mathbf{1}}$ is a $2d$ dimensional vector with first d elements are 1 and the rest are -1. The optimal portfolio weight w^* of the mean-variance problem with transaction cost can be represented by the optimal solution of the above problem v^*:

$$w^* = \tilde{w} + [I, -I]v^*,$$

where I is the d-dimensional identity matrix. One can show that the first d elements of v^* are the positive parts of the weight change, and the rest d elements are the negative parts of the weight change. If $v_k^* > 0$ for some $k = 1, \ldots, d$, then we must have $v_{d+k}^* > 0$, otherwise v^* will not be the optimal solution. The quadratic programming has been thoroughly studied in modern convex optimization theory. Classical algorithm includes the interior-point method and trust-region method, see Nocedal and Wright [49]. Note that $\tilde{\Sigma}$ is not of full rank, this is caused by the non-smoothness of the original problem. One may shrink the eigenvalues of $\tilde{\Sigma}$ a bit to make the problem strictly convex. Thus in practice we usually get an suboptimal solution w^*. If the value of the object function on w^* does not exceed $\tilde{w}'\mu - \frac{\lambda}{2}\tilde{w}'\Sigma\tilde{w}$ then we will keep the portfolio unchanged since the potential benefit of changing the portfolio does not cover the transaction cost.

5.2 On-Line Portfolio Selection

In this section we consider a portfolio allocation framework that is different from the Markowitz's theory. Let $r_{i,t}$ where $i = 1, 2, \ldots, d, t = 1, 2, \ldots, T$ be the log return of the i-th asset at time t, $x_{i,t} = \exp(r_{i,t})$ be the price ratio, $x_t = (x_{1,t}, \ldots, x_{d,t})'$ be the price ratio vector of d assets and $w_t = (w_{1,t}, \ldots, w_{d,t})'$ be the portfolio weights. We assume that the portfolio is long-only; and let $\mathcal{W} = \{w \in \mathbb{R}^d, \text{s.t. } \sum_{i=1}^{d} w_i = 1, w_i \geq 0\}$ be the universe of all long-only portfolio weights. Suppose that the initial wealth is S_0, then the value of a portfolio with strategies: $w_1, w_2, \ldots, w_t \in \mathcal{W}$ is given by:

$$S_t(w_1, \ldots, w_t | x_1, \ldots, x_t) = S_0 \prod_{s=1}^{t} \sum_{i=1}^{d} w_{i,s} x_{i,s}.$$

A general on-line portfolio selection framework proposed by Li and Hoi [34] is as follows:

ALGORITHM: On-line portfolio selection

Input: $x_1, \ldots, x_T$: Historical market sequence

Output: S_T: Final cumulative wealth

Initialize S_0 and w_0

for $t = 1, \ldots, T$ **do**

    Portfolio manager computes a portfolio w_t;

    Market reveals the market price ratio x_t;

    Updates cumulative wealth $S_t = S_{t-1} w_t' x_t$;

    Portfolio manager updates his/her online portfolio selection rules;

end

Here are several examples of on-line portfolio strategies:

5.2.1 Buy and Hold Strategy

The buy and hold strategy simply does not trade anymore once the initial portfolio weight w_0 is given. The dynamic of its portfolio weight is given by:

$$w_{i,t} = \frac{w_{i,t-1} x_{i,t-1}}{\sum_{j=1}^{d} w_{j,t-1} x_{j,t-1}}$$

and the cumulative wealth is:

$$S_t(w_1, \ldots, w_t | x_1, \ldots, x_t) = S_0 \sum_{i=1}^{d} w_{i,0} \prod_{s=1}^{t} x_{i,s}.$$

5.2.2 Constantly Rebalanced Strategy

In contrast to the buy and hold strategy, the constantly rebalanced strategy is to keep rebalancing the portfolio such that $w_0 = w_1 = \cdots = w_t$. Thus the cumulative wealth is

$$S_t(w_1, \ldots, w_t | x_1, \ldots, x_t) = S_0 \prod_{s=1}^{t} \sum_{i=1}^{d} w_{i,0} x_{i,s}.$$

It can be used to replicate the movements of a certain market index. Constantly rebalance and buy and hold are two naïve trading strategies that are often used as benchmarks.

5.2.3 Minimax Strategy

Let $y_1, \ldots, y_T$ be a sequence of integers ranged from 1 to d. Given a sequence of static strategies: $v_1, \ldots, v_T \in \mathcal{W}$, i.e. v_t does not depend on any information prior to t. Then we can define a probability density function of $y_1, \ldots, y_T$:

$$p_T(y_1, \ldots, y_T) = \frac{\sup_{v_1, \ldots, v_T \in \mathcal{W}} \prod_{t=1}^{T} v_{y_t, t}}{\sum_{z_1=1}^{d} \cdots \sum_{z_T=1}^{d} \sup_{v_1, \ldots, v_T \in \mathcal{W}} \prod_{t=1}^{T} v_{z_t, t}}.$$

The marginal density function of $y_1, \ldots, y_t$ for some $t < T$ is given by

$$p_t(y_1, \ldots, y_t) = \sum_{z_{t+1}=1}^{d} \cdots \sum_{z_T=1}^{d} p_T(y_1, \ldots, y_t, z_{t+1}, \ldots, z_T).$$

Given a sequence of price ratio $x_1, \ldots, x_{t-1}$, the minimax strategy on $w_{i,t}$ is defined as

$$w_{i,t} = \frac{\sum_{y_1=1}^{d} \cdots \sum_{y_{t-1}=1}^{d} p_t(y_1, \ldots, y_{t-1}, i) \prod_{s=1}^{t-1} x_{y_s, s}}{\sum_{y_1=1}^{d} \cdots \sum_{y_{t-1}=1}^{d} p_{t-1}(y_1, \ldots, y_{t-1}) \prod_{s=1}^{t-1} x_{y_s, s}}.$$

The minimax strategy is the theoretical best strategy in terms of minimizing the worst-case logarithmic wealth ratio

$$\sup_{x_1, \ldots, x_T} \sup_{v_1, \ldots, v_T \in \mathcal{W}} \log \frac{S_T(v_1, \ldots, v_T | x_1, \ldots, x_T)}{S_T(w_1, \ldots, w_T | x_1, \ldots, x_T)}.$$

This ratio measures the difference between the strategy $w_1, \ldots, w_T$ and the best static strategy with the knowledge of future under the worst case scenario. For detailed proof and the deduction of the minimax strategy we refer readers to Cesa-Bianchi and Lugosi [10].

5.2.4 Universal Portfolio Strategy

The minimax strategy is the theoretical best on-line strategy, but it is hard to achieve in practice. The computation of the densities $p_1, \ldots, p_T$ is often numerically intractable in real market. Cover [13] proposes a computationally efficient strategy called universal portfolio:

$$w_{i,t} = \frac{\int_{\mathcal{W}} u_j S_{t-1}(u, \ldots, u | x_1, \ldots, x_{t-1}) \mu(u) du}{\int_{\mathcal{W}} S_{t-1}(u, \ldots, u | x_1, \ldots, x_{t-1}) \mu(u) du},$$

where $S_{t-1}(u, \ldots, u | x_1, \ldots, x_{t-1})$ is the cumulative wealth of a constantly re-balanced strategy u; and $\mu(u)$ is a density function that can be viewed as a prior distribution of the portfolio weight. At time t the strategy updates the distribution of weight based on the performance of all possible constantly rebalanced strategies. The new strategy is just the expectation of the updated distribution. Cover and Ordentlich [14] show that the worst-case logarithmic wealth ratio of the universal portfolio strategy has an upper bound that increases at the speed of $O(\log T)$ as T increases.

5.2.5 Exponential Gradient (EG) Strategy

The universal portfolio strategy is more practical than the minimax strategy, but still computationally intractable under high dimension; since it involves the calculation of d dimensional integrals. A simple strategy called the EG strategy proposed by Helmbold et al. [30] updates the portfolio weights as follows:

$$w_{i,t} = \frac{w_{i,t-1} \exp\left(\frac{\eta x_{i,t-1}}{\sum_{i=1}^{d} w_{i,t-1} x_{i,t-1}}\right)}{\sum_{j=1}^{d} w_{j,t-1} \exp\left(\frac{\eta x_{j,t-1}}{\sum_{i=1}^{d} w_{i,t-1} x_{i,t-1}}\right)}.$$

The EG strategy is a gradient-based forecaster since the term $\frac{x_{i,t-1}}{\sum_{i=1}^{d} w_{i,t-1} x_{i,t-1}}$ can be viewed as the gradient of logarithmic loss $-\log \sum_{i=1}^{d} w_{i,t-1} x_{i,t-1}$. The upper bound of the worst-case logarithmic wealth ratio of the EG strategy grows with $O(\sqrt{T})$; but in terms of the dimension d it grows only with $O(\sqrt{\log d})$ comparing to the linear growth of universal portfolio.

The above on-line strategies are all based on the assumption that there is no transaction cost. Györfi and Vajda [28] propose an on-line portfolio allocation framework with transaction costs. Suppose that at time $t - 1$ the net wealth of the portfolio is given by N_{t-1}. Given a new strategy w_t and price ratio x_t the gross wealth at time t is given by

$$S_t = N_{t-1} \sum_{i=1}^{d} w_{i,t} x_{i,t}.$$

However, after the rebalancing, the wealth is reduced to $N_t \leq S_t$ because of the transaction costs. Before the rebalancing the weights of each asset are given by

$$\tilde{w}_{i,t} = \frac{w_{i,t} x_{i,t}}{\sum_{j=1}^{d} w_{j,t} x_{j,t}}, i = 1, \ldots, d.$$

In the previous section we simply use $\|w_{t+1} - \tilde{w}_t\|_1$ to approximate the transaction cost. A more precise approximation should be

$$C_t = c_{sell} \sum_{i=1}^{d} \max\{\tilde{w}_{i,t}S_t - w_{i,t+1}N_t, 0\} + c_{buy} \sum_{i=1}^{d} \max\{w_{i,t+1}N_t - \tilde{w}_{i,t}S_t, 0\},$$

where c_{sell} and c_{buy} are the per dollar transaction costs of selling and buying respectively. Using the fact that $N_t = S_t - C_t$ we obtain the following equation

$$1 = \rho_t + c_{sell} \sum_{i=1}^{d} \max\{\tilde{w}_{i,t} - w_{i,t+1}\rho_t, 0\} + c_{buy} \sum_{i=1}^{d} \max\{w_{i,t+1}\rho_t - \tilde{w}_{i,t}, 0\},$$

from with we can solve $\rho_t = N_t/S_t$. Thus instead of S_t we obtain a sequence of net wealth

$$N_t = N_0 \prod_{s=1}^{t} \rho_s \sum_{i=1}^{d} w_{i,s}x_{i,s}.$$

The on-line portfolio allocation with transaction costs can be summarized as:

ALGORITHM: On-line portfolio selection with transaction costs

Input: $x_1, \ldots, x_T$: Historical market sequence, transaction costs c_{sell} and c_{buy}
Output: N_T: Final cumulative net wealth
Initialize ρ_0, S_0 and w_0
for $t = 1, \ldots, T$ **do**
    Portfolio manager computes a portfolio w_t;
    Updates the net wealth $N_{t-1} = \rho_{t-1}S_{t-1}$ after rebalancing;
    Market reveals the market price ratio x_t;
    Updates the gross wealth $S_t = N_{t-1}w_t'x_t$;
    Portfolio manager updates his/her online portfolio selection rules;
end

For more on-line portfolio selection strategies we refer readers to Li and Hoi [34] that provide a review of recent published techniques including some pattern recognition and machine learning strategies.

6 The Future

The rise of big data in financial industry has already been dramatic in the past decade. However we have good reasons to believe that it is just a start; and the adoption of big data technology together with quantitative tools still has a long way to go. Despite of the rapid growth of high frequency industry and systematic trading funds, a number of traditional financial businesses still live in the small data era. A lot of economic data that they collected are weekly, monthly or even quarterly based. Financial analysts may spend several hours on small amount of fundamental data of a single firm; while a large percentage of the work could be done automatically by machine. In addition, there are also more hidden errors in the data that are very difficult to be detected manually, as the information from the data providers such as Bloomberg and Factset grow tremendously. Thus the chances of operational risk made by human analyst who does not have the support of advanced technology increases simultaneously. The most widely used data analyze tool in many financial firms is Microsoft Excel together with Visual Basic for Applications (VBA), which is very inefficient to deal with large datasets. On the other side, although there is a number of professional data analyzing technologies that can process big data in a great efficiency, most of them are not user-friendly and fail to provide a comprehensive visualization of the information for the financial professionals with little technological or mathematical background. Thus the future of big data in finance is likely to be more client-oriented and personalized. This requires a closer connection between the engineers, scientists, financers and bankers [63].

Even in the rapid growing high frequency industry, the technology and theory is far from mature. A unified influential framework such as the classical Black Scholes theory is not discovered yet in high frequency finance. Here we list some potential research topics that might be crucial for the development of quantitative finance.

6.1 Advanced Statistics and Information Theory

In contrast to the classical statistics based on unbiased statistics such as maximum likelihood estimator, biased estimators, shrinkage, Bayesian theory and prior information are getting more and more emphasis in modern statistics in finance. Financial data is highly noisy and inconsistent. And this property would just become more significant as the data size grows bigger. The behavior of financial market also changes over the time. For example, the pattern of some financial instruments is completely changed by the crisis on 2009. New phenomena like the flash crash appears as new technologies are introduced to the market. Simple models fail to capture these changes, and complicated advanced models usually introduce large estimation errors. That is the reason for which the biased estimators often have a better performance than the unbiased ones. However introducing prior information naively could be dangerous. How to shrink the estimators of a distribution? What is the best Bayesian prior? What is the correct way to parameterize a model? All of which are challenging questions in practice. A tool that can address these problems is the information geometry developed by Amari and Nagaoka [3]. By linking probability distributions to differential geometry one can get a better intuition of statistical models and tests. For example, Choi and Mullhaupt [12] investigate the linear time series model on Kähler manifold and construct a Bayesian prior superior than the traditional Jeffers prior. Further researches in different financial econometrics can potentially improve the current models and statistic tests.

6.2 Combination of Machine Learning, Game Theory and Statistics

Markowitz portfolio theory is insightful; but it is clearly not the best strategy that an investor can choose. Given a prediction model and a certain investment period, the theoretical best strategy is provided by dynamic programming, which is numerically unachievable in finance. Machine learning theory provides feasible algorithms that can approximate a dynamic programming strategy. Techniques such as deep learning achieved significant success in different areas such as Chess and Go recently. However unlike the board games, financial market exhibits strong uncertainty; and the information available to each participant is incomplete. Thus machine-learning theory based on modern statistics is necessary for decision making in finance. The on-line portfolio strategies introduced in Sect. 5.2 are just simple examples of the theory. These strategies do not consider stylized facts like mean-reverting of the market, and ignore the transaction costs which are crucial in high frequency trading. Utilizing additional information and signals from the market is an open topic in this area, Li and Hoi [34].

In addition high frequency industry is highly competitive. Buying and selling assets in a short amount of time is approximately a zero-sum game, i.e. someones gain leads to someones loss. Even for the low frequency investors the high frequency traders introduce higher transaction costs that can affect on the long-term profit. Thus an investor may consider opponents actions and the impact of his strategy to the market before executing his strategy. Thus game theory may provide a deeper insight to the high frequency trading than the dynamic programming of a certain utility function.

6.3 Efficient Algorithms in Linear Algebra and Convex Optimization

Linear algebra and convex optimization are the footstones for modern data analysis and financial engineering. Any quantitative model in finance would not be practical without basic tools in linear algebra and optimization, such as matrix inversion, SVD, Cholesky decomposition, QR decomposition, eigenvalue problem, linear and quadratic programming. While most classical algorithms in linear algebra and convex optimization were well developed in the last century, the need of faster and more accurate algorithms keeps increasing as new technologies and new applications appear. First, a number of matrices in financial applications are sparse or structured. Thus

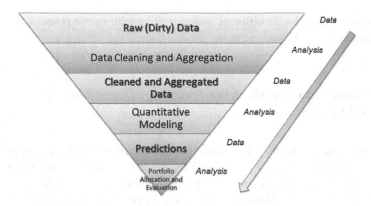

Fig. 8 Inverted pyramid structure of quantitative data analysis in finance

algorithms specificity designed for these matrices can be more efficient than these standard approaches. Second of all, the novel heterogeneous platforms including GPU and MIC [60] has further escalated the computational complexities, although they have improved the computing performance.

7 Conclusion

In this chapter we review the big data concept in quantitative finance. By considering high frequency data as an example, we introduce the basic data cleaning and aggregation approaches, quantitative modeling, portfolio allocation and strategies, which are summarized by Fig. 8.

The inverted pyramid structure illustrated the change of data size after each step. The three topics are also related to the 3V's in Big Data. First of all, raw data is voluminous. Processing and cleaning them requires efficient I/O, ranking and searching techniques. Second, we briefly introduce the typical econometric models but there exist a variety of quantitative models with different degrees of complexity. Different matrix operating and optimization algorithms are needed to deal with different types of the models. Finally, the velocity of model estimation and portfolio allocation is equally important for algorithm trading firms. Even milliseconds difference in speed could make a huge difference for some high frequency investors. However the framework in Fig. 8 is just a coarse summarization of the world of quantitative finance. More researches in market microstructure would be launched in the near future, as more types of data get involved. Appearance of the next Black Scholes theory is just a matter of time.

References

1. I. Aldridge, *High-Frequency Trading: A Practical Guide to Algorithmic Strategies and Trading Systems* (Wiley, Hoboken, 2009)
2. I. Aldridge, Trends: all finance will soon be big data finance (2015). http://www.huffingtonpost.com/irene-aldridge/trends-all-finance-will-s_b_6613138.html
3. S.-I. Amari, H. Nagaoka, *Methods of Information Geometry* (American Mathematical Society, Providence, 2007)
4. T.G. Andersen, T. Bollerslev, Intraday periodicity and volatility persistence in financial markets. J. Empir. Financ. **4**(2), 115–158 (1997)
5. T.G. Andersen, T. Bollerslev et al., Intraday and interday volatility in the Japanese stock market. J. Int. Financ. Mark. Inst. Money **10**(2), 107–130 (2000)
6. A. Beck, Y.S.A. Kim et al., Empirical analysis of ARMA-GARCH models in market risk estimation on high-frequency US data. Stud. Nonlinear Dyn. Econom. **17**(2), 167–177 (2013)
7. F. Black, M. Scholes, The pricing of options and corporate liabilities. J. Polit. Econ. **81**, 637–654 (1973)
8. T. Bollerslev, Generalized autoregressive conditional heteroskedasticity. J. Econom. **31**(3), 307–327 (1986)
9. C.T. Brownlees, G.M. Gallo, Financial econometric analysis at ultra-high frequency: data handling concerns. Comput. Stat. Data Anal. **51**(4), 2232–2245 (2006)
10. N. Cesa-Bianchi, G. Lugosi, *Prediction, Learning, and Games* (Cambridge University Press, Cambridge, 2006)
11. A. Chekhlov, S.P. Uryasev et al., Portfolio optimization with drawdown constraints. Research report 2000-5. Available at SSRN http://dx.doi.org/10.2139/ssrn.223323 (2000)
12. J. Choi, A.P. Mullhaupt, Geometric shrinkage priors for Khlerian signal filters. Entropy **17**(3), 1347–1357 (2015)
13. T.M. Cover, Universal portfolios. Math. Financ. **1**(1), 1–29 (1991)
14. T.M. Cover, E. Ordentlich, Universal portfolios with side information. IEEE Trans. Inform. Theory **42**(2), 348–363 (1996)
15. J.C. Cox, S.A. Ross, The valuation of options for alternative stochastic processes. J. Financ. Econ. **3**(1–2), 145–166 (1976)
16. J.C. Cox, S.A. Ross et al., Option pricing: a simplified approach. J. Financ. Econ. **7**(3), 229–263 (1979)
17. D.W. Diamond, R.E. Verrecchia, Constraints on short-selling and asset price adjustment to private information. J. Financ. Econ. **18**(2), 277–311 (1987)
18. X. Dong, New development on market microstructure and macrostructure: patterns of US high frequency data and a unified factor model framework. Ph.D. Dissertation, State University of New York at Stony Brook (2013)
19. D. Duffie, *Dynamic Asset Pricing Theory* (Princeton University Press, Princeton, 2010)
20. A. Dufour, R.F. Engle, Time and the price impact of a trade. J. Financ. **55**(6), 2467–2498 (2000)
21. D. Easley, M. O'hara, Time and the process of security price adjustment. J. Financ. **47**(2), 577–605 (1992)
22. R.F. Engle, Autoregressive conditional heteroscedasticity with estimates of the variance of United Kingdom inflation. Econom. J. Econom. Soc. **50**, 987–1007 (1982)
23. R.F. Engle, The econometrics of ultra-high-frequency data. Econometrica **68**(1), 1–22 (2000)
24. R.F. Engle, S. Manganelli, CAViaR: conditional autoregressive value at risk by regression quantiles. J. Bus. Econ. Stat. **22**(4), 367–381 (2004)
25. R.F. Engle, J.R. Russell, Autoregressive conditional duration: a new model for irregularly spaced transaction data. Econometrica **66**, 1127–1162 (1998)
26. B. Fang, P. Zhang, in *Big Data in Finance. Big Data Concepts, Theories, and Applications*, ed. by S. Yu, S. Guo (Springer International Publishing, Cham, 2016), pp. 391–412
27. R. Gençay, M. Dacorogna et al., *An Introduction to High-Frequency Finance* (Academic Press, San Diego, 2001)

28. L. Györfi, I. Vajda, Growth optimal investment with transaction costs. *Algorithmic Learning Theory* (Springer, Berlin, 2008)
29. J.M. Harrison, D.M. Kreps, Martingales and arbitrage in multiperiod securities markets. J. Econ. Theory **20**(3), 381–408 (1979)
30. D.P. Helmbold, R.E. Schapire et al., On-line portfolio selection using multiplicative updates. Math. Financ. **8**(4), 325–347 (1998)
31. T. Jia, Algorithms and structures for covariance estimates with application to finance. Ph.D. Dissertation, State University of New York at Stony Brook (2013)
32. Y.S. Kim, Multivariate tempered stable model with long-range dependence and time-varying volatility. Front. Appl. Math. Stat. **1**, 1 (2015)
33. O. Ledoit, M. Wolf, Improved estimation of the covariance matrix of stock returns with an application to portfolio selection. J. Empir. Financ. **10**(5), 603–621 (2003)
34. B. Li, S.C. Hoi, Online portfolio selection: a survey. ACM Comput. Surv. (CSUR) **46**(3), 35 (2014)
35. J. Lintner, The valuation of risk assets and the selection of risky investments in stock portfolios and capital budgets. Rev. Econ. Stat. **47**, 13–37 (1965)
36. C. Liu, D.B. Rubin, The ECME algorithm: a simple extension of EM and ECM with faster monotone convergence. Biometrika **81**(4), 633–648 (1994)
37. H. Markowitz, Portfolio selection. J. Financ. **7**(1), 77–91 (1952)
38. S.L. Marple Jr., *Digital Spectral Analysis with Applications* (Prentice-Hall, Inc, Englewood Cliffs, 1987)
39. Y. Matsuyama, The alpha-EM algorithm: surrogate likelihood maximization using alpha-logarithmic information measures. IEEE Trans. Inform. Theory **49**(3), 692–706 (2003)
40. A.J. McNeil, R. Frey et al., *Quantitative Risk Management: Concepts, Techniques and Tools* (Princeton University Press, Princeton, 2005)
41. X.-L. Meng, D.B. Rubin, Maximum likelihood estimation via the ECM algorithm: a general framework. Biometrika **80**(2), 267–278 (1993)
42. R.C. Merton, Lifetime portfolio selection under uncertainty: the continuous-time case. Rev. Econ. Stat. **51**, 247–257 (1969)
43. A. Meucci, 'P' Versus 'Q': differences and commonalities between the two areas of quantitative finance. GARP Risk Prof., 47–50 (2011)
44. A.M. Mineo, F. Romito, A method to 'clean up' ultra high-frequency data, Vita e pensiero (2007)
45. A.M. Mineo, F. Romito, Different methods to clean up ultra high-frequency data. Atti della XLIV Riunione Scientifica della Societa'Italiana di Statistica (2008)
46. J. Mossin, Equilibrium in a capital asset market. Econom.: J. Econom. Soc. **34**, 768–783 (1966)
47. A.P. Mullhaupt, K.S. Riedel, Band matrix representation of triangular input balanced form. IEEE Trans. Autom. Control (1998)
48. R.M. Neal, G.E. Hinton, A view of the EM algorithm that justifies incremental, sparse, and other variants, *Learning in Graphical Models* (Springer, New York, 1998), pp. 355–368
49. J. Nocedal, S. Wright, *Numerical Optimization* (Springer Science and Business Media, New York, 2006)
50. S.T. Rachev, S. Mittnik et al., *Financial Econometrics: From Basics to Advanced Modeling Techniques* (Wiley, New York, 2007)
51. R.T. Rockafellar, S. Uryasev, Optimization of conditional value-at-risk. J. Risk **2**, 21–42 (2000)
52. D.B. Rubin, D.T. Thayer, EM algorithms for ML factor analysis. Psychometrika **47**(1), 69–76 (1982)
53. J.R. Russell, R. Engle et al., Analysis of high-frequency data. Handb. Financ. Econom. **1**, 383–426 (2009)
54. W.F. Sharpe, Capital asset prices: a theory of market equilibrium under conditions of risk. J. Financ. **19**(3), 425–442 (1964)
55. X. Shi, A. Kim, Coherent risk measure and normal mixture distributions with application in portfolio optimization and risk allocation (2015). Available at SSRN http://dx.doi.org/10.2139/ssrn.2548057

56. W. Sun, S.Z. Rachev et al., Long-range dependence, fractal processes, and intra-daily data, *Handbook on Information Technology in Finance* (Springer, New York, 2008), pp. 543–585
57. S. Tomov, R. Nath et al., Dense linear algebra solvers for multicore with GPU accelerators, in *IEEE International Symposium on Parallel and Distributed Processing, Workshops and PhD Forum (IPDPSW)* (IEEE, 2010)
58. J.L. Treynor, Toward a theory of market value of risky assets. Available at SSRN (1961). doi:10.2139/ssrn.628187
59. Y. Yan, Introduction to TAQ. WRDS Users Conference Presentation (2007)
60. P. Zhang, Y. Gao, Matrix multiplication on high-density multi-GPU architectures: theoretical and experimental investigations, in *High Performance Computing: 30th International Conference, ISC High Performance 2015, Frankfurt, Germany, 12–16 July 2015, Proceedings*, ed. by M.J. Kunkel, T. Ludwig (Springer International Publishing, Cham, 2015), pp. 17–30
61. P. Zhang, Y. Gao et al., A data-oriented method for scheduling dependent tasks on high-density multi-GPU systems, in *IEEE 17th International Conference on High Performance Computing and Communications (HPCC), IEEE 7th International Symposium on Cyberspace Safety and Security (CSS), IEEE 12th International Conference on Embedded Software and Systems (ICESS)* New York, NY, 2015, pp. 694–699
62. P. Zhang, L. Liu et al., A data-driven paradigm for mapping problems. Parallel Comput. **48**, 108–124 (2015)
63. P. Zhang, K. Yu et al., QuantCloud: big data infrastructure for quantitative finance on the cloud. IEEE Trans. Big Data (2016)

Emerging Cost Effective Big Data Architectures

K. Ashwin Kumar

Abstract Volume, velocity and variety of data is increasing at an unprecedented rate. There is a growing consensus that a single system cannot cater to the variety of workloads and real world datasets. As such, different solutions are being researched and developed catering for requirements of different applications. For example, column-stores are optimized specifically for data warehousing applications, whereas row-stores are better suited for transactional workloads. There are also hybrid systems for applications that need support for both transactional workloads and data analytics. Other varied systems are being designed and built to store different types of data, such as document data stores for storing XML or JSON documents, and graph databases for graph-structured or RDF data. Most of these systems focus on minimization of execution time or performance improvement and often ignore optimization of overall cost of data management. A more holistic view of the cost of data management includes energy consumption, and utilization of compute, memory and storage resources which attribute to the cost of data processing especially in a cloud-based pay-as-you-go environments. In this chapter, we discuss a new area of emerging Big Data Architectures that aim at minimization of overall cost of data storage, querying and analysis, while improving performance. We first provide a motivation for the overall problem, with appropriate related work. We then discuss the state-of-the-art and provide key case studies of the emerging cost effective big data architectures that have been recently designed and built with the above mentioned goals in mind. Finally, we enumerate key future directions and conclude.

1 Introduction

The problem of Big Data deals with the exponential increase in the rates of volume, velocity and variety of data. In other words, for organizations with big data projects, data is arriving at an unprecedented speed and volumes, and it has to be managed in a timely fashion—sometimes in real time. They often find it challenging

K. Ashwin Kumar (✉)
Veritas Technologies LLC, Mountain View, CA, USA
e-mail: ashwin.kayyoor@veritas.com

© Springer International Publishing AG 2017
A.Y. Zomaya and S. Sakr (eds.), *Handbook of Big Data Technologies*,
DOI 10.1007/978-3-319-49340-4_22

755

to cope with high-velocity data from internet of things. Also, increasingly, data is arriving in a myriad of formats, from structured ones like traditional databases to unstructured formats like text documents, emails, and videos. Merging, managing, and searching different types of data is one of the challenges associated with big data projects. There is a growing consensus that a single system cannot cater to the variety of workloads and real world datasets. As such, different solutions are being researched and developed catering for requirements of different applications. For example, column-stores [34, 57] are optimized specifically for data warehousing applications, whereas row-stores are better suited for transactional workloads. There are also hybrid systems [24] for applications that need support for both transactional workloads and data analytics. Other varied systems are being designed and built to store different types of data, such as document data stores for storing XML or JSON documents [10], and graph databases for graph-structured or RDF data.

To handle the increasing volumes of data, two solutions are commonly considered. One approach is to use a sufficiently powerful machine that can handle the workload (called the *scale-up approach*), whereas the other approach is to use a cluster of commodity machines to parallelize the compute tasks (*scale-out approach*). The scale-up approach is attractive because it is significantly easier to code for, whereas the scale-out approach requires one to deal with the distributed nature of computation as well as complexities involved in guaranteeing distributed fault-tolerance. However, the scale-up approach is limited in its ability to scale to large volumes of data, and also is typically more expensive. There has been much work on the scale-out approach over the last decade – several high-level programming frameworks and abstractions have been proposed, and numerous systems have been developed for supporting those frameworks over a large number of machines. Popular among these are MapReduce paradigm [12] introduced by Google that makes use of large number of cheap commodity machines to arbitrarily scale the humongous amounts of data. Since, MapReduce is not optimized for iterative algorithms, Spark [69] has been introduced that is specifically optimized for distributed machine learning algorithms (mostly iterative) offers performance up to 10 times faster than previous generation systems like Hadoop MapReduce. On the other hand, to handle and process high-speed streams of data, distributed real-time computation systems such as Storm [60], Apache Samza [49] and Apache Flink [15] has been developed for processing fast, large streams of data. In databases world, variety of database systems (both SQL and NoSQL) [25, 33, 34, 56] have been proposed to process different types of Big Data workloads.

Majority of the systems today focus on minimization of execution time and often ignore optimization of overall cost of data management. Similar to the *performance* that has been central to systems evaluation, energy-efficiency (e.g., completed tasks per unit of energy, or Queries per Joule) is quickly growing in importance for minimizing overall IT costs [62]. A more holistic view of the data management cost includes energy consumption, and utilization of compute, memory and storage resources which attribute to the cost of data processing especially in a cloud-based pay-as-you-go environments.

Energy Efficiency: Fundamentally, energy can be defined as the capacity or power to do work, such as the capacity to move an object of a given mass by the application of force. Energy can exist in a variety of forms, such as electrical, mechanical, chemical, thermal, or nuclear, and can be transformed from one form to another. *Energy efficiency (EE)* can be defined as the ratio of performance to power, in other words, it is the ratio of useful work done to the energy used [62]:

$$EE = \frac{Performance}{Power} = \frac{Work}{Power \times Time} = \frac{Work}{Energy} \tag{1}$$

To understand the importance of this metric, we need to understand the seriousness of the problem resulting from ignoring this metric. Data centers are the backbone of the modern economy – from the server rooms that power small- to medium-sized organizations to the enterprise data centers that support American corporations and the server farms that run cloud computing services hosted by Amazon, Facebook, Google, and others. However, the explosion of digital content, big data, e-commerce, and Internet traffic is also making data centers one of the fastest-growing consumers of electricity in developed countries, and one of the key drivers in the construction of new power plants.

According to a report [4] on data center energy efficiency from the Natural Resources Defense Council (NRDC), an environmental action organization, *"in 2013, U.S. data centers consumed an estimated 91 billion kilowatt-hours of electricity, equivalent to the annual output of 34 large (500-MW) coal-fired power plants. Data center electricity consumption is projected to increase to roughly 140 billion kilowatt-hours annually by 2020, the equivalent annual output of 50 power plants, costing American businesses $13 billion annually in electricity bills and emitting nearly 100 million metric tons of carbon pollution per year"*. Hence, it is critical to save energy by reducing the amount of electricity consumed while increasing the amount of work done within a given time.

There are three key ways to handle the problem of energy consumption, (1) hardware-based, (2) software-based and (3) infrastructural-based. Lets understand each of them:

Hardware-based: One way of tackling the problem of reducing the overall energy consumption in the data centers is by employing more energy efficient hardware solutions. Initial solutions included CPUs with dynamic voltage and frequency settings [43] where processors automatically switch to lower power states when idle. Recently there have been surge in the manufacturing of energy efficient hardware solutions. More comprehensively, today industries are manufacturing compute servers that are ENERGY STAR [14] certified. According to the EPA, computer servers that earn the ENERGY STAR designation will, on average, be 30% more energy efficient than standard servers. The agency also predicts that if all servers sold in the United States were to meet ENERGY STAR specifications, energy cost savings would approach $800 million per year and prevent greenhouse gas emissions equivalent to those from over one million vehicles. On the other hand, because of heterogeneity in the hardware choices, choosing a right variety of hardware for a particular problem can be critical in achieving best energy efficiency. Malik et al. [42] analyze the measurements of performance and power of Big Data applications on two state-of-the-art server platforms, one with $Intel^{TM}$ Xeon; Big cores and the other with $Intel^{TM} Atom$; Little cores. They conclude that that low power embedded core is noticeably more efficient for big data processing across various data sizes. There have been similar studies that make the case for choice of low power embedded cores to improve the energy efficiency of traditional server applications [5, 19, 46]. Similarly, on the storage front, for read-only workloads SSDs are significantly energy efficient than HDDs, for write intensive workloads its vice versa [52].

Software-based: As the hardware choices increase, improved algorithms are needed to make judicious use of underlying hardware to achieve overall energy efficiency. In general, there are three categories of solutions in this area, (a) Algorithms that take advantage of dynamic speed scaling in variable-speed processors. It is common for modern microprocessors to run at variable speed. High speeds result in higher performance but also high energy consumption. Lower speeds save energy but performance degrades. Albers et al. [3] study the problem of scheduling a set of jobs, each specified by a release time, a deadline and a processing volume, on variable speed processors so as to minimize the total energy consumption. (b) power-down mechanisms that conserve energy by shutting down under-utilized server nodes, or transitioning a device into low-power standby or sleep modes. Khuller et al. [26] discuss an algorithm to schedule collection of jobs on set of unrelated machines with a need to decide which subset of machines to activate, while other machines are either shutdown or in low-power standby. Objective is to assign all jobs to machines such that the time when all jobs are complete is minimized. (c) Interestingly, designing algorithms for performance may not be same as designing algorithms for energy efficiency. So, third category consists of optimization algorithms that trade performance with energy. Xu et al. finds opportunity to trade power to performance in current DBMSs and proposes a power aware query optimizer to enable power conservation [66–68]. On the other hand, similar to Xu et al., Lang et al. [35] present a new design for query optimizer that can pick an "energy-enhanced" query plan that also meets any existing response time targets (service level agreements).

Infrastructural: Apart from hardware and software choices, infrastructural choices can also play a key role in improving energy efficiency. One of the dominant costs for data center energy consumption is related to data center cooling. More precisely, cooling consumes 37% of electricity usage. Industries currently employ different strategies for increasing data center cooling efficiency such as, bringing cooling closer to the source of heat essentially reducing the energy required for air movement, placing data centers in the relatively cooler locations, employing cutting edge cooling infrastructures such as variable capacity systems and improved controls.

In this chapter, we will focus mainly on software approaches to cost effective Big Data architectures. We will use the terms "*cost effectiveness*" and "*energy efficiency*" interchangeably as appropriate.

2 Emerging Solutions for Big Data

Issues in energy-efficient computing are being increasingly studied at all layers in today's computing infrastructures. Harizopoulos et al. [62] reported the first results on software-level optimizations to achieve better energy efficiency; they experiment with a system that was configured similarly to an audited TPC-H server and show that making the right physical design decisions can improve energy efficiency. Additionally, they use relational scan operator as a basis to demonstrate that optimizing for performance is different from optimizing for energy efficiency. It is also among the first papers [18, 36, 62] to practically show the importance of energy efficiency in database systems. Graefe [20] also points out various research challenges and promising approaches in energy-efficient database management. In his paper he indicates various promising approaches and techniques to achieve energy efficiency in database systems. He discusses two approaches in this context: processor frequency control and explicit delays.

Overall, in order to minimize the cost of Big Data systems, energy efficiency has to be studied in two key cases, single-node energy efficiency and multi-node energy efficiency. In the case of single-node data systems, objective of maximizing performance and energy efficiency go hand in hand where increasing the performance often improves the energy efficiency as well. Tsirogiannis et al. [62] analyze the energy efficiency of a single-node database server, and argue that the most energy efficient configuration is typically the highest performing one. However, this assertion is valid only for single node database server, and does not hold for scale-out architectures involving multiple machines where parallelization, communication, and startup overheads come into play.

On the other hand, because of the complexity involved in the scale-out architectures, most straightforward and common way of improving energy efficiency or reducing the overall cost is to power down the under utilized machines in a scale-out cluster. Leverich et al. [38] and Lang et al. [37] suggest approaches to conserving energy by powering down Hadoop cluster nodes during periods of low load, and observe that the default replica placement policy is highly inefficient in this regard. In particular, they observe that powering down any three nodes is likely to lead to some data being unavailable, and instead suggest a replication policy such that a small set of cluster nodes cover (contain) at least one replica of each data item. Lang et al. [37] suggest and evaluate an alternative approach where all cluster nodes are powered up (to answer queries), and powered down at the same time, and show that their approach leads to better energy utilization.

In this section, we will study three emerging directions for cost effective Big Data architectures with relatively detailed example systems.

2.1 Workload-Aware Solutions

As the cloud computing industry is maturing, the workloads are becoming stable and somewhat predictable, opening up the possibility of monitoring, capturing, and exploiting workload information to optimize for resource consumption, energy efficiency (or other metrics of interest). Such workload-aware approaches have been shown to be highly successful in making many types of system design decisions in the past work [7]. Recently, Sharov et al. [53] developed a workload-aware optimization framework that dynamically and automatically determines the optimal configuration for leader and quorum based distributed storage systems. Goal of this system was to minimize the overall latency of running jobs and not cost. In the context of cost effective execution of Big Data analytics in a geo-distributed data setup, Vulimiri et al. [63] have proposed a system that can judiciously orchestrate distributed query execution and adjusting data replication across data centers in order to minimize bandwidth usage and reduce overall data transfer costs. On the other hand, Wu et al. [65] propose a novel system SPANStore, which is essentially a cost-effective geo-replicated storage spanning multiple cloud services such as Amazon, Microsoft and Google's cloud services. By understanding the given workload and by exploiting the pricing discrepancies across providers, SPANStore minimizes the overall cost (in terms of price). It also provides an automated way to trade cost with latency by judiciously determining replication policies based on workload properties, and by minimizing the use of compute resources.

Case Study: Minimizing Average Job Span

In this case study, we will discuss the workload-aware approach of minimizing overall cost of Big Data (typically scale-out based) analytics which involves minimizing overall resource consumption while improving performance. To begin with let us take a look at an example of resource inefficiency in scale-out architectures. Consider a job that takes 100 s to execute on a single machine and consumes 100 J of energy. Now consider a situation where the data corresponding to the same job is spread equally on to two machines and job is allowed to execute parallelly on these two machines.

Ideally, this job should finish its execution in exactly half of the time (i.e., 50 s) on each machine and should consume 50 J energy on each machine. But in practice as a result of several overheads and process startup costs the job executes in >50 s time and consumes >50 J of energy on each machine. In summary, a job executing on multiple machines can consume more energy when compared to the same job executing on relatively fewer number of machines. In other words, in the absence of super-linear speedups, more the number of machines a job or a job touches, more energy it consumes. In order to minimize resource consumption of given job, we should minimize the number of machines required for a job for its execution. Also, it can be observed that, for fault tolerance, load balancing, and availability, scale-out based Big Data systems typically maintain several copies of each data item (e.g., Hadoop file system (HDFS) maintains at least 3 copies of each data item by default [64]), and this inherent replication can be exploited to achieve higher colocation by judicious replica creation and placement.

Job Span: A key metric to optimize For a given job or analysis task, its span is defined to be the minimum number of machines that contain the data needed to execute that job or task. Best case is when the data required by a job fits in the memory of a single machine, then a job can be executed on a single machine as efficiently as possible (*job span* = 1). This results in the highest resource efficiency of the underlying system. Minimizing job span has significant advantages that make it an important metric for which to optimize.

• *Minimize the communication overhead*: job span directly impacts the total communication that must be performed to execute a job. This is clearly a concern in distributed setups (e.g., grid systems [58] or multi-datacenter deployments); however even within a datacenter, communication network is oversubscribed, and especially cross-rack communication bandwidth can be a bottleneck [11, 22]. In cloud computing, the total communication directly impacts the total dollar cost of executing a job.

• *Minimize the total amount of resources consumed*: It is well-known that parallelism comes with significant startup and coordination overheads, and we typically see sublinear speedups as a result of these overheads and data skew [44]. Although the response time of a job usually decreases in a parallel setting, the total amount of resources consumed typically increases with increased parallelism. Even in scenarios where we obtain super-linear speedups due to higher aggregate memory across the machines, we expect the total resource consumption to increase with the degree of parallelism.

• *Reduce the energy footprint*: Computing equipment in US costs datacenter operators millions of dollars annually for energy, and also impacts the environment. Energy costs are ever increasing and hardware costs are decreasing – as a result soon the energy costs to operate and cool a datacenter may exceed the cost of the hardware itself. Minimizing the total amount of resources consumed directly reduces the total energy consumption of a task.

In scale-out settings, an effective way to minimize the job span is to co-locate the data items required by the queries on fewer machines. With colocation, data items needed for a particular job or a job can be found in lesser number of machines, consuming lesser resources often also improving execution times. In order to perform data colocation certain information about data access patterns is required. For example, if we know that the data items d_1 and d_2 are being frequently accessed by the queries, then these data items can be colocated and placed in single partition. In other words, we want to understand the history of data access and partition the data items such that frequently co-accessed data items are placed together and when queries access these data items then job span is minimized. Figure 1 gives the detailed pictorial example where it shows that workload-aware smart partitioning and replication minimizes overall job span.

Problem definition We are given a set of data items $\mathcal{D}$ and their sizes – the data items may be files, database relations, vertical or horizontal partitions of database relations,

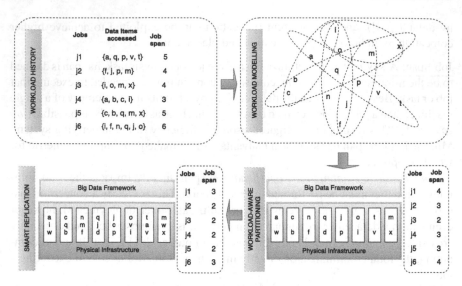

Fig. 1 This figure depicts an overall idea of workload-aware data partitioning and replication to minimize job spans using an example

or tuples. We are also given a set of partitions with associated storage capacities, and an expected job workload in the form of a set of queries over the data items. The queries may be read-only queries, or update transactions. Our goal is to decide which data items to replicate and how to place them on to the partitions so as to minimize the averageshould job span for the queries in the workload. In addition, we may be given a constraint that specifies how much each item should be minimally replicated for fault tolerance and availability.

Modeling workload history Intuitive way to capture or represent a job workload is using hypergraph, $\mathcal{H} = (V, E)$, where the nodes are the data items and each (hyper)edge $e \in E$ corresponds to a job in the workload. Each hyperedge is associated with an edge weight w_e, which represents the frequency of such queries in the workload. Each vertex $v \in V$ is associated with a weight w_v representing either data size or access frequency or a combination of both providing balancing in terms of size and load on each partition. In Fig. 1, block explaining "workload modelling" shows the recording and managing of workload history in terms of list of jobs and data items they have accessed for their execution, then modeling workload history in the form of hypergraph.

Calculating span When there is no replication, calculating the span of a job is straightforward since each data item is associated with a single partition. However, if there is replication, the problem becomes NP-Hard in general. In the simplest case, for read-only queries with strongly consistent replicas (i.e., if all replicas of a data item are kept up-to-date when it is updated), it is identical to the minimum set cover problem [16] – where we are given a collection of subsets of a set (in our case, the partitions) and a job subset, and we are asked to find the minimum number of subsets (partitions) required to cover the job subset.

Hypergraph partitioning Without replication, the problem we defined above is essentially the k-way (balanced) hypergraph partitioning problem that has been very well-studied in the literature. However, the optimization goal of minimizing the average span is unique to this setting; prior work has typically studied how to minimize the number of cut hyperedges instead.

Finding dense subgraphs of a specified size Given a set of nodes S in a graph, the density of the subgraph induced by S is defined to be the ratio of the number of edges in the induced subgraph and $|S|$. The dense subgraph problem is to find the densest subgraph of a given size. To understand the connection to the dense subgraph problem, consider a scenario where we have exactly one "extra" partition for replicating the data items (i.e., $N_e = N - 1$). Further, assume that each job refers to exactly two data items, i.e., the hypergraph $\mathcal{H}$ is just a graph. One approach would then be to first partition the data items into $N - 1$ partitions without replication, and then try to use this extra partition optimally. To do this, we can construct a residual graph, which contains all edges that were cut in this partitioning. The spans of the queries corresponding to these edges is exactly 2. Now, we find the subgraph of size C such that the number of induced edges (among the nodes of the subgraph) is maximized, and we place these data items on the extra partition. The spans of the queries corresponding to these edges are reduced from 2 to 1, and hence this is an optimal way to utilize the extra partition. We can generalize this intuition to hypergraphs and this forms the basis of one of our algorithms. Unfortunately, the problem of finding the most dense subgraph of a specified size is NP-Hard (with no good worst case approximation guarantees), so we have to resort to heuristics. One such heuristic that we adapt in our work is as follows: recursively remove the lowest degree node from the residual graph (and all its incident edges) till the size of the residual graph is exactly C. This heuristic has been analysed by Asahiro et al. [9] who find that this simple greedy algorithm can solve this problem with approximation ratio of approximately $2(\frac{|V|}{C} - 1)$ (when $C \leq |V|/3$).

Smart data placement and replication algorithms Kumar et al. [29, 31] present several algorithms for data placement with replication, with the goal to minimize the average job span. Homogenous scale-out setup is assumed where each machine or a partition has capacity C units. These algorithms are based on existing standard hypergraph partitioning algorithms (denoted HPA) and focus on data placement and smart replication of data items appropriately to reduce the average job span. An HPA algorithm typically tries to find a balanced partitioning (i.e., all partitions are of approximately equal size) that minimizes some optimization goal. Usually, allowing for unbalanced partitions results in better partitioning.

One such algorithm for data placement and replication to minimize the average job spans is known as *local move based replication (LMBR)* algorithm. In this algorithm key idea is to identify small groups of data items that can copied from one partition to another such that we maximize the overall benefit, which is essentially minimization of job span. Algorithm starts with a partitioning returned by HPA. Then algorithm strategically chooses two partitions at a time, more specifically, at each step, we copy a small group of data items from one partition to another. The decisions are made greedily by finding the move that results in the highest decrease in the average job span ("benefit") per data item copied ("cost"). For this purpose, at all times, a priority queue is maintained containing the best moves from $partition_i$ to $partition_j$, for all $i \neq j$. There are clearly many other variations of these algorithms, some of which may work better for some inputs, that can be implemented quickly and efficiently. In practice, taking the best of the solutions produced by running several of these algorithms would guarantee good data placements. The selection of data placement algorithm should primarily be based on the requirements of the application scenario at hand and the granularity of data items. On the other hand, in analytical workloads where the data items are relations or files, LMBR may be more appropriate given it usually results in best colocation. In Fig. 1, blocks explaining "workload-aware data partitioning" and "smart replication" shows that the modeled hypergraph can be min-cut partitioned to perform workload-aware partitioning over cluster of storage nodes such that average job spans can be reduced. Finally, by analyzing the workload and workload-aware partitioning of the data, smart replication is performed to further minimize job spans as well as provide fault-tolerance and data availability. For example in the shown figure: job span for the job j_6 has reduced from 6 to 3 with workload-aware partitioning and smart replication.

Smart data placement and replication to minimize average job spans of the given workload can benefit different types of Big Data stores.

1. **Distributed analytical data stores**: Analytical workloads typically consist of complex distributed join queries. More the number of machines these queries touch for their execution worse the query response time gets. With LMBR-suggested placement (smart data placement and replication) query response times for complex analytical join queries decrease significantly when executed on LMBR-suggested placement. This is because of minimization of overheads caused by distributed analytical processing, e.g., communication overheads in processing complex joins. Colocation of data items and minimization of query spans also reduces the energy consumption significantly both because of better utilization of resources and reduction in distributed overheads.

2. **Distributed transactional data stores**: On the other hand, consider the problem of transparently scaling out transactional (OLTP) workloads on relational data-bases, to support database-as-a-service in cloud computing environment. The primary challenges in supporting such workloads include choosing how to partition the data across a large number of machines, minimizing the number of distributed transactions, providing high data availability, and tolerating failures gracefully. Capturing and modeling the transactional workload over a period of time, and then exploiting that information for data placement and replication has been shown to provide significant benefits in performance, both in terms of transaction latencies and overall throughput. However, such workload-aware data placement approaches can incur very high overheads, and further, may perform worse than naive approaches if the workload changes. Quamar et al. [45] propose a scalable workload-aware data partitioning and placement approach for OLTP

workloads, that incorporates a suite of novel techniques to significantly reduce the overheads incurred both during the initial placement, and during query execution at runtime. Query workload is modelled as a hypergraph over the data items, and propose using a hypergraph compression technique to reduce the overheads of partitioning. To deal with workload changes, an incremental data repartitioning technique is proposed that modifies data placement in small steps without resorting to complete workload repartitioning. A workload-aware active replication mechanism is employed to increase availability and enable load balancing. The use of fine-grained quorums is defined at the level of groups of tuples to control the cost of distributed updates, improve throughput, and provide adaptability to different workloads. Overall, in the case of transactional database workloads it has been shown that workload-aware data partitioning and replication can improve the transactional throughput significantly by reducing the the number of distributed transactions.

3. **Distributed information retrieval systems**: Most prior research on web information retrieval assumes that documents are already assigned to physical partitions [6, 8, 55, 59]. Simplest way of assigning documents to partitions is to distribute these documents randomly across the partitions. Query is sent to all the partitions to retrieve the documents of interest. This approach of distributed exhaustive search increases the query cost significantly [27]. There have been few other studies that have looked into partitioning of a document collection into topical clusters. These studies have shown that search efficiency can be further increased by document cluster selection, that is, by querying a small number of promising document clusters for each query. Kulkarni et al. [27] present the topic-based clustering and partitioning of documents into distributed index or document clusters and show that this approach reduces the search cost significantly when compared to the exhaustive search with no loss of accuracy, on average.

To summarize, workloads will ultimately dictate the resource requirements and performance demands. In order to minimize the overall cost of Big Data analytics, there is a pressing need for Big Data systems to employ workload-aware techniques to optimize dual objectives such as improving performance as well as minimizing the resources consumed. Efforts have already begin in this direction (cost effectiveness), but further impetus is required from larger open source and commercial software community to make *cost effectiveness* or *energy efficiency* to be the key metric to optimize for emerging Big Data systems.

2.2 Scaling-Down Big Data Systems

In the era of Big Data, it has become norm to throw more and more hardware resources in order to solve Big Data problems. Big Data systems today are designed in a way that they can scale linearly with increasing number of machines and application needs. Often, over provisioning or under utilization of the already provisioned servers/machines can lead to massive energy in-efficiencies, increasing the overall cost of data processing. One way to improve the overall energy efficiency is through consolidation of workloads in fewer number of machines, so that all the available machines are fully utilized. Underlying system achieves highest energy efficiency when the data required by a job fits in single machine's memory and the job can be executed on a single machine as fast as possible. In similar essence, Gharaibeh et al. [17] demonstrated that graph processing on single machine using GPUs and CPUs can be more energy efficient than scale-out solutions. More specifically, they show that GPU-acceleration improves both time-to-solution as well as energy consumption for large-scale graph processing, and that this improvement scales when increasing the graph size and adding more GPUs. Further, even within single machine, although the GPUs have one order of magnitude less memory, a hybrid (one CPU and one GPU) system can be more power efficient than a dual-CPU symmetric one.

Scaled-down or single machine solutions, in addition to being cost effective, they often outperform scale-out solutions by eliminating unnecessary distributed overheads. GraphChi [32] is one such popular example of scaled-down Big Data system of GraphLab. GraphChi essentially is a disk-based system for computing efficiently on graphs with billions of edges. By using a well-known method to break large graphs into small parts, and a novel parallel sliding windows method, GraphChi is able to execute several advanced data mining, graph mining, and machine learning algorithms on very large graphs, using just a single consumer-level computer. Hence, because it is extremely efficient and scales well even on single machine, GraphChi can be cost effective than its scale-out counterparts such as Spark, Hadoop, GraphLab. Cost effectiveness of GraphChi can be deduced from Table 2 in [32] by taking into account the amount of resources consumed and execution time by GraphChi when compared to its other scale-out counterparts.

Also, the assumption behind current Big Data systems and the need for distributed processing is that the data to be analyzed cannot be held in memory on a single machine. Today, this assumption needs to be re-evaluated. Although it is true that petabyte-scale datastores are becoming increasingly common, it is unclear whether datasets used in "typical" analytics tasks today are really too large to fit in memory on a single server. Of course, organizations such as Yahoo, Facebook, and Twitter routinely run Pig or Hive jobs that scan terabytes of log data, but these organizations should be considered outliers—they are not representative of data analytics in most enterprise or academic settings. Even still, according to the analysis of Rowstron et al. [48], at least two analytics production clusters (at Microsoft and Yahoo) have median job input sizes under 14 GB and 90% of jobs on a Facebook cluster have input sizes under 100 GB. Holding all data in memory does not seem too far-fetched.

There is one additional issue to consider: over the past several years, the sophistication of data analytics has grown substantially. Whereas yesterday the community was focused on relatively simple tasks such as natural joins and aggregations, there is an increasing trend toward data mining and machine learning. These algorithms usually operate on more refined, and hence, smaller datasets—typically in the range of tens of gigabytes. These factors suggest that it is worthwhile to consider in-memory data analytics on modern servers—but it still leaves open the question of how we orchestrate computations on multi-core, shared-memory machines. Should we go back to multi-threaded programming? That seems like a step backwards because we embraced the simplicity of current Big Data programming paradigms such as MapReduce for good reason—the complexity of concurrent programming with threads is well known. There is a need to scale down Big Data systems to run on shared-memory machines for small to medium sized datasets to improve energy efficiency and reduce the overall cost of analytics.

Case Study: Scaling-Down Hadoop on Shared Memory Systems

In this case study, we will understand about a new system called HONE [30] that attempts to scale down Hadoop on to shared memory systems. We begin by discussing why Hadoop does not perform well on a single machine. To take advantage of multi-core architectures, Hadoop provides pseudo-distributed mode (PDM henceforth), in which all daemon processes run on a single machine (on multiple cores). This serves as a natural point of comparison, and we identify several disadvantages of running Hadoop PDM. (1) *Multi-process overhead*: In PDM, mapper and reducer tasks occupy separate JVM processes. In general, multi-process applications suffer from interprocess communication (IPC) overhead and are typically less efficient than an equivalent multi-threaded implementation that runs in a single process space.break (2) *I/O overhead*: Another disadvantage of Hadoop PDM is the overhead associated with reading from and writing to HDFS. Disk I/O operations using HDFS can be extremely expensive [13, 41] when compared to direct disk access. In Hadoop PDM, mappers read from HDFS and reducers write to HDFS, even though the system is running on a single machine. Thus, Hadoop PDM suffers from these HDFS performance issues. (3) *Framework overhead*: Hadoop is designed for high-throughput processing of massive amounts of data on potentially very large clusters. In this context, startup costs are amortized over long-running jobs and thus do not have a large impact on overall performance. Hadoop PDM inherits this design, and in the context of a single machine running on modest input data sizes, job startup costs become a substantial portion of overall execution time.

HONE attempts to be Hadoop API compatible, so that one can run existing Hadoop jobs on HONE without modification. We describe the Hone system architecture:

Map stage: Analogous to Hadoop, this stage applies the map function on the input dataset to emit intermediate (key, value) pairs. Each mapper is handled by a separate thread, which consumes the supplied InputSplit and processes input records according to the user-specified InputFormat. As with Hadoop, the total number of mappers is determined by the number of input splits. This stage uses a standard thread-pooling technique to control the number of parallel mapper tasks. Mappers in Hone accept input either from disk or from a 'namespace' residing in memory (more below).

Sort stage: Hone sorts intermediate (key, value) pairs emitted by the mappers per the standard contract defined by the MapReduce model. Sorting is handled by a separate thread pool with a builtin load balancer. If the sort streams grow too large then an automatic splitter determines the optimal split size, efficiently splits the streams on the fly, and performs parallel sorting on the split streams. This splitting information is passed on to the reduce stage so that proper stream assignment is performed on the reducers. One can also specify the stream split size in the configuration.

Reduce stage: A reducer in Hone applies a reduce function on intermediate (key, value) pairs emitted by mappers. Depending on how mappers interact with reducers as discussed below, reducers may have to apply the partitioning function on the key to gather the appropriate (key, value) pairs. A reducer either writes output to disk or it can store output in memory for further iterative processing.

MapReduce interaction module: Running MapReduce on multicore shared memory machines creates interesting possibilities in the way mappers interact with reducers. For example, one option is that each mapper emits intermediate (key, value) pairs in a corresponding output stream, and each reducer iterates through the stream corresponding to each mapper and ingests (key, value) pairs as determined by the partitioner. In another possibility, each mapper emits intermediate pairs into r (number of reducers) streams by applying the partitioner to every emitted (key, value) pair. In this case, each reducer only needs to access a single stream. Hone provides a flexible way of controlling these interactions through a user-specified option, by providing three different MapReduce interaction models: pull-based, push-based and hybrid. In the pull-based approach, each mapper emits keys into r streams, where r is the number of reducers. Each mapper applies the partitioning function to assign (key, value) pairs to one of the r corresponding streams. If m is the total number of mappers then there will be a total of $m \times r$ intermediate streams. In the sort stage, these $m \times r$ intermediate streams are sorted in parallel. In the reduce stage, each reducer thread "pulls" the appropriate (key, value) pairs from the appropriate r streams. In the push-based approach, there are only r intermediate streams, each corresponding to a reducer. Each mapper emits intermediate (key, value) pairs directly into these r streams (as determined by the partitioner)—in this way, the (key, value) pairs are "pushed" to the reducers. Because the r streams in this case are being updated by m mappers in parallel, the streams must be synchronized. In the hybrid approach, k ($1 < k < m$) streams are maintained for each reducer and (key, value) pairs emitted by mappers for a particular reducer are distributed among the streams corresponding to that reducer.

Namespace manager: This module manages memory assignment to enable data reading and writing for MapReduce jobs. It converts filesystem paths that are specified in the standard Hadoop API into an abstraction we call a 'namespace': output is directed to an appropriate namespace that resides in memory, and, similarly, input records are directly consumed from memory as appropriate.

Kumar et al. [30] demonstrated that HONE can be significantly faster than scaled-out Hadoop for small to medium sized datasets. As others have suggested, we need to re-think scale-out versus scale-up architectures as the amount of cores and memory on high-end commodity servers continues to increase. There is no doubt that the total amount of data is also growing rapidly, but it is unclear if the datasets used in typical analytical tasks today are increasing as fast. The crux of the scale-out versus scale-up debate hinges on these relative rates of growth: server capacities are (roughly) growing with Moore's Law, which should continue for at least another decade. If dataset sizes are growing at a slower rate, then scale-up architectures will become increasingly attractive.

Ultimately, the datacenter is likely to consist of a mix of scale-out and scale-up systems—we will continue to run large, primarily disk-based jobs to scan petabytes of raw log data to extract interesting features, but this work explores the interesting possibility of switching over to a multi-core, shared-memory system for efficient execution on more refined datasets. With the systems such as HONE, this can all be accomplished without leaving the comforts of Big Data frameworks or paradigms such as MapReduce: we simply select the most appropriate execution environment based on dataset size and other characteristics of the workload. This brings us to the biggest limitation of our current work and the subject of ongoing research—how to a priori determine the best configuration parameters for scaled-down version of Big Data systems in terms of the thread pool sizes etc. In the future, we can imagine an optimizer that is able to examine a Big Data workload and automatically decide what job to run where and the optimal parameter settings.

2.3 *Approximate Computing*

Approximate computing essentially involves computer systems that enables systems to trade-off accuracy for efficiency. In other words, system exposes incorrectness to the application layer in order to minimize the resources consumed. In the context of trading accuracy for energy efficiency, approximation has been applied for hardware technologies such as memory. In general, the idea is to trade energy spent on retaining or accessing data with a very small possibility of data loss in the memory (i.e., bits will flip in the memory). DRAM structures present one such opportunity, where power spent on refresh cycles can be reduced by allowed bit flips [40]. Similarly, SRAM structures represent another opportunity for fidelity trade-offs as they spend significant static power on retaining data [9, 28, 54]. Also, low-power writes to memories like flash can exploit its probabilistic properties while hiding them from software [39, 50, 61].

On the other hand, in the context of software systems, most of the current Big Data systems are meant to process and return the results for terabyte and petabyte amounts of data on clusters of tens, hundreds, or thousands of machines to support near real-time decisions. However, it has been widely observed that many applications can tolerate some degree of inaccuracy. This is especially true for exploratory queries on data, where users are satisfied with "close-enough" answers if they can be provided quickly to the end user. Also, it is not uncommon today for applications to tolerate some degree of inaccuracy. Hence, in order to achieve better cost effectiveness of todays Big Data systems, it is ideal to process sample of data instead of complete data to compute and return approximate results to the queries quickly.

Case Study: **BlinkDB**

BlinkDB [1] is a popular Big Data processing framework built on top of Spark. Given petabytes of data, it creates and maintains a variety of uniform and stratified samples from underlying data. More specifically, stratified samples are created on the most frequently used query column sets to ensure efficient execution for queries on rare values. The term 'stratified' essentially means that rare subgroups are over-represented relative to a uniformly random sample. This ensures that one can answer queries about any subgroup, regardless of its representation in the underlying data. Agarwal et al. [1] have formulated the problem of sample creation as an optimization problem. Given a collection of past query column sets and their historical frequencies, collection of stratified samples are chosen with total storage costs below some user configurable storage threshold. These samples are designed to efficiently answer queries with the same query column sets as past queries, and to provide good coverage for future queries over similar query columns sets.

Once stratified samples are created, BlinkDB returns fast, approximate answers with error bars by executing queries on samples of data. More specifically, based on a query's error/response time constraints, the sample selection module dynamically picks a sample on which to run the query. It does so by running the query on multiple smaller sub-samples to quickly estimate query selectivity and choosing the best sample to satisfy specified response time and error bounds.

Additionally, BlinkDB verifies the correctness of the error bars that it returns at runtime. In other words, it employs bootstrap-based error estimation techniques and diagnostics to validate multiple procedures for generating error bars at runtime. With the diagnostic in place, BlinkDB can answer a range of complex analytic queries on large samples of data (of gigabytes in size) at interactive speeds (i.e., within a couple of seconds), while falling back to non-approximate methods to answer queries whose errors cannot be accurately estimated.

In summary, Big Data systems with approximate computing possess an ability to trade away accuracy for savings in time and energy. In other words, when inaccuracy is not an issue, then we can essentially minimize the overall cost of analytics and improve energy efficiency by retrieving approximate results using sample of data rather than the whole datasets.

3 Future Directions

3.1 Hybrid Big Data Architectures

Previously, it used to be expensive to scale up computing hardware, so most of the large-scale systems used cheap commodity machines. But today scenario is changing where scaling-up of a single machine has become fairly inexpensive. In coming days, the data centers ultimately will consist of a mix of scale-out and scale-up systems. We note that, today most of the systems developed to handle the problem of Big

Data are optimized for scale-out settings. These design decisions often can lead to under utilization of the available resources. In order to judiciously make use of cheap processing power of commodity hardware together with large memories of scaled-up machines, we need systems that can also take advantage of heterogeneity provided by scale-up systems such as large available memories, advanced storage and processor/accelerator types. Key motivation for this direction comes from the fact that data center servers today are notoriously underutilized [47]. Therefore, we need techniques that can adapt dynamically by profiling jobs and assigning right type of resource to them to achieve resource's peak utilization that will maximize the overall energy efficiency and minimize the overall cost.

One idea in this direction is to develop smart data processing techniques that can take advantage of both scale-out and scale-up optimized systems. For example in the scenario of Hadoop, let us say that we have developed a Hadoop compatible system like HONE that is optimized for scaled-up systems. Also let us say we perform a workload-aware data placement and replication in Hadoop. When jobs arrive at Hadoop then based on job profile at hand, it should be automatically routed to either scale-up optimized system like HONE or it can be sent to vanilla scale-out based Hadoop with large number of cheap commodity machines. In summary, in this age of heterogeneity, each Big Data system should have different version of itself optimized for variety of workload profiles, hardware choices and platforms. For particular workload or job profile, appropriate variant (scaled-down version or scaled-out version) of Big Data system must be chosen to minimize the overall cost of analytics.

Essentially, question boils down to this, when should a job run with scale-up rather than scale-out; and for jobs larger than even the largest scale-up machine, should we scale them out with a few large machines or with many small ones? The correct decision depends on job type, job size, job characteristics and pricing and we need an automated way to predict the best architecture and configuration for a given job. Essentially, there is a need for predictive mechanism based on input job sizes and static analysis of the application code. Moreover, the co-existence of scale-up and scale-out machines in a cluster also complicates the management of the cluster, e.g. the design of the scheduler [21, 51].

3.2 Multi-tenancy in Cloud Infrastructures

Cloud computing is gaining popularity where in order to reduce operating costs, multiple businesses or tenants share a common infrastructure owned by a cloud service provider. In coming days, to improve businesses multiple tenants may also share data among themselves while sharing common infrastructure.

In a cloud setting, usually, each tenant has different workload requirements and service level agreements (SLA). Each tenant wants to make profits and also expects cloud service provider to meet his SLAs. Whereas, cloud service provider's goal is to maximize performance for each tenant and minimize the overall cost of their service infrastructure. In other words, cloud service providers objective can be given as: $\max(\frac{performance}{cost})$. In order to meet both tenant and cloud provider's objective, we need techniques to perform multi-tenant workload consolidation.

One open problem in this direction is: given different workloads corresponding to multiple tenants with different SLAs, can we model these workloads together such that the workload-aware data partitioning helps meet multi-tenant SLAs as well as help maximize cloud service provider's objective. Closest work in this direction is by Al-Kiswany et al. [2] where they propose a data sharing framework that hosts a large number of web and mobile applications. This benefits both web and mobile apps by giving them access to rich information within their cost budget, as well as increasing revenue for the cloud provider while saving resources (side effect of data sharing). Clearly more work is needed in this direction since more and more applications are moving to cloud infrastructures and challenges related to the dual problem of performance and cost related issues are soon going to take a center stage.

3.3 Virtualized Environments

Over the past years virtualization has emerged as an ubiquitous technology to increase server utilization. Also, more recently Big Data systems are increasingly run on virtual cloud environments including public clouds such as Amazon EC2. However, performance and virtualization overheads are still open issues in virtualized environments. For performance and energy efficiency reasons, physical clusters currently offer a better choice albeit at the possible cost of utilization. Some of the overheads in virtualization are expected to be alleviated with newer technologies. However, there are still a number of open challenges in trying to determine the performance and energy-efficient configuration of applications running in virtual environments. Very few studies have been performed in this direction, one such work is done by Jin et al. [23] where they empirically investigate the effects of server virtualization on energy usage in physical servers. In addition, they identify a trade-off between the energy saving from server consolidation and the negative effects such as energy overheads and throughput reduction from server virtualization.

4 Conclusion

Big Data management and analytics is one of the most important building blocks of todays modern economy. For the same reason, we are witnessing the emergence of plethora of Big Data systems to handle variety of workloads with extreme volumes, and arriving at tremendous speeds. Most of these systems are optimized for performance and ignore other critical aspects of the overall cost such as resource and power consumption. In extreme scales, such performance-only short sightedness can lead to significant monetary loss, over budgeting, and wastage of resources for the businesses etc. Not only for businesses, these factors can also have ill-effects on global level such as environmental issues. Hence, it has become important to consider overall cost effectiveness that would not only include performance as a key metric but also consider metrics like energy consumption. Research community today is much more aware of these issues than before, and there are few Big Data architecture paradigms that are emerging in this direction. In this chapter, we identified few of these emerging cost effective Big Data architecture paradigms and have discussed it.

References

1. S. Agarwal, B. Mozafari, A. Panda, H. Milner, S. Madden, I. Stoica, BlinkDB: queries with bounded errors and bounded response times on very large data, in *Proceedings of the 8th ACM European Conference on Computer Systems, EuroSys'13* (ACM, New York, 2013), pp. 29–42
2. S. Al-Kiswany, H. Hacıgümüş, Z. Liu, J. Sankaranarayanan, Cost exploration of data sharings in the cloud, in *Proceedings of the 16th International Conference on Extending Database Technology, EDBT'13* (ACM, New York, 2013), pp. 601–612
3. S. Albers, F. Müller, S. Schmelzer, Speed scaling on parallel processors, in *Proceedings of the Nineteenth Annual ACM Symposium on Parallel Algorithms and Architectures, SPAA'07* (ACM, New York, 2007), pp. 289–298
4. America's data centers consuming and wasting growing amounts of energy, http://www.nrdc.org/energy/data-center-efficiency-assessment.asp
5. D.G. Andersen, J. Franklin, M. Kaminsky, A. Phanishayee, L. Tan, V. Vasudevan, FAWN: a fast array of wimpy nodes, in *Proceedings of the ACM SIGOPS 22nd Symposium on Operating Systems Principles, SOSP'09* (ACM, New York, 2009), pp. 1–14
6. J. Arguello, F. Diaz, J. Callan, J.-F. Crespo, Sources of evidence for vertical selection, in *Proceedings of the 32nd International ACM SIGIR Conference on Research and Development in Information Retrieval, SIGIR'09* (ACM, New York, 2009), pp. 315–322
7. N. Bruno, S. Chaudhuri, A.C. König, V.R. Narasayya, R. Ramamurthy, M. Syamala, Autoadmin project at microsoft research: lessons learned. IEEE Data Eng. Bull. **34**(4), 12–19 (2011)
8. J.P. Callan, Z. Lu, W.B. Croft, Searching distributed collections with inference networks, in *Proceedings of the 18th Annual International ACM SIGIR Conference on Research and Development in Information Retrieval, SIGIR'95* (ACM, New York, 1995), pp. 21–28
9. I.J. Chang, D. Mohapatra, K. Roy, A priority-based 6t/8t hybrid SRAM architecture for aggressive voltage scaling in video applications. IEEE Trans. Circuit Syst. Video Technol. **21**(2), 101–112 (2011)
10. K. Chodorow, M. Dirolf, *MongoDB: The Definitive Guide*, 1st edn. (O'Reilly Media, Inc., Sebastopol, 2010)
11. M. Chowdhury, M. Zaharia, J. Ma, M.I. Jordan, I. Stoica, Managing data transfers in computer clusters with orchestra, in *Proceedings of the ACM SIGCOMM 2011 Conference, SIGCOMM'11* (ACM, New York, 2011), pp. 98–109
12. J. Dean, S. Ghemawat, Mapreduce: simplified data processing on large clusters. Commun. ACM **51**(1), 107–113 (2008)
13. B. Dong, J. Qiu, Q. Zheng, X. Zhong, J. Li, Y. Li, A novel approach to improving the efficiency of storing and accessing small files on hadoop: a case study by powerpoint files, in *IEEE International Conference on Services Computing, SCC 2010, Miami, Florida, USA, 5–10 July 2010* (2010), pp. 65–72
14. Energy star program, http://www.energystar.gov/
15. Flink. Apache Software Foundation. http://flink.apache.org
16. M.R. Garey, D.S. Johnson, *Computers and Intractability; A Guide to the Theory of NP-Completeness* (W.H. Freeman & Co., New York, 1990)

17. A. Gharaibeh, E. Santos-Neto, L.B. Costa, M. Ripeanu, The energy case for graph processing on hybrid cpu and gpu systems, in *Proceedings of the 3rd Workshop on Irregular Applications: Architectures and Algorithms, IAAA'13* (ACM, New York, 2013), pp. 2:1–2:8
18. G. Graefe, Database servers tailored to improve energy efficiency, in *Proceedings of the 2008 EDBT Workshop on Software Engineering for Tailor-Made Data Management, SETMDM'08* (ACM, New York, 2008), pp. 24–28
19. S. Harizopoulos, S. Papadimitriou, A case for micro-cellstores: energy-efficient data management on recycled smartphones, in *Proceedings of the Seventh International Workshop on Data Management on New Hardware, DaMoN 2011, Athens, Greece, 13 June 2011* (2011), pp. 50–55
20. S. Harizopoulos, M.A. Shah, J. Meza, P. Ranganathan, Energy efficiency: the new holy grail of data management systems research, in *CoRR* (2009). arXiv:0909.1784
21. B. Hindman, A. Konwinski, M. Zaharia, A. Ghodsi, A.D. Joseph, R. Katz, S. Shenker, I. Stoica, Mesos: a platform for fine-grained resource sharing in the data center, in *Proceedings of the 8th USENIX Conference on Networked Systems Design and Implementation, NSDI'11* (USENIX Association, Berkeley, 2011), pp. 295–308
22. L.-Y. Ho, J.-J. Wu, P. Liu, Optimal algorithms for cross-rack communication optimization in mapreduce framework, in *IEEE International Conference on Cloud Computing (CLOUD)* (2011), pp. 420–427
23. Y. Jin, Y. Wen, Q. Chen, Energy efficiency and server virtualization in data centers: an empirical investigation, in *INFOCOM Workshops* (IEEE, 2012), pp. 133–138
24. A. Jindal, S. Prof, D.J. Dittrich, The mimicking octopus: towards a one-size-fits-all database architecture, in *In VLDB PhD Workshop* (2010)
25. R. Kallman, H. Kimura, J. Natkins, A. Pavlo, A. Rasin, S. Zdonik, E.P.C. Jones, S. Madden, M. Stonebraker, Y. Zhang, J. Hugg, D.J. Abadi, H-Store: a high-performance, distributed main memory transaction processing system. Proc. VLDB Endow. **1**(2), 1496–1499 (2008)
26. S. Khuller, J. Li, B. Saha, Energy efficient scheduling via partial shutdown, in *Proceedings of the Twenty-First Annual ACM-SIAM Symposium on Discrete Algorithms, SODA'10* (Society for Industrial and Applied Mathematics, Philadelphia, 2010), pp. 1360–1372
27. A. Kulkarni, J. Callan, Document allocation policies for selective searching of distributed indexes, in *Proceedings of the 19th ACM International Conference on Information and Knowledge Management, CIKM'10* (ACM, New York, 2010), pp. 449–458
28. A. Kumar, J. Rabaey, K. Ramchandran, SRAM supply voltage scaling: a reliability perspective, in *Proceedings of the 2009 10th International Symposium on Quality of Electronic Design, ISQED'09* (IEEE Computer Society, Washington, 2009), pp. 782–787
29. K.A. Kumar, A. Deshpande, S. Khuller, Data placement and replica selection for improving co-location in distributed environments, in *CoRR* (2013). arXiv:1302.4168
30. K.A. Kumar, J. Gluck, A. Deshpande, J. Lin, Optimization techniques for "scaling down" hadoop on multi-core, shared-memory systems, in *Proceedings of the 17th International Conference on Extending Database Technology, EDBT 2014, Athens, Greece, 24–28 March 2014* (2014), pp. 13–24
31. K.A. Kumar, A. Quamar, A. Deshpande, S. Khuller, SWORD: workload-aware data placement and replica selection for cloud data management systems. VLDB J. **23**(6), 845–870 (2014)
32. A. Kyrola, G. Blelloch, C. Guestrin, GraphChi: large-scale graph computation on just a PC, in *Proceedings of the 10th USENIX Conference on Operating Systems Design and Implementation, OSDI'12* (USENIX Association, Berkeley, 2012), pp. 31–46
33. A. Lakshman, P. Malik, Cassandra: a decentralized structured storage system. SIGOPS Oper. Syst. Rev. **44**(2), 35–40 (2010)
34. A. Lamb, M. Fuller, R. Varadarajan, N. Tran, B. Vandiver, L. Doshi, C. Bear, The vertica analytic database: c-store 7 years later. Proc. VLDB Endow. **5**(12), 1790–1801 (2012)
35. W. Lang, R. Kandhan, J.M. Patel, Rethinking query processing for energy efficiency: slowing down to win the race. IEEE Data Eng. Bull. **34**(1), 12–23 (2011)
36. W. Lang, J.M. Patel, Towards eco-friendly database management systems, in *CIDR 2009, Fourth Biennial Conference on Innovative Data Systems Research, Asilomar, CA, USA, 4–7 January 2009, Online Proceedings* (2009)

37. W. Lang, J.M. Patel, Energy management for mapreduce clusters. Proc. VLDB Endow. **3**(1–2), 129–139 (2010)
38. J. Leverich, C. Kozyrakis, On the energy (in)efficiency of hadoop clusters. SIGOPS Oper. Syst. Rev. **44**(1), 61–65 (2010)
39. R. Liu, C. Yang, W. Wu, Optimizing NAND flash-based SSDs via retention relaxation, in *Proceedings of the 10th USENIX Conference on File and Storage Technologies, FAST 2012, San Jose, CA, USA, 14–17 February 2012* (2012), p. 11
40. S. Liu, K. Pattabiraman, T. Moscibroda, B.G. Zorn, Flikker: saving dram refresh-power through critical data partitioning. SIGPLAN Not. **46**(3), 213–224 (2011)
41. X. Liu, J. Han, Y. Zhong, C. Han, X. He, Implementing WebGIS on Hadoop: a case study of improving small file I/O performance on HDFS, in *CLUSTER* (IEEE Computer Society, 2009), pp. 1–8
42. M. Malik, H. Homayoun, Big data on low power cores: are low power embedded processors a good fit for the big data workloads? in *33rd IEEE International Conference on Computer Design, ICCD 2015, New York City, NY, USA, 18–21 October 2015* (2015), pp. 379–382
43. S. Mittal, A survey of techniques for improving energy efficiency in embedded computing systems. IJCAET **6**(4), 440–459 (2014)
44. A. Pavlo, E. Paulson, A. Rasin, D.J. Abadi, D.J. DeWitt, S. Madden, M. Stonebraker, A comparison of approaches to large-scale data analysis, in *Proceedings of the 2009 ACM SIGMOD International Conference on Management of Data, SIGMOD'09* (ACM, New York, 2009), pp. 165–178
45. A. Quamar, K.A. Kumar, A. Deshpande, Sword: scalable workload-aware data placement for transactional workloads, in *Proceedings of the 16th International Conference on Extending Database Technology, EDBT'13* (ACM, New York, 2013), pp. 430–441
46. V.J. Reddi, B.C. Lee, T.M. Chilimbi, K. Vaid, Web search using mobile cores: quantifying and mitigating the price of efficiency, in *ISCA*, ed. by A. Seznec, U.C. Weiser, R. Ronen (ACM, New York, 2010), pp. 314–325
47. S. Rivoire, M.A. Shah, P. Ranganathan, C. Kozyrakis, J. Meza, Models and metrics to enable energy-efficiency optimizations. IEEE Comput. **40**(12), 39–48 (2007)
48. A. Rowstron, D. Narayanan, A. Donnelly, G. O'Shea, A. Douglas, Nobody ever got fired for using hadoop on a cluster, in *Proceedings of the 1st International Workshop on Hot Topics in Cloud Data Processing, HotCDP'12* (ACM, New York, 2012), pp. 2:1–2:5
49. Samza. Apache Software Foundation. http://samza.apache.org
50. M. Salajegheh, Y. Wang, K. Fu, A. Jiang, E. Learned-Miller, Exploiting half-wits: smarter storage for low-power devices, in *Proceedings of the 9th USENIX Conference on File and Stroage Technologies, FAST'11* (USENIX Association, Berkeley, 2011), pp. 4–4
51. M. Schwarzkopf, A. Konwinski, M. Abd-El-Malek, J. Wilkes, Omega: flexible, scalable schedulers for large compute clusters, in *Proceedings of the 8th ACM European Conference on Computer Systems, EuroSys'13* (ACM, New York, 2013), pp. 351–364
52. E. Seo, S.Y. Park, B. Urgaonkar, Empirical analysis on energy efficiency of flash-based SSDs, in *Proceedings of the 2008 Conference on Power Aware Computing and Systems, HotPower'08* (USENIX Association, Berkeley, 2008), pp. 17–17
53. A. Sharov, A. Shraer, A. Merchant, M. Stokely, Take me to your leader!: online optimization of distributed storage configurations. Proc. VLDB Endow. **8**(12), 1490–1501 (2015)
54. M. Shoushtari, A. BanaiyanMofrad, N. Dutt, Exploiting partially-forgetful memories for approximate computing. IEEE Embed. Syst. Lett. **7**(1), 19–22 (2015)
55. L. Si, J. Callan, Relevant document distribution estimation method for resource selection, in *Proceedings of the 26th Annual International ACM SIGIR Conference on Research and Development in Informaion Retrieval, SIGIR'03* (ACM, New York, 2003), pp. 298–305
56. S. Sivasubramanian, Amazon dynamoDB: a seamlessly scalable non-relational database service, in *Proceedings of the ACM SIGMOD International Conference on Management of Data, SIGMOD 2012, Scottsdale, AZ, USA, 20–24 May 2012*, pp. 729–730
57. M. Stonebraker, D.J. Abadi, A. Batkin, X. Chen, M. Cherniack, M. Ferreira, E. Lau, A. Lin, S. Madden, E. O'Neil, P. O'Neil, A. Rasin, N. Tran, S. Zdonik, C-store: a column-oriented

DBMS, in *Proceedings of the 31st International Conference on Very Large Data Bases, VLDB'05* (VLDB Endowment, 2005), pp. 553–564

58. D. Thain, M. Livny, Building reliable clients and servers, in *The Grid: Blueprint for a New Computing Infrastructure*, ed. by I. Foster, C. Kesselman (Morgan Kaufmann, Amsterdam, 2003)

59. P. Thomas, M. Shokouhi, Sushi: scoring scaled samples for server selection, in *Proceedings of the 32nd International ACM SIGIR Conference on Research and Development in Information Retrieval, SIGIR'09* (ACM, New York, 2009), pp. 419–426

60. A. Toshniwal, S. Taneja, A. Shukla, K. Ramasamy, J.M. Patel, S. Kulkarni, J. Jackson, K. Gade, M. Fu, J. Donham, N. Bhagat, S. Mittal, D. Ryaboy, Storm@twitter, in *Proceedings of the 2014 ACM SIGMOD International Conference on Management of Data, SIGMOD'14* (ACM, New York, 2014), pp. 147–156

61. H.-W. Tseng, L.M. Grupp, S. Swanson, Underpowering NAND flash: profits and perils, in *Proceedings of the 50th Annual Design Automation Conference, DAC'13* (ACM, New York, 2013), pp. 162:1–162:6

62. D. Tsirogiannis, S. Harizopoulos, M.A. Shah, Analyzing the energy efficiency of a database server, in *Proceedings of the 2010 ACM SIGMOD International Conference on Management of Data, SIGMOD'10* (ACM, New York, 2010), pp. 231–242

63. A. Vulimiri, C. Curino, P.B. Godfrey, T. Jungblut, K. Karanasos, J. Padhye, G. Varghese, Wanalytics: geo-distributed analytics for a data intensive world, in *Proceedings of the 2015 ACM SIGMOD International Conference on Management of Data, SIGMOD'15* (ACM, New York, 2015), pp. 1087–1092

64. T. White, *Hadoop: The Definitive Guide*, 1st edn. (O'Reilly Media, Inc., Sebastopol, 2009)

65. Z. Wu, M. Butkiewicz, D. Perkins, E. Katz-Bassett, H.V. Madhyastha, Spanstore: cost-effective geo-replicated storage spanning multiple cloud services, in *Proceedings of the Twenty-Fourth ACM Symposium on Operating Systems Principles, SOSP'13* (ACM, New York, 2013), pp. 292–308

66. Z. Xu, Building a power-aware database management system, in *Proceedings of the Fourth SIGMOD PhD Workshop on Innovative Database Research, IDAR'10* (ACM, New York, 2010), pp. 1–6

67. Z. Xu, Y. Tu, X. Wang, Exploring power-performance tradeoffs in database systems, in *Proceedings of the 26th International Conference on Data Engineering, ICDE 2010, 1–6 March 2010, Long Beach, California, USA* (2010), pp. 485–496

68. Z. Xu, Y. Tu, X. Wang, PET: reducing database energy cost via query optimization. PVLDB 5(12), 1954–1957 (2012)

69. M. Zaharia, M. Chowdhury, M.J. Franklin, S. Shenker, I. Stoica, Spark: cluster computing with working sets, in *Proceedings of the 2nd USENIX Conference on Hot Topics in Cloud Computing, HotCloud'10* (USENIX Association, Berkeley, 2010), pp. 10–10

Bringing High Performance Computing to Big Data Algorithms

H. Anzt, J. Dongarra, M. Gates, J. Kurzak, P. Luszczek, S. Tomov and I. Yamazaki

Abstract Many ideas of High Performance Computing are applicable to Big Data problems. The more so now, that hybrid, GPU computing gains traction in mainstream computing applications. This work discusses the differences between the High Performance Computing software stack and the Big Data software stack and then focuses on two popular computing workloads, the Alternating Least Squares algorithm and the Singular Value Decomposition, and shows how their performance can be maximized using hybrid computing techniques.

1 Introduction

1.1 High Performance Computing Meets Big Data

High Performance Computing (HPC), meaning scientific and engineering computing, with emphasis on simulation, offers decades of experience in crunching numbers

H. Anzt · J. Dongarra · M. Gates · J. Kurzak (✉) · P. Luszczek · S. Tomov · I. Yamazaki
Innovative Computing Laboratory, University of Tennessee,
1122 Volunteer Blvd, Knoxville, TN 37996, USA
e-mail: kurzak@icl.utk.edu

H. Anzt
e-mail: hanzt@icl.utk.edu

J. Dongarra
e-mail: dongarra@icl.utk.edu

M. Gates
e-mail: mgates3@icl.utk.edu

P. Luszczek
e-mail: luszczek@icl.utk.edu

S. Tomov
e-mail: tomov@icl.utk.edu

I. Yamazaki
e-mail: iyamazak@icl.utk.edu

© Springer International Publishing AG 2017
A.Y. Zomaya and S. Sakr (eds.), *Handbook of Big Data Technologies*,
DOI 10.1007/978-3-319-49340-4_23

at the highest speeds, using machines form the high end of the hardware spectrum. Big Data, meaning data analytics, has been shifted more toward the lower end of that spectrum, where the price/performance ratio is more favorable. Now that Big Data problems enter the mainstream of computing, many solutions from HPC can be applied to Big Data.

This chapter opens with a discussion of the main differences between the hardware/software stacks of Big Data and HPC. Then two prominent HPC workloads are introduced, which happen to be in widespread use in the Big Data domain, the Alternating Least Squares (ALS) algorithm and the Singular Value Decomposition (SVD). Then the main techniques for maximizing the performance of the implementations are discussed. A comprehensive discussion of the implementation details of the ALS algorithm follows. Then a thorough presentation of the implementation details of the SVD algorithm is given. The chapter is concluded with the summary of the most important points.

High Performance Computing: In the 1980s, vector supercomputing dominated high-performance computing, as embodied in the eponymously named systems designed by the late Seymour Cray. The 1990s saw the rise of massively parallel processing (MPPs) and shared memory multiprocessors (SMPs) built by Thinking Machines, Silicon Graphics, and others. In turn, clusters of commodity (Intel/AMD x86) and purpose-built processors (such as IBM's BlueGene), dominated the previous decade.

Today, these clusters are augmented with computational accelerators in the form of coprocessors from Intel and graphical processing units (GPUs) from NVIDIA; they also include high-speed, low-latency interconnects (such as InfiniBand). Storage area networks (SANs) are used for persistent data storage, with local disks on each node used only for temporary files. This hardware ecosystem is optimized for performance first, rather than for minimal cost.

Atop the cluster hardware, Linux provides system services, augmented with parallel file systems (such as Lustre) and batch schedulers (such as PBS and SLURM) for parallel job management. MPI and OpenMP are used for internode and intranode parallelism, augmented with libraries and tools (such as CUDA and OpenCL) for coprocessor use. Numerical libraries (such as LAPACK and PETSc) and domain-specific libraries complete the software stack. Applications are typically developed in Fortran, C, or C++. Figure 1 (right) shows the mainstream HPC system stack.

Big Data: Just a few years ago, the very largest data storage systems contained only a few terabytes of secondary disk storage, backed by automated tape libraries. Today, commercial and research cloud-computing systems each contain many petabytes of secondary storage, and individual research laboratories routinely process terabytes of data produced by their own scientific instruments.

As with high-performance computing, a rich ecosystem of hardware and software has emerged for big data analytics. Unlike scientific computing clusters, data-analytics clusters are typically based on commodity Ethernet networks and local storage, with cost and capacity the primary optimization criteria. However, industry is now turning to FPGAs and improved network designs to optimize performance.

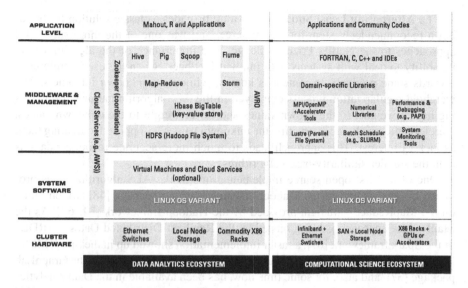

APPLICATION LEVEL	Mahout, R and Applications				Applications and Community Codes		
MIDDLEWARE & MANAGEMENT	Hive	Pig	Sqoop	Flume	FORTRAN, C, C++ and IDEs		
	Map-Reduce		Storm		Domain-specific Libraries		
	Hbase BigTable (key-value store)				MPI/OpenMP +Accelerator Tools	Numerical Libraries	Performance & Debugging (e.g., PAPI)
	HDFS (Hadoop File System)				Lustre (Parallel File System)	Batch Scheduler (e.g., SLURM)	System Monitoring Tools
SYSTEM SOFTWARE	Virtual Machines and Cloud Services (optional)						
	LINUX OS VARIANT				LINUX OS VARIANT		
CLUSTER HARDWARE	Ethernet Switches	Local Node Storage	Commodity X86 Racks		Infiniband + Ethernet Swtiches	SAN + Local Node Storage	X86 Racks + GPUs or Accelerators
	DATA ANALYTICS ECOSYSTEM				COMPUTATIONAL SCIENCE ECOSYSTEM		

(left side vertical labels: Cloud Services (e.g., AWS)), Zookeeper (coordination); right vertical label: AVRO)

Fig. 1 The mainstream big data stack (*left*) versus the mainstream HPC stack (*right*)

Atop this hardware, the Apache Hadoop system implements a MapReduce model for data analytics. Hadoop includes a distributed file system (HDFS) for managing large numbers of large files, distributed (with block replication) across the local storage of the cluster. HDFS and HBase, an open source implementation of Google's BigTable key-value store, are the big data analogs of Lustre for computational science, albeit optimized for different hardware and access patterns.

Atop the Hadoop storage system, tools (such as Pig) provide a high-level programming model for the two-phase MapReduce model. Coupled with streaming data (Storm and Flume), graph (Giraph), and relational data (Sqoop) support, the Hadoop ecosystem is designed for data analysis. Moreover, tools (such as Mahout) enable classification, recommendation, and prediction via supervised and unsupervised learning. Unlike scientific computing, application development for data analytics often relies on Java and Web services tools (such as Ruby on Rails). Figure 1 (left) shows the mainstream Big Data system stack.

1.2 Application Areas

This chapter discusses HPC implementations of two mainstream Big Data algorithms. While the first one, Alternating Least Squares (ALS), has primarily commercial applications, the second one, Singular Value Decomposition (SVD), is uniformly applicable to a wide range of problems in science, engineering, and commerce.

The **Alternating Least Squares** algorithm provides a classic solution for building a recommender system for e-commerce, and was one of the more successful approaches to the Netflix Prize challenge. The importance of the algorithm is in its ability to deal with systems with implicit feedback, when the user's preference towards some products or content is known, while it is unknown for others. The weighted regularization process employed in the ALS algorithm allows for attaching higher weights to the known values and lower weight to the unknown values, therefore effectively reconstructing the unknown values, as opposed to treating them as lack of interest. This approach leads to much more accurate recommendations than the simpler similarity-based algorithms.

One of the first open source implementations of the ALS algorithms was produced in Java, as part of the Mahout machine learning package [48], which relied on the MapReduce paradigm provided by the Hadoop framework [34, 60]. As the MapReduce approach is being ousted by the Resilient Distributed Datasets (RDD) of the Spark framework [67], a faster implementation showed up in the Spark MLlib library [44]. Also, ALS was one of the first algorithms implemented in the GraphLab package [37], and also, for some time now, has been available in the Data Analytics Acceleration Library (DAAL) from Intel [26]. Finally, the first state of the art GPU implementation was produced by the authors of this chapter [16], and followed by similar developments from other groups [57].

The **Singular Value Decomposition** is ubiquitous in statistics and scientific computing and commonly applied to problems where the matrices are large and substantial computational power is required. Prime examples of application areas include astrophysics, genomics, climate data analysis, and information retrieval systems. In astrophysics, the SVD is used on massive datasets from astronomical surveys for spectral classification, e.g., to predict morphological types using galaxy spectra, and to select quasar candidates from sky surveys. In genomics, the SVD is routinely used to analyze genome-wide single-nucleotide polymorphism (SNP) data, for detecting population structure and potential outliers. In climate data analysis, Empirical orthogonal function (EOF) and the SVD are the methods of choice for analyzing spacial and temporal variability of geophysical data. The SVD is also the primary tools for latent semantic indexing (LSI) in information retrieval systems, where it is used to find low-rank approximations to term-document matrices, enabling computation of query-document similarity scores in low-rank representation, as well as automated document categorization.

Randomized algorithms have been developed for the singular value decomposition [36, 42]. Great surveys of recent developments in randomization algorithms were published by Halko [21] and Mahoney [41]. In terms of software, singular value solvers are available in Skylark and Mahout. Skylark is an open-source software project launched by IBM Research with the objective to develop a set of randomized machine learning algorithms that support distributed memory and are accessible through Python interfaces. Skylark uses a number of sketching transforms to implement a few randomized linear algebra solvers, including a singular value solver based on the work by Halko et al. [21]. Mahout is a project of the Apache Software Foundation to produce free implementations of distributed or otherwise scalable machine learning algorithms focused primarily in the areas of collaborative filtering, clustering, and classification [48]. In addition to a classic Lanczos SVD algorithm, Mahout also contains an implementation of a stochastic (randomized) SVD routine [40].

1.3 Tricks of the Trade

Two techniques discussed here and borrowed from the field of High Performance Computing, are automated software tuning and randomization algorithms. The technique of automated software tuning mostly addressed the challenges of programming modern computing devices, such as GPU accelerators, in a way that provides portable performance, i.e., not only allows getting maximum performance from a particular device, but also allows for porting to a new device by retuning rather than rewriting/redesigning the code. The technique of randomization allows dealing with one of the most burning problems of processing Big Data, which is the lagging of IO capabilities behind processing capabilities in modern hardware.

Automated Software Tuning: Although Moore's Law has still been in effect in the last few years, the multicore revolution initiated the trend, in processor design, of going away from architectural features that do not directly contribute to processing throughput. This means preference towards shallow pipelines with in-order execution and cutting down on branch prediction and speculative execution. On top of that, virtually all modern architectures require some form of vectorization to achieve top performance, whether it being short-vector SIMD (Single Instruction Multiple Data) extensions of CPU cores, or SIMT (Single Instruction Multiple Thread) pipelines of GPU accelerators. With the landscape of future High Performance Computing populated with complex, hybrid, vector architectures, automated software tuning may provide a path towards portable performance without heroic programming efforts.

Automated software tuning was pioneered by projects like ATLAS and Spiral, and is the objective of numerous academic projects, and is also practiced by hardware vendors providing libraries like BLAS for their devices. The basic premise is to explore a search space and find the best performers. The search space can be defined by a set of tunable parameters, code transformations, implementation variants, hardware switches, etc. It can then be pruned by applying a set of constraints that eliminate obvious underperformers. Finally, it can be searched to find the winners. Exhaustive search, steepest descent methods, genetic algorithms are all valid approaches.

Randomization Algorithms: The landscape of future High Performance Computing presents an explosive growth in the volume of data, and a relatively dismal growth in the capabilities of communication and IO systems. Under such conditions, it becomes increasingly important to find algorithms that communicate less, and perform IO operations even less. For an important set of problems in numerical computing, a class algorithms emerges that seem to be an answer to these challenges—randomization algorithms.

The new classes of random sampling and random projection algorithms offer numerous advantages when dealing with large datasets coming from both scientific applications (astrophysics, genomics, climate modeling), as well as commercial applications (social networks, information retrieval systems, financial transactions). In many cases, randomized algorithms beat their classical counterparts in terms of accuracy, speed, and robustness. They utilize modern computer architectures better by exposing higher levels of parallelism than traditional numerical methods. At the same time, they often produce more numerically robust solvers by introducing implicit regularization.

2 GPU Acceleration of Alternating Least Squares

Web-based services such as movie databases and online retailers increasingly rely on recommendation systems to suggest products to their customers. Collaborative Filtering (CF) is a class of recommendation systems that recommends products based on what other customers with similar interests have enjoyed [17]. It harvests information collected from a large set of users, which can be either explicit feedback, such as "likes" or product ratings; or implicit feedback, such as purchases, time spent, or search patterns. This yields a large dataset to process, for instance, the Netflix Prize dataset has over 100 million ratings [4].

Collaborative Filtering algorithms are based on observation data in a relation matrix R, where each entry denotes how a user rated or interacted with an item. As each user rates only a small subset of the items, most entries are unknown, i.e., the matrix R is sparse. The goal is to determine the unknown values in R for how a user would hypothetically rate every item. Thus it is an instance of the *matrix completion problem* [9], to determine the unknown entries of a sparsely sampled matrix. In recent years, latent feature models have assumed a small set of features—such as movie genres—drive users' interest. However, these latent features are determined by the algorithm, without any explicit, a priori assigned meaning. This small set of features implies the matrix R is (approximately) low-rank.

Besides providing an algorithm to complete R, an added benefit of the low-rank model is that it determines R in a compact representation, $R = X^T Y$, taking $O(fm + fn)$ space instead of $O(mn)$ space for m users, n items, and rank $f \ll m, n$. For a site with millions of users and millions of products, this compact representation makes storing and accessing the recommendations database tractable.

In addition to recommendation systems, the matrix completion problem occurs in numerous other contexts. Examples include recovery of missing pixels of an image [27], inferring 3D structure from motion of images [10], and determining sensor positions from incomplete distance measurements [8].

Various methods exist for computing the matrix completion. Many CF systems used neighborhood models [30]. For low-rank models, Candès and Recht [9] used convex relaxation, and proved that R can be completed if sufficient entries are known. Stochastic gradient descent [8, 50] and alternating least squares (ALS) [27, 70] are popular methods. We will focus on the ALS method, which has adaptations for both explicit [70] and implicit feedback [23].

We propose both multi-core CPU and GPU implementations that are able to exploit the computing power of state-of-the-art processors and accelerators. We compare performance with the open source implementations available in Mahout [1], GraphLab [12], and Spark MLlib [2, 44, 67], and report significant speedups for selected benchmark datasets.

2.1 Explicit Feedback

For explicit feedback, entry r_{ui} of R denotes how user u rated item i. Since users have not rated all items, the goal is to complete the missing entries of R. We assume that R is approximately low-rank, such that $R \approx X^T Y$, where X is $f \times m$ and Y is $f \times n$ for m users, n items, and rank or feature space size f. This latent feature space is small compared to the number of users and items, e.g., from 10 to 100, depending on the application. Column x_u of X represents user u, and column y_i of Y represents item i, such that their inner product yields the rating, $r_{ui} \approx x_u^T y_i$.

Determining X and Y is commonly expressed as an optimization problem, with a summation over known r_{ui} entries,

$$\min_{X,Y} \sum_{\substack{u,i \\ r_{ui} \text{ is known}}} \left(r_{ui} - x_u^T y_i\right)^2 + \lambda \left(\sum_u \|x_u\|^2 + \sum_i \|y_i\|^2 \right). \tag{1}$$

Here, λ is a regularization term to avoid overfitting. This can be solved with stochastic gradient descent or alternating least squares.

To solve using ALS, we observe that if X or Y is fixed, the cost function (1) becomes a linear least squares problem. ALS iterates two steps: fixing Y and solving for X, then fixing X and solving for Y. In the first step, fixing Y and finding where the gradient is zero yields

$$\left(Y D^u Y^T + \lambda I\right) x_u = Y r_u \qquad \text{for } u = 1, \dots, m$$

to solve for each user-factor x_u. Each of the m user-factors can be solved independently, providing a large amount of parallelism. Here, r_u is row u of the R matrix, and D^u is a binary diagonal matrix that selects columns of Y corresponding to known r_{ui} values. Similarly, in the second step, fixing X yields

$$\left(X D^i X^T + \lambda I\right) y_i = X r_i \qquad \text{for } i = 1, \dots, n$$

to solve for each item-factor y_i, where r_i is column i of the R matrix, and D^i selects columns of X for known r_{ui} values. Experiments have shown that the user- and item-factors typically converge after a few iterations of these two steps [70].

2.2 Implicit Feedback

For implicit feedback, Hu et al. [23] note that a large r_{ui} value does not necessarily indicate a higher preference, but instead gives a higher confidence. For instance, a user may enjoy watching a moderately good TV show every week, yielding a large r_{ui} value, but watch a beloved movie just once or twice, yielding a small r_{ui} value, despite its stronger preference. Therefore, they propose a preference matrix P with binary values,

$$p_{ui} = \begin{cases} 1 & \text{if } r_{ui} > 0, \\ 0 & \text{if } r_{ui} = 0, \end{cases}$$

to indicate whether user u has a preference for item i. Larger r_{ui} values indicate greater confidence in this preference, so a matrix C with entries $c_{ui} = 1 + \alpha r_{ui}$ is introduced that measures the confidence of the preference p_{ui}. Here some minimal confidence is given even to zero entries, while α weights known values more. Hu et al. found $\alpha = 40$ to work well. For implicit feedback, instead of completing the relation matrix R, the goal is to complete the preference matrix as $P \approx X^T Y$. Again, X and Y can be computed by minimizing a cost function,

$$\min_{X,Y} \sum_{u,i} c_{ui} \left(p_{ui} - x_u^T y_i \right)^2 + \lambda \left(\sum_u \|x_u\|^2 + \sum_i \|y_i\|^2 \right). \tag{2}$$

The major difference compared to explicit feedback is that the sum is over all u and i, not just those with nonzero r_{ui} values, since some minimal confidence is given even to zero entries. This means there are mn terms, making stochastic gradient descent prohibitively expensive for implicit feedback, whereas for explicit feedback only the nonzero r_{ui} values have terms in (1). Therefore, we apply the alternating least squares algorithm, similar to the explicit feedback case above, yielding

$$\left(Y C^u Y^T + \lambda I \right) x_u = Y C^u p_u \qquad \text{for } u = 1, \ldots, m;$$
$$\left(X C^i X^T + \lambda I \right) y_i = X C^i p_i \qquad \text{for } i = 1, \ldots, n;$$

to solve for each x_u and for each y_i, where C^u is a diagonal matrix of row u of the confidence matrix C, C^i is a diagonal matrix of column i of C, p_u is row u of the preference matrix P, and p_i is column i of P. Pseudocode is given in Algorithm 1.

Algorithm 1 Pseudocode of alternating least square algorithm iterating user-factors and item-factors.

```
function ALS( input: α, λ, R; output: X, Y )
    set Y to random initial guess
    while not converged
        // update user-factors X
        for u = 1, . . . , m
            solve (YC^u Y^T + λI) x_u = YC^u p_u for x_u
        end
        // update item-factors Y
        for i = 1, . . . , n
            solve (XC^i X^T + λI) y_i = XC^i p_i for y_i
        end
    end
end function
```

The two steps, updating the user-factors and the item-factors, are identical except for swapping the input and output matrices. Therefore, we will subsequently focus on updating the user-factors, and the item-factors will follow similarly. The explicit and implicit feedback ALS algorithms are also very similar; we will concentrate on implicit feedback.

For computational efficiency, the product can be factored as

$$YC^u Y^T = YY^T + \alpha Y R^u Y^T,$$

where R^u is a diagonal matrix of row u of R, as shown schematically in Fig. 2. Since YY^T is the same for all users, it can be computed once per iteration [23], which is done efficiently using the syrk (symmetric rank-k update) BLAS routine. (Explicit feedback lacks the YY^T term.) The remaining term, $\alpha Y R^u Y^T$, involves

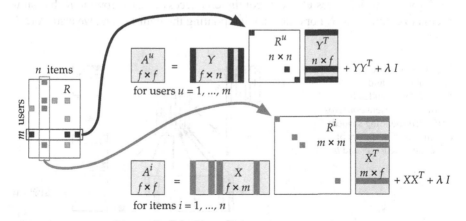

Fig. 2 Diagram of computation of user-factors and item-factors. R is general sparse, R^u and R^i are sparse diagonal, X, Y, A^u, A^i are dense

a dense matrix Y and the sparse diagonal matrix R^u, which will require a custom kernel. Under mild assumptions, $YC^uY^T + \lambda I$ is symmetric positive definite (SPD), allowing us to solve it with the Cholesky factorization.

2.3 CPU Implementation

In the product YR^uY^T, the sparse diagonal matrix R^u selects and scales a few columns of Y, as shown in Fig. 3. Columns of Y corresponding to zeros in R^u are ignored. As k, the number of nonzeros in R^u, is typically much less than n, the number of columns of Y, the kernel should take advantage of this sparsity, reducing the cost from a rank-n update to a rank-k update, with $k \ll n$.

For instance, with the Netflix dataset and $f = 64$, the problem is to generate and solve $m = 480190$ systems, each formed by a 64×64 rank-k update, with the average $k = 209$ (see Fig. 5). There is not enough parallelism in computing a single system for an efficient multi-core implementation. Instead, we do a batched implementation that generates and solves the m systems in parallel. For this, we use OpenMP to parallelize the loops in Algorithm 2.

High efficiency can be attained by relying on optimized Level 3 BLAS routines, which operate on matrices instead of individual vectors, enabling data reuse and optimizations for cache efficiency, improving performance to be compute-bound instead of memory-bound. To use Level 3 BLAS, we copy the relevant columns of Y to workspaces $\hat{Y}$ and V, with the R^u column scaling included in V, as shown in Fig. 3, then use a gemm (general matrix-matrix multiply) BLAS call. Since A is symmetric, work could be reduced by using an extended BLAS routine such as gemmt in Intel MKL [25] or syrkx in NVIDIA cuBLAS [46] instead of gemm.

Updating the item-factors is exactly the same, except it uses columns of R instead of rows of R. For updating the user-factors, we store R in CSR (compressed sparse row) format, which gives efficient, contiguous access to each row of R, but slow access to columns of R. For efficiency in updating the item-factors, we also store R

Fig. 3 Schematic of $A^u = YR^uY^T$ and $b = YC^up_u$. Shaded boxes in row r_u represent nonzeros; only corresponding shaded columns of Y and rows of Y^T contribute to A^u and b

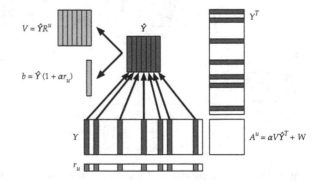

Algorithm 2 Multi-core CPU ALS algorithm.

---
function ALS_CPU(input: α, λ, R; output: X, Y)
   set Y to random initial guess
   while not converged
      // update user-factors X
      $W = YY^T + \lambda I$ using syrk BLAS
      parallel for $u = 1, \ldots, m$
         copy columns of Y corresponding to nonzeros in r_u to $\hat{Y}$
         copy and scale columns of $\hat{Y}$ as $V = \hat{Y} R^u$
         accumulate scaled columns of $\hat{Y}$ as $b_u = \hat{Y}(1 + \alpha r_u)$
         $A^u = \alpha V \hat{Y} + W$ using gemm BLAS (single-threaded)
         solve $A^u x_u = b_u$ using Cholesky (single-threaded)
      end
      // update item-factors Y
      $W = XX^T + \lambda I$ using syrk BLAS
      parallel for $i = 1, \ldots, n$
         copy columns of X corresponding to nonzeros in r_i to $\hat{X}$
         copy and scale columns of $\hat{X}$ as $V = \hat{X} R^i$
         accumulate scaled columns of $\hat{X}$ as $b_i = \hat{X}(1 + \alpha r_i)$
         $A^i = \alpha V \hat{X} + W$ using gemm BLAS (single-threaded)
         solve $A^i y_i = b_i$ using Cholesky (single-threaded)
      end
   end
end function

---

in CSC (compressed sparse column) format, which gives efficient, contiguous access to each column of R.

Because the number of nonzeros per row can vary significantly (see Fig. 5), there will be a load imbalance between different processors. This is easily solved by using the OpenMP dynamic scheduler, adding `schedule(dynamic,NB)`, with a block size NB. We set NB $= 200$, but performance is not sensitive to the exact value.

2.4 GPU Implementation

A brief summary of the GPU architecture will help to understand the GPU implementation. A GPU kernel divides its computation into a grid of thread blocks, and each thread block into a grid of threads. Within each thread block, threads are not independent, but execute the same instructions on different data. Threads can synchronize and communicate via shared memory, which is a kind of fast, user-controlled cache. Each thread's local variables are stored in a large register file. Different thread blocks execute asynchronously, without an easy way to synchronize or communicate. An NVIDIA Kepler GPU contains up to 15 multiprocessors, each with 192 cores.

Due to this GPU architecture, the GPU implementation shown in Algorithm 3 is structured differently than the CPU implementation in Algorithm 2. Each thread block computes one tile of a matrix A^u and its right-hand side b_u. As with the CPU implementation, a single system has insufficient parallelism to fully occupy all the GPU's cores. Filling a GPU requires hundreds of thread blocks and tens of thousands of threads. Therefore, we use a batched implementation, where a single GPU kernel generates a batch of s matrices using the BATCHED_SPARSE_SYRK routine, then a batched Cholesky routine factors them, and finally batched triangular solvers solve the resulting systems. We use the batched Cholesky and triangular solves from the BEAST project [33]. We used a batch size of $s = 4096$ to balance parallelism with GPU memory requirements. However, performance is not sensitive to the exact batch size.

Algorithm 3 GPU implementation of ALS, using batched operations.

---

    function ALS_GPU(input: α, λ, R; output: X, Y)
      // workspaces: A is $f \times f \times s$, B is $f \times s$
      set Y to random initial guess
      while not converged
          // update user-factors X
          $W = YY^T + \lambda I$ using syrk from cuBLAS
          for $k = 1, \ldots, m$ by batch size s
              BATCHED_SPARSE_SYRK computes $A^u = \alpha Y R^u Y^T + W$ and $b_u = Y C^u p_u$
                 for $u = k, \ldots, k + s$
              BATCHED_CHOLESKY factors A^u for $u = k, \ldots, k + s$
              BATCHED_SOLVE solves $A^u x_u = b_u$ for $u = k, \ldots, k + s$
          end
          // update item-factors Y
          $W = XX^T + \lambda I$ using syrk from cuBLAS
          for $i = 1, \ldots, n$ by batch size s
              BATCHED_SPARSE_SYRK computes $A^i = \alpha X R^i X^T + W$ and $b_i = X C^i p_i$
                 for $i = k, \ldots, k + s$
              BATCHED_CHOLESKY factors A^i for $i = k, \ldots, k + s$
              BATCHED_SOLVE solves $A^i y_i = b_i$ for $i = k, \ldots, k + s$
          end
      end
    end function

The implementation of the BATCHED_SPARSE_SYRK GPU kernel is conceptually similar to the CPU kernel. Like the CPU kernel, it copies the relevant columns of Y to a workspace $\hat{Y}$, in this case stored in GPU shared memory. Instead of copying all the relevant columns at once, it copies just one block of kb columns at a time and multiplies these, storing the results in registers, then continues with the next block. Unlike the CPU version, here the copy and multiply are fused into one kernel. The multiply is based an optimized gemm GPU kernel [32], which sub-tiles the output matrix A^u, with each GPU thread computing one entry in each sub-tile (Fig. 4).

A few optimizations can be made. Since A^u is symmetric, only the tiles on or below the diagonal need to be computed; tiles above the diagonal are known by symmetry. Also, since matrix Y is read-only, it is beneficial to bind its memory to GPU *texture memory*, which has optimized caching for read-only data. Texture memory also simplifies the code by dealing with out-of-bounds memory accesses

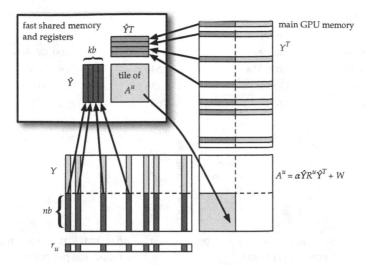

Fig. 4 Schematic of sparse-syrk GPU kernel. A^u is divided into $nb \times nb$ tiles. Block of kb relevant columns are loaded into shared memory and multiplied in registers at a time. At end, tile of A^u in registers is written back to main GPU memory. b_u is also computed (not shown)

in hardware—the software can pretend that Y is bigger than it actually is. This allows for fixed loop bounds and eliminates cleanup code, enabling more compiler optimizations.

2.5 Setup and Datasets

For performance comparison, we chose three ALS implementations from popular data analytics software packages: Mahout version 0.9 [1, 48], GraphLab version 1.3 [12, 37, 38], and Spark MLlib version 1.5 [2]. All results used single precision and were obtained on a two-socket 2.6 GHz Intel Sandy Bridge E5-2670 with 8 cores per socket. CPU implementations were linked with Intel's Math Kernel Library (MKL) version 11.1.2 [25]. Our GPU implementation ran on an NVIDIA Kepler K40c GPU with CUDA version 7.0 [47].

To compare performance, we target several recommendation datasets that are available online: Netflix Prize [4], Million Song [6], and Yahoo! Song [53]. For tuning parameters of the GPU implementation, we employ an autotuning sweep using the BEAST framework [24], with the EachMovie dataset [43, 53], a smaller dataset that permits executing a comprehensive set of kernel configurations in a moderate runtime. Table 1 summarizes properties of the datasets.

For the Netflix Prize dataset, we show histograms in Fig. 5 of the number of nonzeros per row (left) and per column (right). The minimum, median, mean, and maximum number of nonzeros per row and column are annotated in each graph. As

Table 1 Dataset properties

Dataset	# users	# items	# nonzeros
Netflix prize	480,190	17,771	100,480,508
Million song	1,019,318	384,546	48,373,586
Yahoo! song	130,558	136,736	49,770,695
EachMovie	1,623	61,265	2,811,717

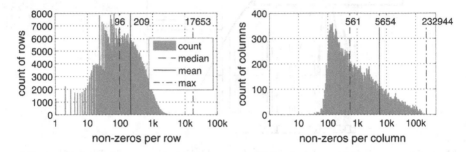

Fig. 5 Nonzero distribution of rows (*left*) and columns (*right*) of Netflix Prize dataset

previously noted, the wide range of nonzeros per row and column means different users and items incur widely different costs in computing YC^uY^T and XC^iX^T, potentially leading to load imbalance.

2.6 Auto Tuning

The sparse-syrk GPU kernel has four tunable parameters: tile size nb, block size kb, and thread block dimensions dx and dy. The kernel is generalized so that any value of nb can be used for any feature space size f. The optimal parameters are not obvious and not easy to derive by an analytical formula. Therefore the factorization calls for a real autotuning sweep. To achieve high performance, classic heuristic automatic software tuning methodology is applied, where a large number of kernels are generated and run, and the fastest ones identified.

The BEAST autotuning framework [39] enumerates and tests all possible kernel configurations. Various constraints are applied to limit the search space. Configurations violating correctness constraints—such as exceeding the maximum shared memory, or nb not divisible by the thread block dimensions—are eliminated. Several heuristic constraints are also applied, for instance, ensuring a compute-intensive kernel by requiring the ratio of multiply-add instructions to load instructions is at least 2. While kernels that violate these soft constraints will run correctly, they will not keep the GPU fully occupied, leading to lower performance.

After applying these constraints, BEAST generated 330 kernel configurations to test. The kernels were tested on the modest sized EachMovie dataset, timing the

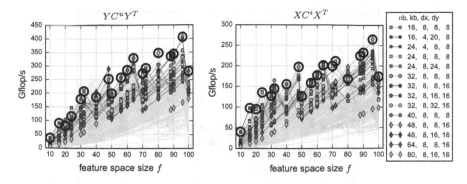

Fig. 6 Performance of all kernels (*gray lines*), highlighting ones that are best for some size. Circled kernel is chosen as best for each size(color figure online)

sparse-syrk for both the user-factor and the item-factor matrix generation. Due to differences in the size of Y and X and the sparsity of R^u and R^i, the performance was not identical between these two. We ran tests for sizes of f that are multiples of 8 and multiples of 10, from 8 to 100.

The performance of all these kernels is plotted in gray in Fig. 6. Kernels that were best for some size are highlighted with colored markers. For each size f, the circled kernel was chosen as the best overall kernel.

Inspecting the data reveals that no one configuration was optimal across all feature space sizes. Taking the yellow diamond (80, 8, 16, 16) kernel as an example: for small f it is a poor performer, but the performance increases as f increases, until it is the best kernel for $f = 80$, where $f = nb$. For the next size, $f = 88$, its performance plummets to less than half the optimal performance. This occurs because it goes from one tile to four tiles covering each matrix A, wasting three large tiles to cover the extra 8 rows and columns. This saw tooth pattern is evident for all the configurations.

While often the best kernel for user-factors (left in Fig. 6) and item-factors (right) is the same, there are several instances where this is not true due to the difference in sparsity patterns. In these cases, the kernel with the best geometric mean performance is chosen as the best compromise between the two.

This analysis highlights the need for autotuning. The performance difference between the best and worst kernels is dramatic—between a factor of 6 and 72 times for a particular f. Also, the optimal kernel configuration depends heavily on the size f, and to a lesser extent on the actual dataset. While some kernel configurations make sense in retrospect, it was infeasible to predict optimal kernels in all cases.

2.7 Performance Evaluation

Execution time of a single ALS iteration (updating user-factors and item-factors once) for the three large benchmark databases—Netflix, Million Song, and Yahoo!

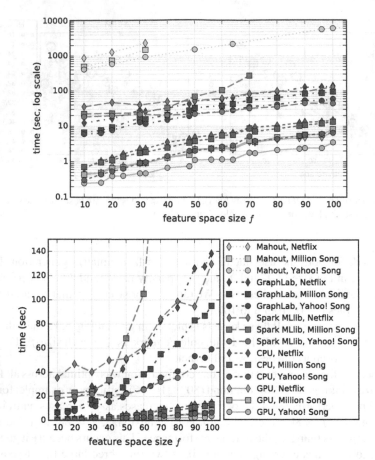

Fig. 7 Time in log scale (*top*) and linear scale (*bottom*) for single ALS iteration, using 16 CPU cores or GPU

Song—is presented in Fig. 7, in both log and linear scale. This covers a range of feature space sizes, all using 16 CPU cores or the GPU. A large performance difference between implements is evident. Mahout is nearly two orders-of-magnitude slower than GraphLab and Spark. This is not surprising, as Mahout is written in Java while GraphLab is a newer implementation written in C++. Spark, while written in Scala/Java, links with native optimized BLAS to achieve good performance. For $f \geq 50$ with the Yahoo and Netflix datasets, Spark had performance comparable to GraphLab. However, with the Million Song dataset, the Spark execution time increased markedly for $f \geq 50$, and it encountered an exception for $f \geq 80$. Our CPU implementation is 10 times faster than GraphLab and 19 times faster than Spark MLlib, on average.

The speedup of our GPU implementation over Mahout, GraphLab, Spark, and our CPU implementation is given in Fig. 8. The GPU achieves an average speedup of 2.1

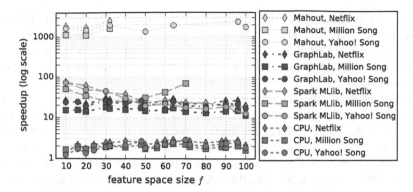

Fig. 8 Speedup in log scale of GPU implementation over Mahout, GraphLab, Spark, and CPU implementations using 16 cores

times over our CPU implementation. Compared to GraphLab, the GPU is on average 20.9 times faster, and compared to Spark it is 35.3 times faster. Mahout performs poorly, taking 1684 times longer, on average, to compute a single ALS iteration.

While speedups are similar across datasets, our GPU implementation consistently gets the best speedups for the Netflix dataset and the least speedups for the Million Song dataset. This may be because the Million Song dataset has the smallest average nonzeros-per-row and nonzeros-per-column, with a mean of 47 nonzeros per row and 126 per column, compared to 209 and 5654 for the Netflix dataset (Fig. 5). Having more nonzeros means a higher floating point operation count in the sparse-syrk routine to amortize memory reads.

We have presented both a multi-core CPU and a GPU implementation for the alternating least-squares algorithm to compute recommendations based on implicit feedback datasets. The central kernel involved is sparse_syrk, an algorithm-specific kernel achieving compute-bound performance for multiplying two dense matrices scaled by a sparse diagonal matrix. Our results demonstrate the advantage of fully exploiting the available parallelism by using a batched implementation, along with using optimized kernels, either from the vendor's BLAS library or custom auto-tuned kernels. This yields good performance over several different datasets and a range of feature space sizes.

3 GPU Acceleration of Singular Value Decomposition

3.1 Introduction

A partial singular value decomposition (SVD) [18] of a sparse matrix is a powerful tool for data analysis, where the data is represented as the sparse matrix. The ability of the SVD to filter out noise and extract the underlying features of the data has been demonstrated in many applications, including Latent Semantic Indexing (LSI) [5, 13], recommendation systems [13, 55], population clustering [49], and subspace tracking [28]. The SVD is also used to compute the leverage scores – statistical measurements for sampling the data in order to reduce the cost of the data analysis [21].

In recent years, the amount of data being generated from the observations, experiments, and simulations has been growing at unprecedented paces in many areas of studies, e.g., science, engineering, medicine, finance, social media, and e-commerce [11, 14]. The algorithmic challenges to analyze such "Big Data" are exacerbated by its massive volume and wide variety as well as its high veracity and velocity [35]. Though the SVD has the potential to address the variety and veracity of the modern data sets, the traditional approaches to computing the partial SVD access the data repeatedly, e.g., block Lanczos [19]. This is a significant drawback on a modern computer, where the data access has become significantly more expensive compared to arithmetic operations, both in terms of time and energy consumptions. The gap between the communication and computation costs is expected to further grow on future computers [15, 20], and this high cost of the communication is exacerbated by the Big Data. This hardware trend is certainly true for the GPU.

3.2 Randomized Algorithms to Compute SVD

To address this hardware trend, a randomized algorithm [21] has been gaining attention since compared to the traditional algorithms, it may require fewer data accesses to compute the SVD of the matrices arising from the modern applications (see Fig. 9 for an illustration of the algorithm). To compare the performance of different algorithms for computing the truncated SVD, we implemented the framework, which encapsulates these algorithms on multicore CPUs with multiple GPUs [64]. This framework not only allows us to develop software whose performance can be tuned based on domain specific knowledge, but it also allows a user from one discipline to test an algorithm from another, or to combine the techniques from different algorithms (see Fig. 10 for the list of the algorithms). For example, we studied the performance of a block Lanczos, combining it with communication-avoiding [22, 62] and thick-restarting [3, 61]; two techniques developed by two different disciplines

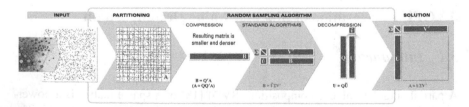

Fig. 9 Randomized algorithm to compute truncated SVD

Fig. 10 Algorithms to compute truncated SVD

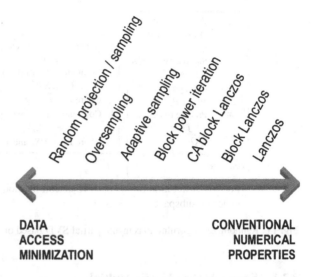

DATA
ACCESS
MINIMIZATION

CONVENTIONAL
NUMERICAL
PROPERTIES

– computer science and numerical linear algebra. These two techniques allow us to build the projection subspace with the minimum data access and accelerate the solution convergence by retaining the useful information when restarting the iteration, respectively. Hence, compared to the randomized algorithm, Lanczos could build a projection subspace of the same dimension, which is richer in useful information with fewer communication phases, and potentially with about the same amount of data access. Unfortunately, this is possible only when the matrix can be partitioned well, while many of the matrices from the modern applications cannot be partitioned in such a way, leading to the significant overheads of the communication-avoiding technique in term of the computation and storage requirements, as well as the communication volume. Hence, there is a growing interest in a novel algorithm that can more efficiently compute the SVD of the massive data that are being generated from many modern applications, and the randomized algorithm is one of such algorithms with the potential.

3.3 Hybrid CPU/GPU Implementation

Figure 11 shows the pseudocode of a randomized algorithm to compute the SVD. Since the computational cost of the randomized algorithm is dominated by the cost of generating the projection basis vectors, $\widehat{P}$ and $\widehat{Q}$, we accelerate this step using GPUs, while the SVD of the projected matrix B is redundantly computed by each MPI process on CPU. To generate the basis vectors, the two main computational kernels of the randomized algorithm are the sparse-matrix dense-matrix multiply (SpMM) and the orthogonalization. In this subsections, we describe our implementations of these two kernels on a hybrid CPU/GPU cluster.

```
for j = 1,2,...,s do
    1.  Orthogonalize Q̂
        QR := Q̂
    2.  Sample range of A
        P̂ := AQ
    3.  Orthogonalize P̂
        PB := P̂
    4.  Prepare to iterate
        if j < s then
            Q̂ := AᵀP
        end if
end for
```

1. Generate $\widehat{P}_{k+\ell}$ and $\widehat{Q}_{k+\ell}$ that approximate $\widehat{A}$,
   $$\widehat{A} \approx \widehat{P}\widehat{Q}^T.$$
2. Compute SVD of the projected matrix B,
   $$B = X\widehat{\Sigma}Y^T,$$
   where $B = \widehat{P}^T\widehat{A}\widehat{Q}$.
3. Compute approximate partial SVD of $\widehat{A}$,
   $$\widehat{A}_k \approx \widehat{U}_k\widehat{\Sigma}_k\widehat{V}_k^T,$$
   where $\widehat{U}_k = \widehat{P}X_k$ and $\widehat{V}_k = \widehat{Q}Y_k$.

(a) Power iteration to generate subspace.   (b) Projection method to compute partial SVD.

Fig. 11 Randomized algorithm to compute partial SVD based on power iteration

3.3.1 Sparse Matrix Matrix Multiply

To perform SpMM with the matrix A on a hybrid CPU/GPU cluster, we distribute A among the GPUs in a 1D block row format (e.g., using a graph or hypergraph partitioning algorithm). The basis vectors $\widehat{P}$ and $\widehat{Q}$ are then distributed in the same formats. Then, to perform SpMM, each GPU first exchanges the required non-local vector elements with its neighboring GPUs. This is done by first copying the required local elements from the GPU to the CPU, then performing the point-to-point communication among the neighbors using the non-blocking MPI (i.e., `MPI_Isend` and `MPI_Irecv`), and finally copying the non-local vector elements back to the GPU. Then, each GPU computes the local part of the next basis vectors using the CuSPARSE SpMM in the compressed sparse row (CSR) format. This was an efficient communication scheme in our previous studies to develop a linear solver [65], where the coefficient matrix A arising from a scientific or engineering simulation is often sparse and structured, e.g., with three-dimensional embedding. Unfortunately, sparse matrices originating from the modern data sets such as social networks and/or commercial applications have irregular sparsity structures, and have wide ranges of nonzero counts per row. In fact, they often exhibit power-law distributions of nonzeros as they result from scale-free graphs. As a result, this point-to-point communication with all the neighbors at once could be inefficient (in term of time and buffer storage). To alleviate the problem, our current implementation is based on a collective communication scheme. For example, using `MPI_Allgatherv`, each process sends its local vector elements, which are needed by at least one of its neighbors, to all the processes. Though this all-to-all approach requires the buffer to store the receiving messages from all the processes at once, it could obtain a significant speedup over the point-to-point communication, especially when the nonzeros of the matrix follows the power-law distribution.

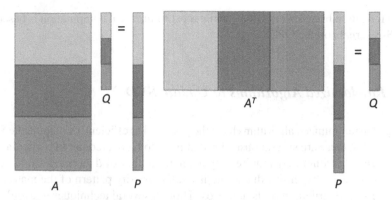

(a) 1DBR with neighborhood-collective before local SpMM.

(b) 1DBC with all-reduce after local SpMM.

Fig. 12 Illustration of matrix and vector distributions for SpMM with A and A^T. The submatrices distributed to the same GPU are colored in the same *color*. In Figure (**a**) or (**b**), the sparse matrices A and A^T are distributed either in 1D block row or block column (1DBR or 1DBC in short), respectively

Sine many matrices of our interests are tall-skinny, to perform SpMM with A^T, our current implementation keeps the input and output vectors, $\widehat{P}$ and $\widehat{Q}$, in the 1D block row distribution, but distribute A^T in the 1D block column (see Fig. 12b). Since the columns of A^T are the same as the rows of A on each GPU, we do not need to separately store A^T and A. In this implementation, each GPU first computes SpMM with its local parts of A^T and $\widehat{P}$, and then copies the partial result to the CPU. Then, the MPI process computes the final result $\widehat{Q}$ by a global all-reduce, and copies its local part back to the GPU. Hence, this requires each MPI process to store the global vectors $\widehat{Q}$. However, when A^T has the power-law distribution, performing SpMM with A^T in the 1D block row requires each GPU to store the much longer global vectors $\widehat{P}$. Our performance results have demonstrated the advantage of this all-reduce communication. Furthermore, partitioning A^T in the 1D block column often led to a higher performance of SpMM on each GPU as the local submatrix becomes more square than tall-skinny.

3.3.2 Orthogonalization

For our experiments in this paper, we used the block classical Gram-Schmidt (CGS) [18] to orthogonalize a set of vectors against another set of vectors (block orthogonalization, or BOrth in short) and the Cholesky QR (CholQR) [56] to orthogonalize the set of vectors against each other. In our previous studies, these algorithms obtained great performance on multiple GPUs on a single compute node [63] or on a hybrid CPU/GPU cluster [65]. This is because these algorithms can orthogonalize the basis vectors with a low communication cost. For example, CholQR requires only one global reduction between the GPUs, while most of the local computation is based on BLAS-3 kernels on the GPU.

3.4   Randomized Algorithms to Update SVD

Though the randomized algorithms have the potential to efficiently compute the SVD on the GPUs, there are several obstacles that need to be overcome. In particular, the randomized algorithm may require only a small number of data accesses, but each data access can be expensive due to the irregular sparsity pattern of the matrix and the power-law distribution of its nonzeros. Though several techniques to avoid such communication have been proposed [22], these techniques may not be effective for computing the SVD of the modern data because they often require a significant computational or communication overhead due to the particular sparsity structure of the matrix [64].

To address this challenge, we studied randomized algorithms to update (rather than recompute) the partial SVD as the changes are made to the data set [66]. This is an attractive approach because compared to recomputing it from scratch, the SVD may be updated more efficiently, while in modern applications, the existing data are being constantly updated and new data is being added. Moreover, in some applications, recomputing the SVD may not be possible because the original data, for which the SVD has been already computed, is no longer available. At the same time, in modern applications, the size of the update is significant even though it is much smaller than the massive data that has been already compressed. Therefore, an efficient updating algorithm is needed to address the large volume and high velocity of the modern data sets. Such applications with the rapidly changing data include the communication and electric grids, transportation and financial systems, personalized services on the internet, particle physics, astrophysics, and genome sequencing [11].

3.4.1   Case Studies

To study the potential of the randomized algorithm, we studied its performance for a popular statistical analysis tool, the principal component analysis (PCA) [7]. In PCA, a multidimensional dataset is projected onto a low-dimensional subspace given by the partial SVD such that related items are close to each other in the projected subspace. Here, we show the results from two particular applications of PCA, Latent Semantic Indexing (LSI) and population clustering.

For information retrieval by text mining [54], a variant of PCA, Latent Semantic Indexing (LSI) [13], has been shown to effectively address the ambiguity caused by the synonymy or polysemy, which are difficult to address using a traditional lexical-matching [31]. Figure 13a compares the average 11-point interpolated precisions [29] after adding different numbers of documents from the MEDLINE matrix. Our test matrices are the term-document matrices generated using the Text to Matrix

Fig. 13 Case studies with randomized algorithms for LSI ($k = 50$)

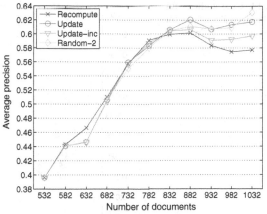

(a) Latent semantic indexing.

Method	Total number of documents, $n + d$							
	700	800	900	1000	1100	1200	1300	1400
Recompute	26.7	30.9	32.0	32.5	32.7	31.3	30.8	29.8
Update	26.7	29.8	30.1	30.7	31.5	30.7	30.4	29.7
Update-inc	26.7	29.8	30.1	30.6	30.9	30.1	29.8	29.5
Random-1	26.7	29.0	29.9	31.9	31.9	30.9	29.5	28.6
Random-2	26.7	29.6	29.6	30.0	31.0	30.1	30.0	29.7
Random-3	26.7	29.6	28.2	28.2	27.9	27.4	26.8	25.8

(b) Average 11-point interpolated precision for 6916-by-1400 CRANFIELD matrix with 225 queries, $n = 700$.

Generator (TMG)[1] and the TREC dataset,[2] and are preprocessed using the lxn.bpx weighing scheme [29]. These are the standard test matrices and were used in the previous studies [59, 68]. For our studies, we first performed 20 power iterations of the randomized algorithm to compute the rank-k approximation of the matrix $\widehat{A}$ representing the first 700 documents Then, the figure shows the average precision after new columns are added (e.g., under the column labeled "1000," 300 documents were added). To recompute the partial SVD of the matrix, we performed 20 power iterations, while the randomized algorithm used the oversampling parameter set to be $\ell = k$ (i.e., $r = 2k$), and performed two iterations that access the matrix three times. Since the basis vectors $\widehat{P}$ and $\widehat{Q}$ approximate the ranges of $\widehat{A}$ and $\widehat{A}^T$, respectively, the randomized algorithm accesses the matrix at least twice. Then, they access the matrix one more time to compute the projected matrix B. We let the incremental update algorithm (Update-inc) add $k + \ell$ columns at a time such that it requires about the same amount of memory as the randomized algorithm. We see that with only three data passes, the randomized algorithm obtained similar precisions as those of the updating algorithm. In some cases, the updating and randomized algorithms obtained higher precisions than recomputing the SVD, while the precisions of

---

[1] http://scgroup20.ceid.upatras.gr:8000/tmg.
[2] http://ir.dcs.gla.ac.uk/resources.

Fig. 14 Case studies with randomized algorithms for population clustering

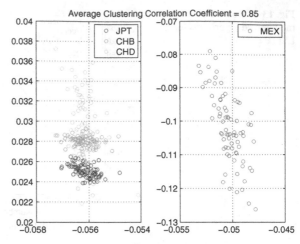

(a) Population clustering.

	JPT+MEX	+ ASW	+ GIH	+ CEU
Recompute	1.00	1.00	1.00	0.97
No update	1.00	0.81	0.84	0.67
Update-inc	1.00	1.00	0.89	0.70
Random-1	1.00	0.95	0.92	0.86

(b) Average correlation coefficients of population clustering based on the five dominant singular vectors, where 83 African ancestry in south west USA (ASW), 88 Gujarati Indian in Houston (GIH), and 165 European ancestry in Utah (CEU) were incrementally added to the 116,565 SNP matrix of 86 Japanese in Tokyo and 77 Mexican ancestry in Los Angeles, USA (JPT and MEX). Random-1 iterated twice with $\ell = k$.

the incremental update slightly deteriorated at the end. Such phenomena were also reported in the previous studies [58, 68].

PCA has been also successfully used to extract the underlying genetic structure of human populations [45, 51, 52]. To study the potential of the randomized algorithm, we used it to update the SVD, when a new population is added to the population dataset from the HapMap project.[3] Figure 14 shows the correlation coefficient of the resulting population cluster, which is computed using the k-mean algorithm of MATLAB in the low-dimensional subspace given by the dominant left singular vectors. We randomly filled in the missing data with either -1, 0, or 1 with the probabilities based on the available information for the SNP. We let the randomized algorithm iterate twice, and with only the three data passes, the randomized algorithm

---

[3]http://hapmap.ncbi.nlm.nih.gov.

improved the clustering results, potentially reducing the number of times the SVD must be recomputed.

3.4.2 Performance Studies

We now study the performance of the randomized algorithm on the Tsubame Computer at the Tokyo Institute of Technology.[4] Each of its compute nodes consists of two six-core Intel Xeon CPUs and three NVIDIA Tesla K20Xm GPUs. We compiled our code using the GNU gcc version 4.3.4 compiler and the CUDA nvcc version 6.0 compiler with the optimization flag -O3, and linked it with Intel's Math Kernel Library (MKL) version xe2013.1.046.

Figure 15a compares the strong parallel scaling of the randomized algorithm with that of the current state-of-the-art updating algorithm [68]. Clearly, the state-of-the-art algorithm can spend significantly longer time in the orthogonalization, leading to a great speedup obtained by the randomized algorithm (i.e., the speedups of up to 14.1). At the same time, the speedup decreased on a larger number of GPUs. This is because the execution time of the randomized algorithm is dominated by SpMM, whose strong parallel scaling suffered from the increasing inter-GPU communication cost for this relatively small-scale matrix that was used for this study. On the other hand, the updating algorithm was still spending a significant amount of its execution time for the orthogonalization which was still compute intensive and scaled over the small number of the GPUs. On a larger number of GPUs, compared to the randomized algorithm, the updating algorithm is expected to suffer from the greater communication latency.

Figure 15b shows the weak parallel scaling results for the document-document matrix used in a previous LSI study [69]. The matrix row contains 2,559,430 documents, and each column contains about $4,176$ nonzero entries. The weak parallel scaling results, in particular, show the advantages of the randomized algorithm due to its ability to compress the desired information into a small projection subspace using a small number of data passes. For the updating algorithm, the accumulated cost of the SVDs of the projected matrices also became significant.

4 Conclusions

In this chapter, two mainstream Big Data algorithms were discussed: the Alternating Least Squares algorithm for solving the matrix completion problem and the Singular Value Decomposition algorithm for computing a low-rank approximation of a matrix, both of which pose significant challenges when offloading to a GPU or a computing cluster with multiple GPUs.

---

[4]http://tsubame.gsic.titech.ac.jp.

Fig. 15 Performance studies with randomized algorithms

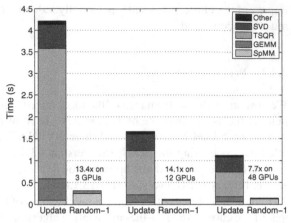

(a) Strong parallel scaling.

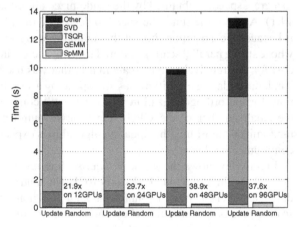

(b) Weak parallel scaling.

In the case of the ALS algorithm, the technique of automatic software tuning was used to achieve top performance, leading to an order of magnitude performance advantage over mainstream open source packages, GraphLab and Spark MLlib, and three orders of magnitude advantage over Mahout (Hadoop), when using a single GPU as opposed to a multicore CPU (16 cores).

In the case of the SVD algorithm, the technique of random projection was applied to implement the algorithm efficiently on a computing cluster with up to 48 GPUs, and also to implement an algorithm for updating a previously computed factorization upon arrival of new data. In this case, the algorithmic innovations also lead to an order of magnitude performance advantage.

Both case studies show the kind of impact that cutting-edge HPC techniques can have on the world of Big Data by enabling efficient use of accelerators, which leads to massive performance improvements.

References

1. Apache, Mahout version 0.9 (2015a). https://mahout.apache.org/
2. Apache, Spark version 1.5 (2015b). http://spark.apache.org/
3. J. Baglama, L. Reichel, Augmented implicitly restarted Lanczos bidiagonalization methods. SIAM J. Sci. Comput. **27**, 19–42 (2005)
4. J. Bennett, S. Lanning, The netflix prize, in *Proceedings of the KDD Cup Workshop 2007* (ACM, New York, 2007), pp 3–6. http://www.cs.uic.edu/~liub/KDD-cup-2007/NetflixPrize-description.pdf
5. M.W. Berry, Large scale sparse singular value computations. Int. J. Supercomput. Appl. **6**, 13–49 (1992)
6. T. Bertin-Mahieux, D.P. Ellis, B. Whitman, P. Lamere, The million song dataset, in *Proceedings of the 12th International Conference on Music Information Retrieval (ISMIR)* (2011)
7. C. Bishop, *Pattern Recognition and Machine Learning* (Springer, New York, 2006)
8. P. Biswas, T.C. Lian, T.C. Wang, Y. Ye, Semidefinite programming based algorithms for sensor network localization. ACM Trans. Sensor Networks (TOSN) **2**(2), 188–220 (2006)
9. E.J. Candès, B. Recht, Exact matrix completion via convex optimization. Found. Comput. Math. **9**(6), 717–772 (2009)
10. P. Chen, D. Suter, Recovering the missing components in a large noisy low-rank matrix: application to SFM. IEEE Trans. Pattern Anal. Mach. Intell. **26**(8), 1051–1063 (2004)
11. Committee on the Analysis of Massive Data, Committee on Applied and Theoretical Statistics, Board on Mathematical Sciences and Their Applications, Division on Engineering and Physical Sciences, National Research Council (2013). Frontiers in Massive Data Analysis. The National Academies Press
12. Dato, GraphLab version 1.3 (2015). https://dato.com/products/create/open_source.html
13. S. Deerwester, S. Dumais, G. Furnas, T. Landauer, R. Harshman, Indexing by latent semantic analysis. J. Am. Soc. Inf. Sci. **41**, 391–407 (1990)
14. DOE Office of Science, Synergistic challenges in data-intensive science and exascale computing. DOE Advanced Scientific Computing Advisory Committee (ASCAC) (2013). Data Subcommittee Report
15. S.H. Fuller, L.I. Millett, *The Future of Computing Performance: Game Over Or Next Level?* (National Academy Press, Washington, DC, 2011)
16. M. Gates, H. Anzt, J. Kurzak, J. Dongarra, Accelerating collaborative filtering using concepts from high performance computing, in *2015 IEEE International Conference on Big Data (Big Data)* (IEEE, 2015), pp. 667–676
17. D. Goldberg, D. Nichols, B.M. Oki, D. Terry, Using collaborative filtering to weave an information tapestry. Commun. ACM **35**(12), 61–70 (1992)
18. G. Golub, C. van Loan, *Matrix Computations*, 4th edn. (The Johns Hopkins University Press, Baltimore, 2012)
19. G. Golub, F. Luk, M. Overton, A block Lanczos method for computing the singular values and corresponding singular vectors of a matrix. ACM Trans. Math. Softw. **7**, 149–169 (1981)
20. S. Graham, M. Snir, C. Patterson, *Getting Up to Speed: The Future of Supercomputing* (The National Academies Press, Washington, DC, 2004)
21. N. Halko, P. Martinsson, J. Tropp, Finding structure with randomness: probabilistic algorithms for constructing approximate matrix decompositions. SIAM Rev. **53**(2), 217–288 (2011)
22. M. Hoemmen, Communication-avoiding Krylov subspace methods. Ph.D. thesis, University of California, Berkeley (2010)

23. Y. Hu, Y. Koren, C. Volinsky, Collaborative filtering for implicit feedback datasets, in *IEEE International Conference on Data Mining (ICDM)* (2008), pp. 263–272
24. Innovative Computing Lab, BEAST (2015). http://icl.utk.edu/beast/
25. Intel Corp, Developer Reference for Intel Math Kernel Library (2015). https://software.intel.com/en-us/articles/mkl-reference-manual
26. Intel Corp, Intel Data Analytics Acceleration Library 2016, Developer Guide (2016)
27. P. Jain, P. Netrapalli, S. Sanghavi, Low-rank matrix completion using alternating minimization, in *Proceedings of the Forty-Fifth annual ACM Symposium on Theory of Computing* (ACM, 2013), pp 665–674
28. I. Karasalo, Estimating the covariance matrix by signal subspace averaging. IEEE Trans. Acoust. Speech Signal Process. **34**(1), 8–12 (1986)
29. T. Kolda, D. O'Leary, A semidiscrete matrix decomposition for latent semantic indexing information retrieval. ACM Trans. Inf. Syst. **16**(4), 322–346 (1998)
30. Y. Koren, Factorization meets the neighborhood: a multifaceted collaborative filtering model, in *Proceedings of the 14th ACM SIGKDD International Conference on Knowledge Discovery and Data Mining, KDD'08* (ACM, New York, 2008), pp. 426–434
31. R. Krovetz, W.B. Croft, Lexical ambiguity and information retrieval. ACM Trans. Inf. Syst. **10**(2), 115–141 (1992)
32. J. Kurzak, S. Tomov, J. Dongarra, Autotuning gemm kernels for the Fermi GPU. IEEE Trans. Parallel Distrib. Syst. **23**(11), 2045–2057 (2012)
33. J. Kurzak, H. Anzt, M. Gates, J. Dongarra, Implementation and tuning of batched Cholesky factorization and solve for NVIDIA GPUs. Trans. Parallel Distrib. Syst. (2015). doi:10.1109/TPDS.2015.2481890
34. C. Lam, *Hadoop in Action* (Manning Publications Co., Stamford, 2010)
35. D. Laney, 3D data management: controlling data volume, velocity, and variety. Application Delivery Strategies by META Group Inc., File: 949 (2001)
36. E. Liberty, F. Woolfe, P.G. Martinsson, V. Rokhlin, M. Tygert, Randomized algorithms for the low-rank approximation of matrices. Proc. National Acad. Sci. **104**(51), 20167–20172 (2007)
37. Y. Low, J. Gonzalez, A. Kyrola, D. Bickson, C. Guestrin, j.M. Hellerstein, GraphLab: a new framework for parallel machine learning. CoRR abs/1006.4990 (2010). http://arxiv.org/abs/1006.4990
38. Y. Low, D. Bickson, J. Gonzalez, C. Guestrin, A. Kyrola, J.M. Hellerstein, Distributed GraphLab: a framework for machine learning and data mining in the cloud. Proc. VLDB Endow. **5**(8), 716–727 (2012)
39. P. Luszczek, M. Gates, J. Kurzak, A. Danalis, J. Dongarra, Search space generation and pruning system for autotuners, in *International Workshop on Automatic Performance Tuning (iWAPT 2016)* (2016, submitted)
40. D. Lyubimov, Command line interface, stochastic SVD. Technical report, The Apache Software Foundation (2014). https://mahout.apache.org/users/dim-reduction/ssvd.page/SSVD-CLI.pdf
41. M.W. Mahoney, Randomized algorithms for matrices and data. Found. Trends® Mach. Learn. **3**(2), 123–224 (2011)
42. P.G. Martinsson, V. Rockhlin, M. Tygert, A randomized algorithm for the approximation of matrices. Technical report, DTIC Document (2006)
43. P. McJones, Eachmovie collaborative filtering data set. DEC Systems Research Center 249 (1997)
44. X. Meng, J. Bradley, B. Yavuz, E. Sparks, S. Venkataraman, D. Liu, J. Freeman, D. Tsai, M. Amde, S. Owen et al., MLlib: Machine learning in Apache Spark (2015). arXiv preprint arXiv:150506807
45. P. Menozzi, A. Piazza, L. C-Sforza, Synthetic maps of human gene frequencies in Europeans. Science **201**, 786–792 (1978)
46. NVIDIA Corp, cuBLAS Library User Guide, v7.0 (2015a)
47. NVIDIA Corp, CUDA C Programming Guide, v7.0 (2015b)

48. S. Owen, R. Anil, T. Dunning, E. Friedman, *Mahout in Action* (Manning Publications Co., Greenwich, 2011)
49. P. Paschou, E. Ziv, E. Burchard, S. Choudhry, W. R-Cintron, M. Mahoney, P. Drineas, PCA-correlated SNPs for structure identification in worldwide human populations. PLoS Genet. **3**, 1672–1686 (2007)
50. A. Paterek, Improving regularized singular value decomposition for collaborative filtering, in *Proceedings of KDD Cup and Workshop* (2007), pp. 39–42
51. N. Patterson, A. Price, D. Reich, Population structure and eigenanalysis. PLoS Genet. **2**(12), 2074–2093 (2006)
52. A. Price, N. Patterson, R. Plenge, M. Weinblatt, N. Shadick, D. Reich, Principal components analysis corrects for stratification in genome-wide association studies. Nature Genet. **38**(8), 904–909 (2006)
53. R.A. Rossi, N.K. Ahmed, The network data repository with interactive graph analytics and visualization, in *Proceedings of the Twenty-Ninth AAAI Conference on Artificial Intelligence* (2015). http://networkrepository.com
54. G. Salton, M. McGill, *Introduction to Modern Information Retrieval* (McGraw-Hill, New York, 1983)
55. B. Sarwar, G. Karypis, J. Konstan, J. Riedl, Analysis of recommendation algorithms for e-commerce, in *Proceedings of the 2nd ACM Conference on Electronic Commerce* (2000), pp 158–167
56. A. Stathopoulos, K. Wu, A block orthogonalization procedure with constant synchronization requirements. SIAM J. Sci. Comput. **23**(6), 2165–2182 (2002)
57. W. Tan, L. Cao, L.L. Fong, Faster and cheaper: Parallelizing large-scale matrix factorization on gpus. CoRR abs/1603.03820 (2016). http://arxiv.org/abs/1603.03820
58. J. Tougas, R. Spiteri, Updating the partial singular value decomposition in latent semantic indexing. Comput. Statist. Data Anal. **52**, 174–183 (2007)
59. E. Vecharynski, Y. Saad, Fast updating algorithms for latent semantic indexing. SIAM J. Matrix Anal. Appl. **35**(3), 1105–1131 (2014)
60. T. White, *Hadoop: The Definitive Guide* (O'Reilly Media, Inc., Sebastopol, 2012)
61. K. Wu, H. Simon, Thick-restart Lanczos method for large symmetric eigenvalue problems. SIAM J. Matrix Anal. Appl. **22**(2), 602–616 (2000)
62. I. Yamazaki, K. Wu, A communication-avoiding thick-restart lanczos method on a distributed-memory system, in *Proceedings of the 2011 International Conference on Parallel Processing, Euro-Par'11* (Springer, Berlin, 2012), pp. 345–354
63. I. Yamazaki, H. Anzt, S. Tomov, M. Hoemmen, J. Dongarra Improving the performance of CA-GMRES on multicores with multiple GPUs, in *Proceedings of the IEEE International Parallel and Distributed Symposium (IPDPS)* (2014a), pp. 382–391
64. I. Yamazaki, T. Mary, J. Kurzak, S. Tomov, Access-averse framework for computing low-rank matrix approximations, in *Proceedings of the International Workshop on High Performance Big Graph Data Management, Analysis, and Minig* (2014b), pp. 70–77
65. I. Yamazaki, S. Rajamanickam, E. Boman, M. Hoemmen, M. Heroux, S. Tomov, Domain decomposition preconditioners for communication-avoiding Krylov methods on a hybrid CPU/GPU cluster, in *Proceedings of the International Conference for High Performance Computing, Networking, Storage and Analysis (SC)* (2014c), pp. 933–944
66. I. Yamazaki, J. Kurzak, P. Luszczek, J. Dongarra, Randomized algorithms to update partial singular value decomposition on a hybrid CPU/GPU cluster, in *Proceedings of the International Conference for High Performance Computing, Networking, Storage and Analysis (SC)* (2015), pp. 345–354
67. M. Zaharia, M. Chowdhury, M.J. Franklin, S. Shenker, I. Stoica, Spark: cluster computing with working sets, in *Proceedings of the 2nd USENIX Conference on Hot Topics in Cloud Computing*, vol. 10 (2010), p.10
68. H. Zha, H. Simon, On updating problems in latent semantic indexing. SIAM J. Sci. Comput. **21**(2), 782–791 (1999)

69. H. Zha, O. Marques, H. Simon, Large-scale SVD and subspace-based methods for information retrieval, in *Solving Irregularly Structured Problems in Parallel*, vol. 1457, Lecture Notes in Computer Science, ed. by A. Ferreira, J. Rolim, H. Simon, S.-H. Teng (Springer, Heidelberg, 1998), pp. 29–42
70. Y. Zhou, D. Wilkinson, R. Schreiber, R. Pan, Large-scale parallel collaborative filtering for the netflix prize in *Proceedings of the 4th International Conference on Algorithmic Aspects in Information and Management, AAIM'08* (Springer, Berlin, 2008), pp. 337–348

Cognitive Computing: Where Big Data Is Driving Us

Ana Paula Appel, Heloisa Candello and Fábio Latuf Gandour

Abstract In this chapter we will discuss the concepts and challenges to design Cognitive Systems. Cognitive Computing is the use of computational learning systems to augment cognitive capabilities in solving real world problems. Cognitive systems are designed to draw inferences from data and pursue the objectives they were given. The era of big data is the basis for innovative cognitive solutions that cannot rely on traditional systems. While traditional computers must be programmed by humans to perform specific tasks, cognitive systems will learn from their interactions with data and humans. Not only is Cognitive Computing a fundamentally new computing paradigm for tackling real world problems, exploiting enormous amounts of data using massively parallel machines, but also it engenders a new form of interaction between humans and computers. As machines start to enhance human cognition and help people make better decisions, new issues arise for research. We will address these questions for Cognitive Systems: What are the needs? Where to apply? Which are the sources of information to relying on?

1 Cognitive Computing: An Alternative Approach for Clear Understanding

Cognitive Computing is a term which became quite popular in a relatively short period of time. Given the origin of this popularity in a contest between man and machine, it was unavoidable the comparison with the former attempts to create artificial intelligence, as coined by John McCarty in 1955 and further explored in the 1980s. Latter in this chapter we will address the similarities and differences between

A.P. Appel (✉) · H. Candello · F.L. Gandour
IBM Research, São Paulo, Brazil
e-mail: apappel@br.ibm.com

H. Candello
e-mail: heloisacandello@br.ibm.com

F.L. Gandour
e-mail: fgandour@br.ibm.com

© Springer International Publishing AG 2017
A.Y. Zomaya and S. Sakr (eds.), *Handbook of Big Data Technologies*,
DOI 10.1007/978-3-319-49340-4_24

807

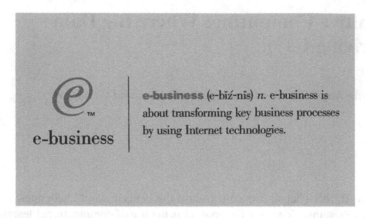

e-business (e-biź-nis) *n.* e-business is about transforming key business processes by using Internet technologies.

e-business

Fig. 1 The original e-Business logo and definition, as it appeared in the WSJ

that AI (Artificial Intelligence), and todays promises of Cognitive Computing. For the moment, we can say that there are more differences than similarities.

The popularity of Cognitive Computing in the 21st century may be understood from 2 perspectives:

- One is the ability of IBM to create global agendas and
- The other is the real need to evolve to a new and innovative level on using data.

A quick review of both view angles may be worthwhile to understand the past, present and eventual future of Cognitive Computing. Lets begin by recovering another successful global agenda, the e-Business. To do so, we suggest you to ask to any person 18 year old or less, if he or she would remember a time when it was impossible to buy goods over the Internet. Very likely, this person will say that he or she has a vague memory of this time or even that this time has never existed. In fact, the action which transformed the Internet in the commercial platform as we see it today was greatly based in the creation made by IBM in the early 90s, made public in an 8 pages e-Business manifesto inserted in the *Wall Street Journal* in October, 1997 (Fig. 1) and highly publicized by Ogilvy and Mather from 1998 onwards.[1] The success of this initiative has transformed the e-Business, an acronym which stands for Electronic Business, into a global agenda and today, using the Internet as a commercial platform may be considered a given element of the human culture [67].

Likewise, in October 2014, IBM has released another manifesto in the same *Wall Street Journal*, now talking about something that became public in an event of great impact: the competition between a machine called **"Watson"** and two human beings at the popular TV show Jeopardy. In fact, the Watson which came up at that impacting media event was made up by several connecting modules and some of them have started to be build many years ago, such as the concepts of UIMA (Unstructured Information Management Architecture). Although UIMA was made popular

---

[1]Figure extract from http://www-03.ibm.com/ibm/history/ibm100/us/en/icons/ebusiness/transform/ last visit 9th March, 2016.

as an Apache component from 2006 on and as an Oasis component from 2009 on
http://uima.apache.org, UIMA concepts started to be created back in 2003 and were
first out together in the IBM Systems Journal [33]. All this historical aspects are
mentioned here to make clear that the Cognitive Computing foundational concepts
have evolved since long ago, in a process of maturation which reached the status we
can appreciate now.

After this retrospect around global agendas, lets focus now on the real need to
evolve to a new and innovative level on using data. We have heard about the evolutionary aspect of IT (Information Technology), moving from DP (Data Processing), to
IM (Information Management) and the contribution to KG (Knowledge Generation).

Figures 2, 3 and 4 illustrates this evolution.

Fig. 2 A set of numbers
consist of pure data

Fig. 3 By adding the title
DOB Date of Birth, the data
set became an information
table

Fig. 4 By adding an
annotation, one can now use
the Information Table to feed
a machine learning process

The addiction of a simple title in this data table will add a new meaning to the data set which will become now an Information Table, as shown in Fig. 3.

From the information table, one already extracts a lot of information such as the average age of this population, higher and lower limits, standard deviation, etc.

A new dimension in the information created by the title DOB, can now be attained by the use of annotation symbols, as show in Fig. 4.

The Information Table properly slashed got the adequate format to be ingested by a machine learning methodology. At this point, when our initial data set has evolved to a new format ready to be understood by a machine, we are getting closer to what has been called Cognitive Computing: the use of computational learning systems to augment cognitive capabilities in solving real world problems.

So far, the evolution from data to information could be attained by a machine but the generation of knowledge on top of the information was essentially the result of a human function.

Running in a parallel track, the Psychology has studied the human mechanism of cognition, defined as "the mental action or process of acquiring knowledge and understanding through thought, experience, and the senses". As a "mental action", cognition has been studied as a process essentially performed by human or, at most, superior living beings, by connecting abilities related to linguistics, neuroscience, psychiatry, psychology, education, philosophy, anthropology and logic.

In another parallel track, due to some reasons including the ever lowering prices of data storage media, the IT facilities ended up storing an astonishing amount of data.

This huge amount of data was in the foundation of the appearance of Big Data and Data Analytics areas. However, the expansion of Social Computing has added another aspect to the data stored: instead of being structured, more and more the data stored is unstructured.

At the end of the first decade of the 21st century, IT world was supported by a robust infrastructure of data storage, a number of algorithms and tools of Data Analytics, stable UIMA protocols developed under the umbrella of Apache Software Foundation since 2006 and Oasis since 2009 https://uima.apache.org/, forming a perfect set of tools and methods to analyze massive unstructured data. At the same time, the prices of high performance hardware came to a reasonable level, including the ever expensive memory chips.

That was the perfect circumstance to face a new challenge: trying to build a machine capable of executing cognition, defined by Psychologists as a mental process [60] but being now, performed by a machine. As a first step to face this challenge, a sophisticated set of modules were put together in the first publicly successful experiment, the Watson System, competing in the Jeopardy TV show in November 2011. Of course, from this experiment on, a lot of improvements have happened but that is the landmark of the beginning of Cognitive Computing as a new era of IT. From then on, we can go from data to information and then, reach a new level of use of the machine as a device with cognitive capability. And this capability will at least, expand the decision making process of the end user, certainly a human being.

As we said, the comparison with the Artificial Intelligence of the 80s is inevitable but the differences are obvious:

- The experiments of AI of the 80s tried to mimic the brain functionality which resulted in a number of knowledge representation forms.
- Under this paradigm, we have had projects based on inference engines operating production rules, like the renowned MYCIN, from Edward Shortliffe [14, 88].
- Todays Cognitive Computing solution do not try to mimic the human brain anymore but rather, work as a complementary expansion of the human way of reasoning.
- To do so, Cognitive Computing foundation will put to work in a collaborative fashion elements of NLP (Natural Language Processing), UIMA and data. An immense amount of data easily available in each semantic field.

Of course, as in any other new wave in the field of IT, the beginning of the Cognitive Computing Era, as it was called in the manifesto of October 2014, will evolve through major changes, peculiar to the onset of any new technology. In any case, we believe that this new era may well change the way see and use data, now more easily transformed in the input for the enhancement of human intelligence [49].

Cognitive Computing has multiple definitions, in a formal way Cognitive Computing was defined by John E. Kelly III, IBM Senior Vice President of Cognitive Solutions and Research in his book [49] as been:

Cognitive computing [is] not just a new computing system or computing paradigm but a whole new era of computing The explosion of data in the world and the rate and pace of change has outstripped our ability to reprogram these systems We have coined this [era] cognitive because it has attributes that are more like human cognition. These are not systems that are programmed; they are systems that learn. These are not systems that require data to be neatly structured in tables or relational databases. They can deal with highly unstructured data, from tweets to signals coming off sensors.

Trying to create a clearer view of this field, we can say that so far, we simple gathered data and put it in a useful format of information:

DATA
INFORMATION

Now, by using Cognitive Computing methods, we will be able to extract knowledge automatically from data: Data Information Cognition Knowledge

If succeeding in this attempt, the extracted knowledge, once acquired, refined and accumulated, may take the us to the next level of sophistication, the generation of wisdom: Data Information Cognition Knowledge Wisdom

If that happens, todays definition of Cognitive Computing can be considered the process to automatically extract knowledge from data and in the future, the same definition may well be expanded to the process of generate wisdom from the accumulated knowledge.

2 Big Data Impulsing Cognitive System

The term **Big Data** was first used at Silicon Graphics (SGI) in the mid 1990s. John Mashey, retired former Chief Scientist at SGI was the responsible to explicit the awareness with this phenomena [26].

Among all the definitions offered for Big Data the one that seams fit better in the context of cognitive system is data thats too big, too fast, or too hard for existing tools to process.

In the last decade the volume of data storage not only from companies but individuals grow exponentially. If in the beginning of Big Data era the World Wide Web (WWW) was the main reason for data volume increase now mobile devices are being the ones responsible for that. Each day the number of videos, photos and text that we storage is bigger that in the previous day. Social networks, which in the beginning was mainly feed with text, is flooding with videos and photos, even in a single "happy B-day" post we find pictures of cakes. Also, much of human communication, whether it is in natural-language text, speech, or images, is unstructured.

How to handle large volumes of data was always a problem in computer science and the era of big data was responsible to push parallel and distribution processing to help handle these large volume of data. Another question with big data is that the volume increased so fast that the development of new technologies to mine these data did not follow this growth. Traditional Database Systems (DBMS) are not useful for this kind of data, since traditional DBMS were best with numbers and small text. As a result, new technologies were developed as distributed processing system such as Map/Reduce models to increase performance and scalability [44].

However, the problem with big data was not only storing and retrieving, mining this large volume of data with noise and unstructured data has being a nightmare for decision making tools. Because most of tools do not support unstructured data from different formats and sources, find a method that can extract useful information about these kind of data is a hard task. On the other hand, this has been a dreaming for data science and research develop new ways to work with these kind of data. This is so true, that in the last couple of years the number of companies looking for data scientist has largely increased.

Doug Laney [52] was the first person to define big data in terms of V's, that are:

- **Volume**: the volume grow each day and the tools do not scale in same proportion;
- **Variety**: we have several types of data as images, videos, long text, sensors, graphs, links and so on
- **Velocity**: data are like streams which means they continuous coming and most of tools are not able to process in real time;

IBM include one more V defining Big Data as Four V's, which is **Veracity** for uncertain in the data. And in [31] more two V's are included, **Variability**: which means that the data structure change and the way people interpret the data changes too; **Value**: the business value that data can bring to organization given them a competitive advantage based on the power to do decision making.

When we move to Cognitive System we can also include a 7th V in this definition, that is **Visualization**: visualize a large amount of data is a tough problem and it is very important to cognitive computing and can attach a enormous power to decision making.

In the business world, companies are still trying to figure out how to use and make decision based on these massive volume of data, and how to use unstructured information to improve their business.

Cognitive System can deal with this massive amount of data amplifying human intelligence, which is not scalable in the way that data is growing.

Cognitive computing is not trying to replicate what the human brain does.

There is a new Moores Law about unstructured data dark data accounts for 80% of all data generated today. Most of that data is dark we cannot make sense of that data. It is noisy or formats that cannot be read by traditional systems. Furthermore, the amount of dark data is expected to grow to over 93% by 2020 as showed in Fig. 5.

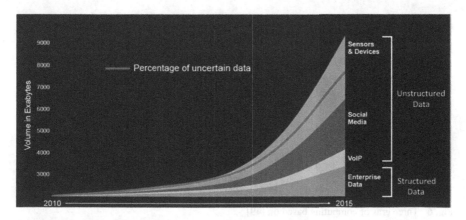

Fig. 5 Data growing expectations

Examples of area where will be a massive unstructured data collection could be found in several industries, as Energy: more than 680 million smart meters will be installed globally by 2017 producing more than 280 PB of new data to be analyzed and acted upon; Security: in 2014, more than 1 billion personal data records were compromised by cyber attacks; Transportation: By 2020, 75% of the worlds cars will be connected and they will produce 350 MB of data per second to be assessed and acted upon. And of course, is healthcare is an enormous industry, prime for disruption and new forms of insight. All these industries will need to make real time decisions about the environment based on learning about the environment and learning about driver behavior.

3   Traditional Systems versus Cognitive Systems?

The history of computing can be divided into three eras as presented in Fig. 6 extract from [49]. The first was the tabulating era, with the early 1900 calculators and tabulating machines made of mechanical systems, and later made of vacuum tubes. In the first era the numbers were fed in on punch cards, and there was no extraction of the data itself.

The second era emerged in the 1940s, programmable machines are still based on a design laid out by Hungarian American mathematician John von Neumann and ranged from vacuum tubes to microprocessors. They are programmed on much way to reach performance while executing restrict tasks, depending totally from the instructions.

The third era is the cognitive computing era, where computing technology represented an intersection between neuroscience, supercomputing and nanotechnology.

Traditional computers must be programmed by humans to perform specific tasks and they are designed to calculate rapidly and have only rudimentary sensing capabilities, such as license-plate-reading systems on toll roads. The development of a traditional system is represented in Fig. 7. First is necessary to have a project def-

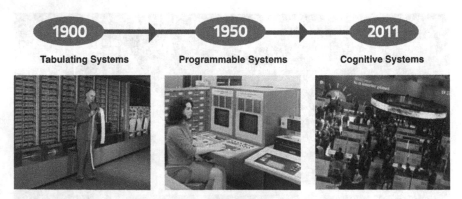

Fig. 6   Three eras of computing based on [49]

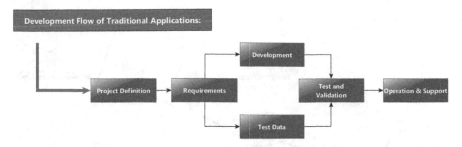

Fig. 7 Focus on functionality: meeting requirements of performance, correctness, scalability, robustness, security, stability, etc

inition to specify the requirements to the system. Them the processes is split in development and test to go to test and validation to finally deploy the system.

On the other hand, cognitive systems will be designed to draw inferences from data and pursue the objectives they were given and will be able to sense more like humans do. Theyll augment our hearing, sight, taste, smell and touch.

Ideally, cognitive systems, humans and machines will collaborate to produce better results each bringing its own skills to the partnership. The machines will be more rational and analytic. People will provide judgment, intuition, empathy, a moral compass, and human creativity (Fig. 8).

Cognitive Computing is poised to transform the IT industry, both in terms of impact and how it works. As we can see the development of cognitive computing solutions follows a very different flow centered on data. We can assure that this will have a strong impact in Knowledge Discovery in Databases (KDD) area, since the Cognitive Computing will change the way we discovery knowledge in this vast amount of unstructured data.

4 Data Mining in the Era of Cognitive Systems

The acronym *KDD* was first define in 1996 by Usama Fayyad, Gregory Piatetsky-Shapiro, and Padhraic Smyth in the paper *From Data Mining to Knowledge Discovery in Databases* [32].

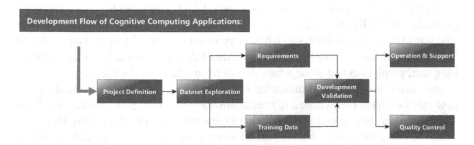

Fig. 8 Focus on data: continuous (re)training and quality data set assessment. Additional challenge: generalization (keep accuracy on new scenarios)

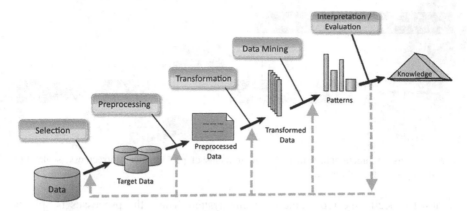

Fig. 9 Overview of the steps that compose KDD Process from [32]

In this paper KDD is defined as: *The nontrivial process of identifying valid, novel, potentially useful, and ultimately understandable patterns in data.*

The KDD process proposed in Fayyad's paper revolutionize a whole era allow that a large volume of data start to bring insight for companies and a new field of research was created.

As we see, even this paper has been written 20 years ago we are in exactly same way as we can see by the beginning of the paper presented bellow:

> Our ability to analyze and understand massive datasets lags far behind our ability to gather and store the data. A new generation of computational techniques and tools is required to support the extraction of useful knowledge from the rapidly growing volumes of data. These techniques and tools are the subject of the emerging field of knowledge discovery in databases (KDD) and data mining. Reference [32]

One could say that the Big Data Era make us go back 20 years but I believe that be in the same way as Fayyad stated in his paper means that now we need to think different and Cognitive system will allow us to do that.

The KDD (as presented in Fig. 9) is composed by following tasks: selection, preprocessing, sub-sampling, and transformations of it; data-mining methods (algorithms) to enumerate patterns from it; and evaluating the products of data mining to identify the subset of the enumerated patterns deemed knowledge.

Over the years the data grew faster than new KDD techniques to mine data. Also, the traditional KDD process and most of the methods proposed are for relational data (the last D comes from database) what today is not a reality. Most of data produced in the last decade, around 80%, is unstructured data composed by video, images, long text, graph data and so on which do not fit in relational databases.

But, even with relational databases not been the storage of these new data and new methods have been created to handle unstructured data, KDD process has been largely used by the community and adapt for these new data and algorithms. However, when we move from tools that support KDD process for system that actually help in KDD process, as cognitive system, this process does not hold completely anymore.

- Procedural Process - a person need to say how she will extract the information, which methods will be used;
- Data Scientist dependent (personal or team experience and knowledge);
- More interested in solving a computational problem than acquisition of knowledge;
- Not sufficient wisdom in using the knowledge;
- User Interaction only with the domain specialist;
- It does not learn with the process.

One of the biggest issue is despite all the tools developed over the years to support KDD process, it is totally dependent of data scientist(s) that are conduct the process. Also, the KDD processes is a procedural processes, where data scientist need to specify not only the task but the methods within the task which makes the processes iterative. This requests data scientist(s) execute and know a lot of techniques to evaluate which one performs better. These techniques are dependent of data scientist (or the team) background and sometimes these are also dependent from which area the data scientist came. There are some works that address the problem of choose the best technique for a dataset as presented in [9, 80]. However, these work are only to find the best data mining method for a couple of dataset, or help find the parameter for a specific type of method as classification as SVM [24].

Recently there is a frenetic search for data scientists and the background requested for them are bigger each day. Not for only for a data scientist, but for any person in mining field, is almost impossible to keep track and updated from all research and development that have been doing in mining area. Not only in data mining traditional but complex networks, videos, images and so on and data process platforms as Spark, Hadoop, NoSQL data management, good reviews about these subjects can be found in [1, 48, 62, 65, 71, 83, 85, 87].

The idea behind a cognitive system is to amplify human's ability and because it is centered on the data, cognitive system could perfectly help data scientist to find the best task and technique to answer hypothesis about the data they were mining. So, instead of be worried about which method, tool, storage and so on the data scientist will be worried in the knowledge acquisition and solve real problems and he or she will guide the system together with the domain specialist. Also, the system will learn about the processes and could replicate or improve that learning that some methods work better with a set of data instead of others (Fig. 10).

Based on this, in our point of view the data mining processes in cognitive age will be:

- Declarative Process - the user specify what he/she want;
- System will learn the process, thus the processes can be replicate;
- Data Scientist will guide the system to extract knowledge;
- Knowledge will be used to improve data domain;
- Interaction with everybody not only data specialists;
- System use the result as a feedback for future use.

In this sense, Cognitive System can encapsulate data mining processes and extend data scientist ability to mining large volume of data in a more efficient way. An example this is "Celia" (Cognitive Environments Laboratory Intelligent Agent) [50]. Celia

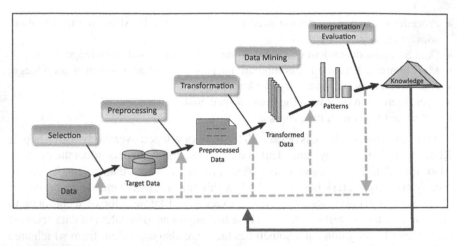

Fig. 10 Data Mining processes being viewed as a system that will be driven by data scientist and will receive the produce results (good or not) as feedback for future use

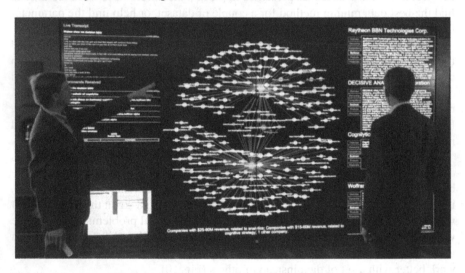

Fig. 11 Scientists working with one another and with the mergers and acquisitions Celia's prototype to discover companies that match desired criteria (Figure extract from [50])

is a prototype of symbiotic cognitive systems built over mergers and acquisitions use case that allows user interact through several cameras, kinects, microphones, input devices such as wands, and displays, including a large 4×4 multi-panel display in the front. Celia communicates through the displays and via speakers that play synthesized speech over the laboratory speakers.

Using Celia specialists are able to discover and visualize companies matching specified characteristics. Figure 11 shows a image of Celia being used. They can also use a decision agent that compares companies and their attributes side by side and receive guidance about which companies are most aligned with their preferences, as inferred through repeated interactions with Celia. A video showing mergers and acquisitions Celia's prototype can be viewed in https://www.youtube.com/watch?v= Sl6ehvwPhKs. Repsol will be the first to apply cognitive technologies for oil and gas applications.[2]

5 Design Methods for Cognitive Systems

Techniques and methods are employed in the design process to facilitate collection, development, evaluation and analysis of data. The choice of a method is related to the purpose of the activity and what is to be achieved using certain techniques. Additionally, cost and time are variables that have an impact on this selection.

Data gathering methods aim to collect data to understand the nature of potential users and context to serve the base for development. Information collected not only comes from user studies, but also other sources such as research literature, competitor analysis, surveys, social networks data and background experience of designers. Design methods are ways to refine, interpret and envision ideas from data collected in the previous phase, in order to create a product and/or a system. Information is transformed into a concrete object tailored to potential users. Evaluation methods elucidate certain issues that were not clear in the previous phases and aim to discover how users will use the system and what can be improved before the final system is fully deployed. Not always this is a linear process, mainly when we are talking about Cognitive systems that depend on data to function.

In this section priority is given to methods and techniques employed in designing cognitive systems.

5.1 Quantitative and Qualitative Methods

It is important to distinguish between the nature of qualitative and quantitative research. It could be suggested that this distinction is not clear. Reference [13] supports this notion stating that the status of the distinction is ambiguous and further suggests that the difference is deeper than the superficial issue of the presence or absence of quantification. Blaxter [11] p.64 concur:

> ... the use of questionnaires as a research technique might be seen as a quantitative strategy, whereas interviews and observations might be thought of as qualitative techniques. In practice, however, it is often more complicated than that. Thus, interviews may be structured and analyzed in a quantitative manner, as when numeric data is collected or when non-numeric answers are categorized and coded in numeric form. Similarly, surveys may allow for open-ended responses and lead to the in-depth study of individual cases.

---

[2]http://www.offshore-technology.com/features/featureturning-the-cogs-ibms-cognitive-environments-lab-takes-on-offshore-exploration-4517222/ last visited in 9th March 2016.

Hence, it could be summarized that qualitative research is more suitable for observing individual behaviors whereas quantitative is better for broader research.

Human-Computer Interaction (HCI) lies at the intersection between the social and behavioral sciences on the one hand, and computer and information technology on the other hand. It is concerned with understanding how people make use of devices and systems that incorporate or embed computation, and how such devices and systems can be more useful and more usable [18]. Therefore, the use of mixed methods is essential to understand how people interact with computer-based systems. The use of quantitative research shows, for example, the numbers of minutes it takes to perform a task or the number of errors made by users. On the other hand, qualitative research focuses on the nature of something and can be represented by themes, patterns, and stories [74] p. 356. Additionally, the use of qualitative methods is motivated by the need to understand user work before design begins, the inadequacy of many traditional forms of data and requirement gathering when applied to interface design problems, the need to involve users in the design process and data analysis [103]. Qualitative strategies were formerly investigated in the field of human-computer interaction, including Grounded Theory [47], Phenomenological research [43], case studies [73] and ethnography [57].

The use of mixed methods provide freedom to apply different approaches in the HCI field. Creswell [20] p. 14 classifies this approach into three general strategies:

- Sequential mixed method procedures are those in which the researcher seeks to elaborate on or expand on the findings of one method with another method.
- Concurrent mixed method procedures are those in which the researcher merges quantitative and qualitative data in order to provide a comprehensive analysis of the research problem.
- Transformative mixed method procedures are those in which the researcher uses a theoretical lens as an overarching perspective within a design that contains both quantitative and qualitative data.

Reference [13] also highlights some approaches to combining qualitative and quantitative research in mixed methods research:

- Triangulation: when the researcher uses more than one method or source of data in the study of social phenomena.
- Context: The application of this approach was for a qualitative study to provide the context for understanding broad-brush quantitative findings.
- Confirm and discover: The inferences that are derived from a qualitative study are then subsequently tested with quantitative research.

We consider that the strategies presented in [20] and the approaches to combining qualitative and quantitative research highlighted in [13] are in concordance. For instance, triangulation is present in both sequential and concurrent mixed methods.

Current projects aimed to enhance human decision-making apply a mixed method approach with qualitative data from interviews [54] and quantitative data from surveys and questionnaires [99, 101]. Some projects aimed to understand users visual preferences applying quantitative techniques such as the study proposed by [104]. In their research, authors used the Amazon Mechanical Turk to understand users comprehension and preferences to composite visualizations under different condition. As a result, they developed taxonomy of participants difficulties in understanding the graphics. Reference [55] describes two cases that use behavioral data to drive requirements to design new services. Although, this data is helpful to generate design insights, still the space of design alternatives is complex, according to the authors, and more knowledge based approaches with their proposal method can improve system design. Therefore, with those methods to gather user information is possible to know WHAT is wrong or not working effectively but its not usually possible to know WHY those behaviors happen without user research methods (contextual inquiry, observation studies).

5.2 Data Gathering Methods

Fieldwork observations, self-reporting methods, including questionnaires and semi-structured interviews, are usually applied in order to provide an overview of user and stakeholders preferences for employing to interface design and better understanding of human reasoning.

Field observations Observational techniques aim to investigate behaviors, interactions and practices in everyday environments. Reference [25] describes observation as a technique that seeks to examine the world through relevant human faculties. He emphasizes that observers make use not only of their visual faculties, but also other senses, from smell to hearing. In addition, [58] mentions some of the observational activities and social interactions that an observer will perform, such as: observing; participating; interrogating; listening and communicating, with a range of other forms of being, doing and thinking. Furthermore, [59] highlight the importance of observational studies as a method for use in investigations of complex interactions in natural social settings.

Observation is one of the key techniques employed in the human-computer interaction (HCI) field. Early in the design process, observation helps designers to understand the user context, task and goals. Observation conducted later in the development process may be used to investigate how well the developing prototype supports these tasks and goals [74]. This method can reinforce findings and is sometimes combined with other methods in order to gather more accurate and rigorous information [25, 59, 74].

There are different kinds of observational studies, one of which is known by the term participant observation. According to [13], ethnography and participant observation are very difficult to define. In his opinion, participant observation is frequently used as a term to describe observation alone. However, ethnography denotes both the observation and the written outcome of the research. Marshall and Rossman [59] state that participant observation researchers immerse themselves in the social setting and also observe everything that they can about it. According to [74], where a particular study falls along this spectrum depends on its goal and on the practical and ethical issues that constrain and shape it. In HCI, the role of the researcher differs from that of traditional ethnographers.

... in addition to seeking understanding of their subjects, user-interface designers focus on the interfaces for the purpose of changing and improving those interfaces. Also, whereas traditional ethnographers immerse themselves in culture for weeks or months, user-interface designers usually need to limit the process to a period of days or even hours to obtain the relevant data needed to influence the redesign ([86] p. 130).

Usually, interactive systems ethnographers use observational studies with a range of other methods such as interviews and focus groups in order to identify the main characteristics of a certain group of people. Reference [46] p. 98 suggest that it is immensely rewarding to make use of ethnographic methods in mobile design projects. They affirm that after a period of observation, the researcher should have a range of sketches, which can be used to create an overview of the field setting. In their words, "the ethnographers job is to portray the action in a vividly colorful way both in responding to design team questions and by providing an account resulting from careful reflection" [46] p. 98.

The observer may participate in the activities through which s/he seeks to observe the users behaviors/tasks either directly or indirectly through records. Overt observation occurs when participants are informed about the objectives of the research and are aware of the researchers presence. In covert observation, observers infiltrate themselves into the activity and do not inform other participants of their presence. According to [13], most research projects use overt observation, but both perspectives are adopted sporadically. In the field, in some cases, it is difficult to avoid coming into contact with people who are unaware of the ethnographers status as a researcher, even though s/he is carrying out overt observation. In addition, it is cumbersome to perform covert observation, as it is difficult to take notes without being noticed by the participants. The researcher cannot take advantage of other methods like interviews, and ethical issues like privacy can arise [13, 25].

There are many issues in the literature regarding covert observation and ethical issues [13, 58, 74]. Among them are issues relating to privacy, deception and the lack of informed consent. In the case of overt observation, the description of what is going to be analyzed has to be clear for the participant. In field work, as well as in the evaluation of systems in laboratories, it is important to clarify the purpose of the study, the kind of information that will be collected, who has the rights to the data and analysis, and how this information will be used. Participants must not only be allowed to withdraw from the study at any time, but must also be permitted to ask questions at any time [46]. It is important to give some kind of payment such as vouchers or a certain amount of money in exchange for the participants views. This should be communicated verbally, or described in a consent form which users are requested to sign before starting the study.

Sometimes is not easy to classify the role that the researcher will play. A number of authors use the classification of participant and observer roles proposed by [39], which is based on the level of involvement with the research, such as: complete observer, observer as a participant, participant as an observer and complete participant

[13, 25]. For instance, as a complete observer, a researcher must not be noticed. In this case, video and/or audio recording or photography may help in the task. An observer as participant carries out short periods of observation and conducts semi-structured interviews. The overt stance is applied. The third role is participant as observer, in which the researcher is a complete participant but the participants know their status. The fourth role is the complete participant; the researcher acts as a covert observer, undertaking the same role as an ordinary participant. The decision about which approach to undertake is affected by the kind of research questions that the study aims to answer.

A broad range of studies in the literature has set the stage and structure of observation studies [13, 25, 58, 59, 86] p. 130. Below is a discussion of the most important issues to be addressed when carrying out observation studies.

Select the setting Researchers should select a setting that contains everything in which they are interested (or as much of the desired material as possible). It is also important to check the availability of the participants and the venue.

Research teams members The researchers may have to train members in what to observe and how to take notes. Sampling According to [13], there are certain types of sampling that are more likely to be used in observational studies, such as:

Convenience sampling: This is a sample that is available to the researcher due to its accessibility;

Snowball sampling: The researcher contacts a small group of people who are relevant to the research and then use these to initiate contact with others;

Theoretical sampling: This is the process of data collection in order to generate theory whereby the analyst jointly collects, codes and analyses the data and decides which data to collect next and where to find them, in order to develop the theory as it emerges ([37] p. 45);

Representative sampling: Reference [46] give advice on selecting a representative sample to observe in the context of mobile design. For instance, a group that characterizes the entire population or a broad section of it should include a diverse range of people;

Defining a broad research question: Normally, the study is guided by a broad research question. After familiarization with the field, new findings will guide the research.

Overall, during the early stages of research, the investigator typically enters the setting with broad areas of interest but without predetermined categories. The value of this is that the researcher is able to discover recurring patterns of behaviors and relationships. After this stage, some patterns are identified and described for an early analysis of fieldwork and records.

There is a broad range of techniques for registering the information identified during observational studies. In the case of direct observation, field notes are the typical choice. The literature provides some general principles on how to take notes [13, 58, 74]. The focus and the type of notes that should be taken are related to the research questions of the study. It is important to be aware that sometimes, the participants may act differently or be self-conscious while the researcher is taking notes. Therefore, most authors suggest carrying out the observations only over short periods of time. Notes should also be clear and legible. Reference [59] also focus on the recording data perspective, saying that detailed, non-judgmental and concrete descriptions of what has been observed should be included in the field notes. After a long day of experiments, researchers should review and reflect on their notes, adding observations where necessary. Frequently, observations of behaviors give researchers cues to validate what participants report in interviews. Likewise in the study by [97], they created ontology to understand participants behaviors in collaborative design meetings that may be applied to create intelligent systems. The researchers used design sessions videos to understand behaviors also giving attention to non-verbal messages. Methodologies and approaches were also used to investigate problems that Cognitive systems might help to solve.

Think aloud technique In interaction design studies, several techniques are applied in order to capture user data, including think aloud protocols, video and audio recording and photography. The think aloud technique, developed by Erikson [30], requires subjects to verbalize their every action or thought during the study. According to [46], this technique has some drawbacks, e.g. it is embarrassing and people forget to speak after a while. A more satisfactory technique may be the constructive interaction technique suggested by [66], in which two participants exchange their opinions with each other, resulting in a more natural task. With this technique is possible to understand the mental maps of target users. For instance, [21] conducted a user qualitative evaluation of an information visualization (RiDeViz) that shows investment alternatives. The aim was to understand the user awareness of risk and uncertainty with bar charts. Observation approach using the think aloud protocol and content analysis were the methods applied with 10 subjects. Participants were asked to choose one investment choice evaluating risk and uncertainty in a bar chat visualization with limited range and a risk explorer table. The system provided different types of information, although participants did not use all for investment decision-making, they focused on small number of salient pieces and concentrated in the perceived consequences of undesirable outcomes.

Multimedia documentation Video and audio recordings are employed at the beginning of the project in order to collect information for the development process and at the end of the project in order to evaluate the prototype design. Video recordings have the advantage of capturing both visual and audio data, but can be intrusive. However, after a while, participants concentrate on the tasks and forget that they are being filmed. Other positive aspects of video recording are that it allows us to capture what users are doing on the system screen, to go back and analyze what happened after or before a specific event and to zoom in on the scene in order to analyze the users face to give some indication of his/her emotional state [46]. Preece [74] mention that is easy to miss things that are outside of the cameras view, which is why the use of more than one camera is an advantage. In addition, the use of more cameras provides a second opportunity to understand data recorded in noisy and windy urban environments.

Photographs also provide contextual information and are an excellent method in addition of gathering audio data and notes [74]. Similarly, sketches of the site, maps, pictures and documents are other resources that can be analyzed [46]. Alternatively, indirect observations can be made when direct observation would be intrusive or participants cannot be present on the day of the study [74]. Diaries and interaction logs are examples of this type of data.

Diaries and interaction logs Diaries are suitable for when the researcher cannot be with the participant when interesting things might happen. With this technique, participants are asked to enter their thoughts about a subject in a diary, through a phone call or on a website [46].

Interaction logs involve the use of software to track and record the users activities in a log that can be analyzed later [46, 74]. In this way, researchers can analyze different aspects of the usability of the activity.

Questionnaires and semi-structured interviews Questionnaires are usually applied in order to request demographic information and to elicit a participants views of a certain system. The questions should be designed to extract the participants opinion efficiently and should also be easy to analyze. Tullis and Albert [95] suggest the use of rating scales such as Likert scales and semantic differential scales for this purpose.

Typical Likert scales consist of positive and negative statements with which participants rate their agreement. Normally, these scales have a five-point scale of agreement, e.g. strongly disagree, disagree, neither agree nor disagree, agree and strongly agree. For instance, in a study integrating fitness and a mobile map-based guide, Buttussi [15] invited 12 participants to test the new system (MOPET). They followed the trails encouraged by an avatar on the mobile guide that reported on their performance as they went. After the test, participants answered a questionnaire rating their motivation and the support offered by the guide in performing the exercises via a Likert scale. As a result, users agreed that MOPET guided them in how to perform the exercises correctly.

On the other hand, semantic differential scales involve the presentation of antithetical words at the extreme points of the scale. As with Likert scales, a five to seven-point scale is normally used. The challenge is to choose suitable and clear opposites so as not to confuse respondents and to obtain dependable data.

Another popular questionnaire is the System Usability Scale (SUS), which was developed by John Brooke [12]. The SUS consists of 10 statements with which users rate their agreement. Half of the questions are positively worded and the other half negatively worded. Reference [5] found that the SUS was highly reliable and useful over a wide range of types of interface. In the same study, they substituted the scores for adjectives and compared them to the school grading scale and acceptability ranges. The scores are classified according to their acceptability. If the mean score is less than 50, the system does not have an acceptable level of usability. A score between 50 and 70 is classified as indicative of marginal acceptability, while a score higher than 70 is acceptable (Fig. 12). The same model was adopted in a mobile phone study. More information on sample questionnaires for usability metrics can be found in [74, 86, 95].

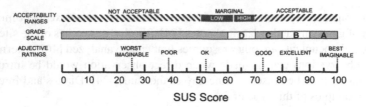

Fig. 12 SUS scale (from [5])

Open-ended questions and interviews are used for qualitative and exploratory research. These may be hard to analyze [46, 95] but lead to interesting findings, as the answers are not predetermined. In a study using a mobile system to track the behavior of young people, this is clear. Reference [76] claim that certain results would not have been available if closed-ended questions were present in their research. In their opinion, when participants are asked to answer closed-ended questions, they have to reframe their thoughts and behavior into the options given, while this does not occur when open-ended questions are used.

Open-ended questions are hard to analyze because the answers may be long and very diverse. The researcher needs to interpret the answers and to find patterns in order to analyze the data. According to [4] this opens the possibilities of misunderstanding and researcher bias. Therefore, the researcher should use a set of codes emerging from the first analysis and try to apply it to the subsequent analyses, with the flexibility to add new codes. It is important to test this method with other researchers, as well as testing questionnaires in pilot studies before using them with a large number of subjects in order to correct possible mistakes and check the time taken to answer the questions [98].

Semi-structured interviews Occasionally, the time available to carry out experiments is not enough for users to write long answers. At other times, users do not feel encouraged to write their own opinions. An alternative is to use semi-structured interviews. Semi-structured interviews are employed in order to support users in answering questionnaires and to give users opportunities to share their experience in a more natural way. According to Bernard [8], this is the best type of interview for questioning someone who will only be interviewed once. It has the freedom of an unstructured interview, but is based on a questionnaire guide. Audio recordings might be used to register the findings, or the researcher might take notes, and therefore users do not have to write their own answers as they would in a questionnaire. For instance, semi-structured interviewing was the method chosen to collect requirements in a project to develop accessible mobile phones for individuals with cognitive disabilities. This method was suitable for interviewing people with disabilities and for developing a contextual and detailed understanding of the role that mobile phones can play in the lives of individuals with cognitive disabilities [22]. As a result, it was found that the mobile phone should be rugged and durable, and should have a simplified menu system; charger input, voicemail access and features targeted at specific remote communication tasks (e.g. sharing ones location).

5.3 Design Methods

Design activities aim to employ the knowledge acquired in the data collection process in designing a product. The main trends that have been identified in the universe in question should be extracted and interpreted in order to give shape to the product and specify requirements. These are applied to the conceptual design, and later to the prototype. Personas and scenarios are inspired by the preliminary research in the field, and used to identify functionality, content and appearance of a prototype.

Requirements and insights elicitation Requirements are statements that define which functional and quality characteristics the system should have. Requirements usually emerge from necessities identified in fieldwork and competitor analysis. Researchers may review the literature in the field and find certain requirements that have already been collected by others to apply to a new product. There are different kinds of requirements: functional and quality requirements. According to the template proposed by [78], eight types of quality requirements can be identified: look and feel requirements; usability and humanity requirements; performance requirements; operational requirements; maintainability and support requirements; security requirements; cultural and political requirements; and legal requirements. The use of this classification helps to identify diverse requirements and to focus not only on functional requirements.

Conceptual design It is not only user requirements, but also factors involved in the activity that emerge from the fieldwork. The environmental context, communication and personal behavior might influence how tasks will be performed with the envisioned product. In addition to a list of requirements, ideas and concepts are put forward about what a system should do and how it should be presented on the interface [74]. Reference [63] explored the design ideas and concepts of place-specific computing based on fieldwork carried out in four countries. The study resulted in 36 concepts, which are available for consultation on their website. In this phase, general specifications of the interface design were developed, and only the main ideas were illustrated by a prototype.

Personas and scenarios Personas are characters that embody certain characteristics of a niche user group. They might be inspired by source documents from field studies. As designs are proposed, they can be checked against personas [46]. The use of personas avoids elastic users facilitating the communication in the development group. Hence, it is the persona who would behave in a certain manner in a specific situation, and not the designer. Personas are actors in scenarios who illustrate interaction with products. According to Carroll [17], every scenario involves at least one agent (persona) and one goal, and if the agent is different, the goal may change. Scenarios are the setting. Design cycles are contemplated in scenarios in order to illustrate interaction and user reactions in a real-life situation. Storyboards are one way to illustrate scenarios and to clarify interaction. From scenarios, it is possible to extract the main necessary elements of the system.

Scenarios to illustrate context to field trails [19] and to envision future use of Cognitive systems is also a common method applied with Protocol analysis and Think aloud techniques [102]. Reference [7] use non-wearable sensors and machine learning algorithms to identify emotions in team meetings using scenarios as a tool. Understanding emotional states of the design team members helps quantify interpersonal interactions and how those interactions might affect resulting design solutions. Participants were invited to a scenario based design meeting and a catalog of 8 body language poses relevant to emotional states was used as data. Their machine learning algorithms identify individuals body language and relates to emotional states to quantify design team interactions.

Prototyping In this stage, system features are specified and transferred to a physical design. Low-tech and high-tech prototypes are normally employed to verify modes of interaction and interface design. Low-tech prototypes are relevant to the resolution of navigational problems and errors that would be complicated to solve after coding. Paper-based prototypes and PowerPoint presentations may serve this purpose. High-tech prototypes are more similar to the final product and are usually ready to be tested by experts and users.

The use of material artefacts to elucidate human thinking is a common trend since getting requirements only with user research to design cognitive systems is not a fixed starting phase. Usually those systems use human parameters and users inputs into technological artefacts for self-improvements, applying machine-learning algorithms. Therefore, some prototypes described into papers were used with the intent to gather parameters for the future systems, as experimental investigations, and not to evaluate a prototype that represents a system. Robins [79] investigated how robotic toys could be used as a play tool to assist in the childrens development. Experimental investigations with artefacts (field trails with children), expert panels and questionnaires (with caretakers) help to develop scenarios for robots to give stimulus for autistic children that may promote further learning.

The question of how close the prototype should be to the final product was investigated by Sauer and Sonderegger [84]. Six experiments were carried out in order to study prototype fidelity. Participants were assigned to diverse user groups such as: paper prototype; computer-based prototype; fully operational appliance, highly appealing and moderately appealing prototype. The task completion time was higher for the computer-based simulation than for the fully operational appliance. On the other hand, the results suggested that perceived usability may be more strongly associated with attractiveness ratings than objectively measured usability parameters. Therefore, aesthetics is as relevant as efficiency and effectiveness for user satisfaction.

5.4 Evaluation Methods

Evaluation can be informal or formal. The first case includes techniques such as self-generated evaluation, peer reviews and casual user testing, which can provide insight into what should be improved in the final product. In the latter, expert evaluations and user testing are applied. Evaluation might occur during the design process (formative evaluation) or in the final stage (summative evaluation) [74]. As Duh [28] and Jeffries [45] both suggest, methods of usability evaluation can be categorized into four main areas: heuristic evaluation, cognitive walkthroughs, usability testing and software guidelines. The latter area is frequently employed as a guide for heuristic evaluators.

Some techniques that are employed in the early stages of development with users may also be applied in the evaluation phase. These include observation of users interacting with the product in the field and/or laboratory settings, questionnaires and interviews. In addition, cognitive walkthroughs and heuristic evaluation are the main techniques used by experts.

Usability testing When people first hear about usability testing, they sometimes assume that it is the same as a focus group. Tullis and Albert [95] in p. 58 clarify the term according to their experience:

> the similarity between the two methods begins and ends with the fact that they both involve representative participants. In a focus group, the participants commonly watch someone demonstrate or describe a potential product, and then react to it. In a usability test, the participants actually try to use some version of the product themselves. Weve seen many cases where a prototype got rave reviews from focus groups and then failed miserably in a usability test.

Giving users a version of the product to test can highlight more interesting issues than demonstrating it and asking for opinions. The same might occur according to the choice of setting (laboratory or field test). Field tests provide a real context for the users experience, despite being more time-consuming and expensive.

Similar approaches for collecting data are also effective during evaluation procedures. Feedback achieved through questionnaires, interviews and thinkaloud techniques allows designers to improve the products based on users opinions. In addition, the use of observational studies helps to clarify the users interaction with the product and the influence of contextual factors on his/her experience

Heuristic evaluation Heuristics evaluation is an expert method of examining system usability. Authors have different opinions of the essence of this technique. Jeffries [45] consider this type of evaluation to be based upon the expertise and experience of the evaluator and that it is not necessary to follow pre-determined design and usability guidelines [66, 74]. The use of a set of heuristics and guidelines is valued by Jones and Marsden [46] in p. 208 as someone who has received these reports on their cherished designs, we appreciate it when the evaluator lists which particular heuristic is being violated, to show that the assessment is based on rational evaluation and not personal opinion. This method is considered as an alternative to formal usability tests with users. It generally requires fewer resources and less time than testing with users [23, 46].

After comparing the results of a heuristic evaluation study and a formal usability evaluation by eye tracking, De Kock [23] identified several differences. First, the purpose of a heuristic evaluation differs from that of usability tests. The former aims to identify usability errors while the second focuses on effectiveness, efficiency and user satisfaction. Second, the resultant data from a heuristic evaluation are highly influenced by the experts experience. In a usability evaluation, it is possible to triangulate results due to a diverse range of methods applied in the same test observation, questionnaires and usability measures. Third, a heuristic evaluation tends to answer questions such as why and when, while usability tests consider what and how information is acquired.

Cognitive walkthrough Cognitive walkthrough is an expert evaluation method that focuses on the steps and goals taken by users in order to predict and solve problems with a future system. It was created by Polson [72] to examine users cognitive activities. Knowledgeable experts are necessary for this method, and sometimes the process is somewhat tedious [45]. The goals and steps taken by users should be well planned in order to facilitate experts reviews and to ensure that they are representative of major interactions [64]. The strengths of this technique are the focus on detailed problems experienced by users and that users do not need to be present; not even a functional prototype is necessary. However, this technique is highly time-consuming. Preece [74] presents a model (p. 703) of how to conduct a cognitive walkthrough. For every task to be accomplished, evaluators should answer yes or no to four questions, described bellow, and add their comments.

- Will the action be sufficiently evident to the user?
- Will users know what to do?
- Will users understand how to do it?
- Will users understand from feedback whether or not the action was correct?

Positive responses to the individual questions support the inference that the interface will be easily learn. Negative responses highlight those steps in an operating procedure that may be difficult to learn [53, 64].

Wizard of Oz A Wizard of OZ technique, where a human (wizard) simulates the intelligent system tasks such as natural language understanding without user awareness, was perceived as one of the main approaches to evaluate cognitive dialogue systems. Steinfeld [91] shows a diagram explaining the combinations of the Wizard of Oz technique. It summarizes these in the context of how close to reality the Wizard and the Oz are within the evaluation. (Fig. 13)

Dow [27] present a study that could be classified in the category of Wizard with OZ (Fig. 13). In their experiment, wizards were human operators type players spoken utterances; then algorithms interpret the players intention, based on a pre-authored dialogue and animate two embodied characters part of a Augmented Reality experience called Façade. Forbes-Riley and Litman [35] also applied the Wizard of Oz technique. The system was a spoken language tutoring system in which the wizard performed speech recognition, natural language understanding, and uncertainty annotation, for each student to answer. 81 students participated in the study. The

Fig. 13 Categories of Wizard of Oz [27]

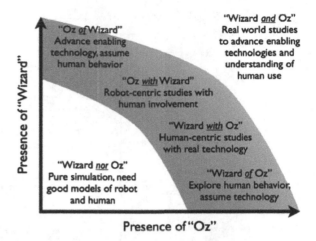

authors also claim it was the first study to show that dynamic responding to student uncertainty can significantly improve learning during computer tutoring. Rieser [77] applied the Wizard of Oz tool to improve information presentation in natural language generation dialogues; humans simulated the intelligent system that provided recommendations of restaurants to other humans. Their aim was to present enough information to users while keeping the utterances short and understandable. Authors identified the adaptive natural language generation, as well the information presentation, affects perceived or objective task success of the system.

5.5 Data Analysis Methods

As previously affirmed, this research has two strands, in order to conduct and analyze the data gathered. Qualitative approaches permeate most of the research process, but quantitative approaches are also present

Qualitative analysis The value of this method is that the researcher is able to discover recurring patterns of behavior and relationship issues in the target group. The study continues until the findings starting to repeat consistently. Different approaches can be applied for data analysis according to the nature of the research. Creswell [20], Marshall and Rossman [59] and Bryman [13] offer an overview of typical procedures in qualitative data analysis: (a) organizing the data; (b) immersion in the data reading through all data; (c) generating codes by theme and/or description; (d) finding relations among themes/description; (e) interpreting the meaning of themes/descriptions; and (f) validating the accuracy of the information.

It is advisable to carry out the first step [13] while the data are being collected. In this phase, data from interviews and video observations are transcribed, field notes are typed up and data are organized according to the source of information. The fol-

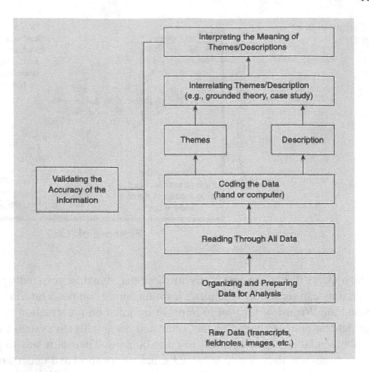

Fig. 14 Validation of data ([20] p. 185)

lowing step involves creating an overview of the issues which have emerged or been examined in the study. With the highlighted issues organized into chunks, codes are generated. Coding is the process of organizing the material into chunks or segments of text before studying the meaning of the information ([81] p. 171 and labeling it with terms [20]. There are certain tools that can make this stage faster, easier and more accurate, such as: software for analyzing videos (NVivo) and spreadsheet applications and tools that support the organization, coding and manipulation of data. After finding the mainstream categories (terms), the researcher looks for relationships, and new categories may emerge during this process. Findings are described through the interpretation of categories, which could be the researchers personal interpretation or/and the meaning derived from a comparison with the literature or theories. Therefore, information could be validated through a comparison to previous research and new questions may emerge ([20] p. 188) (Fig. 14).

The findings are described as contextual trends, and the use of quotations from participants affirms the truthfulness of the research. In addition, initial hypotheses are tested according to the results, and some questions will not have answers [20], perhaps because of the sample size, and therefore more attention should be given to this issue in future research. The use of questions and hypotheses in the initial phase of the study helps the researcher to understand the events, participants and setting [38]. Not only do the recognized patterns, categorizations and hypotheses provide options for analyzing the data, but also the recognition of critical incidents. In interaction design, critical incident analysis has been used in a variety of ways, but the main focus is to identify specific incidents that are significant, and then to focus on these incidents and analyze them in detail, using the rest of the data collected as a context to inform the interpretation ([74] p. 384).

Mixed methods analysis Similar incidents/categories may be seen as emphasizing the strength of issues in the data. Therefore, mixed methods might be applied. Data transformation and content analysis are examples of this approach.

Data transformation is a quantification of qualitative data. This involves creating qualitative codes and themes, and then counting the number of times they occur in the textual data. This enables researchers to compare quantitative results with qualitative data ([20] p. 218).

A more structured and well-known methodology in this line is content analysis. This is considered to be a quantitative approach, despite the fact that in the first steps of the analysis, categories and codes are not always pre-defined. Patterns emerge from the data, key aspects are identified and their frequency is counted. Hence, content analysis reduces qualitative data to a quantified form [56]. It is a flexible, objective and unobtrusive method ([13] p. 289). It is easy to identify the steps of the content analysis approach in Brymans study aimed at analyzing the combined use of qualitative and quantitative research in 232 articles.

First, the rationale given by the authors for combining the two approaches to data collection and/or analysis is coded. For this exercise, the reasons that were given before the findings were presented are typically examined. Then, the ways in which quantitative and qualitative research were actually combined are coded. This coding presents the authors reflections on what they feel has been gleaned from combining quantitative and qualitative research, and any ways in which the two were combined which were not reflected in the authors accounts. The purpose of discriminating between these two ways of thinking about the justification for multi-strategy research was that authors accounts of why they intended to combine quantitative and qualitative research might differ from how they actually combined the methods in practice [13].

Paay [68] describe how content analysis was undertaken in order to analyze and represent peoples understanding of the physical and social aspects of urban settings in order to develop a digital environment. Elements of the physical environment were recorded in the form of photographs and field notes. These elements were coded according to five categories proposed by Lynch [58] landmarks, districts, nodes, edges and paths based on the focal element in each image. Elements were also coded based on 253 patterns, which were investigated by Alexander [2]. Later, maps of the setting Federation Square, Melbourne, Australia were created featuring the same elements. The social context was analyzed using rapid ethnographic methods and contextual interviews. In the opinion of the researchers [69], social and physical elements in the space would not be noticed if the researcher merely examined the original data or visited the space. The results were applied in a pervasive prototype to enrich peoples experiences of Federation Square. For example, the new system identifies peoples previous interactions in the same place, and so it is possible to access information about familiar paths and places that have been visited, the estimated waiting time for a friend, and navigation based on known landmarks.

Quantitative analysis Interaction design data are typically analyzed using simple statistics [74]. The issues that are relevant in the context of the study are denominated as variables. Variables are classified as nominal, ordinal, interval, ratio, dependent and independent. Nominal variables can be defined using categories that are qualitative in nature (e.g. gender). Ordinal ones vary according to degrees (e.g. satisfaction). Interval variables are when the difference between two values is meaningful (e.g. temperature). Ratio variables have the same characteristics as interval variables, but a zero value exists [75, 82]. Dependent variables record the effect provoked by the independent variable, and this is what is measured [4, 82, 93]. For instance, an experimental evaluation of a mobile guide in the field used age as the independent variable and usability measures as dependent variables. The authors [40] suggest a summary of usability measures they found relevant and recommended methods to be employed for evaluating mobile guides based on the study. These included timings, errors, perceived workload, distance traveled and route taken, walking speed and comfort.

The data may be described as descriptive, without saying anything about the large population size. Another type of classification is inferential, from which conclusions can be drawn about a large population [95]. Descriptive statistics covers the measures relating to central tendency, tables and cross-tabulation, which are other ways of investigating data [13, 56, 95].

Measures of central tendency are denominated as the mean, median and mode. The mean is the average score of the dataset, in the everyday sense of the term. The median is the central value of the distribution (half of the values are smaller than or equal to this value). The mode is the value that occurs most frequently within the dataset [56, 75, 82]. When the data have a more limited set of values (such as subjective rating scales), the mode is more useful than when the data are continuous (e.g. completion times) [95].

Tables are used to display counts and percentages for individual variables or to compare one or more variables (cross tabulation). Cross tabulation tables show whether and to what extent two or more nominal-level variables are related. They display the frequency and/or percentage of the categories of one variable cross-tabulated with the frequency and/or percentage of another variable or variables ([82] p. 188). Quantitative data analysis might be assisted by a statistics software package, such as Minitab.

5.6 Using Information Visualization to Understanding Users

Information visualizations are important tools to unveil hidden information in Big Data sets. Information visualization is defined as "The use of computer-supported, interactive, visual representations of data to amplify cognition by Card [16] p.6. Spencer [90] completes affirming that visualization is a process of forming mental model of data, thereby gaining insight into that data. Visualization is a human cognitive activity, not something that a computer does in Spencer [90] views. Thereby, control of information is given over to viewers, not to editors, designers or decorators [94]. To unveil the mental model in every persons mind is a cumbersome activity. Therefore, user experience researchers and designers may dispose what will be formed in peoples mind understanding better their contexts and practices in everyday life. Visualizations have their own purpose give insights on data to solve or clarify certain problems. And it posts a critical question: How best to transform the data into something that people can understand for optimal decision-making? [100]. User cantered design approaches; such as user evaluation studies may enlighten this question.

Designers may employ multimedia elements to built new or redesigned interfaces to improve users interpretation of visual and verbal elements. According to Bertin [10]: A graphic is no longer drawn once and for all; it is constructed and reconstructed until it reveals all the relationships constituted by the interplay of data. In other words, the best graphic operations are those carried out by the decision maker itself. Bertin [10] also created a visual grammar to make designers aware of visual elements characteristics when choosing them to compose a graph. In Tufte [94] p. 8 words When principles of design replicate principles of thought, the act of arranging information is becomes an act of insight. A study and tool proposed by Viegas [96] allows individuals upload data, collaborate and generate visualizations at a large scale in a public website called Many Eyes. This tool also provides a wide range of visualizations types that may help designers to compare and choose appropriate visualizations for different contexts. Baur [6] evaluates visualization techniques with large datasets and recommended-based systems on mobile phones. They evaluated the visualization technique for repeated item selection in the context of music playlist creation. They considered the particularities of the mobile phone devices (orientation) with 12 users to do a user trail. Users selected options and for each option five suggestions were given out, and one should be right selected. Authors measured completion times and error rates. They found that the vertical orientation and interface was faster to interact and had less error rate than the horizontal one. Arshad [3] compared the confidence of expert users and non-expert users varying level of uncertainty presented on a prediction case study of water-pipe failure. Participants did three groups of tasks and received a viewgraph of overlapping and non-overlapping uncertainty presentations as supplementary material for decision-making. Showing this supplementary material improved user confidence and uncertainty with unknown probabilities decreasing user confidence, although uncertainty with known probabilities can increase expert user confidence but the same is not true for non-experts.

As noticed, current methods for designing cognitive system did not differ hugely from traditional design processes. Methods and approaches are applied to understand users to inform the design artefacts and evaluate those with users. The iterative design process is more frequent and necessary when designing and developing Cognitive systems. Cognitive systems use human data as input with the intent to improve and learn in an iterative process. Context is crucial and changes over time. Therefore, design process phases are less linear. Several times prototypes are crucial to gather data and inform the design, and many times prototypes are used as experiments to develop the real system. Primary research, where data gathering methods are applied, usually focuses on context and assembles the basis to the design process. It is not a phase that is isolated as in some traditional design process were ethnography work can be the first phase to inform the overall design process. Evaluation methods are applied in several stages of the design process and not only as the final phase. The evaluation phase never ends, because those systems as much as they learn and are assessed using user information they become well and more user friendly. More studies are necessary to unveil ways of evaluating the huge amount of data available nowadays. In doing so, helping designers, practitioners and researchers to create more effective visual analytics tools and cognitive systems.

6 Cognitive Systems

6.1 IBM Watson

Pretty much like the previous section, any approach about IBM Watson can be made from, at least, 3 different angles

- The media impact of a Deep Qs and As systems which defeated 2 human beings at the popular TV program Jeopardy
- The technology behind the system shown in the TV and
- The consequences of this experiment in terms of science and business

Lets try to do an adequate coverage in each of these possibilities.

The prospective strategy of the IBM company is highly supported by an yearly work intense exercise lead by its Research Division and called GTO Global Technology Outlook. From the GTO, emerge the major opportunities for the foreseeable future and the challenges associated with them, of course.

Back in 2007, the IBM Research Division took the grand challenge of building a computer system that could compete with champions at the game of Jeopardy. In 2011, this system, named Watson, came to life with an excellent result. Beyond any technicality to describe the computer system, one has to have in mind that the Watson competing in the Jeopardy show was an open-domain question-answering system.

In fact, for those interested in the deep technical details of the Watson system, there are several information sources but one of them is essential: the IBM Journal of Research and Development [61], from which most of this chapter has been largely inspired.

In the following lines, we will give an overview from these 3 angles, calling the attention to some important aspect, which can go unnoticed by the reader.

Let's begin by calling the attention to the fact that most of human communication, weather in natural language text, speech or image, is unstructured. Therefore, semantics necessary to interpret unstructured information to solve problems is often implicit and must be derived by using background information and inference. On the other side, with structured information, such as traditional database tables, the data is well-defined, and the semantics is explicit.

To manage unstructured information, "from 2001 through 2006, IBM built the Unstructured Information Management Architecture (UIMA) to facilitate this kind of basic interoperability. UIMA is a software architecture and framework that provides a common platform for integrating diverse collections of text, speech, and image analytics independently of algorithmic approach, programming language, or underlying domain model. UIMA is focused on the general notion of integrating a scalable set of cooperating software programs, called annotators, which assign semantics to some region of text (or image or speech). In 2006, IBM contributed UIMA to Apache,[3] and it is currently in regular use around the world by industry and academia. UIMA provides the essential infrastructure needed to engage large-scale language understanding research".

In an evolutionary perspective, UIMA may well be considered the first foundational roadblock in this journey. But much more were required to progress in towards the great challenge.

In 2007, a baseline solution was used to measure the capability against Jeopardy requirements. The precision performance was nothing better than 16% and this mark was not even enough to qualify for the game. This fact has clearly shown what should be the necessary effort ahead, to face the challenge with some chances to win. After 4 year and many corrections, the project achieved the system called DeepQA framework, with encompassed "more than 100 core algorithmic components. These components were designed to understand questions, search for candidate answers, collect evidence, score evidence and answers, produce confidences, and merge and rank results". As described by Ferrucci [34], the illustration in Fig. 15 summarized the blueprint of DeepQA systems which competed in the Jeopardy TV show.

No question about the success and media impact that the DeepQA system named Watson has caused, in wining two of the top competitors of Jeopardy, in front of a huge TV audience. However, from this success two facts emerged:

- A number of lessons have been learned on how to deal with the complexity in many aspects of what could be considered a cognitive system in the near future and
- The evolutionary aspect of the system design, which in fact has evolved very fast over time.

---

[3]https://uima.apache.org.

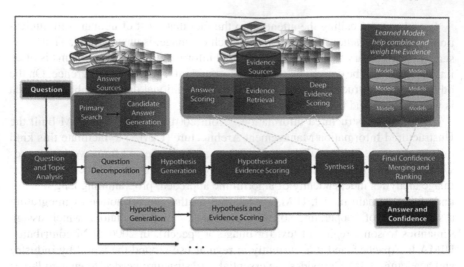

Fig. 15 DeepQA architecture (extract from [34])

The consequences of the experiment went far beyond science and technology! First, after 3 years forming and refining a new set of workable alternatives based on what had been learned in the experiment, IBM created a new line of business named Watson. Second, a totally new and innovative paradigm on how to manage information came up, now fed by unstructured information. At this point, we call the attention to the fact that, for almost 100 years, the IT industry has pushed the market to the direction of structured information but in the last 10 years, with the emergence of Social Computing, unstructured information became the rule. So, finding a way to deal with unstructured information as a tool to support decision making in the business environment was not just a question of transformative innovation. It was a question of survival of the industry!

And third, IBM has once more, fulfilled the tradition of creating every decade, a global agenda as it had made with e-Business in the late 90s.

Today, the evolutionary road took Cognitive Computing to a different format as a number of simplified alternatives came up, closer to the end user and available in a cloud environment. These alternatives do exist under the broad definition of API Application Product Interface, are grouped as services, by function and by industry. The number of APIs grows very fast making impossible to make a list. They are available in the cloud computing architecture of IBM, named BlueMix.[4]

Many advances in analytics and machine learning have been based on our understanding of how the brain works. Deep learning is no exception it takes its inspiration from our understanding of the cortex in the brain. The brain has many regions which form a hierarchy of processing, where sensory data flows from one region to another, being transformed and combined with other information along the way. While it may

---

[4]https://console.ng.bluemix.net/.

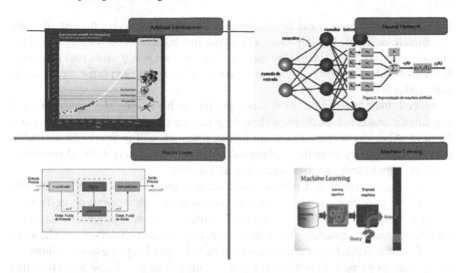

Fig. 16 IT infrastructure evolution

seem instantaneous when we recognize a face or a voice, there are actually many stages of processing between our senses and a set of neurons that we can clearly link to that particular person.

A pictorial review of what has happened in the last 30–35 years and summarized in 4 quadrants is worthwhile to illustrate (Fig. 16) former attempts to create a more intelligent IT infrastructure.

In the early 80s, we started mimicking the brain through what was then called Artificial Intelligence. The most we have achieved was to mimic the brain of a mouse. Then, began the Neural network effort which has created a good foundation to the current Multiple Agents. Then the Fuzzy Logic to support complex mathematical models. And lastly, the Machine Learning projects. Cognitive Computing was the evolution in this sequence.

Today's cognitive computing solutions build on established concepts from artificial intelligence, natural language processing, ontologies, and leverage advances in big data management and analytics. They foreshadow an intelligent infrastructure that enables a new generation of customer and context-aware smart applications in all industries.

6.2 Other Cognitive Systems

To increase the longevity of this chapter, we opted for a different approach to this section because Cognitive Computing drove great motivation in the IT world and new possible solutions are coming up every day in different formats: ideas, proposals and projects some of them already under development. Thus, any attempt to identify the other existing or near-existing Cognitive Systems would carry the risk of being incomplete or simply leaving some of them out just because we have not yet had the chance to get to know them. However, we have not seen any other real Cognitive Systems besides the IBM solutions so far. And real in the former phrase means under full production.

Indeed, the conversation about this topic has to begin with the review of what cognitive is and is not. To do so, we have to get into areas slightly beyond the regular IT field. Also, considering the youth of this area, we will probably go through a period of temporary definitions which will evolve to a more crystallized version of themselves until they are considered final and frozen.

Said that, the simplest definition could be: a cognitive system is a system with the ability to create cognition. Then, we have to define what cognition is. Turning to the dictionary, we find that cognition is The mental action or process of acquiring knowledge and understanding through thought, experience, and the senses. The word mental relates, at this level, the ability to execute the cognition process only to humans or, at least, to living beings that can execute a mental action. These basic thoughts are all useful to support the appearance of new systems that have now some ability to acquire knowledge and understand through thought, experience and, eventually, senses.

One example is Google's DeepMind [5] [41], which is a high performance computer that is inspired by brain's short-term memory properties. The computer is built with a neural network that interacts with external memory. Scientists can use this technology to store experiments, like the brain, and the computer compares to previous data queries and responds more efficiently.

The Zeroth Cognitive Computing Platform[6] relies on visual and auditory cognitive computing to recognize the environment around you so it can capture things that matter to users. This platform allows a device running the platform could recognize objects, read handwriting, identify people and understand the overall context of a setting. One example is that a device could adjust automatically when taking pictures at a sunny beach or a moonlight walk. It could also adjust its microphone settings automatically if there is a lot of background noise to deliver better sound quality. Its ability to understand scenes and context means that partnered with the right devices and image technology, it cloud decipher how people are feeling based off facial expressions, or if their voice reflects levels of stress or anger.

The evolution of IT, mainly in the software arena, has created a number of system features that might suggest they were cognitive. Making this comparison very simple: while I am writing this text, the editor makes a number of corrections. In some cases, the editor decides by itself what needs to be corrected. In other cases, it checks with me if the correction should be made. In the past, somebody could even say that this ability of the text editor is due to knowledge acquisition and understanding through thought and experience and, therefore, the text editor would be considered a cognitive solution. However, today we know that this ability of text editors is related only to

---

[5]https://deepmind.com.

[6]https://www.qualcomm.com/invention/cognitive-technologies/zeroth.

the ability of quick processing of some arithmetic rules, using traditional procedural programming and few heuristic rules.

The conclusion so far is that a cognitive system, as most of the innovative creations in the technology arena, will go through an evolving onset and what is considered cognitive today may not be so cognitive in the near future. With that in mind, we can now try to evaluate the existence of other cognitive systems but rather than naming two or three, lets try to split them into segments we can envision.

From the perspective of the end user:

All Purpose Virtual Assistants: Siri,[7] Alexa,[8] Cortana[9] and others are, undoubtedly, the most popular because of its user-friendliness of the systems said to be cognitive. We will not waste any time here analyzing if they are cognitive or not. Instead, we will accept that their success is, to a great extent, due to a high quality vocal or if you prefer, voice recognition interface. This will be an ever growing segment of cognitive systems in the future and may replace current call center services.

Domain Specific Solutions: less popular than all-purpose assistants, a domain-specific solution can support the decision making process of a person, in their professional role or not, inside a specific domain. In fact, there may be many of these systems already running, but as they are used in connection with some business need (like stock market investment counseling), they are not made public very often.

Public Domain Based Solutions: these solutions are based on information of public interest and will help the user, often a citizen, find an adequate answer to a problem related with public administration at any level. This answer can be related to public services such as the timetable of public transportation and the best alternative to go from A to B , to local regulations such as the requirements to open a new business , or even to safety and security aspects of a community.

From the perspective of the technology and model of operation:

Systems based on existing APIs: such as the ones available in the BlueMix cloud computing environment of IBM. These APIs are focused on a single function, such as relationship extraction, classification, translation, etc. The API will execute only the cognitive function for which it was designed and developed and will usually work on top of a defined segment of cumulated feeding knowledge.

Systems based on very large scale structured content: such as the ones used to guide medical treatment of a particular disease. These systems will be based on clinical protocols, properly supported medical bibliography, consisting of a vast amount of information but already partially or totally curated by the editor of the scientific periodical on which the paper has been published. Given the nature of the input

---

[7]https://www.macstories.net/news/apple-officially-unveils-siri-voice-assistant/ visited 13th March 2016.

[8]http://www.alexa.com/about visited 13th March 2016.

[9]http://research.microsoft.com/en-us/news/features/cortana-041614.aspx visited 13th March 2016.

information and the format of the answer, these systems can benefit from machine learning processes. Gastronomy assistants- such as the Chef Watson,[10] of IBM can also be included in this category.

Systems based on very large amount of unstructured content: such as the ones designed to fulfill the need to solve a specific problem where the solution will be based on a substantial amount of unstructured data. These systems will require good sources of trustable information, permanent curation and the content will be submitted to the process of ingestion, which will generated annotated corpora, after some semantic processing. These are the solutions which will fit properly in cases of legal processes, to help lawyers and eventually judges.

The combination of the 3 categories of users with the 3 categories of technology and operation will result in a 9-alternative matrix of possibilities, which is enough. For now. As much as the technology evolves, there are always needs to be met as new users keep coming, bringing new challenges. That is why the alternatives of the matrix will expand. Aspects regarding this expansion will be discussed in the next section of this chapter.

7   The Future of Cognitive Systems

With different semantic formats, the future of Cognitive Systems has been spread in both scientific and popular publications with the same phrase: AI will be present in our future. This prediction has been based mostly on the history of the IT evolution itself: considering this arena formed basically by 2 segments hardware and software we can say that, periodically, we face disruptive innovation either in the tools to handle information (the hardware) or in the way we explore these tools (the software).

About hardware, a new architecture for cognitive computing is needed, one inspired from human brain. Distributed data processing will be largely used and processing and the memory will be closely integrated in spite of to reduce the shuttling of data and instructions back and forth. Discrete processing tasks will be executed in parallel instead of serially. A cognitive computer employing these systems will respond to inquires more quickly than todays computers; less data movement will be required and less energy will be used.

Evolution in hardware has been based on the Moores Law for a long period of time. Also, we have heard, for quite some time, that Moores Law was hitting in a wall represented by the restriction of materials themselves, which could not cope with the growing need to create smaller, faster and cheaper computers circuitry. Well, in light of the challenging obstacle, the IT hardware industry is now struggling with what will be the new disruptive innovation in this segment and Quantum Computing seems to be the Holy Grail in this case.

---

[10]https://www.ibmchefwatson.com/ visited 13th March 2016.

However, running in another track and maybe in a lower speed , the software technology has evolved to Cognitive Computing, a disruptive innovation by itself. So, its time to enjoy the new rule of the game and be cognitive. Of course, marketing people have already noticed how promising the new wave of Cognitive Computing is and, out of a sudden, everything becomes cognitive. Again, it is very likely that what we will see here is the same pendulum movement that we observe in many other evolutionary and revolutionary events that the technology brings to our lives: on one side, some skepticism and lack of trust in the new technology, on the opposite side, where the new technology seems to be the panacea that heals all wounds. After some time, equilibrium will be established somewhere between these two extremes.

Nowadays, we see a plethora of suggestions on how to use Cognitive Computing to address the human needs to solve problems, using some of the alternatives mentioned in the former section of the chapter. Some of these suggestions will turn into projects and some of these projects may generate cognitive solutions that will be available to the end user. This is the point when they will face their moment of truth will they succeed or succumb? Anyhow, some facts that will undoubtedly outline the future of Cognitive Systems.

Unstructured data will be more and more useful to feed these systems. Also, sensors and actuators evolve, other formats of input will also feed these systems. IoT Internet of Things is a promising field for this.

From the analytics point of view of cognitive computing will contribute to guide data scientist in their journey. Thus, it not will be so dependent of his/her, since cognitive computing will help in the decision making of which technique should be applied to solve the problem and which is the best method for that dataset. This will help companies make sense and take decision using the ocean of data that we are dive. The human-in-the-loop will be each day more present in this type o approach, it was the base of KDD process where the domain specialist and the expert in KDD where the base of this process, know it will help guide and change the decision and parameters during execution time.

Is a common sense that many advances in analytics and machine learning have been based on our understanding of how the brain works. The flavor of the month, deep learning, is inspired in the processes of the cortex in the brain. The brain has many regions which form a hierarchy of processing, where sensory data flows from one region to another, being transformed and combined with other information along the way. While it may seem instantaneous when we recognize a face or a voice, there are actually many stages of processing between our senses and a set of neurons that we can clearly link to that particular person.

Certainly theres a lot of variety to deep learning algorithms, and were likely to see many new variations over the next years as more applications are developed. The current crop of deep learning is drawing on only a fraction of what is known about real neurons and brains, indicating huge potential for this line of scientific exploration.

Not only deep learning, but other machine learning techniques will need to evolve to allow cognitive computing to be better and succeed. In the same way other technologies as data processing with platforms as Spark, data curation, ontologies, reasoning, machine reading, NLP, human computer interaction, visualization and so on. In fact, the hole computer science will need to evolve so cognitive computing can evolve.

After this, inspired by the human cognition, other forms of input will soon be accepted to be ingested by Cognitive Systems. Image utilization is around the corner.

It is fair to say that in a foreseeable future, machines, called by AI or not, will reach a state of of acquiring knowledge and understanding through thought, experience, and the senses. When this happens, we will have reached the end of another evolutionary process for the information technology. The next challenge might be making machines that are able to emulate our emotions, such as love and happiness, and capture the more refined nuances of our intellect, which would make them able to understand irony and, ultimately, to lie in a more humanly fashion [36].

Interdisciplinary studies from computer science, psychology and cognitive science are already helping scientists to emulate emotions. Emotions impact cognitive processes. Reference [70] demonstrated that emotions play an essential role on human creativity, intelligence and also in rational human-thinking and decision making. Empathy is turning out to be one of the most important emotion computers should emulate, to start a conversation and build trust along of the interaction sessions [42].

Scientists are discussing the current social-technical challenges for collaboration with intelligent/autonomous systems and some cooperative work between human and non-human actors are starting to emerge [29]. Discussions about robots and systems as cooperative partners are also in evidence [51, 89]. Additionally, communication modes are still an area to explore, for example human like robots and systems interaction my cause aversive feedback from people [92] and discomfort.[11]

New promises, new ways of interaction and design challenges are part of the everyday life of scientists, designers and consumers of cognitive systems.

8 Final Remarks

The construction of this chapter may summarize the creation of Cognitive Computing itself: challenging, time consuming, working intensive, required multidisciplinary skills and approach and in many points, was not totally conclusive. But the authors really enjoyed the opportunity to impose the discipline required by the scientific writing process to a sort of knowledge which has been sparsely acquired in the last few years and, although available, might not be very well organized. When applying the organization process in a collaborative task, some answers come up but also new questions. Lets begin by the answers:

---

[11] http://arstechnica.com/gadgets/2016/04/how-would-you-feel-if-a-robot-asked-you-to-touch-its-buttocks/.

- Data will be the most efficient fuel to speed up the simpler modalities of Cognitive Systems at the onset of this new era. If the phrase of Clive Humby[12] was a concept still pending to be proved, no question about it anymore: definitely data is the new oil!
- After more than 100 years pushing the world towards structured data, we cannot simply disregard them. Therefore, the data analytics approach will still have a great appeal for a long while.
- Because of the strong appeal of structured data, machine learning methodologies are also very appealing.
- But the Social Computing has been pushing the IT world toward non-structured information and in the mid to long term, they will become the preferred fuel to ignite cognition. Therefore, Cognitive Systems may well become more efficient, more comprehensive and more complex.
- These systems will learn and interact to provide expert assistance to scientists, engineers, lawyers, and other professionals in a fraction of the time it now takes. Far from replacing our thinking, cognitive systems will extend our cognition and free us to think more creatively.

Now, one questions and one requests:

Do you believe that the traditional bibliographic reference can fulfill the curiosity of the interested reader? We bet it cannot and for this reason, we have added other non-traditional sources of information, mainly to reference interesting debates.

Besides this regular reference, we would like to add here some sites with excellent content about Cognitive Computing, most of them are discussions about non-conclusive topics but which are very compelling for all of us:

- Dario Gil's presentation at TED: https://www.ted.com/watch/ted-institute/ted-bcg/dario-gil-the-next-area-of-cognitive-systems
- IBM NYC Cognitive Colloquium: http://www.research.ibm.com/cognitive-computing/index.shtml#fbid=o6epbokzGUL?hashlink=rpi
- Financial Times: AI Can Watson Save IBM? http://www.ft.com/intl/cms/s/2/dced8150-b300-11e5-8358-9a82b43f6b2f.html#axzz3wwMehMjI
- Searching for Eureka: IBMs path back to greatness, and how it could change the world http://qz.com/567658/searching-for-eureka-ibms-path-back-to-greatness-and-how-it-could-change-the-world/

We are pretty curious to know if this chapter will attend the needs of our readers! Please, do not hesitate in contacting us with critics, suggestions or any other feedback.

References

1. V. Abramova, J. Bernardino, P. Furtado, Which nosql database? a performance overview. Open J. Databases (OJDB) 1(2), 17–24 (2014)

---

[12]http://ana.blogs.com/maestros/2006/11/data_is_the_new.html.

2. C. Alexander, S. Ishikawa, M. Silverstein, *A Pattern Language: Towns, Buildings, Construction*, vol. 2 (Oxford University Press, Oxford, 1977)
3. S.Z. Arshad, J. Zhou, C. Bridon, F. Chen, Y. Wang, Investigating user confidence for uncertainty presentation in predictive decision making (2015)
4. E.R. Babbie, *The Practice of Social Research*, vol. 112 (Wadsworth publishing company Belmont, CA, 1998)
5. A. Bangor, P. Kortum, J. Miller, Determining what individual sus scores mean: adding an adjective rating scale. J. Usability Stud. **4**(3), 114–123 (2009)
6. D. Baur, S. Borin, A. Butz, Rush: repeated recommendations on mobile devices, in *Proceedings of the 15th international conference on Intelligent user interfaces* (ACM, 2010), pp. 91–100
7. I. Behoora, C.S. Tucker, Machine learning classification of design team members' body language patterns for real time emotional state detection. Design Stud. **39**, 100–127 (2015)
8. H.R. Bernard, *Social research methods: Qualitative and quantitative approaches* (Sage, 2012)
9. A. Bernstein, F. Provost, S. Hill, Toward intelligent assistance for a data mining process: an ontology-based approach for cost-sensitive classification. IEEE Trans. Knowl. Data Eng. **17**(4), 503–518 (2005)
10. J. Bertin, *Semiology of Graphics: Diagrams, Networks, Maps* (1983)
11. L. Blaxter, *How to Research* (McGraw-Hill Education, New York, 2010)
12. J. Brooke, Sus-a quick and dirty usability scale. Usability Eval. Ind. **189**(194), 4–7 (1996)
13. A. Bryman, *Social Research Methods* (Oxford University Press, Great Britain, 2008)
14. B.G. Buchanan, E.H. Shortliffe, *Rule Based Expert Systems: The Mycin Experiments of the Stanford Heuristic Programming Project (The Addison-Wesley Series in Artificial Intelligence)* (Addison-Wesley Longman Publishing Co. Inc, Boston, MA, USA, 1984)
15. F. Buttussi, L. Chittaro, D. Nadalutti, Bringing mobile guides and fitness activities together: a solution based on an embodied virtual trainer, in *Proceedings of the 8th conference on Human-computer interaction with mobile devices and services* (ACM, 2006), pp. 29–36
16. S.K. Card, J.D. Mackinlay, B. Shneiderman. *Readings in Information Visualization: Using Vision to Think* (Morgan Kaufmann 1999)
17. J.M. Carroll, Five reasons for scenario-based design. Interact. Comput. **13**(1), 43–60 (2000). doi:10.1016/S0953-5438(00)00023-0
18. J.M. Carroll, *HCI Models, Theories, and Frameworks: Toward a Multidisciplinary Science* (Morgan Kaufmann, 2003)
19. A.R. Chatley, K.Dautenhahn, M.L. Walters, D.S. Syrdal, B. Christianson. Theatre as a discussion tool in human-robot interaction experiments - a pilot study, in *Third International Conference on Advances in Computer-Human Interactions, ACHI '10* (2010), pp. 73–78, scenarios theatre
20. J.W. Creswell *Research Design: Qualitative, Quantitative, and Mixed Methods Approaches* (Sage publications, 2013)
21. M. Daradkeh, Exploring the use of an information visualization tool for decision support under uncertainty and risk, in *Proceedings of the The International Conference on Engineering & MIS* (ACM, 2015), pp. 1–7
22. M. Dawe, Understanding mobile phone requirements for young adults with cognitive disabilities, in *Proceedings of the 9th international ACM SIGACCESS conference on Computers and accessibility* (ACM, 2007), pp. 179–186
23. E. De Kock, J. Van Biljon, M. Pretorius, Usability evaluation methods: mind the gaps, in *Proceedings of the 2009 Annual Research Conference of the South African Institute of Computer Scientists and Information Technologists* (ACM, 2009), pp. 122–131
24. P.B.C. de Miranda, R.B.C. Prudêncio, A.C.P.L.F. Carvalho, C. Soares, A hybrid meta-learning architecture for multi-objective optimization of SVM parameters. Neurocomputing **143**, 27–43 (2014)
25. N.K. Denzin, Y.S. Lincoln, *Handbook of Qualitative Research* (Sage Publications Inc, 1994)
26. F.X. Diebold, A personal perspective on the origin (s) and development of 'big data': the phenomenon, the term, and the discipline, second version (2012)

27. S.P. Dow, M. Mehta, B. MacIntyre, M. Mateas, Eliza meets the wizard-of-oz: blending machine and human control of embodied characters, in *Proceedings of the SIGCHI Conference on Human Factors in Computing Systems* (ACM, 2010), pp. 547–556

28. H.B.L. Duh, G.C.B. Tan, V.H. Chen, Usability evaluation for mobile device: a comparison of laboratory and field tests, in *Proceedings of the 8th Conference on Human-Computer Interaction with Mobile Devices and Services* (ACM, 2006), pp. 181–186

29. L. Emanuel, J. Fischer, W. Ju, S. Savage, Innovations in autonomous systems: Challenges and opportunities for human-agent collaboration, in *Proceedings of the 19th ACM Conference on Computer Supported Cooperative Work and Social Computing Companion* (ACM, 2016), pp. 193–196

30. T. Erikson, H. Simon, Protocol analysis: Verbal reports as data. Technical report (MIT Press, 1985)

31. W. Fan, A. Bifet, Mining big data: current status, and forecast to the future. SIGKDD Explor. Newsl. **14**(2), 1–5 (2013)

32. U.M. Fayyad, G. Piatetsky-Shapiro, P. Smyth, Advances in knowledge discovery and data mining. Chapter From Data Mining to Knowledge Discovery: An Overview (American Association for Artificial Intelligence, Menlo Park, CA, USA, 1996), pp. 1–34

33. D. Ferrucci, A. Lally, Building an example application with the unstructured information management architecture. IBM Syst. J. **43**(3), 455–475 (2004)

34. D.A. Ferrucci, Introduction to this is watson. IBM J. Res. Dev. **56**(3.4), 1:1–1:15 (2012)

35. K. Forbes-Riley, D. Litman, Designing and evaluating a wizarded uncertainty-adaptive spoken dialogue tutoring system. Comput. Speech Lang. **25**(1), 105–126 (2011)

36. W. Gibson, *Neuromancer: Roman.* (Heyne, 1992)

37. B.G. Glaser, A. Strauss, *Discovery of Grounded Theory* (Aldine, London, 1967)

38. J.P. Goetz, M.D. Lecompte, *Ethnography and Qualitative Design in Educational Research* (Academic Press, Orlando, Fl, 1984)

39. R.L. Gold, Roles in sociological fieldwork. Soc. Forces **36**, 217–223 (1958)

40. J. Goodman, S. Brewster, P. Gray, Using field experiments to evaluate mobile guides, in *Proceedings of HCI in Mobile Guides, workshop at Mobile HCI*, vol. 2004 (Citeseer, 2004)

41. A. Graves, G. Wayne, I. Danihelka, Neural turing machines. *CoRR*. arXiv:1410.5401 (2014)

42. T.D. Huynh, N.R. Jennings, N.R. Shadbolt, An integrated trust and reputation model for open multi-agent systems. Auton. Agents and Multi-Agent Syst. **13**(2), 119–154 (2006)

43. G. Iacucci, K. Kuutti, R. Mervi, On the move with a magic thing: role playing in concept design of mobile services and devices, in *Proceedings of the 3rd Conference on Designing Interactive Systems: Processes, Practices, Methods, and Techniques* (ACM, 2000), pp. 193–202 347715 193-202

44. H.V. Jagadish, J. Gehrke, A. Labrinidis, Y. Papakonstantinou, J.M. Patel, R. Ramakrishnan, C. Shahabi, Big data and its technical challenges. Commun. ACM **57**(7), 86–94 (2014)

45. R. Jeffries, J.R. Miller, C. Wharton, K. Uyeda, User interface evaluation in the real world: a comparison of four techniques, in *Proceedings of the SIGCHI conference on Human factors in computing systems* (ACM, 2013), pp. 119–124

46. M. Jones, G. Marsden, *Mobile Interaction Design* (Wiley, Glasgow, 2006)

47. J. Joo, Adoption of semantic web from the perspective of technology innovation: a grounded theory approach. Int. J. Hum. - Comput. Stud. **69**(3), 139–154 (2011)

48. A.K. Karun, K. Chitharanjan, A review on hadoop - hdfs infrastructure extensions, in *2013 IEEE Conference on Information Communication Technologies (ICT)* (2013), pp. 132–137

49. J.E. Kelly, S. Hamm, *Smart Machines: IBM's Watson and the Era of Cognitive Computing* (Columbia University Press, New York, NY, USA, 2013)

50. J.O. Kephart, J. Lenchner, A symbiotic cognitive computing perspective on autonomic computing, in *2015 IEEE International Conference on Autonomic Computing (ICAC)* (2015), pp. 109–114

51. C. Lampe, B. Bauer, H. Evans, D. Robson, T. Lau, L. Takayama, Robots as cooperative partners... we hope.., in *Proceedings of the 19th ACM Conference on Computer Supported Cooperative Work and Social Computing Companion* (ACM, 2016), pp. 188–192

52. D. Laney, *3D data management: Controlling data volume, velocity, and variety* (Technical report, META Group, 2001)
53. C. Lewis, P. G. Polson, C. Wharton, J. Rieman, Testing a walkthrough methodology for theory-based design of walk-up-and-use interfaces, in *Proceedings of the SIGCHI Conference on Human Factors in Computing Systems, CHI '90* (ACM, New York, NY, USA, 1990), pp. 235–242
54. M. Li, J. Mao, Hedonic or utilitarian? exploring the impact of communication style alignment on user's perception of virtual health advisory services. Int. J. Inf. Manag. **35**(2), 229–243 (2015)
55. L. Liu, Q. Zhou, J. Liu, Z. Cao, Requirements cybernetics: Elicitation based on user behavioral data. J. Syst. Softw. (2015)
56. S. Love, *Understanding Mobile Human-Computer Interaction* (Elsevier, Oxford, 2005)
57. J. Lundberg, R. Gustavsson, Challenges and opportunities of sensor based user empowerment, in *2011 IEEE International Conference on Networking, Sensing and Control (ICNSC)* (2011), pp. 463–468
58. J. Manson, *Qualitative Researching* (Sage Publications, Great Britain, 2002)
59. C. Marshall, G.B. Rossman, *Designing Qualitative Research* (Sage, 2011)
60. M. Matlin, *Cognitive Psychology* (Wiley, 2009)
61. M.C. McCord, J.W. Murdock, B.K. Boguraev, Deep parsing in watson. IBM J. Res. Dev. **56**(3.4), 3:1–3:15 (2012)
62. X. Meng, J.K. Bradley, B. Yavuz, E.R. Sparks, S. Venkataraman, D. Liu, J. Freeman, D.B. Tsai, M. Amde, S. Owen, D. Xin, R. Xin, M.J. Franklin, R. Zadeh, M. Zaharia, A. Talwalkar. Mllib: Machine learning in apache spark. *CoRR*. arXiv:1505.06807 (2015)
63. J. Messeter, M. Johansson, Place-specific computing: conceptual design cases from urban contexts in four countries, in *Proceedings of the 7th ACM conference on Designing interactive systems* (ACM, 2008), pp. 99–108
64. J. Mowat, Cognitive walkthroughs: where they came from, what they have become, and their application to epss design. *The Herridge Group Inc* (2002)
65. W. Naheman, J. Wei, Review of nosql databases and performance testing on hbase, in *Proceedings 2013 International Conference on Mechatronic Sciences, Electric Engineering and Computer (MEC)* (2013), pp. 2304–2309
66. J. Nielsen, R. Mack, *Usability Inspection Methods* (Wiley, 1994)
67. N. O'Leary, Ogilvy & mather and big blue - a new york agency gives ibm a fresh new look. Commun. Arts Mag. **41**(8), 98–107 (2000)
68. J. Paay, J. Kjeldskov, Understanding and modelling built environments for mobile guide interface design. Behav. Inf. Technol. **24**(1), 21–35 (2005)
69. J. Paay, J. Kjeldskov, S. Howard, B. Dave, Out on the town: a socio-physical approach to the design of a context-aware urban guide. ACM Trans. Comput.-Hum. Interact. (TOCHI) **16**(2), 7 (2009)
70. R.W. Picard, *Affective Computing* (MIT Press, Cambridge, MA, USA, 1997)
71. I. Polato, R.R.A. Goldman, F. Kon, A comprehensive view of hadoop researcha systematic literature review. J. Netw. Comput. Appl. **46**, 1–25 (2014)
72. P.G. Polson, C. Lewis, J. Rieman, C. Wharton, Cognitive walkthroughs: a method for theory-based evaluation of user interfaces. Int. J. Man-Mach. Stud. **36**(5), 741–773 (1992)
73. A. Pommeranz, J. Broekens, P. Wiggers, W.-P. Brinkman, C.M. Jonker, Designing interfaces for explicit preference elicitation: a user-centered investigation of preference representation and elicitation process. User Model. User - Adap.Interact. **22**(4–5), 357 (2012)
74. J.R. Preece, Y. Rogers, *Sharp (2002): Interaction Design: Beyond Human-Computer Interaction* (Wiley, Answers. com Technology, Crawfordsville, 2007)
75. D.G. Rees, *Essential Statistics*, vol. 50 (CRC Press, Boca Raton, 2000)
76. S.C. Reid, S.D. Kauer, P. Dudgeon, L.A. Sanci, L.A. Shrier, G.C. Patton, A mobile phone program to track young peoples experiences of mood, stress and coping. Soc. Psych. Psych. Epidemiol. **44**(6), 501–507 (2009)

77. V. Rieser, O. Lemon, S. Keizer, Natural language generation as incremental planning under uncertainty: adaptive information presentation for statistical dialogue systems. IEEE/ACM Trans. Audio, Speech Lang. Process. **22**(5), 979–994 (2014)
78. J. Robertson, S. Robertson, Volere requirements specification template. *Atlantic System Guild.* www.systemguild.com (2000)
79. B. Robins, E. Ferrari, K. Dautenhahn, G. Kronreif, B. Prazak-Aram, G.-J. Gelderblom, B. Tanja, F. Caprino, E. Laudanna, P. Marti, *human-centred design methods: developing scenarios for robot assisted play informed by user panels and field trials. Int. J. Hum.-Comput. Stud. **68**(12), 873–898 (2010)
80. A.L.D. Rossi, A.C.P. de Leon, Ferreira de Carvalho, C. Soares, B.F. de Souza, Metastream: a meta-learning based method for periodic algorithm selection in time-changing data. Neurocomputing **127**, 52–64 (2014)
81. G.B. Rossman, S.F. Rallis, *Learning in the Field: An introduction to Qualitative Research* (Sage, 2003)
82. A. Rubin, *Statistics for Evidence-Based Practice and Evaluation* (Cengage Learning, 2012)
83. S. Sagiroglu, D. Sinanc, Big data: a review, in *2013 International Conference on Collaboration Technologies and Systems (CTS)* (2013), pp. 42–47
84. J. Sauer, A. Sonderegger, The influence of prototype fidelity and aesthetics of design in usability tests: effects on user behaviour, subjective evaluation and emotion. Appl. Ergon. **40**(4), 670–677 (2009)
85. J. Schindler, Profiling and analyzing the i/o performance of nosql dbs. SIGMETRICS Perform. Eval. Rev. **41**(1), 389–390 (2013)
86. B. Schneiderman, C. Plaisant, *Designing the User Interface: Strategies for Effective Human-Computer Interaction* (Pearson higher education, USA, 2010)
87. A.G. Shoro, T.R. Soomro, Big data analysis: apache spark perspective. Global J. Comput. Sci. Technol. 15(1) (2015)
88. E.H. Shortliffe, B.G. Buchanan, A model of inexact reasoning in medicine. Math. Biosci. **23**, 351–379 (1975)
89. D. Sirkin, B. Mok, S. Yang, W. Ju, Oh, i love trash: personality of a robotic trash barrel, in *Proceedings of the 19th ACM Conference on Computer Supported Cooperative Work and Social Computing Companion* (ACM, 2016), pp. 102–105
90. R. Spencer, *Information Visualization*, vol. 1. (Springer, 2001)
91. A. Steinfeld, O.C. Jenkins, B. Scassellati, The oz of wizard: simulating the human for interaction research, in *2009 4th ACM/IEEE International Conference on Human-Robot Interaction (HRI)* (IEEE, 2009), pp. 101–107
92. M. Strait, L. Vujovic, V. Floerke, M. Scheutz, H. Urry, Too much humanness for human-robot interaction: exposure to highly humanlike robots elicits aversive responding in observers, in *Proceedings of the 33rd Annual ACM Conference on Human Factors in Computing Systems* (ACM, 2015), pp. 3593–3602
93. J.R. Thomas, S. Silverman, J. Nelson, *Research Methods in Physical Activity, 7E* (Human Kinetics, 2015)
94. E.R. Tufte, Envisioning information. Optom. Vis. Sci. **68**(4), 322–324 (1991)
95. T. Tullis, B. Albert, *Measuring the user experience: Collecting, Analysing, and Presenting Usability Metrics* (2008)
96. F.B. Viegas, M. Wattenberg, F. Van Ham, J. Kriss, M. McKeon, Manyeyes: a site for visualization at internet scale. IEEE Trans. Vis. Comput. Gr. **13**(6), 1121–1128 (2007)
97. A.S. Vivacqua, A.C.B. Garcia, A. Gomes, Boo: behavior-oriented ontology to describe participant dynamics in collocated design meetings. Expert Syst. Appl. **38**(2), 1139–1147 (2011). Knowledge acquisition meetings to create domain representations
98. N. Walliman, *Social Research Methods* (Sage, 2006)
99. T. Walsh, P. Nurkka, T. Koponen, J. Varsaluoma, S. Kujala, S. Belt. Collecting cross-cultural user data with internationalized storyboard survey, in *Proceedings of the 23rd Australian Computer-Human Interaction Conference* (ACM, 2011), pp. 301–310
100. C. Ware, *Information Visualization: Perception for Design* (Elsevier, 2012)

101. F. Weber, C. Haering, R. Thomaschke, Improving the human computer dialogue with increased temporal predictability. Hum. Factors: J. Hum. Factors Ergonom. Soc. **55**(5), 881–892 (2013)
102. C.R. Wilkinson, A. De Angeli, Applying user centred and participatory design approaches to commercial product development. Des. Stud. **35**(6), 614–631 (2014)
103. D. Wixon, Qualitative research methods in design and development. Interactions **2**(4), 19–26 (1995)
104. H. Yang, Y. Li, M.X. Zhou, Understand users comprehension and preferences for composing information visualizations. ACM Trans. Comput.-Hum. Interact.(TOCHI) **21**(1), 6 (2014)

Privacy-Preserving Record Linkage for Big Data: Current Approaches and Research Challenges

Dinusha Vatsalan, Ziad Sehili, Peter Christen and Erhard Rahm

Abstract The growth of Big Data, especially personal data dispersed in multiple data sources, presents enormous opportunities and insights for businesses to explore and leverage the value of linked and integrated data. However, privacy concerns impede sharing or exchanging data for linkage across different organizations. Privacy-preserving record linkage (PPRL) aims to address this problem by identifying and linking records that correspond to the same real-world entity across several data sources held by different parties without revealing any sensitive information about these entities. PPRL is increasingly being required in many real-world application areas. Examples range from public health surveillance to crime and fraud detection, and national security. PPRL for Big Data poses several challenges, with the three major ones being (1) scalability to multiple large databases, due to their massive volume and the flow of data within Big Data applications, (2) achieving high quality results of the linkage in the presence of variety and veracity of Big Data, and (3) preserving privacy and confidentiality of the entities represented in Big Data collections. In this chapter, we describe the challenges of PPRL in the context of Big Data, survey existing techniques for PPRL, and provide directions for future research.

Keywords Record linkage · Privacy · Big data · Scalability

D. Vatsalan · P. Christen
Research School of Computer Science, The Australian National University,
Acton, ACT 2601, Australia
e-mail: dinusha.vatsalan@anu.edu.au

P. Christen
e-mail: peter.christen@anu.edu.au

Z. Sehili · E. Rahm (✉)
Database Group, University of Leipzig, 04109 Leipzig, Germany
e-mail: rahm@informatik.uni-leipzig.de

Z. Sehili
e-mail: sehili@informatik.uni-leipzig.de

© Springer International Publishing AG 2017   851
A.Y. Zomaya and S. Sakr (eds.), *Handbook of Big Data Technologies*,
DOI 10.1007/978-3-319-49340-4_25

1 Introduction

With the Big Data revolution, many organizations collect and process datasets that contain many millions of records to analyze and mine interesting patterns and knowledge in order to empower efficient and quality decision making [28, 53]. Analyzing and mining such large datasets often require data from multiple sources to be linked and aggregated. Linking records from different data sources with the aim to improve data quality or enrich data for further analysis is occurring in an increasing number of application areas, such as healthcare, government services, crime and fraud detection, national security, and business applications [28, 52]. Effective ways of linking data from different sources have also played an increasingly important role in generating new insights for population informatics in the health and social sciences [99].

For example, linking health databases from different organizations facilitates quality health data mining and analytics in applications such as epidemiological studies (outbreak detection of infectious diseases) or adverse drug reaction studies [20, 116]. These applications require data from several organizations to be linked, for example human health data, travel data, consumed drug data, and even animal health data [38]. Linked health databases can also be used for the development of health policies in a more efficient and effective way compared to traditionally used time-consuming survey studies [37, 88].

Record linkage techniques are also being used by national security agencies and crime investigators for effective identification of fraud, crime, or terrorism suspects [73, 125, 168]. Such applications require data from law enforcement agencies, immigration departments, Internet service providers, businesses, as well as financial institutions [125].

In recent time, record linkage is increasingly being required by social scientists in the field of population informatics to study insights into our society from 'social genome' data, the digital traces that contain person-level data about social beings [99]. The 'Beyond 2011' program by the Office for National Statistics in the UK, for example, has carried out research to study different possible approaches to producing population and socio-demographics statistics for England and Wales by linking data from several sources [121].

Record linkage within a single organization does not generally involve privacy and confidentiality concerns (assuming there are no internal threats within the organization and the linked data are not being revealed outside the organization). An example application is the deduplication of a customer database by a business using record linkage techniques for conducting effective marketing activities. However, in many countries record linkage across several organizations, as required in the above example applications, might not allow the exchange or the sharing of database records between organizations due to laws or regulations. Some example Acts that describe the legal restrictions of disclosing personal or sensitive data are: (1) the Data-Matching Program Act in Australia,[1] (2) the European Union (EU) Per-

---

[1] https://www.oaic.gov.au/privacy-law/other-legislation/government-data-matching   [Accessed: 15/06/2016].

sonal Data Protection Act in Europe,[2] and (3) the Health Insurance Portability and Accountability Act (HIPAA) in the USA.[3]

The privacy requirements in the record linkage process have been addressed by developing 'privacy-preserving record linkage' (PPRL) techniques, which aim to identify matching records that refer to the same entities in different databases without compromising privacy and confidentiality of these entities. In a PPRL project, the database owners (or data custodians) agree to reveal only selected information about records that have been classified as matches among each other, or to an external party, such as a researcher [164]. However, record linkage requires access to the actual values of certain attributes.

Known as quasi-identifiers (QIDs), these attributes need to be common in all databases to be linked and represent identifying characteristics of entities to allow matching of records. Examples of QIDs are first and last names, addresses, telephone numbers, or dates of birth. Such QIDs often contain private and confidential information of entities that cannot be revealed, and therefore the linkage has to be conducted on masked (encoded) versions of the QIDs to preserve the privacy of entities. Several masking techniques have been developed (as we will describe in Sect. 3.4), using two different types of general approaches: (1) secure multi-party computation (SMC) [111] and (2) data perturbation [87].

Leveraging the tremendous opportunities that Big Data can provide for businesses comes with the challenges that PPRL poses, including scalability, quality, and privacy. Big Data implies enormous data *volume* as well as massive flows (*velocity*) of data, leading to scalability challenges even with advanced computing technology. The *variety* and *veracity* aspects of Big Data require biases, noise, variations and abnormalities in data to be considered, which makes the linkage process more challenging. With Big Data containing massive amounts of personal data, linking and mining data may breach the privacy of those represented by the data. A practical PPRL solution that can be used in real-world applications should therefore address these challenges of scalability, linkage quality, and privacy. A variety of PPRL techniques has been developed over the past two decades, as surveyed in [154, 164]. However, these existing approaches for PPRL fall short in providing a sound solution in the Big Data era by not addressing all of the Big Data challenges. Therefore, more research is required to leverage the huge potential that linking databases in the era of Big Data can provide for businesses, government agencies, and research organizations.

In this chapter, we review the existing challenges and techniques, and discuss research directions of PPRL for Big Data. We provide the preliminaries in Sect. 2 and review existing privacy techniques for PPRL in Sect. 3. We then discuss the scalability challenge and existing approaches that address scalability of PPRL in Sect. 4. In Sect. 5, we describe the challenges and existing techniques of PPRL on multiple databases, which is an emerging research avenue that is being increasingly required in many Big Data applications. In Sect. 6 we discuss research directions in

---

[2]http://ec.europa.eu/justice/data-protection/index_en.htm [Accessed: 15/06/2016].
[3]http://www.hhs.gov/ocr/privacy/ [Accessed: 15/06/2016].

PPRL for Big Data, and in Sect. 7 we conclude this chapter with a brief summary of the topic covered.

2 Background

Building on the introduction to record linkage and privacy-preserving record linkage (PPRL) in Sect. 1, we now present background material that contributes to the understanding of the preliminaries. We describe the basic concepts and challenges in Sect. 2.1, and then describe the process of PPRL in Sect. 2.2.

2.1 Overview and Challenges of PPRL

Record linkage is a widely used data pre-processing and data cleaning task where the aim is to link and integrate records that refer to the same entity from two or multiple disparate databases. The record pairs (when linking two databases) or record sets (when linking more than two databases) are compared and classified as 'matches' by a linkage model if they are assumed to refer to the same entity, or as 'non-matches' if they are assumed to refer to different entities [26, 54]. The frequent absence of unique entity identifiers across the databases to be linked makes it impossible to use a simple SQL-join [30], and therefore linkage requires sophisticated comparisons between a set of QIDs (such as names and addresses) that are commonly available in the records to be linked. However, these QIDs often contain personal information and therefore revealing or exchanging them for linkage is not possible due to privacy and confidentiality concerns.

As an example scenario, assume a demographer who aims to investigate how mortgage stress (having to pay large sums of money on a regular basis to pay off a house) is affecting people with regard to their mental and physical health. This research will require data from financial institutions as well as hospitals as shown in Tables 1 and 2. Neither of these organizations is likely willing or allowed by law to provide their databases to the researcher. The researcher only requires access to some attributes of the records (such as loan type, balance amount, blood pressure, and stress level) that are linked across these databases, but not the actual identities of the individuals that were linked. However, personal details (such as name, age or date of birth, gender, and address) are needed as QIDs to conduct the linkage due to the absence of unique identifiers across the databases.

As illustrated in the above example application (shown in Tables 1 and 2), linking records in a privacy-preserving context is important, as sharing or exchanging sensitive and confidential personal data (contained in QIDs of records) between organizations is often not feasible due to privacy concerns, legal restrictions, or commercial interests. Therefore, databases need to be linked in such ways that no sensitive information is being revealed to any of the organizations involved in a cross-

Table 1 Example bank database

ID	Given_name	Surname	DOB	Gender	Address	Loan_type	Balance
6723	Peter	Robert	20.06.72	M	16 Main Street 2617	Mortgage	230,000
8345	Smith	Roberts	11.10.79	M	645 Reader Ave 2602	Personal	8,100
9241	Amelia	Millar	06.01.74	F	49E Apple-cross Rd 2415	Mortgage	320,750

Table 2 Example health database

PID	Last_name	First_name	Age	Address	Sex	Pressure	Stress	Reason
P1209	Roberts	Peter	41	16 Main St 2617	m	140/90	High	Chest pain
P4204	Miller	Amelia	39	49 Apple-cross Road 2415	f	120/80	High	Headache
P4894	Sieman	Jeff	30	123 Norcross Blvd 2602	m	110/80	Normal	Checkup

organizational linkage project, and no adversary is able to learn anything about these sensitive data. This problem has been addressed by the emerging research area of PPRL [164].

The basic ideas of PPRL techniques are to mask (encode) the databases at their sources and to conduct the linkage using only these masked data. This means no sensitive data are ever exchanged between the organizations involved in a PPRL protocol, or revealed to any other party. At the end of such a PPRL process, the database owners only learn which of their own records match with a high similarity with records from the other database(s). The next steps would be exchanging the values in certain attributes of the matched records (such as loan type, balance amount, blood pressure, and stress level in the above example) between the database owners, or sending selected attribute values to a third party, such as a researcher who requires the linked data for their project [164]. Recent research outcomes and experiments conducted in real health data linkage validate that PPRL can achieve linkage quality with only small loss compared to traditional record linkage using unencoded QIDs [134, 135].

Using PPRL for Big Data involves many challenges, among them the following three key challenges need to be addressed to make PPRL viable for Big Data applications:

1. **Scalability**: The number of comparisons required for classifying record pairs or sets equals to the product of the size of the databases that are linked. This is a performance bottleneck in the record linkage process since it potentially requires comparison of all record pairs/sets using expensive comparison functions [9, 31]. Due to the increasing size of Big Data (*volume*), comparing all records is not feasible in most real-world applications. Blocking and filtering techniques have been used to overcome this challenge by eliminating as many comparisons between non-matching records as possible [9, 29, 150].
2. **Linkage quality**: The emergence of Big Data brings with it the challenge of dealing with typographical errors and other variations in data (*variety* and *veracity*) making the linkage more challenging. The exact matching of QID values, which would classify pairs or sets of records as matches if their QIDs are exactly the same and as non-matches otherwise, will likely lead to low linkage accuracy in the presence of real-world data errors. In addition, the classification models used in record linkage should be effective and accurate in classifying matches and non-matches [31]. Therefore, for practical record linkage applications, techniques are required that facilitate both approximate matching of QID values for comparison, as well as effective classification of record pairs/sets for high linkage accuracy.
3. **Privacy**: The privacy-preserving requirement in the record linkage process adds a third challenge, privacy, to the two main challenges of scalability and linkage quality [164]. Linking Big Data containing massive amounts of personal data generally involves privacy and confidentiality issues. Privacy needs to be considered in all steps of the record linkage process as only the masked (or encoded) records can be used, making the task of linking databases across organizations more challenging. Several masking techniques have been used for PPRL, as we will discuss in detail in Sect. 3.4.

2.2   *The PPRL Process and Techniques Used*

In this section we discuss the steps and the techniques used in the PPRL process, as shown in Fig. 1.

Data Pre-processing and Masking: The first important step for quality linkage outcomes is data pre-processing. Real-world data are often noisy, incomplete and inconsistent [8, 128], and they need to be cleaned in this step by filling in missing data, removing unwanted values, transforming data into well-defined and consistent forms, and resolving inconsistencies in data representations and encodings [28, 36]. In PPRL, data masking (encoding) is an additional step. Data pre-processing and masking can be conducted independently at each data source. However, it is important that all database owners (or parties) who participate in a PPRL project conduct the same data pre-processing and masking steps on the data they will use for linking. Some exchange of information between the parties about what data pre-processing

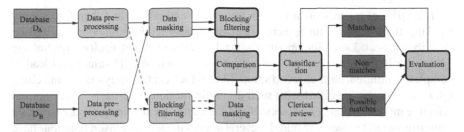

Fig. 1 Outline of the general PPRL process as discussed in Sect. 2.2. The steps shown in *dark outlined boxes* need to be conducted on masked database records, while *dotted arrows* show alternative data flows between steps

and masking approaches they use, as well as which attributes to be used as QIDs, is therefore required [164].

Blocking/filtering: Blocking/filtering is the second step of the process, which is aimed at reducing the number of comparisons that need to be conducted between records by pruning as many pairs or sets of records as possible that unlikely correspond to matches [9, 29]. Blocking groups records according to a blocking criteria (blocking key) such that comparisons are limited to records in the same (or similar) blocks, while filtering prunes potential non-matches based on their properties (e.g. length differences of QIDs) [29]. The output of this step are candidate record pairs (or sets) that contain records that are potentially matching, which need to be compared in more detail. Blocking/filtering can either be conducted on masked records or locally by the database owners on unmasked records. The scalability challenge of PPRL has been addressed by several recent approaches using private blocking and filtering techniques [46, 78, 131, 133, 149, 150, 159, 163], as will be described in Sects. 4 and 5.1.

Comparison: Candidate record pairs (or sets) are compared in detail in the comparison step using comparison (or similarity) functions [32]. Various comparison functions have been used in record linkage including Levenshtein edit distance, Jaro-Winkler comparison, Soft-TFIDF string comparison, and token-based comparison using the Overlap, Dice, or Jaccard coefficient [28]. These comparison functions provide a numerical value representing the similarity of the compared QID values, often normalized into the [0, 1] interval where a similarity of 1 corresponds to two values being exactly the same, and 0 means two values being completely different. Several QIDs are normally used when comparing records, resulting in one weight vector for each compared record pair that contains the numerical similarity values of all compared QIDs.

The QIDs of records often contain variations and errors, and therefore simply masking these values with a secure one-way hash-encoding function (as will be described in Sect. 3.4) and comparing the masked values will not result in high linkage quality for PPRL [35, 122]. A small variation in a pair of QID values will lead to completely different hash-encoded values [35], which enables only exactly matching QID values to be identified with such a simple masking approach. Therefore, an effective masking approach for securely and accurately calculating the approximate similarity of QID values is required. Several approximate comparison functions have been adapted into a PPRL context, including the Levenshtein edit distance [76] and the Overlap, Dice, and Jaccard coefficients [164].

Classification: In the classification step, the weight vectors of the compared candidate record pairs (or sets) are given as input to a decision model which will classify them into matches, non-matches, and possible matches [31, 54, 63], where the latter class is for cases where the classification model cannot make a clear decision. A widely used classification approach for record linkage is the probabilistic method developed by Fellegi and Sunter in the 1960s [54]. In this model, the likelihood that a pair (or set) of records corresponds to a match or a non-match is modelled based on a-priori error estimates on the data, frequency distributions of QID values, as well as their similarity calculated in the comparison step [28]. Other classification techniques include simple threshold-based and rule-based classifiers [28]. Most PPRL techniques developed so far employ a simple threshold-based classification [164].

Supervised machine learning approaches, such as support vector machines and decision trees [14, 25, 51, 52], can be used for more effective and accurate classification results. These require training data with known class labels for matches and non-matches to train the decision model. Once trained, the model can be used to classify the remaining unlabelled pairs/sets of records. Such training data, however, are often not available in real record linkage applications, especially in privacy-preserving settings [28]. Alternatively, semi-supervised techniques (such as active learning-based techniques [3, 11, 169]), that actively use examples manually labeled by experts to train and improve the decision model, need to be developed for PPRL. Recently developed advanced classification models, such as (a) collective linkage [13, 74] that considers relational similarities with other records in addition to QID similarities, (b) group linkage [123] that calculates group of records' similarities based on pair-wise similarities, and (c) graph-based linkage [58, 66, 74] that considers the structure between groups of records, can achieve high linkage quality at the cost of higher computational complexity. However, these advanced classification techniques have not been explored for PPRL so far.

Clerical review: The record pairs/sets that are classified as possible matches require a clerical review process, where these pairs are manually assessed and classified into matches or non-matches [171]. This is usually a time-consuming and error-prone process which depends upon experience of the experts who conduct the review. Active learning-based approaches can be used for clerical review [3, 11, 169]. However, clerical review in its current form is not possible in a PPRL scenario since the actual QID values of records cannot be inspected because this would reveal sensitive private information. Recent work in PPRL suggests an interactive approach with human-machine interaction to improve the quality of linkage results without compromising privacy [100].

Evaluation: The final step in the process is the evaluation of the complexity, quality, and privacy of the linkage to measure the applicability of a linkage project in an application before implementing it into an operational system. A variety of evaluation measures have been proposed [29, 31]. Given in a practical record linkage application the true match status of the compared record pairs are unlikely to be known, measuring linkage quality is difficult [6, 31]. How to evaluate the amount of privacy protection using a set of standard measures is still an immature aspect in the PPRL literature. Vatsalan et al. recently proposed a set of evaluation measures for privacy using probability of suspicion [165]. Entropy and information gain, between unmasked and masked QID values, have also been used as privacy measures [46].

Tools: Different record linkage approaches have been implemented within a number of tools. Koepcke and Rahm provided a detailed overview of eleven such tools in [95] both categories of with and without the use of learning-based (supervised) classification. The comparative evaluation study [96] benchmarks selected tools from both categories for four real-life test cases. It is found that learning-based approaches achieve generally better linkage quality especially for complex tasks requiring the combination of several attribute similarities. Current tools for *link discovery*, i.e., matching entities between sources of linked open data web, are surveyed in [119]. A web-based tool was recently developed to demonstrate several multi-party PPRL approaches (as will be described in Sect. 5) [132].

3 Privacy Aspects and Techniques for PPRL

Several dimensions of privacy need to be considered for PPRL, the four main ones being: (1) the number of parties and their roles, (2) adversary models, (3) privacy attacks, and (4) data masking or encoding techniques. In Sects. 3.1–3.4, we describe these four privacy dimensions, and in Sect. 3.5 we provide an overview of Bloom filter-based data masking, a technique widely used in PPRL.

3.1 PPRL Scenarios

PPRL techniques for linking two databases can be classified into those that require a linkage unit for performing the linkage and those that do not. The former are known as 'three-party protocols' and the latter as 'two-party protocols' [24, 27, 167]. In three-party protocols, a (trusted) third party acts as the linkage unit to conduct the linkage of masked data received from the two database owners, while in two-party

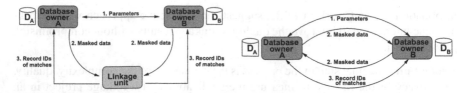

Fig. 2 Outline of PPRL protocols with (*left*) and without (*right*) a linkage unit (also known as three-party and two-party protocols, respectively)

protocols only the two database owners participate in the PPRL process. A conceptual illustration and the main communication steps involved in these protocols are shown in Fig. 2.

A further characterization of PPRL techniques is if they allow the linking of data from more than two data sources (multi-party) or not. Multi-party PPRL techniques identify matching record sets (instead of record pairs) from all parties (more than two) involved in a linkage, or from sub-sets of parties. Only limited work has been so far conducted on multi-party PPRL due to its increased challenges, as we will describe in Sect. 5. Similar to linking two databases, multi-party PPRL may or may not use a linkage unit to perform the linkage.

Protocols that do not require a linkage unit are more secure in terms of collusion (described below) between one of the database owners and the linkage unit. However, they generally require more complex techniques to ensure that the database owners cannot infer any sensitive information about each other's data during the linkage process.

3.2 Adversary Models

Different adversary models are assumed in PPRL techniques, including the most commonly used honest-but-curious (HBC) and malicious models [164].

1. **Honest-but-curious (HBC) or semi-honest** parties are curious in that they try to find out as much as possible about another party's input to a protocol while following the protocol steps [65, 111]. If all parties involved in a PPRL protocol have no new knowledge at the end of the protocol above what they would have learned from the output, which is generally the record pairs (certain attributes) classified as matches, then the protocol is considered to be secure in the HBC model. However, it is important to note that HBC does not prevent parties from colluding with each other with the aim to learn about another party's sensitive information [111]. Most of the existing PPRL solutions assume the HBC adversary model.
2. **Malicious** parties can behave arbitrarily in terms of refusing to participate in a protocol, not following the protocol in the specified way, choosing arbitrary

values for their data input, or aborting the protocol at any time [110]. Limited work has been done in PPRL for the malicious adversary model [57, 105, 118]. Evaluating privacy under this model is very difficult, because there exist potentially unpredictable ways for malicious parties to deviate from the protocol that cannot be identified by an observer [21, 62, 111].

3. **Covert and accountable computing** models are advanced adversary models developed to overcome the problems associated with the HBC and malicious models. The HBC model is not sufficient in many real-world applications because it is suitable only when the parties essentially trust each other. On the other hand, the solutions that can be used with malicious adversaries are generally more complex and have high computation and communication complexities, making their applications not scalable to large databases. The covert model guarantees that the honest parties can detect the misbehavior of an adversary with high probability [4], while the accountable computing model provides accountability for privacy compromises by the adversaries without excessive complexity and cost that incur with the malicious model [72]. Research is required towards transforming existing HBC or malicious PPRL protocols into these models and proving privacy of solutions under these models.

3.3   Attacks

The privacy attacks or vulnerabilities that a PPRL technique is susceptible to allow theoretical and empirical analysis of the privacy guarantees provided by the PPRL technique. The main privacy attacks of PPRL are:

1. **Dictionary attack** is possible with masking functions, where an adversary masks a large list of known values using various existing masking functions until a matching masked value is identified [164]. A keyed masking approach, such as the Hashed Message Authentication Code (HMAC) [97], can be used to prevent dictionary attacks. With HMAC the database owners exchange a secret code (string) that is added to all database values before they are masked. Without knowing the secret key, a dictionary attack is unlikely to be successful.
2. **Frequency attack** is still possible even with a keyed masking approach [164], where the frequency distribution of a set of masked values is matched with the distribution of known unmasked values in order to infer the original values of the masked values [112].
3. **Cryptanalysis attack** is a special category of frequency attack that is applicable to Bloom filter-based data masking techniques. As Kuzu et al. [101] have shown, depending upon certain parameters of Bloom filter masking, such as the number of hash functions employed and the number of bits in a Bloom filter, using a constrained satisfaction solver allows the iterative mapping of individual masked values back to their original values.

4. **Composition attack** can be successful by combining knowledge from more than one independent masked datasets to learn sensitive values of certain records [60]. An attack on distance-preserving perturbation techniques [155], for example, allows the original values to be re-identified with high level of confidence if knowledge about mutual distances between values is available.

5. **Collusion** is another vulnerability associated with multi-party or three-party PPRL techniques, where some of the parties involved in the protocol work together to find out about another database owner's data. For example, one or several database owners collude with the linkage unit or a sub-set of database owners collude among them to learn about other parties' data.

3.4 Data Masking or Encoding

In PPRL, the linkage has to be conducted on a masked or encoded version of the QIDs to preserve the privacy of entities. Data masking (encoding) transforms original data in such a way that there exists a specific functional relationship between the original data and the masked data [55]. Several data masking functions have been used to preserve privacy while allowing the linkage. We categorize existing data masking techniques into three: (1) auxiliary, (2) blocking, and (3) matching techniques. Auxiliary techniques are the ones used as helper functions in PPRL, while blocking and matching categories are used for private blocking and matching (comparison and classification), respectively. In the following we describe key techniques in each of these three categories.

- **Auxiliary**:

1. **Pseudo random function** (PRF) is a deterministic secure function that, when given an n-bit seed k and an n-bit argument x, returns an n-bit string $f_k(x)$ such that it is infeasible to distinguish $f_k(x)$ for different random k from a truly random function [114]. In PPRL, PRFs have been used to generate random secret values to be shared by a group of parties [57, 122, 151].

2. **Reference values** constructed either with random faked values, or values that for example are taken from a public telephone directory, such as all unique surnames and town names, have been used in several PPRL approaches [77, 124, 141, 173]. Such lists of reference values can be used to calculate the distances or similarities between QID values in terms of the distances or similarities between QID and reference values.

3. **Noise addition** in the form of extra records or QID values that are added to the databases to be linked is a data perturbation technique [86] that can be used to overcome the problem of frequency attacks on PPRL protocols [44, 100]. An example is shown in Fig. 3. Adding extra records, however, incurs a cost of lower linkage quality (due to false matches) and scalability (due to extra records that need to be processed and linked) [79]. Section 4.1 discusses noise addition for private blocking.

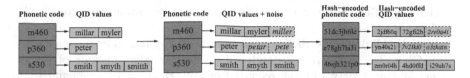

Phonetic code	QID values			Phonetic code	QID values + noise			Hash-encoded phonetic code	Hash-encoded QID values		
m460	millar	myler		m460	millar	myler	*miller*	51dc3jh6le	2jdf60q	72gfi2b	*2re0a4t*
p360	peter			p360	peter	*petar*	*pete*	e78gh7la3i	yn40s21	*7v2lki0*	*o3zkatn*
s530	smith	smyth	smitth	s530	smith	smyth	smitth	46sjb321p0	zm0r04h	4hd0ffd	i29uh7s

Fig. 3 An example of phonetic encoding, noise addition, and secure hash-encoding (adapted from [164]). Values represented with *dotted outlines* are added noise to overcome frequency attacks [165] (as will be discussed in detail in Sect. 4.1)

4. **Differential privacy** [50] has emerged as an alternative to random noise addition in PPRL. Only the perturbed results (with noise) of a set of statistical queries are disclosed to other parties, such that the probability of holding any property on the results is approximately the same whether or not an individual value is present in the database [50]. In recent times, differential privacy has been used in statistical (e.g. counts or frequencies) microdata publication as well as in PPRL [16, 68, 103].

- **Blocking**:

1. **Phonetic encoding**, such as Soundex, NYSIIS or Double-Metaphone, groups values together that have a similar pronunciation [23] in a one-to-many mapping, as shown in Fig. 3. The main advantage of using a phonetic encoding for PPRL is that it inherently provides privacy [79], reduces the number of comparisons, and thus increases scalability [23], and supports approximate matching [23, 79]. Two drawbacks of phonetic encodings are that they are language dependent [126, 146] and are vulnerable to frequency attacks [165]. Section 4.1 discusses phonetic encoding-based blocking in more details.
2. **Generalization techniques** overcome the problem of frequency attacks on records by generalizing records in such a way that re-identification of individual records from the masked data is not feasible [106, 115, 152]. k-anonymity is a widely used generalization technique for PPRL [75, 77, 118], where a database is said to be k-anonymous if every combination of masked QID values (or blocking key values) is shared by at least k records in the database [152]. Other generalization techniques include value generalization hierarchies [67], top-down specialization [118], and binning [108, 162].

- **Matching**:

1. **Secure hash-encoding** is one of the first techniques used for PPRL [17, 49, 127]. The widely known Message Digest (like MD5) and Secure Hash Algorithms (like SHA-1 and SHA-2) [143] are one-way hash-encoding functions [97, 143] that can be used to mask values into hash-codes (as shown in Fig. 3) such that having access to only hash-codes will make it nearly impossible with current computing technology to infer their original values. A major problem with this masking technique is, however, that only exact matches can be found [49]. Even a single

character difference between a pair of original values will result in two completely different hash-codes.

2. **Statistical linkage key (SLK)** is a derived variable generated from components of QIDs. The SLK-581 was developed by the Australian Institute of Health and Welfare (AIHW) to link records from the Home and Community Care datasets [140]. The SLK-581 consists of the second and third letters of first name, the second, third and fifth letters of surname, full date of birth, and sex. Similarly, SLK consisting of month and year of birth, sex, zipcode, and initial of first name was used for linking the Belgian national cancer registry [157]. However, as a recent study has shown these SLK-based masking provides limited privacy protection and poor sensitivity [136].

3. **Embedding space** embeds QID values into a multi-dimensional metric space (such as Euclidean [16, 141, 173] or Hamming [81]) while preserving the distances between these values using a set of pivot values that span the multi-dimensional space. A drawback with this approach is that it is often difficult to determine a good dimension of the metric space and select suitable pivot values.

4. **Encryption schemes**, such as commutative [1] and homomorphic [92] encryption, are used in PPRL techniques to allow secure multi-party computation (SMC) in such a way that at the end of the computation no party knows anything except its own input and the final results of the computation [38, 62, 111]. The secure set union, secure set intersection, and secure scalar product, are the most commonly used SMC techniques for PPRL [38, 143]. A drawback of these cryptographic encryption schemes for SMC, however, is that they are computationally expensive.

5. **Bloom filter** is a bit vector data structure into which values are mapped by using a set of hash functions. Bloom filters have been widely used in PPRL for private matching of records as they provide a means of privacy assurance [46, 47, 76, 104, 147, 170], if effectively used [102]. We will discuss Bloom filter masking in more detail in the following section.

6. **Count-min sketches** are probabilistic data structures (similar to Bloom filters) that can be used to hash-map values along with their frequencies in a sub-linear space [41]. Count-min sketches have been used in PPRL where the frequency of occurrences of a matching pair/set also needs to be identified [80, 139]. However, these approaches only support exact matching of categorical values.

Other privacy aspects in a PPRL project are the secure generation and exchange of public/private key pairs, employee confidentiality agreements to reduce internal threats, as well as encrypted communication, secure connections, and secure servers to reduce external threats.

3.5 Bloom Filters

Bloom filter encoding has been used as an efficient masking technique in a variety of PPRL approaches [46, 104, 130, 148, 150, 158, 160]. A Bloom filter b_i is a bit vector data structure of length l bits where all bits are initially set to 0. k independent hash functions, $h_1, h_2, \ldots, h_k$, each with range $1, \ldots l$, are used to map each of the elements s in a set S into the Bloom filter by setting the bit positions $h_1(s), h_2(s), \ldots, h_k(s)$ to 1. The Bloom filter was originally proposed by Bloom [15] for efficiently checking set membership [19]. Lai et al. [104] first adopted the concept of using Bloom filters in PPRL for identifying exactly matching records across multiple databases, as will be described in Sect. 5.2.

Schnell et al. [148] were the first to propose a method for approximate matching in PPRL using Bloom filters. In their approach, the character q-grams (sub-strings of length q) of QID values of each record in the databases to be linked are hash-mapped into a Bloom filter using k independent hash functions. The resulting Bloom filters are sent to a linkage unit that calculates the similarity between Bloom filters using a set-based similarity function, such as the Dice-coefficient [28]. The Dice-coefficient of two Bloom filters (b_1 and b_2) is calculated as:

$$Dice\_sim(b_1, b_2) = \frac{2 \times c}{(x_1 + x_2)}, \tag{1}$$

where c is the number of common bit positions that are set to 1 in both Bloom filters (common 1-bits), and x_i is the number of bit positions set to 1 in b_i (1-bits), $i \in \{1, 2\}$. An example similarity calculation is illustrated in Fig. 4.

Bloom filters are susceptible to cryptanalysis attacks, as shown by Kuzu et al. [101]. Using a constrained satisfaction solver, such attacks allow the iterative mapping of individual hash-encoded values back to their original values depending upon the number of hash functions employed and the length of a Bloom filter. Different Bloom filter encoding methods have been proposed in the literature to overcome such cryptanalysis attacks and improve linkage quality. Schnell et al.'s proposed method of hash-mapping all QID values of a record into one composite Bloom filter is known as Cryptographic Long-term Key (CLK) encoding [148].

Fig. 4 An example similarity (Dice-coefficient) calculation of Bloom filters for approximate matching using Schnell et al.'s approach [147] (taken from [164])

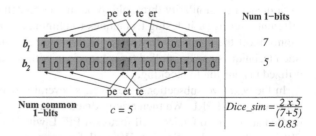

Durham et al. [48] investigated composite Bloom filter encoding in detail by first hash-mapping different attributes into attribute-level Bloom filters of different lengths. These lengths depend upon the weights [54] of QID attributes that are calculated using the discriminatory power of attributes in separating matches from non-matches using a statistical approach. These attribute-level Bloom filters are then combined into one record-level Bloom filter (known as RBF) by sampling bits from each attribute-level Bloom filter. Vatsalan et al. [165] recently introduced a hybrid method of CLK and RBF (known as CLKRBF) where the Bloom filter length is kept to be the same (as in CLK) while using different numbers of hash functions to map different attributes into the Bloom filter based on their weights (to improve matching quality as in RBF).

Several non-linkage unit-based approaches have also been proposed for PPRL using Bloom filter masking, where the database owners (without a linkage unit) collaboratively (or distributively) calculate the similarity of Bloom filters [104, 158, 160]. A recent work proposed novel Bloom filter-based masking techniques that allow approximate matching of numerical data in PPRL [161]. Instead of hash-mapping q-grams of a string, the proposed approaches hash-map a set of neighbouring numerical values to allow approximate matching.

4 Scalability Techniques for PPRL

PPRL for Big Data needs to scale to very large data volumes of many millions of records from multiple sources. As for standard record linkage, the main techniques for high efficiency are to reduce the search space by blocking and filtering approaches and to perform record linkage in parallel on many processors.

These three kinds of optimization are largely orthogonal so that they may be combined to achieve maximal efficiency. Blocking is defined on selected attributes (blocking keys) that may be different from the QID attributes used for comparison, for example zip code. It partitions the records in a database into several blocks or clusters such that comparison can be restricted to the records of the same block, for example persons with the same zip code. Other blocking approaches like sorted neighborhood work differently but are similar in spirit. Filtering is an optimization for the particular comparison approach which optimizes the evaluation of a specific similarity measure for a predefined similarity threshold to be met by matching records. It thus utilizes different filtering or indexing techniques to eliminate pairs (or sets) of records that cannot meet the similarity threshold for the selected similarity measures [28, 43]. Such techniques can be applied for comparison within blocks, i.e., filtering could be utilized in addition to blocking.

In the next two subsections, we discuss several proposed blocking and filtering approaches for PPRL. We then briefly discuss parallel PPRL which has found only limited attention so far. We will focus on PPRL with two data sources (multi-party PPRL is discussed in Sect. 5). We will furthermore mostly assume the use of a dedicated linkage unit (as shown on the left-hand side in Fig. 2) as well as the masking of records using Bloom filters (as described in Sect. 3.5). Note that a linkage unit is ideally suited for high performance as it requires minimal communication between the database owners, and it can utilize a high performance cluster for parallel PPRL as well as blocking and filtering techniques.

4.1 Blocking Techniques

Blocking aims at reducing the search space for linkage by avoiding the comparison between every pair of records and its associated quadratic complexity. There are numerous blocking strategies [28] that mostly group records into disjoint or overlapping blocks such that only records within the same block need to be compared with each other. *Standard blocking* uses the values of a so-called blocking key to partition all records into disjoint blocks. The blocking key values (BKVs) may be the values of a selected attribute or the result of a function on one or several attribute values (e.g. the concatenation of the first two letters of last name and year of birth). Other blocking approaches include *canopy clustering* that results in overlapping clusters or blocks, and *sorted neighborhood* that sorts records according to a sorting key and only compares neighboring records within a certain window [28]. Comparing records only within the predetermined blocks may result in the loss of some matches especially if some BKVs are incorrect or missing. To improve recall a multi-pass blocking approach can be utilized, where records are blocked according to different blocking keys, at the cost of an increased number of additional comparisons.

Blocking for PPRL is based on the known approaches for regular record linkage but aims at improving privacy. A general approach with a central linkage unit is to apply a previously agreed on blocking strategy by the database owners on the original records. Then all records within the different blocks are masked (encoded), e.g. using Bloom filters, and sent to the linkage unit. The linkage unit can then compare the masked records block-wise with each other. In the following, we present selected blocking approaches for PPRL and discuss results from a comparative evaluation of different blocking schemes.

Phonetic Blocking: Blocking records based on their phonetic code is a widely used technique in record linkage [28]. The basic idea is to encode the values of a blocking key attribute (e.g. last name) with a phonetic function (as discussed in Sect. 3.4) such as Soundex or Metaphone [28]. All records with the same phonetic code, i.e. with a similar pronunciation, are then assigned to the same block. The phonetic blocking has been used in several PPRL approaches, in particular in [76, 79]. Karakasidis et al. in [76] use a multi-pass approach with both Soundex and Metaphone encodings to achieve a good recall. Furthermore, they add fake records to the blocks for improved privacy.

As discussed in Sect. 3.4, adding fake records improves privacy but adds overhead in the form of extra comparisons between records and can reduce linkage quality due to the introduction of false matches. A theoretical analysis of the impact of adding fake records for Soundex-based blocking is presented in [79]. The authors study the effect of fake records on the so-called *relative information gain* which is related to the *entropy* measure. A high entropy within blocks caused by fake records introduces a high uncertainty and thus a low information gain [165]. The authors also study different methods to inject fake records to increase entropy. The most flexible of the approaches is based on the concept of k-anonymity and adds as many fake records as required to ensure that each block has at least k records. The approach typically requires only the addition of a modest number of fake records; the choice of k also supports finding a good trade-off between privacy and quality.

Blocking with Reference Values: An alternative to adding fake records for improving the privacy of blocking is the use of reference values (as discussed in Sect. 3.4). The reference values can be used by the database owners for clustering their database records. Comparison can then be restricted to the clusters (blocks) of the same or similar reference records. Such an approach has been proposed in [77] based on k *nearest neighbor (kNN) clustering*. This approach first clusters the reference values identically at each database owner such that each cluster contains at least k reference values to ensure k-anonymity; clustering is based on the Dice-coefficient similarity between values. In the next step, each database owner assigns its records to the nearest cluster based on the Dice-coefficient between records and reference values. Finally each database owner sends its clusters (encoded reference values and records) to the linkage unit which then compares the records between corresponding clusters.

An alternate proposal utilizes a local *sorted neighborhood clustering* (*SNC-3P*) for improved performance in the blocking phase while retaining the use of reference values and support for k-anonymity [159]. Each database owner sorts a shared set of reference values and then inserts its records into the sorted list according to their sorting key. From the sorted list of reference values and records the initial Sorted Neighborhood Clusters (SNCs) are determined such that each cluster contains one reference value and a set of database records. To ensure k-anonymity, the initial clusters are merged into larger blocks containing at least k database records. This differs from kNN clustering where k reference records are needed per cluster. The merging of the initial clusters can be based on similarity or size constraints. The remaining protocol with sending the encoded records to the linkage unit for comparison works as for kNN clustering.

An adaptation of SNC-3P for two parties without a linkage unit (*SNC-2P*) was presented in [163]. In this approach, the two database owners generate their reference values independently, so that they end up with two different sets of reference values. As for SNC-3P, each database owner sorts its reference values, inserts its records into the sorted list, builds initial SNCs (with one reference value and its associated records), and merges these clusters to guarantee k-anonymity. Afterwards the database owners exchange their reference values. These values are then merged with the local reference values and sorted. To find candidate pairs between the sources a sorted neighborhood method with a sliding window w is applied on these reference values. The window size w determines the number of reference values originating from each data source, e.g. for $w = 2$ the sliding window includes 2 reference values from each data source. In the last step, the encoded records falling into a window are exchanged for comparison.

LSH-based blocking: Locality-sensitive hashing (LSH) has been proposed to solve the problem of nearest neighbor search in high dimensional spaces [61, 69]. For blocking, LSH uses a family of hash functions to generate keys used to partition the records in a database, so that similar records are grouped into the same block [89]. Durham investigated the use of LSH for private blocking of records masked as Bloom filters [46]. She considered two families of hash functions depending on the used distance function (Jaccard or Hamming distance). For the Jaccard distance, she proposed the use of *MinHash* functions. A MinHash function h_i permutes the bits of a Bloom filter b_i and selects from the permuted Bloom filter the first index position with a set bit (1-bit). By applying ϕ MinHash functions we obtain ϕ index positions which are concatenated to generate the final MinHash key. So for Bloom filter b_i and the family of hash functions H, we determine $key(b_i)_H = concat(h_1(b_i), h_2(b_i), \ldots, h_\phi(b_i))$, where $h_j \in H$ with $1 \leq j \leq \phi$ and function $concat()$ concatenates a set of input values. For the Hamming distance, Durham proposed the use of *HLSH* hash functions that select the bit value of a Bloom filter at a random position ρ. In the same way as MinHash, ϕ HLSH functions are applied on a Bloom filter b_i and the values of the ϕ selected bits are concatenated to obtain the final hash key of b_i.

Example Consider two Bloom filters $b_1 = 1100100011$ and $b_2 = 1100100111$, two permutations $p_1 = (10, 7, 1, 3, 6, 4, 2, 9, 5, 8)$ and $p_2 = (4, 3, 6, 7, 2, 1, 5, 10, 9, 8)$, and the MinHash family H_1 with two functions $h_1 = Min(p_1(\cdot))$ and $h_2 = Min(p_2(\cdot))$, where $Min(\cdot)$ returns the first position of a set bit in the input bit vector, and $p_i(\cdot)$ returns the input bit vector permuted using p_i. The application of h_1 and h_2 on b_1 results in $h_1(b_1) = Min(p_1(b_1)) = Min(1010001110) = 1$ and $h_2(b_1) = Min(p_2(b_1)) = Min(0000111110) = 5$. Hence the key of b_1 in H_1 is $key(b_1)_{H_1} = concat(h_1(b_1), h_2(b_1)) = (1, 5)$. In the same way we determine the key of b_2, i.e. $key(b_2)_{H_1} = (1, 5)$. Hence, for MinHash family H_1 records b_1 and b_2 are put into the same block and will be compared with each other.

Both families, MinHash and HLSH, depend on two parameters: the number of hash functions ϕ as well as the number of passes or iterations μ. Since the final hash key of a record concatenates ϕ values, using a high ϕ leads to more homogeneous blocks and better precision (i.e., blocks containing similar records with higher probability). However a high ϕ also increases the probability of missing true matches (reduced recall). This problem is addressed by applying μ iterations, each with a different set of hash functions. Therefore each record b_i will be hashed to several blocks to allow identifying more true matches.

In [83] the authors present a theoretical analysis of the use of MinHash functions to identify good values for parameters ϕ_{opt} and μ_{opt} to efficiently achieve a good precision and recall. The naïve approach to improve recall is to increase the number of iterations μ and thus the number of blocks to which records are assigned. The drawbacks of this method are the high runtime caused by the computation of the permutations, increased number of record pairs to compare, and the large space needed to store intermediate results. This observation was experimentally confirmed in [46]. The choice of the ϕ_{opt} is also complex and depends on the expected running time (for details see [83]).

Evaluation of Private Blocking Approaches: The relatively large number of possible blocking approaches requires detailed evaluations regarding their relative scalability, blocking quality and privacy for different kinds of workloads. One of the few studies in this respect has been presented by Vatsalan et al. [165]. For scalability they considered runtime and the so-called *reduction ratio* (RR), a value indicating the number of pruned candidate pairs compared to all possible record pairs (which thus evaluates to what degree the search space is reduced). For blocking quality they considered the recall and precision metrics *pair completeness* (PC) and *pair quality* (PQ), respectively [29]. For privacy they estimated the so-called *disclosure risk* (DR) measures, which represent the probability that masked records/QID values can be linked with records or values in a publicly available dataset.

The evaluation of [165] considers six simulated blocking strategies including kNN [77], SNC-3P [159], SNC-2P [163] and LSH blocking [46]. Regarding blocking runtime, the SNC and LSH schemes performed best. All strategies except SNC-2P achieved a very high RR of almost 1. On the other hand, SNC-2P achieved the best PC. The best trade-off between RR and PC was observed for LSH. Considering the privacy aspect, SNC-2P was found to have a low DR while kNN and LSH generally expose the highest DR.

While this study provides interesting results, we see a need for additional benchmark studies given that further blocking schemes have been developed more recently and that the relative behavior of each approach depends on numerous parameter settings as well as characteristics of the chosen datasets.

4.2 Filtering Techniques

Almost all proposed PPRL schemes based on Bloom filters aim at identifying the pairs of bit vectors with a similarity above a threshold. For regular record linkage, such a threshold-based comparison of record pairs is known as a *similarity join* [39]. The efficient processing of such similarity joins for different kinds of similarity measures has been the focus of much research in the past, e.g. [2, 64, 138, 172]. Several approaches utilize the characteristics of the considered similarity measure and the prespecified similarity threshold to reduce the search space thereby speeding up the linkage process. This holds especially for the broad class of token-based similarity joins where the comparison of records is based on the set of tokens (e.g. q-grams) of QIDs. In this case, one can exclude all pairs of records that do not share at least one token. Further proposed optimizations for such similarity joins include the use of different kinds of filters (for example, length and prefix filters) and dynamically created inverted indexes [10]. PPJoin [172] is an efficient approach that includes these kinds of optimizations. Several filtering approaches also utilize the characteristics of similarity functions for metric spaces to reduce the search space, in particular the so-called triangle inequality (see below) [174].

For PPRL, similar filtering (similarity join) approaches are usable but need to be adapted to the comparison of masked records such as Bloom filters and the associated similarity measures. For Bloom filters it is relatively easy to apply the known token-based similarity measures by considering the set bit positions (1-bits) in the bit vectors as the "tokens". This has already been shown in Sect. 3.5 for the Dice-coefficient similarity which is based on the degree of overlapping bit positions. This is also the case for the related Jaccard similarity. For two bit vectors b_1 and b_2 it is defined as follows:

$$Jacc\_sim(b_1, b_2) = \frac{|b_1 \wedge b_2|}{|b_1 \vee b_2|} = \frac{|b_1 \wedge b_2|}{|b_1| + |b_2| - |b_1 \wedge b_2|}, \tag{2}$$

where $|b_i|$ denotes the number of set bits in bit vector b_i which is also called its *length* or cardinality. For the example Bloom filter pair shown in Fig. 4, the Jaccard similarity is $5/7 = 0.71$. In the following, we outline several filtering approaches that have been proposed for PPRL.

Length Filter: The similarity function $Jacc\_sim$ (as well as $Dice\_sim$) allows the application of a simple *length filter* to reduce the search space. This is because the minimal similarity (overlap of set bits) can only be achieved if the lengths (number of set bits) of the two input records do not deviate too much. Formally, for two records r_i and r_j with $|r_i| \leq |r_j|$, it holds that

$$Jacc\_sim(r_i, r_j) \geq s_t \Rightarrow |r_i| \geq \lceil s_t \cdot |r_j| \rceil \tag{3}$$

For example, two records cannot satisfy a (Jaccard) similarity threshold $s_t = 0.8$ if their lengths differ by more than 20%. Hence for a similarity threshold of 0.8, the length filter would already avoid the two records from the example Bloom filter pair shown in Fig. 4 without comparing in detail, since Eq. 3 ($5 \geq \lceil 0.8 \cdot 7 \rceil = 6$) does not hold. The two-party PPRL approach proposed by Vatsalan and Christen uses such a length filter for Dice-coefficient similarity [158].

PPJoin for PPRL: The privacy-preserving version of PPJoin (called P4Join) utilizes three filters to reduce the search space: the length filter as well as a prefix filter and a position filter [150]. The *prefix filter* is based on the fact that matching bit vectors need a high degree of overlap in their set bit positions in order to satisfy a predefined threshold. Pairs of records can thus be excluded from comparison if they have an insufficient overlap. This overlap test can be limited to a relatively small sub-set of bit positions, e.g. in the beginning (or prefix) of the vectors. To maximize this filter idea, P4Join applies a pre-processing to count for each bit position the number of records where this bit is set to 1 and reorders the positions of all bit vectors in ascending order of these frequency counts. This way the prefixes of bit vectors contain infrequently set bit positions reducing the likelihood of an overlap with other bit vectors. The *position filter* of P4Join can avoid the comparison of two records even if their prefixes overlap depending on the prefix positions where the overlap occurs. For more details of this filter we refer to [150].

As we will see in the comparative evaluation below, the filtering approaches achieve only a relatively small improvement for PPRL since the filter tests imply already a certain overhead which is not much less than for the match tests (which are relatively cheaper for Bloom filters). In addition, Bloom filter masking for PPRL should ideally have 50% of their bits set to 1 in order to make them less vulnerable to frequency attacks [117], making P4Join less effective.

Multi-bit Trees: The use of multi-bit trees was originally proposed for fast similarity search in large databases of chemical *fingerprints* (masked into Bloom filters) [98]. A query Bloom filter b_q is being searched for in a database to retrieve all elements whose similarity with b_q is above the threshold s_t. A multi-bit tree is a binary tree that iteratively assigns fingerprints to its nodes based on so-called match bits. A match bit refers to a specific position of the bit vector and can be 1 or 0: it indicates that all fingerprints in the associated subtree share the specified match bit. When building up the multi-bit tree, one match bit or multiple such bits are selected in each step so that the number of unassigned fingerprints can be roughly split by half. The split is continued as long as the number of fingerprints per node does not fall under a limit ([98] recommends a limit of 6). The match bits can then be used for a query fingerprint to determine the maximal possible similarity for subtrees when traversing the tree and can thereby eliminate many fingerprints to compare.

As suggested in [5], multi-bit trees can easily be applied for PPRL using Bloom filters and Jaccard similarity. For two datasets, the larger input dataset is used to build the multi-bit trees while each record (fingerprint) of the second dataset is used for searching similar records. The multi-bit approach of [5] partitions the fingerprints according to their lengths such that all fingerprints with the same length belong to the same partition (or bucket). To apply the length filter, we can then restrict the search for similar fingerprints to the partitions meeting the length criterion of Eq. 3. Query efficiency is further improved by organizing all fingerprints of a partition within a multi-bit tree.

In evaluations of [5, 145] the multi-bit tree approach was found to be very effective and even better or similarly effective than blocking approaches such as canopy clustering and sorted neighborhood.

PPRL for Metric Space Similarity Measures: A metric space consists of a set of data objects and a metric to compute the distance between the objects. The main property of interest that a metric or distance function d for metric spaces has to satisfy is the so-called triangle inequality. It requires that for any objects x, y and z it holds

$$d(x, z) \leq d(x, y) + d(y, z) \tag{4}$$

Fig. 5 Triangle inequality:
Object y cannot lie within
the search radius of query
object q since the difference
between $d(p, q)$ and $d(p, y)$
exceeds $rad(q)$ (taken
from [149])

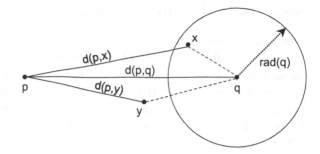

Distance functions for metric spaces satisfying this property include the
Minkowski distances (for example, Euclidean distance), edit distance, Hamming distance and Jaccard-coefficient (but not Dice-coefficient) [174]. The triangle inequality has been used for private comparison and classification in PPRL using reference
values [124, 162].

The triangle inequality has also been used to reduce the search space for similarity
search and record linkage [7, 12]. In both cases we have to find for a query object
q those similar objects x with a distance $d(q, x)$ lower than or equal to a maximal
distance threshold (or above a minimal similarity threshold) which can be seen as a
radius $rad(q)$ around q in Fig. 5. The triangle equality allows one to avoid computing the distance between two objects based on their precomputed distances to a third
reference object or pivot, such as object p in Fig. 5. Utilizing the precomputed distances $d(p, q)$ and $d(p, x)$ we only have to compute the distance $d(q, x)$ for objects
x that satisfy the triangle inequality $d(p, q) - d(p, x) \leq rad(q)$. In all other cases,
comparison can be avoided such as for object y in Fig. 5.

Several alternatives to utilize the triangle inequality to reduce the search space for
PPRL have been studied in [149], in particular for the Hamming distance which has
been shown to be equivalent to the Jaccard similarity [172]. The best performance
was achieved for a pivot-based approach that selects a certain number of data objects
from a sample of the first dataset as pivots and assigns each other object of the
first dataset to its closest pivot. For each pivot, the maximal distance (radius) for
its objects is also recorded. Pivots are iteratively determined from the sample set of
objects such that the object with the greatest distance to all previously determined
pivots becomes the next pivot. The rational behind this selection strategy is to have
a relatively large distance between pivots so that searching for similar objects can be
restricted to objects of relatively few pivots. Determining the pivots from a sample
rather than from all objects limits the overhead of pivot selection. The search for
similar (matching) objects can be restricted to the pivots for which there is a possible
overlap with the radius of the query objects. For the objects of the relevant pivots the
triangle inequality is further used to prune objects from the comparison.

Comparative Evaluation: The performance of pivot-based approaches for metric
similarity measures has been evaluated in [149] and compared with the use of P4Join
and multi-bit trees. The evaluation has been done for synthetically generated datasets

Table 3 PPRL runtime in minutes for different dataset sizes and filtering approaches (taken from [149])

Algorithms	Datasets				
	100,000	200,000	300,000	400,000	500,000
NestedLoop	3.8	20.8	52.1	96.8	152.6
Multi-bit Tree	2.6	11.3	26.5	50.0	75.9
P4Join	1.4	7.4	24.1	52.3	87.9
Pivots (Metric Space)	0.2	0.4	0.9	1.3	1.7

of 100,000–500,000 records such that 80% of the records are in the first dataset and 20% in the second. Bloom filters of length 1,000 bits are used to mask the QIDs of records, and the comparison is based on a Jaccard similarity threshold of 0.8 or the corresponding Hamming distance for the metric-space approach.

Table 3 summarizes the runtimes of the different approaches as well as for a naïve nested loop approach without any filtering (all implemented using Java) on a standard PC (Intel i7-4770, 3.4 GHz CPU with 16 GB main memory). The results show that both multi-bit trees and P4Join perform similarly but achieve only modest improvements (less than a factor of 2) compared to the naïve nested loop scheme. By contrast the pivot-based metric space approach achieves order-of-magnitude improvements. For the largest dataset it only needs 1.7 min and is 40 times faster than using multi-bit trees. A general observation for all approaches is that the runtimes increase more than linearly (almost quadratically) with the size of datasets, indicating a potential scalability problem despite the reduction of the search space. Substantially larger datasets would thus create performance problems even for the best filtering approach indicating that additional runtime improvements are necessary, e.g. by the use of parallel PPRL.

4.3 Parallel PPRL

PPRL in Big Data applications involves the comparison of a large number of masked records as the main part of the overall execution pipeline. Parallel linkage on multiple processors aims at improving the execution time proportionally to the number of processors [42, 90, 91]. This can be achieved by partitioning the set of all record pairs to be compared, and conducting the comparison of the different partitions in parallel on different processors. A special case would be to utilize a blocking approach to compare the records in different blocks in parallel. In the following we discuss two approaches for parallel PPRL that have been proposed: one utilizes graphics processors or GPUs for parallel processing within a single machine, and the other one is based on Hadoop and its MapReduce framework. Both approaches have also been used for general record linkage.

Parallel PPRL with GPUs: The utilization of Graphical Processing Units (GPUs) to speed-up similarity computations is a comparatively new approach [56, 120]. Modern GPUs provide thousands of cores that allow for a massively-parallel application of the same instruction set to disjoint data partitions. The availability of frameworks like OpenCL and CUDA simplify the utilization of GPUs to parallelize general purpose algorithms. The GPU programs (called *kernels*) are typically written in a dialect of the general programming language C. Kernel execution requires the input and output data to be transferred between the main memory of the host system and the memory of the GPU, and it is important to minimize the amount of data to be transferred. Further limitations are that there is no dynamic memory allocation on GPUs (all resources required by a program need to be allocated a priori) and that only basic data types (e.g., int, long, float) and fixed-length data structures (e.g., arrays) can be used.

Despite such limitations, the utilization of GPUs is a promising approach to speed-up PPRL. This is especially the case for Bloom filter masking where all records are represented as bit vectors of equal length. These vectors can easily be stored in array data structures on the GPU. Furthermore, similarity computations can be broken down into simple bit operations which are easily processed by GPUs.

A GPU-based implementation for PPRL using the P4Join filtering is described in [150]. It sorts the bit vectors of the two input datasets initially according to their number of set bits (1-bits) and partitions the set of bit vectors into equal-sized blocks such that multiple of such blocks fit into the GPU memory. Pairs of blocks are then continuously loaded into the GPU for parallel comparison. To limit unnecessary data transfers, the length filter (described in Sect. 4.2) is applied to avoid transferring pairs of blocks that do not meet the length filter restriction. The kernel programs also apply the prefix filter to save comparisons.

The evaluation in [150] showed that the GPU implementation is highly efficient and improves runtimes by a factor of 20, even for a low-profile graphics card (Nvidia GeForce GT 540 M with 96 CUDA cores@672 MHz, 1 GB memory). It would be interesting to realize GPU versions of other PPRL approaches and to utilize more powerful graphics cards with thousands of cores for improved performance.

Hadoop-based Parallel PPRL: Many Big Data applications are based on local Shared Nothing clusters driven by software from the open-source Hadoop ecosystem for parallel processing. Depending on the data volume and needed degree of parallelism up-to thousands of multi-processor nodes are utilized. A main reason for the success of Hadoop is that its programming frameworks, in particular MapReduce and newer platforms such as Apache Spark[4] or Apache Flink,[5] make it easy to develop programs that can be automatically executed in parallel on Hadoop clusters.

Several approaches and implementations have utilized MapReduce for parallel record linkage [93, 166]. In its simplest form, the Map tasks read the input data in parallel and apply a blocking key to assign each record to a block. Then the data records are dynamically redistributed among the Reduce tasks such that all records

---

[4]http://spark.apache.org [Accessed: 15/06/2016].
[5]https://flink.apache.org/ [Accessed: 15/06/2016].

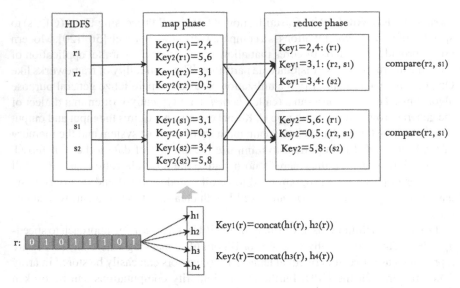

Fig. 6 Parallel PPRL with MapReduce using LSH blocking [82], where the MinHash keys for Bloom filters are computed in the Map phase and records with the same MinHash signature will be sent to the same Reduce task for matching

with the same blocking key are sent to the same Reduce task. Comparison is then performed block-wise and in parallel by the Reduce tasks. For highly skewed block sizes this simple approach can result in load balancing problems for the Reduce tasks; approaches to solve this data skew or load balancing problem are proposed in [94].

The sketched approach can in principle also be applied for parallelizing PPRL, e.g., if the linkage unit utilizes a Hadoop cluster. One such approach for using MapReduce to speed-up PPRL has been proposed in [82]. The authors apply a LSH-based blocking method using the MinHash approach (see Sect. 4.1). The use of MapReduce is rather straightforward and illustrated in Fig. 6. The Bloom filters of both sources are initially stored in the distributed file system (HDFS) as chunks. In the Map phase, records are read sequentially and for each Bloom filter r a set of j MinHash keys are computed. The records are then redistributed so that records with the same key are sent to the same Reduce task for comparison. The main drawback of this strategy is that records may be compared several times by different Reduce tasks because they could share many keys (as shown in Fig. 6 for records r_2 and s_1). To overcome this problem the authors proposed another strategy by chaining two MapReduce jobs, where the first one is similar to the described method except that the Reduce phase only outputs the pairs of records' identifiers instead of comparing the records. In the second MapReduce job, duplicate records pairs are grouped at the same Reducer to be compared only once. In this process, the Bloom filters are not redistributed (but only their identifiers) by storing the Bloom filters in a relational database from where they are read when needed. The evaluation of this parallel LSH approach in [82] was limited to only 2 and 4 nodes and small datasets (about 300,000 records) so that the overall scalability of the approach remains unclear.

For future work, it would be valuable to investigate and compare different parallel PPRL approaches utilizing the Hadoop ecosystem. The approaches could also utilize the Spark or Flink frameworks which support more operators than only Map and Reduce, and support efficient distributed in-memory processing.

5 Multi-party PPRL

While there have been many different approaches proposed for PPRL [164], most work thus far has concentrated on linking records from only two databases (or parties). Only some approaches have investigated linking records from three or more databases [75, 104, 118, 122, 127, 130, 131, 160], with most of these being limited to exact matching or matching of categorical data only, as will be discussed below. However, as the example applications described in Sect. 1 have shown, linking data from multiple databases is increasingly being required for several Big Data applications. In the following, we describe existing techniques of multi-party private blocking and private comparison and classification for multi-party PPRL (MP-PPRL).

5.1 Multi-party Private Blocking Techniques

Private blocking for MP-PPRL is crucial due to the exponential growth of the comparison space with the number of databases linked. However, this has not been studied until recently, making MP-PPRL not practical in real applications.

Tree-based approaches: The first approach [130] is based on a single-bit tree (adapted from multi-bit tree [98]) data structure, which is constructed iteratively to arrange records (masked into Bloom filters) such that similar records are placed into the same tree leaf while non-similar records are placed into different leaf nodes in the tree. At each iteration, the set of Bloom filters in a tree node is recursively split based on selected (according to a privacy criteria) bit positions which are agreed upon by all parties. A drawback with this approach, however, is that it might miss true matches due to the recursive splitting of Bloom filters. Furthermore, a communication step is required among all parties for each iteration.

This limitation of missing true matches in the single-bit tree-based approach [130] has been addressed in [131] using a multi-bit tree [98] data structure (as we discussed in Sect. 4.2) that is combined with canopy clustering. Multi-bit tree-based filtering for PPRL of two databases was first introduced by Schnell [144]. In [131] the concept of multi-bit trees was used to split the databases (masked into Bloom filters) individually by the parties into small mini-blocks, which are then merged into larger blocks

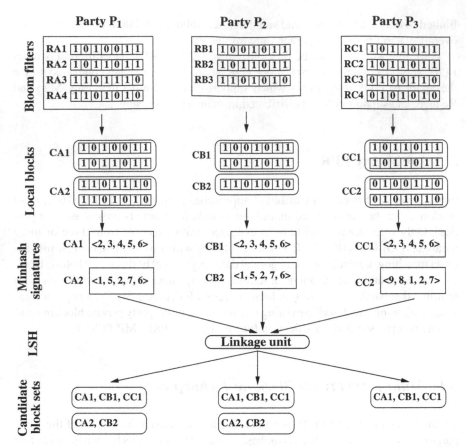

Fig. 7 Multi-party private blocking approach as proposed by Ranbaduge et al. [133] (adapted from [133]). Candidate block sets from all three parties ($CA1$, $CB1$, $CC1$) and sub-set of two parties ($CA2$, $CB2$) are identified to be compared and classified

according to privacy and computational requirements based on their similarity using a canopy clustering technique [40].

Linkage unit-based approaches: A communication-efficient approach for multi-party private blocking by using a linkage unit was recently proposed [133], as illustrated in Fig. 7. In the first step of this approach, local blocks are generated individually by each party using a private blocking technique (which is considered to be a black box). For example, the private blocking approach based on multi-bit tree and canopy clustering [131] (described above) can be used for local blocking. A block representative in the form of a min-hash signature [18] is then generated for each block and sent to a linkage unit. The linkage unit applies global blocking using locality sensitive hashing (LSH) to identify the candidate block sets from all parties or from sub-sets of parties based on the similarity between block representatives. Local blocking provides the database owners with more flexibility and control over their blocks while eliminating all communications among them. This approach outperforms existing multi-party private blocking approaches in terms of scalability, privacy, and blocking quality, as validated by a set of experiments conducted by the authors [133].

Karapiperis and Verykios recently proposed a multi-party private blocking approach based on LSH [84]. This approach uses L independent hash tables (or blocking groups), each of which consists of key-bucket pairs where keys represent the blocking keys and buckets host a linked list aimed at grouping similar records that were previously masked into Bloom filters. Each hash table is assigned with a set of K hash functions which is generated by a linkage unit and sent to all the database owners to populate their set of blocks accordingly. The same authors extended this approach by proposing a frequent pairs scheme (FPS) [85] for further reducing the number of comparisons while maintaining a high level of recall. This approach achieves high blocking quality by identifying similar record pairs that exhibit a number of LSH collisions above a given threshold, and then performs distance calculations only for those similar pairs. Empirical results showed significant improvement in running time due to a drastic reduction of candidate pairs by the FPS, while achieving high blocking quality [85].

A major drawback of these multi-party private blocking techniques is that they still result in an exponential comparison space with an increasing number of databases to be linked, especially when the databases are large. Therefore, efficient communication patterns, such as ring-based or tree-based [113, 142], as well as advanced filtering techniques, such as those discussed in Sect. 4.2, need to be investigated for multi-party PPRL in order to make PPRL scalable and viable in Big Data applications.

5.2 Multi-party Private Comparison and Classification Techniques

Several private comparison and classification techniques for MP-PPRL have been developed in the literature. However, they fall short in providing a practical solution either because they allow exact matching only or they are computationally not feasible with the size and number of multiple databases. In the following we describe these approaches and their drawbacks.

Secure Multi-party Computation (SMC)-based approach: An approach based on SMC using an oblivious transfer protocol was proposed in [122] for multi-party private comparison and classification. While provably secure, the approach only performs exact matching of masked records and it is computationally expensive compared to efficient perturbation-based privacy techniques such as Bloom filters and k-anonymity [164].

Generalization-based approaches: A multi-party private comparison and classification approach was introduced in [75] to perform secure equi-join of masked records
from multiple k-anonymous databases by using a linkage unit. The database records
are k-anonymised by the database owners and sent to a linkage unit. The linkage
unit then compares and classifies records by applying secure equi-join, which allows
exact matching only.

Another multi-party private comparison and classification approach based on k-
anonymity for categorical values was proposed in [118]. In this approach, a top-
down generalization is performed on the QIDs to provide k-anonymous privacy (as
discussed in Sect. 3.4) and the generalized blocks are then classified into matches
and non-matches using the C4.5 decision tree classifier.

Probabilistic data structure-based approaches: An efficient multi-party private
comparison and classification approach for exact matching of masked records using
Bloom filters was introduced by Lai et al. [104], as illustrated in Fig. 8. Each party
hash-maps their record values into a single Bloom filter and then partitions its Bloom
filter into segments according to the number of parties involved in the linkage. The
segments are exchanged among the parties such that each party receives a corresponding Bloom filter segment from all other parties. The segments received by a
party are combined using a conjunction (logical AND) operation. The resulting conjuncted Bloom filter segments are then exchanged among the parties to generate the
full conjuncted Bloom filter. Each party compares its Bloom filter of each record with
the final conjuncted Bloom filter. If the membership test of a record's Bloom filter is
successful then the record is considered to be a match across all databases. Though
the computation cost of this approach is low since the computation is completely
distributed among the parties without a linkage unit and the creation and processing
of Bloom filters are very fast, the approach can only perform exact matching.

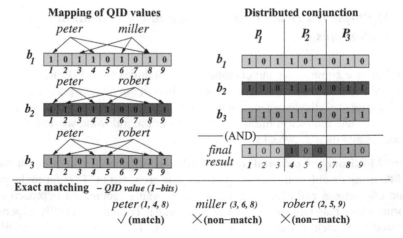

Fig. 8 Bloom filter masking-based exact matching approach for MP-PPRL as proposed by Lai et
al. [104] (adapted from [164])

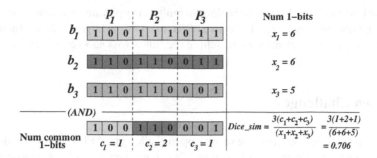

Fig. 9 Bloom filter masking-based approximate matching approach for MP-PPRL proposed by Vatsalan and Christen [160] (adapted from [160])

Another efficient multi-party approach for private comparison and classification of categorical data was recently proposed [80] using a Count-Min sketch data structure (as described in Sect. 3.4). Sketches are used to summarize records individually by each database owner, followed by a secure intersection of these sketches to provide a global synopsis that contains the common records across parties and their frequencies. The approach uses homomorphic operations, secure summation, and symmetric noise addition privacy techniques.

Developing privacy-preserving approximate string comparison functions for multiple (more than two) values has only recently been considered [160]. This MP-PPRL approach adapts Lai et al.'s Bloom filter-based exact matching approach [104] (as described above) for approximate matching to distributively calculate the Dice-coefficient similarity of a set of Bloom filters from different parties using a secure summation protocol. This approach is illustrated in Fig. 9. The Dice-coefficient of P Bloom filters $(b_1, \ldots, b_P)$ is calculated as:

$$Dice\_sim(b_1, \ldots, b_P) = \frac{P \times c}{\sum_{i=1}^{P} x_i} = \frac{P \times \sum_{i=1}^{P} c_i}{\sum_{i=1}^{P} x_i}, \tag{5}$$

where c_i is the number of common bit positions that are set to 1 in ith Bloom filter segment from all P parties such that $c = \sum_{i=1}^{P} c_i$, and x_i is the number of bit positions set to 1 in b_i (1-bits), where $x = \sum_{i=1}^{P} x_i$ and $1 \leq i \leq P$.

Similar to Lai et al.'s approach [104], the Bloom filters are split into segments such that each party receives a certain segment of the Bloom filters from all other parties. A logical conjunction is applied to calculate c_i individually by each party P_i (with $1 \leq i \leq P$) which are then summed to calculate c using a secure summation protocol. A secure summation of x_i is also performed to calculate x. These two sums are then used to calculate the Dice-coefficient similarity of the Bloom filters using Eq. 5. A limitation of this approach is that it can only be used to link a small number of databases due to its large number of logical conjunction calculations (even when a private blocking technique is used).

Therefore, more work needs to be done in multi-party private comparison and classification to enable efficient and effective PPRL on multiple large databases including sub-set matching (i.e. identifying matching records across sub-set of parties).

6 Open Challenges

In this section we first describe the various open challenges of PPRL, and then discuss these challenges in the context of the four V's *volume*, *variety*, *velocity*, and *veracity* of Big Data.

6.1 Improving Scalability

The trend of Big Data growth dispersed in multiple sources challenges PPRL in terms of complexity (*volume*), which increases exponentially with multiple large databases. Much research in recent years has focused on improving the scalability of the PPRL process, both with regard to the sizes of the databases to be linked, as well as with the number of databases to be linked. While significant progress has been made in both these directions, further efforts are required to make all aspects of the PPRL process scalable. Both directions are highly relevant for Big Data applications.

Even small blocks can still lead to a large number of record pair (or set) comparisons that are required in the comparison step, especially when databases from multiple (more than two) sources are to be linked. For each set of blocks across several parties, potentially all combinations of record sets need to be compared. For a block that contains B records from each of P parties, B^P comparisons are required. Crucial are efficient adaptive comparison techniques that stop the comparison of records across parties once a pair of records has been classified to be a non-match between two parties. For example, assume the record set $\langle r_A, r_B, r_C, r_D \rangle$, where r_A is from party A, r_B is from party B, and so on. Once the pair r_A and r_B are compared and classified as a non-match, there is no need to compare all other possible record pairs (r_A with r_C, r_A with r_D, r_B with r_C, and so on) if the aim of the linkage is to identify sets of records that match across all parties involved in a PPRL.

A very challenging aspect is the task of identifying sub-sets of records that match across only a sub-set of parties. An example is to find all patients that have medical records in the databases of any three out of a group of five hospitals. In this situation, all potential sub-sets of records need to be compared and classified. This is a challenging problem with regard to the number of comparisons required and has not been studied in the literature so far.

6.2 Improving Linkage Quality

The *veracity* and *variety* aspects (errors and variations) of Big Data need to be addressed in PPRL by developing accurate and effective comparison and classification techniques for high linkage quality. How to efficiently calculate the similarity of more than two values using approximate comparison functions in PPRL is an important challenge with multi-source linking. Most existing PPRL solutions for multiple parties only support exact matching [80, 104] or they are applicable to QIDs of only categorical data [75, 118]. Thus far only one recent approach supports approximate matching of string data for PPRL on multiple databases [160] (as described in Sect. 5.2).

In the area of non-PPRL, advanced collective [13] and graph-based [58, 74] classification techniques have been developed in recent times. These techniques are able to achieve high linkage quality compared to the basic pair-wise comparison and threshold-based classification approach that is often employed in most PPRL techniques. Group linkage [123] is the only advanced classification technique that has so far been considered for PPRL [105].

For classification techniques that require training data (i.e. supervised classifiers), a major challenge in PPRL is how such training data can be generated. Because of privacy and confidentiality concerns, in PPRL it is generally not possible to gain access to the actual sensitive QID values (to decide if they refer to a true match or a true non-match). The advantage of certain collective and graph-based approaches [13, 74] is that they are unsupervised and therefore do not require training data. However, their disadvantage is their high computational complexities (quadratic or even higher) [137]. Investigating and adapting advanced classification techniques for PPRL will be a crucial step towards making PPRL useful for practical Big Data applications, where training data are commonly not available, or are expensive to generate.

6.3 Dynamic Data and Real-Time Matching

All PPRL techniques developed so far, in line with most non-PPRL techniques, only consider the batch linkage of static databases. However, a major aspect of Big Data is the dynamic nature of data (*velocity*) that requires adaptive systems to link data as they arrive at an organization, ideally in (near) real-time. Limited work has so far investigated temporal data [33, 107] and real-time [32, 70, 129] matching in the context of record linkage. Temporal aspects can be considered by adapting the similarities between records depending upon the time difference between them, while real-time matching can be achieved using sophisticated adaptive indexing techniques. Several works have been done on dynamic privacy-preserving data publishing on the cloud by developing an efficient and adaptive QID index-based approach over incremental datasets [175, 176].

Linking dynamic databases in a PPRL context opens various challenging research questions. Existing masking (encoding) methods used in PPRL assume static databases that allow parameter settings to be calculated a-priori leading to secure masking of QID values. For example, Bloom filters in average should have 50% of their bits set to 1, making frequency attacks more difficult [117]. Such masking might not stay secure as the characteristics of data are changing over time. Dynamic databases also require novel comparison functions that can adapt to changing data as well as adaptive masking techniques.

6.4   Improving Security and Privacy

In addition to the four V's of Big Data, another challenging aspect that needs to be considered for Big Data applications is *security and privacy*. As we discussed in Sect. 3.2, most work in PPRL assumes the *honest-but-curious* (HBC) adversary model [65, 111]. Most PPRL protocols also assume that the parties do not collude with each other (i.e. a sub-set of two or more parties do not collaborate with the aim to learn sensitive information of another party) [111]. However, in a commercial environment and in PPRL scenarios where many parties are involved, such as is likely in Big Data applications, collusion is a real possibility that needs to be prevented. Only few PPRL techniques consider the *malicious* adversary model [164]. The techniques developed based on this security model commonly have high computational complexities and are therefore currently not practical for the linkage of large databases. Therefore, because the HBC model might not be strong enough while the malicious model is computationally too expensive, novel security models that lie between those two need to be investigated for PPRL. Two of these are the *covert* adversary model [4] and *accountable computing* [71], which have been discussed in Sect. 3.2. Research directions are required to develop new protocols that are practical and at the same time more secure than protocols based on the HBC model.

With regard to privacy, most PPRL techniques are known to leak some information during the exchange of data between the parties (such as the number and sizes of blocks, or the similarities between compared records). How sensitive such revealed information is for a certain dataset heavily depends upon the parameter settings used by a protocol. Sophisticated attack methods [101] have been developed that exploit the subtle pieces of information revealed by certain PPRL protocols to iteratively gather information about sensitive values. Therefore, there is a need to harden existing PPRL techniques to ensure they are not vulnerable to such attacks. Preserving privacy of individual entities is more challenging with multi-party PPRL due to the increasing risk of collusion between a sub-set of parties which aim to learn about another (sub-set of) party's private data. Distributing computations among pairs or groups of parties can reduce the likelihood of collusion between parties if individual pairs or groups can use different secret keys (known only to them) for masking their values.

Most PPRL techniques have mainly been focusing on the privacy of the individual records that are to be linked [165]. However, besides individual record privacy, the privacy of a group of individuals also needs to be considered. Often the outcomes of a PPRL project are sets of linked records that represent people with certain characteristics (such as certain illnesses, or particular financial circumstances). While the names, addresses and other personal details of these people are not revealed during or after the PPRL process, their overall characteristics as a group could potentially lead to the discrimination of individuals in this group if these characteristics are being revealed. The research areas of privacy-preserving data publishing [59] and statistical confidentiality [45] have been addressing these issues from different directions.

PPRL is only one component in the management and analysis of sensitive, person-related information by linking different datasets in a privacy-preserving manner. However, achieving an effective overall privacy preservation needs a comprehensive strategy regarding the whole data life cycle including collection, management, publishing, exchange and analysis of data to be protected ('privacy-by-design') [22]. Hence, it is necessary to better understand the role of PPRL in the life cycle for sensitive data to ensure that it can be applied and that the match results are both useful and privacy-preserving.

In research, the different technical aspects to preserve privacy have partially been addressed by different communities with little interaction. For example, there is a large body of research on privacy-preserving data publishing [59] and on privacy-preserving data mining [109, 156] that have been largely decoupled from the research on PPRL. It is well known that data analysis may identify individuals despite the masking of QID values [152]. Hence, there is similar risk that the combined information of matched records together with some background information could lead to the identification of individuals (known as *re-identification*). Such risks must be evaluated and addressed within a comprehensive privacy strategy including a closely aligned PPRL and privacy-preserving data analysis/mining approach.

6.5 Evaluation, Frameworks, and Benchmarks

How to assess the quality (how many classified matches are true matches) and completeness (how many true matches have been classified as matches) of the records linked in a PPRL project is very challenging because it is generally not possible to inspect linked records due to privacy concerns. Manual assessment of individual records would reveal sensitive information which is in contradiction to the objective of PPRL. Not knowing how accurate and complete linked data are is however a major issue that will render any PPRL protocol impractical in applications where linkage completeness and quality are crucial, as is the case in many Big Data applications such as in the health or security domains.

Recent initial work has proposed ideas and concepts for interactive PPRL [100] where parts of sensitive values are revealed for manual assessment. How to actually implement such approaches in real applications, while ensuring the revealed information is limited to a certain level of detail (for example providing k-anonymous privacy for a certain value of $k > 1$ [152]) is an open research question that must be solved. Interactive manual evaluation might also not be feasible in Big Data applications where the size and dynamic nature of data, as well as real-time processing requirements, prohibit any manual inspection.

With regard to evaluating the privacy protection that a given PPRL technique provides, unlike for measuring linkage quality and completeness (where standard measurements such as runtime, reduction ratio, pairs completeness, pairs quality, precision, recall, or accuracy are available [28]), there are currently no standard measurements for assessing privacy in PPRL. Different measurements have been proposed and used [46, 164, 165], making the comparison of different PPRL techniques difficult. How to assess linkage quality and completeness, as well as privacy, are must-solve problems as otherwise it will not be possible to evaluate the efficiency, effectiveness, and privacy protection of PPRL techniques in real-world applications, leaving these techniques non-practical.

An important direction of future work for PPRL is the development of frameworks that allow the experimental comparison of different PPRL techniques with regard to their scalability, linkage quality, and privacy preservation. No such framework currently exists. Ideally, such frameworks allow researchers to easily 'plug-in' their own algorithms such that over time a collection of PPRL algorithms is compiled that can be tested and evaluated by researchers, as well as by practitioners to allow them to identify the best technique to use for their application scenario.

An issue related to frameworks is the availability of publicly available benchmark datasets for PPRL. While this is not a challenge limited to PPRL but to record linkage research in general [28, 95], it is particularly prominent for PPRL as it deals with sensitive and confidential data. While for record linkage techniques publicly available data from bibliographic or consumer product databases might be used [95], such data are less useful for PPRL research as they have different characteristics compared to personal data. The nature of the datasets to be linked using PPRL techniques is obviously in strong contradiction to them being made public. Ideally researchers working in PPRL are able to collaborate with practitioners that do have access to real sensitive and confidential databases to allow them to evaluate their techniques on such data.

A possible alternative to using benchmark datasets is the use of synthetic data that are generated based on the characteristics of real data using data generators [34, 153]. Such generators must be able to generate data with similar distribution of values, variations, and errors as would be expected in real datasets from the same domain. Several such data generators have been developed and are used by researchers working in PPRL as well as record linkage in general.

6.6 Discussion

As we have discussed in this section, there are various challenges that need to be addressed in order to make PPRL practical for applications in a variety of domains.

Some of these challenges are general and not just affect PPRL for Big Data, others are specific to certain types of applications, including those in the Big Data space.

The challenge of scalability of PPRL towards very large databases is highly relevant to the *volume* of Big Data, while the challenge of linkage quality of PPRL is highly relevant to the *veracity* and *variety* of Big Data. The dynamic nature of data in many Big Data applications, and the requirement of being able to link data in real-time, are challenging all aspects of PPRL, as well as record linkage in general [129]. This challenge corresponds to the *velocity* of Big Data and it requires the development of novel techniques that are adaptive to changing data characteristics, and that are highly efficient with regard to fast linking of streams of query records. While the *volume*, *variety*, and *veracity* aspects of Big Data have been studied for PPRL to some extent, the *velocity* aspect has so far not been addressed in a PPRL context.

Making PPRL more secure and more private is challenged by all four V's of Big Data. Larger data *volume* likely means that only encoding techniques that require little computational efforts per record can be employed, while dynamic data (*velocity*) means such techniques have to be adaptable to changing data characteristics. *Variety* means PPRL techniques have to be made more secure and private for various types of data, while *veracity* requires them to also take data uncertainties into account. The challenge of integrating PPRL into an overall privacy-preserving approach has also not seen any work so far. All four V's of Big Data will affect the overall efficiency and effectiveness of systems that enable the management and analysis of sensitive and confidential information in a privacy-preserving manner. The more basic challenges of improving scalability, linkage quality, privacy and evaluation need to solved first before this more complex challenge of an overall privacy-preserving system can be addressed.

The final challenge of evaluation is affected by all aspects of Big Data. Improved evaluation of PPRL systems requires that databases that are large, heterogeneous, dynamic, and that contain uncertain data, can be handled and evaluated efficiently and accurately. So far no research in PPRL has investigated evaluation specifically for Big Data. While the lack of general benchmarks and frameworks is already a gap in PPRL and record linkage research in general, Big Data will make this challenge even more pronounced. Compared to frameworks that can handle small and medium sized static datasets only, it is even more difficult to develop frameworks that enable privacy-preserving linking of very large and dynamic databases, as is making such datasets publicly available. No work addressing this challenge in the context of Big Data has been published.

7 Conclusions

Privacy-preserving record linkage (PPRL) is an emerging research field that is being required by many different applications to enable effective and efficient linkage of databases across different organizations without compromising privacy and confidentiality of the entities in these databases. In the Big Data era, tremendous opportunities can be realized by linking data at the cost of additional challenges. In this chapter, we have provided background material required to understand the applications, process, and challenges of PPRL, and we have reviewed existing PPRL approaches to understand the literature. Based on the analysis of existing techniques, we have discussed several interesting and challenging directions for future work in PPRL for Big Data.

With the increasing trend of Big Data in organizations, more research is required towards the development of techniques that allow for multiple large databases to be linked in privacy-preserving, effective, and efficient ways, thereby facilitating novel ways of data analysis and mining that currently are not feasible due to scalability, quality, and privacy-preserving challenges.

Acknowledgements This work was partially funded by the Australian Research Council under Discovery Project DP130101801, the German Academic Exchange Service (DAAD) and Universities Australia (UA) under the Joint Research Co-operation Scheme, and also funded by the German Federal Ministry of Education and Research within the project Competence Center for Scalable Data Services and Solutions (ScaDS) Dresden/Leipzig (BMBF 01IS14014B).

References

1. R. Agrawal, A. Evfimievski, R. Srikant, Information sharing across private databases, in *ACM SIGMOD* (2003), pp. 86–97
2. A. Arasu, V. Ganti, R. Kaushik, Efficient exact set-similarity joins, in *PVLDB* (2006), pp. 918–929
3. A. Arasu, M. Götz, R. Kaushik, On active learning of record matching packages, in *ACM SIGMOD* (2010), pp. 783–794
4. Y. Aumann, Y. Lindell, Security against covert adversaries: efficient protocols for realistic adversaries. J. Cryptol. **23**(2), 281–343 (2010)
5. T. Bachteler, J. Reiher, and R. Schnell. Similarity Filtering with Multibit Trees for Record Linkage. Technical Report WP-GRLC-2013-01, German Record Linkage Center, 2013
6. D. Barone, A. Maurino, F. Stella, C. Batini, A privacy-preserving framework for accuracy and completeness quality assessment, in *Emerging Paradigms in Informatics, Systems and Communication* (2009), pp. 83–87
7. J.E. Barros, J.C. French, W.N. Martin, P.M. Kelly, T.M. Cannon, Using the triangle inequality to reduce the number of comparisons required for similarity-based retrieval, in *Electronic Imaging Science and Technology* (1996), pp. 392–403
8. C. Batini, M. Scannapieca, *Data quality: Concepts, Methodologies And Techniques. Data-Centric Systems and Applications* (Springer, Berlin, 2006)
9. R. Baxter, P. Christen, T. Churches, A comparison of fast blocking methods for record linkage, in *SIGKDD Workshop on Data Cleaning, Record Linkage and Object Consolidation* (2003), pp. 25–27
10. R.J. Bayardo, Y. Ma, R. Srikant, Scaling Up All Pairs Similarity Search, in *WWW* (2007), pp. 131–140
11. K. Bellare, S. Iyengar, A.G. Parameswaran, V. Rastogi, Active sampling for entity matching, in *ACM SIGKDD* (2012), pp. 1131–1139
12. A. Berman, L.G. Shapiro, Selecting good keys for triangle-inequality-based pruning algorithms, in *IEEE Workshop on Content-Based Access of Image and Video Database* (1998), pp. 12–19
13. I. Bhattacharya, L. Getoor, Collective entity resolution in relational data. ACM TKDD **1**(1), 1–35 (2007)

14. M. Bilenko, R.J. Mooney, Adaptive duplicate detection using learnable string similarity measures, in *ACM SIGKDD* (2003), pp. 39–48
15. B. Bloom, Space/time trade-offs in hash coding with allowable errors. Commun. ACM **13**(7), 422–426 (1970)
16. L. Bonomi, L. Xiong, R. Chen, B. Fung, Frequent grams based embedding for privacy preserving record linkage, in *ACM CIKM* (2012), pp. 1597–1601
17. H. Bouzelat, C. Quantin, L. Dusserre, Extraction and anonymity protocol of medical file, in *AMIA Fall Symposium* (1996), pp. 323–327
18. A.Z. Broder, On the resemblance and containment of documents, in *Compression and Complexity of Sequences*. IEEE (1997), pp. 21–29
19. A. Broder, M. Mitzenmacher, A. Mitzenmacher, Network applications of Bloom filters: a survey. Internet Math. **1**(4), 485–509 (2004)
20. E. Brook, D. Rosman, C. Holman, Public good through data linkage: measuring research outputs from the Western Australian data linkage system. Aust. NZ J. Public Health **32**, 19–23 (2008)
21. R. Canetti, Security and composition of multiparty cryptographic protocols. J. Cryptol. **13**(1), 143–202 (2000)
22. A. Cavoukian, J. Jonas, Privacy by design in the age of Big Data. Technical report, TR Information and privacy commissioner, Ontario (2012)
23. P. Christen, A comparison of personal name matching: techniques and practical issues, in *IEEE ICDM Workshop on Mining Complex Data* (2006), pp. 290–294
24. P. Christen, Privacy-preserving data linkage and geocoding: current approaches and research directions, in *IEEE ICDM Workshop on Privacy Aspects of Data Mining* (2006), pp. 497–501
25. P. Christen, Automatic record linkage using seeded nearest neighbour and support vector machine classification, in *ACM SIGKDD* (2008), pp. 151–159
26. P. Christen, Febrl: an open source data cleaning, deduplication and record linkage system with a graphical user interface, in *ACM SIGKDD* (2008), pp. 1065–1068
27. P. Christen, Geocode matching and privacy preservation, in *Workshop on Privacy, Security, and Trust in KDD* (Springer, Berlin, 2009), pp. 7–24
28. P. Christen, *Data Matching - Concepts and Techniques for Record Linkage, Entity Resolution, and Duplicate Detection* (Springer, Berlin, 2012)
29. P. Christen, A survey of indexing techniques for scalable record linkage and deduplication. IEEE TKDE **24**(9), 1537–1555 (2012)
30. P. Christen, T. Churches, M. Hegland, Febrl – a parallel open source data linkage system, in *Springer PAKDD* (2004), pp. 638–647
31. P. Christen, K. Goiser, Quality and complexity measures for data linkage and deduplication, in *Quality Measures in Data Mining*, vol. 43. Studies in Computational Intelligence (Springer, Berlin, 2007), pp. 127–151
32. P. Christen, R. Gayler, D. Hawking, Similarity-aware indexing for real-time entity resolution, in *ACM CIKM* (2009), pp. 1565–1568
33. P. Christen, R.W. Gayler, Adaptive temporal entity resolution on dynamic databases, in *PAKDD* (2013), pp. 558–569
34. P. Christen, D. Vatsalan, Flexible and extensible generation and corruption of personal data, in *ACM CIKM* (2013), pp. 1165–1168
35. T. Churches, P. Christen, Some methods for blindfolded record linkage. BioMed Cent. Med. Inf. Decision Mak. **4**(9), (2004)
36. T. Churches, P. Christen, K. Lim, J.X. Zhu, Preparation of name and address data for record linkage using hidden Markov models. BioMed Cent. Med. Inf. Decision Mak. **2**(9), (2002)
37. D.E. Clark, Practical introduction to record linkage for injury research. Inj. Prev. **10**, 186–191 (2004)
38. C. Clifton, M. Kantarcioglu, J. Vaidya, X. Lin, M. Zhu, Tools for privacy preserving distributed data mining. SIGKDD Explor. **4**(2), 28–34 (2002)
39. W.W. Cohen, Data integration using similarity joins and a word-based information representation language. ACM TOIS **18**(3), 288–321 (2000)

40. W.W. Cohen, J. Richman, Learning to match and cluster large high-dimensional data sets for data integration, in *ACM SIGKDD* (2002), pp. 475–480
41. G. Cormode, S. Muthukrishnan, An improved data stream summary: the count-min sketch and its applications. J. Algorithms **55**(1), 58–75 (2005)
42. G. Dal Bianco, R. Galante, C.A. Heuser, A fast approach for parallel deduplication on multi-core processors, in *ACM Symposium on Applied Computing* (2011), pp. 1027–1032
43. D. Dey, V. Mookerjee, D. Liu, Efficient techniques for online record linkage. IEEE TKDE **23**(3), 373–387 (2010)
44. W. Du, M. Atallah, Protocols for secure remote database access with approximate matching, in *ACM WSPEC* (Springer, Berlin, 2000), pp. 87–111
45. G.T. Duncan, M. Elliot, J.-J. Salazar-González, *Statistical Confidentiality: Principles and Practice* (Springer, New York, 2011)
46. E. Durham, A framework for accurate, efficient private record linkage. Ph.D. thesis, Faculty of the Graduate School of Vanderbilt University, Nashville, TN, 2012
47. E. Durham, Y. Xue, M. Kantarcioglu, B. Malin, Private medical record linkage with approximate matching, in *AMIA Annual Symposium* (2010), pp. 182–186
48. E.A. Durham, C. Toth, M. Kuzu, M. Kantarcioglu, Y. Xue, B. Malin, Composite Bloom filters for secure record linkage. IEEE TKDE **26**(12), pp. 2956–2968 (2013)
49. L. Dusserre, C. Quantin, H. Bouzelat, A one way public key cryptosystem for the linkage of nominal files in epidemiological studies. Medinfo **8**, 644–647 (1995)
50. C. Dwork, Differential privacy, in *ICALP* (2006), pp. 1–12
51. M.G. Elfeky, V.S. Verykios, A.K. Elmagarmid, TAILOR: a record linkage toolbox, in *IEEE ICDE* (2002), pp. 17–28
52. A. Elmagarmid, P. Ipeirotis, V.S. Verykios, Duplicate record detection: a survey. IEEE TKDE **19**(1), 1–16 (2007)
53. U. Fayyad, G. Piatetsky-Shapiro, P. Smyth, R. Uthurusamy, *Advances in Knowledge Discovery and Data Mining* (The MIT Press, Cambridge, 1996)
54. I.P. Fellegi, A.B. Sunter, A theory for record linkage. J. Am. Stat. Soc. **64**(328), 1183–1210 (1969)
55. S.E. Fienberg, Confidentiality and disclosure limitation. Encycl. Soc. Meas. **1**, 463–469 (2005)
56. B. Forchhammer, T. Papenbrock, T. Stening, S. Viehmeier, U. Draisbach, F. Naumann, Duplicate detection on GPUs, in *BTW* (2013), pp. 165–184
57. M. Freedman, Y. Ishai, B. Pinkas, O. Reingold, Keyword search and oblivious pseudorandom functions, in *Theory of Cryptography* (2005), pp. 303–324
58. Z. Fu, J. Zhou, P. Christen, M. Boot, Multiple instance learning for group record linkage, in *PAKDD, Springer LNAI* (2012), pp. 171–182
59. B. Fung, K. Wang, R. Chen, P.S. Yu, Privacy-preserving data publishing: a survey of recent developments. ACM Comput. Surv. **42**(4), 14 (2010)
60. S.R. Ganta, S.P. Kasiviswanathan, A. Smith, Composition attacks and auxiliary information in data privacy, in *ACM SIGKDD* (2008), pp. 265–273
61. A. Gionis, P. Indyk, R. Motwani, Similarity search in high dimensions via hashing, in *VLDB* (1999), pp. 518–529
62. O. Goldreich, *Foundations of Cryptography: Basic Applications*, vol. 2. (Cambridge University Press, Cambridge, 2004)
63. L. Gu, R. Baxter, Decision models for record linkage, in *Selected Papers from AusDM*. LNCS, vol. 3755 (Springer, Berlin, 2006), pp. 146–160
64. M. Hadjieleftheriou, A. Chandel, N. Koudas, D. Srivastava, Fast indexes and algorithms for set similarity selection queries, in *IEEE ICDE* (2008), pp. 267–276
65. R. Hall, S. Fienberg, Privacy-preserving record linkage, in *PSD* (2010), pp. 269–283
66. M. Herschel, F. Naumann, S. Szott, M. Taubert, Scalable iterative graph duplicate detection. IEEE TKDE **24**(11), 2094–2108 (2012)
67. A. Inan, M. Kantarcioglu, E. Bertino, M. Scannapieco, A hybrid approach to private record linkage, in *IEEE ICDE* (2008), pp. 496–505

68. A. Inan, M. Kantarcioglu, G. Ghinita, E. Bertino. Private record matching using differential privacy, in *EDBT* (2010), pp. 123–134
69. P. Indyk, R. Motwani, *Approximate nearest neighbors: Towards removing the curse of dimensionality*, in *ACM Symposium on the Theory of Computing* (1998), pp. 604–613
70. E. Ioannou, W. Nejdl, C. Niederée, Y. Velegrakis, On-the-fly entity-aware query processing in the presence of linkage. PVLDB 3(1–2), 429–438 (2010)
71. W. Jiang, C. Clifton, Ac-framework for privacy-preserving collaboration, in *SDM* SIAM (2007), pp. 47–56
72. W. Jiang, C. Clifton, M. Kantarcıoğlu, Transforming semi-honest protocols to ensure accountability. Elsevier DKE 65(1), 57–74 (2008)
73. J. Jonas, J. Harper, Effective counterterrorism and the limited role of predictive data mining. Policy Anal. 584, 1–12 (2006)
74. D. Kalashnikov, S. Mehrotra, Domain-independent data cleaning via analysis of entity-relationship graph. ACM TODS 31(2), 716–767 (2006)
75. M. Kantarcioglu, W. Jiang, B. Malin, A privacy-preserving framework for integrating person-specific databases, in *PSD* (2008), pp. 298–314
76. A. Karakasidis, V.S. Verykios, Secure blocking+secure matching = secure record linkage. JCSE 5, 223–235 (2011)
77. A. Karakasidis, V.S. Verykios, Reference table based k-anonymous private blocking, in *ACM SAC* (2012), pp. 859–864
78. A. Karakasidis, V.S. Verykios, A sorted neighborhood approach to multidimensional privacy preserving blocking, in *IEEE ICDMW* (2012), pp. 937–944
79. A. Karakasidis, V.S. Verykios, P. Christen, Fake injection strategies for private phonetic matching. DPM Springer 7122, 9–24 (2012)
80. D. Karapiperis, D. Vatsalan, V.S. Verykios, Large-scale multi-party counting set intersection using a space efficient global synopsis, in *DASFAA* (2015), pp. 329–345
81. D. Karapiperis, D. Vatsalan, V.S. Verykios, P. Christen, Efficient record linkage using a compact hamming space, in *EDBT* (2016), pp. 209–220
82. D. Karapiperis, V.S. Verykios, A distributed framework for scaling up LSH-based computations in privacy preserving record linkage, in *ACM BCI* (2013), pp. 102–109
83. D. Karapiperis, V.S. Verykios, A distributed near-optimal LSH-based framework for privacy-preserving record linkage. ComSIS 11(2), 745–763 (2014)
84. D. Karapiperis, V.S. Verykios, An LSH-based blocking approach with a homomorphic matching technique for privacy-preserving record linkage. IEEE TKDE 27(4), 909–921 (2015)
85. D. Karapiperis, V.S. Verykios, A fast and efficient hamming LSH-based scheme for accurate linkage, in *Springer KAIS* (2016), pp. 1–24
86. H. Kargupta, S. Datta, Q. Wang, K. Sivakumar, On the privacy preserving properties of random data perturbation techniques, in *IEEE ICDM* (2003), p. 99
87. H. Kargupta, S. Datta, Q. Wang, K. Sivakumar, Random-data perturbation techniques and privacy-preserving data mining, Springer KAIS 7(4), 387–414 (2005)
88. C.W. Kelman, J. Bass, D. Holman, Research use of linked health data - a best practice protocol. Aust. NZ J. Public Health 26, 251–255 (2002)
89. H. Kim, D. Lee, Harra: fast iterative hashed record linkage for large-scale data collections, in *EDBT* (2010), pp. 525–536
90. H.-s. Kim, D. Lee, Parallel linkage, in *ACM CIKM* (2007), pp. 283–292
91. T. Kirsten, L. Kolb, M. Hartung, A. Groß, H. Köpcke, E. Rahm, Data partitioning for parallel entity matching, in *QDB* (2010)
92. L. Kissner, D. Song, Private and threshold set-intersection, in *Technical Report*. Carnegie Mellon University, 2004
93. L. Kolb, A. Thor, E. Rahm, Dedoop: efficient deduplication with Hadoop. PVLDB 5(12), 1878–1881 (2012)
94. L. Kolb, A. Thor, E. Rahm, Load balancing for mapreduce-based entity resolution, in *IEEE ICDE* (2012), pp. 618–629

95. H. Köpcke, E. Rahm, Frameworks for entity matching: a comparison. Elsevier DKE **69**(2), 197–210 (2010)
96. H. Köpcke, A. Thor, E. Rahm, Evaluation of entity resolution approaches on real-world match problems. PVLDB **3**(1), 484–493 (2010)
97. H. Krawczyk, M. Bellare, R. Canetti, HMAC: keyed-hashing for message authentication, in *Internet RFCs* (1997)
98. T.G. Kristensen, J. Nielsen, C.N. Pedersen, A tree-based method for the rapid screening of chemical fingerprints. Algorithms Mol. Biol. **5**(1), 9 (2010)
99. H. Kum, A. Krishnamurthy, A. Machanavajjhala, S. Ahalt, Population informatics: tapping the social genome to advance society: a vision for putting "big data" to work for population informatics. Computer (2013)
100. H.-C. Kum, A. Krishnamurthy, A. Machanavajjhala, M.K. Reiter, S. Ahalt, Privacy preserving interactive record linkage. JAMIA **21**(2), 212–220 (2014)
101. M. Kuzu, M. Kantarcioglu, E. Durham, B. Malin, A constraint satisfaction cryptanalysis of Bloom filters in private record linkage. PETS Springer LNCS **6794**, 226–245 (2011)
102. M. Kuzu, M. Kantarcioglu, E.A. Durham, C. Toth, B. Malin, A practical approach to achieve private medical record linkage in light of public resources. JAMIA **20**(2), 285–292 (2013)
103. M. Kuzu, M. Kantarcioglu, A. Inan, E. Bertino, E. Durham, B. Malin, Efficient privacy-aware record integration, in *ACM EDBT* (2013), pp. 167–178
104. P. Lai, S. Yiu, K. Chow, C. Chong, L. Hui, An efficient Bloom filter based solution for multiparty private matching, in *SAM* (2006)
105. F. Li, Y. Chen, B. Luo, D. Lee, P. Liu, Privacy preserving group linkage, in *Scientific and Statistical Database Management* (Springer, Berlin, 2011), pp. 432–450
106. N. Li, T. Li, S. Venkatasubramanian, T-closeness: privacy beyond k-anonymity and l-diversity, in *IEEE ICDE* (2007), pp. 106–115
107. P. Li, X. Dong, A. Maurino, D. Srivastava, Linking temporal records. PVLDB **4**(11), 956–967 (2011)
108. Z. Lin, M. Hewett, R.B. Altman, Using binning to maintain confidentiality of medical data, in *AMIA Symposium* (2002), p. 454
109. Y. Lindell, B. Pinkas, Privacy preserving data mining, in *CRYPTO* (Springer, Berlin, 2000), pp. 36–54
110. Y. Lindell, B. Pinkas, An efficient protocol for secure two-party computation in the presence of malicious adversaries, in *EUROCRYPT* (2007), pp. 52–78
111. Y. Lindell, B. Pinkas, Secure multiparty computation for privacy-preserving data mining. JPC **1**(1), 5 (2009), pp. 59–98
112. H. Liu, H. Wang, Y. Chen, Ensuring data storage security against frequency-based attacks in wireless networks, in *DCOSS, Springer LNCS*, vol. 6131 (2010), pp. 201–215
113. H. Lu, M.-C. Shan, K.-L. Tan, Optimization of multi-way join queries for parallel execution, in *VLDB* (1991), pp. 549–560
114. M. Luby, C. Rackoff, How to construct pseudo-random permutations from pseudo-random functions, in *CRYPTO*, vol. 85 (1986), p. 447
115. A. Machanavajjhala, D. Kifer, J. Gehrke, M. Venkitasubramaniam, l-diversity: privacy beyond k-anonymity. ACM TKDD **1**(1), 3 (2007)
116. B.A. Malin, K. El Emam, C.M. O'Keefe, Biomedical data privacy: problems, perspectives, and recent advances. JAMIA **20**(1), 2–6 (2013)
117. M. Mitzenmacher, E. Upfal, *Probability and Computing: Randomized Algorithms and Probabilistic Analysis* (Cambridge University Press, Cambridge, 2005)
118. N. Mohammed, B. Fung, M. Debbabi, Anonymity meets game theory: secure data integration with malicious participants. PVLDB **20**(4), 567–588 (2011)
119. M. Nentwig, M. Hartung, A.-C. Ngonga Ngomo, E. Rahm, A survey of current link discovery frameworks. *Semantic Web Journal* (2016)
120. A.N. Ngomo, L. Kolb, N. Heino, M. Hartung, S. Auer, E. Rahm, When to reach for the cloud: using parallel hardware for link discovery, in *ESWC* (2013), pp. 275–289
121. Office for National Statistics, Beyond 2011 matching anonymous data (2013)

122. C. O'Keefe, M. Yung, L. Gu, R. Baxter, Privacy-preserving data linkage protocols, in *ACM WPES* (2004), pp. 94–102
123. B. On, N. Koudas, D. Lee, D. Srivastava, Group linkage, in *IEEE ICDE* (2007), pp. 496–505
124. C. Pang, L. Gu, D. Hansen, A. Maeder, Privacy-preserving fuzzy matching using a public reference table, in *Intelligent Patient Management*, vol. 189. Studies in Computational Intelligence (Springer, Berlin, 2009), pp. 71–89
125. C. Phua, K. Smith-Miles, V. Lee, R. Gayler, Resilient identity crime detection. IEEE TKDE **24**(3), 533–546 (2012)
126. C. Quantin, H. Bouzelat, L. Dusserre, Irreversible encryption method by generation of polynomials. Med. Inf. Internet Med. **21**(2), 113–121 (1996)
127. C. Quantin, H. Bouzelat, F. Allaert, A. Benhamiche, J. Faivre, L. Dusserre, How to ensure data security of an epidemiological follow-up: quality assessment of an anonymous record linkage procedure. IJMI **49**(1), 117–122 (1998)
128. E. Rahm, H.H. Do, Data cleaning: problems and current approaches. IEEE Data Eng. Bull. **23**(4), 3–13 (2000)
129. B. Ramadan, P. Christen, H. Liang, R.W. Gayler, Dynamic sorted neighborhood indexing for real-time entity resolution. ACM JDIQ **6**(4), 15 (2015)
130. T. Ranbaduge, P. Christen, D. Vatsalan, Tree based scalable indexing for multi-party privacy-preserving record linkage, in *AusDM* (2014)
131. T. Ranbaduge, D. Vatsalan, P. Christen, Clustering-based scalable indexing for multi-party privacy-preserving record linkage, in *Springer PAKDD* (2015), pp. 549–561
132. T. Ranbaduge, D. Vatsalan, P. Christen, Merlin–a tool for multi-party privacy-preserving record linkage, in *IEEE ICDMW* (2015), pp. 1640–1643
133. T. Ranbaduge, D. Vatsalan, P. Christen, Hashing-based distributed multi-party blocking for privacy-preserving record linkage, in *Springer PAKDD* (2016), pp. 415–427
134. T. Ranbaduge, D. Vatsalan, S. Randall, P. Christen, Evaluation of advanced techniques for multi-party privacy-preserving record linkage on real-world health databases, in *IPDLN* (2016)
135. S.M. Randall, A.M. Ferrante, J.H. Boyd, J.B. Semmens, Privacy-preserving record linkage on large real world datasets, in *Elsevier JBI* (2014) volume 50, pp. 205–212
136. S.M. Randall, A.M. Ferrante, J.H. Boyd, A.P. Brown, J.B. Semmens, Limited privacy protection and poor sensitivity is it time to move on from the statistical linkage key-581? Health Inf. Manag. J. **37**, 60–62 (2016)
137. V. Rastogi, N. Dalvi, M. Garofalakis, Large-scale collective entity matching. in VLDB **4**, 208–218 (2011)
138. C. Rong, W. Lu, X. Wang, X. Du, Y. Chen, A.K.H. Tung, Efficient and scalable processing of string similarity join. IEEE TKDE **25**(10), 2217–2230 (2013)
139. M. Roughan, Y. Zhang, Secure distributed data-mining and its application to large-scale network measurements. ACM SIGCOMM Comput. Commun. Rev. **36**(1), 7–14 (2006)
140. T. Ryan, D. Gibson, B. Holmes, A national minimum data set for home and community care, in *Australian Institute of Health and Welfare* (1999)
141. M. Scannapieco, I. Figotin, E. Bertino, A. Elmagarmid, Privacy preserving schema and data matching, in *ACM SIGMOD* (2007), pp. 653–664
142. D.A. Schneider, D.J. DeWitt, Tradeoffs in processing complex join queries via hashing in multiprocessor database machines, in *VLDB* (1990), pp. 469–480
143. B. Schneier, *Applied Cryptography: Protocols, Algorithms, and Source Code in C*, 2nd edn. (Wiley, New York, 1996)
144. R. Schnell, Privacy-preserving record linkage and privacy-preserving blocking for large files with cryptographic keys using multibit trees, in *JSM* (2013), pp. 187–194
145. R. Schnell, An efficient privacy-preserving record linkage technique for administrative data and censuses. Stat. J. IAOS **30**(3), 263–270 (2014)
146. R. Schnell, T. Bachteler, S. Bender, A toolbox for record linkage. Aust. J. Stat. **33**(1–2), 125–133 (2004)

147. R. Schnell, T. Bachteler, J. Reiher, Privacy-preserving record linkage using Bloom filters. BMC Medi. Inf. Decision Mak. **9**(1), 41 (2009)
148. R. Schnell, T. Bachteler, J. Reiher, A novel error-tolerant anonymous linking code, in *German Record Linkage Center, WP-GRLC-2011-02* (2011)
149. Z. Sehili, E. Rahm, Speeding up privacy preserving record linkage for metric space similarity measures, in *Datenbank-Spektrum* (2016), pp. 1–10
150. Z. Sehili, L. Kolb, C. Borgs, R. Schnell, E. Rahm, Privacy preserving record linkage with PP Join, in *BTW Conference* (2015)
151. D. Song, D. Wagner, A. Perrig, Practical techniques for searches on encrypted data, in *IEEE Symposium on Security and Privacy* (2000), pp. 44–55
152. L. Sweeney, K-anonymity: a model for protecting privacy. Int. J. Uncertaint. Fuzziness Knowl. Based Syst. **10**(5), 557–570 (2002)
153. K.-N. Tran, D. Vatsalan, P. Christen, GeCo: an online personal data generator and corruptor, in *ACM CIKM* (2013), pp. 2473–2476
154. S. Trepetin, Privacy-preserving string comparisons in record linkage systems: a review. Inf. Secur. J.: A Global Perspect. **17**(5), 253–266 (2008)
155. E. Turgay, T. Pedersen, Y. Saygın, E. Savaş, A. Levi, Disclosure risks of distance preserving data transformations, in *Springer SSDBM* (2008), pp. 79–94
156. J. Vaidya, Y. Zhu, C.W. Clifton, *Privacy Preserving Data Mining*, vol. 19. *Advances in Information Security* (Springer, Berlin, 2006)
157. E. Van Eycken, K. Haustermans, F. Buntinx et al., Evaluation of the encryption procedure and record linkage in the Belgian national cancer registry. Archiv. Public Health **58**(6), 281–294 (2000)
158. D. Vatsalan, P. Christen, An iterative two-party protocol for scalable privacy-preserving record linkage, in *AusDM, CRPIT* (2012), pp. 127–138
159. D. Vatsalan, P. Christen, Sorted nearest neighborhood clustering for efficient private blocking, in *Springer PAKDD*, vol. 7819 (2013), pp. 341–352
160. D. Vatsalan, P. Christen, Scalable privacy-preserving record linkage for multiple databases, in *ACM CIKM* (2014), pp. 1795–1798
161. D. Vatsalan, P. Christen, Privacy-preserving matching of similar patients. Elsevier JBI **59**, 285–298 (2016)
162. D. Vatsalan, P. Christen, V.S. Verykios, An efficient two-party protocol for approximate matching in private record linkage, in *AusDM* (2011), pp. 125–136
163. D. Vatsalan, P. Christen, V.S. Verykios, Efficient two-party private blocking based on sorted nearest neighborhood clustering, in *ACM CIKM* (2013), pp. 1949–1958
164. D. Vatsalan, P. Christen, V.S. Verykios, A taxonomy of privacy-preserving record linkage techniques. Elsevier JIS **38**(6), 946–969 (2013)
165. D. Vatsalan, P. Christen, C.M. O'Keefe, V.S. Verykios, An evaluation framework for privacy-preserving record linkage. JPC **6**(1), 3 (2014), pp. 35–75
166. R. Vernica, M.J. Carey, C. Li, Efficient parallel set-similarity joins using MapReduce, in *ACM SIGMOD* (2010), pp. 495–506
167. V.S. Verykios, A. Karakasidis, V. Mitrogiannis, Privacy preserving record linkage approaches. IJDMMM **1**(2), 206–221 (2009)
168. G. Wang, H. Chen, H. Atabakhsh, Automatically detecting deceptive criminal identities. Commun. ACM **47**(3), 70–76 (2004)
169. Q. Wang, D. Vatsalan, P. Christen, Efficient interactive training selection for large-scale entity resolution, in *PAKDD* (2015), pp. 562–573
170. Z. Wen, C. Dong, Efficient protocols for private record linkage, in *ACM Symposium on Applied Computing* (2014), pp. 1688–1694
171. W.E. Winkler, Methods for evaluating and creating data quality. Elsevier JIS **29**(7), 531–550 (2004)
172. C. Xiao, W. Wang, X. Lin, J.X. Yu, Efficient similarity joins for near duplicate detection, in *WWW* (2008), pp. 131–140

173. M. Yakout, M. Atallah, A. Elmagarmid, Efficient private record linkage, in *IEEE ICDE* (2009), pp. 1283–1286
174. P. Zezula, G. Amato, V. Dohnal, M. Batko, *Similarity Search: The Metric Space Approach*, vol. 32 (Springer, Berlin, 2006)
175. X. Zhang, C. Liu, S. Nepal, J. Chen, An efficient quasi-identifier index based approach for privacy preservation over incremental data sets on cloud. J. Comput. Syst. Sci. **79**(5), 542–555 (2013)
176. X. Zhang, C. Liu, S. Nepal, S. Pandey, J. Chen, A privacy leakage upper bound constraint-based approach for cost-effective privacy preserving of intermediate data sets in cloud. IEEE TPDS **24**(6), 1192–1202 (2013)

Printed in the United States
By Bookmasters